Elementary
Classical
Analysis

Elementary
Classical
Analysis

JERROLD E. MARSDEN

University of California, Berkeley

with the assistance of

Michael Buchner
Amy Erickson
Adam Hausknecht
Dennis Heifetz
Janet Macrae
William Wilson

and with contributions by

Paul Chernoff
István Fáry
Robert Gulliver

W. H. Freeman and Company
San Francisco

The cover design is based on an original drawing
by Richard Bassein.

Library of Congress Cataloging in Publication Data

Marsden, Jerrold E.
    Elementary classical analysis.

    Bibliography: p.
    1. Mathematical analysis.   I.   Title.
QA300.M2868   515   74-5764
ISBN 0-7167-0452-8

Printed in the United States of America

AMS 1970 subject classifications:
26-01; 35C10; 42-01.

10 9 8 7 6 5 4 3

to Anne Murray
a good Canadian

Contents

Preface

This book is intended for a two-quarter or one- or two-semester course in advanced calculus and introductory real analysis. The book is classical in the sense that it deals with calculus and Fourier series in Euclidean space. Only a few brief references are made to "modern" topics such as Lebesgue integration, distributions, and quantum mechanics. We resisted the temptation to include vector analysis (the Stokes theorem and so forth). In most curricula, this topic comes earlier in the second year at a more informal level (see, for example, J. Marsden and A. Tromba, *Vector Calculus*, W. H. Freeman and Company, 1975) and possibly later in the context of manifold theory for students who are so inclined.

In presenting the material, we have been deliberately concrete—aiming at a solid understanding of the Euclidean case and introducing abstraction only through examples. For instance, if Euclidean spaces are properly understood, it is a small jump to other spaces such as the space of continuous functions and abstract metric spaces. In the context of the space of continuous functions, we can see the power of abstract metric space methods. When the general theory is presented too soon, the student is confused about its relevance; consequently, much teaching time can be wasted.

The book assumes that the reader has had some calculus; that is, that he or she knows how to differentiate and integrate standard functions. Strictly speaking, the theory is developed logically and requires few prerequisites, although a knowledge of calculus is needed for an understanding of examples and exercises. Also, some brief contact with partial derivatives and multiple integrals is desirable but not essential. Chapter 6, on differentiation, requires the rudiments of linear algebra; specifically, the student should know what a linear transformation and its representing matrix is.

Each chapter is organized as follows. There are numerous sections containing the definitions, statements of the theorems, examples, and fairly easy problems. Once the student masters the theorems and is able to handle the easy problems, he can move on to the end of the chapter to master the technical proofs. Here, numerous further examples and exercises are given. The easier exercises following each section enable the student to master the material as he goes along. The exercises at the end of the chapter then often require an integrated knowledge of the whole chapter or previous chapters including theorem proofs. This plan has worked out well in lectures. When the lectures are devoted to explaining the theorems with only selected proofs given, it is much easier for the student to see what is going on. We found that using this approach one or two sections can be covered in each lecture.

The introductory chapter contains essential terminological material. The student interested in the intricacies of set theory can consult the Appendix, which has been kindly supplied by Professor I. Fáry.

Chapter 1 contains material on the basic structure of the real line needed for later developments. We spend a minimum amount of time on the algebraic axioms and concentrate on the completeness property. The algebraic axioms are usually covered in basic algebra courses, and since the student is used to working with real numbers, it seems logical to accept the basic algebraic skills as valid.

Chapters 2 and 3 treat the topology of $\mathbb{R}^n$ in such a way as to just use the basic metric structure of $\mathbb{R}^n$. This is done to make the transition to other metric spaces, such as the space of continuous functions treated later, almost automatic.

A complete and early introduction of abstract metric spaces is avoided here. Experience has shown that at this level almost two extra weeks are required to achieve this abstraction because, for one thing, one has to go through the usual "bizarre" metric spaces, which students find confusing. The time saved can be used later for more useful topics like the Ascoli theorem, the Stone-Weierstrass theorem, fixed-point theorems, and differential or integral equations.

Chapter 4 continues the development, treating the basic facts about continuity. Chapter 5 gives the more detailed properties of continuous functions related to uniform convergence. A number of more specialized topics are presented in Sections 5.5–5.9, from which a selection can be made.

Chapter 6 deals with differentiation, making some use of linear algebra. All of the usual topics of differential calculus for functions of several variables are treated. A fairly thorough treatment of maxima and minima is given, including an optional discussion of the Morse lemma in Chapter 7. Chapter 7 has as its main topic a complete discussion of the inverse and implicit function theorems. Existence theorems for ordinary differential equations and Lagrange multipliers (constrained extrema) are also given.

Chapter 8 treats the basics of integration. Some may wish to teach this material before Chapter 7. In this chapter, we deal with the Riemann integral but do include Lebesgue's theorem and sets of measure zero. An optional section gives a quick look at distributions, illustrated by the δ-function.

The next chapter proves the two fundamental theorems concerning multiple integrals: the reduction to iterated integrals and the change of variables formula. Numerous applications are given.

The last chapter, Chapter 10, gives a fairly thorough treatment of Fourier series from the point of view of inner product spaces. Some topics such as this are useful to students in introductory analysis courses, since it goes well beyond just "rigorizing" many topics they already knew. One unusual feature of our presentation is the inclusion of some applications to differential equations and quantum mechanics.

Of course, teachers have different tastes concerning rigor, the role of intuition, the choice of subject matter, and so forth. Perhaps a few remarks on variations in the manner of presentation of the material in this book will aid those who wish to adapt it to their own personal style.

First of all, in Chapters 2 through 4, it is possible to lay more emphasis on abstract metric spaces without materially changing the text. It is, in fact, a good exercise to have students do this adaptation themselves, because once they see the "correct" proof in $\mathbb{R}^n$, it becomes rather enjoyable, and rewarding, to make the generalizations. In this regard, there is a table, supplied by R. Gulliver, at the end of Chapter 5; the table indicates which theorems hold for general metric spaces.

Some material in Chapter 5 is a bit more advanced and can be deferred. Also, if a complete logical development is desired, differentiation and integration of functions of one variable should precede Chapter 5; this depends on the background of the students. This material, in its most basic aspects, is used in Sections 5.3, 5.6, and 7.5. In practice, we have found that it offends only the best students to have to use some calculus before it is "correctly" presented in the course. We find this healthy, but some may wish to switch the order of presentation.

At the beginning of Chapter 6, it is good to review a little linear algebra; specifically, the definition of the matrix of a linear transformation. This is also a good time to look over Example 4 at the end of Chapter 4.

For a semester course, some topics have to be cut in order to reach Chapter 10 (such as Sections 5.5–5.9 and 7.3–7.7). In a two-quarter course, there is time to complete the entire text (perhaps omitting Sections 5.8, 5.9, 7.3, 7.4, 7.6, 7.7, 10.7, and 10.8).

The symbols used in this text are standard except possibly for the following: $\mathbb{R}$ denotes the real number line, $\mathbb{C}$ denotes the complex numbers, $\mathbb{R}^n$ denotes Euclidean n-space, "iff" stands for "if and only if," and ∎ denotes the end of a proof. The notation $]a,b[$ is used to indicate the open interval consisting of all real numbers x satisfying $a < x < b$. This European convention avoids confusion with the ordered pair (a,b). The notation $x \mapsto f(x)$ indicates that x is mapped to $f(x)$ by f. The notation $f: A \subset \mathbb{R}^n \to \mathbb{R}^m$ means that f maps the domain A into $\mathbb{R}^m$. Occasionally $\Rightarrow$ is used to denote "implies." The symbol $A \backslash B$ denotes the members of the set A that are not members of B, and $x \in A$ means that x is a member of A.

Sections, theorems, and definitions are numbered consecutively within each chapter. A reference such as "Theorem 24" or "Exercise 3" applies to material within the present chapter or section; otherwise, the chapter or section number is cited.

We thank M. Buchner and W. Wilson who helped with the first draft of the book, and I. Fáry and R. Gulliver for the appendices. We also thank the students of Math 104A-B at Berkeley, especially E. Wong, J. Lim, J. Wing, and J. Seitz, for catching numerous small errors and stylistic points. We thank our colleagues from whose old examinations many of the problems are derived. Several colleagues deserve special mention, especially P. Chernoff, I. Fáry, R. Gulliver, and M. Mayer for reading portions of the manuscript and suggesting several improvements. Sections 5.9 and 10.4 and a number of problems were adapted from class notes of P. Chernoff. The remaining assistants, A. Erickson, A. Hausknecht, D. Heifetz, and J. Macrae helped with portions of the manuscript, eliminated many errors, and checked and prepared answers for most of the problems. Help was also received from M. McCracken, W. A. J. Luxemburg, and R. Graff. We thank I. Workman for a fine job of typing the manuscript and N. Lee for her moral support.

Finally, thanks are extended to R. Abraham, K. McAloon, A. Tromba, and M. O'Nan, officers of Eagle Mathematics Incorporated (an organization of mathematics authors), for their suggestion that this book be written and for their subsequent encouragement.

October 1973   JERROLD E. MARSDEN

Elementary
Classical
Analysis

Introduction

Prerequisites;
Sets and Functions

The student who wishes to use this book successfully should have a sound background in elementary calculus, as well as some knowledge of linear algebra (mostly for use in Chapters 6, 7, and 10) and a little multi-variable calculus. Adequate preparation is normally obtained from two years of undergraduate mathematics. Also required is a basic knowledge of sets and functions, for which the necessary concepts are summarized below. This material should be read briefly and then consulted as needed.

Set theory is the starting point of much of mathematics and is in itself a vast and complicated subject. For brevity and better understanding, we begin our study somewhat intuitively. The reader interested in the subtleties of the subject can consult Appendix A at the end of the book for further information.

A *set* is a collection of "objects" or "things" called *members* of the set. For example, the integers $1, 2, 3, \ldots$ form a set. Likewise the set of all rational numbers (fractions) p/q form a set. If S is a set, and x is a member of S, we write $x \in S$. A *subset* of the set S is a set A such that every element of A is also a member of S; using symbols, $(x \in A) \Rightarrow (x \in S)$. The symbol $\Rightarrow$ denotes "implies." When A is a subset of S, we write $A \subset S$. Sometimes the symbol $A \subseteq S$ is used for what we denote as $A \subset S$. We can also define equality of sets by stating that $A = B$ means that $A \subset B$ and $B \subset A$; that is, A and B have the same elements. The *empty set*, denoted $\varnothing$, is a set with no members. For example, the set of integers n such that $n^2 = -1$ is empty. The empty set $\varnothing$ often mystifies students and it is indeed a strange concept—don't overplay its importance at this stage.

When specifying a set we often list the members in braces. Thus we write

$\mathbb{N} = \{1,2,3,\ldots\}$ to denote the set of positive integers and $\mathbb{Z} = \{\ldots,-3,-2,-1,0,1,2,3,\ldots\}$ for the set of all integers. An example of a subset of $\mathbb{N}$ is the set of even numbers and is written as

$$A = \{2,4,6,\ldots\} = \{x \in \mathbb{N} \mid x \text{ is even}\} \subset \mathbb{N}.$$

We read $\{x \in \mathbb{N} \mid x \text{ is even}\}$ as "the set of all members x of $\mathbb{N}$ such that x is even."

At this point, there is a notational distinction of which we should be aware. Let S be a set. For $a \in S$, $\{a\}$ denotes the subset of S whose members consist of the single element a. Thus $\{a\} \subset S$ while $a \in S$.

For a general set S and for $A \subset S$ and $B \subset S$, we define $A \cup B = \{x \in S \mid x \in A \text{ or } x \in B\}$, which is read "the set of all $x \in S$ which are members of A or B (or both)." The set $A \cup B$ is called the *union* of A and B. Similarly, one can form the union of families of sets. For example, let $A_1, A_2, \ldots$ be subsets of S. Then we define $\bigcup_{i=1}^{\infty} A_i = \{x \in S \mid x \in A_i \text{ for some } i\}$. This union is also written $\bigcup \{A_1, A_2, A_3, \ldots\}$. Note that $A \cup B$ is the special case with $A_1 = A$, $A_2 = B$, and $A_i = \varnothing$ for $i > 2$.

Similarly, one can form the intersections $A \cap B = \{x \in S \mid x \in A \text{ and } x \in B\}$, and $\bigcap_{i=1}^{\infty} A_i = \{x \in S \mid x \in A_i \text{ for all } i\}$. Figure 0-1 presents these operations diagrammatically.

For $A, B \subset S$ we form the *complement* of A relative to B by defining

$$B \backslash A = \{x \in B \mid x \notin A\},$$

where $x \notin A$ means x is *not* contained in A. See Figure 0-2.

The reader can prove (as is done in Example 1 below) that $B \backslash (A_1 \cup A_2) = (B \backslash A_1) \cap (B \backslash A_2)$ and that $B \backslash (A_1 \cap A_2) = (B \backslash A_1) \cup (B \backslash A_2)$ for any sets $A_1, A_2, B \subset S$. This is an example of a "set identity." Other examples are given in the problems.

For sets A, B define the *Cartesian product* of A and B by $A \times B = \{(a,b) \mid a \in A \text{ and } b \in B\}$. It consists of the set of all *ordered pairs* (a,b) with $a \in A$ and $b \in B$. See Figure 0-3.

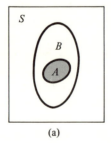 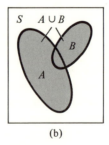 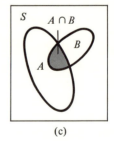

(a) (b) (c)

FIGURE 0-1 (a) Subset. (b) Union. (c) Intersection.

A *function* $f: A \to B$ is a "rule" which assigns to each $a \in A$ a specific element of B, denoted $f(a)$. One often writes $a \mapsto f(a)$ to denote that a is mapped to the element $f(a)$. For example (Figure 0-4), the function $f(x) = x^2$ may be specified by saying $x \mapsto x^2$. Here $A = B$ is the set of all real numbers.

Note: In this book the terms "mapping," "map," "function," and "transformation" are all synonymous.

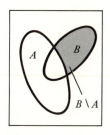

FIGURE 0-2   Complement.

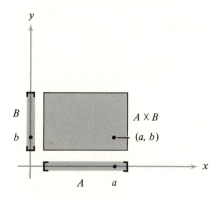

FIGURE 0-3   Cartesian product.

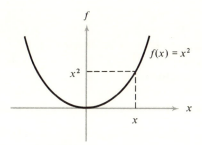

FIGURE 0-4   Function.

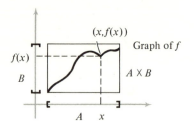

FIGURE 0-5   Graph of a function.

For a function $f: A \to B$, the set A is called the *domain* of f and B is called the *target* of f. The *range* of f is the set $f(A) = \{f(x) \in B \mid x \in A\}$ which is a subset of B. The *graph* of f is the set $\{(x, f(x)) \in A \times B \mid x \in A\}$, as in Figure 0-5.

Someone paying careful attention to logical foundations may object to using colloquial language such as "rule" and would be happier to define a function from A to B as a subset of $A \times B$ with the property that any two members of the set with the same first element are identical; that is, the first element x determines the second, $f(x)$. See Figure 0-5.

A function $f: A \to B$ is called *one-to-one* (also called an *injection*) if* whenever $a_1 \neq a_2$ then $f(a_1) \neq f(a_2)$. Thus a function is one-to-one when no two distinct elements are mapped to the same element.

An extreme example of a function which is not one-to-one is a constant function, a function $f: A \to B$ such that $f(a_1) = f(a_2)$ for all $a_1, a_2 \in A$. See Figure 0-6.

We say $f: A \to B$ is *onto* or is a *surjection* when, for every $b \in B$, there is an $a \in A$ such that $f(a) = b$, in other words when the range equals the target. It should be noted that the choice of A and B is part of the definition of f, and whether or not f is one-to-one or onto depends on this choice. For example, $f(x) = x^2$ is one-to-one and onto when $A = B$ and consists of all real numbers x such that $x \geqslant 0$, is one-to-one but *not* onto when A is all those x such that $x \geqslant 0$ and B is all x, and is *neither* when both A and B are all real numbers x.

For $f: A \to B$ and $D \subset A$, we let $f(D) = \{f(d) \in B \mid d \in D\}$, and for $C \subset B$, define $f^{-1}(C)$ to be the set $\{a \in A \mid f(a) \in C\}$. We call $f(D)$ the *image* of D and $f^{-1}(C)$ the *inverse image* or *pre-image* of C.

If $f: A \to B$ is one-to-one *and* onto, then from the definition it is not hard to see that there is a unique function, denoted $f^{-1}: B \to A$ (not to be confused with $f^{-1}(C)$ above or $1/f$) such that $f(f^{-1}(b)) = b$ for all $b \in B$ and $f^{-1}(f(a)) = a$ for all $a \in A$. We call f^{-1} the *inverse function* of f. A one-to-one

---

* It is a convention that in definitions, "if" stands for "if and only if." The latter is often written "iff," or ⇔. Of course in theorems it is absolutely necessary to distinguish between "if," "only if," and "iff."

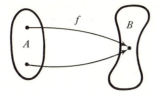

FIGURE 0-6   Constant function.

and onto map is also called a *bijection* or a *one-to-one correspondence.* [*Warning*: We can form $f^{-1}(C)$ for a set $C \subset B$ even though f might not be one-to-one or onto. For practice with these operations, see Exercise 3.]

The map $f: A \rightarrow A$ such that $f(x) = x$ for all $x \in A$ is called the *identity mapping* on A. One should distinguish the identity mappings for different sets. For example, one sometimes uses notation like I_A for the identity mapping on A. Clearly, I_A is one-to-one and onto.

Now consider two functions $f: A \rightarrow B$ and $g: B \rightarrow C$. The *composition* $g \circ f: A \rightarrow C$ is defined by $g \circ f(a) = g(f(a))$. See Figure 0-7. For example, if $f: x \mapsto x^2$ and $g: x \mapsto x + 3$, then $g \circ f: x \mapsto x^2 + 3$ and $f \circ g: x \mapsto (x + 3)^2$ (here A, B, and C consist of all real numbers x).

Sometimes we wish to restrict our attention to just some elements on which a function is defined. This is called *restriction* of a function. More formally, if we have a mapping $f: A \rightarrow B$ and $D \subset A$, we consider a new function denoted $f \mid D: D \rightarrow B$ defined by $(f \mid D)(x) = f(x)$ for all $x \in D$. We call $f \mid D$ the *restriction* of f to D, and also say that f is an *extension* of $f \mid D$. The importance of these notions will become obvious in our later discussions.

A set A is called *finite* if we can display all its elements as follows: $A = \{a_1, a_2, \ldots, a_n\}$ for some integer n. A set which is not finite is called *infinite*. For example, the set of all positive integers $\mathbb{N} = \{1, 2, \ldots\}$ is an infinite set.

In examining examples it may be difficult to decide if one infinite set has more elements than another infinite set. For instance it is not clear at first if there are more rational or irrational numbers. To make this notion precise,

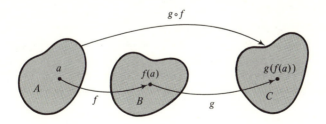

FIGURE 0-7   Composition of mappings.

we say that two sets A and B have the *same number of elements* (or have the same *cardinality*) if there exists a mapping $f\colon A \to B$ which is one-to-one *and* onto.

If an infinite set has the same number of elements as the set of integers $\{1,2,\ldots\}$, it is called *denumerable*. A set that is either finite or denumerable is called *countable*. Otherwise, a set is called uncountably infinite, or just *uncountable*. An example of an uncountable set is the set of all numbers between 0 and 1. (We shall prove this in Chapter 1).

Let S be a set. A *sequence* in S may be viewed as a mapping $f\colon \mathbb{N} \to S$, where $\mathbb{N} = \{1,2,\ldots\}$. Thus we have associated to each integer n an element of S, namely $f(n)$. One often suppresses the fact that we have a function by simply considering a sequence as the image elements, say, $x_1, x_2, x_3, \ldots$ or alternatively, just writes "the sequence x_n" or $\{x_n\}_{n=1}^{\infty}$. By a *subsequence* of $x_1, x_2, \ldots$ we mean a sequence $y_1, y_2, \ldots$ such that each y_n occurs in the set $\{x_1, x_2, \ldots\}$ and if $i < j$ then $y_i = x_l$, $y_j = x_m$ where $l < m$. In other words, a subsequence is obtained by "throwing out" elements of the original sequence and ordering naturally the elements which remain.

Worked Examples for Introductory Chapter

1. For sets $A, B, C \subset S$, show that

$$A \cap (B \cup C) = (A \cap B) \cup (A \cap C).$$

(Distributive law.)

Solution: The method is to show that each side is a subset of the other. So first take $x \in A \cap (B \cup C)$. This means x is a member of both A and $B \cup C$. Therefore, x is in A and x is in either B or C. If $x \in B$, then $x \in A \cap B$, while if $x \in C$, then $x \in A \cap C$. Hence x is in either $A \cap B$ or $A \cap C$; that is $x \in (A \cap B) \cup (A \cap C)$, so $A \cap (B \cup C) \subset (A \cap B) \cup (A \cap C)$. Now let $x \in (A \cap B) \cup (A \cap C)$; thus x is in either $A \cap B$ or is in $A \cap C$. If $x \in A \cap B$, then x is in A and B, and in particular, x is in A and $B \cup C$, so $x \in A \cap (B \cup C)$. Similarly, if $x \in A \cap C$, we conclude that $x \in A \cap (B \cup C)$. Hence $(A \cap B) \cup (A \cap C) \subset A \cap (B \cup C)$, and so we now have equality. This can also be verified diagrammatically as in Figure 0-8.

2. Show that for $A, B \subset S$,

$$A \subset B \Leftrightarrow S \backslash A \supset S \backslash B.$$

Solution: First we prove that $A \subset B$ implies $S \backslash B \subset S \backslash A$. Assume $A \subset B$ and $x \in S \backslash B$. Then $x \notin B$ and therefore $x \notin A$ (for $x \in A \Rightarrow x \in B$), hence $x \in S \backslash A$, proving that $S \backslash B \subset S \backslash A$. To prove the converse, suppose $S \backslash B \subset S \backslash A$ and $x \in A$. Then $x \notin B$ implies $x \in S \backslash B$ which in turn implies $x \in S \backslash A$ and hence $x \notin A$, contradicting the hypothesis; therefore $x \in B$, and $A \subset B$.

3. Let $f(x) = x^2$ (defined on the set of all real numbers) and $B = \{y \mid y \geqslant 1\}$. Compute $f^{-1}(B)$.

$$A \cap (B \cup C) = (A \cap B) \cup (A \cap C)$$

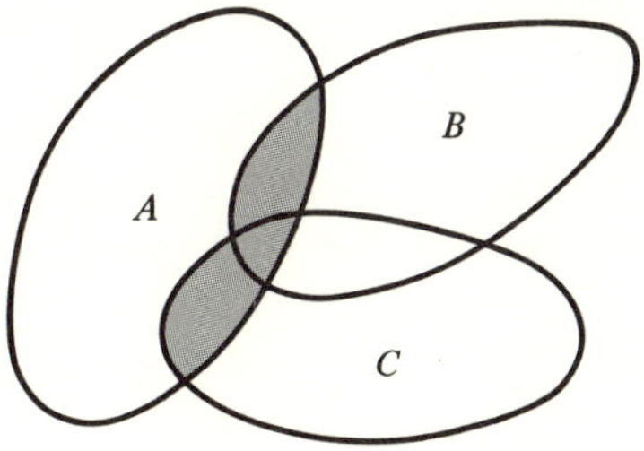

FIGURE 0-8   Distributive law.

Solution: By definition, $f^{-1}(B)$ consists of all x such that $f(x) \in B$; that is, all x such that $x^2 \geq 1$. This happens iff $x \geq 1$ or $x \leq -1$. Thus $f^{-1}(B) = \{x \mid x \geq 1\} \cup \{x \mid x \leq -1\}$.

4. Let A be a set and let $\mathscr{P}(A)$ denote the set of all subsets of A. Prove that A and $\mathscr{P}(A)$ do not have the same cardinality.

Solution: The reasoning here is a little tricky and is similar to various "paradoxes" one finds in set theory (see Appendix A for further details). The result here is due to the work of G. Cantor. Suppose that we have a *bijection* $f: A \to \mathscr{P}(A)$; we shall then derive a contradiction. Let $B = \{x \in A \mid x \notin f(x)\}$. There exists a $y \in A$ such that $f(y) = B$ since f is onto. If $y \in B$, then by definition of B we conclude that $y \notin B$. Similarly if $y \notin B$, then we conclude that $y \in B$. In either case we get a contradiction. Actually the argument shows that there does not exist a function $f: A \to \mathscr{P}(A)$ which is onto.

Exercises for Introductory Chapter

1. The following mappings are defined by stating $f(x)$, the *domain* A, and the *range* B. For $A_0 \subset A$ and $B_0 \subset B$, as given, compute $f(A_0)$ and $f^{-1}(B_0)$.
   (a) $f(x) = x^2$, $A = \{-1,0,1\}$, B = all real numbers,
                       $A_0 = \{-1,1\}$, $B_0 = \{0,1\}$
   (b) $f(x) = \begin{cases} x^2, & \text{if } x \geq 0 \\ -x^2, & \text{if } x < 0 \end{cases}$
       A = all real numbers = B,
       A_0 = all $x > 0$, $B_0 = \{0\}$
   (c) $f(x) = \begin{cases} 1, & \text{if } x > 0 \\ 0, & \text{if } x = 0 \\ -1, & \text{if } x < 0 \end{cases}$
       $A = B$ = all real numbers,
       $A_0 = B_0$ = all x with $-2 < x < 1$.

2. For the functions listed in Exercise 1, determine if they are one-to-one or onto (or both).

3. Let $f: A \to B$ be a function, $C_1, C_2 \subset B$, and $D_1, D_2 \subset A$. Prove
   (a) $f^{-1}(C_1 \cup C_2) = f^{-1}(C_1) \cup f^{-1}(C_2)$
   (b) $f(D_1 \cup D_2) = f(D_1) \cup f(D_2)$
   (c) $f^{-1}(C_1 \cap C_2) = f^{-1}(C_1) \cap f^{-1}(C_2)$
   (d) $f(D_1 \cap D_2) \subset f(D_1) \cap f(D_2)$.

4. Verify the relations in Exercise 3 for the functions in Exercise 1 and the following sets:
   (a) $C_1 = $ all $x > 0$, $D_1 = \{-1,1\}$,
       $C_2 = $ all $x \leqslant 0$, $D_2 = \{0,1\}$;
   (b) $C_1 = $ all $x \geqslant 0$, $D_1 = $ all $x > 0$,
       $C_2 = $ all $x \leqslant 2$, $D_2 = $ all $x \geqslant -1$;
   (c) $C_1 = $ all $x \geqslant 0$, $D_1 = $ all x,
       $C_2 = $ all $x > -1$, $D_2 = $ all $x > 0$;

5. Prove a function $f: A \to B$ is one-to-one iff for all $y \in B$, $f^{-1}(\{y\})$ contains at most one point iff $f(D_1 \cap D_2) = f(D_1) \cap f(D_2)$ for all subsets $D_1, D_2 \subset A$. Develop similar criteria for "ontoness."

6. Show that the open interval* $]0,1[= \{x \mid 0 < x < 1\}$ has as many elements as there are real numbers, by setting up a one-to-one correspondence between $]0,1[$ and the real numbers $\mathbb{R}$.

7. Let A be a finite set with N elements, and let $\mathscr{P}(A)$ denote the collection of all subsets of A, including the empty set. Prove that $\mathscr{P}(A)$ has 2^N elements.

8. Prove that the set $\{\ldots, -2, -1, 0, 1, 2, 3, \ldots\}$ is countable.

9. Show that if $A_1, A_2, \ldots$ are countable sets, so is $A_1 \cup A_2 \cup \cdots$.

10. Let $\mathscr{A}$ be a family of subsets of a set S. Write $\bigcup \mathscr{A}$ for the union of all members of $\mathscr{A}$ and similarly, define $\bigcap \mathscr{A}$. Suppose $\mathscr{B} \supset \mathscr{A}$. Then show $\bigcup \mathscr{A} \subset \bigcup \mathscr{B}$ and $\bigcap \mathscr{B} \subset \bigcap \mathscr{A}$.

11. Let $f: A \to B$, $g: B \to C$, and $h: C \to D$ be mappings. Prove that $h \circ (f \circ g) = (h \circ f) \circ g$ (that is, composition is associative).

12. Prove that a map $f: A \to B$ is a bijection iff there is a map $g: B \to A$ such that $f \circ g = $ identity and $g \circ f = $ identity. Show also that $g = f^{-1}$ and is uniquely determined.

13. Let $f: A \to B$ and $g: B \to C$ be bijections. Then $(g \circ f)$ is a bijection and $(g \circ f)^{-1} = f^{-1} \circ g^{-1}$. [Hint: Use Exercise 12.]

14. Let $\mathscr{A}$ be a collection of subsets of a set S and $\mathscr{B}$ the collection of complementary sets; that is, $B \in \mathscr{B}$ iff $S \backslash B \in \mathscr{A}$. Prove *de Morgan's laws*:
    (a) $S \backslash \bigcup \mathscr{A} = \bigcap \mathscr{B}$
    (b) $S \backslash \bigcap \mathscr{A} = \bigcup \mathscr{B}$.
    Here $\bigcup \mathscr{A}$ denotes the union of all sets in $\mathscr{A}$ (see Exercise 10 and page 2). For

---

* In this text, open intervals are denoted as $]a,b[$ rather than (a,b). This European convention avoids confusion with ordered pairs.

example, if $\mathscr{A} = \{A_1, A_2\}$, then (a) reads $S\backslash(A_1 \cup A_2) = (S\backslash A_1) \cap (S\backslash A_2)$ and (b) reads $S\backslash(A_1 \cap A_2) = (S\backslash A_1) \cup (S\backslash A_2)$.

15. Let $A, B \subset S$. Show that

$$A \times B = \varnothing \Leftrightarrow A = \varnothing \quad \text{or} \quad B = \varnothing.$$

16. Show
(a) $(A \times B) \cup (A' \times B) = (A \cup A') \times B$
(b) $(A \times B) \cap (A' \times B') = (A \cap A') \times (B \cap B')$.

17. Let $f \colon A \to B, g \colon B \to C$ be given mappings. Show that for $C' \subset C, (g \circ f)^{-1}(C') = f^{-1}(g^{-1}(C'))$.

Chapter 1

The Real Line and Euclidean *n*-Space

$\mathbf{A}$ thorough knowledge of the real line and *n*-space is indispensable for a precise treatment of the calculus of functions of several variables as well as for a clear understanding of it. Much of this chapter may appear to be review, the material perhaps having been covered in previous mathematics courses. However, our discussion will be more rigorous and will give some further properties in preparation for later work.

1.1 The Real Line $\mathbb{R}$

Let us begin with the main properties of real numbers. The reader should be familiar with the heuristic (that is, intuitive) arguments which justify the real numbers. Begin with the positive integers 0, 1, 2, 3, . . . , and then adjoin negative integers and non-integral rationals. The system of reals is obtained by adjoining to the rationals all the non-rational limits of rational numbers. For example, the irrational number $\sqrt{2}$ is obtained as the limit of an increasing (or monotone) sequence x_n with $x_n^2 < 2$ and x_n rational. One might use a decimal sequence such as 1, 1.4, 1.41, 1.414, ... It is a well-known fact first proven by Euclid that $\sqrt{2}$ is not rational (see Exercise 2 at the end of this chapter).

Now the question becomes, how do we carry out the above program in a formal manner? Actually, the process is a little long but not difficult, so we shall just provide an outline here. The first thing to do is to isolate the important characteristics which we want the reals to possess. These are as follows:

(I) *Addition axioms.* There is an addition operation "$+$" such that for all numbers x, y, z, we have

   (i) $x + y = y + x$ (commutativity)
   (ii) $x + (y + z) = (x + y) + z$ (associativity)
   (iii) there is a number 0 such that $x + 0 = x$ (existence of zero)
   (iv) for each x there is a number w denoted $-x$ such that $x + w = 0$ (existence of additive inverses).

(II) *Multiplication axioms.* There is a multiplication operation "$\cdot$" such that

   (i) $x \cdot y = y \cdot x$ (commutativity)
   (ii) $x \cdot (y \cdot z) = (x \cdot y) \cdot z$ (associativity)
   (iii) there is a number $1 \neq 0$ such that $1 \cdot x = x$ (existence of unity)
   (iv) for each $x \neq 0$ there exists a number v such that $x \cdot v = 1$ (existence of reciprocals), one writes $v = x^{-1}$ and $yx^{-1} = y/x$
   (v) $x \cdot (y + z) = x \cdot y + x \cdot z$ (distributive law).

Any set or "number system" with operations $+$ and $\cdot$ obeying these rules is called a *field*. For example, the rationals are a field but the integers are not. From now on, we will just write xy for $x \cdot y$.

(III) *Order axioms.* There is an ordering "$\leqslant$" (more precisely, a relation) such that

   (i) if $x \leqslant y$ and $y \leqslant z$, then $x \leqslant z$ (transitivity)
   (ii) $(x \leqslant y$ and $y \leqslant x) \Leftrightarrow (x = y)$ (reflexivity)
   (iii) for any two elements x, y, either $x \leqslant y$ or $y \leqslant x$ (trichotomy)
   (iv) if $x \leqslant y$, then $x + z \leqslant y + z$
   (v) $0 \leqslant x$ and $0 \leqslant y$ implies $0 \leqslant xy$.

A system obeying characteristics (I), (II), and (III) is called an *ordered field*. By definition, $x < y$ shall mean $x \leqslant y$ and $x \neq y$. Other familiar symbols may also be introduced. For example, the *magnitude* of a number x is $|x|$, defined to be x if $x \geqslant 0$ and $-x$ if $x \leqslant 0$. The *distance* between x and y is $|x - y|$. The magnitude obeys the *triangle inequality:* $|x + y| \leqslant |x| + |y|$ as verified in Example 1 at the end of the chapter.

From these axioms follow all the usual manipulative rules that we have lived with since high school. For example one can use the axioms to *prove* that $0 < 1$ (see Example 4 at the end of the chapter). The full details of the above axioms are not important for us to work out at this time and we shall just accept as valid without proof the usual rules of algebra with which we are familiar.

Now, it should be obvious that these axioms cannot be enough to uniquely characterize the reals because the rationals also obey these axioms. Thus we require another condition to ensure that limits of rationals are included in the system.

In order to state this condition, a few additional definitions concerning sequences are needed. Let x_n be a given sequence of numbers. We say x_n *converges to* x if for any number $\varepsilon > 0$ there is an integer N such that $|x_n - x| < \varepsilon$ for all integers $n \geqslant N$. This is written as $\lim_{n \to \infty} x_n = x$ or $x_n \to x$ as $n \to \infty$.

The student has probably encountered convergence of sequences before; intuitively it means that x_n becomes arbitrarily close to x as n gets sufficiently large. Later in Chapter 2 we shall study convergence systematically. For now, it is just used to study the following completeness axiom.

The sequence x_n is *increasing* (or *non-decreasing*) if $x_n \leqslant x_{n+1}$ for all n. A sequence x_n is *bounded* if there is a number M such that $|x_n| \leqslant M$ for all $n = 1, 2, 3, \ldots$.

It is not hard to see that a sequence x_n can converge to, at most, one point. Indeed suppose x_n converges to both x and y. Then $|x - y| = |x - x_n + x_n - y| \leqslant |x - x_n| + |x_n - y|$ by the triangle inequality. If $|x - y| > 0$ then using $|x - y|/2$ as our ε, we can choose N so large that $|x - x_n| < |x - y|/2$ and $|x_n - y| < |x - y|/2$ if $n \geqslant N$. Thus we would conclude $|x - y| < |x - y|$ which cannot be. Hence $|x - y| = 0$ and so $x = y$.

We now state the completeness axiom.

(IV) *Completeness axiom.* If x_n is an increasing sequence which is bounded above, then x_n converges to some number x.

The plausibility of condition (IV) is seen by considering the increasing sequence of decimal approximations: $1, 1.4, 1.41, 1.414, \ldots$, which converge to $\sqrt{2}$.

A number system satisfying axioms (I) through (IV) is called a *complete ordered field*. Condition (IV) is equivalent to the condition that a decreasing sequence bounded below converges. We see this by noting that $(x_n \to x) \Leftrightarrow (-x_n \to -x)$ (see Exercise 18 at the end of the chapter). We are now ready for the statement which coordinates the previous discussion.

Theorem 1. *There is a "unique" number system called the real number system which is a complete ordered field.*

The real number system is denoted $\mathbb{R}$. For the moment, $\pm\infty$ are *not* included in $\mathbb{R}$. In Theorem 1, "uniqueness" means that any two systems satisfying (I)–(IV) can be put into a one-to-one correspondence which is compatible with $+$, $\cdot$, and $\leqslant$. By compatibility with $+$, for example, we mean that the number in the second system corresponding to the sum of the two numbers from the first system is the sum of the corresponding two numbers in the second system. We omit the proof of Theorem 1,* and rather

---

* The interested reader can find a proof outlined in, for example, L. J. Goldstein, *Abstract Algebra*, Prentice-Hall (1973), Chapter IV.

use it as our starting point. The proof is not difficult but is slightly laborious. Existence of ℝ can be done by verifying that the usual decimal expansions have the required properties.

As mentioned above, we do not wish to take too much time to work out all the detailed consequences of the axioms. However one of the "obvious" consequences deserves special mention. Namely, the *Archimedian property:* given any real number x there is an integer N such that $N > x$. (Here the integers may be defined by $2 = 1 + 1$, $3 = 2 + 1$, $4 = 3 + 1$, ...) It is curious to note that this result depends on the completeness axiom and cannot be deduced from the other axioms alone. The reader is asked to prove the Archimedian property in Exercise 30 at the end of the chapter.

The completeness axiom can be put into several other very important equivalent forms. In order to state these, we shall need some further basic terminology.

Definition 1. Let $S \subset \mathbb{R}$ be a subset of ℝ. Thus S is just some collection of real numbers (for example, all the rationals between 0 and 1). A number b is called an *upper bound* for S if for all $x \in S$, we have $x \leqslant b$.

A number b is called a *least upper bound* of S if *first*, b is an upper bound of S and *second*, b is less than or equal to every other upper bound of S. See Figure 1-1.

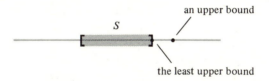

FIGURE 1-1   Least upper bound.

The set $]a,b[= (a,b) = \{x \in \mathbb{R} \mid a < x < b\}$ is called an *open interval* and $[a,b] = \{x \in \mathbb{R} \mid a \leqslant x \leqslant b\}$ is called a *closed interval*.

For example, the closed interval $[0,1]$, the open interval $]0,1[$, and all the rationals less than 1 all have a least upper bound of 1.

Note: The least upper bound of S (also called the *supremum* of S) is denoted sup(S) or lub(S).

There can be at most one least upper bound for S. Indeed, if b and b' are both least upper bounds and since b is less than or equal to every other upper bound, $b \leqslant b'$ and similarly $b' \leqslant b$, so we conclude that $b = b'$. Thus, we may speak of *the* least upper bound.

A set need not have any upper bound. For example, the whole real number system has no upper bound, and the positive integers have no upper bound. In the "degenerate" case of the empty set $\varnothing$, we regard any number as an upper bound.

Observe that if b is an upper bound for the set S and $b \in S$, then b is the least upper bound. The proof of this is very simple. It must be shown that if d is any upper bound for S, then $b \leqslant d$. But $b \in S$ and d is an upper bound, so $b \leqslant d$ as required.

A useful alternative to the definition of least upper bound is stated in Theorem 2 and is sometimes easier to apply.

> **Theorem 2.**   Let $S \subset \mathbb{R}$. Then $b \in \mathbb{R}$ is the least upper bound of S iff b is an upper bound and for every $\varepsilon > 0$ there is an $x \in S$ such that $x > b - \varepsilon$.

The proof is found at the end of this chapter. But the theorem should be pretty obvious because b sits just at the "top" (that is, to the "right") of the set S and there are no "gaps" between it and the set S, so for any $\varepsilon > 0$ we can take x just below b within a distance ε. [*Warning:* This sort of argument is a plausibility argument intended to give you a feel for the statement—do not confuse it with a rigorous proof.]

If S is not bounded above (has no upper bound), we shall say that $\sup(S)$ is infinite and write $\sup(S) = +\infty$. Similarly, a *lower bound* for a set S is a number b such that $b \leqslant x$ for all $x \in S$. Also, b is called a *greatest lower bound* iff it is a lower bound and for any lower bound c of S, $c \leqslant b$. As with least upper bounds, greatest lower bounds are unique if they exist. The greatest lower bound is sometimes called the *infimum* and is denoted $\inf(S)$ or $\text{glb}(S)$. As in Theorem 2, a number c is the greatest lower bound for a set S iff c is a lower bound and for every $\varepsilon > 0$ there is an $x \in S$ such that $x < c + \varepsilon$. Also, if S is not bounded below, we write $\inf(S) = -\infty$.

Another notion we need is that of a Cauchy sequence.

> **Definition 2.**   A sequence x_n in $\mathbb{R}$ is called a *Cauchy sequence* if for every number $\varepsilon > 0$ there is an integer N (depending on ε), such that $|x_n - x_m| < \varepsilon$ whenever $n \geqslant N$ and $m \geqslant N$.

This condition means intuitively that the sequence "bunches up"; that is, all the elements of the sequence are arbitrarily close to one another sufficiently far out in the sequence.

If it is true that x_n converges to x, then x_n is a Cauchy sequence. Indeed, given $\varepsilon > 0$ choose N so that $|x_n - x| < \varepsilon/2$ if $n \geqslant N$. Then, for $n, m \geqslant N$, we have $|x_n - x_m| = |x_n - x + x - x_m| \leqslant |x_n - x| + |x - x_m| < \varepsilon/2 + \varepsilon/2 = \varepsilon$, which proves our assertion. The converse of this statement appears in Theorem 3. Here we have used the triangle inequality $|y + z| \leqslant |y| + |z|$. The special case $|a - b| \leqslant |a - c| + |c - b|$ is very useful, as in the above instance. The next theorem gives some basic properties of real numbers.

Theorem 3.

(i) *Let S be a non-empty set in* ℝ *which has an upper bound. Then S has a least upper bound in* ℝ.

(ii) *Let P be a non-empty set in* ℝ *which has a lower bound. Then P has a greatest lower bound in* ℝ.

(iii) *Every Cauchy sequence x_n in* ℝ *converges to a number x in* ℝ.

This result should also be fairly apparent. Indeed, if a bounded subset of ℝ had no least upper bound, there would be a "hole" at the top of the set and a sequence of members of S increasing toward that hole would not converge to an element in ℝ. Similarly, we must have (ii). Condition (iii) is seen as follows: If we ignore the first N terms of a Cauchy sequence, we know that the remaining terms will be bunched together. As we disregard more and more terms, the remainder of the sequence becomes more tightly grouped and squeezes down to some limiting number, the limit of the sequence. To see more precisely how this is done requires more care and so the actual proof is our only recourse.

Using the methods of the proof we give, it is not too difficult to show that conditions (i), (ii), (iii) are each equivalent to the completeness axiom for an ordered field.

This concludes our brief discussion and review of the real line. Further properties and practice are found in the worked examples which follow and at the end of the chapter.

EXAMPLE 1.   Let $S = \{x \in \mathbb{R} \mid x^2 + x < 3\}$. Find sup(S), inf(S).

Solution: Consider the graph of $y = x^2 + x$ (Figure 1-2). From elementary calculus we see that for $x = -1/2$, y is a minimum. Thus S may be pictured as shown in Figure 1-2. The sup and inf clearly occur when

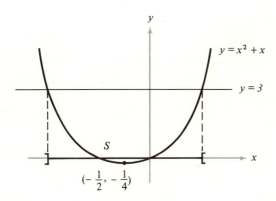

FIGURE 1-2

$x^2 + x = 3$, or from the quadratic formula when

$$x = \frac{-1 \pm \sqrt{1 + 12}}{2} = \frac{(-1 \pm \sqrt{13})}{2}$$

Thus

$$\sup(S) = \frac{(\sqrt{13} - 1)}{2}, \qquad \inf(S) = \frac{-(\sqrt{13} + 1)}{2}.$$

EXAMPLE 2.   Let $x_0 = 0, x_1 = \sqrt{2}, x_2 = \sqrt{2 + x_1}, \ldots, x_n = \sqrt{2 + x_{n-1}}, \ldots$ Show that x_n converges.

Solution: We shall show that x_n is increasing and bounded above and this will prove the assertion. Note that each x_n is non-negative. First, then, we must show $r_n = x_{n+1} - x_n \geq 0$. Let us do this by induction. Clearly, it holds for $n = 0$. Suppose it is true for $n - 1$; then

$$r_n = x_{n+1} - x_n = \sqrt{2 + x_n} - \sqrt{2 + x_{n-1}} = \frac{x_n - x_{n-1}}{\sqrt{2 + x_n} + \sqrt{2 + x_{n-1}}}$$

$$= \frac{r_{n-1}}{(\sqrt{2 + x_n} + \sqrt{2 + x_{n-1}})},$$

so $r_{n-1} \geq 0$ implies $r_n \geq 0$ and therefore x_n is increasing. Now we want to show that x_n is bounded above. For example, one can prove by induction that $x_n \leq 5$. Clearly, $x_0, x_1 \leq 5$. Suppose $x_{n-1} \leq 5$. Then

$$x_n = \sqrt{2 + x_{n-1}} \leq \sqrt{2 + 5} \leq \sqrt{7} \leq 5,$$

and therefore x_n is increasing and bounded above, so it converges.

EXAMPLE 3.   Let x_n be a sequence of real numbers such that $|x_n - x_{n+1}| \leq 1/2^n$. Show that x_n converges.

Solution: We shall show that x_n is a Cauchy sequence and the result then will follow from Theorem 3 (iii). We can write by the triangle inequality,

$$|x_n - x_{n+k}| \leq |x_n - x_{n+1}| + |x_{n+1} - x_{n+2}| + \cdots + |x_{n+k-1} - x_{n+k}|$$

$$\leq \frac{1}{2^n} + \frac{1}{2^{n+1}} + \cdots + \frac{1}{2^{n+k}}$$

$$\leq \frac{2}{2^n}.$$

(since $a + ar + ar^2 + \cdots = a/(1 - r)$ if $0 < r < 1$).
Thus $|x_n - x_m| \leq 1/2^{n-1}$ if $m \geq n$, and given $\varepsilon > 0$, just choose N so that $1/2^{N-1} < \varepsilon$. Hence we get a Cauchy sequence.

Example 4 is inserted to back up our claim that the usual rules of algebra all follow from the axioms. In the exercises, conclusions like these may be taken for granted.

EXAMPLE 4.   Use the axioms for an ordered field to prove
(a) Negatives are unique;
(b) $0x = 0$ for all x;
(c) $(-x)(-y) = xy$;
(d) $0 < 1$.

Solution: For (a), we note that if $x + w = 0$ and $x + y = 0$, then (adding y to $x + w = 0$), $y + (x + w) = y + 0 = y$. By condition I(ii) the left side is $(y + x) + w = 0 + w = w$, so $y = w$. Thus the symbol $-x$ is unambiguous.

For (b), we have $0 + 0 = 0$ and so by II(i) and II(v) we obtain $0 \cdot x = (0 + 0)x = 0 \cdot x + 0 \cdot x$. Adding $-(0 \cdot x)$ to each side gives $0 \cdot x = 0$.

For (c) we first claim $(-x)y = -(xy)$. Indeed, by using II(i) and II(v), $(-x)y + xy = (-x + x)y = 0 \cdot y = 0$ by (b). Next, $(-1)(-1) = 1$ for $(1 - 1)(-1) = 0 \cdot (-1) = 0$ and the left side is $(1)(-1) + (-1)(-1) = -1 + (-1)(-1)$, and by adding 1 to each side we get $(-1)(-1) = 1$. Then since we have proved $(-1)(x) = -(1x) = -x$, we get

$$(-x)(-y) = (-1)x(-1)y = (-1)(-1)xy = 1xy = xy \,.$$

Finally, for (d), by III(iii) the only other possibility for $0 < 1$ is $1 \leqslant 0$. Adding -1 gives $0 \leqslant -1$ (using III(iv)). Then using $x = -1$, $y = -1$ in III(v) gives $0 \leqslant 1$ since $(-1)(-1) = 1$. Hence we must have $0 < 1$, since $0 \neq 1$.

EXAMPLE 5.   Prove that $1/n \to 0$ as $n \to \infty$.

Solution: According to the definition, given any number $\varepsilon > 0$, we must prove that there is an integer N such that if $n \geqslant N$ then $|1/n - 0| < \varepsilon$. It will be so provided that $1/N < \varepsilon$, so it is only necessary to choose $N > 1/\varepsilon$, which is possible by the Archimedian property.

EXAMPLE 6.   Show that $\dfrac{\sqrt{n^2 + 1}}{n!} \to 0$ as $n \to \infty$.

Solution: We must show that $\dfrac{\sqrt{n^2 + 1}}{n!}$ gets small as n gets large.

We can estimate how big $\dfrac{\sqrt{n^2 + 1}}{n!}$ is as follows:

$$\frac{\sqrt{n^2 + 1}}{n!} \leqslant \frac{\sqrt{2n^2}}{n!} = \frac{\sqrt{2}n}{n!} = \frac{\sqrt{2}}{(n - 1)!} \leqslant \frac{\sqrt{2}}{n - 1} \,.$$

Thus given $\varepsilon > 0$ choose N such that $N > \dfrac{\sqrt{2}}{\varepsilon} + 1$. Then $n \geq N$ implies

$$0 \leq \frac{\sqrt{n^2 + 1}}{n!} \leq \frac{\sqrt{2}}{n - 1} \leq \frac{\sqrt{2}}{N - 1} < \varepsilon.$$ This proves the assertion.

Exercises for Section 1.1

1. Let $S = \{x \mid x^3 < 1\}$. Find $\sup(S)$. Is S bounded below?

2. In Example 2, let $\lambda = \lim\limits_{n \to \infty} x_n$. Argue that $\lambda = \sqrt{2 + \lambda}$, that is, that λ is a root of $\lambda^2 - \lambda - 2 = 0$. Find $\lim\limits_{n \to \infty} x_n$.

3. Show that $3^n/n!$ converges to 0.

4. Consider an increasing sequence x_n bounded above and converging to x. Let $S = \{x_n \mid n = 1,2,3,\ldots\}$. Argue that $x = \sup(S)$.

5. Let $x_n = \sqrt{n^2 + 1} - n$. Compute $\lim\limits_{n \to \infty} x_n$.

6. Let x_n be a sequence such that $|x_n - x_{n+1}| \leq 1/n$. Do you think x_n has to converge?

7. If $P \subset Q \subset \mathbb{R}$ and P and Q are bounded above, show that $\sup(P) \leq \sup(Q)$.

1.2   Euclidean *n*-Space $\mathbb{R}^n$

Throughout this book we shall be working with one-, two-, or three-dimensional Euclidean space. However, in many important applications, higher dimensional spaces arise as well. Therefore, it is important to treat the general case, but we usually fall back on the case of one-, two-, or three-space for visualization and intuition.

Let us begin with a formal definition.

> **Definition 3.**   Euclidean *n*-space consists of all ordered *n*-tuples of real numbers and is denoted $\mathbb{R}^n$. Symbolically,
>
> $$\mathbb{R}^n = \{(x_1, \ldots, x_n) \mid x_1, \ldots, x_n \in \mathbb{R}\}.$$

Thus $\mathbb{R}^n$ is the cartesian product of $\mathbb{R}$ by itself n times, and can be written $\mathbb{R}^n = \mathbb{R} \times \cdots \times \mathbb{R}$.

Elements of $\mathbb{R}^n$ are generally denoted by single letters which stand for *n*-tuples $x = (x_1, \ldots, x_n)$, and we speak of x as a *point* in $\mathbb{R}^n$.

Addition and scalar multiplication are defined in the usual way:

$$(x_1, \ldots, x_n) + (y_1, \ldots, y_n) = (x_1 + y_1, \ldots, x_n + y_n)$$

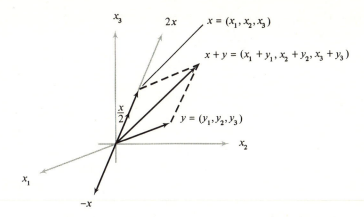

FIGURE 1-3   Addition and scalar multiplication.

and

$$\alpha(x_1,\dots,x_n) = (\alpha x_1,\dots,\alpha x_n) \qquad \text{for } \alpha \in \mathbb{R}.$$

The geometric meaning of these operations are reviewed in Figure 1-3 in the case of three-space, $n = 3$.

For the next theorem the reader should recall the definition of a vector space.

Theorem 4. *Euclidean n-space with the operations of addition and scalar multiplication previously defined is a vector space of dimension n.*

The proof is a straightforward check of the axioms for a vector space, which we shall leave for the student in Exercise 16, p. 30. This theorem should be no surprise. After all, a vector space is an abstraction of the basic properties of vectors in euclidean space. We can show that $\mathbb{R}^n$ has dimension n by exhibiting a basis with n vectors, for example, the *standard basis* $\{e_1 = (1,0,\dots,0), e_2 = (0,1,0,\dots,0), \dots, e_n = (0,0,\dots,0,1)\}$.

In the standard basis, the components of $x = (x_1,\dots,x_n)$ are just $x_1,\dots,x_n$. In another basis for $\mathbb{R}^n$, the components would be different. This means that if $e_1,\dots,e_n$ denotes the standard basis, $x = \sum_{i=1}^n x_i e_i$, but if $f_1,\dots,f_n$ is another basis, $x = \sum_{i=1}^n y_i f_i$ for possibly different numbers $y_1,\dots,y_n$.

Following are some fundamental operations in $\mathbb{R}^n$.

Definition 4.   The *length* or *norm* of a vector x in $\mathbb{R}^n$ is defined by

$$\|x\| = \left(\sum_{i=1}^n x_i^2\right)^{1/2},$$

where $x = (x_1, \ldots, x_n)$. The *distance* between two vectors x and y is the real number defined by

$$d(x,y) = \|x - y\| = \left\{ \sum_{i=1}^{n} (x_i - y_i)^2 \right\}^{1/2}.$$

The *inner product* of x and y is defined by

$$\langle x,y \rangle = \sum_{i=1}^{n} x_i y_i.$$

Thus we have $\|x\|^2 = \langle x,y \rangle$. In $\mathbb{R}^3$, the reader is familiar with another expression for $\langle x,y \rangle$, namely, $\langle x,y \rangle = \|x\| \, \|y\| \cos \theta$, where $\cos \theta$ is the cosine of the angle formed by x and y. See Figure 1-4.

Now let us summarize the basic properties of these operations:

Theorem 5.   *For vectors in* $\mathbb{R}^n$, *we have*

(*I*)   *Properties of the inner product*

   (*i*) $\langle x, y_1 + y_2 \rangle = \langle x, y_1 \rangle + \langle x, y_2 \rangle$

   (*ii*) $\langle x, \alpha y \rangle = \alpha \langle x,y \rangle$ *for* α *real*

   (*iii*) $\langle x,y \rangle = \langle y,x \rangle$

   (*iv*) $\langle x,x \rangle \geqslant 0$ *and* $\langle x,x \rangle = 0$ *iff* $x = 0$

   (*v*) $|\langle x,y \rangle| \leqslant \|x\| \, \|y\|$ *(Cauchy-Schwarz inequality).*
   Note: (*v*) *follows from* (*i*)–(*iv*).

(*II*)   *Properties of the norm*

   (*i*) $\|x\| \geqslant 0$

   (*ii*) $\|x\| = 0$ *iff* $x = 0$

   (*iii*) $\|\alpha x\| = |\alpha| \, \|x\|$ *for real* α

   (*iv*) $\|x + y\| \leqslant \|x\| + \|y\|$ *(triangle inequality).*

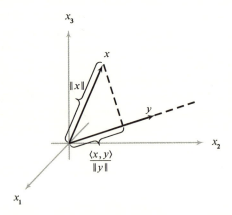

FIGURE 1-4   Length and inner product.

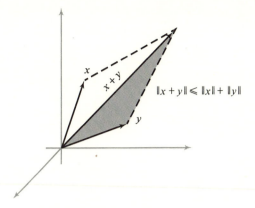

$\|x + y\| \leqslant \|x\| + \|y\|$

FIGURE 1-5   Triangle inequality.

(*III*)   *Properties of the distance*
   (i) $d(x,y) = d(y,x)$
   (ii) $d(x,y) \geqslant 0$
   (iii) $d(x,y) = 0$ *iff* $x = y$
   (iv) $d(x,y) \leqslant d(x,z) + d(z,y)$ (*also called the triangle inequality*).

Each of these properties should be pretty obvious geometrically. For example, (iv) in (II) and (III) just expresses the fact that the length of one side of a triangle is less than or equal to the sum of the lengths of the other sides (Figure 1-5).

A set with a function d obeying rules (III) is called a *metric space*. A vector space with a norm obeying rules (II) is called a *normed space*, and a vector space with an inner product obeying rules (I) is an *inner product* space. As we shall see in the proof, each of these sets (II) and (III) of properties follows from the set of properties above it.*

The reader will recall from linear algebra the notion of a linear subspace. In particular, an $(n - 1)$-dimensional linear subspace of $\mathbb{R}^n$ is called a *hyperplane*. An *affine hyperplane* is a set $x + H$, where H is a hyperplane and $x \in \mathbb{R}^n$; $x + H$ means the set of all $x + y$ as y ranges through H; thus $x + H = \{x + y \mid y \in H\}$. See Figure 1-6.

Finally, generalizing the concepts from $\mathbb{R}^3$, we call $x, y \in \mathbb{R}^n$ *orthogonal* iff $\langle x, y \rangle = 0$. Two subspaces S and T are *orthogonal* iff $\langle x, y \rangle = 0$ for all $x \in S$ and $y \in T$. Furthermore, if in addition, S and T span $\mathbb{R}^n$, they are

---

* The famous inequality of I(v) should, for historical reasons, be called the Cauchy-Bunyakowski-Schwarz inequality, although it is not uncommon to omit the Russian name in English writings and to omit Schwarz's name in Russian works.

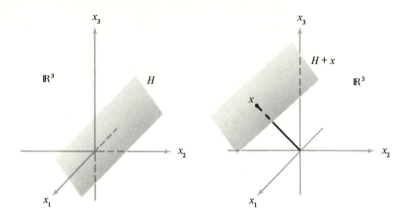

FIGURE 1-6   Hyperplane and affine hyperplane.

called *orthogonal complements*. This will occur iff S and T are orthogonal and the sum of their dimensions equals n (Exercise 20). We define $S^{\perp} = \{ y \in \mathbb{R}^n \mid \langle x,y \rangle = 0 \text{ for all } x \in S \}$. Then it is not difficult to see that S and $S^{\perp}$ are orthogonal complements. We shall not require too much of this linear algebra of $\mathbb{R}^n$ in our work in addition to these basic concepts, so further discussion is not necessary here.

EXAMPLE 1.   Find the length of the line segment joining $(1,1,1)$ to $(3,2,0)$.

Solution: This length is the length of the vector $(3,2,0) - (1,1,1) = (2,1,-1)$ which represents the vector from $(1,1,1)$ to $(3,2,0)$. The length is

$$\|(2,1,-1)\| = \sqrt{2^2 + 1^2 + (-1)^2} = \sqrt{6} .$$

EXAMPLE 2.   In $\mathbb{R}^3$, find the orthogonal complement of the line $x = y = z/2$ (or $x_1 = x_2 = x_3/2$ in different notation).

Solution: This line, call it l, is the one-dimensional subspace spanned by the vector $(1,1,2)$ (see Figure 1-7). The orthogonal complement is a plane (through the origin since it is a subspace) and so has an equation of the form

$$Ax + By + Cz = 0$$

that is,

$$\langle (A,B,C),(x,y,z) \rangle = 0 ,$$

that is, (A,B,C) is normal to the plane; but $(1,1,2)$ is a vector perpendicular to the plane so the orthogonal complement sought is the plane

$$x + y + 2z = 0 .$$

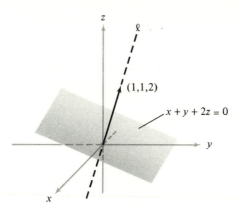

FIGURE 1-7

Exercises for Section 1.2

1. If $\|x + y\| = \|x\| + \|y\|$, argue geometrically that x and y should lie on some line through the origin.

2. What is the angle between $(3,2,2)$ and $(0,1,0)$?

3. Find the orthogonal complement of the plane spanned by $(3,2,2)$ and $(0,1,0)$ in $\mathbb{R}^3$.

4. Describe the sets $B = \{x \in \mathbb{R}^3 \mid \|x\| \leqslant 3\}$ and $Q = \{x \in \mathbb{R}^3 \mid \|x\| < 3\}$.

5. Find the equation of the line through $(1,1,1)$ and $(2,3,4)$. Is it a linear subspace?

Theorem Proofs for Chapter 1

Theorem 2. *Let $S \subset \mathbb{R}$. Then $b \in \mathbb{R}$ is the least upper bound of S iff b is an upper bound and for every $\varepsilon > 0$ there is an $x \in S$ such that $x > b - \varepsilon$.*

Proof: First, suppose $b = \text{lub}(S) = \sup(S)$ and $\varepsilon > 0$. We must produce an $x \in S$, such that $b < x + \varepsilon$. If there were no such x, we would have $b \geqslant x + \varepsilon$ for every $x \in S$, that is, $b - \varepsilon \geqslant x$. Thus $b - \varepsilon$ is an upper bound strictly less than b and therefore b is not the least upper bound, which contradicts our hypothesis.

Conversely, suppose b satisfies the given condition. Let d be an upper bound of S. According to the definition of $\sup(S)$, we must show that $b \leqslant d$. Suppose in fact, $b > d$. Let $\varepsilon = b - d$. Then $d = b - \varepsilon$ and $d \geqslant x$ for all $x \in S$ implies $b - \varepsilon \geqslant x$ or $b \geqslant x + \varepsilon$, and so our condition fails. Thus the supposition that $b > d$ is wrong, and we may then conclude that $b \leqslant d$ as required. This completes the argument. ∎

Note: In this proof, we found it convenient to use the following basic principle of logic: showing that a statement P implies a statement Q (in symbols $P \Rightarrow Q$) is equivalent to showing $\sim Q \Rightarrow \sim P$ where $\sim Q$ is the negation of Q. We call $\sim Q \Rightarrow \sim P$ the *contrapositive* of $P \Rightarrow Q$, whereas $Q \Rightarrow P$ is the *converse*.

Theorem 3.

(i) *Let S be a non-empty set in $\mathbb{R}$ which has an upper bound. Then S has a least upper bound in $\mathbb{R}$.*

(ii) *Let P be a non-empty set in $\mathbb{R}$ which has a lower bound. Then P has a greatest lower bound in $\mathbb{R}$.*

(iii) *Every Cauchy sequence x_n in $\mathbb{R}$ converges to a number x in $\mathbb{R}$.*

*Proof.** (i) Since $S \neq \varnothing$ we can choose some $x_0 \in S$. Let us write $y \geqslant S$ if y is an upper bound of S. Now pick the smallest integer N such that $N \geqslant 1$ and $x_0 + N \geqslant S$. Such an integer exists because S is bounded above. Let $x_1 = x_0 + N - 1$. Thus $x_1 \geqslant x_0$ and there are elements of S greater than x_1 but none greater than $x_1 + 1$. Similarly choose the smallest integer $N_1 \geqslant 1$ such that $x_1 + N_1/2 \geqslant S$ and let $x_2 = x_1 + (N_1 - 1)/2$. If the reader will draw a picture of x_1 and x_2 everything should become clear. Note from your picture that N_1 is either 1 or 2. Inductively define $x_n = x_{n-1} + (N_{n-1} - 1)/2n$ where N_{n-1} is the smallest integer such that $x_{n-1} + N_{n-1}/n \geqslant S$; thus N_{n-1} is $1, 2, \ldots$ or n. Thus there are elements of $S \geqslant x_n$ and no elements of S are $> x_n + 1/n$. Furthermore $x_0 \leqslant x_1 \leqslant x_2 \leqslant \cdots$ so that x_n is an increasing sequence bounded above.

Now we can apply the completeness property of $\mathbb{R}$ to deduce that $x_n \to y$ for some $y \in \mathbb{R}$. We shall show that y is the least upper bound for S. First, let us demonstrate that it is an upper bound. Suppose that $x \in S$ and $x > y$. Select n so that $0 < 1/n \leqslant x - y$, which is possible, since $1/n \to 0$ as $n \to \infty$. Thus, x is an element of S greater than $x_n + 1/n$, which cannot happen by the way we chose x_n above. So $x \leqslant y$, and y is an upper bound. By Theorem 2, it remains to prove that for any given $\varepsilon > 0$ there is an $x \in S$ so that $y < x + \varepsilon$. Choose n such that $y < x_n + \varepsilon$, which is possible as $x_n \to y$. By construction, there is an $x \in S$, with $x \geqslant x_n$. Thus $y < x_n + \varepsilon \leqslant x + \varepsilon$, and the proof of (i) is complete.

(ii) Consider the set $-P = \{-x \mid x \in P\}$. By (i), $-P$ has a least upper bound $c \in \mathbb{R}$ ($-P$ is bounded above because P is bounded below). Also, one easily sees from the definition that $-c$ is the greatest lower bound required. (See Exercise 17 for another proof.)

(iii) Since the completeness axiom implies (i) and (ii), as we just demonstrated, we can make use of them to prove (iii). Thus let x_n be a Cauchy sequence in $\mathbb{R}$. For any integer $M \geqslant 1$, consider the set

$$\{x_M, x_{M+1}, x_{M+2}, \ldots\}$$

(the "tail" of the sequence).

First, we show that this set is bounded above and below. Choose $\varepsilon = 1$. There is an N so that $n, m \geqslant N$ implies $|x_n - x_m| < 1$. Thus all members x_m are a distance $\leqslant 1$ from x_N for $m \geqslant N$. Since this omits only a finite number of terms $(x_1, x_2, \ldots, x_N)$, we obtain our result (drawing a picture may help here).

Now, from what we showed in (i), $\sup\{x_M, x_{M+1}, \ldots\}$ exists; call it A_M. This sequence $\{A_M, A_{M+1}, \ldots\}$ is a decreasing sequence bounded below; $A_{M+1} \leqslant A_M$ since A_{M+1} is the sup of the set $\{x_{M+1}, x_{M+2}, \ldots\} \subset \{x_M, x_{M+1}, \ldots\}$; see Exercise 7, p. 18. Thus A_M converges to a point, say $a \in \mathbb{R}$. We shall prove that $x_n \to a$ as well.

* This proof is a little difficult on first reading, and requires some time and experience to master. If it is not clear now, come back to it after completing Chapter 2.

Given $\varepsilon > 0$, we can choose N_1 so that $0 \leqslant A_n - a < \varepsilon/3$ for $n \geqslant N_1$ since $A_n \to a$. Because x_n is a Cauchy sequence, there is N_2 such that $m,n \geqslant N_2$ implies $|x_m - x_n| < \varepsilon/3$. Because $A_n = \sup\{x_n, x_{n+1}, \ldots\}$, there is, by Theorem 2, an N_4 so that $0 \leqslant A_{N_3} - x_{N_4} < \varepsilon/3$, where N_3 is the maximum of N_1 and N_2. If N is the largest of N_3, N_4, and $n \geqslant N$, we have

$$|x_n - a| \leqslant |x_n - x_{N_4}| + |A_{N_3} - x_{N_4}| + |A_{N_3} - a| < \varepsilon/3 + \varepsilon/3 + \varepsilon/3 = \varepsilon,$$

which proves the assertion. ∎

We remarked earlier that (i), (ii), and (iii) are each equivalent to the completeness axiom for an ordered field. We have shown one-half of the implication, namely, that the completeness axiom implies (i), (ii), and (iii) for an ordered field. Exercise 11 will outline the proof that (i), (ii), and (iii) each imply the completeness axiom.

Theorem 5. *For vectors in $\mathbb{R}^n$, we have*

(I) *Properties of the inner product*
   (i) $\langle x, y_1 + y_2 \rangle = \langle x, y_1 \rangle + \langle x, y_2 \rangle$
   (ii) $\langle x, \alpha y \rangle = \alpha \langle x, y \rangle$ *for α real*
   (iii) $\langle x, y \rangle = \langle y, x \rangle$
   (iv) $\langle x, x \rangle \geqslant 0$ *and* $\langle x, x \rangle = 0$ *iff $x = 0$*
   (v) $|\langle x, y \rangle| \leqslant \|x\| \, \|y\|$ *(Cauchy-Schwarz inequality).*

(II) *Properties of the norm*
   (i) $\|x\| \geqslant 0$
   (ii) $\|x\| = 0$ *iff $x = 0$*
   (iii) $\|\alpha x\| = |\alpha| \, \|x\|$ *for real α*
   (iv) $\|x + y\| \leqslant \|x\| + \|y\|$ *(triangle inequality).*

(III) *Properties of the distance*
   (i) $d(x,y) = d(y,x)$
   (ii) $d(x,y) \geqslant 0$
   (iii) $d(x,y) = 0$ *iff $x = y$*
   (iv) $d(x,y) \leqslant d(x,z) + d(z,y)$ *(also called the triangle inequality).*

Proof: (I) Properties (i) through (iv) are easily verified from the definition of $\langle , \rangle$. We shall deduce property (v) from (i)–(iv). Now, for any $\lambda \in \mathbb{R}$, $0 \leqslant \|\lambda x + y\|^2 = \langle \lambda x + y, \lambda x + y \rangle = \lambda^2 \langle x, x \rangle + 2\lambda \langle x, y \rangle + \langle y, y \rangle$. Considered as a polynomial in λ, we may locate its minimum at $\lambda = -\langle x, y \rangle / \|x\|^2$ (if $x = 0$, the assertion (v) reduces to $0 \leqslant 0$, so we can assume $x \neq 0$). Thus, in particular,

$$0 \leqslant \left(-\frac{\langle x, y \rangle}{\|x\|^2} \right)^2 \langle x, x \rangle + 2 \left(-\frac{\langle x, y \rangle}{\|x\|^2} \right) \langle x, y \rangle + \langle y, y \rangle$$

that is

$$0 \leqslant \left(-\frac{\langle x, y \rangle^2}{\|x\|^2} \right) + \|y\|^2,$$

and $\langle x, y \rangle^2 \leqslant \|x\|^2 \, \|y\|^2$. Taking the square roots gives the desired inequality since $\sqrt{\alpha^2} = |\alpha|$.

(II) (i) and (ii) follow directly from I(iv), and (iii) from I(ii). For (iv) we have, using I(v),

$$\|x + y\|^2 = \langle x + y, x + y \rangle = \langle x,x \rangle + 2\langle x,y \rangle + \langle y,y \rangle$$
$$\leqslant \|x\|^2 + 2|\langle x,y \rangle| + \|y\|^2$$
$$\leqslant \|x\|^2 + 2\|x\|\,\|y\| + \|y\|^2$$
$$= (\|x\| + \|y\|)^2$$

giving the required result.

For III, (i) holds, since $\|x - y\| = \|y - x\|$ using II(iii). Also, (ii) follows from II(i). For (iii) we use II(ii). Finally for (iv) we use III(iii) as follows

$$d(x,y) = \|x - y\| = \|(x - z) + (z - y)\|$$
$$\leqslant \|x - z\| + \|z - y\|$$
$$= d(x,z) + d(z,y) \, .$$

Notice how each succeeding set of properties is deduced from the previous set. ∎

Worked Examples for Chapter 1

1. For real numbers, prove that
   (i) $x \leqslant |x|, \; -|x| \leqslant x$
   (ii) $|x| \leqslant a \Leftrightarrow -a \leqslant x \leqslant a$, where $a \geqslant 0$
   (iii) $|x + y| \leqslant |x| + |y|$.
   Solution:
   (i) If $x \geqslant 0$, then $|x| = x$, while if $x < 0$, $|x| \geqslant x$, since $|x| \geqslant 0$. In any case, $x \leqslant |x|$. The other assertion is similar.
   (ii) If $x \geqslant 0$, then we must show that $0 \leqslant x \leqslant a \Leftrightarrow -a \leqslant x \leqslant a$ which is obvious. Similarly, if $x < 0$, the assertion becomes $(0 \leqslant -x \leqslant a) \Leftrightarrow (-a \leqslant x \leqslant a)$, which is again obvious. Here the fact is used that if $c \leqslant 0$, $(0 \leqslant x \leqslant y) \Leftrightarrow (0 \geqslant cx \geqslant cy)$.
   (iii) By (i), $-|x| \leqslant x \leqslant |x|$ and $-|y| \leqslant y \leqslant |y|$. Adding, we obtain $-(|x| + |y|) \leqslant x + y \leqslant |x| + |y|$. Then, by (ii), $|x + y| \leqslant |x| + |y|$. In addition, this can be proven by cases as we did (ii). Note that this is also a special case of Theorem 5, II(iv).

2. Let S be a set in $\mathbb{R}$ and $x = \sup(S)$. Show that there is a sequence $x_1, x_2, \ldots$ such that $x_k \to x$, and $x_k \in S$.
   Solution: For each k, use Theorem 2 to find an x_k such that $x_k \leqslant x < x_k + 1/k$. Then $x_k \to x$, since for a given $\varepsilon > 0$, we choose $N \geqslant 1/\varepsilon$; then $k \geqslant N$ implies $x_k \leqslant x < x_k + \varepsilon$ or $|x - x_k| < \varepsilon$.

3. For numbers $x_1, \ldots, x_n, y_1, \ldots, y_n$, and $z_1, \ldots, z_n$, show that

$$\left(\sum_{i=1}^{n} x_i y_i z_i \right)^4 \leqslant \left(\sum_{i=1}^{n} x_i^4 \right)\left(\sum_{i=1}^{n} y_i^2 \right)^2\left(\sum_{i=1}^{n} z_i^4 \right).$$

Solution: The CBS inequality (Theorem 5, I(v)) says that

$$\left(\sum w_i y_i\right)^2 \leqslant \left(\sum w_i^2\right)\left(\sum y_i^2\right).$$

Applying this to the numbers $w_i = x_i z_i$ and y_i gives

$$\left(\sum x_i y_i z_i\right)^2 \leqslant \left(\sum (x_i z_i)^2\right)\left(\sum y_i^2\right).$$

Applying this again to x_i^2, z_i^2 gives

$$\left(\sum x_i^2 z_i^2\right)^2 \leqslant \left(\sum x_i^4\right)\left(\sum z_i^4\right)$$

or

$$\left(\sum (x_i z_i)^2\right) \leqslant \left(\sum x_i^4\right)^{1/2}\left(\sum z_i^4\right)^{1/2}$$

and so

$$\left(\sum x_i y_i z_i\right)^2 \leqslant \left(\sum x_i^4\right)^{1/2}\left(\sum z_i^4\right)^{1/2}\left(\sum y_i^2\right).$$

Squaring both sides, the result is obtained. (We have used the fact that if $a, b \geqslant 0$, then $a \leqslant b$ iff $a^2 \leqslant b^2$.)

4. Suppose $x \in \mathbb{R}$ and $x > 0$; show that there is an irrational number between 0 and x.

Solution: If x is rational, then since $\sqrt{2}$ is irrational, so is $x/\sqrt{2}$ (why?) and is between 0 and x. On the other hand, if x is irrational, then $x/2$ is irrational (why?) and lies between 0 and x.

5. Recall that one may define e^x by $e^x = 1 + x + x^2/2! + x^3/3! + \cdots$. (By the ratio test, this series converges for all $x \in \mathbb{R}$. Hence this definition of e^x makes sense.) Show that $e = e^1$ is an irrational number.

Solution: Suppose that $e = a/b$ for integers a and b. Let k be an integer, $k > b$, and let $\alpha = k!\,(e - 1 - 1/2! - 1/3! - \cdots 1/k!)$ so that α is a non-zero integer as well. However, since $e = 1 + 1/2! + 1/3! + \cdots$, we have

$$\alpha = \frac{1}{k+1} + \frac{1}{(k+1)(k+2)} + \cdots$$

$$\leqslant \frac{1}{k+1} + \frac{1}{(k+1)^2} + \cdots$$

$$= \frac{1}{k}.$$

(The last equality follows using the geometric series $y + y^2 + \cdots = y/(1-y)$, $0 \leqslant y < 1$.) But $\alpha < 1/k$ is impossible if α is an integer $\neq 0$. Thus $e = a/b$ is also impossible, and so e is irrational.

Surprisingly, to prove that e^r is irrational for r rational is not at all simple, and the proof that π is irrational is even harder.*

---

* See for example G. H. Hardy and E. M. Wright, *An Introduction to the Theory of Numbers*, New York, Oxford University Press, Fourth edition, 1960. In fact, e and π are transcendental numbers, which means they are not the roots of any polynomial with rational coefficients. This was discovered by Hermite and Lindemann in 1873 and 1882. For an elementary account, see M. Spivak, *Calculus*, W. A. Benjamin Co.

6. Let A and B be sets in $\mathbb{R}$ bounded from above. Let $a = \sup(A)$, $b = \sup(B)$ and let the set C be defined by $C = \{xy \mid x \in A, y \in B\}$. Show that, in general, $ab \neq \sup(C)$. If $a < 0$ and $b < 0$, then prove that $ab = \inf(C)$. If $a > 0$ and $b > 0$, and A, B have only positive elements, then also prove that $ab = \sup(C)$.

Solution: As a specific instance, let $A = \{x \in \mathbb{R} \mid -10 < x < -1\} = \,]-10,-1[$ and $B = \,]0,1/2[$, so that $a = -1$, $b = 1/2$, and $ab = -1/2$. But $C = \,]-5,0[$ and $\sup(C) = 0$.

Now we prove that if $a < 0$ and $b < 0$, then $ab = \inf(C)$. For this, we use the analogue of Theorem 2 for greatest lower bounds. First, let $x \in A$ and $y \in B$. We want to show $xy \geqslant ab$. But, $x \leqslant a$, $y \leqslant b$ or $-x \geqslant -a \geqslant 0$ and $-y \geqslant -b \geqslant 0$, so (using Axiom III(v) for $\mathbb{R}$), $(-x)(-y) \geqslant (-a)(-b)$ or $xy \geqslant ab$. Given $\varepsilon > 0$, we want to find $x \in A$ and $y \in B$ so that $ab > xy - \varepsilon$, or $|ab - xy| < \varepsilon$. Choose x and y so that $a < x + \varepsilon/2(|b| + 1)$, $b < y + \varepsilon/2\,|a|$, and $b < y + 1$. Then, since $|uv| = |u|\,|v|$ and $|y| < |b| + 1$, we get (using the triangle inequality) $|ab - xy| \leqslant |ab - ay| + |ay - xy| = |a|\,|b - y| + |a - x|\,|y| < |a|\,(\varepsilon/2\,|a|) + (\varepsilon/2(|b| + 1))(|b| + 1) = \varepsilon$. The last assertion can be proven in an analogous way.

Exercises for Chapter 1

1. For each of the following sets S, find $\sup(S)$ and $\inf(S)$:
   (a) $\{x \in \mathbb{R} \mid x^2 < 5\}$
   (b) $\{x \in \mathbb{R} \mid x^2 > 7\}$
   (c) $\{1/n \mid n \text{ an integer}, n > 0\}$
   (d) $\{-1/n \mid n \text{ an integer}, n > 0\}$
   (e) $\{.3,.33,.333,\ldots\}$
   (f) the intervals $[a,b]$, $[a,b[$, $]a,b]$, or $]a,b[$.

2. Review the proof that $\sqrt{2}$ is irrational. [Hint: If there were a rational number m/n, where m and n have no common factor, such that $(m/n)^2 = 2$, would m be even or odd?] Generalize this to $\sqrt{k}$ for k a positive integer which is not a perfect square.

3. (a) Let $x \geqslant 0$ be a real number such that for any $\varepsilon > 0$, $x \leqslant \varepsilon$. Show that $x = 0$.
   (b) Let $S = \,]0,1[$. Show that for any $\varepsilon > 0$ there exists $x \in S$, such that $x < \varepsilon$, $x \neq 0$.

4. Show that $d = \inf(S)$ iff d is a lower bound for S and for any $\varepsilon > 0$ there is an $x \in S$, such that $d \geqslant x - \varepsilon$.

5. Let x_n be a monotone increasing sequence bounded above and consider the set $S = \{x_1, x_2, \ldots\}$. Using Theorem 2, show that x_n converges to $\sup(S)$. Make a similar statement for decreasing sequences.

6. Let A and B be two non-empty sets of real numbers with the property that $x \leqslant y$ for all $x \in A$, $y \in B$. Show that there exists a number $c \in \mathbb{R}$ such that $x \leqslant c \leqslant y$ for all $x \in A, y \in B$. Give an example of this statement being false for rational numbers (it is, in fact, equivalent to the completeness axiom and is at the basis for another way of formulating the completeness axiom known as *Dedekind cuts*).

7. For sets $A, B \subset \mathbb{R}$, let $A + B = \{x + y \mid x \in A \text{ and } y \in B\}$. Show that $\sup(A + B) = \sup(A) + \sup(B)$. Make a similar statement for inf's.

8. For sets $A, B \subset \mathbb{R}$, determine which of the following statements are true. Prove the true statements and give a counter-example for those which are false:
   (a) $\sup(A \cap B) \leqslant \inf\{\sup(A), \sup(B)\}$
   (b) $\sup(A \cap B) = \inf\{\sup(A), \sup(B)\}$
   (c) $\sup(A \cup B) \geqslant \sup\{\sup(A), \sup(B)\}$
   (d) $\sup(A \cup B) = \sup\{\sup(A), \sup(B)\}$.

9. Demonstrate that if a subsequence of a Cauchy sequence converges to a point, then the whole sequence converges to that point. Give a counter-example if the original sequence is not a Cauchy sequence.

10. For a given sequence a_n, we define the numbers
$$\lim \sup(a_n) = \inf\{\sup\{a_n, a_{n+1}, \ldots\} \mid n = 1,2,\ldots\}$$
and
$$\lim \inf(a_n) = \sup\{\inf\{a_n, a_{n+1}, \ldots\} \mid n = 1,2,\ldots\}$$
Show that
   (a) $\lim \inf(a_n) \leqslant \lim \sup(a_n)$
   (b) $\lim \sup(a_n) = b$ iff for all $\varepsilon > 0$, there is an N so that $b + \varepsilon > a_n$ for all $n \geqslant N$ and $b - \varepsilon < a_n$ for some $n \geqslant N$
   (c) $a_n \to b$ iff $\lim \sup(a_n) = \lim \inf(a_n) = b$
   (d) let $a_n = (-1)^n$. Compute $\lim \inf(a_n)$, $\lim \sup(a_n)$.

   Note: $\lim \sup(a_n)$ and $\lim \inf(a_n)$ always are defined (but could be $\pm\infty$) although $\lim(a_n)$ need not exist. Also, lim sup is short for *limit superior* and lim inf for *limit inferior*, and these are sometimes written as $\overline{\lim}$ and $\underline{\lim}$, respectively.

11. Show that (i), (ii), and (iii) of Theorem 3 each implies the completeness axiom for an ordered field. [Hint: (i) $\Rightarrow$ completeness axiom is almost immediate. (ii) implies (i) in much the same way as we showed in the proof of Theorem 3 that (i) implies (ii). Therefore (ii) $\Rightarrow$ completeness axiom. To show (iii) $\Rightarrow$ completeness axiom, it is sufficient to show (iii) $\Rightarrow$ (i). To do this, define the sequence x_n as in the proof of completeness axiom $\Rightarrow$ (i) and argue that x_n is a Cauchy sequence. Show that its limit is the sup of the set in question, following the proof that the completeness axiom $\Rightarrow$ (i).]

12. In $\mathbb{R}^n$ show that
   (a) $2\|x\|^2 + 2\|y\|^2 = \|x + y\|^2 + \|x - y\|^2$ (parallelogram law)
   (b) $\|x + y\| \|x - y\| \leqslant \|x\|^2 + \|y\|^2$
   (c) $4\langle x,y \rangle = \|x + y\|^2 - \|x - y\|^2$ (polarization identity).
   Interpret these results geometrically in terms of the parallelogram formed by x and y.

13. What is the orthogonal complement in $\mathbb{R}^4$ of the space spanned by $(1,0,1,1)$ and $(-1,2,0,0)$?

14. (a) Prove Lagrange's identity

$$\left(\sum_{i=1}^{n} x_i y_i\right)^2 = \left(\sum_{i=1}^{n} x_i^2\right)\left(\sum_{i=1}^{n} y_i^2\right) - \sum_{1 \leqslant i < j \leqslant n} (x_i y_j - x_j y_i)^2$$

using algebra techniques and use this to give another proof of the Schwarz inequality.

(b) Show that

$$\left\{\sum_{i=1}^{n} (x_i + y_i)^2\right\}^{1/2} \leqslant \left(\sum_{i=1}^{n} x_i^2\right)^{1/2} + \left(\sum_{i=1}^{n} y_i^2\right)^{1/2}.$$

15. Let x_n be a sequence in $\mathbb{R}$ such that $d(x_n, x_{n+1}) \leqslant d(x_{n-1}, x_n)/2$. Then show that x_n is a Cauchy sequence.

16. Prove Theorem 4. In fact, for vector spaces $V_1, \ldots, V_n$, show that $V = V_1 \times \cdots \times V_n$ is a vector space.

17. Let $S \subset \mathbb{R}$ be bounded below and non-empty. Then show that $\inf(S) = \sup\{x \in \mathbb{R} \mid x$ is a lower bound for $S\}$.

18. Show that in $\mathbb{R}$, $x_n \to x$ iff $-x_n \to -x$. Hence prove that the completeness axiom is equivalent to the statement that every decreasing sequence $x_1 \geqslant x_2 \geqslant x_3 \cdots$ bounded below converges. Prove that the limit of the sequence is $\inf\{x_1, x_2, \ldots\}$.

19. Let $x = (1,1,1) \in \mathbb{R}^3$ be written $x = \sum_{i=1}^{3} y_i f_i$, where $f_1 = (1,0,1)$, $f_2 = (0,1,1)$, and $f_3 = (1,1,0)$. Compute the components y_i.

20. Let S and T be non-zero orthogonal subspaces of $\mathbb{R}^n$. Prove that if S and T are orthogonal complements (that is, S and T span all of $\mathbb{R}^n$) then $S \cap T = \{0\}$ and $\dim(S) + \dim(T) = n$, where $\dim(S)$ denotes dimension of S. Give examples in $\mathbb{R}^3$, where the condition $\dim(S) + \dim(T) = n$ holds, and examples where it fails. Can it fail in $\mathbb{R}^2$?

21. Show that the sequence in Example 2 can be chosen to be increasing.

22. (a) Prove: if in $\mathbb{R}$, $x_k \to x$, then $ax_k \to ax$ for any number a.
    (b) If $x_k \to x$ and $y_k \to y$, then prove $s_k = x_k + y_k$ converges to $x + y$.

23. Let $P \subset \mathbb{R}$ be a set such that $x \geqslant 0$ for all $x \in P$ and for any integer k there is an $x_k \in P$ such that $kx_k \leqslant 1$. Then prove that $0 = \inf(P)$.

24. If $\sup(P) = \sup(Q)$ and $\inf(P) = \inf(Q)$, does $P = Q$?

25. We say that $P \leqslant Q$ if for each $x \in P$, there is a $y \in Q$ with $x \leqslant y$. If $P \leqslant Q$, then prove $\sup(P) \leqslant \sup(Q)$. Is it true that $\inf(P) \leqslant \inf(Q)$? If $P \leqslant Q$ and $Q \leqslant P$, does $P = Q$?

26. Prove that the real numbers form an uncountable set, but the rationals form a countable set. [Hint: First recall how any number $x, 0 \leqslant x \leqslant 1$ can be written as a decimal and that any decimal represents a real number. If the numbers $x, 0 \leqslant x \leqslant 1$ were countable, we could arrange them as $s_n = 0.a_{n,1} a_{n,2} \cdots$. Let $x = 0.b_1 b_2 b_3 \cdots$, where $b_n = 1$ if $a_{n,n} \neq 1$ and $b_n = 2$ if $a_{n,n} = 1$. Show that $x \neq s_n$ for all n. For the rationals, employ Exercise 9 in Introductory chapter.]

27. Suppose $a_n \geqslant 0$ and $a_n \to 0$ as $n \to \infty$. Given any $\varepsilon > 0$, show that there is a subsequence b_n of a_n such that $\sum_{n=1}^{\infty} b_n < \varepsilon$.

28. Let x_n be a Cauchy sequence in $\mathbb{R}$ and let $A_n = \sup\{x_n, x_{n+1}, \ldots\}$ and $B_n = \inf\{x_n, x_{n+1}, \ldots\}$. Prove A_n converges to the same limit as B_n, which in turn is the same as the limit of x_n.

29. For any $x \in \mathbb{R}, x \geqslant 0$ use the axioms for $\mathbb{R}$ to deduce the existence of $y \in \mathbb{R}$ such that $y^2 = x$.

30. Use the axioms for $\mathbb{R}$ to prove the Archimedian property: for every $x \in \mathbb{R}$ there exists an integer N such that $N > x$. [Hint: If $n \leqslant x$ for all $n = 1, 2, 3, \ldots$ use the completeness axiom to prove that $x_n = n$ converges.]

31. Let $A, B \subset \mathbb{R}$ and let $f \colon A \times B \to \mathbb{R}$. Is it true that

$$\sup\{f(x,y) \mid (x,y) \in A \times B\} = \sup\{\sup\{f(x,y) \mid x \in A\} \mid y \in B\}$$

or, the same thing in different notation,

$$\sup_{(x,y) \in A \times B} f(x,y) = \sup_{y \in B}(\sup_{x \in A} f(x,y))?$$

32. (a) Give a reasonable definition for when limit $x_n = \infty$.
    $\qquad n \to \infty$

    (b) Let $x_1 = 1$ and define inductively $x_{n+1} = (x_1 + \cdots + x_n)/2$. Prove that $x_n \to \infty$.

33. (a) Show that $(\log x)/x \to 0$ as $x \to \infty$. (You may consult your calculus text and use for example l'Hopital's rule).

    (b) Show that $n^{1/n} \to 1$ as $n \to \infty$.

Chapter 2

Topology of $\mathbb{R}^n$

In this chapter we begin our study of those basic properties of $\mathbb{R}^n$ which are important for the notion of a continuous function. We will study open sets, which generalize open intervals on $\mathbb{R}$, and closed sets, which generalize closed intervals. The study of open and closed sets constitutes the beginnings of topology. This study will be continued in Chapter 3.

Most of the material in this chapter depends only on the basic properties of the distance function given in Theorem 5, Chapter 1. Recall that the distance function d is given by

$$d(x,y) = \left\{ \sum_{i=1}^{n} (x_i - y_i)^2 \right\}^{1/2},$$

and that the basic properties of d are

    (i) $d(x,y) \geqslant 0$
    (ii) $d(x,y) = 0$ iff $x = y$
    (iii) $d(x,y) = d(y,x)$
    (iv) $d(x,y) \leqslant d(x,z) + d(z,y)$ (triangle inequality).

2.1 Open Sets

In order to define open sets, we first shall introduce the notion of an ε-disc.

Definition 1. For each fixed $x \in \mathbb{R}^n$ and $\varepsilon > 0$, the set
$$D(x,\varepsilon) = \{y \in \mathbb{R}^n \mid d(x,y) < \varepsilon\}$$

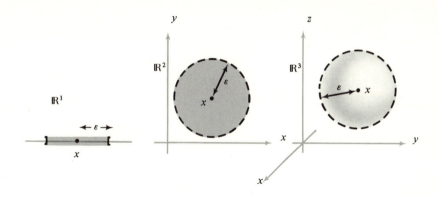

FIGURE 2-1   The ε-disc.

is called the *ε-disc* about x (also called the *ε-neighborhood* or *ε-ball* about x). See Figure 2-1. A set $A \subset \mathbb{R}^n$ is said to be *open* if for each $x \in A$, there exists an $\varepsilon > 0$ such that $D(x,\varepsilon) \subset A$.

It is important to realize that the ε required may depend on x. For example, the unit square in $\mathbb{R}^2$ not including the "boundary" is open, but the ε's needed get smaller as we approach the boundary. However, notice that the ε cannot be zero for any x. See Figure 2-2.

Consider an open interval in $\mathbb{R} = \mathbb{R}^1$, say, $]0,1[$. Indeed, this is an open set (see Figure 2-3). However, if we look upon the set as being in $\mathbb{R}^2$ (as a subset of the x axis), it is no longer open. Thus for a set to be open it is essential to specify which $\mathbb{R}^n$ we are using.

There are numerous examples of sets which are not open. The closed unit disc in $\mathbb{R}^2$, $\{x \in \mathbb{R}^2 \mid \|x\| \leqslant 1\}$, is such an example. This set is not open because for a point on the "boundary" (that is, points x with $\|x\| = 1$), every ε-disc contains points which do not lie in the set. See Figure 2-4.

FIGURE 2-2   An open set.

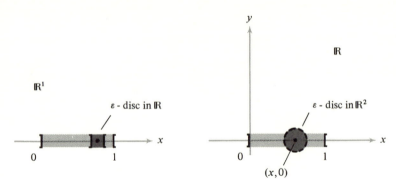

FIGURE 2-3

Theorem 1. *In $\mathbb{R}^n$, for every $\varepsilon > 0$ and $x \in \mathbb{R}^n$, the set $D(x,\varepsilon)$ is open.*

The main idea for the proof is contained in Figure 2-5. Notice in this figure that the size of the disc about the point $y \in D(x,\varepsilon)$ gets smaller as y gets closer to the boundary. The theorem should be "intuitively clear" from this picture.

Some basic laws which open sets obey are the following.

Theorem 2.
(i) *The intersection of a finite number of open subsets of $\mathbb{R}^n$ is an open subset of $\mathbb{R}^n$.*
(ii) *The union of an arbitrary collection of open subsets of $\mathbb{R}^n$ is an open subset of $\mathbb{R}^n$.*

This result is perhaps not entirely clear intuitively. Some idea about the difference between assertions (i) and (ii) may be obtained if we realize that it is not true that the intersection of an arbitrary family of open sets is open. For example, in $\mathbb{R}^1$, a single point (which is not an open set) is the intersection of all open intervals containing it (why?). The remainder of this chapter will rely heavily on the basic properties of open sets which were given in Theorem 2.

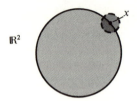

FIGURE 2-4   A non-open set.

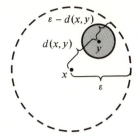

FIGURE 2-5 ε-discs are open.

Note: A set with a specified collection of subsets (called, by definition, open sets) obeying the rules in Theorem 2 and containing the empty set $\varnothing$ and the whole space is called a *topological space*. We shall not deal with general topological spaces in this book, but rather with the case of $\mathbb{R}^n$. However, much of what is said below does apply to the more general setting.

EXAMPLE 1. Let $S = \{(x,y) \in \mathbb{R}^2 \mid 0 < x < 1\}$. Show that S is open.

Solution: In Figure 2-6 we see that about each point $(x,y) \in S$ we can draw the disc of radius $r = \min\{x, 1 - x\}$ and it is entirely contained in S. Hence, by definition, S is open.

EXAMPLE 2. Let $S = \{(x,y) \in \mathbb{R}^2 \mid 0 < x \leqslant 1\}$. Is S open?

Solution: No, because any disc about $(1,0) \in S$ contains points $(x,0)$ with $x > 1$.

EXAMPLE 3. Let $A \subset \mathbb{R}^n$ be open and $B \subset \mathbb{R}^n$. Define

$$A + B = \{x + y \in \mathbb{R}^n \mid x \in A \text{ and } y \in B\}.$$

Prove $A + B$ is open.

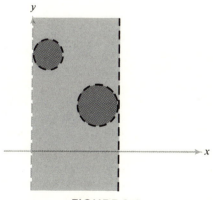

FIGURE 2-6

Solution: Let $x \in A$, $y \in B$ so that $x + y \in A + B$. By definition, there is an $\varepsilon > 0$ so $D(x,\varepsilon) \subset A$. We claim $D(x + y,\varepsilon) \subset A + B$. Indeed, let $z \in D(x + y,\varepsilon)$ so that $d(x + y,z) < \varepsilon$. But, $d(x + y,z) = d(x,z - y)$ (why?) so $z - y \in A$, and then $z = (z - y) + y \in A + B$. Thus $D(x + y,\varepsilon) \subset A + B$, so $A + B$ is open.

Exercises for Section 2.1

1. Show that $\mathbb{R}^2 \backslash \{(0,0)\}$ is open in $\mathbb{R}^2$.

2. Let $S = \{(x,y) \in \mathbb{R}^2 \mid xy > 1\}$. Show that S is open.

3. Let $A \subset \mathbb{R}$ be open and $B \subset \mathbb{R}^2$ be defined by

$$B = \{(x,y) \in \mathbb{R}^2 \mid x \in A\} \ .$$

Show that B is open.

4. Let $B \subset \mathbb{R}^n$ be any set. Define

$$C = \{x \in \mathbb{R}^n \mid d(x,y) < 1 \text{ for some } y \in B\} \ .$$

Show that C is open [Hint: Show that $C = \bigcup_{y \in B} D(y,1)$.]

5. Let $A \subset \mathbb{R}$ be open and $B \subset \mathbb{R}$. Define $AB = \{xy \in \mathbb{R} \mid x \in A \text{ and } y \in B\}$. Is AB necessarily open?

2.2   Interior of a Set

Definition 2.   For any set $A \subset \mathbb{R}^n$, a point $x \in A$ is called an *interior point* of A if there is an open set U such that $x \in U \subset A$. (It should be clear that this is equivalent to the following: there is an $\varepsilon > 0$ such that $D(x,\varepsilon) \subset A$.) The *interior of A* is the collection of all interior points of A and is denoted int(A). This set might be empty.

For example, the interior of a single point is empty. The interior of the unit disc, including its boundary, is the unit disc without its boundary.

We can describe the interior of a set in a somewhat different manner. The interior of A is in fact the union of all open subsets of A (the reader is asked to show this in Exercise 22, p. 58). Thus by Theorem 2, or directly, int(A) is open. Hence int(A) *is the largest open subset of A*. Therefore if there are no open subsets of A, int(A) $= \varnothing$. Also, it is evident that A *is open iff* int(A) $= A$ (again, see Exercise 22).

EXAMPLE 1.   Let $S = \{(x,y) \in \mathbb{R}^2 \mid 0 < x \leqslant 1\}$. Find int($S$).

Solution: To determine the interior points, we just need to locate points about which it is possible to draw an ε-disc entirely contained in S. By considering Figure 2-6, we see that these are points (x,y) where $0 < x < 1$. Thus $\text{int}(S) = \{(x,y) \mid 0 < x < 1\}$.

EXAMPLE 2.   Is it true that $\text{int}(A) \cup \text{int}(B) = \text{int}(A \cup B)$?

Solution: No. Consider in the real line, $A = [0,1]$, $B = [1,2]$. Then $\text{int}(A) =]0,1[$ (why?) and $\text{int}(B) =]1,2[$, so $\text{int}(A) \cup \text{int}(B) =]0,1[\cup]1,2[=]0,2[\backslash\{1\}$, while $\text{int}(A \cup B) = \text{int}[0,2] =]0,2[$.

Exercises for Section 2.2

1. Let $S = \{(x,y) \in \mathbb{R}^2 \mid xy \geqslant 1\}$. Find $\text{int}(S)$.

2. Let $S = \{(x,y,z) \in \mathbb{R}^3 \mid 0 \leqslant x < 1, y^2 + z^2 \leqslant 1\}$. Find $\text{int}(S)$.

3. If $A \subset B$, is $\text{int}(A) \subset \text{int}(B)$?

4. Do you think it is true that $\text{int}(A) \cap \text{int}(B) = \text{int}(A \cap B)$? Try some examples.

2.3   Closed Sets

Definition 3.   A set B in $\mathbb{R}^n$ is said to be *closed* if its complement in $\mathbb{R}^n$ (that is, the set $\mathbb{R}^n \backslash B$) is open.

For example, a single point is a closed set. The set consisting of the unit circle with boundary is closed. Roughly speaking, a set is closed when it contains its "boundary points" (this intuition will be made precise in Section 2.6). See Figure 2-7.

It is entirely possible to have a set which is neither open nor closed. For example, in $\mathbb{R}^1$, a half-open interval $]0,1]$ is neither open nor closed. Thus even if we know A is not open, we *cannot* conclude that it is closed or not closed. The next theorem is analogous to Theorem 2.

FIGURE 2-7   Closed sets.

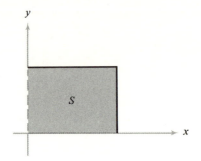

FIGURE 2-8

Theorem 3.

(i) *The union of a finite number of closed subsets of* $\mathbb{R}^n$ *is closed.*

(ii) *The intersection of an arbitrary family of closed subsets of* $\mathbb{R}^n$ *is closed.*

This theorem follows directly from Theorem 2 by noting that unions and intersections are interchanged when we take complements (see Exercise 14 of the Introductory chapter). The proof is left to the reader (Exercise 23) who should also show that (i) cannot be replaced by arbitrary unions.

EXAMPLE 1.   Let $S = \{(x,y) \in \mathbb{R}^2 \mid 0 < x \leqslant 1, 0 \leqslant y \leqslant 1\}$. Is S closed?

Solution: See Figure 2-8. Intuitively, S is not closed because the portion of its boundary on the y-axis is not in S. Also, the complement is not open because any ε-disc about a point on the y-axis, say $(0,1/2)$, will intersect S (and hence not be in $\mathbb{R}^n \backslash S$).

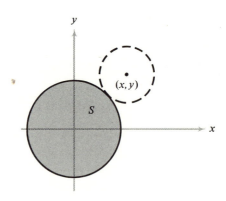

FIGURE 2-9

EXAMPLE 2.   Let $S = \{(x,y) \in \mathbb{R}^2 \mid x^2 + y^2 \leqslant 1\}$. Is S closed?

Solution: Yes. S is just the unit disc, including its boundary. The complement is clearly an open set, because for $(x,y) \in \mathbb{R}^2 \backslash S$, the disc of radius $\varepsilon = \sqrt{x^2 + y^2} - 1$ will be entirely contained in $\mathbb{R}^2 \backslash S$ (Figure 2-9).

EXAMPLE 3.   Show that any finite set in $\mathbb{R}^n$ is closed.

Solution: Single points are closed, and so we may apply Theorem 3(i).

Exercises for Section 2.3

1. Let $S = \{(x,y) \in \mathbb{R}^2 \mid x, y \geqslant 1\}$. Is S closed?

2. Let $S = \{(x,y) \in \mathbb{R}^2 \mid x = 0, 0 < y < 1\}$. Is S closed?

3. Redo Example 3 directly, this time showing that the complement is open.

4. Let $A \subset \mathbb{R}^n$ be arbitrary. Show $\mathbb{R}^n \backslash (\text{int } A)$ is closed.

5. Let $S = \{x \in \mathbb{R} \mid x \text{ is irrational}\}$. Is S closed?

2.4   Accumulation Points

There is another very useful way to determine whether or not a set is closed which depends upon the important concept of an accumulation point.

> **Definition 4.**   A point $x \in \mathbb{R}^n$ is called an *accumulation point* of a set A if every open set U containing x contains some point of A other than x.

That is to say, an accumulation point of a set A is a point such that there are other points of A arbitrarily close by. Accumulation points are also referred to as *cluster points*.

Using Theorem 1, our definition that x be an accumulation point of A is equivalent to the statement that *for every $\varepsilon > 0$, $D(x,\varepsilon)$ contains some point y of A with $y \neq x$.*

For example, in $\mathbb{R}^1$, a set consisting of a single point has no accumulation points and the open interval $]0,1[$ has all points of $[0,1]$ as accumulation points. Note that an accumulation point of a set need *not* lie in that set. The definitions of accumulation points and closed sets are closely related as shown by the next theorem.

> **Theorem 4.**   *A set $A \subset \mathbb{R}^n$ is closed iff all the accumulation points of A belong to A.*

Notice that a set need not have any accumulation points (a single point or the set of integers in $\mathbb{R}^1$ are examples), in which case Theorem 4 still applies and we can conclude that the set is closed. Another useful way to prove that a set is closed is given in Theorem 9 which follows later.

Theorem 4 is intuitively clear because a set being closed means, roughly speaking, that it contains all points on its "boundary," and such points are accumulation points. This sort of rough argument has a pitfall and one has to, in fact, be more careful as some sets are sufficiently complicated that our intuition may fail us. For example, consider $A = \{1/n \in \mathbb{R} \mid n = 1,2,3,\ldots\} \cup \{0\}$. This is a closed set (verify!) and its only accumulation point is $\{0\}$ which lies in A. But our intuition about "boundary" mentioned above is not very clear for this set, hence the need for more careful arguments.

EXAMPLE 1. Let $S = \{x \in \mathbb{R} \mid x \in [0,1]$ and x is rational$\}$. Find the accumulation points of S.

Solution: The set of accumulation points consists of all points in $[0,1]$. Indeed, let $y \in [0,1]$ and $D(y,\varepsilon) = \,]y - \varepsilon, y + \varepsilon[$ be a neighborhood of y. Now we know we can find rational points in $[0,1]$ arbitrarily close to y (other than y) and in particular in $D(y,\varepsilon)$. Hence y is an accumulation point. Any point $y \notin [0,1]$ is not an accumulation point because y has an ε-disc containing it which does not meet $[0,1]$ and therefore S.

EXAMPLE 2. Verify Theorem 4 for the set $A = \{(x,y) \in \mathbb{R}^2 \mid 0 \leqslant x \leqslant 1$ or $x = 2\}$.

Solution: A is shown in Figure 2-10. Clearly, A is closed. The accumulation points of A consist exactly of A itself which lie in A. Note that on $\mathbb{R}$, $[0,1] \cup \{2\}$ has as accumulation points $[0,1]$ *without* the point $\{2\}$.

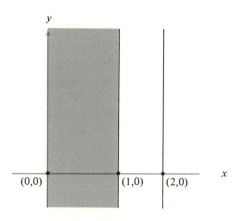

FIGURE 2-10

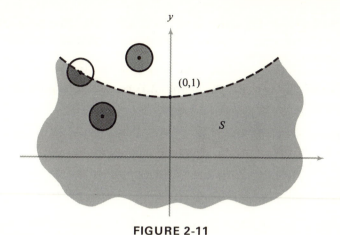

FIGURE 2-11

EXAMPLE 3.   Let $S = \{(x,y) \in \mathbb{R}^2 \mid y < x^2 + 1\}$. Find the accumulation points of S.

Solution: S is sketched in Figure 2-11. The accumulation points constitute the set $\{(x,y) \mid y \leqslant x^2 + 1\}$ as is evident from the figure.

Exercises for Section 2.4

1. Find the accumulation points of $A = \{(x,y) \in \mathbb{R}^2 \mid y = 0 \text{ and } 0 < x < 1\}$.

2. If $A \subset B$ and x is an accumulation point of A, is x an accumulation point of B as well?

3. Find the accumulation points of the following sets in $\mathbb{R}^2$.
    (a) $\{(m,n) \mid m,n \text{ integers}\}$
    (b) $\{(p,q) \mid p,q \text{ rational}\}$
    (c) $\{(m/n, 1/n) \mid m,n \text{ integers}, n \neq 0\}$
    (d) $\{(1/n + 1/m, 0) \mid n,m \text{ integers}, n \neq 0, m \neq 0\}$.

4. Let $A \subset \mathbb{R}$ and $x = \sup(A)$. Must x be an accumulation point of A?

5. Verify Theorem 4 for the set $A = \{(x,y) \in \mathbb{R}^2 \mid x^2 + y + 2x = 3\}$.

2.5   Closure of a Set

The interior of a set A is the largest open subset of A. Similarly, we can form the smallest closed set containing a set A. This set is called the *closure of A* and is denoted cl(A) or sometimes $\bar{A}$.

> **Definition 5.**   Let $A \subset \mathbb{R}^n$. The set cl(A) is defined to be the intersection of all closed sets containing A, (and so cl(A) is *closed* by Theorem 3 (ii)).

For example, on $\mathbb{R}^1$, $\mathrm{cl}(]0,1]) = [0,1]$. Also, note that *A is closed iff* $\mathrm{cl}(A) = A$ (why?). The connection between closure and accumulation points is the following theorem.

> **Theorem 5.** *Let* $A \subset \mathbb{R}^n$. *Then* $\mathrm{cl}(A)$ *consists of* A *plus all the accumulation points of* A.

In other words, to find the closure of a set A, we add to A all the accumulation points not already in A. Theorem 5 should be intuitively clear from the examples presented earlier.

EXAMPLE 1. Find the closure of $A = [0,1[\,\cup\, \{2\}$ in $\mathbb{R}$.

Solution: The accumulation points are $[0,1]$, so the closure is $[0,1] \cup \{2\}$. This is clearly also the smallest closed set we could find containing A.

EXAMPLE 2. For any $A \subset \mathbb{R}^n$, show that $\mathbb{R}^n \backslash \mathrm{cl}(A)$ is open.

Solution: $\mathrm{cl}(A)$ is a closed set and, by definition of a closed set, its complement is open.

EXAMPLE 3. Is it true that $\mathrm{cl}(A \cap B) = \mathrm{cl}(A) \cap \mathrm{cl}(B)$?

Solution: No. Take, for example, $A = [0,1]$, $B =]1,2]$. Then, $A \cap B = \varnothing$ and $\mathrm{cl}(A) \cap \mathrm{cl}(B) = \{1\}$.

Exercises for Section 2.5

1. Find the closure of $S = \{(x,y) \in \mathbb{R}^2 \mid x > y^2\}$.

2. Find the closure of $\{1/n \mid n = 1,2,3,\ldots\}$ in $\mathbb{R}$.

3. Let $A = \{(x,y) \in \mathbb{R}^2 \mid x$ is rational$\}$. Find $\mathrm{cl}(A)$.

4. (a) For $A \subset \mathbb{R}^n$, show $\mathrm{cl}(A) \backslash A$ consists of accumulation points of A.
    (b) Is it all of them?

5. Let $A \subset \mathbb{R}$ and $x = \sup(A)$. Show $x \in \mathrm{cl}(A)$.

2.6 Boundary of a Set

If we consider the unit disc in $\mathbb{R}^2$, we know what we would like to call the boundary—the obvious choice is the unit circle. But, for more complicated sets, such as the rationals, it is not as intuitively clear what the boundary should be. Therefore a precise definition is needed.

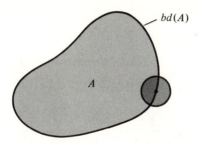

FIGURE 2-12   Boundary of a set.

Definition 6.   For a given set A in $\mathbb{R}^n$, the *boundary* is defined to be the set

$$\mathrm{bd}(A) = \mathrm{cl}(A) \cap \mathrm{cl}(\mathbb{R}^n \backslash A).$$

Sometimes the notation $\partial A = \mathrm{bd}(A)$ is used.

Thus by Theorem 3(ii), $\mathrm{bd}(A)$ is a closed set. Also, note that $\mathrm{bd}(A) = \mathrm{bd}(\mathbb{R}^n \backslash A)$. From Theorem 5, we can deduce that the boundary is also described as follows.

> **Theorem 6.**   *Let* $A \subset \mathbb{R}^n$. *Then* $x \in \mathrm{bd}(A)$ *iff for every* $\varepsilon > 0$, $D(x,\varepsilon)$ *contains points of A and of* $\mathbb{R}^n \backslash A$ *(these points might be x itself).* *See Figure 2-12.*

The original definition states that $\mathrm{bd}(A)$ is the border between A and $\mathbb{R}^n \backslash A$. This is also what Theorem 6 is asserting and therefore Theorem 6 should be intuitively clear.

EXAMPLE 1.   Let $A = \{x \in \mathbb{R} \mid x \in [0,1] \text{ and } x \text{ is rational}\}$. Find $\mathrm{bd}(A)$.

Solution: $\mathrm{bd}(A) = [0,1]$ since, for any $\varepsilon > 0$ and $x \in [0,1]$, $D(x,\varepsilon) =]x - \varepsilon, x + \varepsilon[$ contains both rational and irrational points. The reader should also verify that $\mathrm{bd}(A) = [0,1]$ using the original definition of $\mathrm{bd}(A)$. This example shows that if $A \subset B$ it does *not* necessarily follow that $\mathrm{bd}(A) \subset \mathrm{bd}(B)$ (let A be as above and $B = [0,1]$ in $\mathbb{R}$).

EXAMPLE 2.   If $x \in \mathrm{bd}(A)$, must x be an accumulation point?

Solution: No. Let $A = \{0\} \subset \mathbb{R}$. Then A has no accumulation points, but $\mathrm{bd}(A) = \{0\}$.

EXAMPLE 3.   Let $S = \{(x,y) \in \mathbb{R}^2 \mid x^2 - y^2 > 1\}$. Find $\mathrm{bd}(S)$.

Solution: S is sketched in Figure 2-13. Clearly, $\mathrm{bd}(S)$ consists of the hyperbola $x^2 - y^2 = 1$.

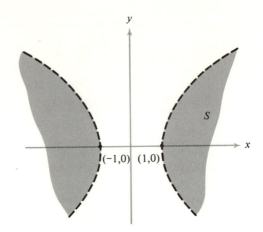

FIGURE 2-13

Exercises for Section 2.6

1. Find bd(A) where $A = \{1/n \in \mathbb{R} \mid n = 1,2,3,\ldots\}$.

2. If $x \in \mathrm{cl}(A)\backslash A$, then show $x \in \mathrm{bd}(A)$. Is the converse true?

3. Find bd(A) where $A = \{(x,y) \in \mathbb{R}^2 \mid x \leqslant y\}$.

4. Is bd(A) = bd(int A)?

5. Let $A \subset \mathbb{R}$ be bounded and $x = \sup(A)$. Is $x \in \mathrm{bd}(A)$?

2.7   Sequences

Let us now consider some aspects of sequences. The definition of convergence in $\mathbb{R}^n$ is very similar to that for real numbers.

> **Definition 7.**   Let x_k be a sequence of points in $\mathbb{R}^n$. We say that x_k *converges to a limit* x in $\mathbb{R}^n$ if for every open set U containing x (also called a neighborhood of x), there is an N (depending on U) such that $x_k \in U$ whenever $k \geqslant N$. See Figure 2-14.

This definition coincides with the usual ε definition as the next theorem shows.

> **Theorem 7.**   *A sequence x_k in $\mathbb{R}^n$ converges to $x \in \mathbb{R}^n$ iff for every $\varepsilon > 0$ there is an N such that $k \geqslant N$ implies $\|x - x_k\| < \varepsilon$.*

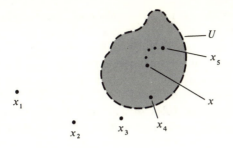

FIGURE 2-14   Convergence of a sequence.

This theorem is entirely analogous to what we know about convergent sequences of real numbers. See Section 1.1. There is another result which is closely allied to the one above. We can show that:

Theorem 8. $x_k \to x$ iff the components of x_k converge to the components of x as sequences of real numbers.

As the proof on page 50 shows, this follows easily from Theorem 7 and the explicit formula for $\|x_k - x\|$.

We can use sequences to determine whether or not a set is closed. The method is as follows:

Theorem 9.
(i) A set $A \subset \mathbb{R}^n$ is closed iff for every sequence $x_k \in A$ which converges, the limit lies in A.
(ii) For a set $B \subset \mathbb{R}^n$, $x \in \mathrm{cl}(B)$ iff there is a sequence $x_k \in B$ with $x_k \to x$.

The intuition behind this theorem is the same as that for Theorems 4 and 5. One should note that these sequences in (i) and (ii) could be trivial, that is $x_k = x$ for all k.

As in the case of $\mathbb{R}^1$, we can define a Cauchy sequence in $\mathbb{R}^n$. (The concepts of monotone sequence and least upper bound also make sense if interpreted coordinate-wise, but these are not very useful in $\mathbb{R}^n$ for $n \neq 1$.)

Definition 8.   A sequence $x_k \in \mathbb{R}^n$ is called a *Cauchy sequence* if for every $\varepsilon > 0$ there is an N such that $l, k \geqslant N$ implies $\|x_k - x_l\| < \varepsilon$.

Theorem 10.   A sequence x_k in $\mathbb{R}^n$ converges to a point in $\mathbb{R}^n$ iff it is a Cauchy sequence.

This is a straightforward generalization of, and follows from, the corresponding theorem for $\mathbb{R}$ (see Theorem 3 of Chapter 1).

As with $\mathbb{R}$, this theorem provides an important test for convergence since the Cauchy condition does not involve the limit point explicitly. Thus we can often tell if a sequence converges even though we do not know the limit.

Note: In a general metric space (a set S and a real-valued distance function d satisfying the rules of Theorem 5, III, Chapter 1) a Cauchy sequence is a sequence $x_k \in S$ such that for all $\varepsilon > 0$, there is an N such that $k, l \geq N$ implies $d(x_k, x_l) < \varepsilon$. The space is called *complete* iff every Cauchy sequence converges to a point in the space. An example of an incomplete space is the rational numbers with $d(x,y) = |x - y|$. Theorem 10 asserts then that $\mathbb{R}^n$ *is a complete metric space.*

EXAMPLE 1.   Show that the sequence $(1/n, 1/n^2)$ converges to $(0,0)$ as $n \to \infty$.

Solution: Each component sequence $1/n$, and $1/n^2$ converges to 0, so by Theorem 8, $x_n = (1/n, 1/n^2)$ converges to $(0,0)$.

EXAMPLE 2.   Let $x_n \in \mathbb{R}^m$ be a convergent sequence with $\|x_n\| \leq 1$ for all n. Then show that the limit x also satisfies $\|x\| \leq 1$. Is this true if $\leq$ is replaced with $<$?

Solution: The unit ball $B = \{y \in \mathbb{R}^m \mid \|y\| \leq 1\}$ is closed. Hence, by Theorem 9(i), $x_n \in B$ implies $x \in B$. This is not true if $\leq$ is replaced by $<$. For example, on $\mathbb{R}$ consider $x_n = 1 - 1/n$.

EXAMPLE 3.   Find the closure of $A = \{1/n \in \mathbb{R} \mid n = 1,2,\ldots\}$.

Solution: We can use, for example, Theorem 9(ii). The sequence $1/n \to 0$ so $0 \in \mathrm{cl}(A)$. Taking other sequences from A will not yield any new points, so

$$\mathrm{cl}(A) = A \cup \{0\}.$$

Exercises for Section 2.7

1. Find the limit of the sequence $[(\sin n)^n/n, 1/n^2]$ in $\mathbb{R}^2$.

2. Let $x_n \to x$ in $\mathbb{R}^m$. Show that $A = \{x_n \mid n = 1,2,\ldots\} \cup \{x\}$ is closed.

3. Let $A \subset \mathbb{R}^m$, $x_n \in A$, and $x_n \to x$. Show that $x \in \mathrm{cl}(A)$.

4. Verify Theorem 9 (ii) for the set $B = \{(x,y) \in \mathbb{R}^2 \mid x < y\}$.

5. Let $S = \{x \in \mathbb{R} \mid x \text{ is rational and } x^2 < 2\}$. Compute $\mathrm{cl}(S)$.

2.8   Series in $\mathbb{R}$ and $\mathbb{R}^n$

Just as in $\mathbb{R}^1$, we can consider series in $\mathbb{R}^n$.

Definition 9.   A series $\sum_{k=0}^{\infty} x_k$, where $x_k \in \mathbb{R}^n$, is said to *converge to* $x \in \mathbb{R}^n$ if the sequence of partial sums $s_k = \sum_{i=0}^{k} x_i$ converges to x, and if so we write $\sum_{k=0}^{\infty} x_k = x$.

As in Theorem 8, $\sum_{k=0}^{\infty} x_k = x$ is equivalent to the corresponding component series converging to the components of x.
Applying Theorem 10 to s_k yields Theorem 11.

Theorem 11.   *A series $\sum x_k$ in $\mathbb{R}^n$ converges iff for every $\varepsilon > 0$, there is an N such that $k \geqslant N$ implies $\|x_k + x_{k+1} + \cdots + x_{k+p}\| < \varepsilon$ for all integers $p = 0, 1, 2, \ldots$.*

In particular, taking $p = 0$ we see that if $\sum x_k$ converges then $x_k \to 0$ as $k \to \infty$ (Exercise 2).
A series $\sum x_k$ is said to be *absolutely convergent* iff the real series $\sum \|x_k\|$ converges.

Theorem 12.   *If $\sum x_k$ converges absolutely, then $\sum x_k$ converges.*

This theorem is useful because it allows us to apply the usual tests for real series (such as the ratio test) to the series $\sum \|x_k\|$ to test for convergence of $\sum x_k$. Of course, it could happen that a particular test fails even though $\sum x_k$ is convergent, in which case some other method is needed.

Now we shall review the most important tests for the convergence of a real series. Some of the main facts are presented in the following theorem. Some other tests for convergence will occur in the exercises and later in Chapter 5.

Theorem 13.
(i) *If $|r| < 1$, then $\sum_{n=0}^{\infty} r^n$ converges to $1/(1 - r)$ and diverges (does not converge) if $|r| \geqslant 1$.*
(ii) *Comparison test: If $\sum_{k=1}^{\infty} a_k$ converges, $a_k \geqslant 0$, and $0 \leqslant b_k \leqslant a_k$, then $\sum_{k=1}^{\infty} b_k$ converges; if $\sum_{k=1}^{\infty} c_k$ diverges, $c_k \geqslant 0$, and $0 \leqslant c_k \leqslant d_k$, then $\sum_{k=1}^{\infty} d_k$ diverges.*
(iii) *p-series test: $\sum_{n=1}^{\infty} n^{-p}$ converges if $p > 1$ and diverges to ∞ (that is, the partial sums increase without bound) if $p \leqslant 1$.*
(iv) *Ratio test: Suppose that $\lim_{n \to \infty} |(a_{n+1}/a_n)|$ exists and is strictly less than 1. Then $\sum_{n=1}^{\infty} a_n$ converges absolutely. If the limit is*

strictly greater than 1, *the series diverges. If the limit equals* 1, *the test is inconclusive.*

(v) *Root test: Suppose that* $\underset{n\to\infty}{\text{limit}}(|a_n|)^{1/n}$ *exists and is strictly less than* 1. *Then* $\sum_{n=1}^{\infty} a_n$ *converges absolutely. If the limit is strictly greater than* 1, *the series diverges; if the limit equals* 1, *the test is inconclusive.*

(vi) *The integral test: If* f *is continuous, non-negative, and monotone decreasing on* $[1, +\infty[$, *then* $\sum_{n=1}^{\infty} f(n)$ *and* $\int_1^{\infty} f(x)\,dx$ *converge or diverge together.*

EXAMPLE 1.   Let $x_n = (1/n^2, 1/n)$. Does $\sum x_n$ converge?

Solution: No, because the harmonic series $\sum 1/n$ diverges by (iii).

EXAMPLE 2.   Let $\|x_n\| \leqslant 1/2^n$; prove $\sum x_n$ converges and $\|\sum_0^{\infty} x_n\| \leqslant 2$.

Solution: Verify the conditions of Theorem 11. Now

$$\|x_k + \cdots + x_{k+p}\| \leqslant \|x_k\| + \cdots + \|x_{k+p}\| \leqslant \frac{1}{2^k} + \cdots + \frac{1}{2^{k+p}}$$

$$\leqslant \sum_{j=k}^{\infty} \frac{1}{2^j} = \frac{1}{2^{k-1}}$$

(by the formula $\sum_0^{\infty} ar^n = a/(1-r)$ for the sum of a geometric series). Thus, given $\varepsilon > 0$, choose N so that $1/2^{N-1} < \varepsilon$. Hence $\sum x_k$ converges. Moreover, the partial sums satisfy

$$\|s_n\| \leqslant \sum_{k=0}^{n} \|x_k\| \leqslant \sum_{k=0}^{n} \frac{1}{2^k} \leqslant 2\,.$$

Thus the limit s also satisfies $\|s\| \leqslant 2$ by Example 2 of Section 2.7. We could also show $\sum \|x_n\|$ converges by direct comparison with the geometric series $\sum 1/2^n$.

EXAMPLE 3.   Test for convergence: $\sum_{n=1}^{\infty} n/3^n$.

Solution: The ratio test is applicable;

$$\left| \frac{a_{n+1}}{a_n} \right| = \frac{n+1}{n}\frac{1}{3} \to \frac{1}{3}$$

so the series converges.

EXAMPLE 4.   Determine whether or not $\sum_{n=1}^{\infty} n/(n^2 + 1)$ converges.

THEOREM PROOFS FOR CHAPTER 2

Solution: Observe that for $x \geq 1$, $f(x) = x/(x^2 + 1)$ is positive and continuous. Since $f'(x) = (-x^2 + 1)/(x^2 + 1)^2 \leq 0$, f is monotone decreasing.

$$\int_1^\infty \frac{x\,dx}{x^2 + 1} = \lim_{b \to \infty} \int_1^b \frac{x\,dx}{x^2 + 1}$$

$$= \lim_{b \to \infty} \left[\tfrac{1}{2} \log(x^2 + 1)\right]_1^b$$

$$= \lim_{b \to \infty} \tfrac{1}{2} \log((b^2 + 1)/2)$$

But, as $b \to \infty$, $\tfrac{1}{2} \log((b^2 + 1)/2) \to \infty$, and so the series diverges by the integral test. One can also proceed as follows: $n/(n^2 + 1) \geq n/(n^2 + n^2) = 1/2n$, so by comparison with the divergent series $(1/2) \sum 1/n$ we get divergence.

Exercises for Section 2.8

1. Determine if $\sum_{n=1}^\infty ((\sin n)/n^2, 1/n^2)$ converges.

2. Show that the series in Example 2 converges absolutely.

3. Let $\sum x_k$ converge in $\mathbb{R}^n$. Show that $x_k \to 0 = (0,\ldots,0) \in \mathbb{R}^n$.

4. Test for convergence $\sum_{n=3}^\infty (2^n + n)/(3^n - n)$.

5. Test for convergence $\sum_{n=0}^\infty n!/3^n$.

Theorem Proofs for Chapter 2

Theorem 1. *In $\mathbb{R}^n$, for every $\varepsilon > 0$ and $x \in \mathbb{R}^n$, the set $D(x, \varepsilon)$ is open.*

Proof: Choose $y \in D(x,\varepsilon)$. We must produce an ε' such that $D(y,\varepsilon') \subset D(x,\varepsilon)$. Figure 2-5 suggests that we try $\varepsilon' = \varepsilon - d(x,y)$, which is strictly positive as $d(x,y) < \varepsilon$. With this choice (which depends on y), we shall show $D(y,\varepsilon') \subset D(x,\varepsilon)$. Let $z \in D(y,\varepsilon')$, so $d(z,y) < \varepsilon'$. We need to prove that $d(z,x) < \varepsilon$. But, by the triangle inequality, $d(z,x) \leq d(z,y) + d(y,x) < \varepsilon' + d(y,x)$, and by the choice of ε', $\varepsilon' + d(y,x) = \varepsilon$. The result follows. ∎

Theorem 2.
(i) *The intersection of a finite number of open subsets of $\mathbb{R}^n$ is an open subset of $\mathbb{R}^n$.*
(ii) *The union of an arbitrary collection of open subsets of $\mathbb{R}^n$ is an open subset of $\mathbb{R}^n$.*

Proof: (i) It suffices to prove that the intersection of two open sets is open, since we can then use induction to get the general result by writing $A_1 \cap \cdots \cap A_n = (A_1 \cap \cdots \cap A_{n-1}) \cap A_n$.

Let A, B be open and $C = A \cap B$; if $C = \varnothing$, C is open by a degenerate case of the definition. Therefore, suppose $x \in C$. Since A, B are open, there are $\varepsilon, \varepsilon' > 0$, such that

$$D(x,\varepsilon) \subset A \quad \text{and} \quad D(x,\varepsilon') \subset B .$$

Let ε'' be the smaller of ε and ε'. Then $D(x,\varepsilon'') \subset D(x,\varepsilon)$ and so $D(x,\varepsilon'') \subset A$ and, similarly, $D(x,\varepsilon'') \subset B$, so $D(x,\varepsilon'') \subset C$ as required.

(ii) The proof for unions is easier. Let $U, V, \ldots$ be the open sets with union A. For $x \in A$, $x \in U$ for some U in the collection. Hence, as U is open, $D(x,\varepsilon) \subset U \subset A$ for some $\varepsilon > 0$, proving that A is open. ∎

Theorem 4. *A set $A \subset \mathbb{R}^n$ is closed iff all the accumulation points of A belong to A.*

Proof: First, suppose A is closed. Let $x \in \mathbb{R}^n$ be an accumulation point and suppose $x \notin A$. Set $U = \mathbb{R}^n \backslash A$, the complement of A. Now, by definition, U is open, contains x, and is hence a neighborhood of x; but $U \cap A = \varnothing$, contradicting the fact that x is an accumulation point. Therefore $x \in A$. Conversely, suppose A contains all its accumulation points. Let $U = \mathbb{R}^n \backslash A$ be the complement of A. We must show U is open. Let $x \in U$. Since x is not an accumulation point of A, there is an $\varepsilon > 0$ such that $D(x,\varepsilon) \cap A = \varnothing$. Hence $D(x,\varepsilon) \subset U$ and, by definition, U is open. ∎

Theorem 5. *Let $A \subset \mathbb{R}^n$. Then $\mathrm{cl}(A)$ consists of A plus all the accumulation points of A.*

Proof: Let B be the union of A and the accumulation points of A. Any closed set containing A contains B by Theorem 4. Therefore, it suffices to prove that B is closed, for B will then be the smallest closed set containing A. Let x be an accumulation point of B. We want to show that $x \in B$. Suppose that $x \notin A$ (or else $x \in B$ trivially). Now it will be shown that x is an accumulation point of A, which will complete the proof (B will be closed, by Theorem 4). Let U be an open set containing x. There exists, by definition, $y \in U \cap B$. Now, either $y \in A$, or y is an accumulation point of A. In the latter case, there exists $z \in U \cap A$. In any case, U contains some element of A (different from x, since $x \notin A$), so x is an accumulation point of A as required. ∎

Theorem 6. *Let $A \subset \mathbb{R}^n$. Then $x \in \mathrm{bd}(A)$ iff for every $\varepsilon > 0$, $D(x, \varepsilon)$ contains points of A and of $\mathbb{R}^n \backslash A$ (these points might consist of x itself).*

Proof: Let $x \in \mathrm{bd}(A) = \mathrm{cl}(A) \cap \mathrm{cl}(\mathbb{R}^n \backslash A)$. Now, either $x \in A$ or $x \in \mathbb{R}^n \backslash A$. If $x \in A$, then, by Theorem 5, x is an accumulation point of $\mathbb{R}^n \backslash A$, and the conclusion follows. The case $x \in \mathbb{R}^n \backslash A$ and the converse are similar. ∎

Theorem 7. *A sequence x_k in $\mathbb{R}^n$ converges to $x \in \mathbb{R}^n$ iff for every $\varepsilon > 0$, there is an N such that $n \geqslant N$ implies $\|x - x_n\| < \varepsilon$.*

Proof: Suppose $x_k \to x$, and $\varepsilon > 0$. Since $D(x,\varepsilon)$ is open, there is an N so $k \geqslant N$ implies $x_k \in D(x,\varepsilon)$, or $d(x,x_k) = \|x - x_k\| < \varepsilon$ as required. Conversely, suppose the condition holds and U is a neighborhood of x. Find $\varepsilon > 0$ so $D(x,\varepsilon) \subset U$. Then there is an N so $k \geqslant N$ implies $\|x_k - x\| < \varepsilon$, that is, $x_k \in D(x,\varepsilon) \subset U$. ∎

Theorem 8. *$x_k \to x$ iff the components of x_k converge to the components of x as sequences of real numbers.*

Proof: Let $x_k = (x_k^1, \ldots, x_k^n)$ (we use superscripts for the components to avoid confusion with the k). Suppose $x_k \to x = (x^1, \ldots, x^n)$. Then, given $\varepsilon > 0$, choose N so

$k \geqslant N$ implies $\|x_k - x\| < \varepsilon$. But,

$$|x_k^1 - x^1| \leqslant \|x_k - x\| = \left(\sum_{i=1}^{n} (x_k^i - x^i)^2 \right)^{1/2}$$

(why?), so that $k \geqslant N$ also implies $|x_k^1 - x^1| < \varepsilon$. Thus $x_k^1 \to x^1$ and similarly, $x_k^i \to x^i$.

Conversely, suppose $x_k^i \to x^i$, for all i. Given $\varepsilon > 0$, choose N so that $|x_k^i - x^i| < \varepsilon/\sqrt{n}$ for $k \geqslant N$ and all $i = 1, \ldots, n$ (where N is the maximum of the N's required for each i). Then

$$\|x_k - x\| = \left(\sum_{i=1}^{n} (x_k^i - x^i)^2 \right)^{1/2} < \left(\sum_{i=1}^{n} \frac{\varepsilon^2}{n} \right)^{1/2} = \varepsilon$$

for $k \geqslant N$, so $x_k \to x$. ∎

Theorem 9.

(i) *A set $A \subset \mathbb{R}^n$ is closed iff for every sequence $x_k \in A$ which converges, the limit lies in A.*

(ii) *For a set $B \subset \mathbb{R}^n$, $x \in \mathrm{cl}(B)$ iff there is a sequence $x_k \in B$ with $x_k \to x$.*

Proof:

(i) First, suppose A is closed. Suppose $x_k \to x$ and $x \notin A$. Then, x is an accumulation point of A, for any neighborhood of x contains $x_k \in A$ for k large. Hence $x \in A$, by Theorem 4.

Conversely, we shall use Theorem 4 to show that A is closed. Let x be an accumulation point of A, and choose $x_k \in D(x, 1/k) \cap A$. Then $x_k \to x$, since for any $\varepsilon > 0$, we can choose $N \geqslant 1/\varepsilon$; then $k \geqslant N$ implies $x_k \in D(x, \varepsilon)$; see Figure 2-15. Hence, by hypothesis, $x \in A$, and so A is closed.

(ii) The argument here is similar and we shall leave it as an exercise (Exercise 7). ∎

Theorem 10. *A sequence x_k in $\mathbb{R}^n$ converges to a point in $\mathbb{R}^n$ iff it is a Cauchy sequence.*

Proof: If x_k converges to x, then for $\varepsilon > 0$, choose N so that $k \geqslant N$ implies $\|x_k - x\| < \varepsilon/2$. Then, for $k, l \geqslant N$, $\|x_k - x_l\| = \|(x_k - x) + (x - x_l)\| \leqslant \|x_k - x\| + \|x - x_l\| < \varepsilon/2 + \varepsilon/2 = \varepsilon$, by the triangle inequality.

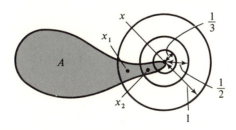

FIGURE 2-15   Accumulation points of a set.

52 TOPOLOGY OF $\mathbb{R}^n$

Conversely, suppose x_k is a Cauchy sequence. Then, since $|x_k^i - x_l^i| \leq \|x_k - x_l\|$, the components are also Cauchy sequences on the real line. By completeness of $\mathbb{R}$ and Theorem 3 of Chapter 1, x_k^i converges to, say, x^i. Then, by Theorem 8, x_k converges to $x = (x^1,\ldots,x^n)$. ∎

Theorem 11. *A series $\sum x_k$ in $\mathbb{R}^n$ converges iff for every $\varepsilon > 0$, there is an N such that $k \geq N$ implies $\|x_k + x_{k+1} + \cdots + x_{k+p}\| < \varepsilon$ for all integers $p = 0, 1, 2, \ldots$.*

Proof: Let $s_k = \sum_{i=1}^k x_i$. Then, by Theorem 10, x_k converges iff s_k is a Cauchy sequence. This is true iff for every $\varepsilon > 0$ there is an N such that $l \geq N$ implies $\|s_l - s_{l+q}\| < \varepsilon$ for all $q = 1, 2, \ldots$. But, $\|s_{l+q} - s_l\| = \|x_{l+1} + \cdots + x_{l+q}\|$, so the result follows with $k = l + 1$ and $p = q - 1$. ∎

Theorem 12. *If $\sum x_k$ converges absolutely, then $\sum x_k$ converges.*

Proof: This follows at once from Theorem 11 with the use of the triangle inequality $\|x_k + \cdots + x_{k+p}\| \leq \|x_k\| + \cdots + \|x_{k+p}\|$. ∎

Theorem 13.

(i) If $|r| < 1$, then $\sum_{n=0}^\infty r^n$ converges to $1/1 - r)$ and diverges (does not converge) if $|r| \geq 1$.

(ii) *Comparison test:* If $\sum_{k=1}^\infty a_k$ converges, $a_k \geq 0$, and $0 \leq b_k \leq a_k$, then $\sum_{k=1}^\infty b_k$ converges; if $\sum_{k=1}^\infty c_k$ diverges, $c_k \geq 0$, and $0 \leq c_k \leq d_k$, then $\sum_{k=1}^\infty d_k$ diverges.

(iii) *p-series test:* $\sum_{n=1}^\infty n^{-p}$ converges if $p > 1$ and diverges to ∞ (that is, the partial sums increase without bound) if $p \leq 1$.

(iv) *Ratio test:* Suppose that $\lim_{n \to \infty} |a_{n+1}/a_n|$ exists and is strictly less than 1. Then $\sum_{n=1}^\infty a_n$ converges absolutely. If the limit is strictly greater than 1, the series diverges. If the limit equals 1, the test is inconclusive.

(v) *Root test:* Suppose that $\lim_{n \to \infty} (|a_n|)^{1/n}$ exists and is strictly less than 1. Then $\sum_{n=1}^\infty a_n$ converges absolutely. If the limit is strictly greater than 1, the series diverges; if the limit equals 1, the test is inconclusive.

(vi) *The integral test:* If f is continuous, non-negative, and monotone decreasing on $[1, +\infty[$, then $\sum_{n=1}^\infty f(n)$ and $\int_1^\infty f(x)\,dx$ converge or diverge together.

Proof:
(i) We have, by elementary algebra, that

$$1 + r + r^2 + \cdots + r^n = \frac{1 - r^{n+1}}{1 - r}$$

if $r \neq 1$. Clearly, $r^{n+1} \to 0$ as $n \to \infty$ if $|r| < 1$, and $|r|^{n+1} \to \infty$ if $|r| > 1$. Thus we have convergence if $|r| < 1$ and divergence if $|r| > 1$. Obviously, $\sum_{n=0}^\infty r^n$ diverges if $|r| = 1$, since $r^n \not\to 0$.

(ii) The partial sums of the series $\sum_{k=1}^\infty a_k$ form a Cauchy sequence and thus the partial sums of the series $\sum_{k=1}^\infty b_k$ also form a Cauchy sequence, since for any k and p we have $b_k + b_{k+1} + \cdots + b_{k+p} \leq a_k + a_{k+1} + \cdots + a_{k+p}$. Hence $\sum_{k=1}^\infty b_k$ converges. A positive series can diverge only to $+\infty$, so given $M > 0$, we can find k_0

such that $k \geqslant k_0$ implies that $c_1 + c_2 + \cdots + c_k \geqslant M$. Therefore, for $k \geqslant k_0$, $d_1 + d_2 + \cdots + d_k \geqslant M$, so $\sum_{k=1}^{\infty} d_k$ also diverges to ∞.

(iii) First suppose that $p \leqslant 1$; in this case $1/n^p \geqslant 1/n$ for all $n = 1, 2, \ldots$. Therefore, by (ii), $\sum_{n=1}^{\infty} 1/n^p$ will diverge if $\sum_{n=1}^{\infty} 1/n$ diverges. We recall the proof of this from calculus.* If $s_k = 1/1 + 1/2 + \cdots + 1/k$, then s_k is a strictly increasing sequence of positive real numbers. Write s_{2^k} as follows:

$$s_{2^k} = 1 + \frac{1}{2} + \left(\frac{1}{3} + \frac{1}{4}\right) + \left(\frac{1}{5} + \frac{1}{6} + \frac{1}{7} + \frac{1}{8}\right)$$

$$+ \cdots + \left(\frac{1}{2^{k-1}+1} + \cdots + \frac{1}{2^k}\right) \geqslant 1 + \frac{1}{2} + \left(\frac{1}{4} + \frac{1}{4}\right) + \left(\frac{1}{8} + \frac{1}{8} + \frac{1}{8} + \frac{1}{8}\right) + \cdots$$

$$= 1 + \frac{1}{2} + \left(\frac{1}{2}\right) + \left(\frac{1}{2}\right) + \cdots + \left(\frac{1}{2}\right) = 1 + \frac{k}{2}.$$

Hence s_k can be made arbitrarily large if k is made sufficiently large; thus $\sum_{n=1}^{\infty} 1/n$ diverges.

Now suppose that $p > 1$. If we let

$$s_k = \frac{1}{1^p} + \frac{1}{2^p} + \frac{1}{3^p} + \cdots + \frac{1}{k^p},$$

then s_k is an increasing sequence of positive real numbers. On the other hand,

$$s_{2^k - 1} = \frac{1}{1^p} + \left(\frac{1}{2^p} + \frac{1}{3^p}\right) + \left(\frac{1}{4^p} + \frac{1}{5^p} + \frac{1}{6^p} + \frac{1}{7^p}\right) \cdots$$

$$+ \left(\frac{1}{(2^{k-1})^p} + \cdots + \frac{1}{(2^k - 1)^p}\right) \leqslant \frac{1}{1^p} + \frac{2}{2^p} + \frac{4}{4^p} + \frac{2^{k-1}}{(2^{k-1})^p}$$

$$= \frac{1}{1^{p-1}} + \frac{1}{2^{p-1}} + \frac{1}{4^{p-1}} + \cdots + \frac{1}{(2^{k-1})^{p-1}} < \frac{1}{1 - \dfrac{1}{2^{p-1}}}$$

(why?). Thus the sequence $\{s_k\}$ is bounded from above by $1/(1 - 1/2^{p-1})$; hence $\sum_{n=1}^{\infty} 1/n^p$ converges.

(iv) Suppose that $\lim_{n \to \infty} |a_{n+1}/a_n| = r < 1$. Choose r' such that $r < r' < 1$ and let N be such that $n \geqslant N$ implies that

$$\left|\frac{a_{n+1}}{a_n}\right| < r'.$$

Then $|a_{N+p}| < |a_N| (r')^p$. Consider the series $|a_1| + \cdots + |a_N| + |a_N| r' + |a_N| (r')^2 +$

---

* We can also prove (iii) by using the integral test for positive series (see vi of the theorem). The demonstration given here however also proves the *Cauchy Condensation Test*: Let $\sum a_n$ be a series of positive terms with $a_{n+1} \leqslant a_n$. Then $\sum a_n$ converges iff $\sum_{j=1}^{\infty} 2^j a_{2^j}$ converges (see G. J. Porter, "An Alternative to the Integral Test for Infinite Series," *American Mathematical Monthly* **79** (1972), page 634).

$|a_N|\,(r')^3 + \cdots$. This converges to

$$|a_1| + \cdots + |a_{N-1}| + \frac{|a_N|}{1 - r'}\,.$$

By (ii) we can conclude that $\sum_{k=1}^{\infty} |a_k|$ converges. If $\lim_{n \to \infty} |a_{n+1}/a_n| = r > 1$, choose r' such that $1 < r' < r$ and let N be such that $n \geqslant N$ implies that $|a_{n+1}/a_n| > r'$. Hence $|a_{N+p}| > (r')^p\, |a_N|$, so $\lim_{n \to \infty} |a_N| = \infty$, whereas the limit would have to be zero if the sum converged. Thus $\sum_{k=1}^{\infty} a_k$ diverges. To see that the test fails if $\lim_{n \to \infty} |a_{n+1}/a_n| = 1$, consider the series $1 + 1 + 1 + \cdots$, and $\sum_{n=1}^{\infty} 1/n^p$ for $p > 1$. In both cases $\lim_{n \to \infty} |a_{n+1}/a_n| = 1$, but the first series diverges and the second converges.

(v) Suppose that $\lim_{n \to \infty} (|a_n|)^{1/n} = r < 1$. Choose r' such that $r < r' < 1$ and N such that $n \geqslant N$ implies that $|a_n|^{1/n} < r'$; in other words, $|a_n| < (r')^n$. The series $|a_1| + |a_2| + \cdots + |a_{N-1}| + (r')^N + (r')^{N+1} + \cdots$ converges to $|a_1| + |a_2| + \cdots + |a_{N-1}| + (r')^N/(1 - r')$, so by (ii), $\sum_{k=1}^{\infty} a_k$ converges. If $\lim_{n \to \infty} (|a_n|)^{1/n} = r > 1$, choose $1 < r' < r$ and N such that $n \geqslant N$ implies that $|a_n|^{1/n} > r'$ or, in other words, $|a_n| > (r')^n$. Hence $\lim_{n \to \infty} |a_n| = \infty$ and therefore, $\sum_{k=1}^{\infty} a_k$ diverges.

To show that the test fails when $\lim_{n \to \infty} (|a_n|)^{1/n} = 1$, observe that, by elementary analysis,

$$\lim_{n \to \infty} \left(\frac{1}{n}\right)^{1/n} = 1 \qquad \text{and} \qquad \lim_{n \to \infty} \left(\frac{1}{n^2}\right)^{1/n} = 1$$

(take logarithms and use the fact that $(\log x)/x \to 0$ as $x \to \infty$.) But $\sum_{n=1}^{\infty} 1/n$ diverges and $\sum_{n=1}^{\infty} 1/n^2$ converges.

(vi) For this part we accept some elementary facts about integrals from calculus. In Figure 2-16a the rectangles of areas $a_1, a_2, \ldots, a_n$ enclose more area than that under the curve from $x = 1$ to $x = n + 1$. Therefore, we have

$$a_1 + a_2 + \cdots + a_n \geqslant \int_1^{n+1} f(x)\, dx$$

If we now consider Figure 2-16(b) and take the area from $x = 1$ to $x = n$ we have

$$a_2 + a_3 + \cdots + a_n \leqslant \int_1^{n} f(x)\, dx$$

Adding a_1 to both sides

$$a_1 + a_2 + a_3 + \cdots + a_n \leqslant a_1 + \int_1^{n} f(x)\, dx$$

Combining our two results, yields

$$\int_1^{n+1} f(x)\, dx \leqslant a_1 + a_2 + \cdots + a_n \leqslant a_1 + \int_1^{n} f(x)\, dx$$

If the integral $\int_1^{\infty} f(x)\, dx$ is finite, then the right-hand inequality implies that the series $\sum_{n=1}^{\infty} a_n$ is also finite by the completeness property of $\mathbb{R}$; p. 12. But if $\int_1^{\infty} f(x)\, dx$ is

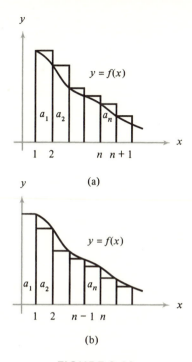

FIGURE 2-16

infinite, the left-hand inequality shows that the series is also infinite. Hence, the series and the integral converge or diverge together. ∎

Worked Examples for Chapter 2

1. Let $S = \{(x_1, x_2) \in \mathbb{R}^2 \mid |x_1| \leqslant 1, |x_2| < 1\}$. Is S open or closed or neither? What is the interior of S?

Solution: S is not open, since there is no neighborhood around any point of S with $x_1 = 1$ which is entirely contained in S. See Figure 2-17. On the other hand, S is not closed, since

$$\mathbb{R}^2 \backslash S = \{(x_1, x_2) \in \mathbb{R}^2 \mid |x_1| > 1, |x_2| \geqslant 1\}$$

and no neighborhood around a point of $\mathbb{R}^2 \backslash S$ with $x_2 = 1$ is contained in $\mathbb{R}^2 \backslash S$.

Alternatively, we see that S is not closed by noting that the sequence $(0, 1 - 1/n)$ converges, but the limit point $(0, 1)$ does not lie in S (see Theorem 9).

We assert that $\text{int}(S) = \{(x_1, x_2) \in \mathbb{R}^2 \mid |x_1| < 1, |x_2| < 1\}$. We check this by showing that the members of this set are the interior points of S. If $|x_1| < 1, |x_2| < 1$, then the disc of center (x_1, x_2) and radius $r = \text{minimum}\{1 - |x_1|, 1 - |x_2|\}$ lies in S. The other points of S are not interior points as we have seen.

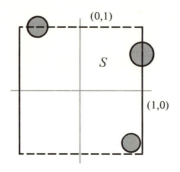

FIGURE 2-17

As the student becomes more familiar with this type of argument, some of the details may be omitted.

2. Show that if x is an accumulation point of a set $S \subset \mathbb{R}^n$ then every open set containing x contains infinitely many points of S.

Solution: We use proof by contradiction. Suppose there were an open set U around x containing only finitely many points of S. Let $x_1, x_2, \ldots, x_m$ be the points of S in U other than x. Let ε be the minimum of the numbers $d(x,x_1), d(x,x_2), \ldots, d(x,x_m)$, so that $\varepsilon > 0$. Then $D(x,\varepsilon)$ contains no points of S other than x, which contradicts the fact that x is an accumulation point of S.

3. If $x = \sup(S)$ for $S \subset \mathbb{R}$, then $x \in \mathrm{cl}(S)$.

Solution: By Theorem 5, it suffices to show that either $x \in S$, or that x is an accumulation point of S. By Theorem 2 in Chapter 1, for any $\varepsilon > 0$, there is a $y \in S$ with $d(x,y) < \varepsilon$. This means that if $x \notin S$, x is an accumulation point of S.

4. A sequence can converge to at most one point (limits are unique).

Solution: Let $x_k \to x$ and $x_k \to y$. Given $\varepsilon > 0$, choose N such that $k \geq N$ implies $\|x_k - x\| < \varepsilon/2$, and M such that $k \geq M$ implies $\|x_k - y\| < \varepsilon/2$. Then, if $k \geq N$ $k \geq M$, we have $\|x - y\| \leq \|x - x_k\| + \|x_k - y\| < \varepsilon$ (by the triangle inequality). Since $0 \leq \|x - y\| < \varepsilon$ holds for every $\varepsilon > 0$, $\|x - y\| = 0$ and so $x = y$.

5. Write $f = O(g)$ if $g(x) > 0$ for x sufficiently large and if $f(x)/g(x)$ is bounded for x sufficiently large. Write $f = o(g)$ if f/g goes to zero as x goes to $+\infty$. Also write $f \sim g$ (read f is asymptotic to g) if $f/g \to 1$ as $x \to \infty$. Prove the following:
(a) $x^2 + x = O(x^2)$
(b) $x^2 + x \sim x^2$
(c) $e^{\sqrt{\log x}} = o(x)$.

Solution: We note that if f is asymptotic to g then it will follow automatically that $f = O(g)$ (why?). Thus (a) will follow from (b). But (b) is easy since we know that $(x^2 + x)/x^2 = 1 + 1/x$ goes to 1 as x goes to infinity. To prove (c) we note that $e^{\log x} = x$, so $e^{(\sqrt{\log x})}/x = e^{(\sqrt{\log x} - \log x)}$. But since $\log x \to \infty$ as $x \to \infty$, $\sqrt{\log x}/\log x \to 0$ as $x \to \infty$ so for x large, $\sqrt{\log x} \leq (\log x)/2$ and hence for x large, $e^{(\sqrt{\log x})}/x \leq e^{-(\log x)/2}$ which goes to zero as $x \to \infty$.

Exercises for Chapter 2

1. Discuss whether the following sets are open or closed:
   (a) $]1,2[$ in $\mathbb{R}^1 = \mathbb{R}$
   (b) $[2,3]$ in $\mathbb{R}$
   (c) $\bigcap_{n=1}^{\infty} [-1,1/n[$ in $\mathbb{R}$
   (d) $\mathbb{R}^n$ in $\mathbb{R}^n$
   (e) A hyperplane in $\mathbb{R}^n$
   (f) $\{r \in]0,1[\mid r$ is rational$\}$ in $\mathbb{R}$
   (g) $\{(x,y) \in \mathbb{R}^2 \mid 0 < x \leqslant 1\}$ in $\mathbb{R}^2$
   (h) $\{x \in \mathbb{R}^n \mid \|x\| = 1\}$ in $\mathbb{R}^n$.

2. Determine the interiors, closures, and boundaries of the sets in Exercise 1.

3. Let U be open in $\mathbb{R}^n$ and $U \subset A$. Then show that $U \subset \text{int}(A)$. What is the corresponding statement for closed sets?

4. (a) Show that if $x_n \to x$ in $\mathbb{R}^m$, then $x \in \text{cl}\{x_1, x_2, \ldots\}$. When is x an accumulation point?
   (b) Can a sequence have more than one accumulation point?
   (c) If x is an accumulation point of a set A, prove that there is a sequence of *distinct* points of A converging to x.

5. Show that $x \in \text{int}(A)$ iff there is an $\varepsilon > 0$ so that $D(x,\varepsilon) \subset A$.

6. Find the limits, if they exist, of these sequences in $\mathbb{R}^2$.
   (a) $\left((-1)^n, \dfrac{1}{n}\right)$
   (b) $\left(1, \dfrac{1}{n}\right)$
   (c) $\left(\left(\dfrac{1}{n}\right)(\cos n\pi), \left(\dfrac{1}{n}\right)\left(\sin\left(n\pi + \dfrac{\pi}{2}\right)\right)\right)$
   (d) $\left(\dfrac{1}{n}, n^{-n}\right)$.

7. Let U be open in $\mathbb{R}^n$. Show that $U = \text{cl}(U)\backslash\text{bd}(U)$. Is this true for any set in $\mathbb{R}^n$?

8. Let $S \subset \mathbb{R}$ and S be bounded above. Show that $\sup(S) \in S$ is closed.

9. Show that $\text{cl}(A) = \mathbb{R}^n\backslash(\text{int}(\mathbb{R}^n\backslash A))$.

10. Determine which of the following statements are true.
    (a) $\text{int}(\text{cl}(A)) = \text{int}(A)$
    (b) $\text{cl}(A) \cap A = A$
    (c) $\text{cl}(\text{int}(A)) = A$
    (d) $\text{bd}(\text{cl}(A)) = \text{bd}(A)$
    (e) If A is open, $\text{bd}(A) \subset \mathbb{R}^n\backslash A$.

11. Show that in $\mathbb{R}^n$, $x_m \to x$ iff for every $\varepsilon > 0$, there is an N such that $m \geqslant N$ implies $\|x_m - x\| \leqslant \varepsilon$ (this differs from Theorem 7 in that here "$<\varepsilon$" is replaced by "$\leqslant\varepsilon$").

12. Prove the following properties (for subsets of $\mathbb{R}^n$).
    (a) $\text{int}(\text{int}(A)) = \text{int}(A)$
    (b) $\text{int}(A \cup B) \supset \text{int}(A) \cup \text{int}(B)$
    (c) $\text{int}(A \cap B) = \text{int}(A) \cap \text{int}(B)$.

13. Show that $\text{cl}(A) = A \cup \text{bd}(A)$ and $\text{int}(A) = A\backslash\text{bd}(A)$.

14. Prove the following (for subsets of $\mathbb{R}^n$).
    (a) $\mathrm{cl}(\mathrm{cl}(A)) = \mathrm{cl}(A)$
    (b) $\mathrm{cl}(A \cup B) = \mathrm{cl}(A) \cup \mathrm{cl}(B)$
    (c) $\mathrm{cl}(A \cap B) \subset \mathrm{cl}(A) \cap \mathrm{cl}(B)$.

15. Prove the following (for subsets of $\mathbb{R}^n$).
    (a) $\mathrm{bd}(A) = \mathrm{bd}(\mathbb{R}^n \backslash A)$
    (b) $\mathrm{bd}(\mathrm{bd}(A)) \subset \mathrm{bd}(A)$
    (c) $\mathrm{bd}(A \cup B) \subset \mathrm{bd}(A) \cup \mathrm{bd}(B) \subset \mathrm{bd}(A \cup B) \cup A \cup B$
    (d) $\mathrm{bd}(\mathrm{bd}(\mathrm{bd}(A))) = \mathrm{bd}(\mathrm{bd}(A))$.

16. Let $a_1 = \sqrt{2}, a_2 = (\sqrt{2})^{\sqrt{2}}, \ldots, a_{n+1} = (\sqrt{2})^{a_n}$. Show $\lim\limits_{n \to \infty} a_n$ exists and compute the limit.

17. If $\sum x_m$ converges absolutely in $\mathbb{R}^n$, then show that $\sum x_m \sin m$ converges.

18. If $x, y \in \mathbb{R}^n$ and $x \neq y$, then prove that there exist open sets U and V such that $x \in U, y \in V$, and $U \cap V = \varnothing$.

19. Define a *limit point* of a set A to be a point $x \in \mathbb{R}^n$, such that if U is any neighborhood of x, then $U \cap A \neq \varnothing$.
    (a) What is the difference between limit points and accumulation points? Give examples.
    (b) If x is a limit point of A, then show that there is a sequence $x_n \in A$ with $x_n \to x$.
    (c) If x is an accumulation point of A, then show that x is a limit point of A. Is the converse true?
    (d) If x is a limit point of A and $x \notin A$, then show that x is an accumulation point.
    (e) Prove: a set is closed iff it contains all of its limit points.

20. For a set A and $x \in \mathbb{R}^n$, let $d(x,A) = \inf\{d(x,y) \mid y \in A\}$, and for $\varepsilon > 0$, let $D(A,\varepsilon) = \{x \mid d(x,A) < \varepsilon\}$.
    (a) Show that $D(A,\varepsilon)$ is open.
    (b) Let $A \subset \mathbb{R}^n$ and $N_\varepsilon = \{x \in \mathbb{R}^n \mid d(x,A) \leqslant \varepsilon\}$, where $\varepsilon > 0$. Show that N_ε is closed and that A is closed iff $A = \bigcap \{N_\varepsilon \mid \varepsilon > 0\}$.
    (c) Give some examples.

21. Prove that a sequence x_k is a Cauchy sequence in $\mathbb{R}^n$ iff for every neighborhood U of 0, there is an N such that $k, l \geqslant N$ implies $x_k - x_l \in U$.

22. Prove that the interior of a set $A \subset \mathbb{R}^n$ is the union of all the subsets of A which are open. Deduce that A is open iff $A = \mathrm{int}(A)$. Also, give a direct proof of the latter fact using the definitions.

23. Prove Theorem 3. [Hint: Use Exercise 14 of the Introductory chapter.]

24. Identify $\mathbb{R}^{n+m}$ with $\mathbb{R}^n \times \mathbb{R}^m$. Show that $A \subset \mathbb{R}^{n+m}$ is open iff for each $(x,y) \in A$, with $x \in \mathbb{R}^n, y \in \mathbb{R}^m$ there exist open sets $U \subset \mathbb{R}^n, V \subset \mathbb{R}^m$ with $x \in U, y \in V$, such that $U \times V \subset A$. Deduce that the product of open sets is open.

25. Prove that a set $A \subset \mathbb{R}^n$ is open iff we can write A as the union of some family of ε-discs.

26. Define the sequence of numbers a_n by

$$a_0 = 1, a_1 = 1 + \frac{1}{1 + a_0}, \ldots, a_n = 1 + \frac{1}{1 + a_{n-1}}.$$

Show a_n is a convergent sequence. Find the limit.

27. Let $S = \{(x,y) \in \mathbb{R}^2 \mid xy > 1\}$ and $B = \{d((x,y),(0,0)) \mid (x,y) \in S\}$. Find inf($B$).

28. Give examples of :
    (a) an infinite set in $\mathbb{R}$ with no accumulation points
    (b) a non-empty subset of $\mathbb{R}$ which is contained in its set of accumulation points
    (c) a subset of $\mathbb{R}$ which has infinitely many accumulation points but which contains none of them
    (d) a set A such that bd(A) = cl(A).

29. Let $A, B \subset \mathbb{R}^n$ and x be an accumulation point of $A \cup B$. Must x be an accumulation point of either A or B?

30. Show that any open set in $\mathbb{R}$ is a union of disjoint open intervals. Is this sort of result true in $\mathbb{R}^n$ for $n > 1$ where we define an open interval as the Cartesian product of n open intervals, $]a_1,b_1[\times \cdots \times]a_n,b_n[$?

31. Let A' denote the set of accumulation points of a set A. Prove that A' is closed. Is $(A')' = A'$ true for all A?

32. Let $A \subset \mathbb{R}^n$ be closed and $x_n \in A$ a Cauchy sequence. Prove x_n converges to a point in A.

33. Let s_n be a bounded sequence of real numbers. Assume $2s_n \leqslant s_{n-1} + s_{n+1}$. Show $\underset{n \to \infty}{\text{limit}}(s_{n+1} - s_n) = 0$.

34. Let $x_n \in \mathbb{R}^k$ and $d(x_{n+1},x_n) \leqslant rd(x_n,x_{n-1})$ where $0 \leqslant r < 1$. Show x_n converges.

35. Show that any family of disjoint non-empty open sets of real numbers is countable.

36. Let $A, B \subset \mathbb{R}^n$ be closed sets. Does $A + B = \{x + y \mid x \in A \text{ and } y \in B\}$ have to be closed?

37. For $A \subset \mathbb{R}^n$, prove bd(A) = $[A \cap \text{cl}(\mathbb{R}^n\backslash A)] \cup [\text{cl}(A)\backslash A]$.

38. Let $x_k \in \mathbb{R}^n$ satisfy $\|x_k - x_l\| \leqslant 1/k + 1/l$. Prove that x_k converges.

39. Let $S \subset \mathbb{R}$ be bounded above and below. Prove sup(S) $-$ inf(S) = sup$\{x - y \mid x \in S \text{ and } y \in S\}$.

40. Suppose in $\mathbb{R}$ that for all n, $a_n \leqslant b_n$, $a_n \leqslant a_{n+1}$, and $b_{n+1} \leqslant b_n$. Prove that a_n converges.

41. Let A_n be subsets of $\mathbb{R}^n$, $A_{n+1} \subset A_n$, $A_n \neq \varnothing$, but $\bigcap_{n=1}^{\infty} A_n = \varnothing$. Suppose $x \in \bigcap_{n=1}^{\infty} \text{cl}(A_n)$. Show x is an accumulation point of A_1.

42. Let $A \subset \mathbb{R}^n$ and $x \in \mathbb{R}^n$. Define $d(x,A) = \inf\{d(x,y) \mid y \in A\}$ (See also Exercise 20). Must there be a $z \in A$ such that $d(x,A) = d(x,z)$?

43. Let $x_1 = \sqrt{3}, \ldots, x_n = \sqrt{3 + x_{n-1}}$. Compute $\underset{n \to \infty}{\text{limit}} x_n$.

44. A set $A \subset \mathbb{R}^n$ is said to be *dense* in $B \subset \mathbb{R}^n$ if $B \subset \text{cl}(A)$. If A is dense in $\mathbb{R}^n$ and U is open, prove $A \cap U$ is dense in U. Is this true if U is not open?

45. Show that $x^{\log x} = o(e^x)$ (see Example 5, p. 56).

46. If $f = o(g)$ and if $g(x) \to 0$ as $x \to \infty$ then show that $e^f = o(e^g)$.

47. Show that $\text{limit}_{x \to \infty}(x \log x)/e^x = 0$ by showing that $x = o(e^{x/2})$ and that $\log x = o(e^{x/2})$.

48. Prove the following generalizations of the ratio and root tests:
    (a) if $a_n > 0$ and $\lim \sup_{n \to \infty} a_{n+1}/a_n < 1$ then $\sum a_n$ converges and if $\lim \inf_{n \to \infty} a_{n+1}/a_n > 1$ then $\sum a_n$ diverges,
    (b) if $a_n \geqslant 0$ and if $\lim \sup_{n \to \infty} \sqrt[n]{a_n} < 1$ (respectively > 1) then $\sum a_n$ converges (respectively diverges). (See Exercise 10, Chapter 1 for treatment of lim sup and lim inf).

49. Prove *Raabe's test*: if $a_n > 0$ and if $a_{n+1}/a_n \leqslant 1 - A/n$ for $A > 1$ some fixed constant and n sufficiently large, then $\sum a_n$ converges. Similarly show that if $a_{n+1}/a_n \geqslant 1 - 1/n$ then $\sum a_n$ diverges. [Hint: show that $a_n = O(n^{-A})$ by considering $P_n = \prod_{k=1}^n (1 - A/k)$ and establishing that $\log P_n = -A \log n + O(1)$.]
    Use the above result to prove that the hypergeometric series converges whose general term is
$$a_n = \frac{\alpha(\alpha + 1) \cdots (\alpha + n - 1)\beta(\beta + 1) \cdots (\beta + n - 1)}{1 \cdot 2 \cdots n \cdot \gamma(\gamma + 1) \cdots (\gamma + n - 1)}$$

    where α, β, and γ are not negative integers.

50. Show that for x sufficiently large the function $f(x) = (x \cos^2 x + \sin^2 x)e^{x^2}$ is monotonic and tends to $+\infty$, but that neither the ratio $f(x)/x^{1/2}e^{x^2}$ nor its reciprocal is bounded. .

51. (a) If $a_n > 0, n = 1, 2, \ldots$, show that
$$\lim \inf u_{n+1}/u_n \leqslant \lim \inf \sqrt[n]{u_n} \leqslant \lim \sup \sqrt[n]{u_n} \leqslant \lim \sup u_{n+1}/u_n .$$

    (See Exercise 10, Chapter 1 for the definition of lim sup and lim inf.)
    (b) Deduce that if $\lim u_{n+1}/u_n = A$ then $\lim \sqrt[n]{u_n} = A$.
    (c) Show that the converse of part (b) is false by use of the sequence $u_{2n} = u_{2n+1} = 2^{-n}$.
    (d) $\lim_{n \to \infty} \frac{1}{n} \sqrt[n]{n!} = ?$

52. Test the following series for convergence.
    (a) $\sum_{k=0}^{\infty} \frac{e^{-k}}{\sqrt{k+1}}$
    (b) $\sum_{k=0}^{\infty} \frac{k}{k^2 + 1}$
    (c) $\sum_{n=0}^{\infty} \frac{\sqrt{n+1}}{n^2 - 3n + 1}$
    (d) $\sum_{k=1}^{\infty} \frac{\log(k+1) - \log k}{\tan^{-1}(2/k)}$
    (e) $\sum_{n=1}^{\infty} \sin(n^{-\alpha})$, α real, > 0
    (f) $\sum_{n=1}^{\infty} \frac{n^3}{3^n} .$

Chapter **3**

Compact and Connected Sets

In this chapter, we study two of the most important and useful kinds of sets in $\mathbb{R}^n$. Intuitively, we want to say that a set in $\mathbb{R}^n$ is compact when it is closed and is contained in a bounded region, and that a set is connected when it is in "one piece." As usual, it is necessary to turn these ideas into rigorous definitions. Figure 3-1 gives some examples. The fruitfulness of these notions is revealed in Chapter 4, where they are applied to the study of continuous functions.

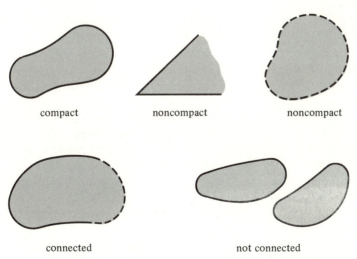

compact   noncompact   noncompact

connected   not connected

FIGURE 3-1   Compact and connected sets in $\mathbb{R}^2$.

3.1   Compact Sets: The Heine-Borel and Bolzano-Weierstrass Theorems

Our first task will be to introduce some terminology prior to giving a precise definition of compactness for sets in $\mathbb{R}^n$. We say that a set $A \subset \mathbb{R}^n$ is *bounded* iff there is a constant $M \geqslant 0$ such that $A \subset D(0,M)$. Thus a set is bounded when it can be enclosed in some (large) disc $D(0,M)$ about the origin; in other words, $\|x\| < M$ for all $x \in A$. A *cover* of a set A is a collection $\{U_i\}$ of sets whose union contains A; it is an *open cover* if each U_i is open. A *subcover* of a given cover is merely a subcollection whose union also contains A or, as we say, *covers A*; it is a *finite subcover* if the subcollection contains only a finite number of sets. For example, the set of discs $\{D((x,0),1) \mid x \in \mathbb{R}\}$ in $\mathbb{R}^2$ covers the real axis, and the subcollection of all discs $D((n,0),1)$ centered at integer points on the real line forms a subcover. Note that the discs $D((n,0),1)$ centered at even integer points on the real line do not form a subcovering (why?).

Note: Open covers are not necessarily countable collections of open sets. We now state the main theorem and an associated definition.

> **Theorem 1.**   *Let $A \subset \mathbb{R}^n$. Then the following conditions are equivalent:*
> (i)   *A is closed and bounded.*
> (ii)   *Every open cover of A has a finite subcover.*
> (iii)   *Every sequence in A has a subsequence which converges to a point of A.*

> **Definition 1.**   A subset of $\mathbb{R}^n$ satisfying one (and hence all) of the conditions (i), (ii), (iii) of Theorem 1 is called *compact*.

The equivalence of (i) and (ii) is often called the *Heine-Borel theorem*, while the assertion that (i) and (iii) are equivalent is called the *Bolzano-Weierstrass theorem*.

Note: For metric spaces, in general (ii) and (iii) are equivalent but (i) is *not* equivalent to (ii) and (iii); for arbitrary metric spaces one *defines* compactness by either of properties (ii) or (iii). The equivalence of (i) with (ii) and (iii) is a *special* and very important property of $\mathbb{R}^n$.

The Bolzano-Weierstrass theorem is very reasonable intuitively. If A is bounded, then any sequence of points in A must "bunch up" somewhere, and the "bunching point" (there can be more than one) must lie in A if A is closed (see Theorem 9, Chapter 2).

The Heine-Borel theorem is less obvious intuitively. Perhaps the best way to see its plausibility is to consider some examples.

EXAMPLE 1.   The entire real line $\mathbb{R}$ is not compact for it is unbounded. Notice that

$$\{D(n,1) =]n - 1, n + 1[\mid n = 0, \pm 1, \pm 2, \ldots\}$$

is an open cover of $\mathbb{R}$ but does not have a finite subcover (why)?

EXAMPLE 2.   Let $A =]0,1]$. Consider the open cover $\{]1/n, 2[\mid n = 1, 2, 3, \ldots\}$. (Why does the union contain all of A?) It, too, cannot have an open subcover. This time, condition (ii) fails because A is not closed; the point 0 is "missing" from A. This collection is not a cover for $[0,1]$ and any open cover for $[0,1]$ must have a finite subcover—the above phenomenon cannot happen.

There is another equivalent way of formulating (iii) which is sometimes useful.

(iii)′   *Every infinite subset of A has an accumulation point in A.*

We shall leave to the student the task of showing that (iii)′ is equivalent to (iii). (See Exercise 3, at the end of the chapter.)

There is an alternative way of stating condition (ii) in terms of closed sets. This is done by means of the "finite intersection property for A." We say that a collection of sets $\{A_i\}$ has the *finite intersection property for A* iff the intersection of any finite number of A_i with A is not void. Then (ii) is equivalent to (ii)′.

(ii)′   *Every collection of closed sets with the finite intersection property for A has a non-empty intersection with A.*

As we shall see in the proof (p. 72), (ii)′ is the same statement as (ii) expressed in terms of the collection of (closed) complements of the open cover in (ii).

EXAMPLE 3.   Determine which of the following are compact.
(a) $\{x \in \mathbb{R} \mid x \geqslant 0\}$.
(b) $[0,1] \cup [2,3]$.
(c) $\{(x,y) \in \mathbb{R}^2 \mid x^2 + y^2 < 1\}$.

Solution: (a) Non-compact because it is unbounded. (b) Compact because it is a closed set and is bounded. (c) Non-compact because it fails to be closed.

EXAMPLE 4.   Let x_k be a sequence of points in $\mathbb{R}^n$ with $\|x_k\| \leqslant 3$ for all k. Show that x_k has a convergent subsequence.

Solution: The set $A = \{x \in \mathbb{R}^n \mid \|x\| \leqslant 3\}$ is closed and bounded, hence compact. Since $x_k \in A$, we can apply Theorem 1 (iii) to obtain the conclusion.

EXAMPLE 5.   In Theorem 1 (ii), can "every" be replaced by "some?"

Solution: No. Let $A = \mathbb{R}$ and consider the open cover consisting of the single open set $\mathbb{R}$. This certainly has a finite subcover, namely itself, but $\mathbb{R}$, being unbounded, is not compact.

EXAMPLE 6.   Let $A = \{0\} \cup \{1, 1/2, \ldots, 1/n, \ldots\}$. Show directly that condition (ii) of Theorem 1 holds.

Solution: Let $\{U_i\}$ be an *arbitrary* open cover of A. We must show that there is a finite subcover. Now 0 lies in one of the open sets, say $0 \in U_1$. Since U_1 is open and $1/n \to 0$, there is an N such that $1/N, 1/(N + 1), \ldots$ lie in U_1. Let $1 \in U_2, \ldots, 1/(N - 1) \in U_N$. Then $U_1, \ldots, U_N$ is a finite subcover since it is a finite subcollection of the $\{U_i\}$ and it includes all of the points of A. Notice that if A were the set $\{1, 1/2, \ldots\}$ then the argument would not work.

Exercises for Section 3.1

1. Which of the following sets are compact?
    (a) $\{x \in \mathbb{R} \mid 0 \leqslant x \leqslant 1 \text{ and } x \text{ is irrational}\}$.
    (b) $\{(x,y) \in \mathbb{R}^2 \mid 0 \leqslant x \leqslant 1\}$.
    (c) $\{(x,y) \in \mathbb{R}^2 \mid xy \geqslant 1\} \cap \{(x,y) \mid x^2 + y^2 < 5\}$.

2. Let $r_1, r_2, r_3, \ldots$ be an enumeration of the rational numbers in $[0,1]$. Show that there is a convergent subsequence.

3. Let $x_k \to x$ be a convergent sequence in $\mathbb{R}^n$.
    (a) Show that $\{x_k\} \cup \{x\}$ is compact.
    (b) Verify explicitly condition (ii) of Theorem 1.

4. Let A be a bounded set. Prove $\mathrm{cl}(A)$ is compact.

5. Let A be an infinite set with a single accumulation point in A. Must A be compact?

3.2   Nested Set Property

There is an important consequence of Theorem 1 called the *nested set property*.

> **Theorem 2.**   *Let F_k be a sequence of compact non-empty sets in $\mathbb{R}^n$ such that $F_{k+1} \subset F_k$ for all $k = 1, 2, \ldots$. Then there is at least one point in $\bigcap_{k=1}^{\infty} F_k$.*

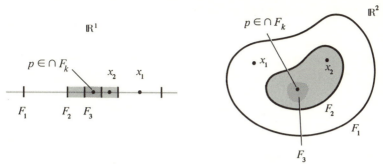

FIGURE 3-2   Nested set property.

Intuitively, the sets F_k are getting smaller and smaller, so it seems very reasonable that there should be a point in all of them. However if the F_k are non-compact the intersection can be empty (see Example 2). Thus the actual proof requires more care.

One can prove this precisely by using the Bolzano-Weierstrass theorem. Pick $x_k \in F_k$ for each k. Then the x_k have a convergent subsequence, as they all lie in F_1, which is compact. The limit point then must lie in all of the sets F_k because they are closed (see Figure 3-2 and Exercise 4).

An even easier proof which uses Theorem 1(ii)′ is found at the end of the chapter.

EXAMPLE 1.   Verify Theorem 2 for $F_k = [0,1/k] \subset \mathbb{R}$.

Solution: Each F_k is compact and clearly, $F_{k+1} \subset F_k$. The intersection is $\{0\}$ which is non-empty.

EXAMPLE 2.   Is Theorem 2 true if "compact non-empty" is replaced by "open non-empty" or "closed non-empty?"

Solution: No. Let $F_k = \,]k,\infty[$ or $[k,\infty[$.

Exercises for Section 3.2

1. Verify Theorem 2 for $F_k = \{x \in \mathbb{R} \mid x \geqslant 0, 2 \leqslant x^2 \leqslant 2 + 1/k\}$.

2. Is Theorem 2 true if "compact non-empty" is replaced by "open bounded non-empty"?

3. Let $x_k \to x$ be convergent in $\mathbb{R}^n$. Verify the validity of Theorem 2 for $F_k = \{x_l \mid l \geqslant k\} \cup \{x\}$. What happens if $F_k = \{x_l \mid l \geqslant k\}$?

4. Let $x_k \to x$ be convergent in $\mathbb{R}^n$. Let $\mathscr{A}$ be a family of closed sets with the property that for each $A \in \mathscr{A}$, there is an N such that $k \geqslant N$ implies $x_k \in A$. Prove that $x \in \cap \mathscr{A}$.

3.3   Path-Connected Sets

The second important topic to be discussed in this chapter is connectedness. We know intuitively to what kind of sets we would like to apply the term "connected." However, our intuition can fail in judging more complicated sets. For example, how do we decide if the set $\{(x,\sin 1/x) \mid x > 0\} \cup \{(0,y) \mid y \in [-1,1]\}$ in $\mathbb{R}^2$ is connected? See Figure 3-3. Therefore, we would like to formulate a sound mathematical definition upon which we can depend.

There are, in fact, two different but closely related, notions of connectedness. The more intuitive and applicable of these is that of path-connectedness, so we begin with it. Our definition must first define what is meant by a curve (or path) joining two points.

> **Definition 2.**   A *continuous path* joining two points x, y in $\mathbb{R}^n$ is a *mapping* $\varphi: [a,b] \to \mathbb{R}^n$ such that $\varphi(a) = x$, $\varphi(b) = y$ and φ is continuous. Here x may or may not equal y and $b \geqslant a$. In Chapter 4, we shall study continuous mappings in detail, but for now, let us call φ *continuous* if
>
> $$(t_k \to t) \Rightarrow (\varphi(t_k) \to \varphi(t))$$
>
> for *every* sequence t_k in $[a,b]$ converging to some $t \in [a,b]$. (The student will recall from earlier courses that, intuitively a continuous path is one with no breaks or jumps in it.) A path φ is said to *lie in a set A* if $\varphi(t) \in A$ for all $t \in [a,b]$. See Figure 3-4.
>
> We say that a set A is *path-connected* if any two points in the set can be joined by a continuous path lying in the set A.

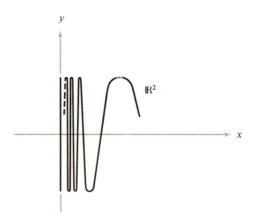

FIGURE 3-3   Connected?

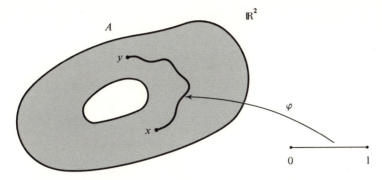

FIGURE 3-4 Curve joining x, y in A.

For example, it is evident that the region A in Figure 3-4 is path-connected. Another path-connected set is the interval $[0,1]$ itself. To prove this, let $x, y \in [0,1]$ and define $\varphi: [0,1] \to \mathbb{R}$ by $\varphi(t) = (y - x)t + x$. This is a path connecting x and y, and it lies in $A = [0,1]$.

Using the above definition of path-connected, a little thought will convince the reader that the set in Figure 3-3 is not path-connected, although the actual proof of this fact is not simple. Most of the time it is rather easy to determine if a set is path-connected. Simply see if any two points can be joined by a continuous curve lying in the set, and this is usually clear geometrically. The second notion of connectedness is harder to check directly but will be very useful. It appears in Section 3.4.

EXAMPLE 1.   Which of the following sets are path-connected?
(a) $[0,3]$.
(b) $[1,2] \cup [3,4]$.
(c) $\{(x,y) \in \mathbb{R}^2 \mid 0 < x \leqslant 1\}$.
(d) $\{(x,y) \in \mathbb{R}^2 \mid 0 < x^2 + y^2 \leqslant 1\}$.

Solution: Only (b) is not path-connected and is clear from a study of Figure 3-5.

EXAMPLE 2.   Must a path-connected set be closed? Or open?

Solution: No; $[0,1]$, $]0,1[$, $[0,1[$ are all path-connected.

EXAMPLE 3.   Let $\varphi: [0,1] \to \mathbb{R}^3$ be a continuous path, and $\mathscr{C} = \varphi([0,1])$. Show that $\mathscr{C}$ is path-connected.

Solution: This is intuitively clear, for we can use the path φ itself to join two points on $\mathscr{C}$. Precisely, if $x = \varphi(a)$, $y = \varphi(b)$, where $0 \leqslant a \leqslant b \leqslant 1$, let $c: [a,b] \to \mathbb{R}^3$, $c(t) = \varphi(t)$. Then, c is a path joining x to y and c lies in $\mathscr{C}$.

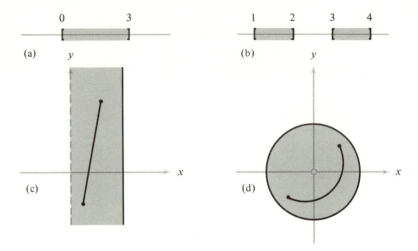

FIGURE 3-5

Exercises for Section 3.3

1. Determine which of the following sets are path-connected.
   (a) $\{x \in [0,1] \mid x \text{ is rational}\}$.
   (b) $\{(x,y) \in \mathbb{R}^2 \mid xy \geqslant 1 \text{ and } x > 1\} \cup \{(x,y) \in \mathbb{R}^2 \mid xy \leqslant 1 \text{ and } x \leqslant 1\}$.
   (c) $\{(x,y,z) \in \mathbb{R}^3 \mid x^2 + y^2 \leqslant z\} \cup \{(x,y,z) \mid x^2 + y^2 + z^2 > 3\}$.
   (d) $\{(x,y) \in \mathbb{R}^2 \mid 0 \leqslant x < 1\} \cup \{(x,0) \mid 1 < x < 2\}$.

2. Let $A \subset \mathbb{R}$ be path-connected. Give plausible arguments that A must be an interval (closed, open, or half-open). Are things as simple in $\mathbb{R}^2$?

3. Let $\varphi: [a,b] \to \mathbb{R}^3$ be a continuous path and $a < c < d < b$. Let $\mathscr{C} = \{\varphi(t) \mid c \leqslant t \leqslant d\}$. Must $\varphi^{-1}(\mathscr{C})$ be path-connected?

3.4   Connected Sets

Definition 3.   A set $A \subset \mathbb{R}^n$ is called *connected* if there do *not* exist two non-empty, open sets U, V such that $A \subset U \cup V$, $A \cap U \neq \varnothing$, $A \cap V \neq \varnothing$, and $A \cap U \cap V = \varnothing$.

Intuitively, the sets U and V would separate A into two pieces, and if this happens, we want to say A is not connected (Figure 3-6).

The set in Figure 3-3 can be shown to be connected but not path-connected; thus the two notions are not the same. However, there is a valid relation between the two ideas, which is presented in the next theorem.

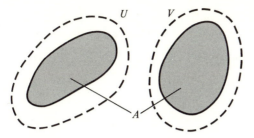

FIGURE 3-6 *A* is neither connected
nor path-connected.

Theorem 3. *If a set A is path-connected, then A is connected.*

Use of this theorem is perhaps the easiest way to identify a connected set. The theorem is reasonable, intuitively. In fact, the (false) converse theorem is also reasonable. Here then is an example of two notions which are closely related, and which are intuitively almost identical, but discerning the true relationship between them requires more care. (The set in Figure 3-3 is connected but not path-connected.)

If a set *A* is not connected (and hence not path-connected), we can divide it up into pieces, or components. More precisely, a *component* of a set *A* is a connected subset $A_0 \subset A$ such that there is no connected set in *A* containing A_0, other than A_0 itself. See Figure 3-6. Thus we see that a component is a maximal connected subset. One can define path-component in a similar way, using path-connectedness instead of connectedness. Some properties of components are found in the exercises of the end of the chapter.

EXAMPLE 1.   Is $\mathbb{Z} = \{\ldots, -2, -1, 0, 1, 2, 3, \ldots\} \subset \mathbb{R}$ connected?

Solution: No, for if $U =]1/2, \infty[$, $V =]-\infty, 1/4[$, then $\mathbb{Z} \subset U \cup V$, $\mathbb{Z} \cap U = \{1, 2, 3, \ldots\} \neq \varnothing$, $\mathbb{Z} \cap V = \{\ldots, -2, -1, 0\} \neq \varnothing$, and $\mathbb{Z} \cap U \cap V = \varnothing$. Hence $\mathbb{Z}$ is not connected. It is also evident that $\mathbb{Z}$ is not path-connected, but observe that this fact cannot be used to conclude that $\mathbb{Z}$ is not connected.

EXAMPLE 2.   Is $\{(x,y) \in \mathbb{R}^2 \mid 0 < x^2 + y^2 \leqslant 1\}$ connected?

Solution: As in Example 1d of Section 3.3, we know that this set is path-connected. Hence, by Theorem 3, it is connected. To prove this directly would be difficult.

Exercises for Section 3.4

1. Is $[0,1] \cup]2,3]$ connected? Prove or disprove.

2. Is $\{(x,y) \in \mathbb{R}^2 \mid 0 \leqslant x \leqslant 1\} \cup \{(x,0) \mid 1 < x < 2\}$ connected? Prove or disprove.

3. Let $A \subset \mathbb{R}^2$ be path-connected. Regarding $A \subset \mathbb{R}^3$ as a subset of the $x - y$ plane, show that A is path-connected. Can you make a similar argument for connected?

4. Discuss the components of:
   (a) $[0,1] \cup [2,3] \subset \mathbb{R}$
   (b) $\mathbb{Z} = \{\ldots, -2, -1, 0, 1, 2, \ldots\}$
   (c) $\{x \in [0,1] \mid x \text{ is rational}\}$.

Theorem Proofs for Chapter 3

Theorem 1. *Let $A \subset \mathbb{R}^n$. Then the following conditions are equivalent:*
(i) *A is closed and bounded.*
(ii) *Every open cover of A has a finite subcover.*
(iii) *Every sequence in A has a subsequence which converges to a point of A.*
(ii)' *Every collection of closed sets with the finite intersection property for A has a non-empty intersection with A.*

Proof: We shall first prove that (i) $\Rightarrow$ (ii) $\Rightarrow$ (iii) $\Rightarrow$ (i), and then (ii) $\Leftrightarrow$ (ii)'. First, let us show that (i) $\Rightarrow$ (ii), which is probably the most difficult. To do this, we begin by proving a special case.

Lemma 1. *The Heine-Borel property (ii) holds for closed intervals $[a,b]$ in $\mathbb{R}$.*

Proof: Let $\mathcal{U} = \{U_i\}$ be an open covering of $[a,b]$. Define $A = \{x \in [a,b] \mid \text{the set } [a,x] \text{ can be covered by a finite collection of the } U_i\}$. We want to show that $A = [a,b]$. To this end, let $c = \sup(A)$. The sup exists because $A \neq \varnothing$ (since $a \in A$), and A is bounded above by b. Also, since $[a,b]$ is closed, $c \in [a,b]$ (see Example 3, Chapter 2). Suppose $c \in U_{i_0}$; such a U_{i_0} exists since the U_i's cover $[a,b]$. Since U_{i_0} is open, there is an $\varepsilon > 0$, such that $]c - \varepsilon, c + \varepsilon[\subset U_{i_0}$. Since $c = \sup(A)$, there exists an $x \in A$, such that $c - \varepsilon < x \leqslant c$ (see Theorem 2, Chapter 1). Because $x \in A$, $[a,x]$ has a finite subcover, say, $U_1, \ldots, U_N$; then $[a, c + \varepsilon/2]$ also has the finite subcover $U_1, \ldots, U_N$, U_{i_0}. Thus we conclude that $c \in A$ and moreover, that $c = b$. Indeed, if $c < b$, we would get a member of A larger than c, since $[a, c + \varepsilon/2]$ has a finite subcover. The latter cannot happen since $c = \sup(A)$. ∎

Question. Why does this fail for $]a,b]$, $[a,b[$, or for $[a,\infty[$?

Lemma 2. *If $A \subset \mathbb{R}^n$ is compact and $x_0 \in \mathbb{R}^m$ then $A \times \{x_0\}$ is compact.*

Proof: Let $\mathcal{U}$ be an open cover of $A \times \{x_0\}$, and $\mathcal{V} = \{V \mid V = \{y \mid (y, x_0) \in U\}$, $U \in \mathcal{U}\}$. Then $\mathcal{V}$ is an open cover of A in $\mathbb{R}^n$, and hence $\mathcal{V}$ has a finite subcover of A, $\mathcal{V}' = \{V_1, \ldots, V_k\}$. Each $V_i \in \mathcal{V}'$ corresponds to a $U_i \in \mathcal{U}$, and $\mathcal{U}' = \{U_1, \ldots, U_k\}$ is then a finite subcover in $\mathcal{U}$ of $A \times \{x_0\}$. ∎

Lemma 3. *If $[-R,R]^{n-1} \subset \mathbb{R}^{n-1}$ has the Heine-Borel property, then $[-R,R]^n \subset \mathbb{R}^n$ has the Heine-Borel property, where $[R,R]^n = [-R,R] \times \cdots \times [-R,R]$, n times.*

Proof: Suppose $[-R,R]^{n-1}$ has the Heine-Borel property and that $\mathscr{U}$ is an open cover of $[-R,R]^n$. Let $S = \{x \in [-R,R] \mid [-R,R]^{n-1} \times [-R,x]$ can be covered by a finite number of sets in $\mathscr{U}\}$. Now $-R \in S$, since $[-R,R]^{n-1}$ satisfies (ii) by hypothesis, and so by Lemma 2 $[-R,R]^{n-1} \times \{-R\}$ has a finite subcover in $\mathscr{U}$. Also, S is bounded above by R and therefore S has a supremum, say x_0. We will show that $x_0 = R$ which will prove the lemma.

Let $\mathscr{U}' \subset \mathscr{U}$ be a finite subcover of $[-R,R]^{n-1} \times \{x_0\}$. For each $(y,x_0) \in [-R,R]^{n-1} \times \{x_0\}$ there exists $\varepsilon_y > 0$ such that $D((y,x_0),\sqrt{2}\,\varepsilon_y)$ is covered by $\mathscr{U}'$. But $V_y = D(y,\varepsilon_y) \times \,]x_0 - \varepsilon_y, x_0 + \varepsilon_y[\, \subset D((y,x_0),\sqrt{2}\,\varepsilon_y)$ so V_y is covered by $\mathscr{U}'$. Consider the open cover $\mathscr{V} = \{V_y \mid y \in [-R,R]^{n-1}\}$ of $[-R,R]^{n-1} \times \{x_0\}$. By Lemma 2, $\mathscr{V}$ has a finite subcover of $[-R,R]^{n-1} \times \{x_0\}$, say $\{V_{y_1}, \ldots, V_{y_N}\}$. Let $\varepsilon = \inf\{\varepsilon_{y_1}, \ldots, \varepsilon_{y_N}\}$. Then $[-R,R]^{n-1} \times \,]x_0 - \varepsilon, x_0 + \varepsilon[\, \subset \bigcup_{i=1}^N V_{y_i}$, and so $[-R,R]^{n-1} \times \,]x_0 - \varepsilon, x_0 + \varepsilon[$ is covered by $\mathscr{U}$.

Now with this ε, there exists $x \in S$ such that $x_0 - \varepsilon < x \leqslant x_0$. Since $x \in S$, there exists a finite subcover $\mathscr{U}'' \subset \mathscr{U}$ which covers $[-R,R]^{n-1} \times [-R,x]$, and $\mathscr{U}' \cup \mathscr{U}''$ is a finite cover of $[-R,R]^{n-1} \times [-R,x_0 + \varepsilon[$. Thus $x_0 \in S$. Suppose $x_0 < R$, then choose δ such that $x_0 + \delta < R$ and $x_0 + \delta < x_0 + \varepsilon$. Thus $[-R,R]^{n-1} \times [-R,x_0 + \delta]$ is covered by $\mathscr{U}' \cup \mathscr{U}''$, and $x_0 + \delta \in S$, a contradiction, and therefore $x_0 = R$. ∎

We will be able to prove our main theorem after we have one more lemma with which to work.

Lemma 4. *If A satisfies (ii), B is closed and $B \subset A$, then B also satisfies (ii).*

Proof: Let $\{U_i\}$ be an open covering of B, and let $V = \mathbb{R}^n \backslash B$. Then $\{U_i, V\}$ is an open cover of A. If $\{U_1, \ldots, U_N, V\}$ is a finite subcover of A, then $\{U_1, \ldots, U_N\}$ is a finite subcover for B. ∎

Theorem 1 Proof that (i) $\Rightarrow$ (ii): Since A is bounded, it lies in some cube $[-R,R]^n$. By Lemma 3 and induction on n, this cube satisfies (ii). By Lemma 4, A does also, since A is closed. ∎

Theorem 1 Proof that (ii) $\Rightarrow$ (iii): Suppose the sequence $x_k \in A$ has *no* convergent subsequences. In particular, this means that x_k has an infinity of distinct points, say, $y_1, y_2, \ldots$. Since there are no convergent subsequences, there is a neighborhood U_k of y_k containing no other y_i. This is because if every neighborhood of y_k contained another y_j we could, by choosing the neighborhoods $D(y_k, 1/m)$, $m = 1, 2, \ldots$ select out a subsequence converging to y_k. Also, we claim that the set $y_1, y_2, \ldots$ is closed. Indeed, it has no accumulation points by the assumption that there are no convergent subsequences (see Exercise 4 at the end of Chapter 2). Now, by Lemma 3 above, $\{y_1, y_2, \ldots\}$ satisfies (ii). But $\{U_k\}$ is an open cover which has no finite subcover, a contradiction. Thus x_k has a convergent subsequence. To show that the limit lies in A amounts to showing that A is closed. We leave this to Exercise 20. ∎

Theorem 1 Proof that (iii) $\Rightarrow$ (i): First, we show that A is closed. For this, we use Theorem 9, Chapter 2. Consider a sequence $x_k \to x$ with $x_k \in A$. By (iii), the limit lies in A, so A is closed.

Next, we prove that A is bounded. Suppose, in fact, A is not bounded. Then there are points $x_k \in A$ with $\|x_k\| \geq k$, $k = 1, 2, 3, \ldots$. This implies that the sequence x_k cannot have any convergent subsequences since, if y was a limit point, $\|y\| = $ limit $_{k \to \infty} \|x_k\| = \infty$ (see Exercise 16). This is impossible if we have $y \in \mathbb{R}^n$. ∎

Theorem 1 Proof that (ii)′ ⇔ (ii): First, we prove that (ii) ⇒ (ii)′. Let $\{F_i\}$ be a collection of closed sets and let $U_i = \mathbb{R}^n \backslash F_i$, so U_i is open. Suppose that $A \cap (\bigcap_{i=1}^{\infty} F_i) = \varnothing$. Then, taking complements, this means that the U_i cover A. Being an open covering, there is a finite subcovering (by assumption (ii)) say, $A \subset U_1 \cup \cdots \cup U_N$. Then $A \cap (F_1 \cap \cdots \cap F_N) = \varnothing$, so $\{F_i\}$ does not have the finite intersection property. Thus if $\{F_i\}$ is a collection of closed sets with the finite intersection property, then $A \cap \{F_i\} \neq \varnothing$.

Finally, we show (ii)′ ⇒ (ii). Indeed, let $\{U_i\}$ be an open covering of A and let $F_i = \mathbb{R}^n \backslash U_i$. Then $A \cap (\bigcap_i F_i) = \varnothing$, and so by (ii)′, $\{F_i\}$ cannot have the finite intersection property for A. Thus $A \cap (F_1 \cap \cdots \cap F_N) = \varnothing$ for some members $F_1, \ldots, F_N$ of the collection. Hence $U_1, \ldots, U_N$ is the required finite subcover. ∎

Theorem 2.   *Let F_k be a sequence of compact nonempty sets in $\mathbb{R}^n$ such that $F_{k+1} \subset F_k$ for all $k = 1, 2, \ldots$. Then $\bigcap_{k=1}^{\infty} F_k \neq \varnothing$.*

Proof: Let us observe that in the compact set $A = F_1$, the sets $F_1, F_2, \ldots$ have the finite intersection property. Indeed, the intersection of any finite collection equals the F_k with the highest index. Thus, since (ii)′ holds for compact sets, we have

$$F_1 \cap \left(\bigcap_{k=1}^{\infty} F_k \right) = \bigcap_{k=1}^{\infty} \{F_k\} \neq \varnothing . ∎$$

Theorem 3.   *If a set A is path- connected, then A is connected.*

Again, we begin by first proving a special case of the theorem.*

Lemma 5.   *The interval $[a,b]$ is connected*

Proof: Suppose the interval were not connected. Then there would be two open sets U and V with $U \cap [a,b] \neq \varnothing$ and $V \cap [a,b] \neq \varnothing$, $[a,b] \cap U \cap V = \varnothing$ and $[a,b] \subset U \cup V$. Further, suppose that $b \in V$. Let $c = \sup(U \cap [a,b])$, which exists as this set is bounded above. Now $U \cap [a,b]$ is closed, since its complement is $V \cup (\mathbb{R} \backslash [a,b])$, which is open. Thus $c \in U \cap [a,b]$ (see Exercise 8, Chapter 2). Now $c \neq b$, since $c \notin V$ and $b \in V$. Any neighborhood of c intersects $V \cap [a,b]$ since $c \neq b$ and no neighborhood of c can be entirely contained in U as $c = \sup(U \cap [a,b])$, so that c is an accumulation point of $V \cap [a,b]$. But as with U, $V \cap [a,b]$ is closed, so $c \in V \cap [a,b]$. This contradicts the fact that $V \cap U \cap [a,b] = \varnothing$. ∎

Proof of Theorem 3: Suppose A is not connected. Then, by definition, there exist open sets U, V such that $A \subset U \cup V$, $A \cap U \cap V = \varnothing$, $U \cap A \neq \varnothing$, and $V \cap A \neq \varnothing$.

---

* It is not necessary to prove Lemma 5 first; one can proceed directly as well, but it seems useful to note that the crux of the argument has to do with connectivity of an interval.

Choose $x \in U \cap A$ and $y \in V \cap A$. Since A is path-connected, there is a path $\varphi \colon [a,b] \to \mathbb{R}^n$ in A joining x and y. Set $U_0 = \varphi^{-1}(U)$ and $V_0 = \varphi^{-1}(V)$, so $U_0, V_0 \subset [a,b]$. Now U_0 is closed, because if we let $t_k \to t$, with $t_k \in U_0$, then, by continuity of φ, $\varphi(t_k) \to \varphi(t)$; but since V is open, $\varphi(t) \notin V$, or else $\varphi(t_k) \in V$ for large k. Hence $\varphi(t) \in U \cap A$ or $t \in U_0$. Thus U_0 is closed. Similarly, V_0 is closed. Let $U' = \;]-\infty,a[\; \cup (\mathbb{R}\backslash V_0)$, and $V' = \;]b,\infty[\; \cup (\mathbb{R}\backslash U_0)$, which are open sets. Observe that $U' \cap [a,b] \neq \varnothing$ (it contains a), $V' \cap [a,b] \neq \varnothing$ (it contains b), $U' \cap V' = \varnothing$, and $U' \cup V' \supset [a,b]$. Thus $[a,b]$ is not connected, contradicting Lemma 5. ∎

Worked Examples for Chapter 3

1. Show that $A = \{x \in \mathbb{R}^n \mid \|x\| \leq 1\}$ is compact and connected.

Solution: To show that A is compact, we show it is closed and bounded. To show that it is closed, consider $\mathbb{R}^n \backslash A = \{x \in \mathbb{R}^n \mid \|x\| > 1\} = B$. For $x \in B$, $D(x, \|x\| - 1) \subset B$ (see Theorem 1, Chapter 2), so B is open, and hence A is closed. It is obvious that A is bounded, since $A \subset D(0,2)$ and therefore A is compact.

To show that A is connected, we show that A is path-connected. Let $x, y \in A$. Then the straight line joining x, y is the required path. Explicitly, use $\varphi \colon [0,1] \to \mathbb{R}^n$, $\varphi(t) = (1 - t)x + ty$. One sees that $\varphi(t) \in A$ since $\|\varphi(t)\| \leq (1 - t)\|x\| + t\|y\| \leq (1 - t) + t = 1$ by the triangle inequality.

2. Let $A \subset \mathbb{R}^n$, $x \in A$, and $y \in \mathbb{R}^n \backslash A$. Let $\varphi \colon [0,1] \to \mathbb{R}^n$ be a (continuous) path joining x and y. Show that there is a t such that $\varphi(t) \in \mathrm{bd}(A)$.

Solution: Intuitively, this result says that a path which joins a set to its complement must pierce the boundary of the set at some point. See Figure 3-7.

Let $B = \{x \in [0,1] \mid \varphi([0,x]) \subset A\} \subset [0,1]$. Now $B \neq \varnothing$, since $0 \in B$. Let $t = \sup(B)$. We shall show that $\varphi(t) \in \mathrm{bd}(A)$. Let U be a neighborhood of $\varphi(t)$. Choose $t_k \in [0,t]$, $t_k \to t$, such that $\varphi(t_k) \in A$. This is possible by definition of B. Then $\varphi(t_k) \in U$ for large k by continuity of φ. By definition of t, there is a point s_k such that $t \leq s_k \leq t + 1/k$ and such that $\varphi(s_k) \notin A$. Now $s_k \to t$, so by continuity of φ, $\varphi(s_k) \in U$ for k large enough. Thus U contains points of A and $\mathbb{R}^n \backslash A$, and so, by Theorem 6 of Chapter 2, $\varphi(t) \in \mathrm{bd}(A)$.

3. Prove: A set $A \subset \mathbb{R}$ is connected iff it is an interval—an interval is a set of the form $[a,b]$, $[a,b[$, $]a,b]$, or $]a,b[$, where a or b can be $\pm\infty$ on an open end of the interval.

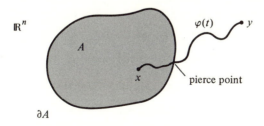

FIGURE 3-7   A path joining A and $\mathbb{R}^n \backslash A$.

Solution: We have already seen that intervals are connected because they are path-connected. Now, for the converse, assume that A is not an interval. We shall show that it is not connected. Saying A is not an interval means that there exist points x, y, z such that $x < y < z$; $x, z \in A$ and $y \notin A$. (Why?). Then $U =]-\infty, y[$ and $V =]y, \infty[$ are open sets such that $A \subset U \cup V, U \cap V \cap A = \varnothing, U \cap A \neq \varnothing$, and $V \cap A \neq \varnothing$. Thus A is not connected.

Exercises for Chapter 3

1. Which of the following sets are compact? Which are connected?
   (a) $\{(x_1, x_2) \in \mathbb{R}^2 \mid |x_1| \leq 1\}$.
   (b) $\{x \in \mathbb{R}^n \mid \|x\| \leq 10\}$.
   (c) $\{x \in \mathbb{R}^n \mid 1 \leq \|x\| \leq 2\}$.
   (d) $\mathbb{Z} = \{$integers in $\mathbb{R}\}$.
   (e) A finite set in $\mathbb{R}^n$.
   (f) $\{x \in \mathbb{R}^n \mid \|x\| = 1\}$ (distinguish between the cases $n = 1$ and $n \geq 2$).
   (g) Perimeter of the unit square in $\mathbb{R}^2$.
   (h) The boundary of a bounded set.
   (i) The rationals in $[0,1]$.
   (j) A closed set in $[0,1]$.

2. Prove that a set $A \subset \mathbb{R}^n$ is not connected iff we can write $A \subset F_1 \cup F_2$, where F_1, F_2 are closed, $A \cap F_1 \cap F_2 = \varnothing, F_1 \cap A \neq \varnothing, F_2 \cap A \neq \varnothing$.

3. Prove the following assertions.
   (a) A set A is compact iff every infinite subset has an accumulation point in A.
   (b) A bounded infinite set A has an accumulation point (not necessarily in A).

4. Show that a set A is bounded iff there is a constant M such that $d(x,y) \leq M$ for all $x, y \in A$. Give a plausible definition of the diameter of a set and reformulate your result.

5. Show that the following sets are not compact by exhibiting an open cover with no finite subcover.
   (a) $\{x \in \mathbb{R}^2 \mid \|x\| < 1\}$.
   (b) $\mathbb{Z}$, the integers in $\mathbb{R}$.

6. Suppose that in Theorem 2 (nested set property), the diameter $(F_k) \to 0$. Show that there is exactly one point in $\cap\{F_k\}$ (by definition, diameter $(F_k) = \sup\{d(x,y) \mid x,y \in F_k\}$).

7. Let x_k be a sequence in $\mathbb{R}^n$ which converges to x. Let $A_k = \{x_k, x_{k+1}, \ldots\}$ and show that $\{x\} = \bigcap_{k=1}^{\infty} \mathrm{cl}(A_k)$.

8. Let $A \subset \mathbb{R}^n$ be compact and x_n a Cauchy sequence in A. Then show that x_n converges to a point in A.

9. Determine (by proof or counterexample) the truth or falsity of the following statements.
   (a) (A compact in $\mathbb{R}^n$) $\Rightarrow$ ($\mathbb{R}^n \backslash A$ connected).

(b) (A connected in $\mathbb{R}^n$) $\Rightarrow$ ($\mathbb{R}^n \backslash A$ connected).

(c) (A connected in $\mathbb{R}^n$) $\Rightarrow$ (A open or closed).

(d) ($A = \{x \in \mathbb{R}^n \mid \|x\| \leqslant 1\}$) $\Rightarrow$ ($\mathbb{R}^n \backslash A$ connected). [Hint: Check the cases $n = 1$ and $n \geqslant 2$.]

10. A set A is said to be *locally path-connected** if each point has a neighborhood U such that $A \cap U$ is path-connected. Show that (A connected and locally path-connected) $\Leftrightarrow$ (A path-connected).

11. (a) Prove that if A is connected in $\mathbb{R}^n$ and $A \subset B \subset \mathrm{cl}(A)$, then B is connected.

    (b) Deduce from (a) that the components of a set A are relatively closed. Give an example in which they are not relatively open. ($C \subset A$ is called *relatively closed in A* if C is the intersection of some closed set in $\mathbb{R}^n$ with A.)

    (c) Show that if sets B_i and B are connected and $B_i \cap B \neq \varnothing$ for all i, then $(\bigcup_i B_i)$ is connected. Give examples.

    (d) Deduce from (c) that every point of a set lies in a unique component.

    (e) Use (c) to show $\mathbb{R}^n$ is connected assuming just that lines in $\mathbb{R}^n$ are connected.

12. Let S be a set of real numbers which is non-empty and bounded above. Let $-S = \{x \in \mathbb{R} \mid -x \in S\}$. Prove that

    (a) $-S$ is bounded below,

    (b) $\sup S = -\inf -S$.

13. Let M be a complete metric space and F_n a collection of closed non-empty subsets of M such that $F_{n+1} \subset F_n$ and diameter $(F_n) \to 0$. Prove that $\bigcap_{n=1}^{\infty} F_n$ consists of a single point.

14. (a) A point $x \in A$ in $\mathbb{R}^n$ is said to be *isolated* in a set A if there is a neighborhood U of x such that $U \cap A = \{x\}$. Show that this is equivalent to saying that there is an $\varepsilon > 0$ such that for all $y \in A$, $y \neq x$, we have $d(x,y) > \varepsilon$.

    (b) A set is called *discrete* if all its points are isolated. Give some examples. Show that a discrete set is compact iff it is finite.

15. Let $K_1 \subset \mathbb{R}^n$ and $K_2 \subset \mathbb{R}^m$ be path-connected (respectively, connected, compact). Show $K_1 \times K_2$ is path-connected (respectively, connected, compact) in $\mathbb{R}^{n+m}$.

16. If $x_k \to x$, then prove that $\|x_k\| \to \|x\|$. Is the converse true? Use this to give a proof that $\{x \in \mathbb{R}^n \mid \|x\| \leqslant 1\}$ is closed, using sequences.

17. Let K be a non-empty closed set in $\mathbb{R}^n$, $x \in \mathbb{R}^n \backslash K$. Show there is a $y \in K$ such that $d(x,y) = \inf\{d(x,z) \mid z \in K\}$. Is this true for open sets?

18. Let $F_n \subset \mathbb{R}$ be defined by $F_n = \{x \mid 2 - 1/n \leqslant x^2 \leqslant 2 + 1/n\}$. Show that $\bigcap_{n=1}^{\infty} F_n \neq \varnothing$. Use this to show the existence of $\sqrt{2}$.

19. Let $V_n \subset \mathbb{R}^m$ be open sets such that $\mathrm{cl}(V_n)$ is compact, $V_n \neq \varnothing$, and $\mathrm{cl}(V_n) \subset V_{n-1}$. Prove $\bigcap_{n=1}^{\infty} V_n \neq \varnothing$.

20. Prove that a set A with property (ii) of Theorem 1 is closed as follows. Let x be an accumulation point of A and suppose $x \notin A$; for each $y \in A$, choose disjoint neighborhoods U_y of y and V_y of x. Consider the open cover $\{U_y\}$.

---

* This terminology differs somewhat from that of standard topology books.

21. (a) Prove: a set $A \subset \mathbb{R}^n$ is connected iff $\varnothing$ and A are the only subsets of A which are open and closed *relative* to A. (A set $U \subset A$ is called open *relative* to A if $U = V \cap A$ for some open set $V \subset \mathbb{R}^n$; closed relative to A is defined similarly).
    (b) Prove that $\varnothing$ and $\mathbb{R}^n$ are the only subsets of $\mathbb{R}^n$ which are both open and closed.

22. Find two subsets $A, B \subset \mathbb{R}^2$ and a point $x_0 \in \mathbb{R}^2$ such that $A \cup B$ is not connected but $A \cup B \cup \{x_0\}$ is connected.

23. Let $\mathbb{Q}$ denote the rationals in $\mathbb{R}$. Show that both $\mathbb{Q}$ and the irrationals $\mathbb{R}\backslash\mathbb{Q}$ are not connected.

24. Prove that a set $A \subset \mathbb{R}^n$ is not connected if we can write A as the disjoint union of two sets B and C with $B \cap A \neq \varnothing$, $C \cap A \neq \varnothing$ and such that neither of the sets B or C has a point of accumulation belonging to the other set.

25. Prove that there is a sequence of distinct integers $n_1, n_2, \ldots \to \infty$ such that $\displaystyle\lim_{k \to \infty} \sin(n_k)$ exists.

26. Show that the completeness property of $\mathbb{R}$ may be replaced by the nested set property.

27. Let $\{x_n\}_{n=1}^{\infty}$ be a bounded sequence in $\mathbb{R}$. Let S be the set of all *limit* points of $\{x_n\}$.
    (a) Prove that S is bounded and non-empty. Let $x^* = \sup S$, $x_* = \inf S$. x^* is called the *limit superior* of $\{x_n\}$ and is denoted by $\limsup x_n$ or $\overline{\lim} x_n$. x_* is called the *limit inferior* and is denoted by $\liminf x_n$ or $\underline{\lim} x_n$. Prove the following.
    (b) The definition coincides with that of Problem 10, Chapter 1.
    (c) $x^* \in S$.
    (d) For each $\varepsilon > 0$ there exists $m \in N$ such that $n \geqslant m \Rightarrow x_n < x^* + \varepsilon$.
    (e) x^* is the only number having both of the properties (c) and (d).
    (f) $\{x_n\}$ converges $\Leftrightarrow x^* = x_*$.
    (g) Let $x_n = (-1)^{n-1}(1 + 1/n)$. Find x^* and x_*.

28. Let $A \subset \mathbb{R}^n$ be connected and contain more than one point. Show that every point of A is an accumulation point of A.

29. Let $A = \{(x,y) \in \mathbb{R}^2 \mid x^4 + y^4 = 1\}$. Show that A is compact. Is it connected?

30. Let U_k be a sequence of open sets in $\mathbb{R}^n$. Prove or disprove that
    (a) $\bigcup_{k=1}^{\infty} U_k$ is open,
    (b) $\bigcap_{k=1}^{\infty} U_k$ is open.

31. Suppose $A \subset \mathbb{R}^n$ is not compact. Show that there exists a sequence $F_1 \supset F_2 \supset F_3 \cdots$ of closed sets such that $F_k \cap A \neq \varnothing$ for all k and

$$\left(\bigcap_{k=1}^{\infty} F_k\right) \cap A = \varnothing .$$

[*Hint:* A set in $\mathbb{R}^n$ is compact iff every *countable* open cover has a finite subcover.]

32. (Baire Category theorem for $\mathbb{R}^n$.) A set $S \subset \mathbb{R}^n$ is called *nowhere dense* if $\text{cl}(S) \cap U \neq U$ for any non-empty open set U. Show that $\mathbb{R}^n$ cannot be written as the countable union of nowhere dense sets, $\mathbb{R}^n = \bigcup_{k=1}^{\infty} S_k$. [Hint: If $\mathbb{R}^n = \bigcup_{k=1}^{\infty} S_k$,

find a non-convergent Cauchy sequence x_k by carefully choosing nested balls $D(x_k, r_k) \subset \mathbb{R}^n \backslash (S_i \cup \cdots \cup S_k)$.]

33. Let x_n be a sequence in $\mathbb{R}^3$ such that $\|x_{n+1} - x_n\| \leqslant 1/(n^2 + n), n \geqslant 1$. Show that x_n converges.

34. Prove that any closed set $A \subset \mathbb{R}^n$ is a countable intersection of open sets. [Hint: Let $U_k = \{y \in \mathbb{R}^n \mid d(x,y) < 1/k \text{ for some } x \in A\}$.]

35. Let the sequence $a_1, a_2, \ldots$ in $\mathbb{R}$ be defined by

$$a_1 = a;$$
$$a_n = a_{n-1}^2 - a_{n-1} + 1, \qquad \text{if } n > 1.$$

For what $a \in \mathbb{R}$ is the sequence (a) monotone? (b) bounded? (c) convergent? Compute the limit in the cases of convergence.

36. Let $A \subset \mathbb{R}^n$ be uncountable. Prove that A has an accumulation point.

37. Let $A, B \subset \mathbb{R}^n$ with A compact, B closed, and $A \cap B = \emptyset$.
    (a) Show there is an $\varepsilon > 0$ so that $d(x,y) > \varepsilon$ for all $x \in A, y \in B$.
    (b) Is (a) true if A, B are merely closed?

38. (Cantor set.) Let $F_1 = [0,1/3] \cup [2/3,1]$ be obtained from $[0,1]$ by removing the middle third. Repeat, obtaining $F_2 = [0,1/9] \cup [2/9,1/3] \cup [2/3,7/9] \cup [8/9,1]$. In general, F_n is a union of intervals and F_{n+1} is obtained by removing the middle third of these intervals. Let $C = \bigcap_1^\infty F_n$, the *Cantor set*. Prove:
    (a) C is compact.
    (b) C has infinitely many points. [Hint: Look at the endpoints of F_n.]
    (c) $\text{int}(C) = \emptyset$.
    (d) Show that C is *perfect*, that is, is closed with no isolated points.

39. Show that $A \subset \mathbb{R}^n$ is not connected iff there exist two disjoint open sets U, V such that $U \cap A \neq \emptyset, V \cap A \neq \emptyset$ and $A \subset U \cup V$. [Hint: Let U_1, V_1 be the open sets from the definition; set $A_1 = A \cap U_1, A_2 = A \cap V_1$ and let $U = \{x \in \mathbb{R}^n \mid d(x,A_1) < d(x,A_2)\}$ and $V = \{x \in \mathbb{R}^n \mid d(x,A_1) > d(x,A_2)\}$.]

40. Let F_k be a nest of compact sets (that is, $F_{k+1} \subset F_k$). Furthermore, suppose each F_k is connected. Prove that $\bigcap \{F_k\}$ is connected. (You may use the result of Exercise 39.) Give an example to show that compactness is an essential condition which cannot be replaced by "F_k a nest of closed connected sets."

Chapter 4

Continuous Mappings

T_o be able to obtain interesting and useful theorems, it is often necessary to make certain restrictions on the mathematical objects one studies. In this chapter we require that the functions studied be continuous, and we will investigate some of the consequences of this restriction. In Chapter 6 we study an even stronger restriction, namely, that of differentiability.

4.1 Continuity

First, let us examine intuitively the notion of continuity for real functions on the real line $\mathbb{R}$. Figure 4-1a shows a continuous function, and 4-1b shows a discontinuous one. A continuous function has the important property that when x is close to x_0, $f(x)$ is close to $f(x_0)$ (as shown in Figure 4-1a). On the other hand, in Figure 4-1b, even if x is very close to x_0, $f(x)$ may not be close to $f(x_0)$. The reader should be familiar with these ideas from basic calculus.

In order to define continuity in precise terms, first the concept of the limit of a function at a point is defined.

> **Definition 1.** Let $A \subset \mathbb{R}^n$, $f: A \to \mathbb{R}^m$, and suppose x_0 is an accumulation point of A. We say that $b \in \mathbb{R}^m$ is the *limit of f at x_0*, written
>
> $$\operatorname*{limit}_{x \to x_0} f(x) = b \, ,$$

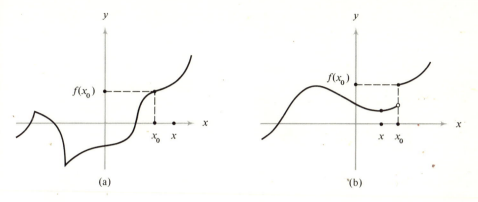

FIGURE 4-1   (a) Continuous function. (b) Discontinuous function.

if given any $\varepsilon > 0$ there exists $\delta > 0$ (depending on f, x_0, and ε) such that for all $x \in A$, $x \neq x_0$, $\|x - x_0\| < \delta$ implies that $\|f(x) - b\| < \varepsilon$.

Intuitively, this says that as x approaches x_0, $f(x)$ approaches b. We also write $f(x) \to b$ as $x \to x_0$. (Compare this with the concept of the limit of a sequence.) Note that if x_0 is not an accumulation point, there will not be any $x \neq x_0$, $x \in A$ near x_0 in which case the condition becomes vacuous.

Note that $\lim_{x \to x_0} f(x)$ may not exist; for example, let $f \colon \mathbb{R}\backslash\{0\} \to \mathbb{R}$ be defined by $f(x) = 1$ if $x < 0$, 2 if $x > 0$. Then 0 is an accumulation point of $\mathbb{R}\backslash\{0\}$ but $\lim f(x)$ doesn't exist. However, if $f(x) = 1$, if $x \neq 0$, and $f(0) = 0$ then $\lim_{x \to 0} f(x) = 1$. Another example is $f \colon \mathbb{R}\backslash\{0\} \to \mathbb{R}$, $f(x) = \sin(1/x)$; this function oscillates faster and faster near 0 and so cannot approach any limit there. However, if $\lim_{x \to x_0} f(x)$ exists, then it is unique, so we are justified in saying *the* limit of f at x_0. To clarify further, suppose $\lim_{x \to x_0} f(x) = b$ and b'. To show $b = b'$, let $\varepsilon > 0$ be given. Then there exist $\delta_1 > 0$ and $\delta_2 > 0$ such that $\|x - x_0\| < \delta_1$ implies $\|f(x) - b\| < \varepsilon/2$ and $\|x - x_0\| < \delta_2$ implies $\|f(x) - b'\| < \varepsilon/2$. Let $\delta = \min\{\delta_1, \delta_2\}$; then $\|x - x_0\| < \delta$ implies $\|b - b'\| \leqslant \|b - f(x)\| + \|f(x) - b'\| < \varepsilon/2 + \varepsilon/2 = \varepsilon$; thus $\|b - b'\| < \varepsilon$ for any ε, and so $\|b - b'\| = 0$, or $b = b'$. (Compare with uniqueness of the limit of a sequence, p. 56.)

We are now ready to define continuity of a function at a point.

Definition 2.   Let $A \subset \mathbb{R}^n$, $f \colon A \to \mathbb{R}^m$, and let $x_0 \in A$. We say that f is *continuous at* x_0 if either x_0 is not an accumulation point of A or $\lim_{x \to x_0} f(x) = f(x_0)$.

Note that this requires the existence of $\lim_{x \to x_0} f(x)$ in addition to specifying its value. Definition 2 can be rephrased as follows: f is continuous at the point x_0 in its domain iff for all $\varepsilon > 0$, there is a $\delta > 0$ such that for all $x \in A$, $\|x - x_0\| < \delta$ implies $\|f(x) - f(x_0)\| < \varepsilon$. In Definition 1 we needed to specify that $x \neq x_0$ because f was not necessarily defined at x_0, but here there is no need to specify $x \neq x_0$ since our condition is certainly valid if $x = x_0$.

There is some additional notation that is useful. Suppose f is defined at least on $]x_0, a] \subset \mathbb{R}$ for some $a > x_0$. Then

$$\lim_{x \to x_0^+} f(x) = b$$

means the limit of f with domain $A =]x_0, a]$. In other words, for every $\varepsilon > 0$ there is a $\delta > 0$ such that $|x - x_0| < \delta$, $x > x_0$ implies $|f(x) - b| < \varepsilon$. Thus we are taking the limit of f as x approaches x_0 *from the right*. Similarly we can define

$$\lim_{x \to x_0^-} f(x) = b \, ,$$

the limit as x approaches x_0 from the left. These are, for obvious reasons, called *one sided limits*. It should now be clear to the reader how to define expressions like $\lim_{x \to \infty} f(x) = \alpha$, and so forth.

Definition 3. A function $f \colon A \to \mathbb{R}^m$ is called *continuous* on the set $B \subset A$ if f is continuous at each point of B. If we just say f is continuous, we mean f is continuous on its domain A.

There are other useful ways of formulating the notion of continuity. One of these is particularly significant because it involves only the topology (that is, the open sets), and so it would be applicable in more general situations.

Theorem 1. *Let $f \colon A \to \mathbb{R}^m$ be a mapping, where $A \subset \mathbb{R}^n$ is any set. Then the following assertions are equivalent.*

(i) *f is continuous on A.*

(ii) *For each convergent sequence $x_k \to x_0$ in A, we have $f(x_k) \to f(x_0)$.*

(iii) *For each open set U in $\mathbb{R}^m$, $f^{-1}(U) \subset A$ is open relative to A; that is, $f^{-1}(U) = V \cap A$ for some open set V.*

(iv) *For each closed set $F \subset \mathbb{R}^m$, $f^{-1}(F) \subset A$ is closed relative to A; that is, $f^{-1}(F) = G \cap A$ for some closed set G.*

Actually condition (ii) in this theorem has an analogous version for limits which can be proved in the same way as (ii) is proved. Namely if $f \colon A \to \mathbb{R}^m$

and x_0 is an accumulation point of A, then

$$\lim_{x \to x_0} f(x) = b$$

if and only if

$$\lim_{k \to \infty} f(x_k) = b$$

for every sequence $x_k \in A$ which converges to x_0.

From this theorem it is evident that our definition of a continuous path, given in Chapter 3, coincides with continuity as we have defined it here. In Section 4.3, we shall establish theorems which will enable us to establish easily the continuity of the more common functions.

We shall now briefly discuss the plausibility of Theorem 1. First of all, that (i) is the same as (ii) should be clear, for (i) means that $f(x)$ is near $f(x_0)$ if x is near x_0, and (ii) is the same except that it lets x approach x_0 via a sequence. The assertions (iii) and (iv) are also the same if we remember that open sets are complements of closed ones.

Let us see what (iii) is telling us. Choose U to be a small open set containing $f(x_0)$. Then $f^{-1}(U)$ being open means that there is a whole open disc around x_0 contained in $f^{-1}(U)$. For x in this disc, x is mapped to U, which represents points near $f(x_0)$. In other words, using U as a measure of closeness of $f(x)$ to $f(x_0)$, if x is near enough to x_0 (namely, $x \in f^{-1}(U)$), $f(x)$ will be near to $f(x_0)$. This therefore represents the same idea expressed in (i).

EXAMPLE 1. Let $f: \mathbb{R}^n \to \mathbb{R}^n$ be the identity function $x \mapsto x$. Show that f is continuous.

Solution: Fix $x_0 \in \mathbb{R}$. By definition we must find $\delta > 0$ for given $\varepsilon > 0$ such that $\|x - x_0\| < \delta$ implies $\|f(x) - f(x_0)\| < \varepsilon$. But, clearly, if we choose $\delta = \varepsilon$, the definition becomes the statement that $\|x - x_0\| < \varepsilon$ implies $\|x - x_0\| < \varepsilon$, which is a tautology. Hence f is continuous.

EXAMPLE 2. Let $f:]0,\infty[\to \mathbb{R}; x \mapsto 1/x$. Show that f is continuous.

Solution: Fix $x_0 \in]0,\infty[$; that is, fix $x_0 > 0$. To determine how to choose δ, we examine the expression

$$|f(x) - f(x_0)| = \left| \frac{1}{x} - \frac{1}{x_0} \right|$$

$$= \frac{|x_0 - x|}{|x \, x_0|}.$$

If $|x - x_0| < \delta$, then we would get

$$|f(x) - f(x_0)| < \frac{\delta}{|x \, x_0|} = \frac{\delta}{x \, x_0}.$$

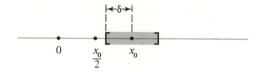

FIGURE 4-2

Now, if we make sure that $\delta < x_0/2$, then we will have $x > x_0/2$ (Figure 4-2) so $\delta/x \, x_0 < (2\delta/x_0^2)$. Now given $\varepsilon > 0$, choose $\delta = \min(x_0/2, \varepsilon x_0^2/2)$. Then the above argument shows that $|f(x) - f(x_0)| < \varepsilon$ if $|x - x_0| < \delta$. Thus f is continuous at x_0.

EXAMPLE 3. Let $f: \mathbb{R}^n \to \mathbb{R}^m$ be continuous. Show that $\{x \in \mathbb{R}^n \mid \|f(x)\| < 1\}$ is open.

Solution: The above set is nothing but $f^{-1}\{y \mid \|y\| < 1\}$ which is the inverse image of an open set, so by Theorem 1 (iii) it is open.

Exercises for Section 4.1

1. (a) Let $f: \mathbb{R} \to \mathbb{R}$, $x \mapsto x^2$. Prove that f is continuous.
   (b) Let $f: \mathbb{R}^2 \to \mathbb{R}$, $(x,y) \mapsto x$. Prove that f is continuous.

2. Use (b) above to show that if $U \subset \mathbb{R}$ is open, then $A = \{(x,y) \in \mathbb{R}^2 \mid x \in U\}$ is open.

3. Let $f: \mathbb{R}^2 \to \mathbb{R}$ be continuous. Prove that $A = \{(x,y) \in \mathbb{R}^2 \mid 0 \leq f(x,y) \leq 1\}$ is closed.

4. Give an example of a continuous function $f: \mathbb{R} \to \mathbb{R}$ and an open set $U \subset \mathbb{R}$ such that $f(U)$ is *not* open.

5. Prove directly that condition (iii) implies condition (iv) in Theorem 1.

4.2   Images of Compact and Connected Sets

Now some important consequences of continuity shall be deduced. The first thing to know is how compact and connected sets behave under continuous mappings. It is crucial to distinguish between the terms *image* and *pre-image* (that is, inverse image) in these theorems; compare the following with Theorem 1 above.

> **Theorem 2.**   *Let $f: A \to \mathbb{R}^m$ be a continuous mapping. Then*
> (i) *if $K \subset A$ and K is connected [respectively, path-connected], then $f(K)$ is connected [respectively, path-connected],*
> (ii) *if $B \subset A$ and B is compact, then $f(B)$ is compact.*

The result of (i) is clearest if we use path-connectedness, that is, if $c(t)$ is a path joining x and y in K, then $f(c(t))$ is a path joining $f(x)$ and $f(y)$ in $f(K)$. (See Theorem 3 below for continuity of $f(x(t))$.) Hence $f(K)$ is path-connected.

The result of (ii) is less obvious intuitively. However, if we use the Bolzano-Weierstrass characterization of compactness it comes fairly easily, for if $f(x_k)$ is a sequence in $f(B)$, then x_k has a convergent subsequence, so we get a corresponding convergent subsequence for $f(x_k)$.

EXAMPLE 1.   Let $K \subset \mathbb{R}^2$ be compact. Prove that $A = \{x \in \mathbb{R} \mid \text{there exists}$ a y such that $(x,y) \in K\}$ is compact.

Solution: Let $f: \mathbb{R}^2 \to \mathbb{R}, (x,y) \to x$. Then f is continuous (see Exercise 1 of Section 4.1). We claim that $A = f(K)$, so A would be compact by Theorem 2. To prove the claim, let $x \in A$. Then $(x,y) \in K$ for some y, so $x = f(x,y) \in f(K)$. Conversely, if $x = f(x,y)$ for some $(x,y) \in K$, then $x \in A$ by definition.

EXAMPLE 2.   Find a continuous map $f: \mathbb{R} \to \mathbb{R}$ and a compact $K \subset \mathbb{R}$ such that $f^{-1}(K)$ is not compact.

Solution: Let $f(x) = 0$ for all $x \in \mathbb{R}$ and let $K = \{0\}$. Then $f^{-1}(K) = \mathbb{R}$ is not compact.

EXAMPLE 3.   Let $f: \mathbb{R}^2 \to \mathbb{R}$ be continuous, and let $A = \{f(x) \mid \|x\| = 1\}$. Show that A is a closed interval.

Solution: Clearly, $A = f(S)$ where $S = \{x \in \mathbb{R}^2 \mid \|x\| = 1\}$ is the unit circle. Now S is connected and compact so A is connected and compact. By Example 3, at the end of Chapter 3, A is an interval. But the only compact intervals are the closed ones.

Exercises for Section 4.2

1. Let $f: \mathbb{R} \to \mathbb{R}$ be continuous. Which sets below are necessarily closed, open, compact, or connected?
   (a) $\{x \in \mathbb{R} \mid f(x) = 0\}$.
   (b) $\{x \in \mathbb{R} \mid f(x) > 1\}$.
   (c) $\{f(x) \in \mathbb{R} \mid x \geqslant 0\}$.
   (d) $\{f(x) \in \mathbb{R} \mid 0 \leqslant x \leqslant 1\}$.

2. Verify Theorem 2 for $f: \mathbb{R}^2 \to \mathbb{R}, f(x,y) = x^2 + y, K = B = \{(x,y) \mid x^2 + y^2 \leqslant 1\}$.

3. Give an example of a continuous map $f: \mathbb{R} \to \mathbb{R}$ and a closed set $B \subset \mathbb{R}$ such that $f(B)$ is not closed. Is it possible if B is bounded as well?

4. Let $A, B \subset \mathbb{R}$, and suppose $A \times B \subset \mathbb{R}^2$ is connected. Prove that A is connected.

5. Let $A, B \subset \mathbb{R}$, and suppose $A \times B \subset \mathbb{R}^2$ is open. Must A be open?

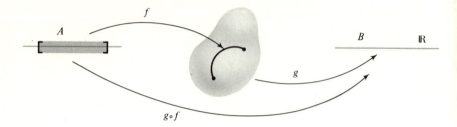

FIGURE 4-3   Composition of mappings.

4.3   Operations on Continuous Mappings

It is intuitively plausible that the composition of continuous functions should be continuous, as we shall now discuss. Recall that for $f: A \to \mathbb{R}^m$ and $g: B \to \mathbb{R}^p$ with $f(A) \subset B$, we define the composition $g \circ f: A \to \mathbb{R}^p$ by $x \mapsto g(f(x))$. If x is close to x_0, then $g \circ f(x)$ is close to $g \circ f(x_0)$ because $f(x)$ is close to $f(x_0)$; hence $g(f(x))$ is close to $g(f(x_0))$. See Figure 4-3.

This indicates the plausibility of the following result.

> **Theorem 3.**   *Suppose $f: A \to \mathbb{R}^m$ and $g: B \to \mathbb{R}^p$ are continuous functions with $f(A) \subset B$. Then $g \circ f: A \to \mathbb{R}^p$ is continuous.*

For example, the function $e^{\sin x}$ is continuous because it is the composition of the two continuous functions $f(x) = \sin x$ and $g(x) = e^x$.

Note: We shall accept as known from calculus the properties of the basic functions such as $\sin x$, e^x, and so forth. These will be used later in several examples.

The following theorem gives some of the fundamental properties concerning the arithmetic of limits.

> **Theorem 4.**   *Let $A \subset \mathbb{R}^n$, and let x_0 be an accumulation point of A.*
> *(i) Let $f: A \to \mathbb{R}^m$ and $g: A \to \mathbb{R}^m$ be functions; assume $\lim\limits_{x \to x_0} f(x)$ and $\lim\limits_{x \to x_0} g(x)$ exist and are equal to a and b respectively. Then $\lim\limits_{x \to x_0} (f + g)(x)$ exists and is equal to $a + b$ (where $f + g: A \to \mathbb{R}^m$ is defined by $(f + g)(x) = f(x) + g(x)$).*
> *(ii) Let $f: A \to \mathbb{R}$ and $g: A \to \mathbb{R}^m$ be functions; assume $\lim\limits_{x \to x_0} f(x)$ and $\lim\limits_{x \to x_0} g(x)$ exist and are equal to a and b respectively. Then $\lim\limits_{x \to x_0} (f \cdot g)(x)$ exists and is equal to ab (where $f \cdot g: A \to \mathbb{R}^m$ is defined by $(f \cdot g)(x) = f(x)g(x)$).*

(iii) Let $f: A \to \mathbb{R}$ and $g: A \to \mathbb{R}^m$ be functions; assume $\lim_{x \to x_0} f(x)$ and $\lim_{x \to x_0} g(x)$ exist and are equal to $a \neq 0$ and b respectively. Then f is non-zero in a neighborhood of x_0 and $\lim_{x \to x_0} (g/f)(x)$ exists and is equal to b/a (where $g/f: A \to R^m$ is defined by $(g/f)(x) = g(x)/f(x)$).

These results are eminently reasonable intuitively. For instance, (i) states that if x is close to x_0 so that $f(x)$ is close to a and $g(x)$ is close to b, then $f(x) + g(x)$ is close to $a + b$. From Theorem 4 we may deduce immediately some basic properties of the arithmetic of continuous functions.

Corollary. Let $A \subset \mathbb{R}^n$, $x_0 \in A$ an accumulation point of A.
(i) Let $f: A \to \mathbb{R}^m$ and $g: A \to \mathbb{R}^m$ be continuous at x_0; then the sum $f + g: A \to \mathbb{R}^m$ is continuous at x_0.
(ii) Let $f: A \to \mathbb{R}$ and $g: A \to \mathbb{R}^m$ be continuous at x_0; then the product $f \cdot g: A \to \mathbb{R}^m$ is continuous at x_0.
(iii) Let $f: A \to \mathbb{R}$ and $g: A \to \mathbb{R}^m$ be continuous at x_0 with $f(x_0) \neq 0$; then f is non-zero in a neighborhood U of x_0 and the quotient $g/f: U \to \mathbb{R}^m$ is continuous at x_0.

For example, we have seen that $f(x) = x$, mapping from $\mathbb{R}$ to $\mathbb{R}$, is continuous, and therefore so is $f(x) = x^n$; also, any polynomial $a_n x^n + a_{n-1} x^{n-1} + \cdots + a_0$.

Now consider $f: \mathbb{R}^2 \to \mathbb{R}$. Think of f as a function of two real variables, $f(x,y)$. It is crucial to distinguish between continuity of f (sometimes called joint continuity) and continuity in each variable separately. For example, consider the function

$$f(x,y) = \begin{cases} 0 & \text{if } x \neq 0 \text{ and } y \neq 0 \,; \\ 1 & \text{if either } x = 0 \text{ or } y = 0 \,. \end{cases}$$

See Figure 4-4. In each individual variable, f is continuous at $(0,0)$ (the

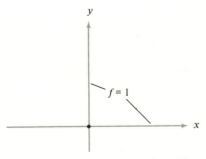

FIGURE 4-4   Separate and joint continuity.

mappings $x \mapsto f(x,0)$ and $y \to f(0,y)$ are constant, and so are continuous), but f itself is not continuous at $(0,0)$ (why?). See Exercise 16 at the end of this chapter for sufficient conditions on separate continuity needed to imply continuity.

EXAMPLE 1.   Let $f: \mathbb{R} \to \mathbb{R}$, $f(x) = x \sin x$. Show that f is continuous.

Solution: We know x and $\sin x$ are continuous and f is the product of two continuous functions and so is continuous.

EXAMPLE 2.   Let $f: \mathbb{R} \to \mathbb{R}^2$ be continuous. Show that $g(x) = f(x^2 + x^3)$ is continuous.

Solution: g is the composition of f on the continuous function $x \mapsto x^2 + x^3$ and so is continuous by Theorem 3.

EXAMPLE 3.   Let $f(x) = x^2/(1 + x)$. Where is f continuous?

Solution: We define f for $x \neq -1$. Then, by Theorem 4(iii), f is continuous at all $x \neq -1$.

Exercises for Section 4.3

1. Where are the following functions continuous?
   (a) $f(x) = x \sin(x^2)$.
   (b) $f(x) = \dfrac{x + x^2}{x^2 - 1}$, $x^2 \neq 1, f(\pm 1) = 0$.
   (c) $f(x) = \dfrac{\sin x}{x}$, $x \neq 0, f(0) = 1$.

2. Let, in $\mathbb{R}$, $a_k \to a$ and $b_k \to b$. Prove $a_k b_k \to ab$ by using Theorems 3 and 4.

3. Let $A = \{x \in \mathbb{R} \mid \sin x = .56\}$. Show that A is a closed set. Is it compact?

4. Show $f: \mathbb{R} \to \mathbb{R}$, $x \mapsto \sqrt{|x|}$ is continuous.

5. Show $f(x) = \sqrt{x^2 + 1}$ is continuous.

4.4   The Boundedness of Continuous Functions on Compact Sets

We are now ready to prove an important property of continuous real-valued functions called the "boundedness theorem." The boundedness theorem says that a continuous function is bounded on a compact set and attains its largest or maximum value and its smallest or minimum value at some point of the set. The precise definitions will be stated in Theorem 5.

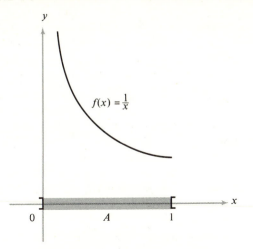

**FIGURE 4-5 An unbounded
continuous function.**

To appreciate this result let us consider what can happen on a noncompact set. First, a continuous function need not be bounded. Figure 4-5 shows the function $f(x) = 1/x$ on the open interval $]0,1[$. As x gets closer to 0, the function becomes arbitrarily large, but f is nevertheless continuous, since f is the quotient of 1 by the continuous function $x \mapsto x$ which does not vanish on $]0,1[$.

Next, we can show that even if a function is bounded and continuous, it might not assume its maximum at any point of its domain. Figure 4-6 shows the function $f(x) = x$ on the interval $[0,1[$. This function never attains a

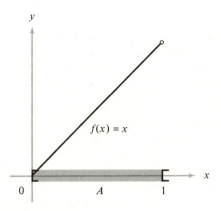

**FIGURE 4-6   A function with no
maximum.**

maximum value because even though there are an infinite number of points as near to 1 as we please, there is no point x for which $f(x) = 1$. From these examples, it should be fairly plausible that, for a continuous function on a compact set, these pathologies cannot occur.

Let us now state the theorem formally.

> **Theorem 5.**   *Let $A \subset \mathbb{R}^n$ and $f: A \to \mathbb{R}$ be continuous. Let $K \subset A$ be a compact set. Then f is bounded on K, that is $B = \{f(x) \mid x \in K\} \subset \mathbb{R}$ is a bounded set. Furthermore, there exist points $x_0, x_1 \in K$ such that $f(x_0) = \inf(B)$ and $f(x_1) = \sup(B)$. We call $\sup(B)$ the (absolute) maximum of f on K and $\inf(B)$ the (absolute) minimum of f on K.*

It should be appreciated that this result is deeper than the usual derivative tests for the location of maxima and minima that we learn from calculus. For example, there are continuous functions on $\mathbb{R}$ which are differentiable at no point; such functions cannot be graphed by a smooth curve so our intuition is not as clear in these cases (see Exercise 19, p. 144).

EXAMPLE 1.   Give an example of a discontinuous function on a compact set which is not bounded.

Solution: Let $f: [0,1] \to \mathbb{R}$ be defined by $f(x) = 1/x$ if $x > 0$ and $f(0) = 0$. Clearly, this function exhibits the same unboundedness property as does $1/x$ on $]0,1]$.

EXAMPLE 2.   Verify Theorem 5 for $f(x) = x/(x^2 + 1)$ on $[0,1]$.

Solution: $f(0) = 0$, $f(1) = 1/2$. We shall verify explicitly that the maximum is at $x = 1$, the minimum is at $x = 0$. (Elementary calculus helps to determine this, but we shall give a direct verification.) First, as $0 \leqslant x \leqslant 1$, $x/(x^2 + 1) \geqslant 0$ since $x \geqslant 0$ and $x^2 + 1 \geqslant 1$, so $f(x) \geqslant f(0)$ for $0 \leqslant x \leqslant 1$. Thus 0 is the minimum. Next, note that $0 \leqslant (x - 1)^2 = x^2 - 2x + 1$, so $x^2 + 1 \geqslant 2x$ and hence for $x \neq 0$,

$$\frac{x}{x^2 + 1} \leqslant \frac{x}{2x} = \frac{1}{2}$$

so $f(x) \leqslant f(1) = 1/2$ and thus $x = 1$ is the maximum point.

EXAMPLE 3.   Show that x_0 and x_1 in Theorem 5 need not be unique.

Solution: Let $f(x) = 1$ for all $x \in [0,1]$. Then any $x_0, x_1 \in [0,1]$ will do.

Exercises for Section 4.4

1. Give an example of a continuous and bounded function on all of ℝ for which Theorem 5 fails.

2. Verify Theorem 5 for $f(x) = x^3 - x$ on $[-1,1]$.

3. Let $f: K \subset \mathbb{R}^n \to \mathbb{R}$ be continuous on a compact K and let $M = \{x \in K \mid f(x)$ is the maximum of f on $K\}$. Show that M is a compact set.

4. Let $f: A \subset \mathbb{R}^n \to \mathbb{R}$ be continuous, $x, y \in A$ and $c: [0,1] \to \mathbb{R}^n$ a curve joining x and y. Show that along this curve f assumes its maximum and minimum values.

5. Study Theorem 5 in the context of $f(x) = (\sin x)/x$ on $]0,\infty[$.

4.5   The Intermediate-Value Theorem

The intermediate-value theorem is perhaps well known from elementary calculus. It states that a continuous function on an interval assumes all values between any two given values. See Figure 4-7a. The noncontinuous function f in Figure 4-7b never assumes the value $1/2$. Briefly, this tells us that while a discontinuous function can jump from one value to another, a continuous function must pass through all intermediate values.

Another way the intermediate-value property can fail is if the domain A is not connected, as illustrated by the continuous function in Figure 4-8.

Thus the crucial assumptions are that f be continuous and f be defined on a connected region. We shall see that the proof of Theorem 6 is quite simple because of the way we have formulated the notion of connectedness (see Example 1 below and the theorem proofs at the end of the chapter).

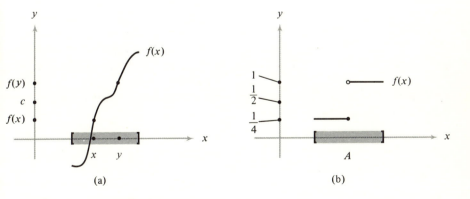

FIGURE 4-7   Intermediate-value theorem.

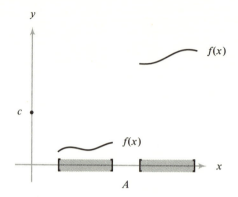

FIGURE 4-8   Continuous function with a disconnected domain.

Theorem 6. *Let* $A \subset \mathbb{R}^n$ *and* $f: A \to \mathbb{R}$ *be continuous. Suppose* $K \subset A$ *is connected and* $x, y \in K$. *For every number* $c \in \mathbb{R}$ *such that* $f(x) \leqslant c \leqslant f(y)$, *there exists a point* $z \in K$ *such that* $f(z) = c$.

Since intervals (open or closed) are connected (this was proven in Example 3 at the end of Chapter 3) the usual intermediate-value theorem becomes a special case of Theorem 6. However, notice that Theorem 6 is more general. It applies, for example, to continuous real-valued functions of several variables $f(x_1, \ldots, x_n)$ defined on all of $\mathbb{R}^n$, which is a connected set.

EXAMPLE 1.   Discuss a possible proof of Theorem 6 using the fact that $f(K)$ is connected.

Solution: That $f(K)$ is connected comes from Theorem 2. Hence $f(K)$ is an interval, possibly infinite. But if $f(x)$, $f(y) \in f(K)$, in particular, $[f(x), f(y)] \subset f(K)$ since $f(K)$ is an interval. So if c is the same as in Theorem 6, $c \in [f(x), f(y)]$ so $c \in f(K)$, so $c = f(z)$ for some z. This is in fact one way of proving Theorem 6. Another is given in the theorem proofs section.

EXAMPLE 2.   Let $f(x)$ be a cubic polynomial. Argue that f has a (real) root x_0 (that is, $f(x_0) = 0$).

Solution: $f(x) = ax^3 + bx^2 + cx + d$, where $a \neq 0$. Suppose that $a > 0$. For $x > 0$, x large, ax^3 is large (and positive) and will be bigger than the other terms so $f(x) > 0$ if x is large, $x > 0$. This requires some exact estimates but should be intuitively clear. Similarly, $f(x) < 0$ if x is large and negative. Hence we can apply the intermediate-value theorem to conclude the existence of a point x_0 where $f(x_0) = 0$.

EXAMPLE 3. Let $f: [1,2] \to [0,3]$ be a continuous function with $f(1) = 0$, $f(2) = 3$. Show that f has a fixed point. That is, show that there is a point $x_0 \in [1,2]$ such that $f(x_0) = x_0$.

Solution: Let $g(x) = f(x) - x$. Then g is continuous, $g(1) = f(1) - 1 = -1$, and $g(2) = f(2) - 2 = 3 - 2 = 1$. Hence by the intermediate-value theorem, g must vanish at some $x_0 \in [1,2]$ and this x_0 is the fixed point for $f(x)$.

Exercises for Section 4.5

1. What happens when you apply the method used in Example 2 to quadratic polynomials? To quintic polynomials?

2. Let f be continuous: $\mathbb{R}^n \to \mathbb{R}^m$. Let $\Gamma = \{(x, f(x)) \mid x \in \mathbb{R}^n\}$ be the graph of f in $\mathbb{R}^n \times \mathbb{R}^m$. Prove that Γ is closed.

3. Let $f: [0,1] \to [0,1]$ be continuous. Prove that f has a fixed point.

4. Let $f: [a,b] \to \mathbb{R}$ be continuous. Show that the range of f is a bounded closed interval.

5. Prove that there exists no continuous map of $[0,1]$ *onto* $]0,1[$.

4.6 Uniform Continuity

Sometimes it is useful to have available a slight variant of the definition of continuity. Often the applications are technical ones, such as labor-saving devices in proofs. Still, the notion of a uniformly continuous function is a basic one and it is used widely. The exact definition is as follows.

> **Definition 3.** Let $f: A \to \mathbb{R}^m$ and $B \subset A$. We say that f is *uniformly continuous* on the set B if for every $\varepsilon > 0$ there is a $\delta > 0$ such that $x, y \in B$ and $d(x,y) < \delta$ implies $d(f(x), f(y)) < \varepsilon$.

The definition is similar to continuity, except that here we must be able to choose δ to work for all x, y once ε is given. For continuity we were only required to choose a δ once we were given $\varepsilon > 0$ and a *particular* x_0. Clearly, if f is uniformly continuous, then f is continuous.

For example, consider $f: \mathbb{R} \to \mathbb{R}, f(x) = x^2$. Then f is certainly continuous, but it is not uniformly continuous. Indeed, for $\varepsilon > 0$ and $x_0 > 0$ given, the $\delta > 0$ we need is at least as small as $\varepsilon/(2x_0)$, so if we choose x_0 large, δ must get smaller. No single δ will do for all x_0. This phenomenon cannot happen on compact sets, as the next theorem shows.

Theorem 7. *Let* $f: A \rightarrow \mathbb{R}^m$ *be continuous and let* $K \subset A$ *be a compact set. Then* f *is uniformly continuous on* K.

The use of merely bounded sets in Theorem 7 will not do, for consider what can happen on the noncompact set $]0,1]$. Let $f(x) = 1/x$. Then if we examine the proof that f is continuous (Example 2, Section 4.1) we see again that f is not uniformly continuous. Of course, we cannot make f continuous on the compact set $[0,1]$ because it is unbounded.

Another very useful criterion for uniform continuity is given in Example 2 below.

EXAMPLE 1.   Let $f:]0,1] \rightarrow \mathbb{R}$, $f(x) = 1/x$. Show f is uniformly continuous on $[a,1]$ for $a > 0$.

Solution: The solution can be immediately drawn from Theorem 7, since $[a,1]$ is a compact set.

EXAMPLE 2.   Let $f:]a,b[\rightarrow \mathbb{R}$ be differentiable and suppose $|f'(x)| \leqslant M$. Here, a or b may be $\pm\infty$, and f' stands for the derivative of f. Show that f is uniformly continuous on $]a,b[$.

Solution: The definition of uniform continuity asks us to estimate $|f(x) - f(y)|$ in terms of $|x - y|$. This suggests using the mean-value theorem (see p. 156 for a review). Indeed,

$$f(x) - f(y) = f'(x_0)(x - y)$$

for some x_0 between x and y. Hence

$$|f(x) - f(y)| \leqslant M |x - y| .$$

Given $\varepsilon > 0$, choose $\delta = \varepsilon/M$. Then $|x - y| < \delta$ implies

$$|f(x) - f(y)| < M \cdot \delta = M \cdot \varepsilon/M = \varepsilon .$$

Hence f is uniformly continuous.

The intuition here may shed some more light on uniform continuity. Namely, this result says that if the slope of the graph of a function is bounded, then it is uniformly continuous. This is often a good guide when examining specific functions or their graphs.

EXAMPLE 3.   Show that $\sin x: \mathbb{R} \rightarrow \mathbb{R}$ is uniformly continuous.

Solution: $d(\sin x)/dx = \cos x$ is bounded in absolute value by 1, so by Example 2, $\sin x$ is uniformly continuous.

Exercises for Section 4.6

1. Demonstrate the conclusion in Example 1 directly from the definition.

2. Prove that $f(x) = 1/x$ is uniformly continuous on $[a,\infty[$ for $a > 0$.

3. Do you think a bounded continuous function on $\mathbb{R}$ has to be uniformly continuous?

4. If f and g are uniformly continuous, $\mathbb{R} \to \mathbb{R}$, must the product $f \cdot g$ be uniformly continuous? What if f and g are bounded?

5. Let $f(x) = |x|$. Show that $f: \mathbb{R} \to \mathbb{R}$ is uniformly continuous.

6. Show that $f: \mathbb{R} \to \mathbb{R}$ is not uniformly continuous iff there exists an $\varepsilon > 0$ and sequences x_n and y_n such that $|x_n - y_n| < 1/n$ and $|f(x_n) - f(y_n)| \geqslant \varepsilon$. Use this to prove that $f(x) = x^2$ is not uniformly continuous.

Theorem Proofs for Chapter 4

Theorem 1. *Let $f: A \to \mathbb{R}^m$ be a mapping, where $A \subset \mathbb{R}^n$ is any set. Then the following assertions are equivalent.*

(i) *f is continuous on A.*

(ii) *For each convergent sequence $x_k \to x_0$ in A, we have $f(x_k) \to f(x_0)$.*

(iii) *For each open set U in $\mathbb{R}^m$, $f^{-1}(U) \subset A$ is open relative to A; that is, $f^{-1}(U) = V \cap A$ for some open set V.*

(iv) *For each closed set $F \subset \mathbb{R}^m$, $f^{-1}(F) \subset A$ is closed relative to A; that is, $f^{-1}(F) = G \cap A$ for some closed set G.*

Proof: The pattern of the proof will be (i) $\Rightarrow$ (ii) $\Rightarrow$ (iv) $\Rightarrow$ (iii) $\Rightarrow$ (i).

Proof of (i) $\Rightarrow$ (ii):   Suppose $x_k \to x_0$. To show that $f(x_k) \to f(x_0)$, let $\varepsilon > 0$; we must find an integer N so that $k \geqslant N$ implies $d(f(x_k), f(x_0)) < \varepsilon$. To accomplish this, choose $\delta > 0$ so that $d(x, x_0) < \delta$ implies $d(f(x), f(x_0)) < \varepsilon$. The existence of a δ is guaranteed by the continuity of f. Then choose N so that $k \geqslant N$ implies $d(x_k, x_0) < \delta$. This choice of N yields the desired conclusion.

Proof of (ii) $\Rightarrow$ (iv):   Let $F \subset \mathbb{R}^m$ be closed. To show $f^{-1}(F)$ is closed in A, we use the fact that a set B is closed relative to A iff for every sequence $x_k \in B$ which converges to a point $x \in A$, then $x \in B$ (see Theorem 9, Chapter 2). The reader should write out the proof of this assertion. Here, let $x_k \in f^{-1}(F)$ and let $x_k \to x$, where $x \in A$. We must show that $x \in f^{-1}(F)$. Now, by (ii), $f(x_k) \to f(x)$, and since $f(x_k) \in F$ and F is closed, we conclude that $f(x) \in F$. Thus $x \in f^{-1}(F)$.

Proof of (iv) $\Rightarrow$ (iii):   If U is open, let $F = \mathbb{R}^m \backslash U$, which is closed. Then, by (iv), $f^{-1}(F) = G \cap A$ for some closed set G. Thus $f^{-1}(U) = A \cap (\mathbb{R}^n \backslash G)$, so $f^{-1}(U)$ is open relative to A.

Proof of (iii) $\Rightarrow$ (i):   Given $x_0 \in A$ and $\varepsilon > 0$, we must find δ so that $d(x, x_0) < \delta$ implies $d(f(x), f(x_0)) < \varepsilon$. Since $D(f(x_0), \varepsilon)$ is an open set, $f^{-1}(D(f(x_0), \varepsilon))$ is open by (iii). Thus by the definition of open set and the fact that $x_0 \in f^{-1}(D(f(x_0), \varepsilon))$, there is a

$\delta > 0$ such that $D(x_0,\delta) \cap A \subset f^{-1}(D(f(x_0),\varepsilon))$. This is another way of saying that $(x \in A$ and $d(x,x_0) < \delta) \Rightarrow (d(f(x),f(x_0)) < \varepsilon)$. ∎

To gain practice with these concepts, one might try proving (directly) other implications above; for example, (i) $\Rightarrow$ (iii), or (ii) $\Rightarrow$ (i).

Theorem 2. *Let* $f: A \to \mathbb{R}^m$ *be a continuous mapping. Then*
(i) *if* $K \subset A$ *is connected, then* $f(K)$ *is connected;*
(ii) *if* $B \subset A$ *is compact, then* $f(B)$ *is compact.*

Proof: (i) Suppose $f(K)$ is not connected. Then, by definition, we can write $f(K) \subset U \cup V$, where $U \cap V \cap f(K) = \varnothing$, $U \cap f(K) \neq \varnothing$, $V \cap f(K) \neq \varnothing$, and U, V are open sets. Now, $f^{-1}(U) = U' \cap A$ for some open set U', and similarly, $f^{-1}(V) = V' \cap A$ for some open set V'. From the conditions on U, V, we see that $U' \cap V' \cap K = \varnothing$, $K \subset U' \cup V'$, $U' \cap K \neq \varnothing$, and $V' \cap K \neq \varnothing$. Thus K is not connected, which proves the assertion.

(ii) Let y_k be a sequence in $f(B)$. By Theorem 1 of Chapter 3, it must be shown that y_k has a subsequence converging to a point in $f(B)$. Let $y_k = f(x_k)$, for $x_k \in B$. Since B is compact, there is a convergent subsequence, say, $x_{k_n} \to x$ for $x \in B$. Now, by Theorem 1(ii), $f(x_{k_n}) \to f(x)$, so $f(x_{k_n})$ is a convergent subsequence of y_k. ∎

The path-connected part of Theorem 2(i) follows as in the text. In the proof of (ii) we used the characterization of a compact set as a set in which every sequence has a convergent subsequence. One may also use the Heine-Borel criterion for compact sets (try it). Note that, in general, the continuous image of a closed set need not be closed. Thus compactness of B is crucial in proving that $f(B)$ is both closed and bounded.

Theorem 3. *Suppose* $f: A \to \mathbb{R}^m$ *and* $g: B \to \mathbb{R}^p$ *are continuous functions with* $f(A) \subset B$. *Then* $g \circ f: A \to \mathbb{R}^p$ *is continuous.*

Proof: Let $U \subset \mathbb{R}^p$ be open. Then $(g \circ f)^{-1}(U) = f^{-1}(g^{-1}(U))$. Now, $g^{-1}(U) = U' \cap B$ for some U' open, and $f^{-1}(U' \cap B) = f^{-1}(U')$, since $f(A) \subset B$. Since f is continuous, $f^{-1}(U') = U'' \cap A$ for U'' open. Thus $g \circ f$ is continuous by Theorem 1. ∎

The other conditions of Theorem 1 can just as easily be used in order to prove Theorem 3. Instead of proving Theorem 4, we shall confine ourselves to proving its Corollary. The general case is similar; the only complexity is in notation.

Corollary. *Let* $A \subset \mathbb{R}^n$.
(i) *Let* $f: A \to \mathbb{R}^m$ *and* $g: A \to \mathbb{R}^m$ *be continuous. Then* $f + g: A \to \mathbb{R}^m$ *defined by* $(f + g)(x) = f(x) + g(x)$ *is continuous.*
(ii) *Let* $f: A \to \mathbb{R}$ *and* $g: A \to \mathbb{R}^m$ *be continuous. Then* $f \cdot g: A \to \mathbb{R}^m$ *defined by* $(f \cdot g)(x) = f(x)g(x)$ *(multiplication of the scalar* $f(x)$ *by the vector* $g(x)$*) is continuous.*
(iii) *Let* $f: A \to \mathbb{R}$ *and* $g: A \to \mathbb{R}^m$ *be continuous for* $A \subset \mathbb{R}^n$. *If* $f(x) \neq 0$ *for all* $x \in A$, *then* g/f *is continuous on* A.

Proof: (i) Let $x_0 \in A$ and suppose $\varepsilon > 0$ is given. Choose $\delta_1 > 0$ such that $d(x,x_0) < \delta_1$ implies $d(f(x),f(x_0)) < \varepsilon/2$ and $\delta_2 > 0$ such that $d(x,x_0) < \delta_2$ implies

$d(g(x),g(x_0)) < \varepsilon/2$. Then let δ be the minimum of δ_1, δ_2. Therefore, if $d(x,x_0) < \delta$, we have by the triangle inequality,

$$\|(f + g)(x) - (f + g)(x_0)\| = \|f(x) - f(x_0) + g(x) - g(x_0)\|$$
$$\leqslant \|f(x) - f(x_0)\| + \|g(x) - g(x_0)\|$$
$$\leqslant \frac{\varepsilon}{2} + \frac{\varepsilon}{2} = \varepsilon .$$

(ii) Let $x_0 \in A$ and suppose $\varepsilon > 0$. Choose δ_1 such that $d(x,x_0) < \delta_1$ implies $|f(x) - f(x_0)| < \varepsilon/2 \|g(x_0)\|$ and $|f(x)| \leqslant |f(x_0)| + 1$ (why is this possible?). Also choose δ_2 such that $d(x,x_0) < \delta_2$ implies that $\|g(x) - g(x_0)\| < \varepsilon/2(|f(x_0)| + 1)$. Then for $\delta = \min(\delta_1,\delta_2)$, $d(x,x_0) < \delta$ implies (by the triangle inequality)

$$\|fg(x) - fg(x_0)\| = \|f(x)g(x) - f(x)g(x_0) + f(x)g(x_0) - f(x_0)g(x_0)\|$$
$$\leqslant |f(x)| \|g(x) - g(x_0)\| + |f(x) - f(x_0)| \|g(x_0)\|$$

(using the fact that $\|\alpha x\| = |\alpha| \|x\|$ for $x \in \mathbb{R}^n$, $\alpha \in \mathbb{R}$). Continuing the above estimate, we get

$$\|fg(x) - fg(x_0)\| < (|f(x_0)| + 1)\varepsilon/2(|f(x_0)| + 1) + \|g(x_0)\| \varepsilon/2 \|g(x_0)\|$$
$$= \frac{\varepsilon}{2} + \frac{\varepsilon}{2} = \varepsilon .$$

(iii) By proof (ii), it suffices to consider the case $1/f$ for $g/f = g \cdot (1/f)$.

To show that $1/f$ is continuous, given $x_0 \in A$, choose δ_1 such that $|f(x) - f(x_0)| \leqslant (|f(x_0)|/2)$ for $\|x - x_0\| < \delta_1$. This is possible by the continuity of f. It follows that $|f(x)| \geqslant (|f(x_0)|/2)$. Now, given $\varepsilon > 0$, choose δ_2 such that $\|x - x_0\| < \delta_2$ implies

$$|f(x) - f(x_0)| < \varepsilon |f(x_0)|^2/2 .$$

Then if $\delta = \min(\delta_1,\delta_2)$, $\|x - x_0\| < \delta$ implies

$$\left| \frac{1}{f(x)} - \frac{1}{f(x_0)} \right| = \left| \frac{f(x_0) - f(x)}{f(x_0)f(x)} \right| \leqslant \frac{|f(x) - f(x_0)|}{|f(x_0)|^2/2} < \varepsilon .$$

This shows that $1/f(x)$ is continuous at x_0, and hence it is continuous on A. ∎

Theorem 5. Let $A \subset \mathbb{R}^n$ and $f: A \to \mathbb{R}$ be continuous. Let $K \subset A$ be a compact set. Then f is bounded on K, that is $B = \{f(x) \mid x \in K\} \subset \mathbb{R}$ is a bounded set. Furthermore, there exist points $x_0, x_1 \in K$ such that $f(x_0) = \inf(B)$ and $f(x_1) = \sup(B)$. We call $\sup(B)$ the (absolute) maximum of f on K and $\inf(B)$ the (absolute) minimum of f on K.

Proof: First, B is bounded above, for $B = f(K)$ is compact by Theorem 2, so it is closed and bounded by the definition of compactness. Second, we want to produce an x_1 such that $x_1 \in K$ and $f(x_1) = \sup(B)$. Now, since B is closed, $\sup(B) \in B = f(K)$ (see Exercise 8, Chapter 2). Thus $\sup(B) = f(x_1)$ for some $x_1 \in K$.

The case of $\inf(B)$ is similar. (The student should write out the details.) ∎

Note: We can also get the case of $\inf(B)$ by applying the above supremum case to $-f$ and observing that the maximum of $-f$ is the minimum of f.

Theorem 6. Let $A \subset \mathbb{R}^n$ and $f\colon A \to \mathbb{R}$ be continuous. Suppose $K \subset A$ is connected and $x, y \in K$. For every number $c \in \mathbb{R}$ such that $f(x) \leqslant c \leqslant f(y)$, there exists a point $z \in K$ such that $f(z) = c$.

Proof: Suppose no such z exists. Then let $U =]-\infty, c[= \{t \in \mathbb{R} \mid t < c\}$ and let $V =]c, \infty[$. Clearly, both U and V are open sets. Since f is continuous, we have $f^{-1}(U) = U_0 \cap K$ for an open set U_0, and similarly, $f^{-1}(V) = V_0 \cap K$. By definition of U and V, we have $U_0 \cap V_0 \cap K = \varnothing$, and by the assumption that $\{z \in K \mid f(z) = c\} = \varnothing$, we have $U_0 \cup V_0 \supset K$. Also, $U_0 \cap K \neq \varnothing$, since $x \in U$; and $V_0 \cap K \neq \varnothing$, since $y \in V_0$. Hence, K is not connected, a contradiction. ∎

Theorem 7. Let $f\colon A \to \mathbb{R}^m$ be continuous and let $K \subset A$ be a compact set. Then f is uniformly continuous on K.

Proof: Given $\varepsilon > 0$, for each $x \in K$, choose δ_x such that $d(x,y) < \delta_x$ implies $d(f(x),f(y)) < \varepsilon/2$. The sets $D(x,\delta_x/2)$ cover K and are open. Therefore, there is a finite covering, say, $D(x_1,\delta_{x_1}/2), \ldots, D(x_N,\delta_{x_N}/2)$. Let $\delta = \text{minimum } \delta_{x_1}/2, \ldots, \delta_{x_N}/2$. Then if $d(x,y) < \delta$, there is an x_i such that $d(x,x_i) < \delta_{x_i}/2$ (since the discs cover K), and therefore $d(x_i,y) \leqslant d(x,x_i) + d(x,y) < \delta_{x_i}$. Thus by the choice of δ_{x_i}, $d(f(x),f(y)) \leqslant d(f(x),f(x_i)) + d(f(x_i),f(y)) < \varepsilon/2 + \varepsilon/2 = \varepsilon$. ∎

Worked Examples for Chapter 4

1. Let $f\colon A \to \mathbb{R}^m$ be written as

$$f(x) = (f_1(x), \ldots, f_m(x)) .$$

Then show that f is continuous iff each component f_i is continuous, $i = 1, \ldots, m$.
Solution: Let f be continuous. If $x_k \to x$ in A, we must show that $f_i(x_k) \to f_i(x)$ for each i. But this is an immediate consequence of the fact that $f(x_k) \to f(x)$, and a sequence in $\mathbb{R}^m$ (here $f(x_k)$) converges iff its component sequences converge (see Section 2.7). The same reasoning proves the converse.

2. Let $f\colon A \to \mathbb{R}^m$ be continuous. For $K \subset A$ a connected set, show that $\{(x,f(x)) \mid x \in K\}$ is connected in $\mathbb{R}^n \times \mathbb{R}^m = \mathbb{R}^{n+m}$. This set is of course just the graph of f.
Solution: Consider the mapping $g\colon K \subset \mathbb{R}^n \to \mathbb{R}^n \times \mathbb{R}^m$ defined by $g(x) = (x,f(x))$. By the previous example, g is continuous. But $g(K) = \{(x,f(x)) \mid x \in K\}$, and the image of a connected set is connected, by Theorem 2.

3. Let $f\colon A \to \mathbb{R}^m$ be continuous at $x_0 \in A$, A open, and $f(x_0) \neq 0$. Then show that f is non-zero in some neighborhood of x_0.
Solution: Given $\varepsilon > 0$ there is a neighborhood U of x_0 such that $\|f(x) - f(x_0)\| < \varepsilon$ for all $x \in U$, by the definition of continuity. For our purpose, choose $\varepsilon = \|f(x_0)\|$. Then $\|f(x) - f(x_0)\| < \|f(x_0)\|$ implies that $f(x) \neq 0$, because it is not true that $\|-f(x_0)\| < \|f(x_0)\|$ (they are equal). Therefore, on the neighborhood U determined by $\varepsilon = \|f(x_0)\|$, f is not zero.

4. Show that a linear map $L\colon \mathbb{R}^n \to \mathbb{R}^m$ is continuous.

Solution: We shall show that for our given linear map $L: \mathbb{R}^n \to \mathbb{R}^m$ we can find a number M such that $\|L(x)\| \leqslant M \|x\|$ for all $x \in \mathbb{R}^n$. Then $\|x - x_0\| < \varepsilon/M$ implies $\|L(x) - L(x_0)\| = \|L(x - x_0)\| \leqslant M \|x - x_0\| < \varepsilon$, which will prove that L is continuous.

Let $M_1 = \sup\{\|L(e_1)\|, \ldots, \|L(e_n)\|\}$, where $e_1, \ldots, e_n$ is the standard basis for $\mathbb{R}^n$. Then for $x = (x_1, \ldots, x_n)$, and using the triangle inequality,

$$\|L(x)\| = \|x_1 L(e_1) + \cdots + x_n L(e_n)\| \leqslant |x_1| \|L(e_1)\| + \cdots + |x_n| \|L(e_n)\|$$
$$\leqslant M_1(|x_1| + \cdots + |x_n|)$$
$$\leqslant M_1 n \|x\|.$$

Thus we can take $M = nM_1$, and we get our result.

5. A multilinear map L from $\mathbb{R}^{n_1} \times \mathbb{R}^{n_2} \times \cdots \times \mathbb{R}^{n_k}$ to $\mathbb{R}^m$ is defined as a mapping such that for each r, $1 \leqslant r \leqslant k$, we have

$$L(a_1, a_2, \ldots, a_r + \lambda b_r, \ldots, a_k) = L(a_1, \ldots, a_r, \ldots, a_k) + \lambda L(a_1, a_2, \ldots, b_r, \ldots, a_k),$$

where the $a_i \in \mathbb{R}^{n_i}$, $b_r \in \mathbb{R}^{n_r}$, and $\lambda \in \mathbb{R}$. Show that a multilinear map is continuous.

Solution: Let $e_1, \ldots, e_n$ be the standard basis of $\mathbb{R}^n$, and for $x \in \mathbb{R}^n$, let $x = (x^1, \ldots, x^n) = \sum_{i=1}^n x^i e_i$. Define fixed elements of $\mathbb{R}^m$, $\alpha_{i_1, \ldots, i_k}$ for integers i_j with $1 \leqslant i_j \leqslant n_j, j = 1, \ldots, k$, by

$$\alpha_{i_1, \ldots, i_k} = L(e_{i_1}, \ldots, e_{i_k}).$$

Then, it is true that

$$L(x_1, \ldots, x_k) = \sum_{i_1 = 1}^{n_1} \cdots \sum_{i_k = 1}^{n_k} \alpha_{i_1, \ldots, i_k} x_1^{i_1} \cdots x_k^{i_k},$$

which is the analogue of writing a linear transformation in terms of a matrix. Indeed, by multilinearity,

$$L(x_1, \ldots, x_k) = L\left(\sum_{i_1 = 1}^{n_1} x_1^{i_1} e_{i_1}, x_2, \ldots, x_k\right)$$
$$= \sum_{i_1 = 1}^{n_1} x_1^{i_1} L(e_{i_1}, x_2, \ldots, x_k).$$

Repeating this k times gives the desired result.

From this formula it is clear that L is continuous since the functions $x_1^{i_1}, \ldots, x_k^{i_k}$ are products of continuous functions and are thus continuous, and L is a sum of these.

Another solution to this problem, which proceeds similarly to Example 4, is to show that there is a constant $M > 0$ such that $L(x_1, \ldots, x_k) \leqslant M \|x_1\| \cdots \|x_k\|$, from which continuity can be deduced directly.

Exercises for Chapter 4

1. (a) Prove directly that the function $1/x^2$ is continuous on $]0, \infty[$.
   (b) A constant function $f: A \to \mathbb{R}^m$ is such that $f(x) = f(y)$ for all $x, y \in A$. Show that f is continuous.
   (c) Is the function $f(y) = \sin(\cos(y^3) \cdot e^y)$ continuous? Justify your answer.

2. (a) Prove that if $f: A \to \mathbb{R}^m$ is continuous and $B \subset A$, then the restriction $f \mid B$ is continuous ($f \mid B$ is the function f but defined only at points of B).

   (b) Find a function $g: A \to \mathbb{R}$ and a set $B \subset A$ such that $g \mid B$ is continuous, but g is continuous at no point of A. [Hint: Let $A = [0,1]$ and B the rationals.]

3. (a) If $f: \mathbb{R} \to \mathbb{R}$ is continuous and $K \subset \mathbb{R}$ is connected, is $f^{-1}(K)$ necessarily connected?

   (b) Show that if $f: \mathbb{R}^n \to \mathbb{R}^m$ is continuous on all of $\mathbb{R}^n$ and $B \subset \mathbb{R}^n$ is bounded, then $f(B)$ is bounded.

4. Discuss why it is necessary to have in the definition of $\lim_{x \to x_0} f(x)$ that $x \neq x_0$ by considering what would happen in the case where $f: \mathbb{R} \to \mathbb{R}$, $f(x) = 0$ if $x \neq 0$ and $f(0) = 1$.

5. Show that $f: A \to \mathbb{R}^m$ is continuous at x_0 iff for every $\varepsilon > 0$ there is a $\delta > 0$ such that $\|x - x_0\| \leq \delta$ implies $\|f(x) - f(x_0)\| \leq \varepsilon$. Can we replace $\varepsilon > 0$ or $\delta > 0$ by $\varepsilon = 0$ or $\delta = 0$?

6. (a) Let $\{c_k\}$ be a sequence in $\mathbb{R}$. Show that $c_k \to c$ iff every subsequence of c_k has a further subsequence which converges to c.

   (b) Let $f: \mathbb{R} \to \mathbb{R}$ be a bounded function. Prove that f is continuous if and only if the graph of f is a closed subset of $\mathbb{R}^2$. What if f is unbounded?

7. Consider a compact set $B \subset \mathbb{R}^n$ and let $f: B \to \mathbb{R}^m$ be continuous and one-to-one. Then prove that $f^{-1}: f(B) \to B$ is continuous. Show by example that this fails if B is not compact. (To find a counterexample it is necessary to take $m > 1$.)

8. Define maps $\oplus: \mathbb{R}^n \times \mathbb{R}^n \to \mathbb{R}^n$ and $\odot: \mathbb{R} \times \mathbb{R}^n \to \mathbb{R}^n$ as addition and scalar multiplication defined by $\oplus(x,y) = x + y$ and $\odot(\lambda,x) = \lambda x$. Show that these mappings are continuous.

9. Prove the following "glueing lemma": Let $f: [a,b] \to \mathbb{R}^m$ and $g: [b,c] \to \mathbb{R}^m$ be continuous. Define $h: [a,c] \to \mathbb{R}^m$ by $h = f$ on $[a,b]$ and $h = g$ on $[b,c]$. If $f(b) = g(b)$, then h is continuous. Generalize this result to sets $A, B \subset \mathbb{R}^n$.

10. Show that $f: A \to \mathbb{R}^m, A \subset \mathbb{R}^n$ is continuous iff for every set $B \subset A, f(\mathrm{cl}(B) \cap A) \subset \mathrm{cl}(f(B))$.

11. (a) For $f:]a,b[\to \mathbb{R}$, show that if f is continuous then its graph Γ is path-connected. Argue intuitively that if the graph of f is path-connected then f is continuous. (The latter fact is true, but is actually more difficult.)

    (b) For $f: A \to \mathbb{R}^m, A \subset \mathbb{R}^n$, show that for $n \geq 2$ connectedness of the graph does not imply continuity. [Hint: For $f: \mathbb{R}^2 \to \mathbb{R}$, cut a slit in the graph.]

    (c) Discuss (b) for $m = n = 1$. [Hint: On $\mathbb{R}$ consider $f(x) = 0$ if $x = 0$, $f(x) = \sin(1/x)$, $x > 0$.]

12. (a) A map $f: A \subset \mathbb{R}^n \to \mathbb{R}^m$ is called *Lipschitz* if there is a constant $L \geq 0$ such that $\|f(x) - f(y)\| \leq L \|x - y\|$, for all $x, y \in A$. Show that a Lipschitz map is uniformly continuous.

    (b) Find a bounded continuous function $f: \mathbb{R} \to \mathbb{R}$ which is not uniformly continuous.

(c) Is the sum (product) of two Lipschitz functions again Lipschitz?

(d) Answer question (c) for uniformly continuous functions.

13. Let f be a bounded continuous function $f: \mathbb{R}^n \to \mathbb{R}$. Prove: $f(U)$ is open for all open sets $U \subset \mathbb{R}^n$ iff for all non-empty open sets $V \subset \mathbb{R}^n$, $\inf_{x \in V} f(x) < f(y) < \sup_{x \in V} f(x)$ for all $y \in V$.

14. (a) Find a function $f: \mathbb{R}^2 \to \mathbb{R}$ such that

$$\lim_{x \to 0} \lim_{y \to 0} f(x,y) \quad \text{and} \quad \lim_{y \to 0} \lim_{x \to 0} f(x,y)$$

exist but are not equal.

(b) Find a function $f: \mathbb{R}^2 \to \mathbb{R}$ such that the two limits in (a) exist and are equal, but f is not continuous. [Hint: $f(x,y) = xy/(x^2 + y^2)$ with $f = 0$ at $(0,0)$.]

(c) Find a function $f: \mathbb{R}^2 \to \mathbb{R}$ which is continuous on every line through the origin but is not continuous. [Hint: Look at the hint in (b), or the function $r \tan(\theta/4)$, $0 \leqslant r < \infty$, $0 \leqslant \theta < 2\pi$ in polar coordinates.]

15. Let $f_1, \ldots, f_N$ be functions from $A \subset \mathbb{R}^n$ to $\mathbb{R}$. Let m_i be the maximum of f_i, that is, $m_i = \sup(f_i(A))$. Let $f = \sum f_i$ and $m = \sup(f(A))$. Show that $m \leqslant \sum m_i$. Give an example where equality fails.

16. Consider a function $f: A \times B \to \mathbb{R}^m$, where $A \subset \mathbb{R}^n$, $B \subset \mathbb{R}^p$. Call f *separately continuous* if for each fixed $x_0 \in A$, the map $g(y) = f(x_0,y)$ is continuous and for $y_0 \in B$, $h(x) = f(x,y_0)$ is continuous. Say f is continuous on A *uniformly with respect to B* if for each $\varepsilon > 0$ and $x_0 \in A$ there is a $\delta > 0$ such that $\|x - x_0\| < \delta$ implies $\|f(x,y) - f(x_0,y)\| < \varepsilon$ for all $y \in B$. Show that if f is separately continuous and is continuous on A uniformly with respect to B, then f is continuous.

17. Demonstrate that multilinear maps on Euclidean space are not necessarily uniformly continuous. [Hint: Try $f(x,y) = xy$.]

18. Let $A \subset \mathbb{R}^n$ be connected and let $f: A \to \mathbb{R}$ be continuous with $f(x) \neq 0$ for all $x \in A$. Then show that $f(x) > 0$ for all $x \in A$ or else $f(x) < 0$ for all $x \in A$.

19. Find a continuous map $f: \mathbb{R}^n \to \mathbb{R}^m$ and a closed set $A \subset \mathbb{R}^n$ such that $f(A)$ is not closed. In fact, do this when $f: \mathbb{R}^2 \to \mathbb{R}$ is the projection on the x axis ($f(x,y) = x$).

20. Give an alternative proof of Theorem 5 using the subsequence characterization of compactness. [Hint: First argue that $\sup(B) < \infty$ as follows. Say $\sup(B) = \infty$ and choose x_k so that $f(x_k) > k$. Then to show $\sup(B) \in f(B)$, choose y_k so that $f(y_k) \leqslant \sup(B) \leqslant f(y_k) + 1/k$ and pass to a convergent subsequence.]

21. Which of the following functions on $\mathbb{R}$ are uniformly continuous?

(a) $f(x) = \dfrac{1}{(x^2 + 1)}$,

(b) $f(x) = \cos^3 x$,

(c) $f(x) = \dfrac{x^2}{(x^2 + 2)}$,

(d) $f(x) = x \sin x$.

22. Give an alternative proof of Theorem 7 using the subsequence characterization of compactness (Bolzano-Weierstrass theorem) as follows. First, show that if f is not uniformly continuous, there is an $\varepsilon > 0$ and sequences x_n, y_n such that $d(x_n, y_n) < 1/n$ and $d(f(x_n), f(y_n)) \geqslant \varepsilon$. Pass to convergent subsequences and obtain a contradiction to the continuity of f.

23. (a) Define the notion of a compact metric space by examining Theorem 1, Chapter 3. Show that all the properties there except (i) are equivalent. Adopt any of these other than (i) as the definition.
    (b) Let X, Y be metric spaces and $f: X \to Y$. Go through Theorem 1, p. 80, and show that it remains valid.
    (c) Let X be a compact metric space and $f: X \to X$ an isometry; that is, $d(f(x), f(y)) = d(x, y)$ for all $x, y \in X$. Show that f is continuous and must be a bijection. [Hint: If $x \in X \setminus f(X)$ show there is a $c > 0$ such that $d(x, y) \geqslant c$ for all $y \in f(X)$. Use the sequence $x, f(x), f(f(x)), \ldots$ to contradict the compactness of X.]

24. Let $f: A \subset \mathbb{R}^n \to \mathbb{R}^m$.
    (a) Prove f is uniformly continuous on A iff for every pair of sequences x_k, y_k of A such that $(x_k - y_k) \to 0$, we have $f(x_k) - f(y_k) \to 0$.
    (b) Let f be uniformly continuous, and x_k be a Cauchy sequence of A. Show $f(x_k)$ is a Cauchy sequence.
    (c) Let f be uniformly continuous. Show f has a unique extension to a continuous function on $\bar{A} = \text{cl}(A)$.

25. Let $f:]0,1[\to \mathbb{R}$ be differentiable and let $f'(x)$ be bounded. Show that $\lim_{x \to 0^+} f(x)$ and $\lim_{x \to 1^-} f(x)$ exist. Do this both (a) directly and (b) by applying Exercise 24(c). Give a counterexample if $f'(x)$ is not bounded.

26. Let $f: [a,b] \to \mathbb{R}$ be continuously differentiable; that is, $f'(x)$ exists and is continuous. Prove f is uniformly continuous.

27. Find the sum of the series $\sum_{k=4}^{\infty} (3/4)^k$.

28. Let $f:]0,1[\to \mathbb{R}$ be uniformly continuous. Must f be bounded?

29. Let $f: \mathbb{R} \to \mathbb{R}$ satisfy $|f(x) - f(y)| \leqslant |x - y|^2$. Prove f is a constant. [Hint: Show that $f'(x) = 0$.]

30. (a) Let $f: [0,\infty[\to \mathbb{R}$, $f(x) = \sqrt{x}$. Prove f is uniformly continuous.
    (b) Let $k > 0$ and $f(x) = (x - x^k)/\log x$ for $0 < x < 1$ and $f(0) = 0, f(1) = 1 - k$. Show $f: [0,1] \to \mathbb{R}$ is continuous. Is f uniformly continuous?

31. Let $f(x) = x^{1/(x-1)}$ for $x \neq 1$. How should $f(1)$ be defined in order to make f continuous at $x = 1$?

32. Let $A \subset \mathbb{R}^n$ be open, $x_0 \in A$, $r_0 > 0$ and $B_{r_0} = \{x \in \mathbb{R}^n \mid \|x - x_0\| \leqslant r_0\}$. Suppose that $B_{r_0} \subset A$. Prove that there is an $r > r_0$ such that $B_r \subset A$.

33. A set $A \subset \mathbb{R}^n$ is called *relatively compact* when $\text{cl}(A)$ is compact. Prove that A is relatively compact iff every sequence in A has a subsequence which converges to a point in $\mathbb{R}^n$.

34. Given that temperature on the surface of the earth is a continuous function, prove that on any great circle of the earth there are two antipodal points with the same temperature. [Hint: Let $T: [0,2\pi] \to \mathbb{R}$ be a continuous function such that $T(0) = T(2\pi)$. Let $f(x) = T(x) - T(x - \pi)$, and show that $f(x) = 0$ for some $x \in [0,2\pi]$.]

35. Let $f: \mathbb{R} \to \mathbb{R}$ be increasing and bounded above. Prove that $\lim\limits_{x \to +\infty} f(x)$ exists.

Chapter 5

Uniform Convergence

In later parts of this book, many of the functions we discuss will be defined by means of infinite sequences or series. To study such functions we shall need to understand the concept of uniform convergence of a sequence or series of (continuous) functions. In order to effectively deal with concrete situations and examples, we will also consider several important tests for uniform convergence. Perhaps the most helpful test in particular examples is the Weierstrass M-test for series. Another test is the Cauchy Criterion which is mainly of theoretical use. We also include the more refined tests of Dirichlet and Abel.

In connection with uniform convergence we introduce a space whose points are functions. On this space we introduce a norm and show that convergence for this norm is exactly uniform convergence. The space is proved to be complete in the sense that Cauchy sequences converge. A second basic property of this space, called the Arzela-Ascoli theorem, establishes compactness of a subset (in the sense of having the Bolzano-Weierstrass property). An important result, called the Stone-Weierstrass theorem, is then proved. This theorem enables one to approximate continuous functions by polynomials, or by functions from other appropriate classes. Finally, some applications of this machinery to differential and integral equations are given.

5.1   Pointwise and Uniform Convergence

The most natural type of convergence for a sequence of functions is probably pointwise (or simple) convergence, defined as follows.

Definition 1.   A sequence of functions $f_k\colon A \to \mathbb{R}^m$, $A \subset \mathbb{R}^n$ is said to *converge pointwise* (or *simply*) to $f\colon A \to \mathbb{R}^m$ if for each $x \in A$, $f_k(x) \to f(x)$ (convergence as a sequence in $\mathbb{R}^m$). We often write $f_k \to f$ (pointwise) if f_k converges pointwise to f.

While this type of convergence is very useful for certain purposes, there are other situations where it is not. The main disadvantage of pointwise convergence is that even if the functions f_k are continuous, f need not be continuous. For example, consider Figure 5-1 in which

$$
f_k(x) = \begin{cases} 0, & x \geqslant \dfrac{1}{k}, \\[2ex] -kx + 1, & 0 \leqslant x < \dfrac{1}{k}. \end{cases}
$$

In this case, for each $x \in [0,1]$, $f_k(x)$ converges. If $x \neq 0$, $f_k(x) \to 0$ (since $f_k(x) = 0$ for k large), while if $x = 0$, $f_k(x) \to 1$ (as $f_k(0) = 1$ for all k). The limit is thus

$$
f(x) = \begin{cases} 0, & x \neq 0, \\[1ex] 1, & x = 0, \end{cases}
$$

which is not a continuous function.

How can we avoid this type of behavior? No matter how large k is, there are points where f_k is not close to f. To remedy this a notion is introduced guaranteeing that f_k will be uniformly close to f (that is, close for all $x \in A$) as follows.

Definition 2.   Let $f_k\colon A \to \mathbb{R}^m$ be a sequence of functions with the property that for every $\varepsilon > 0$ there is an N such that $k \geqslant N$ implies

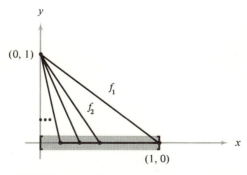

FIGURE 5-1   Pointwise convergence.

$\| f_k(x) - f(x) \| < \varepsilon$ for all $x \in A$. Under these conditions we say f_k *converges uniformly* to f and we write $f_k \to f$ (uniformly).

The condition $\| f_k(x) - f(x) \| < \varepsilon$ means that f_k is within ε of f everywhere. We think of f_k as being within the ε "ribbon" of f. See Figure 5-2.

Perhaps another example will make the idea clearer. On $\mathbb{R}$ consider the sequence

$$f_k(x) = \begin{cases} 0, & x < k, \\ 1, & x \geqslant k, \end{cases}$$

($k = 1, 2, 3, \ldots$). Then $f_k \to 0$ (pointwise) because for each $x \in \mathbb{R}$, $f_k(x) = 0$ for k large ($k > x$). However, f_k does not converge to zero uniformly, for there are points x such that $f_k(x) - 0$ is not small no matter how large k is.

Let us observe that if $f_k \to f$ (uniformly) then $f_k \to f$ (pointwise). This is because for any $x \in A$, and $\varepsilon > 0$ we have an N such that $\| f_k(x) - f(x) \| < \varepsilon$ if $k \geqslant N$, that is $f_k(x) \to f(x)$. We make similar definitions for a series of functions.

Definition 3.   We say the series $\sum_{k=1}^{\infty} g_k$ converges to g *pointwise*, and write $\sum_{k=1}^{\infty} g_k = g$ (pointwise) if the sequence $s_k = \sum_{i=1}^{k} g_i$ converges pointwise to g. Also, we say $\sum_{k=1}^{\infty} g_k = g$ (uniformly) or $\sum g_k$ converges to g *uniformly* if $s_k \to g$ (uniformly). For a sequence f_k (or series $\sum g_k$) we say that f_k (or $\sum g_k$) converges uniformly if there exists a function to which it converges uniformly.

The first basic property of uniform convergence is its connection with continuous functions given in the next theorem.

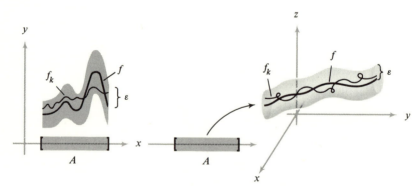

FIGURE 5-2   Uniform closeness. (a) $f: A \subset \mathbb{R} \to \mathbb{R}$. (b) $f: A \subset \mathbb{R} \to \mathbb{R}^3$.

Theorem 1. *Let $f_k: A \to \mathbb{R}^m$ be continuous functions, and suppose that $f_k \to f$ (uniformly). Then f is continuous.*

Thus, uniform convergence is a strong enough condition to guarantee that the limiting function of a sequence of continuous functions is continuous. In view of the preceding examples, this should not be unreasonable.

Corollary 1. *If $g_k: A \to \mathbb{R}^m$ are continuous and $\sum_{k=1}^{\infty} g_k = g$ (uniformly), then g is continuous.*

This follows by applying Theorem 1 to the sequence of partial sums.

EXAMPLE 1.   Let $f_n(x) = (\sin x)/n$, $f_n: \mathbb{R} \to \mathbb{R}$. Show that $f_n \to 0$ uniformly as $n \to \infty$.

Solution: We must show that $|f_n(x) - 0| = |f_n(x)|$ gets small independent of x as $n \to \infty$. But $|f_n(x)| = |\sin x|/n \leqslant 1/n$ which gets small independent of x as $n \to \infty$.

EXAMPLE 2.   Argue that the series for $\sin x$,

$$\sin x = x - \frac{x^3}{3!} + \frac{x^5}{5!} - \cdots$$

converges uniformly, $0 \leqslant x \leqslant r$.

Solution: We must show that

$$s_n(x) = \sum_{k=0}^{n} \frac{(-1)^k x^{2k+1}}{(2k+1)!}$$

converges uniformly to $\sin x$. To do this, estimate the difference:

$$|s_n(x) - \sin x| = \left| \sum_{k=n+1}^{\infty} (-1)^k \frac{x^{2k+1}}{(2k+1)!} \right| \leqslant \sum_{k=n+1}^{\infty} \frac{(r)^{2k+1}}{(2k+1)!}.$$

But this gives a number independent of x which $\to 0$ as $n \to \infty$ since it is the tail of a convergent series. Thus the convergence is uniform. Note that continuity of $\sin x$ follows from this, a result we knew already.

EXAMPLE 3.   Let $f_n(x) = x^n$, $0 \leqslant x \leqslant 1$. Does f_n converge uniformly?

Solution: First we determine the limit point by point. We have $f_n(0) = 0$ for all n and $f_n(x) \to 0$ if $x < 1$, but $f_n(1) = 1$ for all n. Thus f_n converges pointwise to

$$f(x) = \begin{cases} 0, & x \neq 1, \\ 1, & x = 1. \end{cases}$$

It cannot converge uniformly because this limit is not continuous (Figure 5-3).

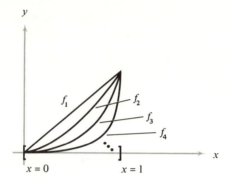

FIGURE 5-3 The sequence $f_n(x) = x^n$.

Exercises for Section 5.1

1. Let $f_n(x) = (x - 1/n)^2, 0 \leqslant x \leqslant 1$. Does f_n converge uniformly?

2. Let $f_n(x) = x - x^n, 0 \leqslant x \leqslant 1$. Does f_n converge uniformly?

3. Let $f_n \colon \mathbb{R} \to \mathbb{R}$ be uniformly continuous and let f_n converge uniformly to f. Do you think that f is uniformly continuous? Discuss.

4. Let $f_n(x) = x^n, 0 \leqslant x \leqslant .999$. Does f_n converge uniformly?

5. Let

$$ f(x) = \sum_{n=0}^{\infty} \frac{x^{n/2}}{n(n!)^2}, \qquad 0 \leqslant x \leqslant 1. $$

Discuss how you might prove f is continuous.

5.2 The Weierstrass M-Test

We shall now consider some tests for uniform convergence. The first is of theoretical use and is entirely analogous to the Cauchy Criterion for a sequence in $\mathbb{R}^n$. It is also called the Cauchy Criterion.

> **Theorem 2.** Let $f_k \colon A \to \mathbb{R}^m$ be a sequence of functions. Then f_k converges uniformly iff for every $\varepsilon > 0$ there is an N such that $l, k \geqslant N$ implies $\| f_k(x) - f_l(x) \| < \varepsilon$ for all $x \in A$.

For the case of series, the Cauchy Criterion takes the following form when applied to the sequence of partial sums (as in Theorem 10, Chapter 2): *The series $\sum_{k=1}^{\infty} g_k$ converges uniformly iff for every $\varepsilon > 0$ there is an N such that $k \geqslant N$ implies $\| g_k(x) + \cdots + g_{k+p}(x) \| < \varepsilon$ for all $x \in A$ and all integers $p = 0, 1, 2, \ldots$.*

Using the above, we can obtain the following important technique for determining the uniform convergence of a series, called the *Weierstrass M-test*.

> **Theorem 3.** *Suppose g_k: $A \to \mathbb{R}^m$ are functions such that there exist constants M_k with $\|g_k(x)\| \leqslant M_k$ for all $x \in A$, and $\sum_{k=1}^{\infty} M_k$ converges. Then $\sum_{k=1}^{\infty} g_k$ converges uniformly (and absolutely).*

It is not always possible to use the *M*-test but it is effective in the majority of cases. For more refined tests, see the Dirichlet and Abel tests in Section 5.8.

Theorem 3 is, in fact, fairly clear intuitively, since the constants M_k give a bound on the "rate of convergence," the point being that the bound is independent of x. (More exactly, the tail of the series $\sum g_k$, which represents the error, is bounded by that of $\sum M_k$, which $\to 0$ independent of x.)

EXAMPLE 1.   Show that

$$\sum_1^{\infty} g_n(x) = \sum_1^{\infty} \frac{(\sin nx)^2}{n^2}$$

converges uniformly.

Solution: Let $M_n = 1/n^2$. Here $|g_n(x)| \leqslant M_n$ since $|\sin nx| \leqslant 1$. Hence by Theorem 3 the convergence is uniform.

EXAMPLE 2.   Prove that

$$f(x) = \sum_0^{\infty} \left(\frac{x^n}{n!}\right)^2$$

is continuous on $\mathbb{R}$.

Solution: Here we cannot choose an M_n for the nth term, because x^n is not bounded. We do not therefore expect uniform convergence on all of $\mathbb{R}$, but we can prove uniform convergence on each interval $[-a,a]$ by letting $M_n = (a^n/n!)^2$ which is an upper bound for the nth term on $[-a,a]$. The ratio test shows $\sum M_n$ converges since

$$\frac{M_{n+1}}{M_n} = \left(\frac{n!}{(n+1)!}\right)^2 \left(\frac{a^{n+1}}{a^n}\right)^2 = \left(\frac{a}{n+1}\right)^2$$

which converges to zero, which is less than one. Hence we have uniform convergence on $[-a,a]$ and so by Theorem 1, we get continuity of f on $[-a,a]$. Since a was arbitrary, we get continuity on all of $\mathbb{R}$.

EXAMPLE 3.   Suppose a sequence $f_n(x)$, $0 \leqslant x \leqslant 1$ converges uniformly and f_n is differentiable. Must $f'_n(x)$ converge uniformly?

Solution: The answer is no. In general, control on the derivatives gives control on the functions via the mean-value theorem, but not vice versa. For example, let $f_n(x) = [\sin(n^2 x)]/n$. Then $f_n \to 0$ uniformly, but $f'_n(x) = n \cos(n^2 x)$ does not converge even pointwise (set $x = 0$, for example).

Exercises for Section 5.2

1. Discuss the convergence and uniform convergence of

   (a) $f_n(x) = \dfrac{x^n}{n + x^n}, x \geqslant 0, n = 1, 2, \dots$

   (b) $f_n(x) = \dfrac{e^{-x^2/n}}{n}, x \in \mathbb{R}, n = 1, 2, \dots$

2. Discuss the uniform convergence of $\displaystyle\sum_1^\infty \frac{x^n}{n^2}, 0 \leqslant x \leqslant 1$.

3. Prove that $f(x) = \displaystyle\sum_1^\infty \frac{x^n}{n^2}$ is continuous on $[0,1]$.

4. Discuss the uniform convergence of $\displaystyle\sum_1^\infty \frac{1}{(x^2 + n^2)}$.

5. If $\displaystyle\sum_1^\infty a_n$ is absolutely convergent, prove $\displaystyle\sum_1^\infty a_n \sin nx$ is uniformly convergent.

5.3   Integration and Differentiation of Series

For a sequence or series converging uniformly, statements can also be made concerning integration and differentiation of the limit function. The question that needs to be answered is whether or not these operations can be performed term by term. For the integration process the answer is yes as can be seen from the next theorem. The general definition of integrability is found in Chapter 8, but the basic properties of integration and differentiation are assumed known from elementary calculus for continuous real-valued functions of a real variable.

    Theorem 4.   *Suppose* $f_k: [a,b] \to \mathbb{R}$ *are continuous functions* $(a,b \in \mathbb{R})$ *and* $f_k \to f$ *uniformly. Then*

$$\int_a^b f_k(x)\, dx \to \int_a^b f(x)\, dx \, .$$

Corollary 2. *Suppose* $g_k: [a,b] \to \mathbb{R}$ *are continuous and* $\sum_{k=1}^{\infty} g_k$ *converges uniformly. Then we may interchange the order of integration and summation*

$$\int_a^b \sum_{k=1}^{\infty} g_k(x)\, dx = \sum_{k=1}^{\infty} \int_a^b g_k(x)\, dx \ .$$

The corollary follows easily from Theorem 4 applied to the sequence of partial sums.

Intuitively, the theorem should be fairly clear, because if f_k is very close to f, then its integral (the area under the curve) should be close to that of f. But be careful here. Indeed, this result is false if f_k only converges pointwise! (See Example 1.)

Note: Actually, there is a theorem with a much wider scope than the above, called Lebesgue's Dominated Convergence theorem. One version of this result states that if f_k converges pointwise to f and the f_k are uniformly bounded (that is, $|f_k(x)| \leqslant M$ for all $k = 1, 2, \ldots, x \in [a,b]$), then the conclusion of Theorem 4 remains valid. We shall be content in this book with the more elementary form of the result in Theorem 4. (See however, Section 8.8.)

Can we take the same liberties with derivatives? The answer to the question of term-by-term differentiation of a uniformly convergent sequence or series is *no* as we saw in Example 3 above. This result is a good illustration of the sort of care that is often needed to turn an intuitively plausible statement into one of actual fact. Thus we need more assumptions than just uniform convergence. Sufficient conditions are given in the following theorem.

Theorem 5. *Let* $f_k: \,]a,b[\,\to \mathbb{R}$ *be a sequence of differentiable functions on the open interval* $]a,b[$ *converging pointwise to* $f: \,]a,b[\,\to \mathbb{R}$*. Suppose the derivatives* f'_k *are continuous and converge uniformly to a function* g*. Then* f *is differentiable and* $f' = g$*.*

Corollary 3. *If the* g_k *are differentiable, the* g'_k *are continuous,* $\sum_{k=1}^{\infty} g_k$ *converges pointwise, and if* $\sum_{k=1}^{\infty} g'_k$ *converges uniformly, then*

$$\left(\sum_{k=1}^{\infty} g_k \right)' = \sum_{k=1}^{\infty} g'_k \ .$$

As usual, the corollary follows by applying the theorem to the sequence of partial sums.

EXAMPLE 1.   Give an example of a sequence $f_k: [0,1] \to \mathbb{R}$ which converges to zero pointwise, but for which $\int_0^1 f_k$ does not converge to zero.

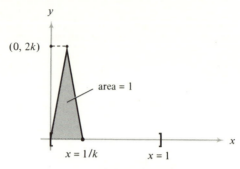

FIGURE 5-4

Solution: Let f_k have the graph in Figure 5-4. Thus, f_k is such that $\int_0^1 f_k = 1$ for all $k = 1, 2, 3, \ldots$. Furthermore, for each x, $f_k(x) \to 0$ as $k \to \infty$ (clearly if $x = 0$ and if $x > 0$, then $f_k(x) = 0$ as soon as $k > 1/x$).

EXAMPLE 2. Let $g_n(x) = nx/(1 + nx)$, $0 \leqslant x \leqslant 1$. Examine Theorem 5 in this case.

Solution: For $x \neq 0$ we see that as $n \to \infty$, $g_n(x) \to 1$, since $g_n(x) = x/(x + 1/n)$. But, for $x = 0$, $g_n(x) = 0$. Thus g_n converges pointwise but not uniformly. The convergence is uniform only on each interval $[\delta, 1]$ where $\delta > 0$.

The derivative is $g_n'(x) = (1/n)/(x + 1/n)^2$. This $\to 0$ uniformly on $[\delta, 1]$, but $g_n'(0) \to \infty$. Thus the conditions of Theorem 5 hold only on $[\delta, 1]$ for $\delta > 0$. The limit function is not differentiable at $x = 0$.

EXAMPLE 3. Verify that $\int_0^x e^t \, dt = e^x - 1$, using $e^x = \sum_0^\infty x^n/n!$ and Theorem 4.

Solution: By the Weierstrass M-test, $e^x = \sum_0^\infty x^n/n!$ converges uniformly on any finite interval. Thus by Corollary 2, applied to the interval $[0, x]$,

$$
\int_0^x e^t \, dt = \sum_0^\infty \int_0^x \frac{t^n}{n!} \, dt
$$

$$
= \sum_{n=0}^\infty \frac{t^{n+1}}{(n+1)!} \Big|_0^x
$$

$$
= \frac{x}{1!} + \frac{x^2}{2!} + \cdots
$$

$$
= e^x - 1 .
$$

Exercises for Section 5.3

1. Investigate the validity of Theorem 4 for the sequence f_n defined by

$$f_n(x) = \frac{nx}{(1 + nx^2)}, \qquad 0 \leqslant x \leqslant 1.$$

2. Show that the sequence $\{f_n\}$ defined by

$$f_n(x) = n^3 x^n (1 - x)$$

converges pointwise to $f = 0$ on $[0,1]$, and then use Theorem 4 to show that the convergence is not uniform.

3. Investigate the validity of Theorems 4 and 5 for $f_n(x) = \sqrt{n}\, x^n(1 - x)$. [Hint: Locate the maximum of $f_n(x)$.]

4. Verify that $\int_0^x \sin t \, dt = 1 - \cos x$, using

$$\sin x = \sum_0^\infty (-1)^n \frac{x^{2n+1}}{(2n + 1)!}.$$

5. Verify that $\sin' x = \cos x$, using the series in Exercise 4 and Corollary 3.

5.4   The Space of Continuous Functions

Fix a set $A \subset \mathbb{R}^n$ and consider the set V of all functions $f: A \to \mathbb{R}^m$. Then V is easily seen to be a vector space. In V, the zero vector is the function which is 0 for all $x \in A$. Also, we define $(f + g)(x) = f(x) + g(x)$ and $(\lambda f)(x) = \lambda(f(x))$ for each $\lambda \in \mathbb{R}$, $f, g \in V$. Now let $\mathscr{C} = \{f \in V \mid f$ is continuous$\}$. If there is danger of confusion we write $\mathscr{C}(A, \mathbb{R}^n)$. Then $\mathscr{C}$ is also a vector space since the sum of two continuous functions is continuous and, for each $\alpha \in \mathbb{R}$ and $f \in \mathscr{C}$, we have $\alpha f \in \mathscr{C}$.

Let $\mathscr{C}_b$ be the vector subspace of $\mathscr{C}$ consisting of bounded functions: $\mathscr{C}_b = \{f \in \mathscr{C} \mid f$ is bounded$\}$. Recall that "f is bounded" means that there is a constant M such that $\| f(x) \| \leqslant M$ for all $x \in A$. If A is compact, then $\mathscr{C}_b = \mathscr{C}$ by Theorem 5, Chapter 4.

For $f \in \mathscr{C}_b$, let $\| f \| = \sup\{\| f(x) \| \mid x \in A\}$, which exists since f is bounded. The number $\| f \|$ is a measure of the size of f and is called the *norm* of f. See Figure 5-5. Note that $\| f \| \leqslant M$ iff $\| f(x) \| \leqslant M$ for all $x \in A$.

What we are trying to do here is to look at the space $\mathscr{C}_b$ in the same way as we look at $\mathbb{R}^n$. Namely, each point (that is, vector) in $\mathscr{C}_b$ (which is a function) has a norm, so we can hope that many of the concepts developed for vectors in $\mathbb{R}^n$ will carry over to $\mathscr{C}_b$. Such a point of view is useful in doing analysis, and some important results (see Section 5.5) can be proved by using the methods of $\mathbb{R}^n$ on the space $\mathscr{C}_b$. For this program to be successful, the

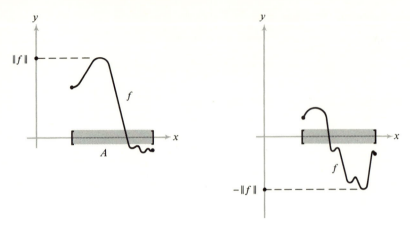

FIGURE 5-5 Norm of a function.

first task is to establish that the basic properties of a norm studied in Chapter 1, Theorem 5 are valid.

Warning: Although we have a norm, we do not have an inner product associated with it such that $\|f\|^2 = \langle f,f \rangle$. Other spaces of functions which we study in Fourier analysis (Chapter 10) do have such an inner product.

Theorem 6. *The function $\|\cdot\|$ on $\mathscr{C}_b(A,\mathbb{R}^m)$ satisfies the properties of a norm:*

(i) $\|f\| \geqslant 0$, *and* $\|f\| = 0$ *iff* $f = 0$,

(ii) $\|\alpha f\| = |\alpha|\,\|f\|$, *for* $\alpha \in \mathbb{R}$, $f \in \mathscr{C}_b$,

(iii) $\|f + g\| \leqslant \|f\| + \|g\|$ *(triangle inequality)*.

These are the basic rules we need to talk about open sets, convergence, and so forth. For example, write $f_k \to f$ in $\mathscr{C}_b$ iff $\|f_k - f\| \to 0$. Recall that a vector space with a norm obeying these rules (i), (ii), (iii) is called a *normed space*. Essentially all of the results of Chapter 2 still hold in the context of normed spaces using the same proofs, and we shall use some of them in the following discussions and proofs. The connection with uniform convergence is simple.

Theorem 7. $(f_k \to f \text{ (uniformly on } A)) \Leftrightarrow (f_k \to f \text{ in } \mathscr{C}_b; \text{ that is,}$ $\|f_k - f\| \to 0).$

Also remember that a sequence f_k is called a *Cauchy sequence* if for any $\varepsilon > 0$ there is an N such that $k, l \geqslant N$ implies $\|f_k - f_l\| < \varepsilon$. A normed space is called *complete* if every Cauchy sequence converges. Another name for a complete normed space is *Banach space*. Completeness is an important technical property for a space, since often we may be able to prove a sequence is Cauchy and we want to deduce its convergence to some element of the space.

Theorem 8. $\mathscr{C}_b$ *is a Banach space.*

The space $\mathscr{C}_b$ is only one of a host of spaces of functions of great importance in analysis. While the rules in Theorem 6 (compare with Theorem 5, Chapter 1) allow us to introduce notions of open sets, convergence, etc. as in $\mathbb{R}^n$, the space $\mathscr{C}_b$ is quite different from $\mathbb{R}^n$ in other respects. For instance, as we have mentioned, $\mathscr{C}_b$ does not have an inner product which gives the norm $\| \cdot \|$, (Exercise 48, at the end of this chapter). Another is that $\mathscr{C}_b$ is not finite-dimensional. In Sections 5.5, 5.6, and 5.7 we shall see some specific problems to which this theory can be applied.

EXAMPLE 1.   Let $B = \{ f \in \mathscr{C}([0,1],\mathbb{R}) \,|\, f(x) > 0 \text{ for all } x \in [0,1] \}$. Show that B is an open set in $\mathscr{C}([0,1],\mathbb{R})$.

Solution: By definition, for $f \in B$ we must find an $\varepsilon > 0$ such that $D(f,\varepsilon) = \{ g \in \mathscr{C} \,|\, \|f - g\| < \varepsilon \} \subset B$. So fix $f \in B$. Now, since $[0,1]$ is compact, f has a minimum value at some point of $[0,1]$, say, m. Thus $f(x) \geqslant m > 0$ for all $x \in [0,1]$. Let $\varepsilon = m/2$. Then if $\|f - g\| < \varepsilon$, the result is that for any x, $|f(x) - g(x)| < \varepsilon = m/2$. Hence $g(x) \geqslant m/2 > 0$, so $g \in B$.

EXAMPLE 2.   What is the closure of the set B in Example 1?

Solution: We assert that the closure is $D = \{ f \in \mathscr{C} \,|\, f(x) \geqslant 0 \text{ for all } x \in [0,1] \}$. This is a closed set because if $f_n(x) \geqslant 0$ and $f_n \to f$ uniformly, and hence pointwise, then $f(x) \geqslant 0$ for all x. To show D is the closure, it suffices to show that for $f \in D$ there is $f_n \in B$ such that $f_n \to f$ (why?). Simply let $f_n = f + 1/n$.

EXAMPLE 3.   Suppose we have a sequence $f_n \in \mathscr{C}_b$ such that $\|f_{n+1} - f_n\| \leqslant r_n$, where $\sum r_n$ is convergent, $r_n \geqslant 0$. Prove that f_n converges.

Solution: We have, by the triangle inequality,

$$\|f_n - f_{n+k}\| \leqslant \|f_n - f_{n+1}\| + \|f_{n+1} - f_{n+2}\| + \cdots + \|f_{n+k-1} - f_{n+k}\|$$

$$\leqslant r_n + r_{n+1} + \cdots + r_{n+k} \,.$$

Since $\sum r_l$ is convergent, this expression $\to 0$ as $n \to \infty$ since it is less than or equal to $s - s_{n-1}$ where s_n is the nth partial sum and s is the sum. Hence f_n is a Cauchy sequence, and so converges.

Exercises for Section 5.4

1. Let $B = \{ f \in \mathscr{C}_b(\mathbb{R},\mathbb{R}) \mid f(x) > 0 \text{ for all } x \in \mathbb{R} \}$. Is B open? If not, what is $\text{int}(B)$?

2. What is the closure of B in Exercise 1?

3. Do you see a connection between Example 3 above and the Weierstrass M-test? Discuss.

4. Let

$$f_n(x) = \frac{1}{n}\left(\frac{nx}{1 + nx} \right), \qquad 0 \leqslant x \leqslant 1 .$$

Show $f_n \to 0$ in $\mathscr{C}([0,1],\mathbb{R})$.

5. Let f_k be a convergent sequence in $\mathscr{C}_b(A,\mathbb{R}^m)$. Prove $\{ f_k \mid k = 1,2,\ldots \}$ is bounded in $\mathscr{C}_b(A,\mathbb{R}^m)$. Is it closed?

5.5 The Arzela-Ascoli Theorem

This theorem is closely related to the notion of compactness in the space $\mathscr{C}_b$ introduced in Section 5.4. As we saw in Chapter 3, in $\mathbb{R}^n$ there are several equivalent ways of formulating the notion of compactness. But in more general spaces, such as $\mathscr{C}_b$, these different ways are not equivalent. Specifically, in Theorem 1 of that chapter, (i) will not be equivalent to the others, but (ii), (ii)', (iii), and (iii)' are all mutually equivalent. An examination of the proof shows this (see also Exercise 21, at the end of this chapter).

In more general spaces, we adopt one of (ii), (ii)', (iii), or (iii)' for the definition of a compact set. The reason for this choice and not (i) is because as we already know in $\mathbb{R}^n$, conditions (ii) through (iii)' are the most useful in proving the key theorems; (see Chapter 4).

The Arzela-Ascoli theorem gives conditions under which a set in $\mathscr{C}$ is compact. Specifically, this is proved in terms of the Bolzano-Weierstrass property. To state the theorem we need a little terminology.

Definition 4. Let $B \subset \mathscr{C}(A,\mathbb{R}^m)$. We say that B is an *equicontinuous* set of functions if for any $\varepsilon > 0$ there is a $\delta > 0$ such that if $x, y \in A$, $d(x,y) < \delta$ implies $d(f(x),f(y)) < \varepsilon$ for all $f \in B$.

This definition is the same as that of uniform continuity except that now we also demand that δ can be chosen independent of f as well as x_0.

The Arzela-Ascoli theorem is as follows.

Theorem 9. *Let $A \subset \mathbb{R}^n$ be compact and let $B \subset \mathscr{C}(A,\mathbb{R}^m)$. If B is bounded and equicontinuous, then any sequence in B has a uniformly convergent subsequence.*

Thus a set in $\mathscr{C}(A,\mathbb{R}^m)$ will be compact if it is closed, bounded, *and* equi-continuous. This result is not really intuitively clear but it is very fundamental for an analysis of the space $\mathscr{C}$ of continuous functions.

EXAMPLE 1.   Let $f_n\colon [0,1] \to \mathbb{R}$ be continuous and be such that $|f_n(x)| \leqslant 100$ and the derivatives f'_n exist and are uniformly bounded on $]0,1[$. Prove f_n has a uniformly convergent subsequence.

Solution: We verify that the set $\{f_n\}$ is equicontinuous and bounded. The hypothesis is that $|f'_n(x)| \leqslant M$ for a constant M. Thus by the mean-value theorem,

$$|f_n(x) - f_n(y)| \leqslant M\,|x - y|\,,$$

so given ε we can choose $\delta = \varepsilon/M$, independent of x, y, and n. Thus $\{f_n\}$ is equicontinuous. It is bounded because $\|f_n\| = \sup_{0 \leqslant x \leqslant 1} |f_n(x)| \leqslant 100$.

EXAMPLE 2.   Is the result of Example 1 valid if we omit "$|f_n(x)|$ bounded?"

Solution: No, for let $f_n(x) = n$. Then $f'_n = 0$ but clearly there is no convergent subsequence.

EXAMPLE 3.   Let $I\colon \mathscr{C}([0,1],\mathbb{R}) \to \mathbb{R}$ be defined by $I(f) = \int_0^1 f(x)\,dx$. Prove I is continuous.

Solution: We must show that $f_n \to f$ in $\mathscr{C}$ implies $I(f_n) \to I(f)$. But this is an immediate consequence of Theorem 4.

Exercises for Section 5.5

1. Show that in Example 1, f_n bounded can be replaced by $f_n(0) = 0$ with the same conclusion.

2. In Theorem 9, need the whole sequence be convergent?

3. (a) Show that
$$\left\{ f \in \mathscr{C}([0,1],\mathbb{R}) \,\middle|\, \int_0^1 f(x)\,dx \in]0,3[\right\}$$
   is open.
   (b) Show that, within the space of all bounded functions on a set A, the space $\mathscr{C}_b$ is closed.

4. Let $B \subset \mathscr{C}([0,1],\mathbb{R})$ be closed, bounded, and equicontinuous. Let $I\colon B \to \mathbb{R}$, $I(f) = \int_0^1 f(x)\,dx$. Show that there is an $f_0 \in B$ at which the value of I is maximized.

5. Let $f_n\colon [a,b] \to \mathbb{R}$ be uniformly bounded continuous functions. Set
$$F_n(x) = \int_a^x f_n(t)\,dt, \qquad a \leqslant x \leqslant b\,.$$

Prove F_n has a uniformly convergent subsequence.

5.6 Fixed Points and Integral Equations

In this section we want to give a brief indication of how analysis on the space $\mathscr{C}(A,\mathbb{R}^m)$ can be used in several applications.

In many physical problems one considers *integral equations*; these have the form

$$f(x) = a + \int_0^x k(x,y)f(y)\, dy \,, \tag{1}$$

where $a = f(0)$ and k are given. We suppose k is continuous.

For example, $f(x) = ae^x$ solves the differential equation $df/dx = f(x)$ which is the same as

$$f(x) = a + \int_0^x f(y)\, dy \,.$$

One can use the Arzela-Ascoli theorem to analyze Eq. 1 (see also Exercise 45, at the end of this chapter). However we shall confine ourselves to studying Eq. 1 under some special hypotheses, for which the following theorem is applicable.

> **Theorem 10** (*Contraction Mapping Principle*). *Let* $T: \mathscr{C}_b(A,\mathbb{R}^m) \to \mathscr{C}_b(A,\mathbb{R}^m)$ *be a given mapping such that there is a constant* $\lambda, 0 \leqslant \lambda < 1$ *with*
>
> $$\| T(f) - T(g)\| \leqslant \lambda \, \|f - g\|$$
>
> *for all* $f, g \in \mathscr{C}_b(A,\mathbb{R}^m)$. *Then* T (*is continuous and*) *has a unique fixed point; that is, there exists a unique point* $f_0 \in \mathscr{C}_b(A,\mathbb{R}^m)$ *such that* $T(f_0) = f_0$.

Note: The proof is actually valid for any complete metric space. The condition on T then reads $d(T(x),T(y)) \leqslant \lambda d(x,y)$. Such a map T is called a *contraction*; it shrinks distances by a factor $\lambda < 1$.

The method of proof is called the method of *successive approximations*. We start with any $f \in \mathscr{C}_b$ and form the sequence

$$f, \quad T(f), \quad T^2(f) = T(T(f)), \quad T^3(f) = T(T(T(f))), \ldots$$

We then show that this sequence is Cauchy, so converges in $\mathscr{C}_b$ and the limit function gives the solution. Observe that the method is constructive. One can successively compute the members of the approximating sequence. Observe that if we started with the solution, or by luck hit it during the iteration, the sequence "stops."

Application of Theorem 10.   *If* $\sup\limits_{x\in[0,r]} \int_0^x |k(x,y)|\,dy = \lambda < 1$ *then Eq. 1 has a unique solution on* $[0,r]$.

Indeed, define $T(f)$ by

$$T(f)(x) = a + \int_0^x k(x,y)f(y)\,dy.$$

Thus a solution of Eq. 1 is a fixed point of T and vice versa. In order to apply Theorem 10 we must verify that T is a contraction: $\|T(f) - T(g)\| \le \lambda \|f - g\|$. Here $A = [0,r]$ and $m = 1$. Now

$$\|T(f) - T(g)\| = \sup_{x\in[0,r]} |T(f)(x) - T(g)(x)|$$

$$= \sup_{x\in[0,r]} \left| \int_0^x k(x,y)[f(y) - g(y)]\,dy \right|$$

$$\le \left(\sup_{x\in[0,r]} \int_0^x |k(x,y)|\,dy \right) \|f - g\|$$

$$= \lambda |f - g|$$

since $|f(y) - g(y)| \le \|f - g\|$, a constant. Hence T is a contraction and so has a unique fixed point, which represents the desired solution.

Later in the book we shall see additional applications of this sort of method. It should be clear that these techniques are very important in the theory of differential and integral equations.

EXAMPLE 1.   Give an example of a complete metric space X and a map $T: X \to X$ with $d(T(x),T(y)) \le d(x,y)$ but with T not having a unique fixed point.

Solution: Let $X = \mathbb{R}$ with the usual distance $d(x,y) = |x - y|$. Let $T(x) = x + 1$. Clearly, there is no x so $x = x + 1$. But $|T(x) - T(y)| = |x - y|$.

This example shows that in Theorem 10, it is essential to have $\lambda < 1$; $\lambda = 1$ will not do.

EXAMPLE 2.   Show that the method of successive approximations applied to $f(x) = 1 + \int_0^x f(y)\,dy$ leads to the usual formula for e^x.

Solution: Begin with the zero function 0. Since $T(g) = 1 + \int_0^x g(y)\,dy$, we get:

$$T(0) = 1 ;$$

$$T^2(0) = T(T(0)) = 1 + \int_0^x dy = 1 + x ;$$

$$T(T^2(0)) = 1 + \int_0^x (1 + y)\, dy = 1 + x + \frac{x^2}{2} ;$$

$$T(T^3(0)) = 1 + \int_0^x \left(1 + y + \frac{y^2}{2}\right) dy = 1 + x + \frac{x^2}{2} + \frac{x^3}{3!} ;$$

$$\cdot$$
$$\cdot$$
$$\cdot$$

$$T^n(0) = 1 + x + \cdots + \frac{x^{n-1}}{(n-1)!} .$$

So this sequence converges to e^x.

EXAMPLE 3. Let $k(x,y) = xe^{-xy}$. On what interval $[0,r]$ does the method of the text guarantee a solution for Eq. 1?

Solution: Evaluate λ and check that $\lambda < 1$. Now

$$\lambda = \sup_{x\in[0,r]} \int_0^x xe^{-xy}\, dy$$

$$= \sup_{x\in[0,r]} (1 - e^{-x^2}) = 1 - e^{-r^2} .$$

Thus we get a unique solution on any interval $[0,r]$.

Exercises for Section 5.6

1. For what α is $T(x) = \alpha x$ a contraction on $\mathbb{R}$?

2. Find a series expression on $[0,\frac{1}{2}]$ for the solution of Exercise 1 if $k(x,y) = x$ and $a = 1$.

3. For what interval $[0,r]$, $r \leqslant 1$ is $f: [0,r] \rightarrow [0,r]$, $x \mapsto x^2$ a contraction?

4. Let $T: \mathscr{C}([0,r],\mathbb{R}) \rightarrow \mathscr{C}([0,r],\mathbb{R})$ be defined by $T(f)(x) = \alpha f(x) + \int_0^x k(x,y)f(y)\, dy$. For what α, k is T a contraction?

5. Convert $dy/dx = 3xy$, $y(0) = 1$ to an integral equation and set up an iteration scheme to solve it.

5.7   The Stone-Weierstrass Theorem

In the study of continuous functions and uniform convergence, two of the most basic results are the Arzela-Ascoli theorem, discussed above, and the Stone-Weierstrass theorem which will be discussed here. This list of theorems is expanded when courses are taken in topology.

The aim of the Stone-Weierstrass theorem is to show that any continuous function can be uniformly approximated by a function which has more easily managed properties, such as a polynomial. Such polynomial approximation techniques are important theoretically and in numerical work.*

We begin by giving the result for the special case of the real line. This was first proved by Weierstrass, but here we present a version due to Bernstein.

Theorem 11.   *Let* $f: [0,1] \to \mathbb{R}$ *be continuous and let* $\varepsilon > 0$*. Then there is a polynomial* $p(x)$ *such that* $\|p - f\| < \varepsilon$*. In fact, the sequence of Bernstein polynomials*

$$p_n(x) = \sum_{k=0}^{n} \binom{n}{k} f\left(\frac{k}{n}\right) x^k (1 - x)^{n-k}$$

converge uniformly to f *as* $n \to \infty$*, where*

$$\binom{n}{k} = \frac{n!}{k!\,(n-k)!}$$

denotes the binomial coefficient.

The first statement here is a consequence of the second. The second can be easily understood if one knows a little probability theory, which is assumed only for the following paragraph of discussion. Needless to say, Bernstein's knowledge of probability theory undoubtedly helped him with the understanding and the proof of this theorem.

Now, an illustration follows. Imagine a "coin" with probability x of getting heads and, consequently, with probability $1 - x$ of getting tails. Then one computes that in n tosses, the probability of getting exactly k heads is

$$\binom{n}{k} x^k (1 - x)^{n-k}.$$

Suppose in a gambling game called "n-tosses," $f(k/n)$ dollars is paid out when exactly k heads turn up when n tosses are made. Then the average amount (after a long evening of playing "n-tosses") paid out when n tosses

* See for instance McAloon, Tromba, *Calculus*, Chapter 12, Harcourt Brace Jovanovich Inc., 1972.

are made is

$$\sum_{k=0}^{n} \binom{n}{k} f(k/n)x^k(1-x)^{n-k} = p_n(x) .$$

Here $f(k/n)$ is f at the fraction of tosses which are heads. Now imagine n very large, that is a great number of tosses. Then we expect that in a typical game of n-tosses, k/n will be very close to $x =$ probability of k heads ($=$ fraction of the time k heads occurs) so our average payout should be very close to $f(x)$. Hence when n is large, we expect $p_n(x)$ to be close to $f(x)$. This is the intuitive reason for the validity of the result. The actual proof is a little complicated, as might be expected from the complexity of the game.

Even for simple f such as $f(x) = \sqrt{x}$, finding an approximating polynomial is not trivial.

We can rephrase the theorem as follows. Let $\mathscr{P}$ denote the set of all polynomials $p: [0,1] \to \mathbb{R}$. Then the first statement of the theorem asserts that $\mathscr{P}$ is *dense* in $\mathscr{C}([0,1],\mathbb{R})$; that is, $\mathrm{cl}(\mathscr{P}) = \mathscr{C}([0,1],\mathbb{R})$.

Stone discovered a very useful generalization of the theorem above by allowing for more general sets than $[0,1]$ and by replacing $\mathscr{P}$ with a general family of functions satisfying certain properties. The proof makes use of the above special case. The theorem is very useful in various branches of analysis (for example, we shall use it in Chapter 10 in our study of Fourier analysis).

> **Theorem 12.**   Let $A \subset \mathbb{R}^n$ be compact and let $\mathscr{B} \subset \mathscr{C}(A,\mathbb{R})$ satisfy
> (i) $\mathscr{B}$ is an algebra; that is, $f, g \in \mathscr{B}, \alpha \in \mathbb{R} \Rightarrow f + g \in \mathscr{B}, f \cdot g \in \mathscr{B}$, and $\alpha f \in \mathscr{B}$;
> (ii) the constant function $x \mapsto 1$ lies in $\mathscr{B}$;
> (iii) $\mathscr{B}$ separates points; that is, for $x, y \in A$, $x \neq y$ there is an $f \in \mathscr{B}$ such that $f(x) \neq f(y)$.
> Then $\mathscr{B}$ is dense in $\mathscr{C}(A,\mathbb{R})$; that is, $\mathrm{cl}(\mathscr{B}) = \mathscr{C}(A,\mathbb{R})$.

EXAMPLE 1.   Let p_n be a uniformly convergent sequence of polynomials and $f = \lim_{n \to \infty} p_n$. Must f be differentiable?

Solution: No, for any continuous function is, by Theorem 11, such a limit and there are plenty of continuous functions which are not differentiable, such as

$$f(x) = \begin{cases} 0, & 0 \leqslant x \leqslant \frac{1}{2}, \\ 2x - 1, & \frac{1}{2} \leqslant x \leqslant 1. \end{cases}$$

EXAMPLE 2.   Prove directly from Theorem 11 or from Theorem 12 that the polynomials on $[a,b]$ are dense in $\mathscr{C}([a,b],\mathbb{R})$.

Solution: (a) We know that if $f: [0,1] \to \mathbb{R}$ is continuous and $\varepsilon > 0$ then there is a polynomial p with $\|f - p\| < \varepsilon$. Now let $g: [a,b] \to \mathbb{R}$ be continuous and let us rescale and set

$$f(x) = g(x(b - a) + a), \qquad 0 \leqslant x \leqslant 1 ,$$

so $f: [0,1] \to \mathbb{R}$. Find p as above and, rescaling backwards let

$$q(x) = p\left(\frac{(x - a)}{(b - a)}\right), \qquad a \leqslant x \leqslant b ,$$

so $q: [a,b] \to \mathbb{R}$. Thus q is a polynomial as well. We claim $\|g - q\| < \varepsilon$. Indeed,

$$|g(x) - q(x)| = \left| f\left(\frac{(x - a)}{(b - a)}\right) - p\left(\frac{(x - a)}{(b - a)}\right)\right| ,$$

so $\|g - q\| < \varepsilon$, since $\|f - p\| < \varepsilon$. Thus the polynomials on $[a,b]$ are dense.

(b) In Theorem 12, we let $A = [a,b]$, and set $\mathcal{B} = \{q \in \mathcal{C}([a,b],\mathbb{R}) \mid q$ is a polynomial$\}$. Then $\mathcal{B}$ clearly satisfies (i) and (ii). It also satisfies (iii) for if $x \neq y$ we can let

$$f(t) = t$$

so $f(x) \neq f(y)$. Thus $\mathcal{B}$ is dense by the theorem.

Exercises for Section 5.7

1. Show that there is a polynomial $p(x)$ such that $|p(x) - \sin x| < 1/100$ for $0 \leqslant x \leqslant 2\pi$.

2. Suppose p_n is a sequence of polynomials converging uniformly to f on $[0,1]$, and f is not a polynomial. Prove that the degrees of the p_n are not bounded. [Hint: An Nth degree polynomial p is uniquely determined by its values at $N + 1$ points $x_0, \ldots, x_N$ via *Lagrange's interpolation formula*

$$p(x) = \sum_{i=1}^{N} \frac{\pi(x)p(x_i)}{\pi'(x_i)(x - x_i)} ,$$

where $\pi(x) = (x - x_0)(x - x_1) \cdots (x - x_N).$]

3. Prove that the polynomials in $\mathcal{C}([a,b],\mathbb{R})$ are not open. Can a subset of a metric space ever be both open and dense?

4. Consider the set of all polynomials $p(x,y)$ in two variables $x, y \in [0,1] \times [0,1]$. Prove this set is dense in $\mathcal{C}([0,1] \times [0,1],\mathbb{R})$.

5. Consider the set of all functions on $[0,1]$ of the form

$$h(x) = \sum_{j=1}^{n} a_j e^{b_j x}, \qquad a_j, b_j \in \mathbb{R} .$$

Is this set dense in $\mathcal{C}([0,1],\mathbb{R})$?

5.8   The Dirichlet and Abel Tests

In some cases in which we would like to determine if we have uniform convergence, the Weierstrass M-test fails. For such instances mathematicians have devised other tests. The first test below was created by the Norwegian mathematician Niels Abel, and the second is credited to P. G. Dirichlet, a German (of French origins) who worked in the first part of the 18th century. These tests are useful in many examples, and are especially useful during the study of Fourier and power series. They are important when we have uniform convergence but not absolute convergence.

> **Theorem 13** (*Abel's Test*).   *Let $A \subset \mathbb{R}^m$ and $\varphi_n \colon A \to \mathbb{R}$ be a sequence of functions which are decreasing; that is, $\varphi_{n+1}(x) \leqslant \varphi_n(x)$ for each $x \in A$. Suppose there is a constant M such that $|\varphi_n(x)| \leqslant M$ for all $x \in A$ and all n. If $\sum_{n=1}^{\infty} f_n(x)$ converges uniformly on A, then so does $\sum_{n=1}^{\infty} \varphi_n(x) f_n(x)$.*

We get useful tests for ordinary series when we take the special case in which φ_n and f_n are constant functions. One has a similar test if the φ_n are increasing, which can be deduced by applying the above to $-\varphi_n$. A related test is the Dirichlet test.

> **Theorem 14** (*Dirichlet Test*).   *Let $s_n(x) = \sum_{m=1}^{n} f_m(x)$ for a sequence $f_n \colon A \subset \mathbb{R}^m \to \mathbb{R}$. Assume there is a constant M such that $|s_n(x)| \leqslant M$ for all $x \in A$ and all n. Let $g_n \colon A \subset \mathbb{R}^m \to \mathbb{R}$ be such that $g_n \to 0$ uniformly, $g_n \geqslant 0$, and $g_{n+1}(x) \leqslant g_n(x)$. Then $\sum_{n=1}^{\infty} f_n(x) g_n(x)$ converges uniformly on A.*

For example, consider the alternating series $\sum (-1)^n g_n(x)$, where $g_n \geqslant 0$, $g_n(x) \to 0$ uniformly, and $g_{n+1} \leqslant g_n$. Let $f_n(x) = (-1)^n$. Then $|s_n(x)| \leqslant 1$ so that $\sum (-1)^n g_n(x)$ converges uniformly. Note that, as a special case, an alternating series whose terms decrease to zero is convergent.

Notice that these theorems are similar but are not the same. The conditions on φ_n in Theorem 13 do not imply that φ_n converges uniformly. Also, in Theorem 13, we do not require $\varphi_n \geqslant 0$. The proofs of these theorems are effected by a device known as Abel's partial summation formula, described in the proofs.

EXAMPLE 1.   Show that $\sum_1^{\infty} (\sin nx)/n$ converges uniformly on $[\delta, \pi - \delta]$, $\delta > 0$.

Solution: We want to apply Theorem 14 with $f_n(x) = \sin nx$ and $g_n(x) = 1/n$. The only hypothesis which is not obvious is $|\sum_{l=1}^{n} f_l(x)| \leqslant M$. To show this requires the use of a technique as follows. Write

$$2 \sin(lx)\sin(\tfrac{1}{2}x) = \cos[(l - \tfrac{1}{2})x] - \cos[(l + \tfrac{1}{2})x]$$

and add from $l = 1, \ldots, n$. We get a collapsing sum so

$$2 \sin(\tfrac{1}{2}x)(\sin x + \cdots + \sin nx) = \cos \tfrac{1}{2}x$$
$$- \cos(n + \tfrac{1}{2})x$$
$$\leqslant 2 .$$

Thus

$$\sin x + \cdots + \sin nx \leqslant \frac{1}{|\sin \tfrac{1}{2}x|} ,$$

which gives a bound on $\sum_{l=1}^{n} f_l(x)$. The bound is good as long as $\sin x/2$ is not zero. For example, on $[\delta, \pi - \delta]$ we get a good bound. Note that the arguments needed here are somewhat more delicate than the M-test.

EXAMPLE 2.   Show that $\sum_{1}^{\infty} (-1)^n e^{-nx}/n$ converges uniformly on $[0,\infty[$.

Solution: This time apply Theorem 13. Let $\varphi_n(x) = e^{-nx}$. For $x \geqslant 0$, φ_n is decreasing and $|e^{-nx}| \leqslant 1$ (why?). We know already that $\sum_{1}^{\infty} (-1)^n/n$ converges, so by Abel's theorem, the series converges uniformly.

EXAMPLE 3.   Let

$$f(x) = \sum_{1}^{\infty} \frac{(-1)^n}{n} e^{-nx} .$$

Show f is continuous.

Solution: The solution is immediate from Example 2 and Corollary 1.

Exercises for Section 5.8

Test the following series for convergence and uniform convergence.

1. $\sum_{1}^{\infty} \frac{x^n}{n!} e^{-nx}, \quad 0 \leqslant x \leqslant 1.$

2. $\sum_{1}^{\infty} \frac{(-1)^n x^n}{n}, \quad 0 \leqslant x \leqslant 1.$

3. $\sum_{1}^{\infty} \frac{(-1)^n}{(n + x)}, \quad 0 \leqslant x < \infty.$

4. $\sum_{1}^{\infty} \frac{\sin nx}{n} e^{-nx}, \quad 0 < \delta \leqslant x \leqslant \pi - \delta.$

5. $\sum_{1}^{\infty} nx^n, \quad 0 \leqslant x < 1.$

5.9 Power Series and Cesaro and Abel Summability

In this section we consider some additional optional topics in the theory of infinite series. We shall begin by studying power series.

> **Definition 5.** A *power series* is a series of the form $\sum_{k=0}^{\infty} a_k x^k$ where the coefficients a_k are fixed real (or complex) numbers. Let
>
> $$\limsup_{k \to \infty} \sqrt[k]{|a_k|} = \frac{1}{R} \; ;$$
>
> R is called the *radius of convergence* of the power series, and $\{x \mid |x| = R\}$ is the *circle of convergence* (x may be real or complex).

See Exercise 10, Chapter 1 for the definition of lim sup and note that $0 \leqslant R \leqslant +\infty$; R may be 0 or $+\infty$. The reason for the terminology in Definition 5 is brought out by the following result.

> **Theorem 15.** $\sum_{k=0}^{\infty} a_k x^k$ *converges absolutely for* $|x| < R$, *converges uniformly for* $|x| \leqslant R'$ *where* $R' < R$ *and diverges if* $|x| > R$. (*The theorem gives no information if* $|x| = R$.)

These convergence properties clearly distinguish R uniquely.

> **Corollary 4.** *The sum of a power series is a* C^{∞} *function inside its circle of convergence. It can be differentiated termwise and the differentiated series has the same radius of convergence.*

The method of proof is to make use of the previous results on termwise differentiation of series.

If $\lim_{n \to \infty} |a_n / a_{n+1}|$ happens to exist, then this limit is R, the radius of convergence. This is easily seen by using Theorem 15 together with the ratio test. We ask the reader to prove this for himself.

Next, the concept of *Cesaro summability* is examined.

> **Definition 6.** Set
>
> $$S_n = \sum_{k=1}^{n} a_k, \qquad \sigma_n = \frac{1}{n} \sum_{k=1}^{n} S_k \; ;$$
>
> thus σ_n is the arithmetic mean of the first n partial sums of the series. Note the formula
>
> $$\sigma_n = \sum_{k=1}^{n} \left(1 - \frac{k-1}{n} \right) a_k \; .$$

The series $\sum_{k=1}^{\infty} a_k$ is called *Cesaro 1-summable* or *(C,1) summable* to A if $\lim_{n \to \infty} \sigma_n = A$. If this is the case, write

$$\sum_{k=1}^{\infty} a_k = A \qquad (C,1) .$$

The idea here is to find some way to attach meaning to otherwise divergent series. For example,

$$\tfrac{1}{2} = 1 - 1 + 1 - 1 + \cdots \qquad (C,1) .$$

Here $S_n = 1, 0, 1, 0, \ldots$

$$T_n = \sum_{k=1}^{n} S_k = 1, 1, 2, 2, 3, 3, \ldots$$

Thus $\sigma_{2n} = n/2n$, $\sigma_{2n+1} = n + 1/(2n + 1)$ and so $\lim_{n \to \infty} \sigma_n = 1/2$.

However, we can introduce a yet *more powerful* method of summation by averaging the σ_n's, just as the (C,1) method was based on averaging the S_n's. That is, we define the (C,2) sum of the given series to be $\lim_{n \to \infty} (\sigma_1 + \sigma_2 + \cdots + \sigma_n)/n$ if the limit exists.

The reader can easily see how to define (C,*r*) summability for arbitrary values $r = 1, 2, \ldots$, obtaining successively more powerful methods of summation. Some basic properties of (C,1) summability follow.

(i) If $\sum a_k = A$ (C,1) and $\sum b_k = B$ (C,1), then $\sum (\alpha a_k + \beta b_k) = \alpha A + \beta B$ (C,1).

(ii) If $\sum_{k=1}^{\infty} a_k = A$ (C,1), then $\sum_{k=1}^{\infty} a_{k+1} = A - a_1$ (C,1) ("decapitation").

(iii) *Regularity*: If $\sum_{k=1}^{\infty} a_k = A$ in the usual sense, then $\sum_{k=1}^{\infty} a_k = A$ (C,1). (Obviously this property is crucial; any self-respecting summation method must have it.)

Proof of (iii):   We have $S_n \to A$. So, given $B < A$, there is an n_0 such that $n \geqslant n_0 \Rightarrow S_n \geqslant B$. Now

$$\sigma_n = \frac{1}{n} (S_1 + \cdots + S_{n_0} + S_{n_0+1} + \cdots + S_n)$$

$$\geqslant \frac{1}{n} (S_1 + \cdots + S_{n_0}) + \frac{n - n_0}{n} B .$$

Hence $\liminf_{n \to \infty} \sigma_n \geqslant B$. Since $B < A$ was arbitrary, $\liminf_{n \to \infty} \sigma_n \geqslant A$. A similar proof shows $\limsup_{n \to \infty} \sigma_n \leqslant A$. Accordingly, $\lim_{n \to \infty} \sigma_n = A$.

Next we turn to another method of summation called *Abel summation* (although it was actually invented by Euler).

Definition 7. $\sum_{k=0}^{\infty} a_k$ is summable to A in the sense of Abel if $\lim_{x \to 1-} \sum_{k=0}^{\infty} a_k x^k = A$. We write $\sum_{k=0}^{\infty} a_k = A$ (Abel).

For instance, we again have

$$\tfrac{1}{2} = 1 - 1 + 1 - 1 + \cdots \qquad \text{(Abel)}$$

since $f(x) = 1 - x + x^2 - \cdots = 1/(1 + x)$ for $|x| < 1$ and this $\to 1/2$ as $x \to 1-$.

Note that (at least in this example) the Abel method gives the same result as the (C,1) method. Actually, this is always the case, as we shall see below. First we shall prove that Abel summability is *regular*.

Theorem 16 (*Abel*).   *If* $\sum_{k=0}^{\infty} a_k = A$ *then* $\sum_{k=0}^{\infty} a_k x^k$ *converges for* $|x| < 1$ *and* $\lim_{x \to 1-} \sum_{k=0}^{\infty} a_k x^k = A$.

Thus, if a power series converges throughout a *closed* interval, its sum is continuous, even at the endpoints.

Actually, Abel's method is more powerful than the (C,1) method.

Theorem 17. $\sum_{k=0}^{\infty} a_k = A$ (C,1) *implies* $\sum_{k=0}^{\infty} a_k = A$ (*Abel*).

It is interesting to ask for conditions under which a Cesaro summable series (or Abel summable series, and so forth) is actually convergent in the usual sense. Along these lines we give a result of G. H. Hardy.

Theorem 18.   *If* $\sum a_n = A(C,1)$ *and if* $a_n = O(1/n)$ *(that is, if* $|a_n| \leqslant M/n$ *for a constant* M *and* n *large), then* $\sum a_n$ *converges (to* A*) in the usual sense.*

Note: Theorems of the above type are known as "Tauberian," after A. Tauber, who proved such a theorem relating Abel summability to ordinary convergence.

EXAMPLE 1.   Find the radius of convergence of $\sum x^k$ and $\sum x^k/k!$

Solution: In these cases we can use the formula

$$R = \lim_{n \to \infty} \left| \frac{a_n}{a_{n+1}} \right|.$$

The first example gives $R = 1$ and the second gives

$$R = \lim_{n \to \infty} \left(\frac{(n + 1)!}{n!} \right) = \lim_{n \to \infty} (n + 1) = \infty.$$

Thus $\sum x^k$ converges (to $1/(1-x)$) if $|x| < 1$ and $\sum x^k/k!$ converges (to e^x) for all x.

EXAMPLE 2.   Show that $\sum_{k=1}^{\infty} (-1)^k k$ is *not* summable (C,1).

Solution: Here a_n: $\quad -1, \quad +2, \quad -3, \quad +4, \quad -5, \quad +6, \ldots$

$\qquad\qquad\qquad S_n$: $\quad -1, \quad +1, \quad -2, \quad +2, \quad -3, \quad +3, \ldots$

$\qquad\qquad\qquad T_n$: $\quad -1, \quad\ \ 0, \quad -2, \quad\ \ 0, \quad -3, \quad\ \ 0, \ldots$

$$\sigma_{2n} = \quad 0, \quad \sigma_{2n-1} = -\frac{n}{2n-1} \to -\frac{1}{2}$$

Thus $\lim_{n \to \infty} \sigma_n$ does *not* exist. However, the (C,2) sum is $-1/4$ (Exercise).

EXAMPLE 3.   Show $\sum_{k=1}^{\infty} (-1)^k k = -1/4$ (Abel). Here

$$\sum_{k=1}^{\infty} (-1)^k k x^k = x \frac{d}{dx} \sum (-1)^k x^k$$

$$= x \frac{d}{dx} \frac{1}{1+x}$$

$$= -\frac{x}{(1+x)^2}, \qquad |x| < 1.$$

This $\to -1/4$ as $x \to 1-$, so

$$\sum_{k=1}^{\infty} (-1)^k k = -\frac{1}{4} \qquad \text{(Abel)}.$$

Exercises for Section 5.9

1. Compute the radius of convergence of

$$\sum x^k/k^2 \qquad \text{and of} \qquad \sum k! \, x^k.$$

2. Show that

$$\frac{1}{(1-x)^2} = \sum_{k=0}^{\infty} k x^{k-1} = \sum_{k=0}^{\infty} (k+1) x^k, \qquad -1 < x < 1$$

by differentiating an appropriate series.

3. Show that $2/3 = 1 + 0 - 1 + 1 + 0 - 1 + 1 + 0 - 1 + \cdots$ (C,1). (Note that insertion of zeros can alter the Cesaro sum.)

4. Show $1 + 0 - 1 + 1 + 0 - \cdots = 2/3$ (Abel).

Theorem Proofs for Chapter 5

Theorem 1. *Let f_k: $A \to \mathbb{R}^m$ be continuous functions, and suppose that $f_k \to f$ (uniformly). Then f is continuous.*

Proof: Since $f_n \to f$ uniformly, given $\varepsilon > 0$, we can find an N such that $k \geqslant N$ implies that $\|f_k(x) - f(x)\| < \varepsilon/3$ for all $x \in A$. Consider a particular point $x_0 \in A$. Since f_N is continuous, there exists a $\delta > 0$ such that $(\|x - x_0\| < \delta, x \in A) \Rightarrow (\|f_N(x) - f_N(x_0)\| < \varepsilon/3)$. Then for $\|x - x_0\| < \delta$, $\|f(x) - f(x_0)\| \leqslant \|f(x) - f_N(x)\| + \|f_N(x) - f_N(x_0)\| + \|f_N(x_0) - f(x_0)\| < \varepsilon/3 + \varepsilon/3 + \varepsilon/3 = \varepsilon$. Since x_0 is arbitrary, f is continuous at each point of A, hence it is continuous. ∎

Theorem 2. *Let f_k: $A \to \mathbb{R}^m$ be a sequence of functions. Then f_k converges uniformly iff for every $\varepsilon > 0$ there is an N such that $l, k \geqslant N$ implies $\|f_k(x) - f_l(x)\| < \varepsilon$ for all $x \in A$.*

Proof: If $f_k \to f$ uniformly, then given $\varepsilon > 0$, we can find an N such that $k \geqslant N$ implies $\|f_k(x) - f(x)\| < \varepsilon/2$ for all x. Then if $k, l \geqslant N$, $\|f_k(x) - f_l(x)\| \leqslant \|f_k(x) - f(x)\| + \|f(x) - f_l(x)\| < \varepsilon/2 + \varepsilon/2 = \varepsilon$.

Conversely, if given $\varepsilon > 0$, we can find an N such that $k, l \geqslant N$ implies $\|f_k(x) - f_l(x)\| < \varepsilon$ for all x, then $f_k(x)$ is a Cauchy sequence at each point x, so $f_k(x)$ certainly converges pointwise to, say, $f(x)$. Moreover, we can find an N such that $k, l \geqslant N$ implies $\|f_k(x) - f_l(x)\| < \varepsilon/2$ for all x. Since $f_k(x) \to f(x)$ at each point x, we can find for each x an N_x such that $l \geqslant N_x \Rightarrow \|f_l(x) - f(x)\| < \varepsilon/2$. Let $l \geqslant \max\{N, N_x\}$. Then $k \geqslant N \Rightarrow \|f_k(x) - f(x)\| \leqslant \|f_k(x) - f_l(x)\| + \|f_l(x) - f(x)\| < \varepsilon/2 + \varepsilon/2 = \varepsilon$. Since this is true for each point x, we have found an N such that $k \geqslant N \Rightarrow \|f_k(x) - f(x)\| < \varepsilon$ for all x. Hence $f_k \to f$ (uniformly). ∎

Theorem 3. *Suppose g_k: $A \to \mathbb{R}^m$ are functions such that there exist constants M_k with $\|g_k(x)\| \leqslant M_k$ for all $x \in A$, and $\sum_{k=1}^{\infty} M_k$ converges. Then $\sum_{k=1}^{\infty} g_k$ converges uniformly (and absolutely).*

Proof: Since $\sum M_k$ converges, for every $\varepsilon > 0$ there is an N such that $k \geqslant N$ implies $|M_k + \cdots + M_{k+p}| < \varepsilon$ for all $p = 1, 2, \ldots$ (see Theorem 11, Chapter 2). For $k \geqslant N$ we have, by the triangle inequality,

$$\|g_k(x) + \cdots + g_{k+p}(x)\| \leqslant \|g_k(x)\| + \cdots + \|g_{k+p}(x)\|$$

$$\leqslant M_k + \cdots + M_{k+p} < \varepsilon$$

for all $x \in A$. Thus by the Cauchy criterion for series, $\sum g_k$ converges uniformly. ∎

Theorem 4. *Suppose f_k: $[a,b] \to \mathbb{R}$ are continuous functions and $f_k \to f$ uniformly. Then*

$$\int_a^b f_k(x)\, dx \to \int_a^b f(x)\, dx \ .$$

Corollary 2.   *Suppose* $g_k: [a,b] \to \mathbb{R}$ *are continuous and* $\sum_{k=1}^{\infty} g_k$ *converges uniformly. Then we may interchange the order of integration and summation*

$$\int_a^b \sum_{k=1}^{\infty} g_k(x)\, dx = \sum_{k=1}^{\infty} \int_a^b g_k(x)\, dx .$$

Proof: For integrals recall that if $|f(x)| \leqslant M$ then

$$\left| \int_a^b f(x)\, dx \right| \leqslant M(b - a) .$$

For $\varepsilon > 0$ choose N such that $k \geqslant N$ implies $|f_k(x) - f(x)| < \varepsilon/(b - a)$. Then

$$\left| \int_a^b f_k(x)\, dx - \int_a^b f(x)\, dx \right| = \left| \int_a^b (f_k(x) - f(x))\, dx \right| \leqslant \varepsilon \cdot \frac{(b - a)}{(b - a)} = \varepsilon$$

as required.

For the corollary, let $f_k = \sum_{i=1}^{k} g_i$; then $f_k \to f = \sum_{k=1}^{\infty} g_k$ (uniformly), and so by the above

$$\int_a^b f_k(x)\, dx \to \int_a^b f(x)\, dx .\quad \blacksquare$$

Theorem 5.   *Let* $f_k: \,]a,b[\,\to \mathbb{R}$ *be a sequence of differentiable functions on the open interval* $]a,b[$ *converging pointwise to* $f: \,]a,b[\,\to \mathbb{R}$. *Suppose the derivatives* f'_k *are continuous and converge uniformly to a function* g. *Then* f *is differentiable and* $f' = g$.

Proof:*   Write $f_k(x) = f_k(x_0) + \int_{x_0}^{x} f'_k(x)\, dt$, where $a < x_0 < b$. This is possible by the fundamental theorem of calculus. Letting $k \to \infty$, we get $f(x) = f(x_0) + \int_{x_0}^{x} g(t)\, dt$ using Theorem 4. Hence $f' = g$ again by the fundamental theorem. Here g is continuous by Theorem 1. $\blacksquare$

Theorem 6.   *The function* $\| \ \|$ *on* $\mathscr{C}_b(A,\mathbb{R}^m)$ *satisfies*

  (i) $\|f\| \geqslant 0$, *and* $\|f\| = 0$ *iff* $f = 0$;
  (ii) $\|\alpha f\| = |\alpha|\, \|f\|$; *for* $\alpha \in \mathbb{R}, f \in \mathscr{C}_b$;
  (iii) $\|f + g\| \leqslant \|f\| + \|g\|$ *(triangle inequality)*.

Proof: (i) and (ii) are clear. For (iii),

$$\|f + g\| = \sup\{\|(f + g)(x)\| \mid x \in A\}$$
$$\leqslant \sup\{\|f(x)\| + \|g(x)\| \mid x \in A\}$$

by the triangle inequality in $\mathbb{R}^m$. Now, since $\sup(P + Q) = \sup(P) + \sup(Q)$ (Exercise 7, at the end of Chapter 1), and

$$\{\|f(x)\| + \|g(x)\| \mid x \in A\} \subset \{\|f(x)\| + \|g(y)\| \mid x, y \in A\}$$

we have

$$\sup\{\|f(x)\| + \|g(x)\| \mid x \in A\} \leqslant \|f\| + \|g\| .\quad \blacksquare$$

Theorem 7. $(f_k \to f$ *(uniformly on* A)) $\Leftrightarrow (f_k \to f$ *in* $\mathscr{C}_b$; *that is,* $\|f_k - f\| \to 0)$.

---

* Note: Actually one can prove the theorem even if f'_k are not continuous, but it requires more work; see Apostol, *Mathematical Analysis*, p. 402.

Proof: This is nothing more than a transcription of the definitions. The student should write it out. ∎

Theorem 8. $\mathscr{C}_b$ *is a Banach space.*

Proof: Let f_k be a Cauchy sequence. By Theorem 2, f_k converges uniformly to f. Since $\|f(x)\| \leqslant \|f_k\| + 1$ for k large, f is bounded, and by Theorem 1, f is continuous. Thus $f \in \mathscr{C}_b$ and therefore f_k converges in $\mathscr{C}_b$. ∎

The proof of the Arzela-Ascoli theorem is a little long and involved. It may be omitted, if desired, in a less ambitious course.

Theorem 9. *Let $A \subset \mathbb{R}^n$ be compact and let $B \subset \mathscr{C}(A, \mathbb{R}^m)$. If B is bounded and equicontinuous, then any sequence in B has a uniformly convergent subsequence.*

To prove this, we first prove a lemma.

Lemma 1. *Let $A \subset \mathbb{R}^n$ be any set. Then there is a countable set $C \subset A$ whose closure contains A.*

Proof: The points in $\mathbb{R}^n$ with rational coordinates are a countable set (see the Introductory chapter). Call them $x_1, x_2, \ldots$. Consider for each integer n the discs

$$D\left(x_1, \frac{1}{n}\right), D\left(x_2, \frac{1}{n}\right), \ldots$$

These clearly cover all of $\mathbb{R}^n$. Whenever one of these, $D(x_l, (1/n))$, meets A, select one point from $D(x_l, (1/n)) \cap A$, and the collection so obtained will define our set C. Now C is countable since this collection $\{D(x_l, (1/n)) \mid l, n \in \mathbb{N}\}$ is countable.

We claim that $\mathrm{cl}(C) \supset A$. Indeed, let $x \in A$, $\varepsilon > 0$. Choose n so $1/n < \varepsilon/2$. Now x lies in some $D(x_l, (1/n))$ for some value of l, so there is a point in $C \cap D(x_l, (1/n))$, say y. Thus $d(x, y) \leqslant d(x, x_l) + d(x_l, y) \leqslant 1/n + 1/n < \varepsilon$. Hence $x \in \mathrm{cl}(C)$, so $\mathrm{cl}(C) \supset A$. ∎

We shall need to exploit compactness of A in the following way.

Lemma 2. *Let A be compact and C be constructed as above. Then for any $\delta > 0$ there is a finite set $C_1 \subset C$ say $C_1 = \{y_1, \ldots, y_k\}$ such that each $x \in A$ is, within δ of some $y_l \in C_1$.*

Proof: Choose n so $1/n < \delta$. Then in Lemma 1, there is a finite number of the collection $D(x_1, (1/n)), D(x_2, (1/n)), \ldots$ which cover A, because A is compact. Then C_1 is defined as those members in this finite collection which were chosen for C. The result then follows as in Lemma 1. ∎

Now we turn to the proof of the theorem. Let C be as constructed in Lemma 1, say $C = \{x_1, x_2, \ldots\}$. Let f_n be our sequence in B. Now $\{f_n\}$ is bounded, so, in particular, the sequence $f_n(x_1)$ is bounded in $\mathbb{R}^m$. It follows from the Bolzano-Weierstrass theorem in $\mathbb{R}^m$ that there is a subsequence of $f_n(x_1)$ which is convergent. Let us denote this subsequence by

$$f_{11}(x_1), f_{12}(x_1), \ldots, f_{1n}(x_1), \ldots.$$

Similarly, the sequence $f_{1k}(x_2)$: $k = 1, 2, \ldots$ is bounded in $\mathbb{R}^m$; hence it has a subsequence

$$f_{21}(x_2), f_{22}(x_2), \ldots, f_{2n}(x_2), \ldots$$

which is convergent. Continuing the process, the sequence $f_{2k}(x_3)$: $k = 1, 2, \ldots$ is bounded in $\mathbb{R}^m$, so some subsequence

$$f_{31}(x_3), f_{32}(x_3), \ldots, f_{3n}(x_3), \ldots$$

is convergent. We proceed in this way and then set $g_n = f_{nn}$ so that g_n is the nth function occurring in the nth subsequence.

Diagramatically, g_n is obtained by picking out the diagonal:

$$
\begin{array}{llllll}
f_{11} & f_{12} & f_{13} & \cdots & f_{1n} & \cdots \quad \text{(first subsequence)} \\
f_{21} & f_{22} & f_{23} & \cdots & f_{2n} & \cdots \quad \text{(second subsequence)} \\
f_{31} & f_{32} & f_{33} & \cdots & f_{3n} & \cdots \quad \text{(third subsequence)} \\
& & & & & \\
f_{n1} & f_{n2} & f_{n3} & \cdots & f_{nn} & \cdots \quad \text{(nth subsequence)} \\
\end{array}
$$

This trick is called the "diagonal process" and is useful in a variety of situations.

From the construction of the sequence g_n, we see the sequence g_n converges at each point of C; indeed g_n is a subsequence of each sequence f_{mk}: $k = 1, 2, \ldots$.

We shall now prove that the sequence g_n converges at each point of A and that the convergence is uniform and this will prove the theorem. To do this, let $\varepsilon > 0$ and let δ be as in the definition of equicontinuity. Let $C_1 = \{y_1, \ldots, y_k\}$ be a finite subset of C such that every point in A is within δ of some point in C_1 (see Lemma 2). Since the sequences

$$(g_n(y_1)), (g_n(y_2)), \ldots, (g_n(y_k))$$

all converge, there is an integer N such that if $m, n \geqslant N$, then

$$\|g_m(y_i) - g_n(y_i)\| < \varepsilon \qquad \text{for } i = 1, 2, \ldots, k \,.$$

For each $x \in A$, there exists a $y_j \in C_1$ such that $\|x - y_j\| < \delta$. Hence, by the assumption of equicontinuity, we have

$$\|g_n(x) - g_n(y_j)\| < \varepsilon$$

for all $n = 1, 2, \ldots$. Therefore, we have

$$\|g_n(x) - g_m(x)\| \leqslant \|g_n(x) - g_n(y_j)\| + \|g_n(y_j) - g_m(y_j)\| + \|g_m(y_j) - g_m(x)\|$$

$$< \varepsilon + \varepsilon + \varepsilon = 3\varepsilon \,,$$

provided $m, n \geqslant N$. This shows that

$$\|g_n - g_m\| \leqslant 3\varepsilon \qquad \text{for } m, n \geqslant N,$$

so the uniform convergence of the sequence g_n on A follows from the Cauchy Criterion (see Theorem 2). ∎

Instead of proving Theorem 10, the following more general result is established.

Theorem 10'. *Let X be a complete metric space and let $T: X \to X$ be a contraction: $d(T(x), T(y)) \leqslant \lambda\, d(x,y)$, where $0 \leqslant \lambda < 1$ is a fixed constant. Then T is continuous and has a unique fixed point.*

Proof: That T is uniformly continuous is immediate, for given $\varepsilon > 0$ we can use $\delta = \varepsilon/\lambda$; $d(x,y) < \delta$ implies $d(T(x),T(y)) < \lambda\delta = \varepsilon$.

Let $x_0 \in X$, and let $x_1 = T(x_0)$, $x_2 = T(x_1)$, ..., $x_{n+1} = T(x_n) = T^{n+1}(x_0)$. We claim x_n is a Cauchy sequence. Note that

$$
\begin{aligned}
d(x_{n+1}, x_n) &= d(T(x_n), T(x_{n-1})) \\
&\leqslant \lambda\, d(x_n, x_{n-1}) \\
&= \lambda\, d(T(x_{n-1}), T(x_{n-2})) \\
&\leqslant \lambda^2\, d(x_{n-1}, x_{n-2}) \\
&\quad\vdots \\
&\leqslant \lambda^n\, d(Tx_0, x_0)\,.
\end{aligned}
$$

Hence

$$
\begin{aligned}
d(x_n, x_{n+k}) &\leqslant d(x_n, x_{n+1}) + d(x_{n+1}, x_{n+2}) + \cdots + d(x_{n+k-1}, x_{n+k}) \\
&\leqslant (\lambda^n + \lambda^{n+1} + \cdots + \lambda^{n+k-1})\, d(Tx_0, x_0)\,.
\end{aligned}
$$

Now since $\lambda < 1$, $\sum \lambda^n$ is a convergent geometric series, so given $\varepsilon > 0$ there is an N such that $n \geqslant N$ implies $(\lambda^n + \cdots + \lambda^{n+k-1}) < \varepsilon/(d(Tx_0, x_0))$. Hence $n \geqslant N$ implies $d(x_n, x_{n+k}) < \varepsilon$. Thus we have a Cauchy sequence, and by the assumption of completeness, $x_n \to x$ for some $x \in X$.

We claim $Tx = x$. Indeed, $x = \lim\limits_{n\to\infty} x_n$, so $Tx = \lim\limits_{n\to\infty} T(x_n)$ by continuity of T. But $Tx_n = x_{n+1}$ so $Tx = \lim\limits_{n\to\infty} x_{n+1} = x$.

Finally, x, the fixed point is unique, for suppose that $Tx = x$ and $Ty = y$. Then

$$d(x,y) = d(Tx, Ty) \leqslant \lambda\, d(x,y)\,.$$

If $d(x,y) \neq 0$ we would get $1 \leqslant \lambda$, a contradiction. Hence $d(x,y) = 0$, so $x = y$.

Theorem 11. *Let $f: [0,1] \to \mathbb{R}$ be continuous and let $\varepsilon > 0$. Then there is a polynomial $p(x)$ such that $\|p - f\| < \varepsilon$. In fact, the sequence of Bernstein polynomials*

$$p_n(x) = \sum_{k=0}^{n} \binom{n}{k} f\!\left(\frac{k}{n}\right) x^k (1 - x)^{n-k}$$

converge uniformly to f as n → ∞ where

$$\binom{n}{k} = \frac{n!}{k!\,(n-k)!}$$

denotes the binomial coefficient.

Proof: The binomial theorem asserts

$$(x + y)^n = \sum_{k=0}^{n} \binom{n}{k} x^k y^{n-k} \, . \tag{1}$$

Differentiating Eq. 1 with respect to x and multiplying by x gives the identity

$$nx(x + y)^{n-1} = \sum_{k=0}^{n} k \binom{n}{k} x^k y^{n-k} \tag{2}$$

Similarly, by differentiating twice,

$$n(n-1)x^2(x+y)^{n-2} = \sum_{k=0}^{n} k(k-1) \binom{n}{k} x^k y^{n-k} \, . \tag{3}$$

Let (for notation) $r_k(x) = \binom{n}{k} x^k (1-x)^{n-k}$. Thus Eqs. 1, 2, and 3 read, with $y = 1 - x$,

$$\sum_{k=0}^{n} r_k(x) = 1, \qquad \sum_{k=0}^{n} k r_k(x) = nx, \qquad \sum_{k=0}^{n} k(k-1) r_k(x) = n(n-1)x^2 \, .$$

It follows that we have the identity

$$\begin{aligned}
\sum_{k=0}^{n} (k - nx)^2 r_k(x) &= n^2 x^2 \sum_{k=0}^{n} r_k(x) - 2nx \sum_{k=0}^{n} k r_k(x) + \sum_{k=0}^{n} k^2 r_k(x) \\
&= n^2 x^2 - 2nx \cdot nx + [nx + n(n-1)x^2] \\
&= nx(1 - x) \, .
\end{aligned} \tag{4}$$

Now choose M such that $|f(x)| \leqslant M$ on $[0,1]$. Since f is uniformly continuous there is, for $\varepsilon > 0$, a $\delta > 0$ such that $|x - y| < \delta$ implies $|f(x) - f(y)| < \varepsilon$.

We want to estimate the expression

$$\begin{aligned}
|f(x) - p_n(x)| &= \left| f(x) - \sum_{k=0}^{n} f(k/n) r_k(x) \right| \\
&= \left| \sum_{k=0}^{n} \left(f(x) - f\left(\frac{k}{n}\right) \right) r_k(x) \right| \, .
\end{aligned}$$

To do this, divide this sum into two parts; those for which $|k - nx| < \delta n$ and those for which $|k - nx| \geqslant \delta n$. If $|k - nx| < \delta n$, then $|x - (k/n)| < \delta$, so $|f(x) - f(k/n)| < \varepsilon$, and therefore, remembering that $r_k(x) \geqslant 0$, these terms give a sum $\leqslant \varepsilon \sum r_k(x) = \varepsilon$. The second type of terms have a sum

$$\leqslant 2M \sum_{|k-nx| \geqslant \delta n} r_k(x) \leqslant \frac{2M}{n^2 \delta^2} \sum_{k=0}^{n} (k - nx)^2 r_k(x)$$

which, by Eq. 4, is

$$\frac{2Mx(1-x)}{n\delta^2} \leqslant \frac{M}{2\delta^2 n},$$

since $x(1-x) \leqslant 1/4$ (why?). Thus we have proven that for any $\varepsilon > 0$ there is a $\delta > 0$ such that

$$|f(x) - p_n(x)| < \varepsilon + \frac{M}{2\delta^2 n}.$$

Thus if n is sufficiently large, $M/(2\delta^2 n) < \varepsilon$ so

$$|f(x) - p_n(x)| < 2\varepsilon$$

if $n \geqslant M/2\delta^2\varepsilon$. Thus $p_n \to f$ uniformly. ∎

Theorem 12.   *Let $A \subset \mathbb{R}^n$ be compact and let $\mathscr{B} \subset \mathscr{C}(A,\mathbb{R})$ satisfy*
 (i) $\mathscr{B}$ is an algebra;
 (ii) the constant function $x \mapsto 1$ lies in $\mathscr{B}$;
 (iii) $\mathscr{B}$ separates points.
Then $\mathscr{B}$ is dense in $\mathscr{C}(A,\mathbb{R})$.

Proof: Let us introduce some notations as follows:

$$(f \vee g)(x) = \max(f(x),g(x)) \quad \text{and} \quad (f \wedge g)(x) = \min(f(x),g(x)).$$

(See Figure 5-6.) Let $\bar{\mathscr{B}}$ be the closure of $\mathscr{B}$. Then by continuity of addition and multiplication, we see that $\bar{\mathscr{B}}$ also satisfies (i). It clearly satisfies (ii), (iii). Thus $\bar{\mathscr{B}}$ is closed and what we then want to show is that $\bar{\mathscr{B}} = \mathscr{C}(A,\mathbb{R})$.

By the preceding theorem and solution (a) to Example 2, Section 5.7 we can find a sequence of polynomials $p_n(t)$ such that

$$||t| - p_n(t)| < \frac{1}{n} \quad \text{for } -n \leqslant t \leqslant n.$$

Thus

$$||f(x)| - p_n(f(x))| < \frac{1}{n} \quad \text{if } -n \leqslant f(x) \leqslant n.$$

This proves that for $f \in \bar{\mathscr{B}}$, $|f| \in \bar{\mathscr{B}}$, because $p_n \circ f \in \bar{\mathscr{B}}$ since $\bar{\mathscr{B}}$ is an algebra.

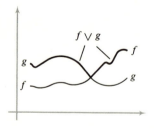
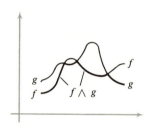

FIGURE 5-6

Now we have the identities

$$f \vee g = \frac{f + g}{2} + \frac{|f - g|}{2} ,$$

$$f \wedge g = \frac{f + g}{2} - \frac{|f - g|}{2}$$

(an exercise for reader), so if $f, g \in \mathscr{B}$, $f \vee g$ and $f \wedge g$ lie in $\mathscr{B}$ as well.

Let $h \in \mathscr{C}(A, \mathbb{R})$ and $x_1, x_2 \in A$ with $x_1 \neq x_2$. Choose $g \in \mathscr{B}$ such that $g(x_1) \neq g(x_2)$ (which is possible by hypothesis (iii)), and let

$$f_{x_1 x_2}(x) = \alpha g(x) + \beta ,$$

where

$$\alpha = \frac{[h(x_1) - h(x_2)]}{[g(x_1) - g(x_2)]} \quad \text{and} \quad \beta = \frac{[g(x_1)h(x_2) - h(x_1)g(x_2)]}{[g(x_1) - g(x_2)]} .$$

The numbers of α, β are chosen so $f_{x_1 x_2}(x_1) = h(x_1)$ and $f_{x_1 x_2}(x_2) = h(x_2)$.

Let $\varepsilon > 0$ and $x \in A$. For $y \in A$ there is a neighborhood $U(y)$ of y such that

$$f_{yx}(z) > h(z) - \varepsilon \quad \text{if } z \in U(y) .$$

This is simply by continuity of h. Let $U(y_1), \ldots, U(y_l)$ be a finite subcover of A, which is possible by the Heine-Borel theorem. Set $f_x = f_{y_1 x} \vee \cdots \vee f_{y_l x}$. Thus, as above, $f_x \in \mathscr{B}$ and $f_x(z) > h(z) - \varepsilon$ for all $z \in A$. Also, $f_x(x) = h(x)$. Thus there is a neighborhood $V(x)$ such that $f_x(y) < h(y) + \varepsilon$ if $y \in V(x)$. Let $V(x_1), \ldots, V(x_k)$ cover A and set

$$f = f_{x_1} \wedge \cdots \wedge f_{x_k} .$$

Then again $f \in \mathscr{B}$. Now $f(z) > h(z) - \varepsilon$ for all $z \in A$ because $f_{x_j}(u) > h(u) - \varepsilon$ for all $u \in A$ and also for $y \in A$, $y \in V(x_j)$ for some x_j so $f(y) \leqslant f_{x_j}(y) < h(y) + \varepsilon$. Thus $|f(z) - h(z)| < \varepsilon$, so $h \in \mathscr{B}$. Thus $\mathscr{B} = \mathscr{C}(A, \mathbb{R})$. ∎

For both Theorems 13 and 14 which follow *Abel's partial summation formula* is employed; this is contained in the next lemma.

Lemma 1. *Consider two sequences $a_1, a_2, \ldots$ and $b_1, b_2, \ldots$ of real numbers. Let $s_n = a_1 + \cdots + a_n$. Then*

$$\sum_{k=1}^{n} a_k b_k = s_n b_{n+1} - \sum_{k=1}^{n} s_k(b_{k+1} - b_k)$$

$$= s_n b_1 + \sum_{k=1}^{n} (s_n - s_k)(b_{k+1} - b_k) .$$

Proof: Note that $a_n = s_n - s_{n-1}$. Then

$$\sum_{k=1}^{n} a_k b_k = \sum_{k=1}^{n} (s_k - s_{k-1}) b_k = \sum_{k=1}^{n} s_k b_k - \sum_{k=1}^{n} s_{k-1} b_k ,$$

where $s_0 = 0$. Now

$$\sum_{k=1}^{n} s_{k-1} b_k = \sum_{k=1}^{n} s_k b_{k+1} - s_n b_{n+1} ,$$

so we obtain the first result. The second equality is obtained by putting

$$b_{n+1} = \sum_{k=1}^{n} (b_{k+1} - b_k) + b_1$$

in the first equality. ∎

Theorem 13 (*Abel's Test*).   *Let* $A \subset \mathbb{R}^m$ *and* $\varphi_n: A \to \mathbb{R}$ *be a sequence of functions which are decreasing; that is,* $\varphi_{n+1}(x) \leqslant \varphi_n(x)$ *for each* $x \in A$. *Suppose there is a constant* M *such that* $|\varphi_n(x)| \leqslant M$ *for all* $x \in A$ *and all* n. *If* $\sum_{n=1}^{\infty} f_n(x)$ *converges uniformly on* A, *then so does* $\sum_{n=1}^{\infty} \varphi_n(x)f_n(x)$.

Proof: Let

$$s_n(x) = \sum_{k=1}^{n} f_k(x) \qquad \text{and} \qquad r_n(x) = \sum_{k=1}^{n} \varphi_k(x)f_k(x) .$$

Then, by the second equality of the lemma, we find that

$$r_n(x) - r_m(x) = (s_n(x) - s_m(x))\varphi_1(x) + \sum_{k=m+1}^{n} (s_n(x) - s_k(x))(\varphi_{k+1}(x) - \varphi_k(x))$$

for $n > m$, so that

$$|r_n(x) - r_m(x)| \leqslant |s_n(x) - s_m(x)| \, |\varphi_1(x)| + \sum_{k=m+1}^{n} |s_n - s_k| \, |\varphi_{k+1}(x) - \varphi_k(x)|$$

since

$$\varphi_{k+1} \leqslant \varphi_k, \qquad |\varphi_{k+1} - \varphi_k| = \varphi_k - \varphi_{k+1} .$$

Given $\varepsilon > 0$, choose N so that $n, m \geqslant N$ implies $|s_n(x) - s_m(x)| < \varepsilon/3M$ for all $x \in A$. Then

$$|r_n(x) - r_m(x)| < \frac{\varepsilon}{3} + \left(\frac{\varepsilon}{3M}\right) \sum_{k=m+1}^{n} [\varphi_k(x) - \varphi_{k+1}(x)]$$

$$= \frac{\varepsilon}{3} + \left(\frac{\varepsilon}{3M}\right)[\varphi_{m+1}(x) - \varphi_{n+1}(x)]$$

$$\leqslant \frac{\varepsilon}{3} + \left(\frac{\varepsilon}{3M}\right)[|\varphi_{m+1}(x)| + |\varphi_{n+1}(x)|]$$

$$\leqslant \frac{\varepsilon}{3} + \frac{\varepsilon}{3} + \frac{\varepsilon}{3} = \varepsilon$$

for all $x \in A$. Hence by the Cauchy Criterion (Theorem 2), $f_n(x)$ converges uniformly. ∎

Theorem 14 (*Dirichlet Test*).   *Let* $s_n(x) = \sum_{m=1}^{n} f_m(x)$ *for a sequence* $f_n: A \subset \mathbb{R}^m \to \mathbb{R}$. *Assume there is a constant* M *such that* $|s_n(x)| \leqslant M$ *for all* $x \in A$ *and all* n. *Let* $g_n: A \subset \mathbb{R}^m \to \mathbb{R}$ *be such that* $g_n \to 0$ *uniformly,* $g_n \geqslant 0$, *and* $g_{n+1}(x) \leqslant g_n(x)$. *Then* $\sum_{n=1}^{\infty} f_n(x)g_n(x)$ *converges uniformly on* A.

Proof: We use the same notation as in the above proof, writing $\varphi_n = g_n$. Now, however, to compute $r_n - r_m$ we use the first equality in the lemma. Namely,

$$r_n(x) - r_m(x) = s_n(x)\varphi_{n+1}(x) - s_m(x)\varphi_{m+1}(x)$$

$$- \sum_{k=m+1}^{n} s_k(x)(\varphi_{k+1}(x) - \varphi_k(x))$$

so that, since $\varphi_k \geqslant 0$ and $\varphi_{k+1} \leqslant \varphi_k$,

$$|r_n(x) - r_m(x)| \leqslant M(\varphi_{n+1}(x) + \varphi_{m+1}(x))$$

$$+ M \sum_{k=m+1}^{n} (\varphi_k(x) - \varphi_{k+1}(x))$$

$$= M(\varphi_{n+1}(x) + \varphi_{m+1}(x) + \varphi_{m+1}(x) - \varphi_{n+1}(x))$$

$$= 2M\varphi_{m+1}(x) .$$

Now, given $\varepsilon > 0$, choose N so that $m > N$ implies $\varphi_m(x) < \varepsilon/2M$ for all x. Then $m, n \geqslant N$ implies $|r_n(x) - r_m(x)| < \varepsilon$, which proves the assertion. ∎

Theorem 15. $\sum_{k=0}^{\infty} a_k x^k$ *converges absolutely for* $|x| < R$, *converges uniformly for* $|x| \leqslant R'$ *where* $R' < R$ *and diverges if* $|x| > R$.

Proof: Let $R' < R$. Choose R'' with $R' < R'' < R$. Then, for n sufficiently large,

$$\sqrt[n]{|a_n|} \leqslant \frac{1}{R''}, \qquad \text{that is, } |a_n| \leqslant \left(\frac{1}{R''}\right)^n .$$

Hence if $|x| \leqslant R'$, we have

$$|a_n x^n| \leqslant \left(\frac{R'}{R''}\right)^n .$$

Since $R'/R'' < 1$, we have uniform absolute convergence in the disk $|x| \leqslant R'$ by the Weierstrass M-test.

On the other hand, suppose $\sum a_n x^n$ converges. Then $a_n x^n \to 0$, so $|a_n x^n| \leqslant 1$ for n large. Thus $\sqrt[n]{|a_n|} \leqslant |x|^{-1}$ for n large. Hence $R^{-1} = \limsup \sqrt[n]{|a_n|} \leqslant |x|^{-1}$, that is, $|x| \leqslant R$. ∎

Corollary 4. *The sum of a power series is a* C^∞ *function inside its circle of convergence. It can be differentiated termwise and the differentiated series has the same radius of convergence.*

Proof: The series obtained by termwise differentiating is $\sum k a_k x^{k-1}$. The radius of convergence is R', where

$$1/R' = \limsup \sqrt[k]{k\,|a_k|} .$$

But $\sqrt[k]{k} \to 1$ (why?), so

$$\frac{1}{R'} = \limsup \sqrt[k]{|a_k|} = \frac{1}{R} \qquad \text{that is, } R' = R .$$

Thus, by Corollary 3, the differentiated series converges uniformly inside any smaller circle, and therefore it *is* the derivative of the sum of the original series. By induction, we see that the original series is differentiable any number of times. ∎

Theorem 16 (*Abel*).   If $\sum_{k=0}^{\infty} a_k = A$ then $\sum_{k=0}^{\infty} a_k x^k$ converges for $|x| < 1$ and $\lim_{x \to 1-} \sum_{k=0}^{\infty} a_k x^k = A$.

Proof: By changing a_0, we can assume $A = 0$. Since a_k is bounded (in fact $a_k \to 0$) the series $\sum a_k x^k$ converges for $|x| < 1$ by Theorem 15 on the radius of convergence.

Write $S_n = \sum_{k=0}^{n} a_k$. Since S_n is bounded as $n \to \infty$, the series $\sum S_k x^k$ likewise converges for $|x| < 1$. Now, since $A = 0$, $S_n \to 0$ as $n \to \infty$. Write $f(x) = \sum_{k=0}^{\infty} a_k x^k$, $|x| < 1$. Then

$$f(x) = S_0 + \sum_{k=1}^{\infty} (S_k - S_{k-1})x^k$$

$$= (1 - x) \sum_{k=0}^{\infty} S_k x^k .$$

Since $S_n \to 0$, given $\varepsilon > 0$ we can find n_0 so that $|S_n| \leqslant \varepsilon$ for $n > n_0$. Then

$$|f(x)| \leqslant (1 - x) \left| \sum_{k=0}^{n_0} S_k x^k \right| + (1 - x) \sum_{k=n_0+1}^{\infty} \varepsilon x^k$$

$$\leqslant (1 - x) \left| \sum_{k=0}^{n_0} S_k x^k \right| + (1 - x) \cdot \varepsilon x^{n_0 + 1}(1 - x)^{-1}$$

$$\leqslant (1 - x) \left| \sum_{k=0}^{n_0} S_k x^k \right| + \varepsilon .$$

Accordingly, $\lim_{x \to 1-} \sup |f(x)| \leqslant \varepsilon$. Since $\varepsilon > 0$ was arbitrary,

$$\lim_{x \to 1-} f(x) = 0 . \quad ∎$$

Theorem 17. $\sum_{k=0}^{\infty} a_k = A$ (C,1) *implies* $\sum_{k=0}^{\infty} a_k = A$ (*Abel*).

Proof: As before, we may suppose $A = 0$. Write $S_n = \sum_{k=0}^{n} a_k$, $T_n = \sum_{k=0}^{n} S_k$. Then, by assumption, $T_n = O(n)$. Hence $S_n = T_n - T_{n-1} = O(n)$ and $a_n = S_n - S_{n-1} = O(n)$. Accordingly all three series $\sum a_k x^k$, $\sum S_k x^k$, and $\sum T_k x^k$ converge if $|x| < 1$. Also,

$$f(x) = \sum a_k x^k = (1 - x) \sum S_k x^k$$

$$= (1 - x)^2 \sum T_k x^k .$$

Now, as $T_n = O(n)$, given $\varepsilon > 0$ we may choose n_0 so that $n \geqslant n_0$ implies $|T_n| \leqslant \varepsilon n$. Accordingly,

$$|f(x)| \leqslant (1 - x)^2 \left| \sum_{k \leqslant n_0} T_k x^k \right| + (1 - x)^2 \sum_{k > n_0} \varepsilon k x^k$$

$$\leqslant (1 - x)^2 \left| \sum_{k \leqslant n_0} T_k x^k \right| + (1 - x)^2 \cdot \varepsilon x(1 - x)^{-2}$$

and we find $\lim_{x \to 1-} \sup |f(x)| \leqslant \varepsilon$. Thus, as in the previous theorem, $\lim_{x \to 1-} f(x) = 0$. ∎

Theorem 18. *If $\sum a_n = A$ (C,1) and if $a_n = O(1/n)$ then $\sum a_n = A$.*

Proof: We can as usual, suppose that $A = 0$. Write $S_n = \sum_1^n a_k$, $T_n = \sum_1^n S_k$. Then the first hypothesis is written as $T_n = o(n)$. The second hypothesis implies there exists a constant C with $|a_n| \leqslant C/n$ for all n.

We want to show $S_n \to 0$. If not, then for some $\delta > 0$, $|S_n| \geqslant \delta$ for infinitely many indices n. It can be assumed (by reversing all signs if need be) that $S_n \geqslant \delta$ for infinitely many values of n. But if $S_n \geqslant \delta$ and $r > S$, we have

$$S_r = S_n + a_{n+1} + a_{n+2} + \cdots + a_r$$

$$\geqslant \delta - C\left(\frac{1}{n+1} + \cdots + \frac{1}{r}\right)$$

$$\geqslant \delta - C \log\left(\frac{r}{n}\right).$$

This will be $\geqslant \delta/2$ provided $C \log(r/n) \leqslant \delta/2$, that is, $r/n \leqslant e^{\delta/2C} = \lambda$. (Note that $\lambda > 1$). Hence we have

$$([\lambda n] - n)\frac{\delta}{2} \leqslant \sum_{r=n+1}^{[\lambda v]} S_r = T_{[\lambda n]} - T_n.$$

(Here $[x]$ means the largest integer $\leqslant x$.) Now the right side of this inequality is $o(n)$, but the left side is of the order $(\lambda - 1)\delta n/2$, a contradiction. Hence S_n must tend to 0.   ∎

Worked Examples for Chapter 5

1.   (i) If $f_k \to f$ (pointwise) and $g_k \to g$ (pointwise), then show that $f_k + g_k \to f + g$ (pointwise) for functions $f, g: A \subset \mathbb{R}^n \to \mathbb{R}^m$.

(ii) Answer the same question for uniform convergence.

Solution:

(i) For $x \in A$, we must show that $(f_k + g_k)(x) \to (f + g)(x)$. Given $\varepsilon > 0$, choose N_1 so that $k \geqslant N_1$ implies $\|f_k(x) - f(x)\| < \varepsilon/2$ and N_2 so that $k \geqslant N_2$ implies $\|g_k(x) - g(x)\| < \varepsilon/2$. Then let $N = \max(N_1, N_2)$ so that $k \geqslant N$ implies (by the triangle inequality)

$$\|(f_k + g_k)(x) - (f + g)(x)\| \leqslant \|f_k(x) - f(x)\| + \|g_k(x) - g(x)\| < \varepsilon.$$

(ii) Repeat the argument in (i) where each statement is to hold for all $x \in A$.

2. Prove that a sequence $f_k: A \to \mathbb{R}^n$ converges pointwise (uniformly) iff its components converge pointwise (uniformly).

Solution: The portion of the example on pointwise convergence follows from the fact that a sequence in $\mathbb{R}^m$ converges iff its components do (see Chapter 2). However, write out the argument again so its validity for uniform convergence can be seen.

Let $x = (x^1, \ldots, x^m) \in \mathbb{R}^m$. Then $|x^i| \leqslant \|x\| \leqslant \sum_{i=1}^m |x|$. Indeed, the first inequality is obvious and the second follows from the triangle inequality if we write $x = (x^1, 0, \ldots, 0) + (0, x^2, 0, \ldots, 0) + \cdots + (0, 0, \ldots, x^m)$.

Applied to $f_k = (f_k^1, \ldots, f_k^m)$, we have

$$|f_k^i(x) - f^i(x)| \leqslant \|f_k(x) - f(x)\| \leqslant \sum_{i=1}^{m} |f_k^i(x) - f^i(x)| .$$

Now if $f_k(x)$ is a Cauchy sequence for all x, so is $f_k^i(x)$ by the first inequality. Hence f_k converging pointwise implies that f_k^i converges pointwise. The same inequality and Theorem 2 show that if f_k converges uniformly, so does f_k^i.

Conversely, suppose $f_k^i(x)$ converges for each i and x. Choose N_i such that $k, l \geqslant N_i$ implies $|f_k^i(x) - f_l^i(x)| < \varepsilon/m$. Then if $N = \max(N_1, \ldots, N_m)$, $k, l \geqslant N$ implies $\|f_k(x) - f_l(x)\| < \varepsilon/m + \cdots + \varepsilon/m = \varepsilon$, so $f_k(x)$ converges.

For uniform convergence we repeat the argument with each statement holding for all $x \in A$.

3. Find an example of a sequence f_k converging uniformly to zero on $[0, \infty[$, with each $\int_0^\infty f_k(x)\, dx$ existing (that is, converging), but $\int_0^\infty f_k(x)\, dx \to +\infty$. Does this contradict Theorem 4?

Solution: Let

$$f_k(x) = \begin{cases} \dfrac{1}{k}, & \text{if } 0 \leqslant x \leqslant k^2, \\[2mm] 0, & \text{if } x > k^2 . \end{cases}$$

Then $f_k \to 0$ uniformly, since $|f_k(x)| \leqslant 1/k$ for all x. However,

$$\int_0^\infty f_k(x)\, dx = \frac{k^2}{k} = k \to \infty .$$

This does not contradict Theorem 4 because that theorem dealt with finite intervals.

4. (Dini's theorem). Let $A \subset \mathbb{R}^n$ be compact and f_k a sequence of continuous functions $f_k : A \to \mathbb{R}$ such that
(a) $f_k(x) > 0$ for $x \in A$;
(b) $f_k \to 0$ pointwise;
(c) $f_k(x) \leqslant f_l(x)$ whenever $k \geqslant l$.
Prove that $f_k \to 0$ uniformly.

Solution: This example requires a little care because we are trying to deduce uniform convergence from pointwise convergence plus some other hypotheses and we know that the result won't be true without these extra ones (study Figure 5-1, where all the hypotheses here are valid, except $f_k(0) \to 0$ as $k \to \infty$).

Given $\varepsilon > 0$ we want to find an N so that $|f_k(x)| < \varepsilon$ for all $k \geqslant N$ and all $x \in A$. For each $x \in A$, find N_x so that $|f_k(x)| < \varepsilon/2$ if $k \geqslant N_x$. We write N_x to emphasize that this number depends on x. Here we have used hypothesis (b). By continuity of $f_k(x)$ there is a neighborhood $U_{x,k}$ of x such that $|f_k(y) - f_k(x)| < \varepsilon/2$ for $y \in U_{x,k}$. The neighborhoods U_{x,N_x} form a covering of A, so by compactness there is a finite subcover, say centered at $x_1, \ldots, x_M$. Let $N = \max(N_{x_1}, \ldots, N_{x_M})$. Now let $x \in A$, $k \geqslant N$. Then $x \in U_{x_l, N_l}$ for some l, so $|f_{N_l}(x) - f_{N_l}(x_l)| < \varepsilon/2$. Thus, using (c),

$$0 \leqslant f_k(x) \leqslant f_N(x) \leqslant f_{N_l}(x) = f_{N_l}(x_l) + [f_{N_l}(x) - f_{N_l}(x_l)] < \frac{\varepsilon}{2} + \frac{\varepsilon}{2} = \varepsilon .$$

Therefore $f_k(x) < \varepsilon$ for $k \geqslant N$, $x \in A$ and so we have uniform convergence.

5. *Preamble.* Consider the alternating series $\sum_{n=1}^{\infty} (-1)^n/n$ which converges (see Theorem 14 above). However, we cannot rearrange the terms of the series, or else we may get divergence. In fact, the series $\sum (-1)^n/n$ can be rearranged to yield any desired sum! This was discovered by Riemann (see Exercise 17).

To be able to rearrange series, it is necessary to have absolute convergence. First, let us define a rearrangement. Let $\sum_{i=1}^{\infty} a_i$ be a series. A rearrangement is the series $\sum_{i=1}^{\infty} a_{\sigma(i)}$, where σ is a permutation of $\{1,2,3,\dots\}$, or more precisely, a bijection $\sigma: \{1,2,3,\dots\} \to \{1,2,3,\dots\}$.

Prove the following theorem.

Theorem. *Let $g_k \in \mathbb{R}^m$ and suppose $\sum_{k=1}^{\infty} g_k$ converges absolutely; that is, $\sum_{k=1}^{\infty} \|g_k\|$ converges. Then any rearrangement of the series $\sum_{k=1}^{\infty} g_k$ also converges absolutely, and to the same limit.*

Solution: Let $g_{\sigma(k)}$ be the rearranged series. Given $\varepsilon > 0$, there is an N such that $n \geq N$ implies

$$\|g_n\| + \cdots + \|g_{n+p}\| < \varepsilon .$$

Now choose an integer N_1 so that $\sigma(n) > N$ whenever $n > N_1$. (We can do this because there are only finitely many integers n for which $\sigma(n) \leq N$ because σ is a bijection). Thus if $n > N_1$, we have $\sigma(n + k) > N$, so by the above,

$$\|g_{\sigma(n)}\| + \cdots + \|g_{\sigma(n+p)}\| < \varepsilon .$$

By the Cauchy Criterion then, $\sum g_{\sigma(n)}$ converges absolutely (Theorem 10, Chapter 2).

To show that the limits are the same, given ε, select $N_2 > N$, where N is as above so that if $1 \leq n \leq N$, then $n = \sigma(k)$ for some k, $1 \leq k \leq N_2$. This is because such k are finite in number and σ is onto. Then let $N_0 = \max(N_1, N_2)$ and so for $m > N_0$,

$$\left\| \sum_{k=1}^{m} g_{\sigma(k)} - \sum_{n=1}^{\infty} g_n \right\| = \left\| \sum_{k=1}^{m} g_{\sigma(k)} - \sum_{n=1}^{N_0} g_n - \sum_{n=N_0+1}^{\infty} g_n \right\|$$

$$\leq \left\| \sum_{k=1}^{m} g_{\sigma(k)} - \sum_{n=1}^{N_0} g_n \right\| + \left\| \sum_{n=N_0+1}^{\infty} g_n \right\|$$

$$= \left\| \sum_{n=N_0+1}^{m} g_{\sigma(n)} \right\| + \left\| \sum_{n=N_0+1}^{\infty} g_n \right\|$$

$$< \varepsilon + \varepsilon = 2\varepsilon .$$

Here we have used the fact that

$$\sum_{n=1}^{\infty} g_n = \sum_{n=1}^{N_0} g_n + \sum_{n=N_0+1}^{\infty} g_n$$

and that

$$\sum_{k=1}^{m} g_{\sigma(k)} - \sum_{n=1}^{N_0} g_n = \sum_{n=N_0+1}^{m} g_{\sigma(n)} ,$$

which holds by construction of N_2.

Thus the series $\sum_{k=1}^{m} g_{\sigma(k)}$ converges to $\sum_{n=0}^{\infty} g_n$, which is the desired conclusion. The result of this example is closely related to important rearrangement theorems for double series (see Exercise 51).

Exercises for Chapter 5

1. (a) Let f_k be a sequence of functions from $A \subset \mathbb{R}^n$ to $\mathbb{R}^m$. Suppose there are constants m_k, such that $\|f_k(x) - f(x)\| \leqslant m_k$ for all $x \in A$ and that $m_k \to 0$. Prove that $f_k \to f$ uniformly.
   (b) If $m_k \to m \in \mathbb{R}$ and $\|f_k(x) - f_l(x)\| \leqslant |m_k - m_l|$ for all $x \in A$, then show that f_k converges uniformly.

2. Determine which of the following sequences converge (pointwise or uniformly). Check the continuity of the limit in each case.
   (a) $\dfrac{(\sin x)}{k}$ on $\mathbb{R}$.
   (b) $\dfrac{1}{(kx + 1)}$ on $]0,1[$.
   (c) $\dfrac{x}{(kx + 1)}$ on $]0,1[$.
   (d) $\dfrac{x}{(1 + kx^2)}$ on $\mathbb{R}$.
   (e) $\left(1, \dfrac{(\cos x)}{k^2}\right)$, a sequence of functions from $\mathbb{R}$ to $\mathbb{R}^2$.

3. Determine which of the following series $\sum_{k=1}^{\infty} g_k$ converge (pointwise or uniformly). Check the continuity of the limit in each case.
   (a) $g_k(x) = \begin{cases} 0, & x \leqslant k, \\ (-1)^k, & x > k. \end{cases} \quad g_k: \mathbb{R} \to \mathbb{R}$
   (b) $g_k(x) = \begin{cases} \dfrac{1}{k^2}, & |x| \leqslant k, \\ \dfrac{1}{x^2}, & |x| > k. \end{cases} \quad g_k: \mathbb{R} \to \mathbb{R}$
   (c) $g_k(x) = \left(\dfrac{(-1)^k}{\sqrt{k}}\right)\cos(kx)$ on $\mathbb{R}$.
   (d) $g_k(x) = x^k$ on $]0,1[$.

4. Let $f_n: [1,2] \to \mathbb{R}$ be defined by $f_n(x) = \dfrac{x}{(1 + x)^n}$.
   (a) Prove that $\sum_1^{\infty} f_n(x)$ is convergent for $x \in [1,2]$.
   (b) Is it uniformly convergent?
   (c) Is $\int_1^2 (\sum_1^{\infty} f_n(x))\, dx = \sum_1^{\infty} \int_1^2 f_n(x)\, dx$?

5. Suppose $f_k \to f$ uniformly, where $f_k: A \to \mathbb{R}$ and $g_k \to g$ uniformly where $g_k: A \to \mathbb{R}$ and there is a constant M_1 such that $\|g(x)\| \leqslant M_1$ for all x, and a constant M_2 such that $\|f(x)\| \leqslant M_2$ for all x. Then show that $f_k g_k \to fg$ uniformly. Find a counterexample if M_1 or M_2 does not exist. Are M_1 and M_2 necessary for pointwise convergence?

6. Prove that the sequence $f_k: A \to \mathbb{R}^m$ converges pointwise iff for each $x \in A$, $f_k(x)$ is a Cauchy sequence.

7. For functions $f: A \to \mathbb{R}$, form $\mathscr{C}_b$ as in the text. Show that we always have $\|fg\| \leqslant \|f\| \cdot \|g\|$. Discuss with examples.

8. Does pointwise convergence of continuous functions on a compact set to a continuous limit imply uniform convergence on that set?

9. Suppose $\sum_{k=1}^{\infty} g_k$ converges uniformly on A. If $x_k \to x_0$ in A, prove that

$$\sum_{n=1}^{\infty} g_n(x_k) \to \sum_{n=1}^{\infty} g_n(x_0) \ .$$

10. For the sequences and series of Exercises 2 and 3, when can we integrate or differentiate term by term?

11. (a) Must a contraction on a metric space have a fixed point? Discuss.
    (b) Let $f: X \to X$, where X is a complete metric space (such as $\mathbb{R}$) satisfying $d(f(x),f(y)) < d(x,y)$ for all $x, y \in X$. Must f have a fixed point? Discuss. What if X is compact?

12. A function $f: A \to \mathbb{R}$, $A \subset \mathbb{R}^n$ is called *lower semicontinuous* if whenever $x_0 \in A$ and $\lambda < f(x_0)$, there is a neighborhood U of x_0 such that $\lambda < f(x)$ for all $x \in U \cap A$. *Upper semicontinuity* is defined similarly.
    (a) Show that f is continuous iff it is both upper and lower semicontinuous.
    (b) If f_k are lower semicontinuous, $f_k \to f$ pointwise and $f_{k+1}(x) \geq f_k(x)$ then prove that f is lower semicontinuous.
    (c) In (b) show that f need not be continuous even if the f_k are continuous.
    (d) Let $f: [0,1] \to \mathbb{R}$ and let $g(x) = \sup_{\delta > 0} \inf_{|y-x| < \delta} f(y)$. Prove that g is lower semicontinuous.

13. In Theorem 5, show that $f_k \to f$ uniformly. [Hint: Use the mean-value theorem].

14. Let $f: X \to X$ be a contraction on a compact metric space X. Show that $\bigcap_{n=1}^{\infty} f^n(X)$ is a single point where $f^n = f \circ f \circ \cdots \circ f$ (n times). Is this true if $X = \mathbb{R}$?

15. Let $g_k \in \mathbb{R}^m$ and let f_k be a subsequence of g_k. Prove that if $\sum g_k$ converges absolutely, then $\sum f_k$ converges absolutely as well. Find a counterexample if $\sum g_k$ is just convergent.

16. Observe that in Example 5, the same argument applies in any normed space. Use this observation and the space $\mathscr{C}_b$ to prove the following:

    Theorem. *Let $g_k: A \subset \mathbb{R}^n \to \mathbb{R}^m$ be bounded, continuous, and suppose $\sum g_k$ converges uniformly and absolutely. Then any rearrangement also converges uniformly and absolutely, and to the same limit.*

17. Let $\sum_{n=0}^{\infty} a_n$ be a convergent, not absolutely convergent, real series. Given any number x, show that there is a rearrangement $\sum b_n$ of the series which converges to x. [Hint: Let p_n denote the nth positive term of a_n and $-q_n$ its nth negative term. Non-absolute convergence implies that both of these series $\sum p_n, \sum q_n$ diverge. Let $x_n = x - 1/n$ and $y_n = x + 1/n$. Choose k_1, so that $s_1 = p_1 + \cdots + p_{k_1} > x_1$ and l_1 so that $r_1 = p_1 + \cdots + p_{k_1} - q_1 - \cdots - q_{l_1} < y_1$. Then choose further terms so that $s_2 = p_1 + \cdots + p_{k_1} - q_1 - \cdots - q_{l_1} + p_{k_1} + \cdots + p_{k_2} > x_2$. Repeat this, obtaining a series with partial sums $s_1, r_1, s_2, r_2, \ldots$. Argue that we can choose, for k large enough, $x_k \leq s_k \leq y_k$ and $x_k \leq r_k \leq y_k$, from the fact that $p_n, q_n \to 0$. Show that this is the desired rearrangement.]

18. Give an example of a sequence of discontinuous functions f_k converging uniformly to a limit function f which is continuous.

19. Construct the function $g(x)$ by $g(x) = |x|$ if $x \in [-1/2, 1/2]$ and extend g so that it becomes periodic. Define

$$f(x) = \sum_{n=1}^{\infty} \frac{g(x4^{n-1})}{4^{n-1}}.$$

(a) Sketch g and the first few terms in the sum.
(b) Use the Weierstrass M-test to show f is continuous.
(c) Prove f is differentiable at *no* point [Hint: It would be helpful to consult Gelbaum and Olmsted, *Counterexamples in Analysis*, p. 38].

20. Prove that $\sum_{n=1}^{\infty} \left(\frac{\sin nx}{n^2} \right) x^3$ defines a continuous function on all of $\mathbb{R}$.

21. (a) Prove that if $A \subset \mathbb{R}^n$ is compact, $B \subset \mathscr{C}(A, \mathbb{R}^m)$ is compact $\Leftrightarrow B$ is closed, bounded, and equicontinuous. *Note:* One half of this, $\Leftarrow$, was proved in the text.
(b) Let $D = \{ f \in \mathscr{C}([0,1], \mathbb{R}) \mid \|f\| \leqslant 1 \}$. Show D is closed and bounded, but is not compact. Construct a sequence in D which is not equicontinuous and then make use of (a).

22. Let $B \subset \mathscr{C}(A, \mathbb{R}^m)$ and A compact. Suppose for each $x_0 \in A$, and $\varepsilon > 0$ there is a $\delta > 0$ such that $d(x, x_0) < \delta$ implies $d(f(x), f(x_0)) < \varepsilon$ for all $f \in B$. Prove B is equicontinuous.

23. Let $f \colon \mathbb{R} \to \mathbb{R}$ and suppose $f \circ f$ is continuous. Then must f be continuous?

24. A metric space X is called *second countable* if there is a countable collection $U_1, U_2, \ldots$ of open sets in X such that every open set in X is the union of members of this collection. Prove that such an X has a countable subset C such that $\mathrm{cl}(C) = X$. (We then say that X is *separable*). Prove conversely that a separable metric space is second countable.

25. Let $g \colon [0,1] \to \mathbb{R}$ be continuous and one-to-one. Show that g is either increasing or decreasing.

26. Let $k(x,y)$ be a continuous real-valued function on the square $U = \{(x,y) \mid 0 \leqslant x \leqslant 1, 0 \leqslant y \leqslant 1\}$ and assume $|k(x,y)| < 1$ for each $(x,y) \in U$. Let $A \colon [0,1] \to \mathbb{R}$ be continuous. Prove that there is a unique continuous real-valued function $f(x)$ on $[0,1]$ such that

$$f(x) = A(x) + \int_0^1 k(x,y) f(y) \, dy.$$

27. Let $f \colon \,]a,b[\,\to \mathbb{R}$ be uniformly continuous, and suppose that $x_n \to b$. Show that $\lim_{n \to \infty} f(x_n)$ exists.

28. Let $f_n(x) = x/n$. Is f_n uniformly convergent on $[0,396]$? On $\mathbb{R}$?

29. Discuss the uniform continuity of the following.

    (a) $f(x) = x^2$, $x \in]-1,1[$.

    (b) $f(x) = x^{1/3}$, $x \in [0,\infty[$.

    (c) $f(x) = e^{-x}$, $x \in [0,\infty[$.

    (d) $f(x) = x \sin\left(\dfrac{1}{x}\right)$, $0 < x \leqslant 1, f(0) = 0$

    (e) $f(x) = \sin[\ln(1 + x^3)]$, $-1 < x \leqslant 1, f(-1) = 0$.

30. Discuss and prove the statement "Every continuous function on a compact metric space is uniformly continuous."

31. Let a_n be a convergent sequence of real numbers, $a_n \to a$. Let $b_n = (a_1 + \cdots + a_n)/n$. Show $b_n \to a$ as well.

32. Discuss and prove the following. Let X and Y be metric spaces and $f : X \to Y$ continuous. Suppose $f(X)$ consists of two distinct points. Then prove X is not connected.

33. Let $f_n : [0,1] \to \mathbb{R}$ be a sequence of increasing functions on $[0,1]$ and suppose $f_n \to 0$ pointwise. Must f_n converge uniformly? What if f_n just converges pointwise to some limit f?

34. Find a sequence $f_n : [0,1] \to \mathbb{R}$ of differentiable functions such that $f_n \to 0$ uniformly, but such that $f'_n(1/2)$ does not converge to 0.

35. Let $f : \mathbb{R} \to \mathbb{R}$ be continuous and bijective. Show that f^{-1} is continuous (see Exercise 7, Chapter 4. For a generalization, see M. Hoffman, *Continuity of Inverse Functions*, Mathematica Magazine (not yet published).)

36. Let $f(x,y) = x^2 y/(x^4 + y^2)$. Discuss the behavior of f near $(0,0)$ with regard to the limits

    (a) $\displaystyle \lim_{(x,y) \to (0,0)} f(x,y)$,

    (b) $\displaystyle \lim_{x \to 0}[\lim_{y \to 0} f(x,y)]$,

    (c) $\displaystyle \lim_{y \to 0}[\lim_{x \to 0} f(x,y)]$.

37. Suppose $f : \mathbb{R} \to \mathbb{R}$ is continuous and $f(1) = 7$. Suppose $f(x)$ is rational for all x. Prove f is constant.

38. Prove $1 + 1/2 + 1/4 + 1/8 + \cdots$ converges and $1 - 1/2 + 1/3 - 1/4 + \cdots$ converges, but not absolutely.

39. A function $g : [0,1] \to \mathbb{R}$ is called *simple* if we can divide up $[0,1]$ into subintervals on which g is constant, except perhaps at the end points. Let $f : [0,1] \to \mathbb{R}$ be continuous and $\varepsilon > 0$. Prove there is a simple function g such that $\|f - g\| < \varepsilon$.

40. (a) Define $\delta : \mathscr{C}([0,1],\mathbb{R}) \to \mathbb{R}, f \mapsto f(0)$. Prove δ is continuous and is linear.

    (b) Let $g : \mathbb{R} \to \mathbb{R}$ be continuous. Define $F : \mathscr{C}([0,1],\mathbb{R}) \to \mathscr{C}([0,1],\mathbb{R})$ by $F(f) = g \circ f$. Prove that F is continuous; prove that if g is uniformly continuous then F is uniformly continuous.

41. Show that there is a polynomial $p(x)$ such that $|p(x) - |x|^3| < 1/10$ for $-1000 \leqslant x \leqslant 1000$.

42. Study the possibility of replacing the sequence of Bernstein polynomials in Theorem 11 by a sequence of Lagrange interpolation polynomials (see Exercise 2, Section 5.7 for the definition and properties) to effect the proof of the theorems in Section 5.7.

43. Let $\mathscr{C}_e([-1,1],\mathbb{R})$ denote the set of even functions in $\mathscr{C}([-1,1],\mathbb{R})$.
    (a) Show $\mathscr{C}_e$ is closed and not dense in $\mathscr{C}$.
    (b) Show the even polynomials are dense in $\mathscr{C}_e$, but not in $\mathscr{C}$.

44. *Projects:* Examine the possibility of extending the Stone-Weierstrass theorem to
    (a) complex valued functions (keep the same hypotheses on $\mathscr{B}$ except add "$f \in \mathscr{B}$ implies $\bar{f} \in \mathscr{B}$" (overbar) denoting the complex conjugate);
    (b) non compact domains (consult Simmons, *Introduction to Topology and Modern Analysis*);
    (c) Use (b) to study the density of the Hermite functions in a suitable space of continuous functions (the Hermite functions are defined and studied in, for example, Courant-Hilbert, *Methods of Mathematical Physics, I*).

45. Let $f(t,x)$ be defined and continuous for $a \leqslant t \leqslant b$ and $x \in \mathbb{R}^n$. The purpose of this exercise is to show that the problem $dx/dt = f(t,x)$, $x(a) = x_0$ has a solution on an interval $t \in [a,c]$ for some $c > a$ (it is unique only under more stringent conditions). Perform the operations as follows: divide $[a,b]$ into n parts $t_0 = a, \ldots, t_n = b$, and define a continuous function x_n by

$$\begin{cases} x_n'(t) = f(t_i,x_n(t_i)), & t_i < t < t_{i+1}, \\ x_n(a) = x_0. \end{cases}$$

Put $\Delta_n(t) = x_n'(t) - f(t,x_n(t))$ so that

$$x_n(t) = x_0 + \int_0^t f(s,x_n(s)) + \Delta_n(s)\, ds.$$

Use the Arzela-Ascoli theorem to pull out a convergent subsequence of the x_n. This method is called *polygonal approximation;* compare with Sections 6 and 7.5.

46. (a) Let $f_n: \mathbb{R}^p \to \mathbb{R}^q$ be a sequence of equicontinuous functions on a compact set K converging pointwise. Prove that the convergence is uniform.
    (b) Let

$$f_n(x) = \frac{x^2}{[x^2 + (1 - nx)^2]}, \qquad 0 \leqslant x \leqslant 1.$$

Show that f_n converges pointwise but not uniformly. What can you conclude from (a)?

47. Let $f_n: K \subset A \to \mathbb{R}^m$ be a sequence of equicontinuous functions. Suppose that f_n converges on a dense subset of A. Prove that the sequence converges on all of A. Does this shed any light on the proof of Theorem 9?

48. Prove that the norm on $\mathscr{C}([0,1],\mathbb{R})$ is not derived from an inner product $\langle,\rangle$ by $\|f\| = \sqrt{\langle f,f\rangle}$, as the norm on $\mathbb{R}^n$ is. (An inner product on a vector space S is a

function $\langle,\rangle: S \times S \to \mathbb{R}$ satisfying the property in Theorem 5, Chapter 1.) [Hint: Show that the property in Exercise 12a, Chapter 1 fails, and note that this property follows only from the fact that the norm on $\mathbb{R}^n$ derives from an inner product satisfying the properties in Theorem 5, Chapter 1.]

49. Let S be a set and let $\mathscr{B}$ denote the set of *all* bounded real valued functions on S; endow $\mathscr{B}$ with the sup norm. Prove that $\mathscr{B}$ is a Banach space.

50. Let $f: \mathbb{R} \to \mathbb{R}$ be a uniform limit of polynomials. Prove that f is a polynomial.

51. Consider a double series

$$\sum_{m,n=0}^{\infty} a_{mn} \quad \text{where } a_{mn} \in \mathbb{R}, m, n = 0, 1, 2, \ldots.$$

Say that it converges to S if for any $\varepsilon > 0$ there is an N such that $n, m \geq N$ implies

$$\left| \sum_{k,l=0}^{m,n} a_{k,l} - S \right| < \varepsilon.$$

Define absolute convergence in the obvious way. Prove that if $\sum_{n,m=0}^{\infty} a_{nm}$ is absolutely convergent, then the sum can be rearranged as follows:

$$\sum_{n,m=0}^{\infty} a_{nm} = \sum_{n=0}^{\infty} \left(\sum_{m=0}^{\infty} a_{nm} \right) = \sum_{m=0}^{\infty} \left(\sum_{n=0}^{\infty} a_{nm} \right).$$

Interpret this result in terms of summing entries in an infinite matrix by rows and columns.

52. Can we differentiate the series

$$x = \sum_{k=1}^{\infty} \left(\frac{x^k}{k} - \frac{x^{k+1}}{k+1} \right), \quad 0 \leq x \leq 1$$

term by term?

53. Evaluate the following limits:

(i) $\lim_{x \to 0} \dfrac{1 - \cos x}{3^x - 2^x}$

(ii) $\lim_{x \to 0+} (1 + \sin 2x)^{1/x}$

(iii) $\lim_{x \to 0+} \dfrac{1}{\sin x} - \dfrac{1}{x}.$

54. Test the following infinite series for convergence or divergence:

(i) $\sum_{k=1}^{\infty} \dfrac{\sqrt{k} \log k}{k^2 + 2k + 3}$

(ii) $\sum_{k=1}^{\infty} \dfrac{k! \, 3^k}{k^k}$

(iii) $\sum_{k=1}^{\infty} \dfrac{(k!)^2}{(2k)!}.$

55. Prove that $\pi/4 = 1 - 1/3 + 1/5 - 1/7 + \cdots$ starting from

$$(1 + x^2)^{-1} = \sum_{k=0}^{\infty} (-1)^k x^{2k}, \qquad |x| < 1 .$$

56. Test the following series for absolute and conditional convergence:

   (a) $\displaystyle\sum_{n=1}^{\infty} (-1)^n n^{-\alpha}, \qquad \alpha$ real

   (b) $\displaystyle\sum_{k=3}^{\infty} \frac{(-1)^k \log k}{k \log \log k}$

   (c) $\displaystyle\sum_{k=2}^{\infty} \frac{(-1)^k}{k^\alpha + (-1)^k}, \qquad \alpha > 0 .$

57. Prove that if
   (a) $f_n(x), g(x)$ continuous, $\quad 0 \leqslant x < \infty$
   (b) $|f_n(x)| \leqslant g(x), \qquad n = 1, 2, 3, \ldots \qquad 0 \leqslant x < \infty$
   (c) $f_n(x) \to f(x)$ uniformly, $\quad 0 \leqslant x \leqslant R$, for any $R < \infty$ and
   (d) $\int_0^\infty g(x)\, dx \quad < \infty$
   then $\displaystyle\lim_{n \to \infty} \int_0^\infty f_n(x)\, dx = \int_0^\infty f(x)\, dx.$

58. Prove the following convergence tests (see, for example, Exercise 49, p. 60).

   (a) $u_n > 0, \dfrac{u_{n+1}}{u_n} \leqslant 1 - \dfrac{1}{n} - \dfrac{\alpha}{n \log n}, \qquad \alpha > 1$
   $\Rightarrow \sum u_n$ converges,

   (b) $u_n > 0, \dfrac{u_{n+1}}{u_n} \geqslant 1 - \dfrac{1}{n} - \dfrac{1}{n \log n}$
   $\Rightarrow \sum u_n$ diverges.

59. (a) Let $p > 1$ with $1/p + 1/q = 1$. For $a, b, t > 0$ prove that

$$ab \leqslant \frac{a^p t^p}{p} + \frac{b^q t^{-q}}{q}$$

and that ab is the minimum value of the right side. (One way to prove this is to use elementary calculus.)

   (b) Prove *Hölder's inequality:* $a_k, b_k \geqslant 0, p > 1, 1/p + 1/q = 1$

$$\Rightarrow \sum_1^n a_k b_k \leqslant \left(\sum_1^n a_k^p \right)^{1/p} \left(\sum_1^n b_k^q \right)^{1/q} .$$

[Hint: Imitate the proof of the Cauchy-Schwarz inequality, using part (a).]

   (c) Prove *Minkowski's inequality:* $a_k, b_k \geqslant 0, p > 1$

$$\Rightarrow \left(\sum_1^n (ca_k + b_k)^p \right)^{1/p} \leqslant \left(\sum_1^n a_k^p \right)^{1/p} + \left(\sum_1^n b_k^p \right)^{1/p} .$$

[Hint: Write

$$\sum_1^n (a_k + b_k)^p = \sum_1^n (a_k + b_k)^{p-1} a_k + \sum_1^n (a_k + b_k)^{p-1} b_k ,$$

and use Hölder's inequality in a clever way.]

60. The series $\sum_{k=1}^{\infty} x^k/k$ converges if $|x| < 1$. For which complex x with $|x| = 1$ does it converge?

61. Let $\sum a_k x^k$ have radius of convergence R. Show $\sum a_k(x - b)^k$ converges inside the disc of center b, radius R.

62. Find the radis of convergence:

$$\sum x^k k/(k + 1), \qquad \sum x^k/\log k .$$

63. (Binomial series.) Consider

$$\sum_{k=0}^{\infty} \frac{\alpha(\alpha - 1) \cdots (\alpha - k + 1)}{k!} x^k .$$

Assume α is not an integer $\geqslant 0$. Show the radius of convergence is $R = 1$. (See Exercise 49, Chapter 2 on the hypergeometric series for behavior of the series at $x = \pm 1$.)

64. Does $1 + 1/2 + 1/3 + \cdots$ converge (C,1) or (Abel)?

65. Let $f(x)$ be continuous, $0 \leqslant x < \infty$. We normally define

$$\int_0^{\infty} f(x) \, dx = \lim_{R \to +\infty} \int_0^R f(x) \, dx ,$$

if the limit exists. By analogy with (C,1) summability, define a notion of "(C,1) *integrability* from 0 to ∞," and prove that your method of integrability is *regular*, that is, agrees with the usual $\int_0^{\infty}$ if the latter converges.

66. Define t_n inductively by $t_1 = 1$, and $t_{n+1} = t_n/(1 + t_n^{\beta})$ where β is fixed, $0 \leqslant \beta < 1$. Prove that $\sum_{n=1}^{\infty} t_n$ converges. [Hint: Try to show that there is a constant C such that $t_n \leqslant C/n^{1/\beta}$.]

67. Let $A = \{j/2^n \in [0,1] \mid n = 1,2,3,\ldots,j = 0,1,2,\ldots,2^n\}$ and let $f: A \to \mathbb{R}$ satisfy the following condition: there is a sequence $\varepsilon_n > 0$ with $\sum_{n=1}^{\infty} \varepsilon_n < \infty$ and

$$\left| f\left(\frac{j-1}{2^n}\right) - f\left(\frac{j}{2^n}\right) \right| < \varepsilon_n \text{ for all } n > 0, \qquad j = 1, 2, \ldots, 2^n.$$

Prove that f has a unique extension to a continuous function from $[0,1]$ to $\mathbb{R}$. [Hint: Show that $|f(t_1) - f(t_2)| \leqslant 2 \sum_{n=N}^{\infty} \varepsilon_n$ if $|t_1 - t_2| \leqslant 1/2^N$ and apply Exercise 24, Chapter 4.]

68. Let $A \subset \mathbb{R}^n$ be compact and let $B \subset \mathscr{C}(A,\mathbb{R}^m)$ be compact. Prove that B is equicontinuous as follows:
    (a) Prove that the map $E: \mathscr{C}(A,\mathbb{R}^m) \times A \to \mathbb{R}^m$ $(f,x) \mapsto f(x)$ is continuous;
    (b) use uniform continuity of E restricted to $B \times A$ to deduce the result.
    (This method of proof is due to J. Allen.)

APPENDIX TO CHAPTER 5:
When can $\mathbb{R}^n$ be replaced by "metric space"?
by R. Gulliver

In this book we have concentrated much of our attention on concrete metric spaces, especially $\mathbb{R}^n$. The question naturally arises, how general are the results we have obtained? In many exercises we have already asked the reader to verify that some results hold in general metric spaces (see for example p. 100). In the table below are gathered together some of the important results, (including some not formally stated as theorems in the text) and the general contexts in which they are valid are stated. The proofs are, in almost every case, the same as those given in the text. The reader should pick out some of these theorems and verify that this generalization is indeed valid.

Theorem	Valid in Metric spaces?
Chapter 2	
Theorem 1: For all $\varepsilon > 0$ and $x \in \mathbb{R}^n$, $D(x,\varepsilon)$ is open.	Yes.
Theorem 2: (i) the intersection of a finite number of open sets is open; (ii) the union of any collection of open sets is open.	Yes.
Theorem 3: (reverse of Theorem 2 for closed sets).	Yes.
Theorem 4: $A \subset \mathbb{R}^n$ is closed iff all accumulation points of A in $\mathbb{R}^n$ belong to A.	Yes.
Theorem 5: $cl(A)$ consists of A plus all its accumulation points in $\mathbb{R}^n$.	Yes.
Theorem 6: $x \in bd(A)$ iff every neighborhood of x in $\mathbb{R}^n$ contains points of A and points of $\mathbb{R}^n \backslash A$.	Yes.
Theorem 7: $x_k \to x$ iff for all $\varepsilon > 0$ there exists N such that if $k > N$ then $\|x_k - x\| < \varepsilon$.	Yes.
Theorem 8: $x_k, x \in \mathbb{R}^n$: $x_k \to x$ iff each sequence of components of x_k converges to the corresponding component of x.	Meaningless in a general metric space.
Theorem 9: $A \subset \mathbb{R}^n$ is closed iff for all sequences $\{x_k\}$, $x_k \in A$ which converge in $\mathbb{R}^n$, the limit is in A.	Yes.
Theorem 10: A sequence $\{x_k\}$ in $\mathbb{R}^n$ converges iff it is a Cauchy sequence.	$\Rightarrow$ yes. $\Leftarrow$ is the definition of a complete metric space;
Theorem 11: For $x_k \in \mathbb{R}^n$: $\sum x_k$ converges iff for all $\varepsilon > 0$ there exists N such that if $k \geqslant N$ and $p \geqslant 0$ then $\|x_k + x_{k+1} + \cdots + x_{k+p}\| < \varepsilon$.	Valid in complete normed space ($=$ Banach space).

Theorem	Valid in Metric spaces?
Theorem 12: $x_k \in \mathbb{R}^n$: If $\sum \|x_k\|$ converges in $\mathbb{R}$ then $\sum x_k$ converges in $\mathbb{R}^n$.	Valid in Banach space.
Theorem 13: (iv) If $\lim\limits_{k \to \infty} \|x_{k+1}\|/\|x_k\|$ exists and is < 1 then $\sum x_k$ converges. (Also (v) is valid).	Valid in Banach space.
Baire Category Theorem: The intersection of a countable number of dense open subsets of $\mathbb{R}^n$ is dense in $\mathbb{R}^n$.	Valid in complete metric space.
Theorem: $\mathbb{R}^n$ has a countable dense subset.	This defines a "separable" metric space; not always true. However, $\mathscr{C}(A,\mathbb{R}^m)$ is separable, for $A \subset \mathbb{R}^n$ compact (prove this using the Stone-Weierstrass theorem).

Chapter 3

Theorem 1: The following are equivalent for $A \subset \mathbb{R}^n$: (i) A is closed and bounded. (ii) A has the Heine-Borel property. (iii) A has the Bolzano-Weierstrass property.	No! However, (ii) and (iii) are equivalent, and each implies (i). If A has (ii), we call it *compact*.
Theorem 2: $\{F_k\}$ a sequence of non-empty compact subsets of $\mathbb{R}^n$ with $F_{k+1} \subset F_k$. Then $\bigcap_{k=1}^{\infty} F_k$ is non-empty.	Yes (using the above definition of compact).
Theorem 3: If A is path-connected then it is connected.	Yes.
Theorem: If A is open $\subset \mathbb{R}^n$ and A is connected, then it is path-connected.	In a normed linear space.
Proposition: $\tilde{A}$ a closed subset of A, A compact $\Rightarrow \tilde{A}$ is compact.	Yes.
Proposition: A a closed subset of $\mathbb{R}^n$, $x \notin A \Rightarrow$ there exists $y \in A$ with $d(x,y) = \inf\{d(x,z) \mid z \in A\}$	No!

Chapter 4

Theorem 1: For $f: A \to \mathbb{R}^m$, $A \subset \mathbb{R}^n$, these are equivalent: (i) f is continuous on A. (ii) For each sequence $x_k \to x$, $x_k \in A$, $x \in A$, there holds $f(x_k) \to f(x)$. (iii) For all open sets $U \subset \mathbb{R}^m$, $f^{-1}(U)$ is a relatively open subset of A. (iv) For all closed sets $K \subset \mathbb{R}^m$, $f^{-1}(K)$ is a relatively closed subset of A.	Yes (replace A by one metric space, $\mathbb{R}^m$ by another metric space)

(continued)

Theorem	Valid in Metric spaces?

Theorem 2: $A \subset \mathbb{R}^n$ and $f: A \to \mathbb{R}^m$ continuous. Then
  (i) If $K \subset A$ is connected, then $f(K)$ is connected.
  (ii) If $K \subset A$ is compact, then $f(K)$ is compact.
	Yes.

Theorem 3: $A \subset \mathbb{R}^n$, $f: A \to \mathbb{R}^m$; $B \subset f(A) \subset \mathbb{R}^m$, $g: B \to \mathbb{R}^p$. If f and g are continuous then $g \circ f: A \to \mathbb{R}^p$ is also continuous.	Yes.
Theorem 4: Sums and scalar products of continuous functions are again continuous.	In a normed space.
Theorem 5: $A \subset \mathbb{R}^n$ compact, $f: A \to \mathbb{R}$ continuous. Then $f(A)$ is bounded and contains its sup and inf.	Yes.
Theorem 6: $A \subset \mathbb{R}^n$ connected, $f: A \to \mathbb{R}$ continuous. For any x, $y \in A$ and $c \in \mathbb{R}$ with $f(x) < c < f(y)$, there exists $z \in A$ such that $f(z) = c$.	Yes.
Theorem 7 (Heine's Theorem): $A \subset \mathbb{R}^n$ compact, $f: A \to \mathbb{R}^m$ continuous. Then f is uniformly continuous on A.	Yes.
Chapter 5	
Theorem 1: $f_k \to f$ uniformly, $f_k, f: A \to \mathbb{R}^m: A \subset \mathbb{R}^n$. If each f_k is continuous then f is continuous.	Yes.
Theorem 3 (Weierstrass M-test): $A \subset \mathbb{R}^n$ $g_k: A \to \mathbb{R}^m$, $\|g_k\|_{\sup} \leqslant M_k$ and $\sum M_k$ converges. Then $\sum g_k$ converges uniformly.	A may be any metric space; $\mathbb{R}^m$ must be replaced by a Banach space.
Theorem 8: For $A \subset \mathbb{R}^n$, $\mathscr{C}_b(A, \mathbb{R}^m)$ is a Banach space.	A any metric space; $\mathbb{R}^m$ must be a Banach space.
Theorem 9 (Arzela-Ascoli): $A \subset \mathbb{R}^n$ compact, $B \subset \mathscr{C}(A, \mathbb{R}^m)$. B is compact iff B is closed, bounded, and equicontinuous.	A may be any compact metric space, but $\mathbb{R}^m$ must be $\mathbb{R}^m$.
Theorem 12 (Stone-Weierstrass): $A \subset \mathbb{R}^n$ compact, $B \subset \mathscr{C}(A, \mathbb{R})$. If B is an algebra which separates points and if the constant functions are included in B, then B is dense.	A may be any compact metric space.

Further results on metric spaces:

Theorem: If X is a complete metric space, A a closed subset of X, then A is a complete metric space.

Definition: A metric space X is *totally bounded* if for all $\varepsilon > 0$ there exists a finite set $\{x_1, \ldots, x_n\} \subset X$ such that $X \subset \bigcup_{i=1}^{n} D(x_i, \varepsilon)$.

Theorem: Let X be a metric space. X is compact iff X is complete and totally bounded.

Chapter 6

Differentiable Mappings

In this chapter we shall discuss the notion of a differentiable map from $\mathbb{R}^n$ to $\mathbb{R}^m$. We shall start right in with the general case since the reader should have some familiarity with the derivative for functions of one variable. Pertinent facts from one variable calculus will be brought in as they are needed.

Starting with this chapter a certain amount of linear algebra will be used. In particular the student should now review the notion of a linear transformation and its matrix representation.* We shall be defining the derivative as a linear mapping; the connection with partial derivatives will be found in Section 6.2. After this we will generalize the usual theorems of calculus to the multivariable case (such as differentiability implies continuity, the chain rule, mean-value theorem, Taylor's theorem, tests for extrema, and so forth).

6.1   Definition of the Derivative

For a function of one variable $f: \,]a,b[\,\to \mathbb{R}$ we recall that f is called *differentiable at* $x_0 \in \,]a,b[$ if the limit

$$f'(x_0) = \lim_{h \to 0} \frac{f(x_0 + h) - f(x_0)}{h}$$

exists. We recall that one also writes df/dx for $f'(x)$. Equivalently, we may

---

* See for example, M. O'Nan, *Linear Algebra*, Harcourt Brace, Jovanovich, (1971).

write the above formula as

$$\lim_{h \to 0} \frac{f(x_0 + h) - f(x_0) - f'(x_0)h}{h} = 0$$

that is,

$$\lim_{x \to x_0} \frac{f(x) - f(x_0) - f'(x_0)(x - x_0)}{x - x_0} = 0$$

or, what is the same,

$$\lim_{x \to x_0} \frac{|f(x) - f(x_0) - f'(x_0)(x - x_0)|}{|x - x_0|} = 0 \ .$$

We recall that this number $f'(x_0)$ represents the slope of the line tangent to the graph of f at the point $(x_0, f(x_0))$. See Figure 6-1.

To generalize this notion to maps $f: A \subset \mathbb{R}^n \to \mathbb{R}^m$ we make the following definition.

Definition 1.   A map $f: A \subset \mathbb{R}^n \to \mathbb{R}^m$ is said to be *differentiable* at $x_0 \in A$ if there is a linear function, denoted $Df(x_0): \mathbb{R}^n \to \mathbb{R}^m$ and called the *derivative* of f at x_0, such that

$$\lim_{x \to x_0} \frac{\| f(x) - f(x_0) - Df(x_0)(x - x_0) \|}{\| x - x_0 \|} = 0 \ .$$

Here, $Df(x_0)(x - x_0)$ denotes the value of the linear map $Df(x_0)$ applied to the vector $x - x_0 \in \mathbb{R}^n$, so $Df(x_0)(x - x_0) \in \mathbb{R}^m$. We shall often write $Df(x_0) \cdot h$ for $Df(x_0)(h)$. (In this definition, as usual,

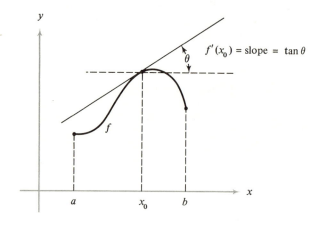

FIGURE 6-1

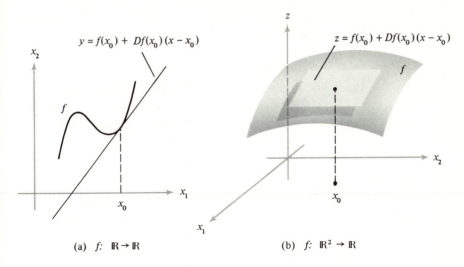

(a) $f: \mathbb{R} \to \mathbb{R}$ (b) $f: \mathbb{R}^2 \to \mathbb{R}$

FIGURE 6-2   (a) $f: \mathbb{R} \to \mathbb{R}$. (b) $f: \mathbb{R}^2 \to \mathbb{R}$.

we exclude $x = x_0$ in taking the limit, since we are dividing by $\|x - x_0\|$, and take the limit through those $x \in A$).

More explicitly, it may be rewritten by saying that for every $\varepsilon > 0$ there is a $\delta > 0$ such that $x \in A$, and $\|x - x_0\| < \delta$ implies

$$\|f(x) - f(x_0) - Df(x_0)(x - x_0)\| \leqslant \varepsilon \|x - x_0\| \; .$$

In this formulation we can allow $x = x_0$ since then both sides reduce to zero.

Intuitively, $x \mapsto f(x_0) + Df(x_0)(x - x_0)$ is supposed to be the *best affine approximation** to f near the point x_0. See Figure 6-2. In this figure we have indicated the equations of the tangent planes to the graph of f.

If f is differentiable at each point of A, we just say f is *differentiable* on A. We expect intuitively (as in Figure 6-2) that there can be only one best linear approximation. This is in fact true if we assume that A is an open set. If we compare the definitions of $Df(x)$ and $df/dx = f'(x)$, we see that $Df(x)(h) = f'(x) \cdot h$ (the product of the numbers $f'(x)$ and $h \in \mathbb{R}$). Thus the linear map $Df(x)$ is just multiplication by df/dx.

Theorem 1. *Let A be an open set in $\mathbb{R}^n$ and suppose $f: A \to \mathbb{R}^m$ is differentiable at x_0. Then $Df(x_0)$ is uniquely determined by f.*

EXAMPLE 1. Let $f: \mathbb{R} \to \mathbb{R}$, $f(x) = x^3$. Compute $Df(x)$ and df/dx.

---

* An affine mapping is a linear mapping plus a constant.

Solution: In this case we know from elementary calculus (or from rules developed below) that $dx^3/dx = 3x^2$. Thus in this example $Df(x)$ is the linear mapping

$$h \mapsto Df(x) \cdot h = 3x^2h .$$

EXAMPLE 2.   Show that, in general, Df is not uniquely determined.

Solution: For example, if $A = \{x_0\}$ is a single point any $Df(x_0)$ will do, because $x \in A$, $\|x - x_0\| < \delta$ holds only when $x = x_0$, in which case the expression

$$\|f(x) - f(x_0) - Df(x_0)(x - x_0)\|$$

is zero. The definition is then fulfilled in a trivial way.

Note: If the proof of Theorem 1 is examined closely one sees that $Df(x)$ is unique (assuming it exists) on a wider range of sets than open sets. For example, the theorem is valid for closed intervals in $\mathbb{R}$ or generally for closed discs in $\mathbb{R}^n$.

At this point it is convenient to recall some facts about derivatives of functions of one variable. Specifically, recall the logical steps leading up to the important mean-value theorem. We shall shortly be generalizing these ideas to functions of several variables.

Fact 1.   *If f: $]a,b[\to \mathbb{R}$ is differentiable at $c \in]a,b[$ and f has a maximum (respectively minimum) at c, then $f'(c) = 0$.*

Proof: Let f have a maximum at c. Then for $h \geqslant 0$, $[f(c + h) - f(c)]/h \leqslant 0$, and so letting $h \to 0$, $h \geqslant 0$ we get $f'(c) \leqslant 0$. Similarly for $h \leqslant 0$ we obtain $f'(c) \geqslant 0$. Hence $f'(c) = 0$. ∎

The reader should be familiar with the geometric significance of this result.

Fact 2.   *(Rolle's Theorem).   If f: $[a,b] \to \mathbb{R}$ is continuous, f is differentiable on $]a,b[$ and $f(b) = f(a) = 0$, then there is a number $c \in]a,b[$ such that $f'(c) = 0$.*

Proof: If $f(x) = 0$ for all $x \in [a,b]$ we can choose any c. So assume f is not identically zero. From Chapter 4, we know that there is a point c_1 where f assumes its maximum and a point c_2 where f assumes its minimum. By our assumption and the fact that $f(a) = f(b) = 0$, at least one of c_1, c_2 lies in $]a,b[$. If $c_1 \in]a,b[$ we get $f'(c_1) = 0$ by Fact 1; similarly for c_2. ∎

Fact 3.   *(Mean-Value Theorem).   If f: $[a,b] \to \mathbb{R}$ is continuous and differentiable on $]a,b[$, there is a point $c \in]a,b[$ such that $f(b) - f(a) = f'(c)(b - a)$.*

Proof: Let $\varphi(x) = f(x) - f(a) - (x - a)[f(b) - f(a)]/(b - a)$ (see Figure 6-3) and apply Rolle's theorem. ∎

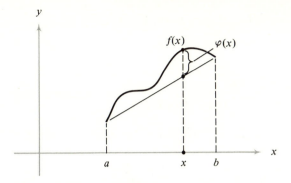

FIGURE 6-3

Corollary. *If, in addition, $f' = 0$ on $]a,b[$, then f is constant.*

Proof: Applying Fact 3 to f on $[a,x]$ we have $f(x) - f(a) = f'(c)(x - a) = 0$, so $f(x) = f(a)$ for all $x \in [a,b]$, and therefore f is constant.

The list of basic theorems continues to include differentiability implies continuity, sum rule for derivatives, quotient rule, chain rule, and Taylor's theorem. These will all be dealt with below in the general case of functions of several variables, but the reader may wish to review the one variable case first.

EXAMPLE 3. Let $f: \ ]a,b[\ \to \mathbb{R}$ be differentiable and $|f'(x)| \leqslant M$. Prove that $|f(x) - f(y)| \leqslant M \, |x - y|$ for all $x, y \in \ ]a,b[$.

Solution: By the mean-value theorem,

$$f(x) - f(y) = f'(c)(x - y)$$

for some $c \in \ ]x,y[$. Taking absolute values gives the result.

Exercises for Section 6.1

1. Compute $Df(x)$ for $f: \mathbb{R} \to \mathbb{R}, \ f(x) = x \sin x$.

2. Prove that $D(f + g) = Df + Dg$.

3. Let $A = \{(x,y) \in \mathbb{R}^2 \mid 0 \leqslant x \leqslant 1, y = 0\}$. Prove that the conclusion of Theorem 1 is false for this A. [Hint: Take, for example, $f(x,y) = 0$ and show that $Df(x,y) = 0$ and $Df(x,y)(h,k) = k$ both satisfy the definition.]

4. Let $f: \mathbb{R}^n \to \mathbb{R}^m$ and suppose there is a constant M such that for $x \in \mathbb{R}^n$, $\|f(x)\| \leqslant M \, \|x\|^2$. Prove f is differentiable at $x_0 = 0$ and that $Df(x_0) = 0$.

5. If $f: \mathbb{R} \to \mathbb{R}$ and $|f(x)| \leqslant |x|$, must $Df(0) = 0$?

6. Does the mean-value theorem apply to $f(x) = \sqrt{x}$ on $[0,1]$? Does it apply to $g(x) = \sqrt{|x|}$ on $[-1,1]$?

6.2   Matrix Representation

In addition to the above, there is another way to differentiate a function f of several variables. We can write it in component form $f(x_1, \ldots, x_n) = (f_1(x_1, \ldots, x_n), \ldots, f_m(x_1, \ldots, x_n))$ and compute the partial derivatives, $\partial f_j / \partial x_i$ for $j = 1, \ldots, m$ and $i = 1, \ldots, n$, where the symbol $\partial f_j / \partial x_i$ means that we compute the usual derivative of f_j with respect to x_i while keeping the other variables $x_1, \ldots, x_{i-1}, x_{i+1}, \ldots, x_n$ fixed. Explicitly, Definition 2 follows.

> **Definition 2.** $\partial f_j / \partial x_i$ is given by the following limit, when the latter exists:
>
> $$\frac{\partial f_j}{\partial x_i}(x_1, \ldots, x_n) = \lim_{h \to 0}\left\{\frac{f_j(x_1, \ldots, x_i + h, \ldots, x_n) - f_j(x_1, \ldots, x_n)}{h}\right\}.$$

In Section 6.1 we saw that $Df(x)$ for $f: \mathbb{R} \to \mathbb{R}$ is just the linear map multiplication by df/dx. This fact, which was obvious from the definitions, can be generalized to the following theorem.

> **Theorem 2.**   Suppose $A \subset \mathbb{R}^n$ is an open set and $f: A \to \mathbb{R}^m$ is differentiable. Then the partial derivatives $\partial f_j / \partial x_i$ exist, and the matrix of the linear map $Df(x)$ with respect to the standard bases in $\mathbb{R}^n$ and $\mathbb{R}^m$ is given by
>
> $$\begin{pmatrix} \dfrac{\partial f_1}{\partial x_1} & \dfrac{\partial f_1}{\partial x_2} & \cdots & \dfrac{\partial f_1}{\partial x_n} \\[2mm] \dfrac{\partial f_2}{\partial x_1} & \dfrac{\partial f_2}{\partial x_2} & \cdots & \dfrac{\partial f_2}{\partial x_n} \\[2mm] \cdot & \cdot & & \cdot \\ \cdot & \cdot & & \cdot \\ \cdot & \cdot & & \cdot \\[2mm] \dfrac{\partial f_m}{\partial x_1} & \dfrac{\partial f_m}{\partial x_2} & \cdots & \dfrac{\partial f_m}{\partial x_n} \end{pmatrix}$$
>
> where each partial derivative is evaluated at $x = (x_1, \ldots, x_n)$. This matrix is called the Jacobian matrix of f.

In doing practical computations one can usually compute the Jacobian matrix easily and Theorem 2 then gives us Df. In some books, Df is called the *differential* or the *total derivative* of f.

One should take special note when $m = 1$, in which case we have a real-valued function of n variables. Then Df has the matrix

$$\left(\frac{\partial f}{\partial x_1} \quad \cdots \quad \frac{\partial f}{\partial x_n} \right)$$

and the derivative applied to a vector $e = (a_1, \ldots, a_n)$ is

$$Df(x) \cdot e = \sum_{i=1}^{n} \frac{\partial f}{\partial x_i} a_i .$$

It should be emphasized that Df is a linear mapping at each $x \in A$ and the definition of $Df(x)$ is independent of the basis used. If we change the basis from the standard basis to another one, the matrix elements will of course change. If one examines the definition of the matrix of a linear transformation* it can be seen that the columns of the matrix relative to the new basis will be the derivative $Df(x)$ applied to the new basis in $\mathbb{R}^n$ with this image vector expressed in the new basis in $\mathbb{R}^m$. Of course, the linear map $Df(x)$ itself does not change from basis to basis. In the case $m = 1$, $Df(x)$ is, in the standard basis, a $1 \times n$ matrix. The vector whose components are the same as those of $Df(x)$ is called the *gradient* of f, and is denoted grad f or ∇f. Thus for

$$f \colon A \subset \mathbb{R}^n \to \mathbb{R}, \text{ grad } f = \left(\frac{\partial f}{\partial x_1}, \ldots, \frac{\partial f}{\partial x_n} \right) .$$

(Sometimes it is said that grad f is just Df with commas inserted!).

An important special case occurs when $f = L$ is already linear. Then from the definition (see Example 2 below) we see that $DL = L$, as expected since the best affine approximation to a linear map is the linear map itself. Thus the Jacobian matrix of L is the matrix of L itself in this case. Another case of interest is a constant map. Indeed one sees that a constant map has derivative zero; zero is the linear map $f \colon \mathbb{R}^n \to \mathbb{R}^m$ such that $f(x) = 0 = (0, \ldots, 0)$ for all $x \in \mathbb{R}^n$.

EXAMPLE 1.   Let $f \colon \mathbb{R}^2 \to \mathbb{R}^3$, $f(x,y) = (x^2, x^3 y, x^4 y^2)$. Compute Df.

Solution: According to Theorem 2, $Df(x,y)$ is the linear map whose matrix is

$$\begin{pmatrix} \dfrac{\partial f_1}{\partial x} & \dfrac{\partial f_1}{\partial y} \\[2mm] \dfrac{\partial f_2}{\partial x} & \dfrac{\partial f_2}{\partial y} \\[2mm] \dfrac{\partial f_3}{\partial x} & \dfrac{\partial f_3}{\partial y} \end{pmatrix} = \begin{pmatrix} 2x & 0 \\[2mm] 3x^2 y & x^3 \\[2mm] 4x^3 y^2 & 2x^4 y \end{pmatrix}$$

where $f_1(x,y) = x^2$, $f_2(x,y) = x^3 y$, $f_3(x,y) = x^4 y^2$.

* See M. O'Nan, *Linear Algebra*, Harcourt Brace Jovanovich, New York, (1971), Chapter 5.

EXAMPLE 2.   Let $L: \mathbb{R}^n \to \mathbb{R}^m$ be a linear map (that is, $L(x + y) = L(x) + L(y)$ and $L(\alpha x) = \alpha L(x)$). Show that $DL(x) = L$.

Solution: Given x_0, and $\varepsilon > 0$ we must find $\delta > 0$ such that $\|x - x_0\| < \delta$ implies

$$\|L(x) - L(x_0) - DL(x) \cdot (x - x_0)\| \leqslant \varepsilon \|x - x_0\| .$$

But with $DL(x) = L$ the left side becomes

$$\|L(x) - L(x_0) - L(x - x_0)\| ,$$

which is zero since $L(x - x_0) = L(x) - L(x_0)$ by linearity of L. Hence $DL(x) = L$ satisfies the definition (with any $\delta > 0$).

EXAMPLE 3.   Let $f(x,y,z) = x(\sin y)/z$. Compute grad f.

Solution: grad $f = (\partial f/\partial x, \partial f/\partial y, \partial f/\partial z)$, and here

$$\frac{\partial f}{\partial x} = \frac{(\sin y)}{z}, \qquad \frac{\partial f}{\partial y} = \frac{x(\cos y)}{z}, \qquad \frac{\partial f}{\partial z} = -\frac{x(\sin y)}{z^2},$$

so

$$\text{grad } f(x,y,z) = \left(\frac{(\sin y)}{z}, \frac{x(\cos y)}{z}, -\frac{x(\sin y)}{z^2} \right) .$$

Exercises for Section 6.2

1. Let $f: \mathbb{R}^3 \to \mathbb{R}^2$, $f(x,y,z) = (x^4 y, xe^z)$. Compute Df.

2. Let $f: \mathbb{R}^3 \to \mathbb{R}$, $(x,y,z) \mapsto e^{x^2 + y^2 + z^2}$. Compute Df and grad f.

3. Let L be a linear map of $\mathbb{R}^n \to \mathbb{R}^m$, $g: \mathbb{R}^n \to \mathbb{R}^m$ such that $\|g(x)\| \leqslant M \|x\|^2$, and let $f(x) = L(x) + g(x)$. Prove $Df(0) = L$.

4. Let $f(x,y) = (xy, y/x)$. Compute Df. Compute the matrix of $Df(x,y)$ with respect to the basis $(1,0)$, $(1,1)$ in $\mathbb{R}^2$.

5. Discuss the possibility of defining Df for f, a mapping from one normed space to another.

6.3   Continuity of Differentiable Mappings; Differentiable Paths

The reader might recall from elementary calculus that a differentiable map is continuous. This is appealing intuitively since having a tangent line (or plane) to the graph is stronger than having no breaks in the graph.

For real functions of a single variable, we recall the proof: let $f:]a,b[\to \mathbb{R}$

be differentiable at x_0. Then

$$\underset{x \to x_0}{\text{limit}}(f(x) - f(x_0)) = \underset{x \to x_0}{\text{limit}}\left(\frac{f(x) - f(x_0)}{x - x_0}\right) \cdot (x - x_0)$$

$$= f'(x_0) \cdot \underset{x \to x_0}{\text{limit}}(x - x_0) = f'(x_0) \cdot 0 = 0$$

so $\underset{x \to x_0}{\text{limit}}(f(x) - f(x_0)) = 0$ which implies f is continuous at x_0.

These ideas are readily generalized to the case of $f \colon A \subset \mathbb{R}^n \to \mathbb{R}^m$ and the next theorem follows.

Theorem 3. *Suppose $A \subset \mathbb{R}^n$ is open and $f \colon A \to \mathbb{R}^m$ is differentiable on A. Then f is continuous. In fact, for each $x_0 \in A$ there is a constant $M > 0$ and a $\delta_0 > 0$ such that $\lVert x - x_0 \rVert < \delta_0$ implies $\lVert f(x) - f(x_0) \rVert \leqslant M \lVert x - x_0 \rVert$. (This is called the Lipschitz property.)*

Earlier we examined the special case of real-valued functions, $f \colon \mathbb{R}^n \to \mathbb{R}$. The case of a function $c \colon \mathbb{R} \to \mathbb{R}^m$ is also important. Here c represents a curve or path in $\mathbb{R}^m$. In this case $Dc(t) \colon \mathbb{R} \to \mathbb{R}^m$ is represented by the vector

$$\begin{pmatrix} \dfrac{dc_1}{dt} \\ \cdot \\ \cdot \\ \cdot \\ \dfrac{dc_m}{dt} \end{pmatrix}$$

where $c(t) = (c_1(t),\ldots,c_m(t))$. This vector is denoted $c'(t)$ and is called the *tangent vector* or *velocity vector* to the curve. If we note that $c'(t) = \underset{h \to 0}{\text{limit}}(c(t + h) - c(t))/h$ and use the fact that $[c(t + h) - c(t)]/h$ is a chord which approximates the tangent line to the curve, we see that $c'(t)$ should represent the exact tangent vector (see Figure 6-4). In terms of a moving particle, $(c(t + h) - c(t))/h$ is an approximation to the velocity since it is displacement/time, so $c'(t)$ is the *instantaneous velocity*.

Strictly speaking we should always represent $c'(t)$ as a column vector, since the matrix of $Dc(t)$ is a 3×1 matrix. However this is typographically awkward, and so we shall write $c'(t)$ as a row vector.

EXAMPLE 1.   Prove that $f \colon \mathbb{R} \to \mathbb{R}$, $x \mapsto |x|$ is continuous but not differentiable at 0.

Solution: $f(x) = x$ for $x \geqslant 0$ and $f(x) = -x$ for $x < 0$ so f is continuous on $]0,\infty[$ and $]-\infty,0[$. Since $\underset{x \to 0}{\text{limit}} f(x) = 0 = f(0)$, f is also continuous at

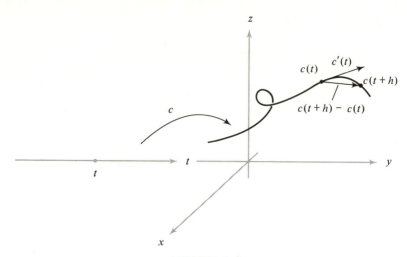

FIGURE 6-4

0, so f is continuous at all points. Finally, f is not differentiable at 0, for if it were,

$$\underset{x \to 0}{\text{limit}} \frac{f(x) - f(0)}{x - 0} = \underset{x \to 0}{\text{limit}} \frac{f(x)}{x}$$

would exist. But for $x > 0$, $f(x)/x$ is $+1$ and for $x < 0$ it is -1. Hence the limit cannot exist.

EXAMPLE 2.   Must the derivative of a function be continuous?

Solution: The answer is no, but an example is not obvious. Perhaps the simplest known example is

$$f(x) = \begin{cases} x^2 \sin\left(\dfrac{1}{x}\right), & x \neq 0, \\ 0, & x = 0. \end{cases}$$

See Figure 6-5.

To demonstrate the differentiability at zero we shall show

$$\frac{f(x)}{x} \to 0 \quad \text{as} \quad x \to 0.$$

Indeed, $|f(x)/x| = |x \sin(1/x)| \leqslant |x| \to 0$ as $x \to 0$. Thus $f'(0)$ exists and is zero. Hence f is differentiable at 0. Now, by elementary calculus,

$$f'(x) = 2x \sin\left(\frac{1}{x}\right) - \cos\left(\frac{1}{x}\right), \quad x \neq 0$$

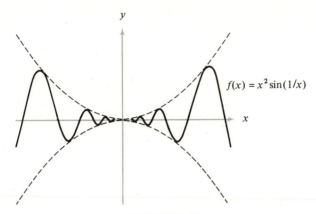

FIGURE 6-5

As $x \to 0$ the first term $\to 0$ but the second term oscillates between $+1$ and -1 so $\lim\limits_{x \to 0} f'(x)$ does not exist. Thus f' exists but is not continuous.

EXAMPLE 3.   Let $c(t) = (t^2, t, \sin t)$. Find the tangent vector to $c(t)$ at $c(0) = (0,0,0)$.

Solution: $c'(t) = (2t, 1, \cos t)$. Setting $t = 0$, $c'(0) = (0,1,1)$ which is the vector tangent to $c(t)$ at $(0,0,0)$.

Exercises for Section 6.3

1. Let
$$f(x) = \begin{cases} x^2, & \text{if } x \text{ is irrational}, \\ 0, & \text{if } x \text{ is rational}. \end{cases}$$

   Show $f'(0)$ exists. Is f continuous at 0?

2. Is the Lipschitz condition in Theorem 3 enough to guarantee differentiability?

3. Must the derivative of a continuous function exist at its maximum?

4. Let $f(x) = x \sin (1/x)$, $x \neq 0$ and $f(0) = 0$. Investigate the continuity and differentiability of f at 0.

5. Find the tangent vector to the curve $c(t) = (3t^2, e^t, t + t^2)$ at $t = 1$.

6.4   Conditions for Differentiability

Since the Jacobian matrix provides an effective computational method, we should like to know if the existence of the usual partial derivatives implies that the derivative Df exists. This is, unfortunately, not true in general.

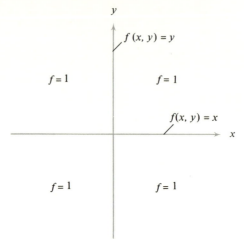

FIGURE 6-6

For example, take $f: \mathbb{R}^2 \to \mathbb{R}$ defined by $f(x,y) = x$ when $y = 0$, $f(x,y) = y$ when $x = 0$, and $f(x,y) = 1$ elsewhere. Then $\partial f/\partial x$ and $\partial f/\partial y$ exist at $(0,0)$ and are equal to 1. However, f is not continuous at $(0,0)$ (why?), so the derivative Df cannot possibly exist at $(0,0)$. See Figure 6-6. (See the Examples and Exercises for more exotic examples.)

It is quite simple to understand such behavior. The partial derivatives depend only on what happens in the directions of the x and y axes, whereas the definition of Df involves the combined behavior of f in a whole neighborhood of a given point.

We can, however, assert the following.

> **Theorem 4.** *Let $A \subset \mathbb{R}^n$ be an open set and $f: A \subset \mathbb{R}^n \to \mathbb{R}^m$. Suppose $f = (f_1, \ldots, f_m)$. If each of the partials $\partial f_j/\partial x_i$ exists and is continuous on A, then f is differentiable on A.*

Let us now discuss the directional derivative.

> **Definition 3.** Let f be real-valued, defined in a neighborhood of $x_0 \in \mathbb{R}^n$ and let $e \in \mathbb{R}^n$ be a unit vector. Then
>
> $$\frac{d}{dt} f(x_0 + te)\bigg|_{t=0} = \lim_{t \to 0} \frac{f(x_0 + te) - f(x_0)}{t}$$
>
> is called the *directional derivative* of f at x_0 in the direction e.

From this definition, the directional derivative is just the rate of change of f in the direction e; see Figure 6-7.

We claim that the directional derivative in the direction of e equals $Df(x_0) \cdot e$. To see this just look at the definition of $Df(x_0)$ with $x = x_0 + te$; we get

$$\left\| \frac{f(x_0 + te) - f(x_0)}{t} - Df(x_0) \cdot e \right\| \leqslant \varepsilon \|e\| \qquad \text{for any } \varepsilon > 0$$

if $|t|$ is sufficiently small. This proves that *if f is differentiable at x_0 then the directional derivatives also exist and are given by*

$$\lim_{t \to 0} \frac{f(x_0 + te) - f(x_0)}{t} = Df(x_0) \cdot e .$$

In particular, observe that $\partial f / \partial x_i$ is the derivative of f in the direction of the ith coordinate axis (with $e = e_i = (0,0,\ldots,0,1,0,\ldots,0)$).

Notice that for a function $f \colon \mathbb{R}^2 \to \mathbb{R}$ the directional derivatives $Df(x_0) \cdot e$ can be used to determine the plane tangent to the graph of f (compare Figure 6-2). Namely, the line l, $z = f(x_0) + Df(x_0) \cdot te$ is tangent to the graph of f since, as in Figure 6-7, $Df(x_0) \cdot e$ is just the rate of change of f in the direction e. Thus the tangent plane to the graph of f at $(x_0, f(x_0))$ may be described by the equation

$$z = f(x_0) + Df(x_0) \cdot (x - x_0) ,$$

(see Figure 6-8). Since we have not defined rigorously the notion of the

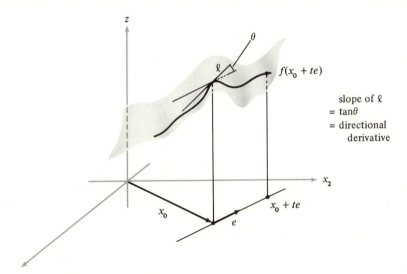

FIGURE 6-7   Slope of $l = \tan \theta = $ directional derivative.

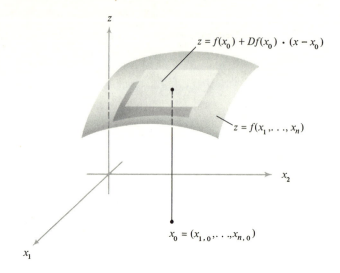

FIGURE 6-8

tangent plane to a surface, we shall adopt the above equation as a *definition* of the tangent plane.

EXAMPLE 1.   Show that the existence of all directional derivatives at a point does not imply differentiability.

Solution: We consider $f: \mathbb{R}^2 \to \mathbb{R}$,

$$f(x,y) = \begin{cases} \dfrac{xy}{(x^2 + y)}, & x^2 \neq -y, \\ 0, & x^2 = -y. \end{cases}$$

Then if $e = (e_1, e_2)$,

$$\frac{1}{t} f(te_1, te_2) = \frac{1}{t} \frac{t^2 e_1 e_2}{t^2 e_1^2 + te_2} = \frac{te_1 e_2}{t^2 e_1^2 + te_2} \to e_1$$

as $t \to 0$ (the case $e_2 = 0$, however, gives zero). Thus each directional derivative exists at $(0,0)$, but f is not continuous at $(0,0)$ since for x^2 near $-y$ with both x, y small, f is very large. (For instance, given δ and M, choose (x,y) such that $x^2 = -y + \varepsilon$ and $\|(x,y)\| < \delta$. Then $f(x,y) = xy/\varepsilon$, which for ε small can be made larger than M. Thus f is not bounded on $D((0,0),\delta)$ for any $\delta > 0$ and so is not continuous at $(0,0)$.) Hence, by Theorem 3, f is not differentiable at $(0,0)$.

Note: This example shows that existence of all directional derivatives would not be a convenient definition of differentiability since it would not even imply continuity. This is the reason one adopts the more restrictive notion in Definition 1.

EXAMPLE 2.   Let $f(x,y) = x^2 + y$. Compute the equation of the plane tangent to the graph of f at $x = 1, y = 2$.

Solution: Here $Df(x,y)$ has matrix

$$\left(\frac{\partial f}{\partial x}, \frac{\partial f}{\partial y}\right) = (2x, 1)$$

so $Df(1,2) = (2,1)$. Thus the equation of the tangent plane becomes

$$z = 3 + (2,1)\binom{x-1}{y-2} = 3 + 2(x-1) + (y-2)$$

that is,

$$2x + y - z = 1 .$$

Exercises for Section 6.4

1. Use Theorem 4 to show that

$$f(x,y) = \begin{cases} \dfrac{(xy)^2}{\sqrt{x^2+y^2}}, & (x,y) \neq (0,0), \\ 0, & (x,y) = (0,0) \end{cases}$$

   is differentiable at $(0,0)$.

2. Investigate the differentiability of

$$f(x,y) = \frac{xy}{\sqrt{x^2+y^2}}$$

   at $(0,0)$ if $f(0,0) = 0$.

3. Find the tangent plane to the graph of $z = x^2 + y^2$ at $(0,0)$.

4. Find the equation of the tangent plane to $z = x^3 + y^4$ at $x = 1, y = 3$.

5. Find a function $f: \mathbb{R}^2 \to \mathbb{R}$ which is differentiable at each point, but the partials are not continuous at $(0,0)$. [Hint: Study Example 2, Section 6.3.]

6.5   The Chain Rule or Composite Mapping Theorem

One of the most important techniques of differentiation is the chain rule ("function of a function rule"). For example, to differentiate $(x^3 + 3)^6$ let $y = x^3 + 3$ and first differentiate y^6, getting $6y^5$, then multiply by the derivative of $x^3 + 3$ to obtain the final answer $6(x^3 + 3)^5 3x^2$. There is a similar process for functions of several variables. For example, if u, v, and f are real-valued functions of two variables then

$$\frac{\partial}{\partial x} f(u(x,y),v(x,y)) = \frac{\partial f}{\partial u}\frac{\partial u}{\partial x} + \frac{\partial f}{\partial v}\frac{\partial v}{\partial x}.$$

The general theorem is now given which includes all of these as special cases.

> **Theorem 5.**   Let $f\colon A \to \mathbb{R}^m$ be differentiable on the open set $A \subset \mathbb{R}^n$ and $g\colon B \to \mathbb{R}^p$ be differentiable on the open set $B \subset \mathbb{R}^m$, and suppose that $f(A) \subset B$. Then the composite $g \circ f$ is differentiable on A and $D(g \circ f)(x_0) = Dg(f(x_0)) \circ Df(x_0)$.

Note that this formula is logical because $Df(x_0)\colon \mathbb{R}^n \to \mathbb{R}^m$ and $Dg(f(x_0))\colon \mathbb{R}^m \to \mathbb{R}^p$ so their composition is defined.

Recall that the product of two matrices corresponds to the composition of the corresponding linear maps they represent. Thus from Theorem 5 we get the important fact that the Jacobian matrix of $g \circ f$ at $x = (x_1, \ldots, x_n)$ is the product of the Jacobian matrix of g evaluated at $f(x)$ with the Jacobian matrix of f evaluated at x (in that order). Thus if $h = g \circ f$ and $y = f(x)$, then

$$Dh(x) = \begin{pmatrix} \dfrac{\partial g_1}{\partial y_1} & \cdots & \dfrac{\partial g_1}{\partial y_m} \\[2mm] \cdot & & \cdot \\ \cdot & & \cdot \\ \cdot & & \cdot \\[1mm] \dfrac{\partial g_p}{\partial y_1} & \cdots & \dfrac{\partial g_p}{\partial y_m} \end{pmatrix} \begin{pmatrix} \dfrac{\partial f_1}{\partial x_1} & \cdots & \dfrac{\partial f_1}{\partial x_n} \\[2mm] \cdot & & \cdot \\ \cdot & & \cdot \\ \cdot & & \cdot \\[1mm] \dfrac{\partial f_m}{\partial x_1} & \cdots & \dfrac{\partial f_m}{\partial x_n} \end{pmatrix}$$

where $\partial g_i/\partial y_j$ are evaluated at $y = f(x)$ and $\partial f_i/\partial x_j$ at x. Writing this out, we obtain, for example,

$$\frac{\partial h_1}{\partial x_1} = \sum_{j=1}^{m} \frac{\partial g_1}{\partial y_j}\frac{\partial f_j}{\partial x_1}.$$

This situation occurs when we "change variables." For example, suppose $f(x,y)$ is a real-valued function, and let $x = r \cos \theta$, $y = r \sin \theta$ for the new variables r, θ (polar coordinates). We form the function

$$h(r,\theta) = f(r \cos \theta, r \sin \theta) .$$

Then

$$\frac{\partial h}{\partial r} = \frac{\partial f}{\partial x} \cos \theta + \frac{\partial f}{\partial y} \sin \theta ,$$

and

$$\frac{\partial h}{\partial \theta} = -\frac{\partial f}{\partial x} r \sin \theta + \frac{\partial f}{\partial y} r \cos \theta .$$

The reader should derive similar formulas for spherical coordinates (r,φ,θ), where $x = r \cos \theta \sin \varphi$, $y = r \sin \theta \sin \varphi$, $z = r \cos \varphi$ (spherical coordinates are discussed in detail in Section 9.5).

The chain rule (Theorem 5) is also called the composite function theorem, since it tells us how to differentiate composite functions.

Another illustration may clarify matters. Suppose we have functions $u(x,y)$, $v(x,y)$, $w(x,y)$, and $f(u,v,w)$, and form the function $h(x,y) = f(u(x,y),v(x,y),w(x,y))$. Then Theorem 5 yields

$$\frac{\partial h}{\partial x} = \frac{\partial f}{\partial u}\frac{\partial u}{\partial x} + \frac{\partial f}{\partial v}\frac{\partial v}{\partial x} + \frac{\partial f}{\partial w}\frac{\partial w}{\partial x} .$$

We can see this formula (as an illustrative case) roughly as follows, write

$$\frac{[h(x + \Delta x,y) - h(x,y)]}{\Delta x}$$

$$= \frac{\begin{aligned}[f(u(x + \Delta x,y),v(x + \Delta x,y),w(x + \Delta x,y)) \\ - f(u(x,y),v(x + \Delta x,y),w(x + \Delta x,y))]\end{aligned}}{\Delta x}$$

$$+ \frac{\begin{aligned}[f(u(x,y),v(x + \Delta x,y),w(x + \Delta x,y)) \\ - f(u(x,y),v(x,y),w(x + \Delta x,y))]\end{aligned}}{\Delta x}$$

$$+ \frac{[f(u(x,y),v(x,y),w(x + \Delta x,y)) - f(u(x,y),v(x,y),w(x,y))]}{\Delta x} .$$

Now this is approximately (using $f(u + \Delta u,v,w) - f(u,v,w) \approx \Delta u\, \partial f/\partial u$),

$$\frac{\partial f}{\partial u}\frac{\Delta u}{\Delta x} + \frac{\partial f}{\partial v}\frac{\Delta v}{\Delta x} + \frac{\partial f}{\partial w}\frac{\Delta w}{\Delta x} .$$

So, letting $\Delta x \to 0$ gives the formula.

EXAMPLE 1. Verify the chain rule for $f(u,v,w) = u^2v + wv^2$ and $u = xy$, $v = \sin x$, $w = e^x$.

Solution: Here $h(x,y) = f(u(x,y),v(x,y),w(x,y))$ is given by

$$h(x,y) = x^2y^2 \sin x + e^x \sin^2 x$$

so, directly,

$$\frac{\partial h}{\partial x} = 2xy^2 \sin x + x^2y^2 \cos x + e^x \sin^2 x + e^x 2 \sin x \cos x \; .$$

On the other hand,

$$\frac{\partial f}{\partial u}\frac{\partial u}{\partial x} + \frac{\partial f}{\partial v}\frac{\partial v}{\partial x} + \frac{\partial f}{\partial w}\frac{\partial w}{\partial x} = 2uv\frac{\partial u}{\partial x} + u^2\frac{\partial v}{\partial x} + 2wv\frac{\partial v}{\partial x} + v^2\frac{\partial w}{\partial x}$$

$$= 2xy^2 \sin x + x^2y^2 \cos x + 2e^x \sin x \cos x$$

$$+ \, e^x \sin^2 x \; ,$$

which is the same result. The formula for $\partial h/\partial y$ can be checked similarly.

EXAMPLE 2.   Let $f: \mathbb{R} \to \mathbb{R}$ and let $F: \mathbb{R}^2 \to \mathbb{R}$ be given by $F(x,y) = f(xy)$. Verify

$$x\frac{\partial F}{\partial x} = y\frac{\partial F}{\partial y} \; .$$

Solution: By the chain rule,

$$\frac{\partial F}{\partial x} = f'(xy)y$$

and

$$\frac{\partial F}{\partial y} = f'(xy)x \; ,$$

so the statement is clear.

Exercises for Section 6.5

1. Write out the chain rule for

$$h(x,y,z) = f(u(x,y,z),v(x,y),w(y,z)) \; .$$

2. Verify the chain rule for

$$u(x,y,z) = xe^y \; ,$$

$$v(x,y,z) = (\sin x)yz \; ,$$

and

$$f(u,v) = u^2 + v \sin u$$

with

$$h(x,y,z) = f(u(x,y,z),v(x,y,z)) \; .$$

3. Let $F(x,y) = f(x^2 + y^2)$. Show that $x(\partial F/\partial y) = y(\partial F/\partial x)$.

4. Write out the chain rule for spherical coordinates, as we did in the text for polar coordinates.

5. Let $f: \mathbb{R} \to \mathbb{R}$ and $F: \mathbb{R}^2 \to \mathbb{R}$ be differentiable and satisfy $F(x, f(x)) = 0$ and $\partial F/\partial y \neq 0$. Prove that $f'(x) = -(\partial F/\partial x)/(\partial F/\partial y)$ where $y = f(x)$.

6.6 Product Rule and Gradients

Another well-known rule of differential calculus is the *product rule* or *Leibnitz rule*.

> **Theorem 6.** Let $A \subset \mathbb{R}^n$ be open and let $f: A \to \mathbb{R}^m$ and $g: A \to \mathbb{R}$ be differentiable functions. Then gf is differentiable and for $x \in A$, $D(gf)(x): \mathbb{R}^n \to \mathbb{R}^m$ is given by $D(gf)(x) \cdot e = g(x)(Df(x) \cdot e) + (Dg(x) \cdot e)f(x)$ for all $e \in \mathbb{R}^n$. (Note that this makes sense since $g(x) \in \mathbb{R}$ and $Dg(x) \cdot e \in \mathbb{R}$).

We sometimes abbreviate this result by saying that

$$D(gf) = gDf + (Dg)f,$$

but the precise meaning is as stated in the theorem.

The reader is undoubtedly familiar with the product rule from elementary calculus. In terms of components, the theorem simply states that

$$\frac{\partial}{\partial x_i}(gf_k) = g\left(\frac{\partial f_k}{\partial x_i}\right) + \left(\frac{\partial g}{\partial x_i}\right)f_k.$$

For quotients, we have a similar result. If $g \neq 0$, then

$$D\left(\frac{f}{g}\right) = \frac{(g \cdot Df - f \cdot Dg)}{g^2}.$$

In order to prove this formula, it suffices, by Theorem 6, to demonstrate it for the case $1/g$. This reduces it to a problem in elementary calculus with which the reader should be acquainted, so we shall omit details.*

Other rules of differentiation are encompassed in the statement that D is linear; that is, $D(f + g) = Df + Dg$ and $D(\lambda f) = \lambda Df$ for $\lambda \in \mathbb{R}$, a constant. The reader will be able to supply the proofs without difficulty.

Let us consider the geometry of gradients a little further. Let $f: A \subset \mathbb{R}^n \to \mathbb{R}$ be differentiable. Then we have the gradient

$$\text{grad } f(x) = \left(\frac{\partial f}{\partial x_1}, \dots, \frac{\partial f}{\partial x_n}\right).$$

* (See McAloon-Tromba, *Calculus*, Harcourt Brace Jovanovich (1972), Section 3.3).

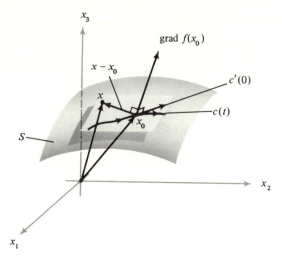

FIGURE 6-9

Hence the directional derivative in direction h is (see Theorem 2 above)

$$Df(x) \cdot h = \langle \text{grad } f(x), h \rangle$$

$$= \text{rate of change of } f \text{ at the point } x \text{ in direction } h \ .$$

Consider now the "surface" S defined by the equation $f(x) =$ constant. We assert that grad $f(x)$ is *orthogonal to this surface* (this is intuitive since we have not been precise about the nature of this surface—see however Section 7.7). To prove this, consider a curve $c(t)$ in S with its tangent vector $c'(0)$ where $c(0) = x_0$. We assert that

$$\langle \text{grad } f(x_0), c'(0) \rangle = 0 \ .$$

Now since $c(t) \in S$, $f(c(t)) =$ constant. Differentiating and using the chain rule, we get

$$Df(c(t)) \cdot c'(t) = 0 \ .$$

Setting $t = 0$, and using $Df(x) \cdot h = \langle \text{grad } f(x), h \rangle$, gives the desired relation. See Figure 6-9.

Note that we may describe the tangent plane to S: $f(x) =$ constant at x_0 by $\langle \text{grad } f(x_0), x - x_0 \rangle = 0$, since grad $f(x_0)$ is orthogonal to S.

It is also evident from the equation

$$\langle \text{grad } f(x_0), h \rangle = \| \text{grad } f(x_0) \| \cos \theta$$

(where $\| h \| = 1$ and θ is the angle between grad $f(x_0)$ and h) that grad $f(x_0)$ is the direction in which f is changing the fastest. It is not unreasonable because if we suppose that f represents the height function of a mountain, then f = constant are the level contours. To climb or descend the mountain as quickly as possible, we should walk perpendicular to the level contours. (Figure 6-10).

These facts are actually of value in practical optimal control problems. In such problems one is given a function $f(x_1,\ldots,x_n)$ and the problem is to maximize or "optimize" f by some practical scheme. A common method is to take a trial point x_0 and proceed along a straight line in the direction of the gradient of f to reach a new point at which f will be larger (at least if we do not go too far), and repeat.

EXAMPLE 1.   Find the normal to the surface $x^2 + y^2 + z^2 = 3$ at $(1,1,1)$.

Solution: Here $f(x,y,z) = x^2 + y^2 + z^2$ has gradient grad $f = (2x,2y,2z)$ which, at $(1,1,1)$, is $(2,2,2)$. Normalizing, the unit normal is $(1/\sqrt{3},1/\sqrt{3},1/\sqrt{3})$.

EXAMPLE 2.   Find the direction of greatest rate of increase of $f(x,y,z) = x^2 y \sin z$ at $(3,2,0)$.

Solution: The direction is that of the gradient vector, which is $(2xy \sin z, x^2 \sin z, x^2 y \cos z)$ which becomes $(0,0,18)$ at $(3,2,0)$.

EXAMPLE 3.   What is the tangent plane to the surface $x^2 - y^2 + xz = 2$ at $(1,0,1)$?

Solution: Here grad $f(1,0,1) = (3,0,1)$ so the tangent plane is $\langle(x - 1,y,z - 1),(3,0,1)\rangle = 0$, that is, $3x + z = 4$.

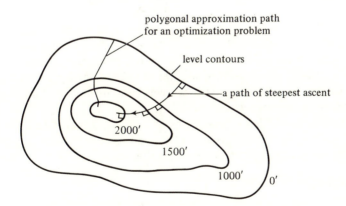

FIGURE 6-10   Direction of steepest ascent is orthogonal to the level contours.

Exercises for Section 6.6

1. Prove

$$\frac{d}{dt} f(x_0 + th)\Big|_{t=0} = Df(x_0) \cdot h$$

   by using the chain rule, where $f: \mathbb{R}^m \to \mathbb{R}^n$.

2. Find the unit normal to the surface $x^2 - y^2 + xyz = 1$ at $(1,0,1)$.

3. Find the equation of the tangent plane to the surface $x^2 - y^2 + xyz = 1$ at $(1,0,1)$.

4. In what direction is $f(x,y) = e^{x^2}y$ increasing the fastest?

5. Let $f: \mathbb{R}^n \to \mathbb{R}$, $g: \mathbb{R}^n \to \mathbb{R}$. Show grad($fg$) $= f$ grad $g + g$ grad f.

6. Show that grad f being the normal to the tangent plane is a more general description of the tangent plane than the description in Section 6.4.

6.7   Mean-Value Theorem

We will now consider two very important theorems. These are the mean-value theorem and Taylor's theorem. First, let us turn our attention to the mean-value theorem. In Fact 3, Section 6.1 we recalled the proof of the mean-value theorem of elementary calculus, which stated that if $f: [a,b] \to \mathbb{R}$ is continuous and if f is differentiable on $]a,b[$, there exists a point $c \in]a,b[$ such that $f(b) - f(a) = f'(c)(b - a)$, where $f' = df/dx$.

Unfortunately, for $f: A \subset \mathbb{R}^n \to \mathbb{R}^m$ this version of the mean-value theorem simply is not true. For example, consider $f: \mathbb{R} \to \mathbb{R}^2$, defined by $f(x) = (x^2,x^3)$. Let us try to find a c such that $0 \leqslant c \leqslant 1$ and $f(1) - f(0) = Df(c)(1 - 0)$. This means that $(1,1) - (0,0) = (2c,3c^2)$, and thus $2c = 1$ and $3c^2 = 1$. It is obvious that there is no c satisfying these equations.

Experience leads us to believe that some restrictive condition might provide a valid theorem. In this case, for the above version to hold, f must be real-valued. In order to give the correct theorem let us first make precise the meaning of "c is between x and y" for $c, x, y \in \mathbb{R}^n$.

We say c is on *the line segment joining* x and y, or is *between* x and y if $c = (1 - \lambda)x + \lambda y$ for some $0 \leqslant \lambda \leqslant 1$. See Figure 6-11.

We are now prepared to state our next theorem.

Theorem 7.

(i) *Suppose $f: A \subset \mathbb{R}^n \to \mathbb{R}$ is differentiable on an open set A. For any $x, y \in A$ such that the line segment joining x and y lies in A (which need not happen for all x, y), there is a point c on*

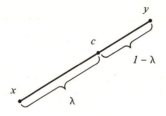

FIGURE 6-11

that segment such that

$$f(y) - f(x) = Df(c)(y - x) .$$

(ii) *Suppose* $f: A \subset \mathbb{R}^n \to \mathbb{R}^m$ *is differentiable on the open set* A. *Suppose the line segment joining* x *and* y *lies in* A *and* $f = (f_1, \ldots, f_m)$. *Then there exist points* $c_1, \ldots, c_m$ *on that segment such that*

$$f_i(y) - f_i(x) = Df_i(c_i)(y - x), \qquad i = 1, \ldots, m .$$

An important alternative formulation of the mean value theorem is given in Example 5, at the end of the chapter.

EXAMPLE 1.   A set $A \subset \mathbb{R}^n$ is said to be *convex* if for each $x, y \in A$ the segment joining x, y also lies in A. See Figure 6-12. Let $A \subset \mathbb{R}^n$ be an open convex set and let $f: A \to \mathbb{R}^m$ be differentiable. If $Df = 0$, then show that f is constant. (Generalizations of this are given in Exercise 9, at the end of the chapter.)

Solution: For $x, y \in A$ we have for each component f_i a vector c_i such that

$$f_i(y) - f_i(x) = Df_i(c_i)(y - x) .$$

Since $Df = 0$, $Df_i = 0$ for each i (why?), and so $f_i(y) = f_i(x)$. It follows that $f(y) = f(x)$, which means that f is constant.

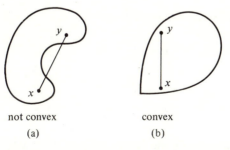

not convex   convex

(a)   (b)

FIGURE 6-12   (a) Not convex. (b) Convex.

EXAMPLE 2.   Suppose $f: [0,\infty[\to \mathbb{R}$ is continuous, $f(0) = 0$, f is differentiable on $]0,\infty[$ and f' is non-decreasing. Prove that $g(x) = f(x)/x$ is non-decreasing for $x > 0$.

Solution: From the mean-value theorem we see that a function $h: \mathbb{R} \to \mathbb{R}$ is non-decreasing if $h'(x) \geqslant 0$, because $x \leqslant y$ implies that

$$h(y) - h(x) = h'(c)(y - x) \geqslant 0 \ .$$

Now

$$g'(x) = \frac{[xf'(x) - f(x)]}{x^2}$$

and

$$f(x) = f(x) - f(0) = f'(c) \cdot x \leqslant xf'(x)$$

since $0 < c < x$ and $f'(x) \geqslant f'(c)$. Thus $xf'(x) - f(x) \geqslant 0$, so $g' \geqslant 0$ which implies that g is non-decreasing.

Exercises for Section 6.7

1. If $f: \mathbb{R} \to \mathbb{R}$ is differentiable and is such that $f'(x) > 0$, prove f is (strictly) increasing. Define your terms.

2. Prove *l'Hôpital's rule:* if f', g' exist at x_0, $g'(x_0) \neq 0$, and if $f(x_0) = 0 = g(x_0)$, then

$$\lim_{x \to x_0} \frac{f(x)}{g(x)} = \frac{f'(x_0)}{g'(x_0)} \ .$$

3. Use Exercise 2 to evaluate

   (a) $\displaystyle \lim_{x \to 0} \frac{\sin x}{x}$,

   (b) $\displaystyle \lim_{x \to 0} \frac{e^x - 1}{x}$.

4. Which of the following sets are convex?

   (a) $\{(x,y) \in \mathbb{R}^2 \mid y \geqslant 0\}$,
   (b) $\{x \in \mathbb{R}^n \mid 0 < \|x\| < 1\}$,
   (c) $\mathbb{R}\setminus\{0\}$.

5. Let $f: A \subset \mathbb{R}^n \to \mathbb{R}$ be differentiable with A convex and suppose $\|\text{grad } f(x)\| \leqslant M$ for $x \in A$. Prove $|f(x) - f(y)| \leqslant M \|x - y\|$ for $x, y \in A$. Do you think this is true if A is not convex?

6. Let $f: \mathbb{R} \to \mathbb{R}$ be differentiable. Assume that for all $x \in \mathbb{R}$, $0 \leqslant f'(x) \leqslant f(x)$. Show that $g(x) = e^{-x}f(x)$ is decreasing. If f vanishes at some point, conclude that f is zero.

6.8 Taylor's Theorem and Higher Derivatives

Next, we would like to discuss Taylor's formula for the general case of functions $f: A \subset \mathbb{R}^n \to \mathbb{R}^m$. To be able to do this, we must first discuss derivatives of higher order. For $f: \mathbb{R}^n \to \mathbb{R}$ there is no problem defining partial derivatives of higher order; we just iterate the process of partial differentiation

$$\frac{\partial^2 f}{\partial x_1 \, \partial x_2} = \frac{\partial}{\partial x_1}\left(\frac{\partial}{\partial x_2} f\right),$$

and so on. However, regarding the derivative as a linear map needs a little more care.

The second derivative is obtained by differentiating Df, if it exists, and is accomplished as follows.

> **Definition 4.** Let $L(\mathbb{R}^n, \mathbb{R}^m)$ denote the space of linear maps from $\mathbb{R}^n$ to $\mathbb{R}^m$. (If we choose a basis in $\mathbb{R}^n$ and $\mathbb{R}^m$, then $L(\mathbb{R}^n, \mathbb{R}^m)$ can be identified with the $m \times n$ matrices and hence with $\mathbb{R}^{nm}$.) Now $Df: A \to L(\mathbb{R}^n, \mathbb{R}^m)$; that is, at each $x \in A$ we get a linear map $Df(x_0)$. If we differentiate Df at x_0 we get a linear map from $\mathbb{R}^n$ to $L(\mathbb{R}^n, \mathbb{R}^m)$ by definition of the derivative. We write $D(Df(x_0)) = D^2 f(x_0)$. We define the map $B_{x_0}: \mathbb{R}^n \times \mathbb{R}^n \to \mathbb{R}^m$ by setting $B_{x_0}(x_1, x_2) = [D^2 f(x_0)(x_1)](x_2)$.

This makes sense because $D^2 f(x_0): \mathbb{R}^n \to L(\mathbb{R}^n, \mathbb{R}^m)$ and so $D^2 f(x_0)(x_1) \in L(\mathbb{R}^n, \mathbb{R}^m)$; therefore it can be applied to x_2. The reason we do this is that B_{x_0} avoids the unnecessary use of the conceptually difficult space $L(\mathbb{R}^n, \mathbb{R}^m) \approx \mathbb{R}^{nm}$.

By definition, a *bilinear map* $B: E \times F \to G$, where E, F, G are vector spaces, is a map which is linear in each variable separately; for example, in the first variable this means $B(\alpha e_1 + \beta e_2, f) = \alpha B(e_1, f) + \beta B(e_2, f)$, where $e_1, e_2 \in E$, $f \in F$, and $\alpha, \beta \in \mathbb{R}$. The map B_{x_0} defined above is easily seen to be a bilinear map of $\mathbb{R}^n \times \mathbb{R}^n \to \mathbb{R}^m$.

Now, with a bilinear map $B: E \times F \to \mathbb{R}$, we can associate a matrix for each basis $e_1, \ldots, e_n$ of E and $f_1, \ldots, f_m$ of F. Namely let

$$a_{ij} = B(e_i, f_j) .$$

Then if

$$x = \sum_{i=1}^{n} x_i e_i \quad \text{and} \quad y = \sum_{j=1}^{m} y_j f_j ,$$

we have

$$B(x,y) = \sum_{i,j} a_{ij}x_i y_j$$

$$= (x_1, x_2, \ldots, x_n) \begin{pmatrix} a_{11} & & a_{1m} \\ & & \\ & & \\ & & \\ a_{n1} & \cdots & a_{nm} \end{pmatrix} \begin{pmatrix} y_1 \\ \\ \\ y_m \end{pmatrix}.$$

Note: For the second derivative, we shall by abuse of notation, still write $D^2f(x_0)$ for the bilinear map B_{x_0} obtained by differentiating Df at x_0 as described above.

Theorem 8. *Let $f: A \subset \mathbb{R}^n \to \mathbb{R}$ be twice differentiable on the open set A. Then the matrix of $D^2f(x): \mathbb{R}^n \times \mathbb{R}^n \to \mathbb{R}$ with respect to the standard basis is given by*

$$\begin{pmatrix} \dfrac{\partial^2 f}{\partial x_1\,\partial x_1} & \cdots & \dfrac{\partial^2 f}{\partial x_1\,\partial x_n} \\ & & \\ \dfrac{\partial^2 f}{\partial x_n\,\partial x_1} & \cdots & \dfrac{\partial^2 f}{\partial x_n\,\partial x_n} \end{pmatrix}$$

where each partial derivative is evaluated at the point $x = (x_1, \ldots, x_n)$.

For higher derivatives, proceed in an analogous manner. For example, D^3f gives a trilinear map for each x, $D^3f(x): \mathbb{R}^n \times \mathbb{R}^n \times \mathbb{R}^n \to \mathbb{R}^m$. We do not associate a matrix with this map, but rather a quantity labeled by three indices; which, as above, is just $\partial^3 f_k/(\partial x_l\,\partial x_j\,\partial x_i)$ for each component f_k. (Such quantities are called tensors.)

Before proceeding with Taylor's theorem, a very important property of the second derivative shall be given: the matrix in Theorem 8 is symmetric, that is,

$$\frac{\partial^2 f}{\partial x_i\,\partial x_j} = \frac{\partial^2 f}{\partial x_j\,\partial x_i}.$$

Theorem 9. *Let $f: A \to \mathbb{R}$ be twice differentiable on the open set A with D^2f continuous (that is, the functions $\partial^2 f/(\partial x_i\,\partial x_j)$ are continuous). Then D^2f is symmetric; that is,*

$$D^2f(x)(x_1, x_2) = D^2f(x)(x_2, x_1)$$

or, in terms of components,

$$\frac{\partial^2 f}{\partial x_i\, \partial x_j} = \frac{\partial^2 f}{\partial x_j\, \partial x_i} .$$

From this, it can be proven that all the higher derivatives are symmetric as well under analagous conditions. The case $f: A \to \mathbb{R}^m$ is handled by applying the above to the components of f.

The symmetry of second derivatives represents a fundamental property not encountered in single variable calculus. Let us verify these principles through an example.

Suppose $f(x,y,z) = e^{xy} \sin x + x^2 y^4 \cos^2 z$, so $f: \mathbb{R}^3 \to \mathbb{R}$. Then

$$\frac{\partial f}{\partial x} = e^{xy} \cos x + y e^{xy} \sin x + 2xy^4 \cos^2 z ,$$

$$\frac{\partial f}{\partial y} = x e^{xy} \sin x + 4x^2 y^3 \cos^2 z ,$$

and

$$\frac{\partial^2 f}{\partial y\, \partial x} = x e^{xy} \cos x + e^{xy} \sin x + xy e^{xy} \sin x + 8xy^3 \cos^2 z$$

which is the same as $\partial^2 f / \partial x\, \partial y$.

Theorem 9 is not as obvious intuitively. However, some intuition can be gained from the proof.

> **Definition 5.** A function is said to be of *class C^r* if the first r derivatives exist and are continuous. (Equivalently, this means that all partial derivatives up to order r exist and are continuous, see Theorems 2, 3, and 4). A function is said to be *smooth* or *of class C^∞* if it is of class C^r for all positive integers r.

Using the formula in Theorem 5 (the coordinate form is easiest) one can show that the composite of C^r functions is also C^r (see Exercise 23).

Taylor's theorem is as follows:

> **Theorem 10.** *Let $f: A \to \mathbb{R}$ be of class C^r for $A \subset \mathbb{R}^n$ an open set. Let $x, y \in A$ and suppose that the segment joining x and y lies in A. Then there is a point c on that segment such that*
>
> $$f(y) - f(x) = \sum_{k=1}^{r-1} \frac{1}{k!} D^k f(x)(y - x,\ldots,y - x)$$
>
> $$+ \frac{1}{r!} D^r f(c)(y - x,\ldots,y - x)$$

where $D^k f(x)(y - x, \ldots, y - x)$ denotes $D^k f(x)$ as a k-linear map applied to the k-tuple $(y - x, \ldots, y - x)$. In coordinates,

$$D^k f(x)(y - x, \ldots, y - x)$$

$$= \sum_{i_1, \ldots, i_k = 1}^{n} \left(\frac{\partial^k f}{\partial x_{i_1} \cdots \partial x_{i_k}} \right) (y_{i_1} - x_{i_1}) \cdots (y_{i_k} - x_{i_k}) \,.$$

Setting $y = x + h$, we can write the Taylor formula as

$$f(x + h) = f(x) + Df(x) \cdot h + \cdots$$

$$+ \frac{1}{(r - 1)!} D^{r-1} f(x) \cdot (h, \ldots, h) + R_{r-1}(x, h)$$

where $R_{r-1}(x, h)$ is the remainder. Furthermore,

$$\frac{R_{r-1}(x, h)}{\|h\|^{r-1}} \to 0 \qquad as\ h \to 0 \,.$$

There are other forms in which the remainder term can be cast which are given in the proof of the theorem. This theorem is a generalization of the mean-value theorem (in which case $r = 1$) and of Taylor's theorem encountered in one variable calculus.*

From Taylor's theorem we are led to form the *Taylor series* about x_0,

$$\sum_{k=0}^{\infty} \frac{1}{k!} D^k f(x_0)(x - x_0, \ldots, x - x_0) \,.$$

This need not converge to $f(x)$ even if f is C^∞. If it does so in a neighborhood of x_0, we say f is *real analytic* at x_0. To show f is real analytic amounts to showing that the remainder term $(1/r!)D^r f(c)(x - x_0, \ldots, x - x_0) \to 0$ as as $r \to \infty$. This then is used to establish the usual power series expressions for $\sin x$, $\cos x$, and so forth. (See Section 5.9 for a discussion of power series).

EXAMPLE 1.   Verify Theorem 9 for $f(x, y) = yx^2(\cos y^2)$.

Solution:

$$\frac{\partial f}{\partial x} = 2xy \cos y^2, \qquad \frac{\partial^2 f}{\partial y\, \partial x} = 2x \cos y^2 - 4xy^2 \sin y^2 \,;$$

$$\frac{\partial f}{\partial y} = x^2 \cos y^2 - 2y^2 x^2 \sin y^2, \qquad \frac{\partial^2 f}{\partial x\, \partial y} = 2x \cos y^2 - 4y^2 x \sin y^2 \,.$$

---

* See McAloon and Tromba, *Calculus*, Harcourt Brace Jovanovich (1972), Section 10.5.

EXAMPLE 2.   If f is C^∞ on $\mathbb{R}$ and for every interval $[a,b]$ there is a constant M such that $|f^{(n)}(x)| \leq M^n$ for all n and $x \in [a,b]$, show that f is analytic at each x_0 and

$$f(x) = \sum_{n=0}^{\infty} \frac{f^{(n)}(x_0)}{n!} (x - x_0)^n .$$

Solution: The remainder is

$$\left| \frac{f^{(n)}(c)}{n!} (x - x_0)^n \right| \leq \frac{M^n |x - x_0|^n}{n!} ,$$

which $\to 0$ as $n \to \infty$, since by the ratio test, the corresponding series converges. Observe that the convergence is uniform on all bounded intervals (why?).

EXAMPLE 3.   Give an example of a C^∞ function which is not analytic.

Solution: Let

$$f(x) = \begin{cases} 0, & x \leq 0 , \\ e^{-1/x}, & x > 0 . \end{cases}$$

The only place where smoothness of f is in doubt is at $x = 0$. But, for $x > 0$,

$$f'(x) = \frac{1}{x^2} e^{-1/x} ,$$

which $\to 0$ as $x \to 0+$ (by l'Hôpital's rule, for example). Similarly, one sees $f^{(n)}(x) \to 0$ as $x \to 0+$. Thus using the mean-value theorem we see that f is C^∞ at 0 and $f^{(n)}(0) = 0$. Hence the Taylor series about $x = 0$ is identically zero, so f is not equal to its Taylor series about $x = 0$, and therefore f is not analytic.

EXAMPLE 4.   Compute the second order Taylor formula for $f(x,y) = \sin(x + 2y)$, around $(0,0)$.

Solution: Here $f(0,0) = 0$,

$$\frac{\partial f}{\partial x} (0,0) = \cos(0 + 2 \cdot 0) = 1 ,$$

$$\frac{\partial f}{\partial y} (0,0) = 2 \cos(0 + 2 \cdot 0) = 2 ,$$

$$\frac{\partial^2 f}{\partial x^2} (0,0) = 0 ,$$

$$\frac{\partial^2 f}{\partial y^2} (0,0) = 0 ,$$

and

$$\frac{\partial^2 f}{\partial x\, \partial y}(0,0) = 0\ .$$

Thus

$$f(h,k) = h + 2k + R_2(h,k), (0,0)\ ,$$

where

$$R_2(h,k), (0,0)/|(h,k)|^2 \to 0 \qquad \text{as } (h,k) \to (0,0)\ .$$

Exercises for Section 6.8

1. Verify Theorem 9 for $f(x,y) = (e^{x^2 + y^2 x})xy^2$.

2. Use Example 2 to establish the Taylor series and analyticity of e^x, $\sin x$, $\cos x$ on all of $\mathbb{R}$.

3. Let

$$f(x) = \begin{cases} x^2 \sin\left(\dfrac{1}{x}\right), & x \in\,]-1,1[, x \ne 0\ , \\[2mm] 0, & x = 0\ . \end{cases}$$

   Investigate Taylor's theorem for f about the point $x = 0$.

4. Find the Taylor series representation about $x = 0$ for $\log(1 - x)$, $-1 < x < 1$ and show that it equals $\log(1 - x)$ on $-1 < x < 1$ and also show that it converges uniformly on closed subintervals of $]-1,1[$.

5. Verify that if the conditions in Example 2 are met then we can differentiate the Taylor series term by term to obtain $f'(x)$.

6. Compute the second order Taylor formula for $f(x,y) = e^x \cos y$ around $(0,0)$.

6.9 Maxima and Minima

There is a very important application of Theorem 10 which provides us with a method for determining the maxima and minima of functions. As we might expect from our knowledge of functions of one variable the criteria involves the second derivative. Let us first recall the real variable case.

If $f\colon \mathbb{R} \to \mathbb{R}$ has a local maximum or minimum at x_0, and f is differentiable at x_0, then $f'(x_0) = 0$. Furthermore if f is twice continuously differentiable and if $f''(x_0) < 0$, x_0 is a local maximum and if $f''(x_0) > 0$, it is a local minimum.

We want to generalize these facts now to functions $f\colon A \subset \mathbb{R}^n \to \mathbb{R}$. Let us begin by giving the relevant definitions.

Definition 6. Let $f\colon A \subset \mathbb{R}^n \to \mathbb{R}$ where A is open. If there is a neighborhood of $x_0 \in A$ on which $f(x_0)$ is a maximum, that is, if $f(x_0) \geqslant f(x)$ for all x in the neighborhood, we say $f(x_0)$ is a *local maximum* for f. Similarly, we can define a *local minimum* of f. A point is called *extreme* if it is either a local minimum or a local maximum for f. A point x_0 is a *critical point* if f is differentiable at x_0 and if $Df(x_0) = 0$.

The first basic fact is presented in the next theorem.[*]

Theorem 11. *If $f\colon A \subset \mathbb{R}^n \to \mathbb{R}$ is differentiable, A is open, and if $x_0 \in A$ is an extreme point of f, then $Df(x_0) = 0$; that is, x_0 is a critical point.*

The proof is much the same as for elementary calculus. The result is intuitively obvious since at an extreme point the graph of f must have a horizontal tangent plane. However, just being a critical point is not sufficient to guarantee that the point is also extreme. For example, consider $f(x) = x^3$. For this function 0 is a critical point, since $Df(0) = 0$. But $x^3 > 0$ for $x > 0$ and $x^3 < 0$ for $x < 0$, so 0 is not extreme. Another example is given by $f(x,y) = y^2 - x^2$. Here $0 = (0,0)$ is a critical point, since $\partial f/\partial x = -2x$, $\partial f/\partial y = 2y$, so $Df(0,0) = 0$. However, in any neighborhood of 0 we can find points where f is greater than 0 and points where f is less than 0. A critical point which is not a local extreme value is called a *saddle point*. Figure 6-13 shows how this terminology originated.

In the case of $f\colon A \subset \mathbb{R} \to \mathbb{R}$ we have already mentioned that $f(x)$ is a local maximum if $f'(x) = 0$ and $f''(x) < 0$. Recall that this is geometrically

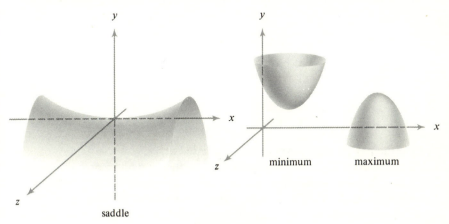

saddle

minimum   maximum

FIGURE 6-13

* The problem of "maximizing" vector functions is important in economics. See S. Smale, "Global Analysis and Economics," p. 537 in *Dynamical Systems*, M. M. Peixoto, ed., Academic Press, N.Y. (1973).

clear if we remember that $f''(x) < 0$ means f is concave downwards. To generalize this the concept of the *Hessian* of a function g at x_0 is introduced.

Definition 7. If $g: B \subset \mathbb{R}^n \to \mathbb{R}$, is of class C^2, the *Hessian* of g at x_0 is defined to be the bilinear function $H_{x_0}(g): \mathbb{R}^n \times \mathbb{R}^n \to \mathbb{R}$ given by $H_{x_0}(g)(x,y) = -D^2 g(x_0)(x,y)$ (note the minus sign).* Thus the Hessian is, as a matrix, just the *negative* of the matrix of second partials.

A bilinear form, that is, a bilinear mapping, $B: \mathbb{R}^n \times \mathbb{R}^n \to \mathbb{R}$ is called *positive definite* if $B(x,x) > 0$ for all $x \neq 0$ in $\mathbb{R}^n$ and is called *positive semidefinite* if $B(x,x) \geqslant 0$ for all $x \in \mathbb{R}^n$. *Negative definite* and *negative semidefinite* bilinear forms are defined similarly.

Now we can make the following generalization to the multi-variable case.

Theorem 12.

(i) If $f: A \subset \mathbb{R}^n \to \mathbb{R}$ is a C^2 function defined on an open set A and x_0 is a critical point of f such that $H_{x_0}(f)$ is positive definite, then f has a local maximum at x_0.

(ii) If f has a local maximum at x_0, then $H_{x_0}(f)$ is positive semidefinite.

The case for minima is covered by changing "positive" to "negative" in the above theorem. Note that a minimum of f is a maximum of $-f$.

As we have noted, the matrix of $H_{x_0}(f)$ with respect to the standard basis is

$$
\begin{pmatrix}
-\dfrac{\partial^2 f}{\partial x_1\, \partial x_1} & \cdots & -\dfrac{\partial^2 f}{\partial x_1\, \partial x_n} \\
\cdot & & \cdot \\
\cdot & & \cdot \\
\cdot & & \cdot \\
-\dfrac{\partial^2 f}{\partial x_n\, \partial x_1} & \cdots & -\dfrac{\partial^2 f}{\partial x_n\, \partial x_n}
\end{pmatrix}
$$

where the partial derivatives are all evaluated at x_0.

When we have $n = 1$, Theorem 12(i) reduces to the one variable test $f''(x_0) < 0$. As in the case $n = 1$, one can have a maximum or a saddle point or a minimum if $f''(x_0) = 0$ (in this case the test fails). For example, $f(x) = -x^4$ has a maximum, x^5 a saddle point, and x^4 a minimum at $x_0 = 0$, although $f''(0) = 0$. For a test in these cases see Exercise 17.

---

* This minus sign is purely conventional and is not of essential importance. The reader who is so inclined can make the appropriate changes in the text if he desires to set $H_{x_0} = +D^2 g(x_0)$.

It might be helpful to mention a few facts from linear algebra, which will be helpful in using the above theorem. Let Δ_k be the determinant of the matrix

$$\begin{pmatrix} -\dfrac{\partial^2 f}{\partial x_1\, \partial x_1} & \cdots & -\dfrac{\partial^2 f}{\partial x_1\, \partial x_k} \\ \cdot & & \cdot \\ \cdot & & \cdot \\ \cdot & & \cdot \\ -\dfrac{\partial^2 f}{\partial x_k\, \partial x_1} & \cdots & -\dfrac{\partial^2 f}{\partial x_k\, \partial x_k} \end{pmatrix}$$

This is the matrix of the Hessian with the last $n - k$ rows and columns removed. Then the symmetric matrix $H_{x_0}(f)$ is positive definite iff $\Delta_k > 0$ for $k = 1, \ldots, n$ and positive semidefinite only if $\Delta_k \geqslant 0$ for $k = 1, \ldots, n$. We shall not prove this here in general. In Example 1 below we prove it for 2×2 matrices, which will often suffice. There is also a criterion for the negative definite case given below. *Thus if $\Delta_k > 0$ for $k = 1, \ldots, n$, then f has a (local) maximum at the critical point x_0.* This is probably the best way to apply Theorem 12. If $\Delta_k < 0$ for any k, f cannot have a maximum at x_0. Similarly, f has a (local) minimum at x_0 if $H_{x_0}(f)$ is negative definite. By changing the sign of $H_{x_0}(f)$ in the above and using properties of determinants, it follows that $H_{x_0}(f)$ is negative definite iff $\Delta_k < 0$ for k odd and $\Delta_k > 0$ for k even, and $H_{x_0}(f)$ is negative semidefinite only if $\Delta_k \leqslant 0$ for k odd and $\Delta_k \geqslant 0$ for k even. Thus f has a minimum at x_0 if $\Delta_k < 0$ for k odd and $\Delta_k > 0$ for k even.

If $\Delta_k > 0$ for some odd k or $\Delta_k < 0$ for some even k, then f cannot have a minimum value at x_0. In fact, if $\Delta_k < 0$ for some even k, f can have neither a maximum nor minimum at x_0, and x_0 must be a saddle point of f (see Exercise 21).

This theorem is also useful in mechanics when f is the potential of a system, for then a minimum corresponds to stability, and the maxima and saddle points correspond to instability.*

EXAMPLE 1.   Show that the matrix

$$\begin{bmatrix} a & b \\ b & d \end{bmatrix}$$

is positive definite iff $a > 0$ and $ad - b^2 > 0$.

* See Marsden and Tromba, *Vector Calculus*, W. H. Freeman Co. (1975), Chapter 4 for details.

Solution: Positive definite means that

$$[x,y]\begin{bmatrix} a & b \\ b & d \end{bmatrix}\begin{bmatrix} x \\ y \end{bmatrix} > 0 \qquad \text{if } (x,y) \neq (0,0),$$

that is, $ax^2 + 2bxy + dy^2 > 0$. First, suppose this is true for all $(x,y) \neq (0,0)$. Setting $y = 0$, $x = 1$ we get $a > 0$. Setting $y = 1$ we have $ax^2 + 2bx + d > 0$ for all x. This function is a parabola with a minimum (since $a > 0$) at $2ax + 2b = 0$, that is, $x = -b/a$. Hence

$$a\left(-\frac{b}{a}\right)^2 + 2b\left(-\frac{b}{a}\right) + d > 0$$

that is, $ad - b^2 > 0$. The converse may be proved in the same way.

EXAMPLE 2.   Investigate the nature of the critical point $(0,0)$ of

$$f(x,y) = x^2 - xy + y^2 .$$

Solution: Here

$$\frac{\partial f}{\partial x} = 2x - y, \quad \frac{\partial^2 f}{\partial x^2} = 2, \quad \frac{\partial^2 f}{\partial x\,\partial y} = -1, \quad \frac{\partial f}{\partial y} = -x + 2y, \quad \frac{\partial^2 f}{\partial y^2} = 2 ,$$

so the Hessian is

$$\begin{bmatrix} -2 & 1 \\ 1 & -2 \end{bmatrix}.$$

Here $\Delta_1 = -2 < 0$ and $\Delta_2 = 4 - 1 = 3 > 0$ so the Hessian is negative definite. Thus we have a local minimum.

Exercises for Section 6.9

1. Prove

$$\begin{bmatrix} a & b \\ b & d \end{bmatrix}$$

is negative definite iff $a < 0$ and $ad - b^2 > 0$.

2. Investigate the nature of the critical point $(0,0)$ of $f(x,y) = x^2 + 2xy + y^2 + 6$.

3. Investigate the nature of the critical point $(0,0,0)$ of $f(x,y,z) = x^2 + y^2 + 2z^2 + xyz$.

4. (This exercise assumes a good knowledge of linear algebra.) Let A be a symmetric matrix. Show that if A is positive definite the eigenvalues of A (which exist and are real since A is symmetric) are positive.

5. Determine the nature of the critical point $(0,0)$ of $x^3 + 2xy^2 - y^4 + x^2 + 3xy + y^2 + 10$.

Theorem Proofs for Chapter 6

Theorem 1. *Let A be an open set in $\mathbb{R}^n$ and suppose $f\colon A \to \mathbb{R}^m$ is differentiable at x_0. Then $Df(x_0)$ is uniquely determined by f.*

Proof: Let L_1 and L_2 be two linear mappings satisfying conditions of Definition 1. We must show that $L_1 = L_2$. Fix $e \in \mathbb{R}^n$, $\|e\| = 1$, and let $x = x_0 + \lambda e$ for $\lambda \in \mathbb{R}$. Then note that

$$|\lambda| = \|x - x_0\| \quad \text{and} \quad \|L_1 \cdot e - L_2 \cdot e\| = \frac{\|L_1 \cdot \lambda e - L_2 \cdot \lambda e\|}{|\lambda|}.$$

Since A is open, $x \in A$ for λ sufficiently small. By the triangle inequality

$$\|L_1 \cdot e - L_2 \cdot e\| = \frac{\|L_1(x - x_0) - L_2(x - x_0)\|}{\|x - x_0\|}$$

$$\leqslant \frac{\|f(x) - f(x_0) - L_1(x - x_0)\|}{\|x - x_0\|}$$

$$+ \frac{\|f(x) - f(x_0) - L_2(x - x_0)\|}{\|x - x_0\|}.$$

As $\lambda \to 0$ these two terms each $\to 0$, so $L_1 \cdot e = L_2 \cdot e$. Our selection of e was arbitrary, except that $\|e\| = 1$. But for any $y \in \mathbb{R}^n$, $y/\|y\| = e$ has length 1 and, by linearity, if $L_1(e) = L_2(e)$, then $L_1(y) = L_2(y)$. ∎

Theorem 2. *Suppose $A \subset \mathbb{R}^n$ is an open set and $f\colon A \to \mathbb{R}^m$ is differentiable. Then the partial derivatives $\partial f_j/\partial x_i$ exist, and the matrix of the linear map $Df(x)$ with respect to the standard bases in $\mathbb{R}^n$ and $\mathbb{R}^m$ is given by*

$$\begin{pmatrix} \dfrac{\partial f_1}{\partial x_1} & \dfrac{\partial f_1}{\partial x_2} & \cdots & \dfrac{\partial f_1}{\partial x_n} \\[2mm] \dfrac{\partial f_2}{\partial x_1} & \dfrac{\partial f_2}{\partial x_2} & \cdots & \dfrac{\partial f_2}{\partial x_n} \\[2mm] \cdot & & & \\ \cdot & & & \\ \cdot & & & \\ \dfrac{\partial f_m}{\partial x_1} & \dfrac{\partial f_m}{\partial x_2} & \cdots & \dfrac{\partial f_m}{\partial x_n} \end{pmatrix}$$

where each partial derivative is evaluated at $x = (x_1, \ldots, x_n)$.

Proof: By definition of the matrix of a linear mapping, the jith matrix element of $Df(x)$ is given by the jth component of the vector $Df(x) \cdot e_i = Df(x)$ applied to the ith standard basis vector, e_i. Call this component a_{ji}. Now let $y = x + he_i$ for $h \in \mathbb{R}$ and note that

$$\frac{\|f(y) - f(x) - Df(x) \cdot (y - x)\|}{\|y - x\|}$$

$$= \frac{\|f(x_1, \ldots, x_i + h, \ldots, x_n) - f(x_1, \ldots, x_n) - h Df(x) \cdot e_i\|}{|h|}.$$

Since this $\to 0$ as $h \to 0$, then so does the jth component of the numerator, which means that as $h \to 0$,

$$\frac{|f_j(x_1, \ldots, x_i + h, \ldots, x_n) - f_j(x_1, \ldots, x_n) - h a_{ji}|}{|h|} \to 0 .$$

Therefore, we have

$$a_{ji} = \lim_{h \to 0} \frac{[f_j(x_1, \ldots, x_i + h, \ldots, x_n) - f_j(x_1, \ldots, x_n)]}{h} = \frac{\partial f_j}{\partial x_i} . \quad \blacksquare$$

Theorem 3.   *Suppose $A \subset \mathbb{R}^n$ is open and $f: A \to \mathbb{R}^m$ is differentiable on A. Then f is continuous. In fact, for each $x_0 \in A$ there is a constant $M > 0$ and a $\delta_0 > 0$ such that $\|x - x_0\| < \delta_0$ implies $\|f(x) - f(x_0)\| \leqslant M \|x - x_0\|$. (This is called the Lipschitz property.)*

For the proof we need to recall that if $L: \mathbb{R}^n \to \mathbb{R}^m$ is a linear transformation, there is a constant M_0 such that $\|Lx\| \leqslant M_0 \|x\|$ for all $x \in \mathbb{R}^n$ (see Example 4 at the end of Chapter 4). Here we shall be taking $L = Df(x_0)$.

Proof: To prove continuity, it suffices to prove the stated Lipschitz property, for given $\varepsilon > 0$ we can choose $\delta = \min(\delta_0, \varepsilon/M)$. To do this, let $\varepsilon = 1$ in Definition 1. Then there is a δ_0 so that $\|x - x_0\| < \delta_0$ implies

$$\|f(x) - f(x_0) - Df(x_0)(x - x_0)\| \leqslant \|x - x_0\| ,$$

which in turn gives

$$\|f(x) - f(x_0)\| \leqslant \|Df(x_0)(x - x_0)\| + \|x - x_0\|$$

(here we use the triangle inequality in the form $\|y\| - \|z\| \leqslant \|y - z\|$, which follows by writing $y = (y - z) + z$ and applying the usual form of the triangle inequality). Let $M = M_0 + 1$ and use the fact that $\|Df(x_0)(x - x_0)\| \leqslant M_0 \|x - x_0\|$ to give the result. $\blacksquare$

Theorem 4.   *Let $A \subset \mathbb{R}^n$ be an open set and $f: A \subset \mathbb{R}^n \to \mathbb{R}^m$. Suppose $f = (f_1, \ldots, f_m)$. If each of the $\partial f_j / \partial x_i$ exists and is continuous on A, then f is differentiable on A.*

Proof: If $Df(x)$ is to exist, its matrix representation must be the Jacobian matrix by Theorem 2. Thus we need to show that with $x \in A$ fixed, for any $\varepsilon > 0$ there is a $\delta > 0$ such that $\|y - x\| < \delta$, $y \in A$ implies

$$\|f(y) - f(x) - Df(x)(y - x)\| < \varepsilon \|y - x\| .$$

To do this, it suffices to prove this for each component of f separately (why?). Therefore we can suppose $m = 1$.

We can write $f(y) - f(x) = f(y_1, \ldots, y_n) - f(x_1, y_2, \ldots, y_n) + f(x_1, y_2, \ldots, y_n) - f(x_1, x_2, y_3, \ldots, y_n) + f(x_1, x_2, y_3, \ldots, y_n) - f(x_1, x_2, x_3, y_4, \ldots, y_n) + \cdots + f(x_1, \ldots, x_{n-1}, y_n) - f(x_1, \ldots, x_n)$. Now we use the mean-value theorem which enables us to write $f(y_1, \ldots, y_n) - f(x_1, y_2, \ldots, y_n) = \partial f / \partial x_1 (u_1, y_2, \ldots, y_n)(y_1 - x_1)$ for some u_1 between x_1 and y_1 ($y_2, \ldots, y_n$ are fixed). We write similar expressions for the other terms and get

$$f(y) - f(x) = \left(\frac{\partial f}{\partial x_1} (u_1, y_2, \ldots, y_n) \right)(y_1 - x_1) + \left(\frac{\partial f}{\partial x_2} (x_1, u_2, y_3, \ldots, y_n) \right)(y_2 - x_2) + \cdots$$

$$+ \left(\frac{\partial f}{\partial x_n} (x_1, x_2, \ldots, x_{n-1}, u_n) \right)(y_n - x_n) .$$

Therefore, since $Df(x)(y - x) = \sum_{i=1}^{n} \frac{\partial f}{\partial x_i} (x_1, \ldots, x_n)(y_i - x_i)$,

$$\| f(y) - f(x) - Df(x)(y - x) \| \leqslant \left\{ \left| \frac{\partial f}{\partial x_1} (u_1, y_2, \ldots, y_n) - \frac{\partial f}{\partial x_1} (x_1, \ldots, x_n) \right| + \cdots \right.$$

$$\left. + \left| \frac{\partial f}{\partial x_n} (x_1, \ldots, x_{n-1}, u_n) - \frac{\partial f}{\partial x_n} (x_1, \ldots, x_n) \right| \right\} \| y - x \|$$

using the triangle inequality and the fact that $|y_i - x_i| \leqslant \| y - x \|$. But since the terms $\partial f / \partial x_i$ are continuous, and u_i lies between y_i and x_i, there is a $\delta > 0$ such that the term in braces is less than ε for $\| y - x \| < \delta$. This estimate proves the assertion. ∎

Theorem 5. *Let $f: A \to \mathbb{R}^m$ be differentiable on the open set $A \subset \mathbb{R}^n$ and $g: B \to \mathbb{R}^p$ be differentiable on the open set $B \subset \mathbb{R}^m$, and suppose that $f(A) \subset B$. Then the composite $g \circ f$ is differentiable on A and $D(g \circ f)(x_0) = Dg(f(x_0)) \circ Df(x_0)$.*

Proof: To show $D(g \circ f)(x_0) \cdot y = Dg(f(x_0)) \cdot (Df(x_0) \cdot y)$, we want to show that

$$\lim_{x \to x_0} \frac{\| g \circ f(x) - g \circ f(x_0) - Dg(f(x_0)) \cdot (Df(x_0)(x - x_0)) \|}{\| x - x_0 \|} = 0 .$$

To do this, estimate the numerator as follows:

$$\| g \circ f(x) - g \circ f(x_0) - Dg(f(x_0)) \cdot (Df(x_0)(x - x_0)) \|$$

$$= \| g(f(x)) - g(f(x_0)) - Dg(f(x_0))(f(x) - f(x_0)) + Dg(f(x_0))(f(x) - f(x_0)) - Df(x_0)(x - x_0)) \|$$

$$\leqslant \| g(f(x)) - g(f(x_0)) - Dg(f(x_0))(f(x) - f(x_0)) \| + \| Dg(f(x_0))(f(x) - f(x_0)) - Df(x_0)(x - x_0)) \|$$

by the triangle inequality. Since f is differentiable, there is a δ_0 and $M > 0$ such that $\| f(x) - f(x_0) \| \leqslant M \| x - x_0 \|$ whenever $\| x - x_0 \| < \delta_0$, by Theorem 3. Now given $\varepsilon > 0$, there is, by the definition of the derivative, a $\delta_1 > 0$ such that $\| y - f(x_0) \| < \delta_1$ implies

$$\| g(y) - g(f(x_0)) - Dg(f(x_0))(y - f(x_0)) \| < \left(\frac{\varepsilon}{2M} \right) \| y - f(x_0) \| .$$

Thus $\|x - x_0\| < \delta_2 = \min\{\delta_0, \delta_1\}$ implies

$$\frac{\|g(f(x)) - g(f(x_0)) - Dg(f(x_0))(f(x) - f(x_0))\|}{\|x - x_0\|} < \frac{\varepsilon}{2}.$$

Since $Dg(f(x_0))$ is a linear map, we know that there is a constant $\tilde{M}$ such that $\|Dg(f(x_0))(y)\| \leqslant \tilde{M} \cdot \|y\|$ for all $y \in \mathbb{R}^m$, where it can be assumed $\tilde{M} \neq 0$. Now by definition of the derivative there is $\delta_3 > 0$ such that $\|x - x_0\| < \delta_3$ implies

$$\frac{\|f(x) - f(x_0) - Df(x_0)(x - x_0)\|}{\|x - x_0\|} < \frac{\varepsilon}{2\tilde{M}}.$$

Then $\|x - x_0\| < \delta_3$ implies

$$\frac{\|Dg(f(x_0))(f(x) - f(x_0) - Df(x_0)(x - x_0))\|}{\|x - x_0\|}$$

$$\leqslant \frac{\tilde{M} \cdot \|f(x) - f(x_0) - Df(x_0)(x - x_0)\|}{\|x - x_0\|} < \frac{\varepsilon}{2}.$$

Let $\delta = \min\{\delta_2, \delta_3\}$. Thus $\|x - x_0\| < \delta$ implies

$$\frac{\|g \circ f(x) - g \circ f(x_0) - Dg(f(x_0)) \cdot Df(x_0)(x - x_0)\|}{\|x - x_0\|}$$

$$\leqslant \frac{\|g(f(x)) - g(f(x_0)) - Dg(f(x_0))(f(x) - f(x_0))\|}{\|x - x_0\|}$$

$$+ \frac{\|Dg(f(x_0))(f(x) - f(x_0) - Df(x_0)(x - x_0))\|}{\|x - x_0\|} < \frac{\varepsilon}{2} + \frac{\varepsilon}{2} = \varepsilon,$$

which proves the formula. ∎

Theorem 6. *Let $A \subset \mathbb{R}^n$ be open and let $f: A \to \mathbb{R}^n$ and $g: A \to \mathbb{R}$ be differentiable functions. Then gf is differentiable and for $x \in A$, $D(gf)(x): \mathbb{R}^n \to \mathbb{R}^m$ is given by $D(gf)(x) \cdot e = g(x)(Df(x) \cdot e) + (Dg(x) \cdot e)f(x)$ for all $e \in \mathbb{R}^n$.*

Proof: Given $\varepsilon > 0$ and $x_0 \in A$, choose $\delta > 0$ such that $\|x - x_0\| < \delta$ implies
(i) $|g(x)| \leqslant |g(x_0)| + 1 = M$;

(ii) $\|f(x) - f(x_0) - Df(x_0)(x - x_0)\| \leqslant \dfrac{\varepsilon}{3M} \|x - x_0\|$;

(iii) $\|g(x) - g(x_0) - Dg(x_0)(x - x_0)\| \leqslant \dfrac{\varepsilon}{3 \|f(x_0)\|} \|x - x_0\|$;

(iv) $\|g(x) - g(x_0)\| \leqslant \dfrac{\varepsilon}{3M}$

where $\|Df(x_0)y\| \leqslant M\|y\|$, ((iii) and (iv) are needed only if $f(x_0) \neq 0$ and $Df(x_0) \neq 0$). Why is the choice of δ possible?

Then we have for $\|x - x_0\| < \delta$, using the triangle inequality,

$$\|g(x)f(x) - g(x_0)f(x_0) - g(x_0)Df(x_0)(x - x_0) - [Dg(x_0)(x - x_0)]f(x_0)\|$$

$$\leqslant \|g(x)f(x) - g(x)f(x_0) - g(x)Df(x_0)(x - x_0)\|$$

$$+ \|g(x)Df(x_0)(x - x_0) - g(x_0)Df(x_0)(x - x_0)\|$$

$$+ \|g(x)f(x_0) - g(x_0)f(x_0) - [Dg(x_0)(x - x_0)]f(x_0)\|$$

$$\leqslant M \cdot \frac{\varepsilon \|x - x_0\|}{3M} + \frac{\varepsilon}{3M} M \|x - x_0\| + \frac{\varepsilon \|x - x_0\|}{3 \|f(x_0)\|} \cdot \|f(x_0)\|$$

$$= \varepsilon \|x - x_0\| . \quad \blacksquare$$

Theorem 7.

(i) *Suppose* $f: A \subset \mathbb{R}^n \to \mathbb{R}$ *is differentiable on the open set* A. *For any* $x, y \in A$ *such that the line segment joining* x *and* y *lies in* A *there is a point* c *on that segment such that*

$$f(y) - f(x) = Df(c)(y - x) .$$

(ii) *Suppose* $f: A \subset \mathbb{R}^n \to \mathbb{R}^m$ *is differentiable on the open set* A. *Suppose the line segment joining* x *and* y *lies in* A *and* $f = (f_1, \ldots, f_m)$. *Then there exist points* $c_1, \ldots, c_m$ *on that segment such that*

$$f_i(y) - f_i(x) = Df_i(c_i)(y - x), \qquad i = 1, \ldots, m .$$

Proof: (i) Consider the function $h: [0,1] \to \mathbb{R}$ defined by $h(t) = f((1 - t)x + ty)$. The function h is differentiable in t on $]0,1[$. There is a $t_0 \in]0,1[$ such that $h(1) - h(0) = h'(t_0)(1 - 0)$ by the ordinary mean-value theorem. Now $h(1) = f(y)$ and $h(0) = f(x)$. Differentiating using the chain rule, we obtain $h'(t_0) = Df((1 - t_0)x + t_0y)(y - x)$, since the derivative of $(1 - t)x + ty$ with respect to t is $y - x$ (explain). Hence we can take $c = (1 - t_0)x + t_0y$.

(ii) This follows by applying (i) to each component of f separately. $\quad \blacksquare$

Theorem 8. *Let* $f: A \subset \mathbb{R}^n \to \mathbb{R}$ *be twice differentiable on the open set* A. *Then the matrix of* $D^2f(x): \mathbb{R}^n \times \mathbb{R}^n \to \mathbb{R}$ *with respect to the standard basis is given by*

$$\begin{pmatrix} \dfrac{\partial^2 f}{\partial x_1 \, \partial x_2} & \cdots & \dfrac{\partial^2 f}{\partial x_1 \, \partial x_n} \\ \cdot & & \cdot \\ \cdot & & \cdot \\ \cdot & & \cdot \\ \dfrac{\partial^2 f}{\partial x_n \, \partial x_1} & \cdots & \dfrac{\partial^2 f}{\partial x_n \, \partial x_n} \end{pmatrix}$$

where each partial derivative is evaluated at the point $x = (x_1, \ldots, x_n)$.

Proof: The matrix representation of $Df: A \to \mathbb{R}^n$ is given by the row vector $(\partial f/\partial x_1, \ldots, \partial f/\partial x_n)$ so that by Theorem 2, in a version suitable for row vectors $D^2f: A \to \mathbb{R}^{n^2}$ is given by

$$
\begin{pmatrix}
\dfrac{\partial^2 f}{\partial x_1\, \partial x_1} & \dfrac{\partial^2 f}{\partial x_1\, \partial x_2} & \cdots & \dfrac{\partial^2 f}{\partial x_1\, \partial x_n} \\[2ex]
\cdot & & & \cdot \\
\cdot & & & \cdot \\
\cdot & & & \cdot \\[1ex]
\dfrac{\partial^2 f}{\partial x_n\, \partial x_1} & \dfrac{\partial^2 f}{\partial x_n\, \partial x_2} & \cdots & \dfrac{\partial^2 f}{\partial x_n\, \partial x_n}
\end{pmatrix}
$$

Regarding D^2f as a bilinear map will not change the matrix representation as a consideration of the definitions shows. ∎

Theorem 9. *Let $f: A \to \mathbb{R}$ be twice differentiable on the open set A with D^2f continuous (that is, the functions $\partial^2 f/\partial x_i\, \partial x_j$ are continuous). Then D^2f is symmetric, that is,*

$$D^2f(x)(x_1, x_2) = D^2f(x)(x_2, x_1)$$

or, in terms of components,

$$\frac{\partial^2 f}{\partial x_i\, \partial x_j} = \frac{\partial^2 f}{\partial x_j\, \partial x_i}$$

Proof: We want to show $D^2f(x) \cdot (y, z) = D^2f(x) \cdot (z, y)$; that is,

$$\frac{\partial^2 f}{\partial x_i\, \partial x_j} = \frac{\partial^2 f}{\partial x_j\, \partial x_i}.$$

By holding all other variables fixed, we are reduced to the two-dimensional case. Thus we can assume f is of class C^2 on $A \subset \mathbb{R}^2$ and is real-valued.

Consider, for fixed $(x, y) \in A$ and small h, k, the quantity (see Figure 6-14)

$$S_{h,k} = [f(x + h, y + k) - f(x, y + k)] - [f(x + h, y) - f(x, y)].$$

Let us define the function g_k by $g_k(u) = f(u, y + k) - f(u, y)$, and observe that the formula for $S_{h,k}$ can be written

$$S_{h,k} = g_k(x + h) - g_k(x).$$

Thus, by the mean-value theorem, $S_{h,k} = g_k'(c_{k,h}) \cdot h$ for some $c_{k,h}$ lying between x and $x + h$. Hence

$$
\begin{aligned}
S_{h,k} &= \left\{ \frac{\partial f}{\partial x}(c_{k,h}, y + k) - \frac{\partial f}{\partial x}(c_{k,h}, y) \right\} \cdot h \\[1ex]
&= \frac{\partial^2 f}{\partial y\, \partial x}(c_{k,h}, d_{k,h}) \cdot hk
\end{aligned}
$$

for some $d_{h,k}$ lying between y and $y + k$.

Now $S_{h,k}$ is "symmetrical" in h, k and x, y. By interchanging the two middle terms

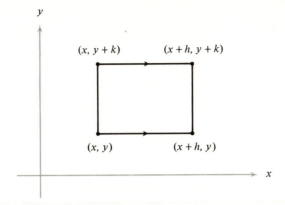

FIGURE 6-14

in $S_{h,k}$, we can derive (in the same way) that

$$S_{h,k} = \frac{\partial^2 f}{\partial x\, \partial y}(\tilde{c}_{h,k}, \tilde{d}_{h,k}) \cdot hk \ .$$

Equating these two formulas for $S_{h,k}$, cancelling h, k, and letting $h \to 0, k \to 0$ (using continuity of $D^2 f$) gives the result. ∎

Note: A refinement is this result: if f is C^1 and $\partial^2 f/\partial x\, \partial y$ exists and is continuous, then $\partial^2 f/\partial y\, \partial x$ exists and these are equal. This requires more work than the above, but the idea is the same (Exercise 24).

Theorem 10. *Let $f: A \to \mathbb{R}$ be of class C^r for $A \subset \mathbb{R}^n$ an open set. Let $x, y \in A$ and suppose that the segment joining x and y lies in A. Then there is a point c on that segment such that*

$$f(y) - f(x) = \sum_{k=1}^{r-1} \frac{1}{k!} D^k f(x)(y-x, \ldots, y-x) + \frac{1}{r!} D^r f(c)(y-x, \ldots, y-x) \ .$$

Proof: If we remember that

$$\frac{d}{dt} f(x + th) = Df(x + th) \cdot h$$

$$= \sum_{i=1}^{n} \frac{\partial f}{\partial x_i}(x + th) h_i$$

from the chain rule, then we can integrate* both sides from $t = 0$ to $t = 1$ to obtain

$$f(x + h) - f(x) = \int_0^1 \sum_{i=1}^{n} \frac{\partial f}{\partial x_i}(x + th) h_i \, dt \ .$$

We now want to integrate the expression on the right-hand side by parts. Remember

* We assume the reader is familiar with the fundamental theorem of calculus. A detailed discussion of integration, including this theorem, is given in Chapter 8.

the general formula,

$$\int_0^1 u \frac{dv}{dt} dt = - \int_0^1 v \frac{du}{dt} dt + uv \Big|_0^1 .$$

In our case, we let $u = (\partial f/\partial x_i)(x + th)h_i$ and let $v = t - 1$. Therefore,

$$\sum_{i=1}^n \int_0^1 \frac{\partial f}{\partial x_i}(x + th)h_i \, dt - \sum_{i,k=1}^n \int_0^1 (1 - t) \frac{\partial^2 f}{\partial x_i \, \partial x_k}(x + th)h_i h_k \, dt + h_i \frac{\partial f}{\partial x_i}(x)$$

since

$$\frac{du}{dt} = \frac{\partial^2 f}{\partial x_i \, \partial x_k}(x + th)h_i h_k$$

from the chain rule, and

$$uv \Big|_0^1 = (t - 1) \frac{\partial f}{\partial x_i}(x + th)h_i \Big|_{t=0} = \frac{\partial f}{\partial x_i}(x)h_i .$$

Thus we have proved the identity

$$f(x + h) - f(x) = \sum_{i=1}^n \frac{\partial f}{\partial x_i}(x) \cdot h_i + R_1(h,x) ,$$

where

$$R_1(h,x) = \sum_{i,k=1}^n \int_0^1 (1 - t) \frac{\partial^2 f}{\partial x_i \, \partial x_k}(x + th)h_i h_k \, dt .$$

Since $|h_i| \leqslant \|h\|$, we have

$$|R_1(h,x_0)| \leqslant \|h\|^2 \left\{ \sum_{i,k=1}^n \int_0^1 (1 - t) \left| \frac{\partial^2 f}{\partial x_i \, \partial x_k}(x_0 + th) \right| dt \right\} .$$

If we integrate $R_1(h,x_0)$ by parts again with

$$u = \frac{\partial^2 f}{\partial x_i \, \partial x_k}(x + th)h_i h_k$$

and

$$v = -(t - 1)^2/2 ,$$

we get

$$R_1(h,x) = \sum_{i,j,k} \int_0^1 \frac{(t - 1)^2}{2} \frac{\partial^3 f}{\partial x_i \, \partial x_j \, \partial x_k}(x + th)h_i h_j h_k \, dt + \sum_{i,j} \frac{1}{2} \frac{\partial^2 f}{\partial x_i \, \partial x_j}(x)h_i h_j .$$

Thus we have proved that

$$f(x + h) = f(x) + \sum_{i=1}^n h_i \frac{\partial f}{\partial x_i}(x) + \sum_{i,j=1}^n \frac{h_i h_j}{2} \frac{\partial^2 f}{\partial x_i \, \partial x_j}(x) + R_2(h,x) ,$$

where

$$R_2(h,x) = \sum_{i,j,k} \int_0^1 \frac{(t - 1)^2}{2} \frac{\partial^3 f}{\partial x_i \, \partial x_j \, \partial x_k}(x + th)h_i h_j h_k \, dt .$$

Now the integrand in the last formula is a continuous function and is therefore bounded on a small neighborhood of x (remember it has to be close to the value of the function at x). Thus for a constant $M \geq 0$ we get, for $\|h\|$ small,

$$|R_2(h,x)| \leq \|h\|^3 \, M .$$

In particular note that $R_2(h,x_0)/\|h\|^2 \leq \|h\| \, M \to 0$ as $h \to 0$. The formulas stated in the theorem for the remainder (Lagrange's form of the remainder) is obtained by applying the second mean-value theorem for integrals. Recall that this states that*

$$\int_a^b f(x)g(x) \, dx = f(c)\int_a^b g(x) \, dx$$

provided f and g are continuous and $g \geq 0$ on $[a,b]$; here c is some number between a and b. Thus we obtain

$$R_1(h,x_0) = \sum_{i,k=1}^n \int_0^1 (1 - t) \frac{\partial^2 f}{\partial x_i \, \partial x_k}(x + th)h_i h_k \, dt$$

$$= \int_0^1 (1 - t)D^2 f(x + th)(h,h) \, dt$$

$$= \frac{1}{2} D^2 f(c)(h,h)$$

where c lies somewhere on the line joining x to $y = x + h$.

Similarly,

$$R_2(h,x_0) = \sum_{i,j,k=1}^n \int_0^1 \frac{(t - 1)^2}{2} \frac{\partial^3 f}{\partial x_i \, \partial x_j \, \partial x_k}(x + th)h_i h_j h_k \, dt$$

$$= \int_0^1 \frac{(t - 1)^2}{2} D^3 f(x + th) \cdot (h,h,h) \, dt$$

$$= \frac{1}{3!} D^3 f(c) \cdot (h,h,h)$$

where c lies somewhere on the line joining x to $y = x + h$. One can proceed by induction using the same method to get the general result. ∎

Remarks: 1. Actually, with more effort one can prove a stronger theorem. Namely if f is C^r, then

$$f(x + h) = f(x) + \sum_{k=1}^r \frac{1}{k!} D^k f(x) \cdot (h, \ldots,h) + R_r(x,h)$$

where $R_r(x,h)/\|h\|^r \to 0$ as $h \to 0$, $h \in \mathbb{R}^n$. We leave the investigation of this point to the interested reader.

2. There is another proof of Theorem 10 which uses Taylor's formula from one variable calculus as follows. Let $g(t) = f(x + t(y - x))$ for $x \in [0,1]$. Applying Taylor's formula on $\mathbb{R}$ we know there is some $\bar{t} \in [0,1]$ such that

$$g(1) - g(0) = \sum_{k=1}^{r-1} \frac{1}{k!} g^{(k)}(0) + \frac{1}{r!} g^{(r)}(\bar{t}) .$$

* See McAloon and Tromba, *Calculus*, Harcourt Brace Jovanovich (1972), p. 280.

Note that $g(1) = f(y)$ and $g(0) = f(x)$. Let $p(t) = x + t(y - x)$. Then $g = f \circ p$ and $Dp(t)(1) = y - x$ for all x, so by Exercise 6b,

$$g^{(k)}(t) = D^k g(t)(1, \ldots, 1) = D^k f(p(t))(Dp(t)(1), \ldots, Dp(t)(1))$$
$$= D^k f(x + t(y - x))(y - x, \ldots, y - x) .$$

Substituting, we get

$$f(y) - f(x) = \sum_{k=1}^{r-1} \frac{1}{k!} D^k f(x)(y - x, \ldots, y - x)$$
$$+ \frac{1}{r!} D^r f(x + \bar{t}(y - x))(y - x, \ldots, y - x)$$

which completes the proof, with $c = x + \bar{t}(y - x)$.

Theorem 11. *If $f: A \subset \mathbb{R}^n \to \mathbb{R}$ is differentiable, A is open, and if $x_0 \in A$ is an extreme point of f, then $Df(x_0) = 0$; that is, x_0 is a critical point.*

Proof: If $Df(x_0) \neq 0$, we can find an $x \in \mathbb{R}^n$ such that $Df(x_0)x = c \neq 0$, say, $c > 0$. Then we can find a $\delta > 0$ such that

$$\|h\| < \delta \Rightarrow \|f(x_0 + h) - f(x_0) - Df(x_0)h\| \leqslant \frac{c}{2\|x\|} \|h\| .$$

Pick $\lambda > 0$ such that $\lambda \|x\| < \delta$. Then $\|f(x_0 + \lambda x) - f(x_0) - Df(x_0)\lambda x\| \leqslant c \lambda \|x\|/2 \|x\| = c\lambda/2$. Now $Df(x_0)\lambda x = \lambda c$. Therefore we must have $f(x_0 + \lambda x) - f(x_0) > 0$. Similarly, $\|f(x_0 - \lambda x) - f(x_0) + Df(x_0)\lambda x\| \leqslant c\lambda/2$ implies $f(x_0 - \lambda x) - f(x_0) < 0$. Since $f(x_0 + \lambda x) > f(x_0)$ and $f(x_0 - \lambda x) < f(x_0)$, we see that $f(x_0)$ is not a local extreme value. That is, we can find points y arbitrarily close to x_0 such that $f(y) > f(x_0)$, and similarly, there are points y arbitrarily close to x_0 such that $f(y) < f(x_0)$. ∎

Theorem 12.

(i) *If $f: A \subset \mathbb{R}^n \to \mathbb{R}$ is a C^2 function defined on an open set A and x_0 is a critical point of f such that $H_{x_0}(f)$ is positive definite, then f has a local maximum at x_0.*

(ii) *If f has a local maximum at x_0, then $H_{x_0}(f)$ is positive semidefinite.*

Proof: (i) $H_{x_0}(f)(x,x) > 0$ for all $x \neq 0$ in $\mathbb{R}^n$ implies $D^2 f(x_0)(x,x) < 0$ for all $x \neq 0$ in $\mathbb{R}^n$. By Example 5, Chapter 4 we know that a bilinear function is continuous. Hence $D^2 f(x_0)(x,x)$ is a continuous function of x. Moreover, $S = \{x \in \mathbb{R}^n \mid \|x\| = 1\}$ is compact, so there is some point $\bar{x} \in S$ such that $0 > D^2 f(x_0)(\bar{x},\bar{x}) \geqslant D^2 f(x_0)(x,x)$ for all $x \in S$. Now let $\varepsilon = -D^2 f(x_0)(\bar{x},\bar{x})$. Then $D^2 f(x_0)(x,x) = \|x\|^2 D^2 f(x_0)(x/\|x\|, x/\|x\|) \leqslant -\varepsilon \|x\|^2$ for any $x \neq 0$ in $\mathbb{R}^n$. Since $D^2 f$ is continuous, there is a $\delta > 0$ such that $\|y - x_0\| < \delta$ implies $\|D^2 f(y) - D^2 f(x_0)\| < \varepsilon/2$ and we may also pick our δ such that $D(x_0, \delta) \subset A$. If $y \in D(x_0, \delta)$, Taylor's theorem may be used to obtain $f(y) - f(x_0) = Df(x_0)(y - x_0) + (1/2)D^2 f(c)(y - x_0, y - x_0)$, where $c \in D(x_0, \delta)$. Thus

$$\|D^2 f(c) - D^2 f(x_0)\| < \varepsilon/2$$

implies

$$D^2f(c)(y - x_0, y - x_0) \leqslant D^2f(x_0)(y - x_0, y - x_0)$$
$$+ \|D^2f(c)(y - x_0, y - x_0) - D^2f(x_0)(y - x_0, y - x_0)\|$$
$$\leqslant -\varepsilon \|y - x_0\|^2 + (\varepsilon/2)\|y - x_0\|^2$$
$$= -(\varepsilon/2)\|y - x_0\|^2 .$$

Remember that $Df(x_0) = 0$, since x_0 is a critical point. Thus Taylor's theorem gives

$$f(y) - f(x_0) = (1/2)D^2f(c)(y - x_0, y - x_0) \leqslant (1/2)(-\varepsilon/2 \|y - x_0\|^2) < 0 .$$

Hence $f(y) < f(x_0)$ for all $y \in D(x_0, \delta)$, $y \neq x_0$ and so f has a local maximum at x_0.

(ii) To prove this part of the theorem we argue by contradiction. Let f have a local maximum at x_0 and suppose $D^2f(x_0)(x,x) > 0$ for some $x \in \mathbb{R}^n$. Now consider $g(t) = -f(x_0 + tx)$. Since f is defined in a neighborhood of x_0, g is defined in a neighborhood of 0. We have $D^2g(0)(1,1) = -D^2f(x_0)(x,x) < 0$. Using the proof of (i), there is a δ such that $|t| < \delta$, $t \neq 0$ implies $g(t) < g(0)$. Thus $|t| < \delta$ implies $f(x_0 + tx) > f(x_0)$, so f does not have a local maximum at x_0. This contradiction implies that $D^2f(x_0)(x,x) \leqslant 0$ for all $x \in \mathbb{R}^n$. Hence $H_{x_0}(f)(x,x) \geqslant 0$ for all $x \in \mathbb{R}^n$. ∎

Worked Examples for Chapter 6

1. Let $f: B \subset \mathbb{R}^n \to \mathbb{R}$ where $B = \{x \in \mathbb{R}^n \mid \|x\| \leqslant 1\}$ be continuous and let f be differentiable on int(B). Suppose $f(x) = 0$ on bd(B). Show that there is a point $x_0 \in$ int(B) for which $Df(x_0) = 0$.

Solution: This is the multidimensional version of Rolle's theorem. If f is identically zero, the theorem is trivial. Therefore, suppose $f(x) \neq 0$ for some $x \in$ int(B). Then f attains a maximum, or minimum, at some interior point, since B is compact. Thus there is an extreme point $x_0 \in$ int(B) and hence by Theorem 11, $Df(x_0) = 0$.

2. Show that for a bilinear map $f: \mathbb{R}^n \times \mathbb{R}^m \to \mathbb{R}^p$, we have $Df(x_0, y_0)(x,y) = f(x_0, y) + f(x, y_0)$. (The map $f(x,y)$ is called bilinear when it is linear in each of x and y separately; see Example 5, Chapter 4.)

Solution: We know that f is differentiable because from its matrix representation, we see that the partial derivatives exist and are continuous. Since $f(x, y_0)$ is a linear function of x, the derivative in direction $(x,0)$ is $Df(x_0, y_0)(x,0) = f(x, y_0)$, as in Example 2, Section 6.2. Similarly, $Df(x_0, y_0)(0,y) = f(x_0, y)$. Thus, since $Df(x_0, y_0)$ is linear, and $(x,y) = (x,0) + (0,y)$, we have $Df(x_0, y_0)(x,y) = f(x_0, y) + f(x, y_0)$.

3. Find the Jacobian of $f(x,y) = (\sin(x \sin y), (x + y)^2)$; $f: \mathbb{R}^2 \to \mathbb{R}^2$.

Solution: We have

$$\frac{\partial f_1}{\partial x} = \frac{\partial}{\partial x}(\sin(x \sin y)) = \cos(x \sin y)\frac{\partial}{\partial x}(x \sin y) = \sin y \cos(x \sin y);$$

$$\frac{\partial f_2}{\partial x} = 2(x + y)\frac{\partial}{\partial x}(x + y) = 2(x + y);$$

$$\frac{\partial f_1}{\partial y} = \cos(x \sin y)\frac{\partial}{\partial y}(x \sin y) = \cos(x \sin y)x \cos y;$$

$$\frac{\partial f_2}{\partial y} = 2(x + y).$$

Thus by Theorem 2, the Jacobian matrix (where $x = x_1$ and $y = x_2$) is

$$\begin{pmatrix} \sin y \cos(x \sin y) & x \cos y \cos(x \sin y) \\ 2(x + y) & 2(x + y) \end{pmatrix}.$$

Jacobian matrices generally are not symmetric and indeed need not be square. Symmetry is only a property of the *second* derivative of a function $f: \mathbb{R}^n \to \mathbb{R}$.

4. Find the critical points of $f(x,y) = x^3 - 3x^2 + y^2$ and determine whether f has a (local) maximum, (local) minimum, or saddle at each of these critical points.

Solution: The critical points are precisely those points (x,y) for which

$$\frac{\partial f}{\partial x} = 3x^2 - 6x = 0$$

and

$$\frac{\partial f}{\partial y} = 2y = 0.$$

Solving for x, we see that $x = 0$ or $x = 2$. Therefore the critical points of f are $(0,0)$ and $(2,0)$. The matrix of the Hessian at (x,y) is

$$\begin{pmatrix} -\dfrac{\partial^2 f}{\partial x \, \partial x} & -\dfrac{\partial^2 f}{\partial x \, \partial y} \\ -\dfrac{\partial^2 f}{\partial y \, \partial x} & -\dfrac{\partial^2 f}{\partial y \, \partial y} \end{pmatrix} = \begin{pmatrix} -6x + 6 & 0 \\ 0 & -2 \end{pmatrix}.$$

At $(0,0)$ the matrix of the Hessian is

$$\begin{pmatrix} 6 & 0 \\ 0 & -2 \end{pmatrix}$$

and $\Delta_1 = +6$, $\Delta_2 = -12$. Hence f has a saddle point at $(0,0)$. At $(2,0)$ the matrix of the Hessian is

$$\begin{pmatrix} -6 & 0 \\ 0 & -2 \end{pmatrix}$$

and $\Delta_1 = -6$, $\Delta_2 = 12$ and so f has a local minimum.

5. Let A be an open convex set in $\mathbb{R}^n$ and let $f\colon \mathbb{R}^n \to \mathbb{R}^m$ be differentiable with a continuous derivative. Suppose $\|Df(x)y\| \leqslant M\,\|y\|$ for all $x \in A$, $y \in \mathbb{R}^n$. Prove the *mean-value inequality*:

$$\|f(x_1) - f(x_2)\| \leqslant M\,\|x_1 - x_2\|\ .$$

Solution: For $n = 1, m = 1$ this follows directly from the mean-value theorem. To get the general case we can proceed as follows. By the chain rule, we have $(d/dt)f(tx_1 + (1 - t)x_2) = Df(tx_1 + (1 - t)x_2) \cdot (x_1 - x_2)$. Integrate both sides in t from $t = 0$ to $t = 1$ to obtain $f(x_1) - f(x_2) = \int_0^1 Df(tx_1 + (1 - t)x_2) \cdot (x_1 - x_2)\, dt$. The integral here is defined as the integral of the component functions. Taking absolute values and using the hypothesis on Df now gives the result desired. We used the fact that the absolute value of an integral is less than or equal to the integral of the absolute value, a fact which will be reviewed in Chapter 8 (the case of vector functions is similar—Exercise 2 at the end of Chapter 8.)

Exercises for Chapter 6

1. If $f\colon A \subset \mathbb{R}^n \to \mathbb{R}^m$ and $g\colon B \subset \mathbb{R}^n \to \mathbb{R}^m$ are differentiable functions on the (open) sets A and B, and α, β are constants, prove that $\alpha f + \beta g\colon A \cap B \subset \mathbb{R}^n \to \mathbb{R}^m$ is differentiable and $D(\alpha f + \beta g)(x) = \alpha Df(x) + \beta Dg(x)$.

2. Show that for $f\colon A \subset \mathbb{R} \to \mathbb{R}^m$, if df_i/dx exists for $i = 1, \ldots, m$, then Df exists.

3. Let $f\colon [0,\infty[\to \mathbb{R}$ be continuous and let f be differentiable on $]0,\infty[$. Assume $f(0) = 0$ and $f(x) \to 0$ as $x \to +\infty$. Show that there is a $c \in]0,\infty[$ such that $f'(c) = 0$.

4. If $f\colon A \subset \mathbb{R}^n \to \mathbb{R}^m$ is a constant function, then show that $Df(x) = 0$ for all $x \in A$.

5. Calculate the Jacobians of the following functions.
   (a) $f(x,y) = \sin(x^2 + y^3)$.   (b) $f(x,y,z) = (z \sin x, z \sin y)$.
   (c) $f(x,y) = xy$.   (d) $f(x,y,z) = x^2 + y^2$.
   (e) $f(x,y) = (\sin(xy),\cos(xy),x^2y^2)$. (f) $f(x,y,z) = x^{y+z}$.
   (g) $f(x,y,z) = xyz$.   (h) $f(x,y,z) = (z^{xy},x^2,\tan(xyz))$.

6. (a) If $f\colon A \subset \mathbb{R}^n \to \mathbb{R}^m$ and $g\colon B \subset \mathbb{R}^m \to \mathbb{R}^p$ are twice differentiable and $f(A) \subset B$, then for $x_0 \in A$, $x, y \in \mathbb{R}^n$, show that

$$D^2(g \circ f(x_0))(x,y) = D^2(g(x_0))(Df(x_0) \cdot x, Df(x_0) \cdot y)$$
$$+ Dg(f(x_0)) \cdot D^2f(x_0)(x,y)\ .$$

   (b) If $p\colon \mathbb{R}^n \to \mathbb{R}^m$ is a linear map plus some constant and $f\colon A \subset \mathbb{R}^m \to \mathbb{R}^s$ is k-times differentiable, prove that

$$D^k(f \circ p)(x_0)(x_1,\ldots,x_k) = D^kf(p(x_0))(Dp(x_0)(x_1),\ldots,Dp(x_0)(x_k))\ .$$

7. Find the critical points of the following functions and determine whether they are local maxima, local minima, or saddle points.
   (a) $f(x,y) = x^3 + 6x^2 + 3y^2 - 12xy + 9x$.
   (b) $f(x,y) = \sin x + y^2 - 2y + 1$.
   (c) $f(x,y,z) = \cos 2x \cdot \sin y + z^2$.
   (d) $f(x,y,z) = (x + y + z)^2$.

8. Show that if $f: A \subset \mathbb{R}^2 \to \mathbb{R}$ has a critical point $x_0 \in A$ and

$$\Delta = \frac{\partial^2 f}{\partial x_1 \, \partial x_1} \cdot \frac{\partial^2 f}{\partial x_2 \, \partial x_2} - \left(\frac{\partial^2 f}{\partial x_1 \, \partial x_2} \right)^2$$

   at x_0, then

   (a) $\Delta > 0$ and $\dfrac{\partial^2 f}{\partial x_1 \, \partial x_1} > 0$ imply f has a local minimum at x_0.

   (b) $\Delta > 0$ and $\dfrac{\partial^2 f}{\partial x_1 \, \partial x_1} < 0$ imply f has a local maximum at x_0.

   (c) $\Delta < 0$ implies x_0 is a saddle point of f.

9. Let $X \subset \mathbb{R}^n$ be an open set with either of the following (non-equivalent) properties.
   (1) For some $x_0 \in X$, each $x \in X$ can be joined to x_0 by a straight line.
   (2) For some $x_0 \in X$, each $x \in X$ can be joined to x_0 by a differentiable path.
   Give some examples of such sets which are not convex. If $f: X \to \mathbb{R}$ with $Df = 0$, then prove that f is constant. Argue that for X open the following are equivalent:
   (a) Condition (2) above,
   (b) path-connectedness,
   (c) connectedness.
   [Hint: See Exercise 11, Chapter 3 for (b) $\Leftrightarrow$ (c). It is easy to show that (a) $\Rightarrow$ (b). For (b) $\Rightarrow$ (a), show first that any two points can be joined by a finite collection of line segments and then "smooth out" the corners.]

10. Prove the analogue of Theorem 12 for minima. [Hint: Apply Theorem 12 to $-f$.]

11. Prove the analogue of Theorem 5, Chapter 5 for $f: A \subset \mathbb{R}^n \to \mathbb{R}^m$.

12. A function $f: \mathbb{R}^n \to \mathbb{R}$ is called homogeneous of degree m if $f(tx) = t^m f(x)$ for all $x \in \mathbb{R}^n$, $t \in \mathbb{R}$. If f is differentiable, show that for $x \in \mathbb{R}^n$,

$$Df(x)x = \cdot m f(x) \,,$$

   that is,

$$\sum_{i=1}^{n} x_i \frac{\partial f}{\partial x_i} = mf(x) \,.$$

   [Hint: Let $g(t) = f(tx)$ and compute $g'(1)$ using the chain rule.] Show that maps multilinear in k variables (see Examples 5, Chapter 4) are homogeneous of degree k. Give other examples.

13. Use the chain rule to find derivatives of the following, where $f(x,y,z) = x^2 + yz$, $g(x,y) = y^3 + xy$, and $h(x) = \sin(x)$.
   (a) $F(x,y,z) = f(h(x),g(x,y),z)$.
   (b) $G(x,y,z) = h(f(x,y,z)g(x,y))$.
   (c) $H(x,y,z) = g(f(x,y,h(x)),g(z,y))$.
   Also find general formulas for the derivatives of F, G, H.

14. (a) Extend Example 2 to multilinear maps.
    (b) By applying the result in (a) to the case of the determinant map det: $\mathbb{R}^{n^2} = \mathbb{R}^n \times \cdots \times \mathbb{R}^n \to \mathbb{R}$, show that $A \in \mathbb{R}^{n^2}$ is a critical point of det iff A has rank $n-2$.

15. Let $f: \mathbb{R} \to \mathbb{R}$ be differentiable. Assume there is no $x \in \mathbb{R}$ such that $f(x) = 0 = f'(x)$. Show that $S = \{x \mid 0 \leqslant x \leqslant 1, f(x) = 0\}$ is finite.

16. If $f: \mathbb{R}^n \to \mathbb{R}^m$ is differentiable and Df is constant, then show that f is linear plus a constant and that the linear part of f is the constant value of Df.

17. If $f: A \subset \mathbb{R}^n \to \mathbb{R}$ is of class C^r and $Df(x_0) = 0$, $D^2 f(x_0) = 0, \ldots, D^{r-1}f(x_0) = 0$ but $D^r f(x_0)(x, \ldots, x) < 0$ for all $x \in \mathbb{R}^n, x \neq 0$, then prove that f has a local maximum at x (use Taylor's formula).

18. Prove that the equation $x^3 + bx + c = 0$ where $b > 0$ has exactly one solution $x \in \mathbb{R}$. [Hint: Use Rolle's theorem.]

19. In each of the following problems, determine the second-order Taylor formula for the given function about the given point (x_0, y_0).
    (a) $f(x,y) = (x + y)^2$, $x_0 = 0$, $y_0 = 0$.
    (b) $f(x,y) = e^{x+y}$, $x_0 = 0$, $y_0 = 0$.
    (c) $f(x,y) = \dfrac{1}{x^2 + y^2 + 1}$, $x_0 = 0$, $y_0 = 0$.
    (d) $f(x,y) = e^{-x^2-y^2} \cos(xy)$, $x_0 = 0$, $y_0 = 0$.
    (e) $f(x,y) = \sin(xy) + \cos(xy)$, $x_0 = 0$, $y_0 = 0$.
    (f) $f(x,y) = e^{(x-1)^2} \cos y$, $x_0 = 1$, $y_0 = 0$.

20. Let $L: \mathbb{R}^n \to \mathbb{R}^m$ be a linear map. Define $\|L\| = \inf\{M \mid \|Lx\| \leqslant M \|x\|$ for all $x \in \mathbb{R}^n\}$. Show that $\|\cdot\|$ is a norm on the space of linear maps of $\mathbb{R}^n$ to $\mathbb{R}^m$.

21. (a) If, for $f: A \subset \mathbb{R}^n \to \mathbb{R}$, at x_0, $\Delta_k > 0$ for k odd, or $\Delta_k < 0$ for k even, then show that f cannot have a (local) minimum at x_0.
    (b) If $\Delta_k < 0$ for k even, prove that f has a saddle point at x_0.

22. Give an example of a continuous map $f:]0,1[\to \mathbb{R}$ whose graph is not closed. Can this happen for $f: A \subset \mathbb{R} \to \mathbb{R}$ where A is closed?

23. Write down the first four terms in the Taylor expansion of $\log(\cos x)$ about $x = 0$.

24. Let $f(x,y)$ be a real-valued function on $\mathbb{R}^2$. Use the proof of Theorem 9 to show that if f is of class C^1 and $\partial^2 f/\partial x\, \partial y$ exists and is continuous, then $\partial^2 f/\partial y\, \partial x$ exists, and

$$\frac{\partial^2 f}{\partial x\, \partial y} = \frac{\partial^2 f}{\partial y\, \partial x}$$

(this is weaker than saying that f is of class C^2). Generalize.

25. Let $f: \mathbb{R}^n \to \mathbb{R}$ and suppose $\partial f/\partial x_i, i = 1, \ldots, n$ exist and $\partial f/\partial x_i, i = 1, \ldots, n-1$ are continuous. Then prove that f is differentiable.

26. (a) If $f: \mathbb{R} \to \mathbb{R}$ and f' exists on a neighborhood of $x = a$ and limit $f'(x) = l$,
    then prove that $f'(a) = l$. [Hint: Use the mean-value theorem.] $_{x \to a+}$

    (b) Can $f(x) = \begin{cases} 1, & x < 0, \\ 0, & x \geqslant 0, \end{cases}$ be the derivative of any function?

27. Let $f: A \to \mathbb{R}$ be continuous, $A \subset \mathbb{R}^n$ open. Assume that all directional derivatives exist and define at each $x_0 \in A$, a linear map $Df(x_0)$. Must f be differentiable? [Hint: Consider $f(x,y) = x^2 y \sqrt{x^2 + y^2}/(x^4 + y^2)$, a function suggested by F. Weisler.]

28. Let f be differentiable on $[a,b]$. Verify that $f'(x)$ satisfies the conclusion of the intermediate-value theorem (remember f' need not be continuous) [Hint: If we seek x_0 such that $f'(x_0) = c$, consider $g(x) = f(x) - cx$ and $\inf g(x)$.]

29. Let $f_n(x) = xe^{-nx}$, $x \in [0, \infty[$, $n = 0, 1, 2, \ldots$.
    (a) Show $f(x) = \sum_{n=0}^{\infty} f_n(x)$ exists. Compute f explicitly.
    (b) Is f continuous?
    (c) Find a suitable set on which the convergence is uniform.
    (d) May we differentiate term by term?

30. Suppose $f: \mathbb{R} \to \mathbb{R}$ is bounded and has a continuous derivative. What is right and what is wrong in the following string of conclusions?

    We want to prove that the set T of all points at which f assumes its (absolute) maximum is closed. Since f is differentiable it is continuous. Hence it assumes its maximum, that is, T is not empty. Denote by S the set of points at which $f'(x) = 0$. Then $T \subset S$. On the other hand, if $x \in S$, then $f'(x) = 0$, hence f either achieves a maximum or a minimum there. If it achieves a maximum, we must have $f(x) \geqslant 0$. Hence $T = S \cap \{x \mid f(x) \geqslant 0\}$. $\{x \mid f(x) \geqslant 0\}$ is closed and so is S; therefore T is closed.

    Is T really closed or not?

31. Let $A \subset \mathbb{R}^n$ be compact and construct the normed space $\mathscr{C}(A, \mathbb{R})$ as in Chapter 5. Define, for $x_0 \in A$, $\delta_{x_0}: \mathscr{C}(A, \mathbb{R}) \to \mathbb{R}; f \mapsto f(x_0)$. Prove δ_{x_0} is continuous.

32. Let $f: \mathbb{R}^2 \to \mathbb{R}$, $f(x,y) = (xy(x^2 - y^2))/(x^2 + y^2)$ if $(x,y) \neq (0,0)$ and $f(0,0) = 0$. Show $\partial^2 f/\partial x\, \partial y$ and $\partial^2 f/\partial y\, \partial x$ exist at $(0,0)$ but are not equal.

33. Use Taylor's theorem to prove the binomial theorem

$$(a + x)^n = \sum_{k=0}^{n} \binom{n}{k} a^k x^{n-k} .$$

34. Consider the sequence of real numbers (a continued fraction)

$$\frac{1}{2}, \frac{1}{2 + (1/2)}, \frac{1}{2 + 1/(2 + 1/2)}, \ldots$$

    Show that it is convergent and find the limit [Hint: Prove that the even terms and the odd terms are monotone.]

35. Let $f:]a,b[\to \mathbb{R}$ be twice differentiable. Suppose f vanishes at three distinct points. Prove there is a $c \in]a,b[$ such that $f''(c) = 0$.

36. Suppose $f: [0,1] \to \mathbb{R}$ is continuous, f is differentiable on $]0,1[$, and $f(0) = 0$. Assume $|f'(x)| \leq |f(x)|, 0 < x < 1$. Prove $f(x) = 0$ for all $x \in [0,1]$.

37. Let $f: \mathbb{R}^2 \to \mathbb{R}$ be C^2. f is called *harmonic* if $\partial^2 f/\partial x^2 + \partial^2 f/\partial y^2 = 0$. Assume (x_0, y_0) is a strict local maximum and f is harmonic. Prove that all second derivatives of f vanish at (x_0, y_0).

38. Find the equation of the plane tangent to the following surfaces at the indicated points.
    (a) $z = x^2 + y^2$, $(0,0)$;
    (b) $z = x^2 - y^2 + x$, $(1,0)$;
    (c) $z = (x + y)^2$, $(3,2)$.

39. Analyze the behavior of the following functions at the indicated points.
    (a) $z = x^2 - y^2 + 3xy$, $(0,0)$;
    (b) $z = Ax^2 - By^2 + Cxy$, $(0,0)$.

40. Find the equation of the tangent plane to the surface S given by the graph of
    (a) $f(x,y) = \sqrt{x^2 + y^2} + (x^2 + y^2)$ at $(1,0,2)$;
    (b) $f(x,y) = \sqrt{x^2 + 2xy - y^2 + 1}$ at $(1,1,\sqrt{3})$.

Chapter 7

The Inverse and Implicit Function Theorems and Related Topics

$\mathbf{W}$e know from linear algebra that a system of linear equations

$$a_{11}x_1 + \cdots + a_{1n}x_n = y_1$$

$$\cdot \qquad\qquad \cdot \qquad \cdot$$
$$\cdot \qquad\qquad \cdot \qquad \cdot$$
$$\cdot \qquad\qquad \cdot \qquad \cdot$$

$$a_{n1}x_1 + \cdots + a_{nn}x_n = y_n$$

can be solved uniquely for $x_1, \ldots, x_n$ if the matrix $A = (a_{ij})$ is non-singular, that is, if $\det(A) \neq 0$, where $\det(A)$ denotes the determinant of A. What about functional equations? When can we solve a system of the form

$$f_1(x_1, \ldots, x_n) = y_1$$

$$\cdot \qquad\qquad \cdot$$
$$\cdot \qquad\qquad \cdot$$
$$\cdot \qquad\qquad \cdot$$

$$f_n(x_1, \ldots, x_n) = y_n$$

for $x_1, \ldots, x_n$? The object which generalizes the determinant is the *Jacobian determinant* defined by $Jf(x) = \det(Df(x))$, where $x = (x_1, \ldots, x_n)$ and

$f = (f_1, \ldots, f_n)$. Written out in coordinates, at $x = (x_1, \ldots, x_n)$,

$$Jf(x) = \begin{vmatrix} \dfrac{\partial f_1}{\partial x_1} & \cdots & \dfrac{\partial f_1}{\partial x_n} \\[2mm] \cdot & & \cdot \\ \cdot & & \cdot \\ \cdot & & \cdot \\[2mm] \dfrac{\partial f_n}{\partial x_1} & \cdots & \dfrac{\partial f_n}{\partial x_n} \end{vmatrix}.$$

Sometimes one writes

$$\frac{\partial(f_1, \ldots, f_n)}{\partial(x_1, \ldots, x_n)}$$

instead of Jf and

$$\frac{\partial(f_1, \ldots, f_n)}{\partial(x_1, \ldots, x_n)} (x_1, \ldots, x_n)$$

instead of $Jf(x)$. If $Jf(x) \neq 0$, one might expect to be able to solve $f(x) = y$ for x. The theorem which justifies such results is the main subject of Section 7.1. We shall also consider the case when we wish to solve $f(x,y) = 0$ for y (implicit function theorem). In the latter sections we shall apply some similar existence theorems to ordinary differential equations and an important theoretical result called the "Morse lemma." The final section is concerned with extremum problems in the presence of constraints.

7.1   Inverse Function Theorem

Notice that $Jf(x) \neq 0$ implies that $Df(x)$: $\mathbb{R}^n \to \mathbb{R}^n$ is a linear isomorphism (that is, its matrix is invertible). Thus, from the fact that the best linear approximation is invertible, we want to conclude that the function itself is invertible.

There are, however, some restrictions. To appreciate these, examine the case f: $\mathbb{R} \to \mathbb{R}$. It is true that if f is C^1 and if $f'(x_0) \neq 0$, then f is invertible (one-to-one) in a neighborhood of x_0. Geometrically this is quite clear, for $f'(x_0) \neq 0$ means f has a non-zero slope at, and consequently near, x_0 (see Figure 7-1).

Thus our main concern will be with *local invertibility*, that is, with invertibility of $f(x)$ for x near x_0 and y near $y_0 = f(x_0)$.

It is easy to compute the derivative of the inverse function $f^{-1}(y)$ from

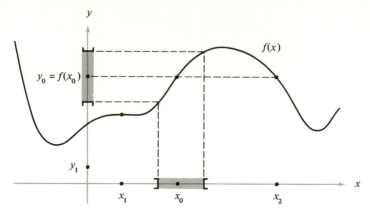

FIGURE 7-1

the chain rule: from $f^{-1}(f(x)) = x$, we get $(df^{-1}/dy) \cdot f'(x) = 1$, so

$$\left.\frac{df^{-1}}{dy}\right|_{y=f(x)} = \frac{1}{df/dx}.$$

To actually check that f^{-1} is differentiable requires a little more care.

If $f'(x_0) = 0$, then f may or may not be invertible near x_0; in Figure 7-1, f is not invertible near x_1, but $f(x) = x^3$ is invertible near $x_0 = 0$. In the case where $f'(x_0) = 0$, then, no conclusion can be drawn (some further analysis would be required). In general, $f'(x_0) \neq 0$ does not guarantee that we can solve $f(x) = y$ for all y. For example, there is no x_3 such that $f(x_3) = y_1$ for y_1 as in Figure 7-1. Also, from the same figure we see that solutions are generally not unique, for $f(x_0) = f(x_2)$. There will be a unique solution only if our attention is restricted to a suitably small neighborhood of x_0.

Therefore, all we can expect is that f is invertible near $f(x_0)$. That is, for y close to $f(x_0)$ we can solve uniquely for some x near x_0 such that $f(x) = y$. The question of "how near?" is a subtle one requiring detailed analysis of the proof. Fortunately, for many purposes, this is not important.

Theorem 1 includes the single variable situation just described as a special case.

> **Theorem 1.** Let $A \subset \mathbb{R}^n$ be an open set and let $f \colon A \subset \mathbb{R}^n \to \mathbb{R}^n$ be of class C^1 (that is, Df exists and is continuous). Let $x_0 \in A$ and suppose $Jf(x_0) \neq 0$. Then there is a neighborhood U of x_0 in A and an open neighborhood W of $f(x_0)$ such that $f(U) = W$ and f has a C^1 inverse $f^{-1} \colon W \to U$. Moreover, for $y \in W$, $x = f^{-1}(y)$, we have
>
> $$Df^{-1}(y) = [Df(x)]^{-1}$$

the inverse of $Df(x)$ meaning the inverse as a linear mapping (corresponding to the inverse matrix). If f is of class C^p, $p \geqslant 1$, then so is f^{-1}.*

Saying that f has an inverse f^{-1} means exactly that we can uniquely solve $f(x) = y$ for $x \in U$ given any $y \in W$.

The proof of the theorem depends on a certain existence argument. That is, for y near y_0 we need to prove the existence of an x such that $f(x) = y$. The basic technical tool which is used is the contraction mapping principle; see Section 5.6. In Section 5.6 we saw how that result could be used to prove the existence of solutions to some simple integral equations. In Section 7.5 we shall use these same sorts of arguments to solve differential equations as well.

EXAMPLE 1.   Consider the equations $(x^4 + y^4)/x = u(x,y)$, $\sin x + \cos y = v(x,y)$. Near which points (x,y) can we solve for x, y in terms of u, v?

Solution: Here the functions are $u(x,y) = f_1(x,y) = (x^4 + y^4)/x$ and $v(x,y) = f_2(x,y) = \sin x + \cos y$. We want to know the points near which we can solve for x, y as functions of u and v and to compute $\partial x/\partial u$, and so forth. According to the inverse function theorem we must first compute $\partial(f_1,f_2)/\partial(x,y)$. Observe that for $f = (f_1,f_2)$ we take its domain to be $A = \{(x,y) \in \mathbb{R}^2 \mid x \neq 0\}$. Now

$$\frac{\partial(f_1,f_2)}{\partial(x,y)} = \begin{vmatrix} \dfrac{\partial f_1}{\partial x} & \dfrac{\partial f_1}{\partial y} \\[2mm] \dfrac{\partial f_2}{\partial x} & \dfrac{\partial f_2}{\partial y} \end{vmatrix} = \begin{vmatrix} \dfrac{3x^4 - y^4}{x^2} & \dfrac{4y^3}{x} \\[2mm] \cos x & -\sin y \end{vmatrix}$$

$$= \frac{(\sin y)}{x^2}(y^4 - 3x^4) - \frac{4y^3}{x}\cos x \ .$$

Therefore, at points where this does not vanish we can solve for x, y in terms of u and v. In other words, we can solve for x, y near those x, y for which $x \neq 0$ and $(\sin y)(y^4 - 3x^4) \neq 4xy^3 \cos x$. Such conditions generally cannot be solved explicitly. For example, if $x_0 = \pi/2$, $y_0 = \pi/2$, we can solve for x, y near x_0, y_0, because there, $\partial(f_1,f_2)/\partial(x,y) \neq 0$.

The derivatives $\partial x/\partial u$, etc., are obtained according to Theorem 1 by inverting the Jacobian matrix. In the 2×2 case this comes down to the

---

* If $f: A \subset \mathbb{R}^n \to \mathbb{R}^m$ is C^1 and $Df(x_0)$ is one-to-one, then f is also locally one-to-one near x_0. Similarly if $Df(x_0)$ is onto, then f is onto some neighborhood of $f(x_0)$. These more general results follow from Theorem 1 by the methods of Section 7.2; see Exercise 11 at the end of this chapter.

following:

$$\frac{\partial x}{\partial u} = \frac{1}{Jf(x,y)} \frac{\partial v}{\partial y}, \qquad \frac{\partial x}{\partial v} = \frac{-1}{Jf(x,y)} \frac{\partial u}{\partial y};$$

$$\frac{\partial y}{\partial u} = \frac{-1}{Jf(x,y)} \frac{\partial v}{\partial x}, \qquad \frac{\partial y}{\partial v} = \frac{1}{Jf(x,y)} \frac{\partial u}{\partial x}$$

(see Example 2, at the end of the chapter for more details).

In this example,

$$\frac{\partial x}{\partial u} = \frac{-(x^2 \sin y)}{\{(\sin y)(y^4 - 3x^4) - 4y^3 x \cos x\}}.$$

Notice that the answer is expressed in terms of x and y and not u, v. Thus $\partial x/\partial u$ is evaluated at the point $u(x,y), v(x,y)$.

The inverse function theorem is useful because it tells us that there are solutions to equations and it explains how to differentiate the solutions, although it may be impossible to solve the equations explicitly.

EXAMPLE 2.   Let $u(x,y) = e^x \cos y$, $v(x,y) = e^x \sin y$. Show that $(x,y) \mapsto (u(x,y),v(x,y))$ is locally invertible, but is not invertible.

Solution: Here

$$\frac{\partial(u,v)}{\partial(x,y)} = \begin{vmatrix} \dfrac{\partial u}{\partial x} & \dfrac{\partial u}{\partial y} \\[2mm] \dfrac{\partial v}{\partial x} & \dfrac{\partial v}{\partial y} \end{vmatrix} = \begin{vmatrix} e^x \cos y & -e^x \sin y \\[2mm] e^x \sin y & e^x \cos y \end{vmatrix}$$

$$= e^{2x}(\cos^2 y + \sin^2 y) = e^{2x} \neq 0.$$

Hence by the inverse function theorem the map is locally invertible. It is not (globally) one-to-one, however, because

$$u(x,y + 2\pi) = u(x,y), \qquad v(x,y + 2\pi) = v(x,y).$$

Notice that for $f: \mathbb{R} \to \mathbb{R}$ if f is differentiable and if $f'(x) \neq 0$ for all x, then $f'(x)$ is either >0 or <0 since f' satisfies the intermediate value theorem (see Exercise 28, Chapter 6), hence f must be (globally) one-to-one as f is always increasing or decreasing. The example above shows that this need not be the case in $\mathbb{R}^2$.

Exercises for Section 7.1

1. Let $u(x,y) = x^2 - y^2$, $v(x,y) = 2xy$. Show that the map $(x,y) \mapsto (u,v)$ is locally invertible at all points $(x,y) \neq (0,0)$.

2. Compute $\dfrac{\partial x}{\partial u}, \dfrac{\partial x}{\partial v}, \dfrac{\partial y}{\partial u}, \dfrac{\partial y}{\partial v}$ in Exercise 1.

3. Let $f(x) = x + 2x^2 \sin(1/x)$, $x \neq 0$, $f(0) = 0$. Show that $f'(0) \neq 0$ but that f is not locally invertible near 0. Why does this not contradict Theorem 1?

4. Let $L: \mathbb{R}^n \to \mathbb{R}^n$ be a linear isomorphism, and $f(x) = L(x) + g(x)$, where $\|g(x)\| \leqslant M \|x\|^2$ and f is C^1. Show f is locally invertible near 0.

5. Investigate whether the system

$$u(x,y,z) = x + xyz \; ;$$
$$v(x,y,z) = y + xy \; ;$$
$$w(x,y,z) = z + 2x + 3z^2$$

can be solved for x, y, z in terms of u, v, w near $(0,0,0)$.

7.2   Implicit Function Theorem

In studying the implicit function theorem we are again interested in the existence and differentiability of certain functions. Undoubtedly, the student has worked with functions defined implicitly before; however, he or she may not know why the manipulations are justified. Possible questions we would like to ask will be more obvious after looking at some examples.

Suppose we consider those x and y related by an equation $F(x,y) = 0$. We would like to say that this defines a function $y = f(x)$ (one says that $y = f(x)$ is *defined implicitly*), and we would like to compute dy/dx. As in the previous section, given such an F, one generally cannot solve for y explicitly, so it is important to know that such a function does indeed exist without having to solve for it.

To motivate the next result, consider the function $F(x,y) = x^2 + y^2 - 1$. We are interested in those x and y related by $F(x,y) = 0$, which is just the unit circle. A function $f(x)$ is a "solution" iff $F(x,f(x)) = 0$ for all x in the domain of f. Clearly, f must be given by $f(x) = \pm\sqrt{1 - x^2}$, and either of these is a solution. We note therefore that f need not be unique. Given (x_0,y_0) such that $F(x_0,y_0) = 0$, we would like to know if we can find $f(x)$ such that $F(x,f(x)) = 0$ and f is differentiable and *unique near* (x_0,y_0). If $x_0 \neq \pm 1$, this is true if f is taken to be the appropriate square root. The given y_0 determines which square root must be selected. See Figure 7-2. The points $x_0 = \pm 1$ are exceptional for several reasons. First, f is not differentiable

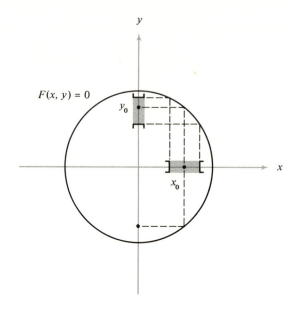

FIGURE 7-2

there and second, near $x_0 = \pm 1$, f could be either square root, so it is not uniquely determined. These exceptional points are exactly the places where $\partial F/\partial y = 0$. Thus, in general, we want some condition like $\partial F/\partial y \neq 0$ to guarantee that, locally at least, we can find a unique differentiable f such that $F(x, f(x)) = 0$.

In the general case we shall have a function $F \colon \mathbb{R}^n \times \mathbb{R}^m \to \mathbb{R}^m$, and consider the relation $F(x, y) = 0$, or written out,

$$F_1(x_1, \ldots, x_n, y_1, \ldots, y_m) = 0$$

$$\cdot \qquad\qquad\qquad \cdot$$
$$\cdot \qquad\qquad\qquad \cdot$$
$$\cdot \qquad\qquad\qquad \cdot$$

$$F_m(x_1, \ldots, x_n, y_1, \ldots, y_m) = 0 \ ,$$

and we want to solve for these m unknowns $y_1, \ldots, y_m$ from the m equations in terms of $x_1, \ldots, x_n$.

The theorem is as follows.

Theorem 2. (Implicit Function theorem).   Let $A \subset \mathbb{R}^n \times \mathbb{R}^m$ be an open set and let $F \colon A \to \mathbb{R}^m$ be a function of class C^p (that is F has p

continuous derivatives where p is a positive integer). Suppose $(x_0, y_0) \in$
A and $F(x_0, y_0) = 0$. Form

$$
\Delta = \begin{vmatrix}
\dfrac{\partial F_1}{\partial y_1} & \cdots & \dfrac{\partial F_1}{\partial y_m} \\
\cdot & & \cdot \\
\cdot & & \cdot \\
\cdot & & \cdot \\
\dfrac{\partial F_m}{\partial y_1} & \cdots & \dfrac{\partial F_m}{\partial y_m}
\end{vmatrix}
$$

evaluated at (x_0, y_0), where $F = (F_1, \ldots, F_m)$. Suppose that $\Delta \neq 0$.
Then there is an open neighborhood $U \subset \mathbb{R}^n$ of x_0 and a neighborhood
V of y_0 in $\mathbb{R}^m$ and a unique function $f: U \to V$ such that

$$F(x, f(x)) = 0$$

for all $x \in U$. Furthermore, f is of class C^p.

Actually, we shall see that this theorem follows fairly easily from the
inverse function theorem. The intuitive reason for the validity of the theorem
and the necessity for the restriction $\Delta \neq 0$ should be clear from the above
example. From the equation $F(x, f(x)) = 0$ one can determine Df using the
chain rule. First, take the case $m = 1$. Then, by the chain rule,

$$0 = \frac{\partial}{\partial x_i} F(x, f(x)) = \frac{\partial F}{\partial x_i} + \frac{\partial F}{\partial y} \frac{\partial f}{\partial x_i}.$$

So we get the important equation (notice the minus sign):

$$\frac{\partial f}{\partial x_i} = -\frac{\partial F / \partial x_i}{\partial F / \partial y}.$$

The reader is especially warned that in

$$\frac{(\partial F / \partial x_i)}{(\partial F / \partial y)}$$

it is *incorrect* to "cancel" the ∂F's to obtain $\partial y / \partial x_i$. Thus, while such memory
devices are sometimes useful, they do have limitations.

We can formulate the general solution analogous to the above.

Corollary 1. *In Theorem 2, $\partial f_j/\partial x_i$ are given by*

$$
\begin{pmatrix}
\dfrac{\partial f_1}{\partial x_1} & \cdots & \dfrac{\partial f_1}{\partial x_n} \\
& & \\
\cdot & & \cdot \\
\cdot & & \cdot \\
\cdot & & \cdot \\
& & \\
\dfrac{\partial f_m}{\partial x_1} & \cdots & \dfrac{\partial f_m}{\partial x_n}
\end{pmatrix}
= -
\begin{pmatrix}
\dfrac{\partial F_1}{\partial y_1} & \cdots & \dfrac{\partial F_1}{\partial y_m} \\
& & \\
\cdot & & \cdot \\
\cdot & & \cdot \\
\cdot & & \cdot \\
& & \\
\dfrac{\partial F_m}{\partial y_1} & \cdots & \dfrac{\partial F_m}{\partial y_m}
\end{pmatrix}^{-1}
\begin{pmatrix}
\dfrac{\partial F_1}{\partial x_1} & \cdots & \dfrac{\partial F_1}{\partial x_n} \\
& & \\
\cdot & & \cdot \\
\cdot & & \cdot \\
\cdot & & \cdot \\
& & \\
\dfrac{\partial F_m}{\partial x_1} & \cdots & \dfrac{\partial F_m}{\partial x_n}
\end{pmatrix}
$$

where $^{-1}$ denotes the inverse matrix.

The proof is similar to the case $m = 1$ given above and will be left as an exercise.

EXAMPLE 1.   Consider the system of equations

$$xu + yv^2 = 0 \; ;$$
$$xv^3 + y^2u^6 = 0 \; .$$

Are they uniquely solvable for u, v in terms of x and y near $x = 0$, $y = 1$, $u = 0$, $v = 0$? Compute $\partial u/\partial x$ at $x = 0$, $y = 1$ if it exists.

Solution: Here we have $F(x,y,u,v) = 0$ where F stands for the left-hand sides of the given equations. We want to see if we can solve for $u(x,y)$, $v(x,y)$. Thus we form

$$
\Delta =
\begin{vmatrix}
\dfrac{\partial F_1}{\partial u} & \dfrac{\partial F_1}{\partial v} \\
& \\
\dfrac{\partial F_2}{\partial u} & \dfrac{\partial F_2}{\partial v}
\end{vmatrix}
=
\begin{vmatrix}
x & 2yv \\
6y^2u^5 & 3xv^2
\end{vmatrix}
$$

which, at the given point is equal to 0. Thus the implicit function theorem states that we cannot expect to uniquely solve for u, v in terms of x and y. To actually determine solvability would require a direct analysis not provided by the implicit function theorem.

Exercises for Section 7.2

1. Check directly where we can solve the equation $F(x,y) = y^2 + y + 3x + 1 = 0$ for y in terms of x.

2. Check that your answer in Exercise 1 agrees with the answer you expect from the implicit function theorem. Compute dy/dx.

3. Consider $(x,y) \mapsto \left(\dfrac{x^2 - y^2}{x^2 + y^2}, \dfrac{xy}{x^2 + y^2} \right)$. Does this have a local inverse near $(0,1)$?

4. Discuss the solvability in the system

$$3x + 2y + z^2 + u + v^2 = 0 \; ;$$
$$4x + 3y + z + u^2 + v + w + 2 = 0 \; ;$$
$$x + z + w + u^2 + 2 = 0 \; ,$$

for u, v, w in terms of x, y, z near $x = y = z = 0, u = v = 0, w = -2$.

5. Discuss the solvability of

$$y + x + uv = 0 \; ;$$
$$uxy + v = 0 \; ,$$

for u, v in terms of x, y near $x = y = u = v = 0$ and check directly.

7.3 Straightening-Out Theorem

We now give another consequence of the implicit function theorem which is an important technical tool in the study of surfaces. This result states, roughly speaking, that if $f : A \subset \mathbb{R}^n \to \mathbb{R}$ has a non-zero derivative at a point x_0, then in a neighborhood of x_0, f can be "straightened out"; in fact, f can be deformed into the map which is the projection onto the coordinate axis x_n by composing it with a "coordinate change," which means (by definition) a smooth function which has a smooth inverse. See Figure 7-3, where the coordinate change is denoted h, and in which h straightens out the surfaces of constant f to be planes. The exact result is stated in the next theorem.

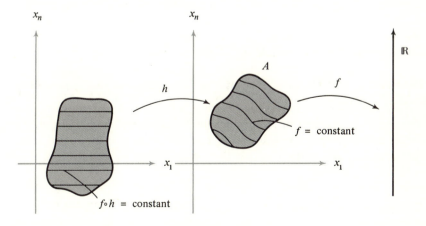

FIGURE 7-3

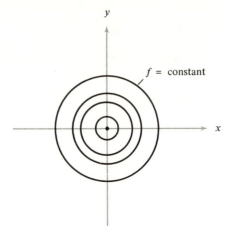

FIGURE 7-4

Theorem 3. *Let $A \subset \mathbb{R}^n$ be an open set and let $f : A \to \mathbb{R}$ be a function of class C^p, $p \geqslant 1$. Let $x_0 \in A$ and suppose $Df(x_0) \neq 0$. Then there is an open set U, an open set V containing x_0, and a function $h: U \to V$ of class C^p, with inverse $h^{-1}: V \to U$ of class C^p, such that*

$$f(h(x_1, \ldots, x_n)) = x_n .$$

This theorem has a generalization to functions $f: A \subset \mathbb{R}^n \to \mathbb{R}^m$, $m \leqslant n$, given in Exercise 3 at the end of the chapter.

The plausibility of the theorem is seen from Figure 7-3. The function of h is to twist things in a way so that the level surfaces of f become planes of dimension $n - 1$. The condition $Df(x_0) \neq 0$ comes in to guarantee that the surfaces $f = $ constant are "non-degenerate" or intuitively, have dimension $n - 1$. An example will clarify this point.

EXAMPLE 1.   Let $f(x,y) = x^2 + y^2$. Can we "straighten out" f near $(0,0)$?

Solution: No, not necessarily, because $Df(0,0) = 0$. Indeed, this is clear intuitively because the surfaces of constant f degenerate at $(0,0)$ from being circles to being a point (Figure 7-4). Clearly, there is no way we can deform the surfaces $f = $ constant near $(0,0)$ to planes. But we can do this at any point $(x_0, y_0) \neq (0,0)$.

EXAMPLE 2.   Let $f(x,y) = x^3 + x + y$. Can f be "straightened out" near $(0,0)$?

Solution: Yes, for $Df(0,0) = (1,1) \neq 0$.

Exercises for Section 7.3

1. At what (x,y) can $f(x,y) = x^2 - y^2$ be "straightened out"?

2. Sketch the graphs of $f =$ constant in Exercise 1 and explain your answer geometrically.

3. Can $f(x,y) = x^3 + y^2 + 1$ be straightened out near $(0,0)$? Near $(0,1)$?

7.4   Further Consequences of the Implicit Function Theorem

Theorem 3 says that we can find a function h which "straightens out" the domain of f so that $f \circ h$ is simply a projection. Analogous to this we can look for a function g which "straightens out" the range of f so that $g \circ f$ looks like a projection.

> **Theorem 4.**   Let $A \subset \mathbb{R}^p$ be an open set and $f: A \to \mathbb{R}^n$ a function of class C^r and $p \leqslant n$. Let $x_0 \in A$ and suppose the rank* of $Df(x_0)$ is p. Then there are open sets U and V in $\mathbb{R}^n$ with $f(x_0) \in U$ and a function $g: U \to V$ of class C^r with inverse $g^{-1}: V \to U$ also of class C^r such that $g \circ f(x_1,. . .,x_p) = (x_1,. . .,x_p,0,. . .,0)$ for all $(x_1,. . .,x_p) \in A$.

The intuition is given in Figure 7-5, which should be compared to Figure 7-3. In the present case the function g flattens out the image of f. Notice that this is intuitively correct; we expect the range of f to be a p-dimensional "surface" so it should be possible to flatten it to a piece of $\mathbb{R}^p$. Note that the range of a linear map of rank p is a linear subspace of dimension exactly p, so this result expresses, in a sense, a generalization of the linear case.

To use Theorems 3 (or 4) we must have the rank of Df equal to the dimension of its image space (or the domain space). However, we can use the inverse function theorem again to tell us that if $Df(x)$ has constant rank m in a neighborhood of x_0, we can straighten out the domain of f with some invertible function h such that $f \circ h$ depends only on $x_1, \ldots, x_m$. Then we can also apply Theorem 4. This is the essence of the following theorem and its corollary. Roughly speaking, the theorem says that if Df has rank m on $\mathbb{R}^n$, then $n - m$ variables are redundant and can be eliminated. For example, if $f(x,y) = x - y$, $f: \mathbb{R}^2 \to \mathbb{R}^1$, Df has rank 1, and so we can express f using just one variable, namely, let $h(x,y) = (x + y,y)$ so that $f \circ h(x,y) = x$, which depends only on x.

---

* Recall that the rank of a linear map is the dimension of its image. Equivalently, by linear algebra, the rank is the size of the largest square submatrix with non-zero determinant (see any linear algebra text, such as O'Nan, *Linear Algebra* for details).

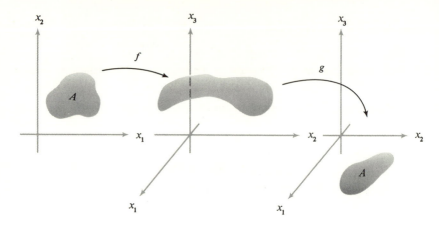

FIGURE 7-5

Theorem 5.   *Let* $f: A \subset \mathbb{R}^n \to \mathbb{R}^N$ *(where A is open in* $\mathbb{R}^n$*) be a* C^r
function such that $Df(x)$ *has rank m for all x in a neighborhood of*
$x_0 \in A$. *Then there is an open set* $U \subset \mathbb{R}^n$ *and an open set* $V \subset \mathbb{R}^n$
with $x_0 \in V$ *and a function* $h: U \to V$ *of class* C^r *with inverse*
$h^{-1}: V \to U$ *of class* C^r *such that* $f \circ h$ *depends only on* $x_1, \dots, x_m$.
That is $f \circ h(x_1, \dots, x_m, x_{m+1}, \dots, x_n) = \tilde{f}(x_1, \dots, x_m)$ *for some* C^r
function $\tilde{f}$. *See Figure* 7-6.

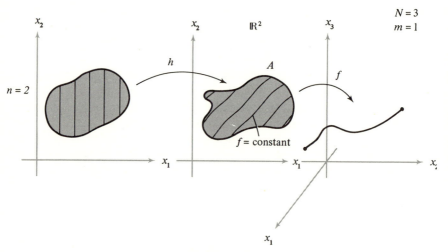

FIGURE 7-6

Corollary 2. Let $f: A \subset \mathbb{R}^n \to \mathbb{R}^N$ (where A is open in $\mathbb{R}^n$) be a function of class C^r such that $Df(x)$ has rank m for all x in a neighborhood of $x_0 \in A$. Then there is an open set $U_1 \subset \mathbb{R}^n$, an open set $U_2 \subset \mathbb{R}^n$ with $x_0 \in U_2$, an open set V_1 around $f(x_0)$, an open set $V_2 \subset \mathbb{R}^N$, and functions $h: U_1 \to U_2$ and $g: V_1 \to V_2$ of class C^r with inverses of class C^r such that $g \circ f \circ h(x_1, \ldots, x_n) = (x_1, \ldots, x_m, 0, \ldots, 0)$.

Some further applications of the implicit function theorem to surface theory and Lagrange multipliers (extremum problems with constraints) are given in Section 7.7. Also, in the sections below some (optional) topics are treated by these same or similar methods.

EXAMPLE 1. Let $f: \mathbb{R}^2 \to \mathbb{R}^3$, $(x,y) \mapsto (x + y^3, xy, y + y^2)$. Can the range of f be "straightened out" near $(0,0)$?

Solution: Here we employ Theorem 4. First, we compute the Jacobian matrix:

$$\begin{pmatrix} 1 & 3y^2 \\ y & x \\ 0 & 1 + 2y \end{pmatrix}$$

which, at $(0,0)$, is

$$\begin{pmatrix} 1 & 0 \\ 0 & 0 \\ 0 & 1 \end{pmatrix}.$$

This matrix has rank 2 (since there is a 2×2 submatrix with non-zero determinant). Hence Theorem 4 applies, so we can straighten out the range. It will be, intuitively, a two-dimensional surface near $(0,0)$.

EXAMPLE 2. Let $f: \mathbb{R}^2 \to \mathbb{R}$, $f(x,y) = x^2 + y$. Can f be expressed as a function of only one variable near $(0,0)$?

Solution: Yes, since (by Theorem 5), $Df(0,0) = (0,1) \neq 0$. Note that this can also be answered by using Theorem 3.

Exercises for Section 7.4

1. Let $f: \mathbb{R}^2 \to \mathbb{R}^3$, $(x,y) \mapsto (x + y^2, xy, y^2)$. Can the range be straightened out near $(0,0)$? Near $(0,1)$?

2. What does Theorem 5 say about $f: \mathbb{R}^3 \to \mathbb{R}^2$, $(x,y,z) \mapsto (x^2 + 2y^2, z^2 + 3xy)$ near $(0,0,0)$? Near $(0,1,0)$?

3. What does Corollary 2 say about $f: \mathbb{R}^3 \to \mathbb{R}^3$, $(x,y,z) \mapsto (x + 2y, 6x + 12y, x + y^3 + z^3)$ near $(0,0,0)$?

4. Examine the statement of Corollary 2 in the case where f is a linear mapping.

7.5 An Existence Theorem for Ordinary Differential Equations

In calculus we learn how to solve simple linear differential equations; for example, one learns that the solution to $d^2x/dt^2 + k^2x = 0$ is $x(t) = A \cos(kt - \omega)$ for constants A and ω. It is interesting to investigate whether or not general differential equations always have solutions. This will be the main concern here. The methods one uses are constructive and suitable for numerical computation; that is, a definite sequence of approximating solutions is constructed.

An example may clarify matters.

EXAMPLE 1. Consider the non-linear equation $dx/dt = x^2$, $x(0) = 1$. Can we compute $x(1)$?

Solution: In this case we can solve the equation explicitly: we have $dx/x^2 = dt$, so integrating, $-1/x = t + C$, that is, $x = -1/(t + C)$. At $t = 0$, $x = 1$, so $C = -1$. Thus $x = 1/(1 - t)$ is our solution. This is the only solution starting at $t = 0$, with $x(0) = 1$. At $t = 1$ the solution $x(t)$ blows up. Thus $x(1)$ is not defined. Note that we cannot find a differentiable solution $x(t)$ defined for all $t \geqslant 0$. (Figure 7-7).

This example points out the important fact that in general our solutions $x(t)$ may be defined and differentiable only for a small t-interval.

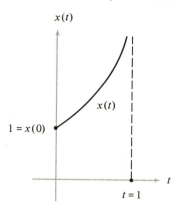

FIGURE 7-7

Another statement is important here. If we allow vector differential equations, then higher order equations may be reduced to first-order ones. Example 2 will illustrate this.

EXAMPLE 2.   Reduce $d^2x/dt^2 + kx = 0$ to a first-order equation.

Solution: We let $y = dx/dt$ and write:

$$\begin{cases} \dfrac{dx}{dt} = y \\[2mm] \dfrac{dy}{dt} = -kx \end{cases}$$

which is first order in the vector (x,y), and is equivalent to the original equation.

The main existence and uniqueness theorem will now be given. In the theorem, we write $\bar{D}(x_0,r)$ for the closed ball of radius r about x_0, so

$$\bar{D}(x_0,r) = \{y \in \mathbb{R}^n \mid \|x_0 - y\| \leqslant r\} \ .$$

Theorem 6.   *Let $f: [-a,a] \times \bar{D}(x_0,r) \to \mathbb{R}^n$ be a given continuous mapping. Let $C = \sup\{\|f(t,x)\| \mid (t,x) \in [-a,a] \times \bar{D}(x_0,r)\}$. Suppose there is a constant K such that*

$$\|f(t,x) - f(t,y)\| \leqslant K \|x - y\|$$

for all $t \in [-a,a]$, $x, y \in \bar{D}(x_0,r)$. Let $b < \min\{a,r/C,1/K\}$. Then there is a unique continuously differentiable map $x: [-b,b] \to \bar{D}(x_0,r) \subset \mathbb{R}^n$ such that

$$\begin{cases} x(0) = x_0 & \textit{(initial condition)} \ , \\[2mm] \dfrac{dx}{dt} = f(t,x(t)) \ . \end{cases}$$

The main condition on f is this *Lipschitz condition:*

$$\|f(t,x) - f(t,y)\| \leqslant K \|x - y\| \ .$$

Here K is called the *Lipschitz constant* and we say f is *Lipschitz* in the variable x. To verify this condition one often uses the following device.

Device: If $D_x f(t,x)$ denotes the derivative of f for fixed t, and $\|D_x f(t,x)y\| \leqslant K \|y\|$ for all $y \in \mathbb{R}^n$ then f is Lipschitz with Lipschitz constant K. For example, if $n = 1$ this holds if $|\partial f(t,x)/\partial x| \leqslant K$ on $-a \leqslant t \leqslant a$, $-r \leqslant x - x_0 \leqslant r$.

One sees this by using the chain rule as follows:

$$\frac{d}{ds} f(t,y + s(x - y)) = D_x f(t,y + s(x - y)) \cdot (x - y) \,,$$

so, integrating between $s = 0$ and $s = 1$,

$$f(t,x) - f(t,y) = \int_0^1 D_x f(t,y + s(x - y)) \cdot (x - y) \, ds \,.$$

Taking absolute values then yields the result. The device is the method we normally would use to determine K. Note that if f is C^1 such a K will always exist (why?).

Often f is independent of t, in which case we say we have an *autonomous system*. If f is merely continuous, the existence (but not uniqueness) of $x(t)$ in Theorem 6 is true; see Exercise 45 at the end of **Chapter 5**.

The idea of the proof of Theorem 6 is to use successive approximations; start with

$$x_1(t) = x_0,$$

and write

$$x_2(t) = x_0 + \int_0^t f(s,x_1(s)) \, ds$$

$$x_3(t) = x_0 + \int_0^t f(s,x_2(s)) \, ds$$

$$\begin{array}{ccc} \cdot & \cdot & \cdot \\ \cdot & \cdot & \cdot \\ \cdot & \cdot & \cdot \end{array}$$

$$x_n(t) = x_0 + \int_0^t f(s,x_{n-1}(s)) \, ds \,.$$

Then one wants to prove $x_n(t)$ converges to a solution $x(t)$ which will satisfy

$$x(t) = x_0 + \int_0^t f(s,x(s)) \, ds$$

(this equation is equivalent to the differential equation plus the initial condition).

If we compare this with Chapter 5, Section 6, we see that what is really going on is the search for a fixed point of the map of one function to another given by

$$y(t) \mapsto x_0 + \int_0^t f(s,y(s)) \, ds$$

and we might expect that we can use the contraction mapping principle. We can indeed and this is how the actual proof goes.

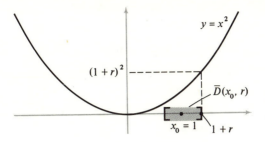

FIGURE 7-8

EXAMPLE 3.   Compute b for Example 1.

Solution: Here $dx/dt = x^2, x(0) = 1$ is our equation. Let, for the moment, a, r be undetermined. Now

$$C = \sup\{|f(t,x)| \mid -a \leqslant t \leqslant a, -r \leqslant x - 1 \leqslant r\}$$
$$= \sup\{x^2 \mid -r \leqslant x - 1 \leqslant r\}$$
$$= (r + 1)^2$$

(see Figure 7-8). Thus $r/C = r/(r + 1)^2$. Also, $\partial f/\partial x = 2x$, so

$$K = \sup\{2 |x| \mid -r \leqslant x - 1 \leqslant r\}$$
$$= 2(r + 1).$$

Since a is not involved we can just choose a large enough so that it does not interfere, say, $a = 100$. Then, by the theorem, we must choose

$$b < \min\left\{\frac{r}{(r + 1)^2}, \frac{1}{2(r + 1)}\right\}.$$

This will work for any choice of r. For example, if we let $r = 1$ we get a time of existence $b < 1/4$. This is not as good as we found directly (a time of existence <1) but one can reapply the theorem to get a new time of existence at $t = 1/4$ and gradually work out to any $t < 1$. But we could never go past $t = 1$.

Exercises for Section 7.5

1. Solve $dx/dt = 1 + x^2$, $x(0) = 0$ by the method of successive approximations. Is $x(t)$ defined for all $t \geqslant 0$?

2. Compute b from Theorem 6 for Exercise 1.

3. Show that $dx/dt = \sqrt{x}$, $x(0) = 0$ has two solutions:

$$x(t) = 0 \quad \text{and} \quad x(t) = \begin{cases} 0, & t \leqslant 0, \\ \dfrac{t^2}{4}, & t > 0. \end{cases}$$

Does this contradict Theorem 6?

4. Consider the equation $dx/dt = te^{x^2} \sin x$, $x(0) = 1$. Obtain an estimate on how long we can define the solution $x(t)$.

5. Let A be an $n \times n$ matrix and consider the linear system

$$\frac{dx}{dt} = A \cdot x(t), \qquad x(t) \in \mathbb{R}^n.$$

(a) Show that a solution is

$$x(t) = e^{tA}x(0), \qquad \text{where } e^B = \sum_{n=0}^{\infty} \frac{B^n}{n!}.$$

(b) The time of existence here extends for all t; can this fact also be derived from Theorem 6?

7.6 The Morse Lemma

In Chapter 6, Section 9 we saw that the Hessian of a function $f: \mathbb{R}^n \to \mathbb{R}$ at a critical point determined the local behavior of f near this point. The Morse lemma carries this result one step further. It states that if, for example, f has a local minimum at x_0, not only does f *look like* a paraboloid but that we can change the coordinates (as in Sections 7.3 and 7.4) so that f really "*is*" *a paraboloid* in the new coordinates. The "lemma" (it really is a "theorem") also applies to saddle surfaces.

The Morse lemma is fundamental in more advanced work in topology and analysis, but even here it helps us understand the shape of functions near a critical point.

> **Theorem 7.** *Let $A \subset \mathbb{R}^n$ be open and $f: A \to \mathbb{R}$ a smooth (that is, f is infinitely differentiable) function. Suppose $Df(x_0) = 0$ and the Hessian of f at x_0 is non-singular. Then there is a neighborhood U of x_0 and a neighborhood V of 0 in $\mathbb{R}^n$ and a smooth map $g: V \to U$ with a smooth inverse such that $f \circ g = h$ has the form*
>
> $$h(y) = f(x_0) - [y_1^2 + y_2^2 + \cdots + y_\lambda^2] + [y_{\lambda+1}^2 + \cdots + y_n^2],$$
>
> *where λ is some fixed integer between 0 and n.*

One calls g a *change of coordinates* and we speak of $y = g^{-1}(x)$ as the *new coordinates*.

A critical point at which the Hessian matrix $\Delta = -\partial^2 f/\partial x_i \, \partial x_j$ is non-singular is called a *non-degenerate critical point*. Thus Theorem 7 gives a rather complete description of functions in the neighborhood of a non-degenerate critical point. The number λ is called the *index* of the critical point. Figure 7-9 illustrates the graphs of the quadratic forms $-y_1^2 - \cdots - y_\lambda^2 + y_{\lambda+1}^2 + \cdots + y_n^2$ for various indices in $\mathbb{R}^2$.

For functions of two variables it is easy to determine the index; namely, if Δ is positive definite (see Section 6.9) the index is 2; if Δ is negative definite the index is zero and otherwise it is one. (Note how this ties together with Theorem 12 of Chapter 6.)

In general, to find the index one needs to know a little more linear algebra. The knowledgable reader can check that the index is exactly the number of positive eigenvalues of Δ.

EXAMPLE 1. What is the shape of the surface $z = x^2 + 2xy + 2y^2 + y^3$ near $(0,0)$?

Solution: We have a critical point at $(0,0)$ and the Hessian is

$$\Delta = \left[-\frac{\partial^2 f}{\partial x_i \, \partial x_j} \right] = \begin{bmatrix} -2 & -2 \\ -2 & -4 \end{bmatrix},$$

which is negative definite since $-2 < 0$, and $\det(\Delta) = (-2)(-4) - (-2)(-2) > 0$. Thus the index is 0 and near $(0,0)$ the surface is approximately a paraboloid and in some other coordinate system it is exactly a paraboloid.

EXAMPLE 2. Compute the index of $x^2 - 3xy + y^2 + 8xy^2 + 6$ at $(0,0)$.

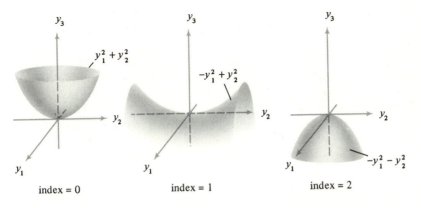

index = 0   index = 1   index = 2

FIGURE 7-9 (a) Index $= 0$. (b) Index $= 1$. (c) Index $= 2$.

Solution: (0,0) is a critical point and the Hessian is

$$\Delta = \begin{bmatrix} -2 & 3 \\ 3 & -2 \end{bmatrix}$$

which is neither positive definite nor negative definite. Thus we have index 1 and hence a saddle point at (0,0).

Exercises for Section 7.6

1. Compute the index of $2x^2 + 6xy - y^2 + y^4$ at (0,0).

2. What is the shape of the surface $x^2 + 3xy - y^2$ at (0,0)?

3. Does Theorem 7 apply to $x^2 - 2xy + y^2$? What happens?

4. Let $f(x,y) = x^2 + y^2 + 3y^3 + 8x^4 + x^2 e^x \sin x + 6$. Show that there exist new coordinates ξ, η, where

$$\xi = \xi(x,y), \qquad \eta = \eta(x,y)$$

for which

$$f(x,y) = \xi^2 + \eta^2 + 6$$

in a whole neighborhood of (0,0).

5. (a) If f has a non-degenerate critical point at $x_0 \in \mathbb{R}^n$ show that there is a neighborhood of x_0 containing no other critical points.
   (b) What are the critical points of the function $f(x,y) = x^2 y^2$?

7.7 Constrained Extrema and Lagrange Multipliers

In some problems we want to maximize a function subject to certain *constraints* or *side conditions*. Such situations arise, for example, in economics. Suppose you are selling two kinds of goods, say, I and II; let x and y represent the quantity of each sold. Then let $f(x,y)$ represent the profit we earn when x amount of I and y amount of II is sold. But our production is limited by our capital, so we are constrained to work subject to a relation, say, $g(x,y) = 0$. Thus we want to maximize $f(x,y)$ among those x, y satisfying $g(x,y) = 0$. The condition $g(x,y) = 0$ is called the *constraint* in the problem.

The purpose of this section is to briefly discuss some methods which will enable us to handle this and similar situations. Theorem 8 is the main result.

> **Theorem 8.** *Let* $f: U \subset \mathbb{R}^n \to \mathbb{R}$ *and* $g: U \subset \mathbb{R}^n \to \mathbb{R}$ *be given* C^1
> *functions. Let* $x_0 \in U$, $g(x_0) = c_0$ *and let* $S = g^{-1}(c_0)$, *the level set*
> *for* g *with value* c_0. *Assume* $\nabla g(x_0) \neq 0$. *If* $f \mid S$, *which denotes* f

restricted to S (that is, to those $x \in U$ satisfying $g(x) = c_0$) has a maximum or minimum at x_0, then there is a real number λ such that

$$\nabla f(x_0) = \lambda \, \nabla g(x_0) \, .$$

The idea of the proof is as follows. Recall that the tangent space to S at x_0 is defined as the space orthogonal to $\nabla g(x_0)$ (see Section 6.6). We motivated this definition by considering tangents to paths $c(t)$ which lie in S, as follows: if $c(t)$ is a path in S, $c(0) = x_0$, then $c'(0)$ is a tangent vector to S at x_0 since

$$\frac{d}{dt} g(c(t)) = \frac{d}{dt} c_0 = 0$$

and, on the other hand, by the chain rule,

$$\frac{d}{dt} g(c(t)) \bigg|_{t=0} = \nabla g(x_0) \cdot c'(0) \, ,$$

so $c'(0)$ is orthogonal to $\nabla g(x_0)$.

Now if $f \mid S$ has a maximum at x_0, then certainly $f(c(t))$ has a maximum at $t = 0$. Hence,

$$0 = \frac{d}{dt} f(c(t)) \bigg|_{t=0} = \nabla f(x_0) \cdot c'(0) \, .$$

Thus $\nabla f(x_0)$ is also perpendicular to the tangent space to S at x_0 and so $\nabla f(x_0)$ and $\nabla g(x_0)$ are parallel. Since $\nabla g(x_0) \neq 0$ it follows that $\nabla f(x_0)$ is a multiple of $\nabla g(x_0)$, which is exactly the conclusion of the theorem.

Let us extract from this proof the geometry of the situation and formulate a corollary as follows.

Corollary 3. *If f, when constrained to a surface S, has a maximum or minimum at x_0, then $\nabla f(x_0)$ is perpendicular to S at x_0 (see Figure 7-10).*

These results tell us that to find the constrained extrema of f we must look among those x_0 satisfying the conclusions of the theorem or the corollary. We shall give several illustrations of how to use each.

When the method in Theorem 8 is used we must look for a point x_0 and a constant λ, called a *Lagrange multiplier*, such that $\nabla f(x_0) = \lambda \, \nabla g(x_0)$. This method is more analytical in nature while the method of Corollary 3 is more geometrical.

Unfortunately, for constrained problems there is no simple test to distinguish maxima from minima as there was in Section 6.9 for unconstrained extrema. Therefore one must examine each x_0 separately using the given data or other geometric arguments.

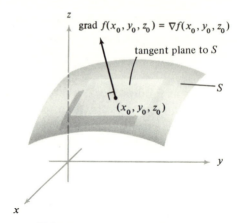

FIGURE 7-10   The geometry of constrained extrema.

EXAMPLE 1.   Let $S \subset \mathbb{R}^2$ be a line through $(-1,0)$ inclined at $45°$, and let $f: \mathbb{R}^2 \to \mathbb{R}$, $(x,t) \mapsto x^2 + t^2$. Find the minimum of f on S.

Solution: Here $S = \{(x,y) \mid y - x - 1 = 0\}$ so we choose $g(x,y) = y - x - 1$. The relative extrema of $f \mid S$ must be found among the points at which ∇f is orthogonal to S, that is, is inclined at $-45°$. But $\nabla f(x,t) = (2x, 2t)$ and has the desired slope whenever $x = -t$, or (x,t) lies in the line L, through the origin inclined at $-45°$. This can occur for a point (x,t) lying in the set S only for the single point at which L and S intersect (see Figure 7-11). Reference to the level curves of f indicates that this point, $(-1/2, 1/2)$, is a relative minimum of $f \mid S$ (but not of f).

EXAMPLE 2.   Let $f: \mathbb{R}^2 \to \mathbb{R}: (x,y) \mapsto x^2 - y^2$, and S be the circle around the origin of radius 1. Find the critical points of f on S.

Solution: Here $S = g^{-1}(1)$, where $g: \mathbb{R}^2 \to \mathbb{R}$, $(x,y) \mapsto x^2 + y^2$. The level curves, tangent spaces, and gradients are shown in Figure 7-12. Clearly, the gradient of f is orthogonal to S at the four points $(0, \pm 1), (\pm 1, 0)$, which are relative minima and maxima, respectively, of $f \mid S$.

This problem can be performed analytically by the method of Lagrange multipliers. Clearly,

$$f(x,y) = \left(\frac{\partial f}{\partial x}, \frac{\partial f}{\partial y}\right) = (2x, -2y)$$

and

$$\nabla g(x,y) = (2x, 2y) .$$

Thus, according to Theorem 1, we seek to find a λ such that

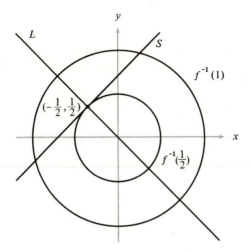

FIGURE 7-11   Locating the critical points of *f* restricted to *S*.

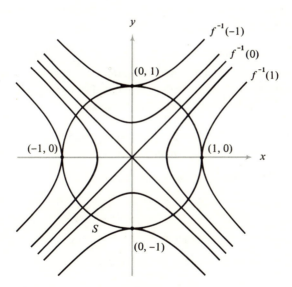

FIGURE 7-12

$$\begin{cases} (2x, -2y) = \lambda(2x, 2y) \\ (x,y) \in S, \qquad \text{or } x^2 + y^2 = 1 \ . \end{cases}$$

These are three equations which can be solved for the three unknowns x, y, and λ. From $2x = \lambda 2x$ we conclude either $x = 0$ or $\lambda = 1$. If $x = 0$ then $y = \pm 1$ and $-2y = \lambda 2y$ implies $\lambda = -1$. If $\lambda = 1$, then $y = 0$ and $x = \pm 1$. Thus we get the same points $(0, \pm 1), (\pm 1, 0)$ as before. As we have mentioned, the method only locates potential extrema; whether they are maxima or minima or neither must be determined by other means.

If the surface S is defined by a number of constraints,

$$g_1(x_1, \ldots, x_n) = c_1$$
$$g_2(x_1, \ldots, x_n) = c_2$$
$$.$$
$$.$$
$$.$$
$$g_k(x_1, \ldots, x_n) = c_k$$

(above we just had one g), then Theorem 8 may be generalized as follows. If f has a maximum or minimum at x_0 on S, there must exist constants $\lambda_1, \ldots, \lambda_k$ such that

$$\nabla f(x_0) = \lambda_1 \, \nabla g_1(x_0) + \cdots + \lambda_k \, \nabla g_k(x_0) \ .$$

This may be proved by generalizing the method used to prove Theorem 8. This argument is left to the interested reader. Let us now give an example of how this more general formulation may be used.

EXAMPLE 3.   Find the extreme points of $f(x,y,z) = x + y + z$ subject to the conditions $x^2 + y^2 = 2$, and $x + z = 1$.

Solution: Here there are two constraints,

$$g_1(x,y,z) = x^2 + y^2 - 2 = 0$$

and

$$g_2(x,y,z) = x + z - 1 = 0 \ .$$

Thus we must find x, y, z and λ_1 and λ_2 such that

$$\nabla f(x,y,z) = \lambda_1 \, \nabla g_1(x,y,z) + \lambda_2 \, \nabla g_2(x,y,z)$$

and

$$\begin{cases} g_1(x,y,z) = 0 \\ g_2(x,y,z) = 0 \ , \end{cases}$$

that is,

$$\begin{cases} 1 = \lambda_1 \cdot 2x + \lambda_2 \cdot 1 \\ 1 = \lambda_1 \cdot 2y + \lambda_2 \cdot 0 \\ 1 = \lambda_1 \cdot 0 + \lambda_2 \cdot 1 \\ x^2 + y^2 = 2 \\ x + z = 1 \ . \end{cases}$$

and

These are five equations for x, y, z, λ_1, and λ_2. From the third, $\lambda_2 = 1$ and so $2x\lambda_1 = 0$, $2y\lambda_1 = 1$. Since the second implies $\lambda_1 \neq 0$, we have $x = 0$. Thus $y = \pm\sqrt{2}$ and $z = 1$. Hence our points are $(0, \pm\sqrt{2}, 1)$. By inspection one can show that $(0, \sqrt{2}, 1)$ gives a maximum, $(0, -\sqrt{2}, 1)$ gives a minimum.

EXAMPLE 4.   Maximize $f(x,y,z) = x + z$ subject to the constraint $x^2 + y^2 + z^2 = 1$.

Solution: Here we use Theorem 1. We seek λ and (x,y,z) such that

$$1 = 2x\lambda$$
$$0 = 2y\lambda$$
$$1 = 2z\lambda$$

and

$$x^2 + y^2 + z^2 = 1 \ .$$

Since $\lambda \neq 0$, we get $y = 0$. From the first and third equations, $x = z$ and $4\lambda^2 x^2 - 4\lambda^2 z^2 = 0$; from the fourth, $4\lambda^2 x^2 + 4\lambda^2 z^2 = 4\lambda^2$, which together imply $8\lambda^2 x^2 = 4\lambda^2$ and so $x = \pm 1/\sqrt{2} = z$. Hence our points are $(1/\sqrt{2}, 0, 1/\sqrt{2})$ and $(-1/\sqrt{2}, 0, -1/\sqrt{2})$. Clearly, the first yields the maximum of f, the second the minimum. Since S is compact, f must achieve a maximum and a minimum on S.

EXAMPLE 5.   Find the largest volume a rectangular box can have subject to the constraint that the surface area be fixed at 10 square meters.

Solution: Here, if x, y, z are the lengths of the sides, the volume is $f(x,y,z) = xyz$. The constraint is that $2(xy + xz + yz) = 10$, that is, $xy + xz + yz = 5$. Thus our conditions are

$$\begin{cases} yz = \lambda(y + z) \\ xz = \lambda(x + z) \\ xy = \lambda(y + x) \\ xy + xz + yz = 5 \ . \end{cases}$$

First of all, $x \neq 0$, for $x = 0$ implies $yz = 5$ and $0 = \lambda z$, so $\lambda = 0$ and $yz = 0$. Similarly, $y \neq 0$, $z \neq 0$, and $x + y \neq 0$, and so forth. Elimination of λ from the first two equations gives $yz/(y + z) = xz/(x + z)$, which gives $x = y$. Similarly, $y = z$. Using the last equation, $3x^2 = 5$, that is, $x = \sqrt{5/3}$. Thus $x = y = z = \sqrt{5/3}$, and so $xyz = (5/3)^{3/2}$. This is the solution. It should be geometrically clear that the maximum occurs when $x = y = z$.

Exercises for Section 7.7

In Exercises 1–5 find the extrema of f subject to the stated constraints.

1. $f(x,y,z) = x - y + z$, $\quad x^2 + y^2 + z^2 = 2$

2. $f(x,y) = x - y$, $\quad x^2 - y^2 = 2$

3. $f(x,y) = z$, $\quad x^2 + 2y^2 = 3$

4. $f(x,y) = 3x + 2y$, $\quad 2x^2 + 3y^2 = 3$

5. $f(x,y,z) = x + y + z$, $\quad x^2 - y^2 = 1$, $\quad 2x + z = 1$

Theorem Proofs for Chapter 7

Theorem 1. *Let $A \subset \mathbb{R}^n$ be an open set and let $f: A \subset \mathbb{R}^n \to \mathbb{R}^n$ be of class C^1. Let $x_0 \in A$ and suppose $Jf(x_0) \neq 0$. Then there is a neighborhood U of x_0 in A and an open neighborhood W of $f(x_0)$ such that $f(U) = W$ and f has a C^1 inverse $f^{-1}: W \to U$. Moreover, for $y \in W$, $x = f^{-1}(y)$, we have*

$$Df^{-1}(y) = [Df(x)]^{-1}$$

If f is of class $C^p, p \geq 1$, then so is f^{-1}.

The proof of the inverse function theorem is not especially easy in its technical details, but this theorem represents one of the most important cornerstones of analysis so should be mastered. A proof will be given based on the contraction lemma (see Section 5.6). This technique is useful as it is applicable to many situations.

We begin by recalling the contraction lemma. Here we use the special case of a closed subset of $\mathbb{R}^n$.

Lemma 1. *Let M be a closed subset of $\mathbb{R}^n$, and d the distance function on $\mathbb{R}^n$. Let f be a mapping of M into M. Assume there exists a constant K, where $0 < K < 1$, such that for any two points x and y in M we have $d(f(x), f(y)) \leq Kd(x,y)$. Then there exists a unique $x \in M$ such that $f(x) = x$ (x is called a fixed point of f).*

Before beginning the proof of the inverse function theorem, it is helpful to have a technical lemma about the set of invertible linear maps (or equivalently, the set of invertible matrices). Now an $m \times n$ matrix (or a linear map from $\mathbb{R}^n \to \mathbb{R}^m$) is simply an mn-tuple of real numbers, since a matrix A with entries (a_{ij}) can be regarded as an

mn-tuple $(a_{11}, \ldots, a_{1n}, a_{21}, \ldots, a_{m1}, \ldots, a_{mn})$. It makes sense then, to say that a certain subset of the set of all matrices is open or that a map from the set of $m \times n$ matrices to the set of $p \times q$ matrices is differentiable. Let $L(\mathbb{R}^n, \mathbb{R}^n)$ denote the set of all $n \times n$ matrices (or linear maps from $\mathbb{R}^n$ to $\mathbb{R}^n$) and let $GL(\mathbb{R}^n, \mathbb{R}^n)$ denote the set of all invertible matrices (or invertible linear maps from $\mathbb{R}^n$ to $\mathbb{R}^n$), which is called the *general linear group*. Thus $GL(\mathbb{R}^n, \mathbb{R}^n) = \{A \in L(\mathbb{R}^n, \mathbb{R}^n) \mid \det A \neq 0\}$. Let $\mathscr{L}^{-1}: GL(\mathbb{R}^n, \mathbb{R}^n) \to GL(\mathbb{R}^n, \mathbb{R}^n)$ denote the map which takes an invertible matrix A to its inverse A^{-1}. The lemma that we need is Lemma 2.

Lemma 2.

    (i) $GL(\mathbb{R}^n, \mathbb{R}^n)$ *is an open subset of* $L(\mathbb{R}^n, \mathbb{R}^n)$.

    (ii) $\mathscr{L}^{-1}$ *is a* C^∞ *mapping*.

    Proof: *   (i) The determinant mapping det: $\mathbb{R}^n \times \cdots \times \mathbb{R}^n$ (n times) $\to \mathbb{R}$ is an n-linear map. (Recall that the determinant is linear in the rows.) Hence by Example 5, Chapter 4, which shows that a multilinear map from $\mathbb{R}^{n_1} \times \cdots \times \mathbb{R}^{n_k}$ to $\mathbb{R}^m$ is continuous, the determinant mapping is continuous, and, by Example 2, or Exercise 14 at the end of Chapter 6, it is differentiable. Because the set consisting of zero $\{0\}$ is closed, we have that $\det^{-1}(\{0\})$ is closed (by Theorem 1, Chapter 4). Hence $L(\mathbb{R}^n, \mathbb{R}^n) \backslash \det^{-1}(\{0\})$ is open. But $L(\mathbb{R}^n, \mathbb{R}^n) \backslash \det^{-1}(\{0\})$ is the set of all those $n \times n$-matrices with non-zero determinant, and these are exactly all the invertible matrices $GL(\mathbb{R}^n, \mathbb{R}^n)$.

    (ii) It is easy to see, from the explicit expression for the inverse of a matrix, that $\mathscr{L}^{-1}$ is C^∞. Indeed, the expression for the inverse of the matrix A is $A^{-1} = (\det A)^{-1} \operatorname{adj} A$, where $\operatorname{adj} A$ is a matrix such that $(\operatorname{adj} A)_{ij} = (-1)^{i+j} \det A(j \mid i)$, where $\det A(j \mid i)$ denotes the determinant of the matrix obtained from A by deleting the jth row and ith column. As $(\det A)^{-1}$ is a real differentiable function of A, we only need to show that the mapping adj: $L(\mathbb{R}^n, \mathbb{R}^n) \to L(\mathbb{R}^n, \mathbb{R}^n)$, which takes a matrix to its adjoint, is C^∞. Regarded as a function from $\mathbb{R}^{n^2}$ to $\mathbb{R}^{n^2}$, the adjoint is simply an n^2-tuple of functions like $(\operatorname{adj} A)_{ij} = (-1)^{i+j} \det A(j \mid i)$. Now as we have mentioned a multilinear map from $\mathbb{R}^{n_1} \times \cdots \times \mathbb{R}^{n_k}$ to $\mathbb{R}^m$ is C^∞. Thus each of the n^2 component functions of adj is C^∞. Hence the adj map is C^∞. ∎

    Proof of Theorem 1:   For the sake of clarity we will now break up the proof of Theorem 1 into a number of steps.

Step 1:   *Simplification to a special case.*

    We will prove the theorem below for the case when $Df(x_0)$ is the identity transformation. Here we show that this is indeed sufficient to prove the general case.

    Let $\lambda = Df(x_0)$; then λ^{-1} exists, and by the chain rule

$$D(\lambda^{-1} \circ f)(x_0) = D(\lambda^{-1})(f(x_0)) \circ Df(x_0) = \lambda^{-1} \circ Df(x_0) = \text{identity transformation}.$$

    Now if the theorem is true for $\lambda^{-1} \circ f$, then the theorem is also true for f. Indeed, if g is an inverse for $\lambda^{-1} \circ f$, the inverse for f will be $g \circ \lambda^{-1}$.

    We can make one further simplifying assumption, namely, that $x_0 = 0$ and $f(x_0) = 0$. To see this, let us suppose we have proven the theorem for the special case $x_0 = 0$ and

---

* For a more "intrinsic" proof see Dieudonné, *Foundations of Modern Analysis*, p. 179.

$f(x_0) = 0$. We want to see how to prove the general case from this. Let $h(x) = f(x + x_0) - f(x_0)$. Then $h(0) = 0$ and $Dh(0) = Df(x_0)$, so $Dh(0)$ is invertible. Then if h has an inverse near $x = 0$, the required inverse for f near x_0 is given by

$$f^{-1}(y) = h^{-1}(y - f(x_0)) + x_0 \ .$$

In summary, Step 1 demonstrates that *it is sufficient to prove the theorem under the assumptions $x_0 = 0$, $f(x_0) = 0$ and $Df(0)$ is the identity.* This will be assumed in the remainder of the discussion.

Step 2:   *Application of the contraction lemma to get a local inverse.*

If we bear the preceding remarks in mind, it follows that what we would like is two neighborhoods of 0 such that given any y from the first neighborhood of 0 there is a unique x from the second neighborhood such that $f(x) = y$. To do this, consider the function g_y defined by $g_y(x) = y + x - f(x)$. If for some closed neighborhood of zero this is a contracting mapping, then it has a unique fixed point, say x, and so $x = y + x - f(x)$ or x is the unique point belonging to the neighborhood such that $f(x) = y$. Now construct this neighborhood: define $g(x) = x - f(x)$; then $Dg(0) = 0$. Assume g to be of class C^p, with $p \geq 1$. This means in particular that Dg is a continuous function, and so by continuity at 0 there exists an $r > 0$ such that $\|x\| < r$ implies $\|Dg_i(x)\| < 1/2n$, where $g = (g_1, \ldots, g_n)$. By the mean-value theorem, given $x \in D(0,r)$ there are points $c_1, c_2, \ldots, c_n$ in $D(0,r)$ such that $g_i(x) = g_i(x) - g_i(0) = Dg_i(c_i)(x - 0) = Dg_i(c_i)(x)$. Therefore

$$\|g(x)\| \leq \sum_{i=1}^{n} \|g_i(x)\| = \sum_{i=1}^{n} \|Dg_i(c_i)(x)\| \leq \sum_{i=1}^{n} \|Dg_i(c_i)\| \, \|x\| < \frac{\|x\|}{2} < \frac{r}{2},$$

using the C.B.S. inequality.

This establishes that g maps the closed r-ball $\bar{D}(0,r)$ into the closed $r/2$-ball $\bar{D}(0,r/2)$. Now let y be any member of $\bar{D}(0,r/2)$. The mapping g_y takes $\bar{D}(0,r)$ into $\bar{D}(0,r)$; for $\|y\| \leq r/2$ and $x \in \bar{D}(0,r)$ implies

$$\|g_y(x)\| = \|y + g(x)\| \leq \|y\| + \|g(x)\| < \frac{r}{2} + \frac{r}{2} = r \ .$$

Let x_1 and x_2 be any two points in $\bar{D}(0,r)$. Then $\|g_y(x_1) - g_y(x_2)\| = \|g(x_1) - g(x_2)\|$, and by the mean-value theorem as above, $\|g(x_1) - g(x_2)\| \leq (1/2)\|x_1 - x_2\|$, and so g_y is a contracting map (with constant $K = 1/2$). Now we apply the contraction lemma, which implies that there is a unique fixed point $x \in \bar{D}(0,r)$ for g_y, and as we observed before, this implies $f(x) = y$. This means that f has an inverse $f^{-1}: \bar{D}(0,r/2) \subset \mathbb{R}^n \to \bar{D}(0,r) \subset \mathbb{R}^n$.

Step 3:   *The inverse is continuous.*

Let x_1 and $x_2 \in \bar{D}(0,r)$; then recalling the definition of g, we get

$$\|x_1 - x_2\| \leq \|f(x_1) - f(x_2)\| + \|g(x_1) - g(x_2)\| \leq \|f(x_1) - f(x_2)\| + (1/2)\|x_1 - x_2\| \ ,$$

and hence $\|x_1 - x_2\| \leq 2\|f(x_1) - f(x_2)\|$. Therefore if y_1 and $y_2 \in \bar{D}(0,1/2)$, we get $\|f^{-1}(y_1) - f^{-1}(y_2)\| \leq 2\|y_1 - y_2\|$, so f^{-1} is continuous.

Step 4:   *For suitably small r, the inverse is differentiable on $D(0, r/2)$.*

We were given that $Df(0)$ is invertible, that $Df: A \subset \mathbb{R}^n \to \mathbb{R}^{n^2}$ is continuous, and we have shown that $GL(\mathbb{R}^n,\mathbb{R}^n)$ is open in $L(\mathbb{R}^n,\mathbb{R}^n)$. Together, these facts show that for all x in some neighborhood around 0, $[Dg(x)]^{-1}$ exists. If this neighborhood does not contain $D(0,r/2)$, r is restricted further until this is the case. Hence we can assume $[Df(x)]^{-1}$ exists for all $x \in D(0,r/2)$. Moreover, we can assume $\|[Df(x)]^{-1}y\| \leqslant M \|y\|$ for all $x \in D(0,r/2)$ and $y \in \mathbb{R}^n$ by continuity of $Df(x)^{-1}$ (see Example 4, Chapter 4).

Now, for $y_1, y_2 \in D(0,r/2)$, $x_1 = f^{-1}(y_1)$ and $x_2 = f^{-1}(y_2)$,

$$\frac{\|f^{-1}(y_1) - f^{-1}(y_2) - [Df(x_2)]^{-1} \cdot (y_1 - y_2)\|}{\|y_1 - y_2\|}$$

$$= \frac{\|x_1 - x_2 - [Df(x_2)]^{-1} \cdot (f(x_1) - f(x_2))\|}{\|f(x_1) - f(x_2)\|}$$

$$= \left[\frac{\|x_1 - x_2\|}{\|f(x_1) - f(x_2)\|} \right] \frac{\|\{[Df(x_2)]^{-1}\}\{Df(x_2)(x_1 - x_2) - (f(x_1) - f(x_2))\}\|}{\|x_1 - x_2\|}.$$

Using $\|x_1 - x_2\| \leqslant 2 \|f(x_1) - f(x_2)\|$ and $\|Df(x_2)^{-1}y\| \leqslant M \|y\|$ gives that the above is

$$\leqslant 2M \frac{\|Df(x_2)(x_1 - x_2) - (f(x_1) - f(x_2))\|}{\|x_1 - x_2\|}.$$

The last expression has a limit zero as $\|x_1 - x_2\| \to 0$, by the differentiability of f at x_2. This shows that f^{-1} is differentiable at y_2 with derivative $[Df(x_2)]^{-1} = [Df(f^{-1}(y_2))]^{-1}$.

In the theorem we set $W = D(0,r/2)$ and $U = f^{-1}(W)$, both open sets.

Step 5: $f^{-1}: D(0,r/2) \to \mathbb{R}^n$ *is of class* C^p.

From Step 4 it follows that $f^{-1}: D(0,r/2) \to \mathbb{R}^n$ is differentiable on $D(0,r/2)$ and that $Df^{-1}(y) = [Df(f^{-1}(y))]^{-1}$. We have shown that $f^{-1}: D(0,r/2) \to \mathbb{R}^n$ is continuous; Df is continuous by assumption; and the inversion mapping from $GL(\mathbb{R}^n,\mathbb{R}^n)$ (the invertible linear maps from $\mathbb{R}^n$ to $\mathbb{R}^n$) to $GL(\mathbb{R}^n,\mathbb{R}^n)$ is continuous and, in fact, C^∞ by Lemma 2. This implies that Df^{-1} is a continuous map from $D(0,r/2)$ into $L(\mathbb{R}^n,\mathbb{R}^n)$. Hence f^{-1} is of class C^1. Again look at $Df^{-1}(y) = [Df(f^{-1}(y))]^{-1}$ and observe that since f^{-1} is of class C^1, Df is of class C^{p-1} and since inversion is C^∞, Df^{-1} is of class C^1. Hence f^{-1} is of class C^2. Continuing in this way by induction we finally conclude that f^{-1} is of class C^p. ∎

Theorem 2. (*Implicit Function theorem.*) *Let* $A \subset \mathbb{R}^n \times \mathbb{R}^m$ *be an open set and let* $F: A \to \mathbb{R}^m$ *be a function of class* C^p. *Suppose that* $(x_0,y_0) \in A$ *and* $F(x_0,y_0) = 0$. *Form*

$$\Delta = \begin{vmatrix} \dfrac{\partial F_1}{\partial y_1} & \cdots & \dfrac{\partial F_1}{\partial y_m} \\ \cdot & & \cdot \\ \cdot & & \cdot \\ \cdot & & \cdot \\ \dfrac{\partial F_m}{\partial y_1} & \cdots & \dfrac{\partial F_m}{\partial y_m} \end{vmatrix}$$

evaluated at (x_0, y_0), *where* $F = (F_1, \ldots, F_m)$, *and suppose that* $\Delta \neq 0$. *Then there is an open neighborhood* $U \subset \mathbb{R}^n$ *of* x_0, *and a neighborhood* V *of* y_0 *in* $\mathbb{R}^m$, *and a unique function* $f: U \to V$ *such that*

$$F(x, f(x)) = 0$$

for all $x \in U$. *Furthermore,* f *is of class* C^p.

Proof: Define the function $G: A \subset \mathbb{R}^n \times \mathbb{R}^m \to \mathbb{R}^n \times \mathbb{R}^m$ by $G(x,y) = (x, F(x,y))$. Since F is of class C^p and the identity mapping is of class C^∞, it follows that G is of class C^p. The matrix of partial derivatives of G (Jacobian matrix) is

$$\begin{pmatrix}
1 & 0 & \cdots & 0 & 0 & \cdots & 0 \\
0 & 1 & & & & & \\
\cdot & & \cdot & & \cdot & & \cdot \\
\cdot & & \cdot & & \cdot & & \cdot \\
\cdot & & \cdot & & \cdot & & \cdot \\
0 & & \cdots & 1 & 0 & \cdots & 0 \\
\dfrac{\partial F_1}{\partial x_1} & & \cdots & \dfrac{\partial F_1}{\partial x_n} & \dfrac{\partial F_1}{\partial y_1} & \cdots & \dfrac{\partial F_1}{\partial y_m} \\
\cdot & & \cdot & & \cdot & & \cdot \\
\cdot & & \cdot & & \cdot & & \cdot \\
\cdot & & \cdot & & \cdot & & \cdot \\
\dfrac{\partial F_m}{\partial x_1} & & \cdots & \dfrac{\partial F_m}{\partial x_n} & \dfrac{\partial F_m}{\partial y_1} & \cdots & \dfrac{\partial F_m}{\partial y_m}
\end{pmatrix}$$

The determinant of this matrix evaluated at (x_0, y_0) is equal to

$$\Delta = \begin{vmatrix}
\dfrac{\partial F_1}{\partial y_1} & \cdots & \dfrac{\partial F_1}{\partial y_m} \\
\cdot & & \cdot \\
\cdot & & \cdot \\
\cdot & & \cdot \\
\dfrac{\partial F_m}{\partial y_1} & \cdots & \dfrac{\partial F_m}{\partial y_m}
\end{vmatrix}$$

evaluated at (x_0, y_0). Therefore, by hypothesis, $JG(x_0, y_0) \neq 0$ and thus by the inverse function theorem, there is an open set W containing $(x_0, 0)$ and an open set S containing (x_0, y_0) such that $G(S) = W$ and G has a C^p-inverse $G^{-1}: W \to S$. From the definition of an open set we see that there are open sets $U \subset \mathbb{R}^n$ and $V \subset \mathbb{R}^m$ with $x_0 \in U$ and $y_0 \in V$ such that $U \times V \subset S$ (see Exercise 24, Chapter 2). Let $G(U \times V) = Y \subset W$. Thus $G: U \times V \to Y$ is a C^p-diffeomorphism (this means that G is of class C^p and has inverse $G^{-1}: Y \to U \times V$ also of class C^p). Now G^{-1} is of the form $G^{-1}(x,w) = (x, H(x,w))$, where H is a C^p function from Y to V, since G is of this form, as is easy to see.

Let $\pi: \mathbb{R}^n \times \mathbb{R}^m \to \mathbb{R}^m$ be defined by $\pi(x,y) = y$, so $F(x,H(x,w)) = \pi \circ G(x,H(x,w)) = \pi \circ G \circ G^{-1}(x,w) = w$. Also observe that because G^{-1} is of the form $G^{-1}(x,w) = (x,H(x,w))$, if $(x,w) \in Y$, then $x \in U$. Define $f: U \to V$ by $f(x) = H(x,0)$. Then, as $F(x,H(x,w)) = w$, we get $F(x,f(x)) = 0$. Also, as H is of class C^p, f must also be of class C^p. By Theorem 1, $H(x,w)$ is uniquely determined. Since f must be given by $H(x,0)$, f is seen to be unique as well. ∎

Theorem 3. *Let $A \subset \mathbb{R}^n$ be an open set and let $f: A \to \mathbb{R}$ be a function of class C^p, $p \geqslant 1$. Let $x_0 \in A$ and suppose $f(x_0) = 0$ and $Df(x_0) \neq 0$. Then there is an open set U, an open set V containing x_0, and a function $h: U \to V$ of class C^p, with inverse $h^{-1}: V \to U$ of class C^p, such that*

$$f(h(x_1, \ldots, x_n)) = x_n .$$

Proof: Since $Df(x_0) \neq 0$, there must exist some i such that $(\partial f/\partial x_i)(x_0) \neq 0$. Define $g: \mathbb{R}^n \to \mathbb{R}^n$ by $(x_1, \ldots, x_n) \mapsto (x_1, \ldots, x_{i-1}, x_n, x_{i+1}, \ldots, x_{n-1}, x_i)$. The permutation map g is linear and hence C^∞ and because f is C^p we have by the chain rule that $f \circ g$ is of class C^p. So $(\partial(f \circ g)/\partial x_n)(g^{-1})(x_0) = \partial f(x_0)/\partial x_i \neq 0$, which implies that $f \circ g$ is a function of the type described in the hypotheses of Theorem 2, with $m = 1$. Hence, just as in the proof of Theorem 2, if we define $G: A \subset \mathbb{R}^{n-1} \times \mathbb{R} \to \mathbb{R}^{n-1} \times \mathbb{R}$ by $G(x,y) = (x, f \circ g(x,y))$, there are open sets $W \subset \mathbb{R}^n$ and $U \subset \mathbb{R}^n$ with $x_0 \in W$ and $(x_0^1, \ldots, x_0^{n-1}, 0) \in U$ (where $x_0 = (x_0^1, \ldots, x_0^n)$) such that $G: W \to U$ has an inverse $G^{-1}: U \to W$ of class C^p. Now, $(f \circ g) \circ G^{-1}(x_1, \ldots, x_n) = (\pi \circ G) \circ G^{-1}(x_1, \ldots, x_n) = x_n$, where $\pi: \mathbb{R}^{n-1} \times \mathbb{R} \to \mathbb{R}$ is the projection on the last coordinate. Define $V = g(W)$ and $h: U \to V$ by $h = g \circ G^{-1}$. Then h is a C^p function with C^p inverse, since both g and G^{-1} have this property; and $f(h(x_1, \ldots, x_n)) = x_n$. ∎

It is possible to prove a theorem that is more general than the one above, using a similar technique. That is, if $f: A \subset \mathbb{R}^n \to \mathbb{R}^m$, $n \geqslant m$, and $Df(x_0)$ as a linear map has rank m, then f can locally be made to look like a projection on the last m factors by composing it after a smooth function with smooth inverse. In Exercise 3, we state this exactly and give a hint as to the proof. Note that here the range has dimension less than or equal to that of the domain. In the following theorem the opposite is the case.

Theorem 4. *Let $A \subset \mathbb{R}^p$ be an open set and $f: A \to \mathbb{R}^n$ a function of class C^r and $p \leqslant n$. Let $x_0 \in A$ and suppose the rank of $Df(x_0)$ is p. Then there are open sets U and V in $\mathbb{R}^n$ with $f(x_0) \in U$ and a function $g: U \to V$ of class C^r with inverse $g^{-1}: V \to U$ of class C^r such that $g \circ f(x_1, \ldots, x_p) = (x_1, \ldots, x_p, 0, \ldots, 0)$ for all $(x_1, \ldots, x_p) \in A$.*

Proof: Since $Df(x_0)$ has rank p, some $p \times p$ submatrix of $Df(x_0)$ has non-zero determinant. By relabeling, if necessary, we may assume

$$\begin{vmatrix} \dfrac{\partial f^1}{\partial x_1} & \cdots & \dfrac{\partial f^1}{\partial x_p} \\ \cdot & & \cdot \\ \cdot & & \cdot \\ \cdot & & \cdot \\ \dfrac{\partial f^p}{\partial x_1} & \cdots & \dfrac{\partial f^p}{\partial x_p} \end{vmatrix} \neq 0 ,$$

where $f = (f^1, \ldots, f^n)$. Define $\varphi: A \times \mathbb{R}^{n-p} \to \mathbb{R}^n$ by $\varphi(x,y) = f(x) + (0,y)$. Then the matrix of $D\varphi$ is

$$
\begin{pmatrix}
\dfrac{\partial f^1}{\partial x_1} & \cdots & \dfrac{\partial f^1}{\partial x_p}, & 0,\ldots,0 \\[2mm]
\cdot & & \cdot \\
\cdot & & \cdot \\
\cdot & & \cdot \\
\dfrac{\partial f^p}{\partial x_1} & \cdots & \dfrac{\partial f^p}{\partial x_p}, & 0,\ldots,0 \\[2mm]
\dfrac{\partial f^{p+1}}{\partial x_1} & \cdots & \dfrac{\partial f^{p+1}}{\partial x_p}, & 1,\ldots,0 \\[2mm]
\cdot & & \cdot \\
\cdot & & \cdot \\
\cdot & & \cdot \\
\dfrac{\partial f^n}{\partial x_1} & \cdots & \dfrac{\partial f^n}{\partial x_p}, & 0,\ldots,1
\end{pmatrix}
$$

and

$$
J\varphi(x_0,0) =
\begin{vmatrix}
\dfrac{\partial f^1}{\partial x_1} & \cdots & \dfrac{\partial f^1}{\partial x_p} \\[2mm]
\cdot & & \cdot \\
\cdot & & \cdot \\
\cdot & & \cdot \\
\dfrac{\partial f^p}{\partial x_1} & \cdots & \dfrac{\partial f^p}{\partial x_p}
\end{vmatrix}
\neq 0 .
$$

Hence by the inverse function theorem there is an open set U around $f(x_0)$, an open set V around $(x_0,0)$, and a function $g: U \to V$ of class C^r such that $g = \varphi^{-1}$. Then $g(f(x)) = g(f(x) + (0,0)) = (x,0)$ as desired. ∎

Theorem 5. *Let $f: A \subset \mathbb{R}^n \to \mathbb{R}^N$ (where A is open in $\mathbb{R}^n$) be a C^r function such that $Df(x)$ has rank m for all x in a neighborhood of $x_0 \in A$. Then there is an open set $U \subset \mathbb{R}^n$ and an open set $V \subset \mathbb{R}^n$ with $x_0 \in V$ and a function $h: U \to V$ of class C^r with inverse $h^{-1}: V \to U$ of class C^r such that $f \circ h$ depends only on $x_1, \ldots, x_m$. That is, $f \circ h(x_1, \ldots, x_m, x_{m+1}, \ldots, x_n) = \tilde{f}(x_1, \ldots, x_m)$ for some C^r function $\tilde{f}$.*

Proof: Let N_0 be the kernel of $Df(x_0)$; that is, let $N_0 = \{y \in \mathbb{R}^n \mid Df(x_0) \cdot y = 0\}$ (a subspace of $\mathbb{R}^n$ of dimension $n - m$) and let M be an m-dimensional complement of N_0 in $\mathbb{R}^n$, that is, $M \cap N_0 = \{0\}$ and $\{x + y \mid x \in M, y \in N_0\} = \mathbb{R}^n$. Let $c_1, \ldots, c_m$ be a basis for M and $c_{m+1}, \ldots, c_n$ be a basis for N_0. Now each $x \in \mathbb{R}^n$ can be written uniquely as $x = \psi_1(x)c_1 + \cdots + \psi_n(x)c_n$. Define $G(x) = (0, \ldots, 0, \psi_{m+1}(x), \ldots, \psi_n(x))$. Then G is linear and hence smooth. Now $Df(x_0)$ has rank m, so $Df(x_0)(\mathbb{R}^n)$ is an m-dimensional subspace P of $\mathbb{R}^N$. Moreover, the set $\{d_i = Df(x_0)c_i \mid 1 \leqslant i \leqslant m\}$ is a basis for P. Any $x \in \mathbb{R}^N$ may be written uniquely as

$$x = \varphi_1(x)\, d_1 + \cdots + \varphi_m(x)\, d_m + \varphi_{m+1}(x)\, d_{m+1} + \cdots + \varphi_N(x)\, d_N \,,$$

where $d_1, \ldots, d_N$ is a basis for $\mathbb{R}^N$, with $d_1, \ldots, d_m$ the basis for P given above. Define $H: \mathbb{R}^N \to \mathbb{R}^n$ by $H(x) = (\varphi_1(x), \ldots, \varphi_m(x), 0, \ldots, 0)$.

Now let $g(x) = H(f(x)) + G(x)$. Then g maps $\mathbb{R}^n$ to $\mathbb{R}^n$, and as H and G are linear, we have $Dg(x_0) \cdot s = DH(f(x_0)) \circ Df(x_0)(s) + DG(x_0)(s) = H(Df(x_0)(s)) + G(s)$. If we write the matrix of the linear transformation $Dg(x_0)$ in terms of the bases $c_1, \ldots, c_n$ and the standard basis, we get the identity matrix. Hence $Dg(x_0)$ is invertible. We may use the inverse function theorem to find an open set U around $H(f(x_0)) + G(x_0)$ and an open set V around x_0 and a smooth inverse function $g^{-1}: U \to V$. Now for each $x \in V$, $Dg(x)$ is invertible. That is, $Dg(x)$ must be a one-to-one linear map of $\mathbb{R}^n$ onto $\mathbb{R}^n$. We may assume rank $\{Df(x)\} = m$ for all $x \in A$ (otherwise restrict f to an even smaller neighborhood of x_0). For $x \in A$, $Df(x)(\mathbb{R}^n)$ is an m-dimensional subspace, say, P_x of $\mathbb{R}^N$. Now if $s \in M$, $Dg(x) \cdot s = H(Df(x) \cdot s) + G(s) = H(Df(x) \cdot s)$. Thus if $x \in V$, $Df(x)$ restricted to M must be a one-to-one linear map of M onto P_x. That the mapping is onto follows from the fact that M and P_x both have dimension m. Similarly, H must be a one-to-one linear map of P_x onto $\mathbb{R}^m$. Denote the inverse of this map by $L_x: \mathbb{R}^m \to P_x$.

Let $h = g^{-1}: U \to V$; we shall show that $f \circ h(x_1, \ldots, x_n)$ does not depend on $x_{m+1}, \ldots, x_n$. To do this we may assume that U is a ball. It suffices to show that $D_2 f_1 = 0$, where $D_2 f_1$ is the derivative of $f_1 = f \circ h$ restricted to $\{0\} \times \mathbb{R}^{n-m}$, that is, we are showing $\partial f_1 / \partial x_i = 0$, $i = m+1, \ldots, n$. It of course follows that f_1 is constant with respect to $x_{m+1}, \ldots, x_n$. Now $f = f_1 \circ g$, so

$$Df(x) \cdot y = Df_1(g(x)) \cdot Dg(x) \cdot y = D_1 f_1(g(x)) \cdot H(Df(x) \cdot y) + D_2 f_1(g(x)) \cdot G(y) \,. \quad (1)$$

Since G is a mapping of $\mathbb{R}^n$ onto $\{0\} \times \mathbb{R}^{n-m}$, it suffices to show that $D_2 f_1(g(x)) \circ G(y) = 0$ for $y \in \mathbb{R}^n$. Returning to Eq. 1 and using $L_x \circ H = $ identity, we obtain

$$D_2 f_1(g(x)) \circ G(y) = L_x \circ H(Df(x) \cdot y) - D_1 f_1(g(x)) \circ H(Df(x) \cdot y)$$
$$= (L_x - D_1 f_1(g(x))) \circ H(Df(x) \cdot y) \qquad (2)$$

for all $y \in \mathbb{R}^n$. Now $L_x - D_1 f_1(g(x))$ is defined on $\mathbb{R}^m \times \{0\}$ and $H \circ Df(x)$ maps M onto $\mathbb{R}^m \times \{0\}$. Hence to show $L_x - D_1 f_1(g(x)) = 0$, it suffices to show

$$(L_x - D_1 f_1(g(x))) \circ H(Df(x) \cdot y) = 0$$

for $y \in M$. But this follows because for $y \in M$, $G(y) = 0$, and so $D_2 f_1(g(x)) \circ G(y) = 0$. Therefore $L_x - D_1 f_1(g(x))$ is identically zero and thus $D_2 f_1(g(x)) = 0$. ∎

Corollary 2. Let $f: A \subset \mathbb{R}^n \to \mathbb{R}^N$ (where A is open in $\mathbb{R}^n$) be a function of class C^r such that $Df(x)$ has rank m for all x in a neighborhood of $x_0 \in A$. Then there is an open set $U_1 \subset \mathbb{R}^n$, an open set $U_2 \subset \mathbb{R}^n$ with $x_0 \in U_2$, an open set V_1 around $f(x_0)$, an open set $V_2 \subset \mathbb{R}^N$, and functions $h: U_1 \to U_2$ and $g: V_1 \to V_2$ of class C^r with inverses of class C^r such that $g \circ f \circ h(x_1, \ldots, x_n) = (x_1, \ldots, x_m, 0, \ldots, 0)$.

Proof: By Theorem 5 there is a C^r function $h: U \to V$ with C^r inverse, such that

$$f \circ h(x_1, \ldots, x_m, x_{m+1}, \ldots, x_n) = \tilde{f}(x_1, \ldots, x_m)$$

for some $\tilde{f}: W \subset \mathbb{R}^m \to \mathbb{R}^N$. Now $D\tilde{f}$ has rank m (since h is invertible), so by Theorem 4, there is an invertible C^r function g_0 such that

$$g_0 \circ \tilde{f}(x_1, \ldots, x_m) = (x_1, \ldots, x_m, 0, \ldots, 0) \ .$$

Define g on $\mathbb{R}^n$ by $g(x_1, \ldots, x_n) = (g_0(x_1, \ldots, x_m), x_{m+1}, \ldots, x_n)$. Then g is also C^r and invertible and we have

$$g \circ f \circ h(x_1, \ldots, x_n) = (x_1, \ldots, x_m, 0, \ldots, 0) \ . \ \blacksquare$$

Theorem 6. *Let $f: [-a,a] \times \bar{D}(x_0, r) \to \mathbb{R}^n$ be a given continuous mapping. Let $C = \sup\{\|f(t,x)\| \mid -a \leqslant t \leqslant a, \ x \in \bar{D}(x_0, r)\}$. Suppose there exists a $K \in \mathbb{R}$ such that for all $t \in [-a,a]$ and $x, y \in \bar{D}(x_0, r)$,*

$$\|f(t,x) - f(t,y)\| \leqslant K \|x - y\|$$

and that $b < \min\{a, r/C, 1/K\}$. Then there is a unique continuously differentiable map $x: [-b,b] \to \bar{D}(x_0, r)$ such that

$$x(0) = x_0 \qquad and \qquad \frac{dx}{dt} = f(t, x(t)) \ .$$

Proof: The differential equation and the initial condition $x(0) = x_0$ is clearly equivalent to the condition

$$x(t) = x_0 + \int_0^t f(s, x(s)) \, ds \ .$$

Consider $\mathscr{C}([-b,b], \mathbb{R}^n)$ which we know (from Chapter 5, Section 4) is a complete metric space. Let

$$A = \{\varphi \in \mathscr{C}([-b,b], \mathbb{R}^n) \mid \varphi(0) = x_0 \text{ and } \varphi(t) \in \bar{D}(x_0, r)\} \ .$$

Then $A \subset \mathscr{C}([-b,b], \mathbb{R}^n)$ is closed (why?) and therefore A is also a complete metric space. We will apply the contraction mapping principle proved in Section 5.6 to this space A.

Define $F: A \to A$ by*

$$F(\varphi)(t) = x_0 + \int_0^t f(s, \varphi(s)) \, ds \ .$$

First we must show $F(\varphi) \in A$. Clearly, $F(\varphi) \in \mathscr{C}([-b,b], \mathbb{R}^n)$. Also, $F(\varphi)(0) = x_0$, and for all $t \in [-b,b]$,

$$\|F(\varphi)(t) - x_0\| = \left\| \int_0^t f(s, \varphi(s)) \, ds \right\| \leqslant \int_0^t \|f(s, \varphi(s))\| \, ds \leqslant b \cdot C < r$$

since $b < r/C$. Thus $F(\varphi)(t) \in \bar{D}(x_0, r)$, so $F(\varphi) \in A$.

_____

* $\int_0^t f(s, \varphi(s)) \, ds$ is obtained by integrating each component of f; the result is a vector. The inequality

$$\left\| \int_0^t f(s, \varphi(s)) \, ds \right\| \leqslant \int_0^t \|f(s, \varphi(s))\| \, ds$$

is analogous to the similar result for the case of real functions—we accept it here; see Chapter 8 for a detailed discussion.

Next, for $\varphi, \psi \in A$,

$$\|F(\varphi) - F(\psi)\| = \sup_{-b \leq t \leq b} \|F(\varphi)(t) - F(\psi)(t)\|$$

$$= \sup_{-b \leq t \leq b} \left\| \int_0^t (f(s,\varphi(s)) - f(s,\psi(s))) \, ds \right\|$$

$$\leq \sup_{-b \leq t \leq b} \int_0^t \|f(s,\varphi(s)) - f(s,\psi(s))\| \, ds$$

$$\leq \sup_{-b \leq t \leq b} \int_0^t K \|\varphi(s) - \psi(s)\| \, ds$$

$$\leq \sup_{-b \leq t \leq b} K \int_0^t \|\varphi - \psi\| \, ds \leq Kb \|\varphi - \psi\|$$

where $Kb < 1$.

Therefore if we let $k = b \cdot K < 1$, $d(F(\varphi),F(\psi)) \leq kd(\varphi,\psi)$ and so F is a contraction and thus has a unique fixed point: $x = F(x)$. This fixed point $x(t)$ is the unique solution we were seeking. ∎

The iteration scheme mentioned in the text comes about because, as we saw in the proof of the contraction mapping theorem, the unique fixed point is the limit $F^n(\varphi)$ as $n \to \infty$ for any $\varphi \in A$. We chose $\varphi(t) \equiv x_0$.

Theorem 7.   *Let $A \subset \mathbb{R}^n$ be open and $f: A \to \mathbb{R}$ a smooth function. Suppose $Df(x_0) = 0$ and $\Delta = [-\partial^2 f / \partial x_i \, \partial x_j]$ is non-singular. Then there is a neighborhood U of x_0 and a neighborhood V of 0 in $\mathbb{R}^n$ and a smooth map $g: V \to U$ with smooth inverse such that*

$$f \circ g(y) = f(x_0) - [y_1^2 + \cdots + y_\lambda^2] + [y_{\lambda+1}^2 + \cdots + y_n^2]$$

for all $y \in V$. Here λ is a fixed integer $0 \leq \lambda \leq n$.

Proof:*   It is easy to see that we lose no generality if we assume $x_0 = 0$ and $f(x_0) = 0$.

Write

$$f(x_1,\ldots,x_n) = \int_0^1 \frac{df(tx_1,\ldots,tx_n)}{dt} \, dt$$

$$= \int_0^1 \sum_{i=1}^n x_i \frac{\partial f}{\partial x_i} (tx_1,\ldots,tx_n) \, dt \ .$$

Thus we see that if we set

$$g_i(x_1,\ldots,x_n) = \int_0^1 \frac{\partial f}{\partial x_i} (tx_1,\ldots,tx_n) \, dt \ ,$$

then

$$f(x_1,\ldots,x_n) = \sum_{i=1}^n x_i g_i(x_1,\ldots,x_n) \ .$$

Since $x_0 = 0$ is a critical point, $\partial f / \partial x_i (0) = g_i(0) = 0$. Also, g_i are smooth functions—one only needs to justify differentiating under the integral sign—you may accept it now, or refer ahead to Example 2 at the end of Chapter 9 for detailed justification.

---

* The proof makes use of some facts on quadratic forms; see O'Nan, *Linear Algebra*, Chapter 7. An alternative, perhaps simpler proof, kindly supplied by A. Tromba, is given in Exercise 33.

Since $g_i(0) = 0$ we can apply the same procedure as above to write

$$g_i(x_1,\ldots,x_n) = \sum_{j=1}^{n} x_j h_{ij}(x_1,\ldots,x_n)$$

for certain smooth functions h_{ij}, and therefore

$$f(x_1,\ldots,x_n) = \sum_{i,j=1}^{n} x_i x_j h_{ij}(x_1,\ldots,x_n) .$$

We can assume $h_{ij} = h_{ji}$ by replacing h_{ij} by $\tilde{h}_{ij} = 1/2(h_{ij} + h_{ji})$ if necessary, which does not alter the expression for f. Note that at zero $\partial^2 f/\partial x_i\, \partial x_j = 2h_{ij}(0)$, so $h_{ij}(0)$ is non-singular.

Now f is written in a way analogous to a quadratic form. What we want to do is to "diagonalize" it. Proceed by induction. Suppose there exist coordinates $u_1, \ldots, u_n$ in a neighborhood U_1 of 0 such that

$$f = \pm(u_1)^2 \pm \cdots \pm (u_{r-1})^2 + \sum_{i,j \geqslant r} u_i u_j H_{ij}(u_1,\ldots,u_n) \qquad (1)$$

on U_1 where $r \geqslant 1$ and H_{ij} are symmetric. We have this as above for $r = 1$. (Coordinates $u_1, \ldots, u_n$ means, as in the text, that $(u_1,\ldots,u_n)$ are invertible functions of $(x_1,\ldots,x_n)$.)

We can make a linear coordinate change in $u_r, \ldots, u_n$ in order to diagonalize

$$\sum_{i,j \geqslant r} u_i u_j H_{ij}(0) .$$

In particular, since $H_{ij}(0)$ is non-singular the diagonal terms are non-zero. Thus we can assume $H_{rr}(0) \neq 0$. Let $g(u_1,\ldots,u_n) = |H_{rr}(u_1,\ldots,u_n)|^{1/2}$; in some smaller neighborhood $U_2 \subset U_1$ of 0, g will be a C^∞ non-zero function. Define

$$\begin{cases} V_i = u_i & i \neq r \\ V_r(u_1,\ldots,u_n) = g(u_1,\ldots,u_n)\left[u_r + \sum_{i>r} \dfrac{u_i H_{ir}(u_1,\ldots,u_n)}{H_{rr}(u_1,\ldots,u_n)} \right]. \end{cases} \qquad (2)$$

The Jacobian at 0 is

$$\frac{\partial(V_1,\ldots,V_n)}{\partial(u_1,\ldots,u_n)} = \begin{pmatrix} 1 & 0 & \cdots & & & & 0 \\ 0 & 1 & & & & & \\ \cdot & & \cdot & & & & \cdot \\ \cdot & & & \cdot & & & \cdot \\ \cdot & & & & \cdot & & \cdot \\ & & & & 1 & & 0 \\ \dfrac{\partial V_r}{\partial x_1} & \cdots & & g(0) & \cdots & & \dfrac{\partial V_r}{\partial x_n} \\ & & & & 1 & & \\ \cdot & & & & & \cdot & \cdot \\ \cdot & & & & & & \cdot \\ \cdot & & & & & & \cdot \\ 0 & & & & & & 0 \\ 0 & & \cdots & & & 0 & 1 \end{pmatrix}$$

which is non-singular. Therefore, by the inverse-function theorem $(u_1, \ldots, u_n) \mapsto (V_1, \ldots, V_n)$ is a C^∞ map with a C^∞ inverse on some smaller neighborhood U_3 of 0. In other words, $(V_1, \ldots, V_n)$ will serve as coordinates.

Now consider

$$u_r u_r H_{rr}(u_1, \ldots, u_n) + 2 \sum_{j=r+1}^{n} u_j u_r H_{jr}(u_1, \ldots, u_n) , \tag{3}$$

which are the terms in Eq. 1 with either i or $j = r$. Here we have used symmetry of H_{ij}. Comparing Eq. 1 with Eq. 2 we see that Eq. 3 equals

$$\pm \, V_r V_r - \frac{1}{H_{rr}} \left[\sum_{i > r} u_i H_{ir}(u_1, \ldots, u_n) \right]^2 ,$$

the plus or minus coming about because $H_{rr} = \pm g^2$, where we use $+$ if H_{rr} is positive and $-$ if H_{rr} is negative.

From this we see that Eq. 1 becomes

$$f = \sum_{i \leqslant r} \pm \, (V_i)^2 + \sum_{i,j > r} V_i V_j \tilde{H}_{ij}(V_1, \ldots, V_n)$$

for new symmetric $\tilde{H}_{ij}$. Thus we have inductively gone from r to $r + 1$ in Eq. 1. Hence it is true for $r = n + 1$, which proves the theorem.* ∎

Theorem 8. *Let* $f: U \subset \mathbb{R}^n \to \mathbb{R}$ *and* $g: U \subset \mathbb{R}^n \to \mathbb{R}$ *be given* C^1 *functions. Let* $x_0 \in U$, $g(x_0) = c_0$ *and let* $S = g^{-1}(c_0)$ *the level set for* g *with value* c_0. *Assume* $\nabla g(x_0) \neq 0$. *If* $f \,|\, S$ *has a maximum or minimum at* x_0 *then there is a real number* λ *such that*
$$\nabla f(x_0) = \lambda \, \nabla g(x_0)$$

Proof: The only thing not complete about the sketch of the proof given in Section 7.7 is that we need to know that if $v \perp \nabla g(x_0)$ then $v = c'(0)$ for a C^1 curve $c(t)$ in S, with $c(0) = x_0$.

This can be established as follows. By Theorem 3 there is a change of coordinates h such that $g(h(x_1, \ldots, x_n)) = x_n$. Thus $h^{-1}(S)$ is the coordinate plane $x_n = c_0$. Let $w = Dh^{-1}(x_0) \cdot v$. We claim that the last coordinate of w is zero, that is, w lies in the plane $x_n = c_0$. Indeed let $e_n = (0, 0, \ldots, 1)$. We shall show that $\langle w, e_n \rangle = 0$. But from the chain rule, $g(h(x_1, \ldots, x_n)) = x_n$ implies

$$\langle \nabla g(x_0), Dh(y_0) \cdot w \rangle = \langle w, e_n \rangle$$

where $h(y_0) = x_0$. But the left side is $\langle \nabla g(x_0), v \rangle = 0$. Now let $c(t) = h(y_0 + tw)$. This lies in S, $c(0) = x_0$, and from the chain rule, $c'(0) = v$.

The proof may now be completed as in the text. ∎

Worked Examples for Chapter 7

1. (Product rule for Jacobians.) Let $f: A \subset \mathbb{R}^n \to \mathbb{R}^n$, $g: B \subset \mathbb{R}^n \to \mathbb{R}^n$, and $f(A) \subset B$. Then show that for $x \in A$,
$$J_{g \circ f}(x) = J_g(f(x)) \cdot J_f(x)$$
(product of real numbers).

* Although the applications of this theorem to topology are fairly advanced, the reader interested in this material may consult J. Milnor, *Morse Theory*, Princeton University Press, 1963.

Solution: By the chain rule,

$$D(g \circ f)(x) = Dg(f(x)) \circ Df(x) \,,$$

which may be interpreted either as composition of linear maps, or as a matrix product. Since the determinant of a matrix product is the product of the determinants we immediately get the required result.

2. Consider equations $u = f_1(x,y)$ and $v = f_2(x,y)$. Show that they are invertible near (x_0,y_0) iff

$$\Delta = \frac{\partial f_1}{\partial x}\frac{\partial f_2}{\partial y} - \frac{\partial f_1}{\partial y}\frac{\partial f_2}{\partial x}$$

does not vanish at (x_0,y_0). If $x(u,v)$, $y(u,v)$ are the solutions, show that

$$\frac{\partial x}{\partial u} = \frac{1}{\Delta}\frac{\partial v}{\partial y}\,, \qquad \frac{\partial x}{\partial v} = -\frac{1}{\Delta}\frac{\partial u}{\partial y}\,,$$

$$\frac{\partial y}{\partial u} = -\frac{1}{\Delta}\frac{\partial v}{\partial x}\,, \qquad \frac{\partial y}{\partial v} = \frac{1}{\Delta}\frac{\partial u}{\partial x}\,.$$

Solution: This is just a special case of Theorem 1 for $n = 2$. Here Δ is exactly the Jacobian determinant. The matrix of derivatives of the solutions is, by Theorem 1, the inverse of the matrix of derivatives of f_1, f_2. Since the inverse of the matrix $\begin{pmatrix} a & b \\ c & d \end{pmatrix}$ is $1/\Delta\begin{pmatrix} d & -b \\ -c & a \end{pmatrix}$, where $\Delta = ad - bc$, we get the stated result.

3. Let $A \subset \mathbb{R}^n$ be an open set and $f: A \subset \mathbb{R}^n \to \mathbb{R}^n$ a one-to-one continuously differentiable function such that $Jf(x) = \det(Df(x)) \neq 0$ for all $x \in A$. Show that $f(A)$ is an open set and $f^{-1}: f(A) \to A$ is differentiable.

Solution: Let $y \in f(A)$ and suppose $y = f(x)$. Since f is continuously differentiable and $Df(x)$ has non-zero determinant, the inverse function theorem tells us that there exist open neighborhoods U of x and V of y such that $f \mid U$ (the restriction of f to U) is a C^1 diffeomorphism (that is, it has a C^1 inverse) of U onto V. Hence $V \subset f(A)$, so $f(A)$ is open. Now $(f \mid U)^{-1} = f^{-1} \mid f(U)$ and $(f \mid U)^{-1}$ is differentiable at y, and so f^{-1} is differentiable at y. Hence, f^{-1} is differentiable on $f(A)$.

4. Consider the following equations:

$$\begin{cases} x^2 - yu = 0 \,, \\ xy + uv = 0 \,. \end{cases}$$

Using the implicit function theorem describe under what conditions these equations can be solved for u and v. Then solve the equations directly and check these conditions.

Solution: Define $f_1: \mathbb{R}^4 \to \mathbb{R}$ by $f_1(x,y,u,v) = x^2 - yu$ and define $f_2: \mathbb{R}^4 \to \mathbb{R}$ by $f_2(x,y,u,v) = xy + uv$. Let $f: \mathbb{R}^4 \to \mathbb{R}^2$ be defined by $f = (f_1,f_2)$; then f is a smooth function. The matrix

$$\begin{pmatrix} \dfrac{\partial f_1}{\partial u} & \dfrac{\partial f_1}{\partial v} \\[2mm] \dfrac{\partial f_2}{\partial u} & \dfrac{\partial f_2}{\partial v} \end{pmatrix} = \begin{pmatrix} -y & 0 \\ v & u \end{pmatrix}$$

has determinant $-yu$. If (x_0,y_0,u_0,v_0) is such that $y_0u_0 \neq 0$, then the hypotheses of the implicit-function theorem are satisfied, and so there are neighborhoods A of (x_0,y_0) and B of (u_0,v_0) and a unique continuously differentiable function $g: A \to B$ such that $f(x,y,g(x,y)) = 0$ for all $(x,y) \in A$. If we let $u = g_1$ and $v = g_2$ (where $g = (g_1,g_2)$), then u and v are the solutions to the simultaneous equations. Thus these equations can be solved uniquely for u and v in neighborhoods around (x_0,y_0) and (u_0,v_0) satisfying the equations provided that $y_0u_0 \neq 0$, which is equivalent to requiring x_0 and $y_0 \neq 0$ since $f(x_0,y_0,u_0,v_0) = 0$, or $x_0^2 - y_0u_0 = 0$ and $x_0y_0 + u_0v_0 = 0$.

By direct computation, the solutions are $u = x^2/y$ and $v = -y^2/x$, which are valid except when $x_0 = 0$ or $y_0 = 0$.

5. (Functional dependence.) Let $A \subset \mathbb{R}^n$ be an open set and let the functions $f_1, \ldots, f_n: A \to \mathbb{R}$ be smooth. The functions $f_1, \ldots, f_n$ are said to be *functionally dependent* at $x_0 \in A$ if there is a neighborhood U of the point $(f_1(x_0),\ldots,f_n(x_0)) \in \mathbb{R}^n$ and a smooth function $F: U \to \mathbb{R}$ such that $DF \neq 0$ on a neighborhood of $(f_1(x_0),\ldots,f_n(x_0))$, and

$$F(f_1(x),\ldots,f_n(x)) = 0$$

for all x in some neighborhood of x_0.
(i) Show that if $f_1, \ldots, f_n$ are functionally dependent at x_0, then

$$\frac{\partial(f_1,\ldots,f_n)}{\partial(x_1,\ldots,x_n)} = 0 \text{ at } x_0 .$$

(ii) If

$$\frac{\partial(f_1,\ldots,f_{n-1})}{\partial(x_1,\ldots,x_{n-1})} \neq 0 \quad \text{and} \quad \frac{\partial(f_1,\ldots,f_n)}{\partial(x_1,\ldots,x_n)} = 0$$

on a neighborhood of x_0, then show that $f_1, \ldots, f_n$ are functionally dependent, and further,

$$f_n = G(f_1,\ldots,f_{n-1})$$

for some G.

Solution: Let $f = (f_1,\ldots,f_n)$.
(i) We have $F \circ f = 0$, so $DF(f(x)) \circ Df(x) = 0$. Now if $Jf(x_0) \neq 0$, $Df(x)$ would be invertible in a neighborhood of x_0, implying $DF(f(x)) = 0$. By the inverse function theorem, this implies $DF(y) = 0$ on a whole neighborhood of $f(x_0)$.
(ii) The conditions of (ii) imply that Df has rank $n - 1$. Hence by Corollary 2 there are functions g, h such that

$$g \circ f \circ h(x_1,\ldots,x_n) = (x_1,\ldots,x_{n-1},0) .$$

Let F be the last component of g. Then

$$F(f_1, \ldots, f_n) = 0$$

Since g is invertible, $DF \neq 0$.

It follows from the implicit-function theorem that $f_n = G(f_1, \ldots, f_{n-1})$, that is, we can locally solve

$$F(f_1, \ldots, f_n) = 0$$

for $f_n = G(f_1, \ldots, f_{n-1})$, provided we can show $\Delta = \partial F / \partial y_n \neq 0$. Now, as we saw above,

$$DF(f(x)) \cdot Df(x) = 0 \,,$$

or, in components, if $y = f(x)$,

$$\left(\frac{\partial F}{\partial y_1}, \ldots, \frac{\partial F}{\partial y_n} \right) \begin{pmatrix} \dfrac{\partial f_1}{\partial x_1} & \cdots & \dfrac{\partial f_1}{\partial x_n} \\ \cdot & & \cdot \\ \cdot & & \cdot \\ \cdot & & \cdot \\ \dfrac{\partial f_n}{\partial x_1} & \cdots & \dfrac{\partial f_n}{\partial x_n} \end{pmatrix} = 0 \,.$$

If $\partial F / \partial y_n = 0$, we would have

$$\left(\frac{\partial F}{\partial y_1}, \ldots, \frac{\partial F}{\partial y_{n-1}} \right) \begin{pmatrix} \dfrac{\partial f_1}{\partial x_1} & \cdots & \dfrac{\partial f_1}{\partial x_{n-1}} \\ \cdot & & \\ \cdot & & \\ \cdot & & \\ \dfrac{\partial f_{n-1}}{\partial x_1} & \cdots & \dfrac{\partial f_{n-1}}{\partial x_{n-1}} \end{pmatrix} = 0 \,,$$

or

$$\left(\frac{\partial F}{\partial y_1}, \ldots, \frac{\partial F}{\partial y_{n-1}} \right) = 0 \,,$$

since the square matrix is invertible by the assumption that

$$\frac{\partial(f_1, \ldots, f_{n-1})}{\partial(x_1, \ldots, x_{n-1})} \neq 0 \,.$$

This implies $DF = 0$, which is not true. Hence $\partial F / \partial y_n \neq 0$, and we have the desired result.

The reader should note the analogy between linear dependence and functional dependence, where rank or determinant conditions are replaced by the analogous conditions on the Jacobian matrix.

Exercises for Chapter 7

1. Write an expression for $\partial f/\partial x$ if $f(x,y) = g(x,h(x,y))$ where $g, h\colon \mathbb{R}^2 \to \mathbb{R}$.

2. Consider the following set of p equations in $n + p$ unknowns.

$$a_{11}x_1 + \cdots + a_{1n}x_n + a_{1,n+1}x_{n+1} + \cdots + a_{1,n+p}x_{n+p} = 0$$

.
.
.

$$a_{p1}x_1 + \cdots + a_{pn}x_n + a_{p,n+1}x_{n+1} + \cdots + a_{p,n+p}x_{n+p} = 0$$

What does the implicit function theorem say about the solution of these equations for the unknowns $x_{n+1}, \ldots, x_{n+p}$? Does it reduce to a theorem you know from linear algebra?

3. Prove the following generalization of Theorem 3. Let $A \subset \mathbb{R}^n$ be an open set and $f\colon A \subset \mathbb{R}^n \to \mathbb{R}^m$, $m \leqslant n$, a function of class C^p. Let $x_0 \in A$ and suppose $f(x_0) = 0$ and rank $Df(x_0) = m$. Then there is an open set U, an open set V containing x_0, and a function $h\colon U \to V$ of class C^p, with inverse $h^{-1}\colon V \to U$ of class C^p (that is, a C^p-diffeomorphism), such that $f(h(x_1, \ldots, x_n)) = (x_{n-m+1}, \ldots, x_n)$. [Hint: If $Df(x_0)$ has rank m there must exist $j_i, \ldots, j_m$ such that the matrix $(D_{j_k} f_i)$, $1 \leqslant i, k \leqslant m$, is invertible. Define the permutation map $g\colon \mathbb{R}^n \to \mathbb{R}^n$ by

$$g(x_1, \ldots, x_n) = (x_1, \ldots, x_{j_1-1}, x_{n-m+1}, x_{j_1+1}, \ldots, x_{j_m-1}, x_n, x_{j_m+1}, \ldots, x_{n-m}, x_{j_1}, \ldots, x_{j_m})$$

and make appropriate modifications to the proof of Theorem 3.]

4. Let $f\colon \mathbb{R}^n \to \mathbb{R}^n$ and $g\colon \mathbb{R}^n \to \mathbb{R}^n$ be functions of class C^1. Define $h\colon \mathbb{R}^n \to \mathbb{R}^n$ by $h(x) = f(g_1(x_1), \ldots, g_n(x_n))$, where $g = (g_1, \ldots, g_n)$ and $x = (x_1, \ldots, x_n)$. Show that

$$Dh(x) = Df(g_1(x_1), \ldots, g_n(x_n)) \begin{pmatrix} g_1'(x_1) & & 0 \\ & \ddots & \\ 0 & & g_n'(x_n) \end{pmatrix}.$$

5. (a) Define $x\colon \mathbb{R}^2 \to \mathbb{R}$ by $x(r,\theta) = r \cos \theta$ and define $y\colon \mathbb{R}^2 \to \mathbb{R}$ by $y(r,\theta) = r \sin \theta$. Show that

$$\frac{\partial(x,y)}{\partial(r,\theta)}(r_0,\theta_0) = r_0 .$$

(b) When can we form a smooth inverse function $r(x,y)$, $\theta(x,y)$? Check directly and with the inverse function theorem.

(c) Consider the following transformations for spherical coordinates:

$$x(r,\varphi,\theta) = r \sin \varphi \cos \theta ;$$
$$y(r,\varphi,\theta) = r \sin \varphi \sin \theta ;$$
$$z(r,\varphi,\theta) = r \cos \varphi .$$

Show that

$$\frac{\partial(x,y,z)}{\partial(r,\varphi,\theta)} = r^2 \sin \varphi .$$

(d) When can we solve for (r,φ,θ) in terms of (x,y,z)?

6. Let f satisfy the conditions of the inverse-function theorem and let g be the local inverse $g = f^{-1}: W \to U$. Let $x_0 \in U$ and let $y_0 = f(x_0)$. Consider the case $n = 3$ and show that

$$Jf(x_0)D_1g_i(y_0) = \begin{vmatrix} \delta_{i,1} & D_1 f_2(x_0) & D_1 f_3(x_0) \\ \delta_{i,2} & D_2 f_2(x_0) & D_2 f_3(x_0) \\ \delta_{i,3} & D_3 f_2(x_0) & D_3 f_3(x_0) \end{vmatrix},$$

where $\delta_{i,j} = 1$ if $i = j$ and 0 if $i \neq j$. From this deduce the following expression for $D_1 g_1$:

$$D_1 g_1 = \frac{\partial(f_2,f_3)/\partial(x_2,x_3)}{\partial(f_1,f_2,f_3)/\partial(x_1,x_2,x_3)} .$$

Also, obtain expressions for the other eight partial derivatives $D_j g_i$.

7. Determine whether the "curve" described by the equation $x^2 + y + \sin(xy) = 0$ can be written in the form $y = f(x)$ in a neighborhood of $(0,0)$. Does the implicit function theorem allow you to say whether the equation can be written in the form $x = h(y)$ in a neighborhood of $(0,0)$?

8. Let (x_0,y_0,z_0) be a point of the locus defined by $z^2 + xy - a = 0$, $z^2 + x^2 - y^2 - b = 0$.
   (a) Under what sufficient conditions may the part of the locus near (x_0,y_0,z_0) be represented in the form $x = f(z), y = g(z)$?
   (b) Compute $f'(z)$ and $g'(z)$.

9. Let f_1, f_2, f_3 be continuously differentiable functions from $\mathbb{R}^4$ to $\mathbb{R}$. Give sufficient conditions so that the equations

$$f_1(x,y,z,t) = 0 , \qquad f_2(x,y,z,t) = 0 , \qquad f_3(x,y,z,t) = 0$$

can be solved for x, y, z in terms of t.

10. (a) Let $f: \mathbb{R}^2 \to \mathbb{R}^2$ be smooth and suppose that

$$\frac{\partial f_1}{\partial x} = \frac{\partial f_2}{\partial y} , \qquad \frac{\partial f_1}{\partial y} = -\frac{\partial f_2}{\partial x} .$$

(These equations are called the *Cauchy-Riemann Equations* and arise naturally in complex variable theory.*) Show that $Jf(x,y) = 0$ iff $Df(x,y) = 0$; hence f is locally invertible iff $Df(x,y) \neq 0$. Prove that the inverse function also satisfies the Cauchy-Riemann equations.

(b) Show that the conclusion of (a) is false (by giving an example) if f does not satisfy the Cauchy-Riemann equations.

* See for example, J. Marsden, *Basic Complex Analysis*, W. H. Freeman Co. (1973).

11. (a) Suppose that $f: \mathbb{R}^n \to \mathbb{R}^m$ is of class C^1 and $Df(x_0)$ has rank m. This means that $Df(x_0)$ as a linear map is onto. Then show that there is a whole neighborhood of $f(x_0)$ lying in the image of f.
    (b) Suppose $f: \mathbb{R}^n \to \mathbb{R}^m$ is C^1 and $Df(x_0)$ is one-to-one. Show f is one-to-one on a neighborhood of x_0.

12. Show that the implicit function theorem implies the inverse function theorem.

13. Prove Corollary 1.

14. (Based on Example 5.) Prove that if $f_1, \ldots, f_k$ are functionally independent on $\mathbb{R}^n$ (that is, Df has rank k for $f = (f_1, \ldots, f_k)$, $k \leqslant n$) and $g, f_1, \ldots, f_k$ are functionally dependent, then locally we can write $g = F(f_1, \ldots, f_k)$.

15. Consider the map $\mathscr{L}^{-1}: GL(\mathbb{R}^n, \mathbb{R}^n) \to GL(\mathbb{R}^n, \mathbb{R}^n)$, $A \mapsto A^{-1}$, taking a matrix to its inverse. Show that the derivative of this map is given by

$$D\mathscr{L}^{-1}(A) \cdot B = -A^{-1} \circ B \circ A^{-1}$$

(consult Lemma 2 in the proof of Theorem 1). [Hint: Differentiate the relation $\mathscr{L}^{-1}(A) \circ A = $ identity with respect to A.]

16. Does the function h in Theorem 3 have to be unique? Discuss.

17. Give a direct proof of the Morse lemma for functions $f: \mathbb{R} \to \mathbb{R}$. Does it apply to

    (a) $f(x) = x^3$, or   (b) $f(x) = x \sin\left(\dfrac{1}{x}\right)$?

18. Let $f: \mathbb{R}^4 \to \mathbb{R}^2$ be $F(x,y,u,v) = (u^3 + vx + y, uy + v^3 - x)$. At what points can we solve for $F(x,y,u,v) = 0$ for u, v in terms of x, y? Compute $\partial u / \partial x$.

19. Let $f: \mathbb{R} \to \mathbb{R}$ be C^1 and

$$u = f(x)\,,$$
$$v = -y + xf(x)\,.$$

If $f'(x_0) \neq 0$ show that this transformation is invertible near (x_0, y) and has the form

$$x = f^{-1}(u)\,,$$
$$y = -v + uf^{-1}(u)\,.$$

20. Show that the equations

$$x^2 - y^2 - u^3 + v^2 + 4 = 0$$
$$2xy + y^2 - 2u^2 + 3v^4 + 8 = 0$$

determine functions $u(x,y)$, $v(x,y)$ near $x = 2$, $y = -1$ such that $u(2,-1) = 2$, $v(2,-1) = 1$. Compute $\partial u / \partial x$.

21. "If $f(x,y,z) = 0$ then $\partial z / \partial y \cdot \partial y / \partial x \cdot \partial x / \partial z = -1$." What do you think this really means.*

* Thermodynamics books are notorious for such mystifying statements.

22. Let $f(x,y) = (xy(x^2 - y^2))/(x^2 + y^2)$ for $(x,y) \neq (0,0)$ and $f(0,0) = 0$. Is f of class C^2? [Hint: See Exercise 32 at the end of Chapter 6.]

23. Let $C \subset \mathbb{R}^n$ be a closed subset such that $x \in C \Rightarrow \alpha x \in C$ for $\alpha \geqslant 0$.
    (a) Discuss what C "looks like."
    (b) Let $f: C \to \mathbb{R}^n$ be continuous and $f(\alpha x) = \alpha f(x)$ for $x \in C$, $\alpha \geqslant 0$. Show that there is an M such that
    $$\| f(x) \| \leqslant M \|x\|$$
    for all $x \in C$.

24. Let $f(x,y,z) = x^2 - yz - \sin(xz)$ and $g(x,y,z) = (x \cos y, x \sin y \cos z, x \sin y \sin z)$. Compute the derivative of $f \circ g$.

25. Let $\bar{D}(0,r) = \{x \in \mathbb{R}^n \mid \|x\| \leqslant r\}$. Let $f: \bar{D}(0,r) \to \mathbb{R}^n$ be a map with
    (a) $\| f(x) - f(y) \| \leqslant 1/3 \|x - y\|$,   (b) $\| f(0) \| \leqslant 2/3r$.
    Prove that there is a unique $x \in \bar{D}(0,r)$ such that $f(x) = x$.

26. Show that there exist positive numbers $p > 0$, $q > 0$ such that there are unique functions u, v from $]-1 - p, -1 + p[$ into $]1 - q, 1 + q[$ for which
    $$xe^{u(x)} + u(x)e^{v(x)} = 0 = xe^{v(x)} + v(x)e^{u(x)}$$
    for all $x \in]-1 - p, -1 + p[$ and $u(-1) = 1 = v(-1)$.

27. Obtain an estimate on the length of time the solution of $dx/dt = t^2 x^3 e^{tx}$, $x(0) = 1$ exists.

28. Let $A \subset \mathbb{R}^n$ be compact and let $B \subset \mathscr{C}(A,\mathbb{R})$ be compact (see Section 5.5). Show that there is an $f_0 \in B$ and an $x_0 \in A$ such that $g(x) \leqslant f_0(x_0)$ for all $g \in B$ and $x \in A$.

29. Let $a_n \geqslant a_{n+1} \geqslant 0$ and $a_n \to 0$. Let $f(x) = \sum_{n=0}^{\infty} a_n x^n$. Show that $f(x)$ is continuous on $[-1,0]$.

30. Is it possible to solve
    $$xy^2 + xzu + yv^2 = 3$$
    $$u^3 yz + 2xv - u^2 v^2 = 2$$
    for $u(x,y,z)$, $v(x,y,z)$ near $(x,y,z) = (1,1,1)$, $(u,v) = (1,1)$? Compute $\partial v/\partial y$.

31. Consider the equation $dx/dt = 1 + tx$, $x(0) = 0$. Examine the iteration scheme given in the text to obtain a power series expression for the solution. Examine the radius of convergence.

32. Compute the index of the function $x^2 + y^2 - 7x - 8y + xy + 16 + (x - 2)^3$ at its critical point $x = 2$, $y = 3$. Discuss the nature of the function near this point.

33. Give another proof of Theorem 7 as follows. Assume $x_0 = 0$ and $f(x_0) = 0$. Use Taylors theorem to write $f(x) = 1/2 D^2 f(0) \cdot (x,x) + 1/2 R_x(x,x) = 1/2 \langle A_x x, x \rangle$ so that for each x, A_x is a symmetric linear transformation of $\mathbb{R}^n$. By assumption, A_0 is an isomorphism. By Lemma 2, p. 231, A_x is an isomorphism if x is near to 0. Let $Q_x = A_0 A_x^{-1}$ so that $Q_0 = I$. Using a power series, we can define the square root T_x of Q_x for x close to 0, that is, $T_x^2 = Q_x$. Show that $Q_x A_x = A_x Q_x^T$, where T

means the transpose matrix, and using the power series for T_x show that the same
equation holds for T_x. Let $S_x = T_x^{-1}$ and conclude that $A_x = S_x A_0 S_x^T$. Let
$h(x) = S_x^T x$ and show that $Dh(0) = I$; now apply the inverse function theorem to
conclude that h is locally invertible. Let $g = h^{-1}$. Show that

$$f(x) = (1/2)\langle A_0 h(x), h(x)\rangle$$

and deduce that $f \circ g(x) = (1/2)D^2 f(0)(x,x)$. Finally, use a linear change of coordinates to diagonalize the quadratic form $(1/2)D^2 f$.

Find the relative extrema of $f \mid S$ in Exercises 34–37: use both Theorem 8 and
Corollary 3.

34. $f: \mathbb{R}^2 \to \mathbb{R}$, $(x,y) \mapsto x^2 + y^2$, $S = \{(x,2) \mid x \in \mathbb{R}\}$.

35. $f: \mathbb{R}^2 \to \mathbb{R}$, $(x,y) \mapsto x^2 + y^2$, $S = \{(x,y) \mid x^2 - y^2 = 1\}$.

36. $f: \mathbb{R}^2 \to \mathbb{R}$, $(x,y) \mapsto x^2 - y^2$, $S = \{(x,\cos x) \mid x \in \mathbb{R}\}$.

37. $f: \mathbb{R}^3 \to \mathbb{R}$, $(x,y,z) \mapsto x^2 + y^2 + z^2$, $S = \{(x,y,z) \mid z \geqslant -2 + x^2 + y^2\}$.

38. A rectangular box with no top is to have a surface area of 16 square meters. Find
the dimensions that maximize the volume.

39. Design a cylindrical can to contain 1 liter of water, but uses the minimum amount
of metal.

40. Let f_n be monotone increasing continuous functions on $[0,1]$. Suppose $F(x) =
\sum_{n=1}^\infty f_n(x)$ converges for each $x \in [0,1]$. Prove that F is continuous.

ter **8**

egration

T he reader is undoubtedly familiar with the integration process for functions of one variable and how to apply this to practical problems involving area, volume, arc length, and so on. Some familiarity with simple situations involving multiple integrals would be useful but is not essential. The purpose of the next two chapters is to review, solidify, and extend this knowledge. In this chapter we will formulate the basic definitions for a general theory of integration. The connection with the usual method of evaluating integrals by antiderivatives is made by the fundamental theorem of calculus.

The powerful computational theorems for multiple integrals will be given in the next chapter. These are Fubini's theorem, which enables us to reduce a multiple integral to iterated single integrals, and the change of variables formula, which enables us to change to a more convenient system of coordinates such as polar or spherical coordinates. To obtain a satisfactory theory of multiple integrals, even for continuous functions, it is convenient to introduce the notion of a set of "measure zero." We shall see that one of the main theorems states that a function is integrable iff its discontinuities form a set of measure zero. As a result, a function with a finite or countable number of discontinuities will be integrable.

Although the manipulations which are required for integration in dimensions larger than one are considerably more complicated, the basic idea of integration remains the same. We begin by recalling these ideas in one and two dimensions.

8.1   Review of Integration in $\mathbb{R}$ and $\mathbb{R}^2$

First, let us look briefly at the basic ideas involved in integration in one and two dimensions. These cases provide the clues on how to generalize to functions of several variables. In our discussion, let f be a non-negative real-valued bounded function defined on some bounded subset $A \subset \mathbb{R}$.

When we say that we want to integrate the function f over the set A we mean that we would like to *find the area* under the graph of f (see Figure 8-1). To do this, note first, as A is bounded, that there is a closed interval $[a,b] \supset A$. We consider f to be *defined* over the whole interval $[a,b]$ by letting f be zero on $[a,b] \backslash A$. Next, partition $[a,b]$, which means that we pick points $x_0 = a, x_1, \ldots, x_{n-1}, x_n = b$ in such a way that $a = x_0 < x_1 < \cdots < x_{n-1} < x_n = b$. Denote such a partition by P, that is, $P = \{x_0, \ldots, x_n\}$. Then, form the two sums

$$U(f,P) = \sum_{i=0}^{n-1} [\sup\{f(x) \mid x \in [x_i, x_{i+1}]\}](x_{i+1} - x_i)$$

and

$$L(f,P) = \sum_{i=0}^{n-1} [\inf\{f(x) \mid x \in [x_i, x_{i+1}]\}](x_{i+1} - x_i)$$

called the *upper* and *lower sums*, respectively. The first sum is the sum over all intervals $[x_i, x_{i+1}]$ of the maximum $(= \sup)$ of f in that interval times the length of the interval and has value equal to the area of the shaded region shown in Figure 8-1. Since f is assumed to be bounded, the sup exists in

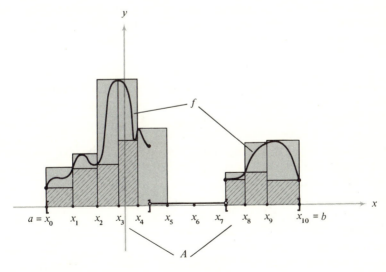

FIGURE 8-1

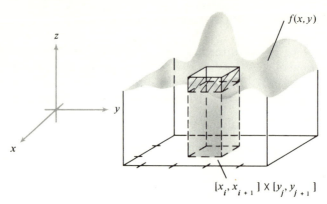

z

y

x

$f(x,y)$

$[x_i, x_{i+1}] \times [y_j, y_{j+1}]$

FIGURE 8-2

each interval. The second sum is the sum over all intervals $[x_i, x_{i+1}]$ of the minimum or inf of f in that interval times the length of the interval and is the hatched region shown in Figure 8-1. The boundedness of the functions again guarantees that the inf exists.

Since f is bounded, say, $-M \leqslant f \leqslant M$, we see that $-(b-a)M \leqslant L(f,P) \leqslant U(f,P) \leqslant (b-a)M$ for any partition P of $[a,b]$. Let

$$S = \inf\{U(f,P) \mid P \text{ is any partition}\}$$

and

$$s = \sup\{L(f,P) \mid P \text{ is any partition}\} .$$

If we again look at Figure 8-1, it seems reasonable to expect that as the size of the intervals in P get smaller, $U(f,P)$ decreases while $L(f,P)$ increases, and in the limit of decreasing size of the intervals of P, the numbers $U(f,P)$ and $L(f,P)$ should converge to a common value. This leads us to the following definition.

> **Definition 1.** We say that f is *Riemann integrable* (or just *integrable* or the *integral exists*, for short) if $s = S$. The common value $s = S$ is denoted by $\int_A f$ or by $\int_A f(x)\, dx$.

It should be noted that integrability does not really involve smoothness or continuity properties of f. In fact, some badly discontinuous functions can still be integrable.

Now suppose $f \colon A \subset \mathbb{R}^2 \to \mathbb{R}$ is a bounded non-negative function (see Figure 8-2), where A is a bounded set.

The graph of the function f is a surface in $\mathbb{R}^3$, and the integration process is used to find the volume under this surface. We enclose A in some rectangle $[a_1, b_1] \times [a_2, b_2]$ and extend f to the whole rectangle by defining it to be zero outside of A. Then we divide $[a_1, b_1] \times [a_2, b_2]$ into smaller rectangles by partitioning $[a_1, b_1]$ by, for example, $a_1 = x_0 < x_1 < \cdots < x_{n-1} < b_1 = x_n$ and partitioning $[a_2, b_2]$ by, say, $a_2 = y_0 < y_1 < \cdots < y_{m-1} < b_2 = y_m$, thus forming mn rectangles $[x_i, x_{i+1}] \times [y_j, y_{j+1}]$. Then we form volumes $\inf\{f(z) \mid z \in [x_i, x_{i+1}] \times [y_j, y_{j+1}]\} \cdot (x_{i+1} - x_i)(y_{j+1} - y_j)$ (the shaded block of Figure 8-2) and $\sup\{f(z) \mid z \in [x_i, x_{i+1}] \times [y_j, y_{j+1}]\} \cdot (x_{i+1} - x_i)(y_{j+1} - y_j)$ (the shaded block plus the cross-hatched block of Figure 8-2). Next, we sum these volumes over all i and j (that is, all the mn "subrectangles" of the rectangle $[a_1, b_1] \times [a_2, b_2]$), and get two values $L(f, P)$ and $U(f, P)$, where P stands for the partitioning. Again, if $\sup\{L(f, P) \mid P$ is a partition$\} = \inf\{U(f, P) \mid P$ is a partition$\}$, we say f is *(Riemann) integrable* over A and define the *(Riemann) integral* of f over the set A, written $\int_A f$, or $\int_A \int f(x, y) \, dx \, dy$, by

$$\int_A f = \sup\{L(f, P)\} = \inf\{U(f, P)\} .$$

One thing about this procedure may seem puzzling. Why do we insist that $\inf\{U(f, P)\} = \sup\{L(f, P)\}$? At first, we might think that this relation will always hold. However, this is not always the case, as the next example shows.

EXAMPLE 1. Consider $\inf\{U(f, P)\}$ and $\sup\{L(f, P)\}$ for the following function

$$f: [0, 1] \subset \mathbb{R} \to \mathbb{R}$$

defined by

$$f(x) = \begin{cases} 1, & x \text{ irrational}, \\ 0, & x \text{ rational} . \end{cases}$$

It is not difficult to see that $\inf\{U(f, P)\} = 1$ and $\sup\{L(f, P)\} = 0$ (because on any interval f is always one at some points and zero at others, so the inf on any interval is zero and the sup is one). Therefore, the integral of this function over the set $[0, 1]$ does not exist for our purposes. In more advanced work the integral of such a pathological function can be defined, but we shall be dealing mostly with "decent functions" for which the integral exists.

EXAMPLE 2. Suppose $f: [a, b] \to \mathbb{R}$ is (Riemann) integrable and $f \geqslant 0$. Show $\int_a^b f(x) \, dx \geqslant 0$.

Solution: By definition, the integral is the infimum of sums of the form

$$\sum_{i=0}^{n-1} \left(\sup_{x \in [x_i, x_{i+1}]} f(x) \right) \cdot (x_{i+1} - x_i)$$

over all partitions. But each of these sums $U(f, P)$ is non-negative since $f \geqslant 0$. Hence the integral is $\geqslant 0$ since the inf of a set of non-negative numbers is also non-negative.

Exercises for Section 8.1

1. Show directly from the definition that $\int_a^b dx = (b - a)$.

2. If f and g are integrable on $[a,b]$ and if $f \geqslant g$ on $[a,b]$, show $\int_a^b f \geqslant \int_a^b g$.

3. Show that for $f: [a,b] \to \mathbb{R}$ and P any partition of $[a,b]$, $U(f,P) \geqslant L(f,P)$.

4. Let $f: [a,b] \subset \mathbb{R} \to \mathbb{R}$ be integrable and $f \leqslant M$. Prove that $\int_a^b f(x)\, dx \leqslant (b - a)M$.

8.2 Integrable Functions

In essence we have already introduced all the ideas needed for a theory of integration of bounded functions over bounded sets for arbitrary dimensions. Most of what remains is to formalize the statements for the case $\mathbb{R}^n$.

Let $f: A \subset \mathbb{R}^n \to \mathbb{R}$ be a bounded function with domain a bounded set A. Let $[a_1,b_1] \times \cdots \times [a_n,b_n]$ be a rectangle which encloses A. Furthermore, let f be defined over the whole rectangle by setting it equal to 0 at points not contained in A. Let P be a partition of $[a_1,b_1] \times \cdots \times [a_n,b_n]$ obtained by dividing each $[a_i,b_i]$ by points $x_0^i, \ldots, x_{m_i}^i$ and forming the $m_1 m_2 \cdots m_n$ *rectangles*

$$[x_{j_1}^1, x_{j_1+1}^1] \times \cdots \times [x_{j_n}^n, x_{j_n+1}^n], \qquad \text{where } 0 \leqslant j_i \leqslant m_i.$$

Define the volume of the rectangle $B = [a_1,b_1] \times \cdots \times [a_n,b_n]$ by $v(B) = (b_1 - a_1)(b_2 - a_2) \cdots (b_n - a_n)$. Let $L(f,P)$ denote the lower sum of f for P, defined by

$$L(f,P) = \sum_{R \in P} [\inf\{f(x) \mid x \in R\}] v(R),$$

the sum being over all subrectangles R of the partition P, and let $U(f,P)$ denote the upper sum for P; $U(f,P) = \sum_{R \in P} [\sup\{f(x) \mid x \in R\}] v(R)$. Now we observe some properties of $L(f,P)$ and $U(f,P)$. From the definition we see that for any partition P, $L(f,P) \leqslant U(f,P)$. Now suppose P' is any partition which is a *refinement of* or is *finer than* P; this means that each subrectangle belonging to P' is *contained* in a subrectangle belonging to P. Then we see that $L(f,P) \leqslant L(f,P')$. Indeed, we can observe that the minimum of f on a rectangle is less than or equal to the minimum on any rectangle contained in it. Similarly, $U(f,P') \leqslant U(f,P)$. This has the following consequence. *If P' and P'' are any two partitions of $[a_1,b_1] \times \cdots \times [a_n,b_n]$, then $L(f,P') \leqslant U(f,P'')$.* To clarify this, let P be a partition of the rectangle which refines both P' and P'', which we can always arrange by using all the subdivision points of P' and P''; then $L(f,P') \leqslant L(f,P) \leqslant U(f,P) \leqslant U(f,P'')$.

As before, the set $\{L(f,P) \mid P$ is any partition$\}$ is bounded from above and thus has a sup. The set $\{U(f,P) \mid P$ is any partition$\}$ is bounded from below and thus has an inf. Let $s = \sup\{L(f,P) \mid P$ is a partition$\}$ and $S = \inf\{U(f,P) \mid P$ is a partition$\}$; then $s \leqslant S$. With this notation we can make another definition.

Definition 2. Let $\overline{\int}_A f = S$, called the *upper integral* of f and let $\underline{\int}_A f = s$, called the *lower integral* of f. If $s = S$, we say f is *Riemann integrable* (from now on we will just use the word integrable) and define the *integral* of f over the set A by

$$\int_A f = \sup\{L(f,P) \mid P \text{ is a partition}\} = \inf\{U(f,P) \mid P \text{ is a partition}\} .$$

Instead of $\int_A f$, the notation $\int_A f(x) \, dx$ or $\int \cdots \int_A f(x_1,\ldots,x_n) \, dx_1 \cdots dx_n$ is frequently employed. If $f: [a,b] \to \mathbb{R}$, the notation $\int_a^b f$ or $\int_a^b f(x) \, dx$ is also used.

There is an important equivalent characterization of the Riemann integral as presented in the next theorem.

Theorem 1 (*Darboux's Theorem*). *Let $A \subset \mathbb{R}^n$ be bounded and lie in some rectangle S. Let $f: A \to \mathbb{R}$ be bounded and be extended to S by defining $f = 0$ outside A. Then f is integrable with integral I iff for any $\varepsilon > 0$ there is a $\delta > 0$ such that if P is any partition into rectangles $S_1, \ldots, S_N$ with sides $< \delta$ and if $x_1 \in S_1, \ldots, x_N \in S_N$, we have*

$$\left| \sum_{i=1}^N f(x_i)v(S_i) - I \right| < \varepsilon .$$

We call $\sum_{i=1}^n f(x_i)v(S_i)$ a Riemann sum.

This theorem is an important tool for proving many properties of the integral. In Example 1 the theorem was rewritten for the special case $n = 1$ in order to gain some insight into the meaning of the theorem. There it is shown why the theorem is intuitively plausible.

A condition closely related to Theorem 1 follows.

Riemann's condition: *f is integrable iff for any $\varepsilon > 0$ there is a partition P_ε of S such that $0 \leqslant U(f,P_\varepsilon) - L(f,P_\varepsilon) < \varepsilon$.*

The proof of Riemann's condition will be given along with the proof of Theorem 1 at the end of the chapter.

Notice that if f is continuous we can realize the upper and lower sums as special Riemann sums since f assumes its maximum and minimum at some point of the interval. If f is continuous on the whole rectangle S ($=$ interval if $n = 1$) then it follows easily from uniform continuity of f (see Section 4.6)

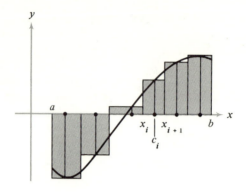

FIGURE 8-3

and Riemann's condition that f is integrable. We shall, in fact, prove a more general result in Theorem 3 below.

EXAMPLE 1. Interpret the Riemann sums geometrically, for $f: [a,b] \to \mathbb{R}$.

Solution: Let $P: a = x_0 < x_1 < \cdots < x_n = b$ be a partition and let $c_i \in [x_i, x_{i+1}]$. By definition the Riemann sum is

$$R = \sum_{i=0}^{n-1} f(c_i)(x_{i+1} - x_i) ,$$

which is the total area of the rectangles represented in Figure 8-3, with the area of rectangles below the x-axis counted with a negative sign. We observe that $L(f,P) \leqslant R \leqslant U(f,P)$, so the result in Theorem 1 is plausible.

EXAMPLE 2. Show $\int_0^1 x \, dx = 1/2$ using the definition of the integral. Compare with a geometrical computation.

Solution: Break up $[0,1]$ into n equal parts

$$\left[0, \frac{1}{n}\right], \left[\frac{1}{n}, \frac{2}{n}\right], \ldots, \left[\frac{n-1}{n}, 1\right].$$

Using this as a partition, note that on $[i/n, (i+1)/n]$, $f(x) = x$ has inf $= i/n$ and sup $= (i+1)/n$. Thus, calling this partition P,

$$U(f,P) = \sum_{i=0}^{n-1} \left[\frac{i+1}{n}\right] \cdot \left[\frac{1}{n}\right]$$

$$= \frac{1}{n^2} \cdot \sum_{i=0}^{n-1} (i+1) = \frac{1}{n^2}(1 + 2 + \cdots + n)$$

$$= \frac{1}{n^2} \cdot \frac{1}{2} n \cdot (n+1)$$

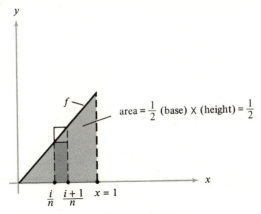

FIGURE 8-4

and

$$L(f,P) = \sum_{i=0}^{n-1} \frac{i}{n} \cdot \frac{1}{n} = \frac{1}{n^2}(0 + 1 + \cdots + (n-1))$$

$$= \frac{1}{n^2}\frac{1}{2}(n-1)(n)$$

where, we recall, $1 + 2 + \cdots + k = \frac{1}{2}k(k+1)$. Thus

$$U(f,P) = \frac{1}{2}\left(1 + \frac{1}{n}\right) \qquad \text{and} \qquad L(f,P) = \frac{1}{2}\left(1 - \frac{1}{n}\right).$$

These both converge to 1/2 as $n \to \infty$. Thus from Riemann's condition (or Darboux's theorem) we see that f is integrable with integral = 1/2. This is also geometrically obvious from Figure 8-4.

Exercises for Section 8.2

1. Give a formal proof that if R is any Riemann sum for a function f and partition P, then $L(f,P) \leqslant R \leqslant U(f,P)$.

2. Let $f: [0,1] \to \mathbb{R}$,
$$f(x) = 0 \qquad \text{if } x \neq 1/2 ;$$
$$f(1/2) = 1 .$$

   Prove f is integrable and $\int_0^1 f(x)\, dx = 0$.

3. Let $f: [0,2] \to \mathbb{R}$, $f(x) = 0$, $0 \leqslant x \leqslant 1$, and $f(x) = 1$, $1 < x \leqslant 2$. Compute, using the definition, $\int_0^2 f(x)\, dx$.

4. Let $A \subset \mathbb{R}^n$ and let $f(x) = 1$ for $x \in A$. What do you think $\int_A f$ should be?

5. Evaluate $\int_0^1 (3x + 4)\, dx$ using the definition and compare the answer with a geometrical computation of area.

6. Let $f: [a,b] \to \mathbb{R}$ be continuous. Use Riemann's condition and uniform continuity of f to prove that f is integrable.

8.3   Volume and Sets of Measure Zero

On the real line one usually integrates over intervals. However, in $\mathbb{R}^n$ we wish to integrate over more complicated sets. We must be sure that the sets we are dealing with are restricted in such a way that the partitioning in the definition of integrability is reasonable. Here, "reasonable" means, roughly speaking, that the boundary of the set is not too complicated. Our immediate goal is to develop enough machinery so that we can make these ideas precise. First, let us define the volume of a set.

Definition 3.   If $A \subset \mathbb{R}^n$, define the *characteristic function* 1_A of A by

$$1_A: \mathbb{R}^n \to \mathbb{R}, \; 1_A(x) = \begin{cases} 1, & x \in A , \\ 0, & x \notin A . \end{cases}$$

We say that A *has volume* if 1_A is integrable, and the *volume* of A is the number

$$\int_A 1_A(x)\, dx = v(A) .$$

(If A is a bounded set, it makes sense to talk about integrability of 1_A.)

This definition is natural because the region under the graph of 1_A is just "cylindrical" with height one and base A (Figure 8-5). We shall also use the phrase "A has content" to mean the same as "A has volume." Sometimes a set which has volume is called *Jordan measurable*.

Notice that in the case of $n = 1$ when $A \subset \mathbb{R}$, we speak of $v(A)$ as the *length* of A and when $A \subset \mathbb{R}^2$, we use the term *area* of A for $v(A)$.

We say A has *volume zero* (or *content zero*) if $v(A) = 0$. From the definition of the integral this is equivalent to the statement that for every $\varepsilon > 0$ there is a finite covering of A by rectangles, say, $S_1, \ldots, S_m$ such that the total volume is $< \varepsilon$; that is,

$$\sum_{i=1}^{m} v(S_i) < \varepsilon,$$

where $v(S_i)$ is computed for rectangles as before. (The details are worked out in Example 1 at the end of the chapter.)

It is useful to allow countable coverings as well as finite ones. These ideas were systematically introduced for the first time by Henri Lebesgue around 1900.

Definition 4.   A set $A \subset \mathbb{R}^n$ (not necessarily bounded) is said to have *measure zero* if for every $\varepsilon > 0$ there is a covering of A, say, S_1, $S_2, \ldots$, by a countable (or finite) number of rectangles such that the total volume $\sum_{i=1}^{\infty} v(S_i) < \varepsilon$. Recall that $S_1, S_2, \ldots$ are said to *cover A* when $\bigcup_{i=1}^{\infty} S_i \supset A$.

It is important to realize that these concepts depend on the space in which we are working. To illustrate the point, consider an example.

EXAMPLE 1.   Show that, regarded as a subset of $\mathbb{R}^2$, the real line has measure zero, but as a subset of $\mathbb{R}$ it does not.

Solution: To prove the first assertion, given $\varepsilon > 0$, we want to find rectangles $S_1, S_2, \ldots$ which enclose the x-axis and have total area $< \varepsilon$. Let

$$S_i = [-i, i] \times \left[-\frac{\varepsilon}{(2i \cdot 2^{i+1})}, \frac{\varepsilon}{(2i \cdot 2^{i+1})} \right].$$

See Figure 8-6. Now

$$v(S_i) = (2i) \cdot \frac{2\varepsilon}{(2i \cdot 2^{i+1})} = \frac{\varepsilon}{2^i}.$$

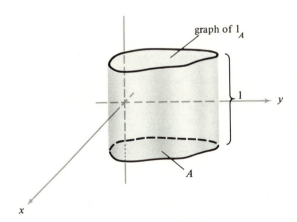

graph of 1_A

y

1

A

x

FIGURE 8-5

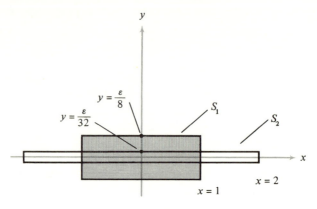

FIGURE 8-6

Thus

$$\sum_{i=1}^{\infty} v(S_i) < \sum_{i=1}^{\infty} \frac{\varepsilon}{2^i} = \varepsilon$$

since $1/2 + 1/4 + 1/8 + \cdots = 1$. It is clear that the real line as a subset of itself cannot have measure zero because in covering it by intervals the total length of the intervals will be $+\infty$.

This demonstration is typical of the way one proves a set has measure zero. Another example of a set of measure zero is the sphere

$$S = \{x \in \mathbb{R}^n \mid \|x\| = 1\} .$$

From the definition of volume it is clear that if A has volume zero, then A has measure zero. Indeed, if A has volume zero and $\varepsilon > 0$, we can even find a finite covering by rectangles for A with total volume $< \varepsilon$. Also, note that if A has measure zero and $B \subset A$ then B has measure zero as well.

The main advantage of measure zero over volume zero is indicated in the following theorem.

Theorem 2. *Suppose $A_1, A_2, \ldots$ have measure zero in $\mathbb{R}^n$. Then $A_1 \cup A_2 \cup \cdots$ has measure zero in $\mathbb{R}^n$.*

From this we conclude, for example, that any set comprised of a countable number of points has measure zero.

EXAMPLE 2.   Consider the set A of rationals in $[0,1] \subset \mathbb{R}$. The set A does not have volume, that is, 1_A is not integrable. Indeed, the function that has the value 1 on rationals, 0 on the irrationals, is not integrable as we have seen in Example 1, Section 8.1. Nevertheless, the set A does have measure zero because a point has volume and measure zero, and A consists of countably many points, so Theorem 2 applies.

Exercises for Section 8.3

1. Argue that $\{(x,y) \in \mathbb{R}^2 \mid x^2 + y^2 = 1\}$ has volume zero.

2. Does the x-axis in $\mathbb{R}^2$ have volume zero? Can you even cover it by finitely many rectangles?

3. If $A \subset [a,b]$ has measure zero in $\mathbb{R}$ prove that $[a,b]\backslash A$ does not have measure zero in $\mathbb{R}$. (Exercise 10, at the end of the chapter shows that $[a,b]$ does not have measure zero.)

4. Use Exercise 3 to show that the irrationals in $[0,1]$ do not have measure zero.

5. Must the boundary of a set have measure zero?

8.4   Lebesgue's Theorem

We now consider a theorem which is probably one of the most important results in integration theory. We feel intuitively that most "decent" functions, like continuous ones, ought to be integrable since the area under their graphs should be definable. To settle the question of exactly how decent is "decent" we have the theorem of H. Lebesgue. With this theorem Lebesgue opened up new advances in integration theory by stressing the measure zero concept. It led to the success of the fundamental subject of measure theory. One learns this subject in more advanced courses.*

> **Theorem 3.** Let $A \subset \mathbb{R}^n$ be bounded and let $f: A \to \mathbb{R}$ be a bounded function. Extend f to all of $\mathbb{R}^n$ by letting it be zero at points not contained in A. Then f is (Riemann) integrable iff the points at which the extended f is discontinuous form a set of measure zero.

We can draw two important conclusions from this result as stated in the following two corollaries.

> **Corollary 1.** A bounded set $A \subset \mathbb{R}^n$ has volume iff the boundary of A has measure zero.

> **Corollary 2.** Let $A \subset \mathbb{R}^n$ be bounded and have volume. A bounded function $f: A \to \mathbb{R}$ with a finite or countable number of points of discontinuity is integrable.

* For further discussion, see Section 9.7 and Royden, *Real Analysis*, Macmillan.

This result includes most functions one meets in practice. For example, a continuous function on an interval $[a,b]$ is integrable because $[a,b]$ has volume (the boundary consists of two points). Piecewise continuous functions are integrable for the same reason. (A function is called *piecewise continuous* if it has a finite number of points of discontinuity.)

Notice that integrability of f in Theorem 3 depends on the extension of f. For instance, if A is the set of all rationals in $[0,1]$ and f is identically one, f restricted to A is continuous on A but the extended f is nowhere continuous and in fact is not integrable. In Corollary 2 it is *not* necessary to extend f. This is accounted for by the fact that A is assumed to have volume, having regard for Corollary 1.

Another useful result is as follows.

Theorem 4.
 (i) *Let $A \subset \mathbb{R}^n$ be bounded and have measure zero and let $f: A \to \mathbb{R}$ be any (bounded) integrable function. Then $\int_A f(x)\,dx = 0$.*
 (ii) *If $f: A \to \mathbb{R}$ is integrable and $f(x) \geq 0$ for all x and $\int_A f(x)\,dx = 0$, then the set $\{x \in A \mid f(x) \neq 0\}$ has measure zero.*

This theorem is not unreasonable. Indeed, a set of measure zero is "small" with, essentially, zero volume so the integral of any function over it ought to be zero. The second part is likewise reasonable.

EXAMPLE 1.   Let
$$f(x) = \begin{cases} x, & -1 \leq x \leq 0 \\ 3x + 8, & 0 < x \leq 1 \end{cases}.$$
Show f is integrable on $[-1,1]$.

Solution: The set $[-1,1]$ has volume and f has only one discontinuity at $x = 0$. Thus by Corollary 2, f is integrable, since f is bounded.

EXAMPLE 2.   Let $f(x) = \sin(1/x)$, $x > 0$, $f(0) = 0$. Show f is integrable on $[0,1]$.

Solution: Here f has one point of discontinuity at $x = 0$. Also, $|f(x)| \leq 1$ so f is bounded. Thus by Corollary 2, f is integrable.

EXAMPLE 3.   Let $f(x,y) = x^2 + \sin(1/y)$, $y \neq 0$ and $f(x,0) = x^2$. Show f is integrable on $A = \{(x,y) \mid x^2 + y^2 < 1\}$.

Solution: Here f is bounded on $A =$ interior of unit disc in $\mathbb{R}^2$, and has discontinuities on the line $y = 0$ which is a set of zero measure in $\mathbb{R}^2$. Also, A has volume (its boundary has zero volume). Hence, by Corollary 2, f is integrable.

Exercises for Section 8.4

1. Let $f(x) = x^3$ on $[-1,1]$. Prove f is integrable.

2. Let $f(x,y) = 1$ if $x \neq 0$, $f(0,y) = 0$. Prove f is integrable on $A = [0,1] \times [0,1] \subset \mathbb{R}^2$.

3. Compute $\int_A f$ where f, A are as in Exercise 2.

4. Let $A \subset \mathbb{R}^n$ be open and have volume, and let $f: A \to \mathbb{R}$ be continuous, $f(x) \geqslant 0$ and $f(x_0) > 0$ for some $x_0 \in A$. Show $\int_A f > 0$.

5. Let $r_1, r_2, \ldots$ be an enumeration of the rationals in $[0,1]$ and let

$$U = \bigcup_{k=1}^{\infty} D\left(r_k, \frac{1}{2^k}\right),$$

an open set. Discuss whether or not U has volume.

8.5 Properties of the Integral

We now present some of the elementary properties of the integral. For the case of functions on an interval, the reader is probably familiar with some of these properties.

Theorem 5. *Let $A, B \subset \mathbb{R}^n$, $c \in \mathbb{R}$ and $f, g: A \to \mathbb{R}$ be integrable. Then*

(i) *$f + g$ is integrable and $\int_A f + g = \int_A f + \int_A g$.*
(ii) *cf is integrable and $\int_A cf = c \int_A f$.*
(iii) *$|f|$ is integrable and $|\int_A f| \leqslant \int_A |f|$.*
(iv) *If $f \leqslant g$, then $\int_A f \leqslant \int_A g$.*
(v) *If A has volume, and $|f| \leqslant M$, then $|\int_A f| \leqslant Mv(A)$.*
(vi) *(Mean-Value Theorem for Integrals.) If $f: A \to \mathbb{R}$ is continuous, A has volume and is compact and connected, then there is an $x_0 \in A$ such that $\int_A f(x)\,dx = f(x_0)v(A)$.*
    The quantity $\int_A f/v(A)$ is called the average of f over A.
(vii) *Let $f: A \cup B \to \mathbb{R}$. If A and B are such that $A \cap B$ has measure zero, and $f \mid A \cap B$, $f \mid A$, and $f \mid B$ are integrable, then f is integrable and $\int_{A \cup B} f = \int_A f + \int_B f$.*

This last conclusion is quite useful. For example, if $a < b < c$ on $\mathbb{R}$, (vii) implies that

$$\int_a^c f(x)\,dx = \int_a^b f(x)\,dx + \int_b^c f(x)\,dx .$$

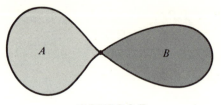

FIGURE 8-7

In the plane, if A and B are as depicted in Figure 8-7, the integral over their union is the sum of the individual integrals because the intersection has zero measure (it is a point).

EXAMPLE 1. If A, B have volume show directly (without Theorem 5(vii)) that $A \cup B$ has volume.

Solution: We must show $\mathrm{bd}(A \cup B)$ has measure zero (see Corollary 1). But

$$\mathrm{bd}(A \cup B) \subset \mathrm{bd}(A) \cup \mathrm{bd}(B)$$

(see Exercise 15, Chapter 2) so that as the right side has measure zero, so does the left.

EXAMPLE 2. Give a geometrical interpretation of property (iii) above for $f\colon [a,b] \to \mathbb{R}$.

Solution: $\int_a^b f(x)\, dx$ represents the area under the graph of f with the portion below the x-axis counted negatively. The magnitude of this is clearly less than (or equal to) the area under the graph of $|f|$; see Figure 8-8.

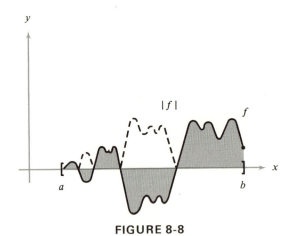

FIGURE 8-8

Exercises for Section 8.5

1. If $A_1, A_2, \ldots$ have volume and $A = A_1 \cup A_2 \cup \cdots$ is bounded, does A have volume?

2. Give a geometrical motivation for properties (iv) and (v) in Theorem 5.

3. Set $\int_a^b f(x)\, dx = -\int_b^a f(x)\, dx$ if $a > b$. Establish

$$\int_a^c f(x)\, dx = \int_a^b f(x)\, dx + \int_b^c f(x)\, dx$$

for all a, b, c (not assuming $a < b < c$ as we did in the text).

4. Let A, B have volume and $A \cap B$ have zero volume. Show that

$$v(A \cup B) = v(A) + v(B)\,.$$

8.6   Fundamental Theorem of Calculus

Now that we have characterized a large class of integrable functions, we may still ask what is a practical way to compute integrals. The answer in one dimension is, of course, that we use antiderivatives and the usual techniques of integration. The techniques for higher dimensions are given in the next chapter.

For $f\colon [a,b] \to \mathbb{R}$, an *antiderivative* of f is a continuous function $F\colon [a,b] \to \mathbb{R}$ such that F is differentiable on $]a,b[$ and $F'(x) = f(x)$ for $a < x < b$. The following theorem provides an effective method for computing integrals of a wide class of functions.

> **Theorem 6** (*Fundamental Theorem of Calculus*).   *Let $f\colon [a,b] \to \mathbb{R}$ be continuous. Then f has an antiderivative F and*
>
> $$\int_a^b f(x)\, dx = F(b) - F(a)\,.$$
>
> *If G is any other antiderivative of f, we also have $\int_a^b f(x)\, dx = G(b) - G(a)$.*

EXAMPLE   1. $\int_0^{\pi/2} \sin x\, dx = 1$, because $d(-\cos x)/dx = \sin x$ and $-\cos(\pi/2) - (-\cos(0)) = 1$. The reader should be familiar with these ideas.

Recall the basic intuition concerning Theorem 6. Namely, one sets $F(x) = \int_a^x f(y)\, dy$. Then suppose $f \geq 0$ for simplicity. F represents the area under the graph of f from a to x. The fact that $F' = f$ comes about because $f(x)$ is the rate at which this area is increasing. Indeed, this ought to be clear because $F(x + \Delta x) - F(x) \approx f(x)\, \Delta x$ (Figure 8-9).

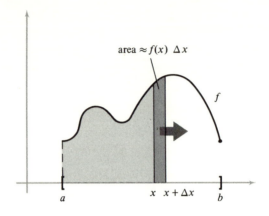

area $\approx f(x)\ \Delta x$

f

$x\ \ x + \Delta x$

a

b

FIGURE 8-9

We assume the reader knows, or is willing to review, the basic integration techniques which are obtained from this theorem, such as the method of substitution (chain rule) and integration by parts. These shall be taken for granted in some discussions to follow.

EXAMPLE 2.   Let $F(x) = \int_a^x f(t)\,dt$. Is F differentiable if f is merely (Riemann) integrable?

Solution: No, continuity in Theorem 6 is essential. For example, let

$$f(x) = \begin{cases} 0, & 0 \leqslant x \leqslant 1, \\ 1, & 1 < x \leqslant 2. \end{cases}$$

Then

$$F(x) = \begin{cases} 0, & 0 \leqslant x \leqslant 1, \\ x - 1, & 1 < x \leqslant 2. \end{cases}$$

Thus, F is continuous but not differentiable at $x = 1$ (see Figure 8-10).

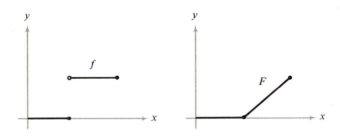

FIGURE 8-10

Exercises for Section 8.6

1. Evaluate $\int_0^3 (x + 5)\, dx$.

2. Evaluate $\int_0^1 (x + 2)^9\, dx$.

3. Evaluate $\int_0^5 x e^{3x^2}\, dx$.

4. Let $f: [a,b] \to \mathbb{R}$ be Riemann integrable and $|f(x)| \leqslant M$. Let $F(x) = \int_a^x f(t)\, dt$. Prove that $|F(y) - F(x)| \leqslant M\,|y - x|$. Deduce that F is continuous. Does this check with Example 2?

5. Let $f: [0,1] \to \mathbb{R}$, $f(x) = 1$ if $x = 1/n$, n an integer, and $f(x) = 0$ otherwise.
   (a) Prove f is integrable.   (b) Show $\int_0^1 f(x)\, dx = 0$.

8.7   Improper Integrals

We often find it is necessary to integrate unbounded functions or to integrate over unbounded regions. Integrals of unbounded functions or integrals over unbounded regions are called *improper integrals*. These lead to convergence problems quite analogous to those for an infinite series.

One usually defines improper integrals by

$$\int_0^\infty f(x)\, dx = \underset{k \to \infty}{\text{limit}} \int_0^k f(x)\, dx \, ,$$

or if the function h is unbounded near 0, by

$$\int_0^b h(x)\, dx = \underset{\varepsilon \to 0}{\text{limit}} \int_\varepsilon^b h(x)\, dx \, ,$$

and so forth. Our definitions conform to these notions as will be explained below. However, a word of caution is advisable at this point. Namely, we do *not* define

$$\int_{-\infty}^\infty f(x)\, dx = \underset{k \to \infty}{\text{limit}} \int_{-k}^k f(x)\, dx \, .$$

If we did, consider what would happen for $f(x) = x$; $\int_{-\infty}^\infty x\, dx$ would be zero, while $\int_0^\infty x\, dx$ and $\int_{-\infty}^0 x\, dx$ would not exist; they would be $+\infty$ and $-\infty$, respectively. Thus if one wishes to retain additivity of integrals, one must proceed more carefully. One possible procedure to avoid this "cancelling of infinities" is to break up f into positive and negative parts, as indeed we shall do below.

Generally, improper integrals are of two types, depending on whether it is the function or the domain which is unbounded. First, we shall consider the case of unbounded regions.

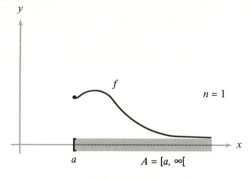

FIGURE 8-11

Note. The case of $\mathbb{R}^1$ is probably the most important, so if this is the primary concern of the reader, proceed directly to the computational methods given in Theorem 8 and use that theorem as the definition of improper integrals.

To start the discussion, suppose $f \geqslant 0$ is bounded and $A \subset \mathbb{R}^n$ is arbitrary, possibly unbounded. Extend f to the whole space as usual by setting $f = 0$ outside A. See Figure 8-11.

> **Definition 5.** Define $\int_A f$ to be $\underset{a \to \infty}{\text{limit}} \int_{[-a,a]^n} f$ if this limit exists, where $[-a,a]^n = [-a,a] \times \cdots \times [-a,a]$ (a cube with side of length a). Here f should be bounded and integrable on each $[-a,a]^n$. If $\int_A f$ exists (and is finite), we say f is *integrable*.

> **Theorem 7.** *For $f \geqslant 0$ and bounded and integrable on any cube $[-a,a]^n$, f is integrable iff the following condition holds: for any sequence B_k of bounded sets with volume such that (i) $B_k \subset B_{k+1}$ and (ii) for any cube C we have $C \subset B_k$ for sufficiently large k, then $\underset{k \to \infty}{\text{limit}} \int_{B_k} f$ exists. In this case $\underset{k \to \infty}{\text{limit}} \int_{B_k} f = \int_A f$.*

This theorem is reasonable in that we get $\int_A f$ no matter how we expand out to infinity. See Figure 8-12.

Observe that if $f \geqslant 0$ is integrable and $0 \leqslant g \leqslant f$, then g is integrable as well for $\int_{[-a,a]^n} g$ is increasing with a and is bounded by the integral of f, so it converges as $a \to \infty$.

Next, let us treat the case of an arbitrary function $f \geqslant 0$ that is unbounded and defined on an unbounded region. The significance of these conditions in the case of $\mathbb{R}$ will be given shortly. Unfortunately, one cannot just drop the requirement that f is bounded and use the definition in Theorem 1 because that actually would imply that f is bounded (this is not obvious). Another way to proceed is as follows.

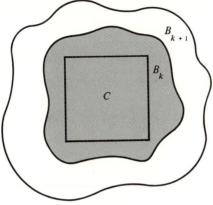

FIGURE 8-12

Definition 6.   For each positive real number M, let

$$f_M(x) = \begin{cases} f(x), & f(x) \leqslant M , \\ 0, & f(x) > M , \end{cases}$$

(see Figure 8-13).

Thus f_M is bounded by M and $f_M \geqslant 0$. Hence we can define $\int_A f_M$ as in Definition 5. Note that $\int_A f_M$ increases as M increases, and $0 \leqslant f_M \leqslant f$. We then define

$$\int_A f = \underset{M \to \infty}{\text{limit}} \int_A f_M$$

if this limit is finite, and in that case we say f is *integrable*.

As before, if $f \geqslant 0$ is integrable and $0 \leqslant g \leqslant f$, then g is also integrable (this is called the *comparison test*).

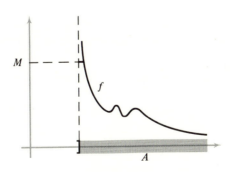

FIGURE 8-13

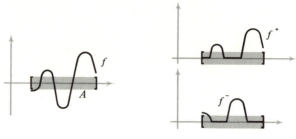

FIGURE 8-14

Definition 7.   For a general function $f: A \to \mathbb{R}$ (that is, f might not be positive), we let

$$f^+(x) = \begin{cases} f(x), & f(x) \geqslant 0, \\ 0, & f(x) < 0, \end{cases}$$

and

$$f^-(x) = \begin{cases} -f(x), & f(x) \leqslant 0, \\ 0, & f(x) > 0, \end{cases}$$

(see Figure 8-14).

Then if both $\int_A f^+$ and $\int_A f^-$ exist, we let $\int_A f = \int_A f^+ - \int_A f^-$ and say f is *integrable*.

Observe that $|f| = f^+ + f^-$. Then if f is integrable, so is $|f|$ and $\int_A |f| = \int_A f^+ + \int_A f^- \geqslant |\int_A f|$.

Conversely, if $\int |f|$ exists, then both $\int f^+$ and $\int f^-$ must be finite, since they are non-negative and $0 \leqslant f^+ \leqslant |f|, 0 \leqslant f^- \leqslant |f|$. Thus f is integrable iff $|f|$ is integrable.

While integration in higher dimensions is important and does occur in practice, the case of the real line deserves special attention. In this case there is a method for computing integrals which is particularly simple to use.

Theorem 8.

(i) *Suppose* $f: [a, \infty[\to \mathbb{R}$ *is continuous and* $f \geqslant 0$. *Let F be an antiderivative of f. Then f is integrable iff* $\underset{x \to \infty}{\text{limit}}\, F(x)$ *exists. In this case*

$$\int_{[a, \infty[} f = \int_a^\infty f(x)\, dx = \left\{ \underset{x \to \infty}{\text{limit}}\, F(x) \right\} - F(a).$$

(ii) *Suppose* $f: \,]a, b] \to \mathbb{R}$ *is continuous and* $f \geqslant 0$. *Then f is integrable iff*

$$\underset{\varepsilon \to 0+}{\text{limit}} \int_{a+\varepsilon}^b f(x)\, dx$$

exists. This limit equals $\int_a^b f(x)\, dx$.

The generalization of this statement to $\mathbb{R}^n$ is given in Exercise 26 below. Instead of "$\int_A f$ exists," one often says "$\int_A f$ converges." As with infinite series, it is necessary to be able to test an improper integral for convergence or divergence. Some of the tests closely resemble those used for series.

One of the most useful tests is the *comparison test*. If $\int_a^\infty f(x)\,dx$ converges and $f \geqslant 0$ and $0 \leqslant g \leqslant f$, then $\int_a^\infty g(x)\,dx$ converges. The reason, as stated earlier, is simply that $\int_a^b g(x)\,dx$ increases as $b \to \infty$ and is bounded above by $\int_a^\infty f(x)\,dx$, so it converges.

In our treatment of improper integrals, we have used what amounts to the most natural approach on $\mathbb{R}^n$. Our method is particularly desirable because most of Theorem 8 remains valid (this is outlined in Exercise 26). However, in the special case of $\mathbb{R}^1 = \mathbb{R}$, it is also useful to consider a weaker type of convergence, called *conditional convergence*. Here we define

$$\int_a^\infty f(x)\,dx \text{ (conditional)} = \lim_{b\to\infty}\int_a^b f(x)\,dx$$

if the limit exists. This is not the same as our earlier definition (called *absolute convergence*) because we demanded that the limit hold separately for f^+ and f^-. An example will serve to point up the difference.

EXAMPLE 1. Let $f(x) = (\sin x)/x$. Show $\int_1^\infty f$ is conditionally but not absolutely convergent.

Solution: If f were integrable on $[1,\infty[$, then $|f|$ would be also. But then

$$\int_1^\infty \frac{|(\sin x)|}{x}\,dx \geqslant \int_\pi^{n\pi}\frac{|(\sin x)|}{x}\,dx = \sum_{k=2}^n \int_{(k-1)\pi}^{k\pi}\frac{|(\sin x)|}{x}\,dx \geqslant \frac{2}{\pi}\sum_{k=2}^n\frac{1}{k}$$

since on the interval $[(k-1)\pi, k\pi]$, $1/x \geqslant 1/k\pi$ and $\int_{(k-1)\pi}^{k\pi}|(\sin x)|\,dx = 2$. But $\sum_{k=2}^n 1/k \to \infty$ as $n \to \infty$, so $\int_1^\infty |\sin x|/x\,dx = +\infty$.

However, $\lim_{b\to\infty}\int_1^b (\sin x)/x\,dx$ exists. To see this, note that an integration by parts gives

$$\int_1^b \frac{\sin x}{x}\,dx = -\int_1^b\frac{d(\cos x)}{x} = -\frac{\cos b}{b} + \cos 1 - \int_1^b\frac{\cos x}{x^2}\,dx ,$$

and $\int_1^\infty \cos x/x^2\,dx$ exists because

$$\int_1^b\frac{|\cos x|}{x^2}\,dx \leqslant \int_1^b\frac{1}{x^2}\,dx = 1 - \frac{1}{b} ,$$

which converges as $b \to \infty$. So $\int_1^\infty (\sin x/x)\,dx$ is conditionally convergent.

One can give refined tests like the Dirichlet test for series to obtain conditional convergence when absolute convergence fails. See Exercise 31 at the end of this chapter.

EXAMPLE 2.   We give some standard improper integrals which are useful in conjunction with the comparison test. They may all be proved by direct integration, successive integration by parts, or by other tricks.

(a) $\displaystyle\int_1^\infty x^p\,dx \begin{cases} \text{converges if } p < -1\,, \\ \text{diverges if } p \geqslant -1\,; \end{cases}$

(b) $\displaystyle\int_0^a x^p\,dx \begin{cases} \text{converges if } p > -1\,, \\ \text{diverges if } p \leqslant -1\,; \end{cases}$

(c) $\displaystyle\int_1^\infty e^{-x}x^p\,dx$ converges for all p ;

(d) $\displaystyle\int_0^a e^{1/x}x^p\,dx$ diverges for all p ;

(e) $\displaystyle\int_0^a \log x\,dx$ converges ;

(f) $\displaystyle\int_1^\infty \left(\frac{1}{\log x}\right) dx$ diverges .

For instance,

$$\int_1^c x^p\,dx = \begin{cases} \dfrac{x^{p+1}}{p+1}\Big|_1^c, & p \neq -1\,, \\[2ex] \log x\,\big|_1^c, & p = -1\,. \end{cases}$$

Now $\log c \to \infty$ as $c \to \infty$, and $c^{p+1} \to \infty$ as $c \to \infty$ if $p + 1 > 0$, that is, $p > -1$. This gives (a). Part (b) is similar and (e) and (f) can be proved in the same way. We outline (c) in Exercise 2 below and (d) is similar to that.

EXAMPLE 3.   Show

$$\int_1^\infty \frac{1}{\sqrt{x^3 + 1}}\,dx \text{ converges .}$$

Solution: This is improper at $x \to \infty$. Now for $x \geqslant 0$,

$$\frac{1}{\sqrt{x^3 + 1}} \leqslant \frac{1}{\sqrt{x^3}} = x^{-3/2}$$

and $\int_1^\infty x^{-3/2}\,dx$ converges by (a) of Example 2. Hence, by comparison, this integral converges as well.

Exercises for Section 8.7

1. Let $f: [a,\infty[\to \mathbb{R}$ be Riemann integrable on bounded intervals. Show $\int_a^\infty f$ exists iff for every $\varepsilon > 0$ there is a T such that $t_1, t_2 \geqslant T$ implies $\int_{t_1}^{t_2} f(x)\, dx < \varepsilon$ (this is called the *Cauchy criterion*).

2. Establish formula (c) of Example 2 as follows. Prove $e^{-x}x^{p+2} \to 0$ as $x \to \infty$ and then compare the integral with $\displaystyle\int_1^\infty \frac{1}{x^2}\, dx$.

3. Show $\int_0^1 e^{-x}x^p\, dx$ converges if $-1 < p$.

4. Show $\displaystyle\int_0^\infty \frac{\sin x}{x^2 + 1}\, dx$ is convergent.

5. For what α is $\displaystyle\int_1^\infty \frac{x^\alpha}{1 + x^\alpha}\, dx$ convergent?

8.8 Some Convergence Theorems

In Chapter 5 we saw that uniform convergence is sufficient to allow us to interchange limit and integration operations (see also Example 2 at the end of this chapter). In this section we shall refine that result.

A course at this level is not the proper place for an exhaustive treatment of convergence theorems, since they fit in more naturally in advanced courses in measure and integration. Therefore we confine ourselves to an illustrative theorem—the monotone convergence theorem. The result will be needed in Chapter 10 for Fourier series.*

> **Theorem 9** *(Lebesgue's Monotone Convergence Theorem). Let* $g_n: [0,1] \to \mathbb{R}$ *be a sequence of non-negative functions such that each improper integral* $\int_0^1 g_n(x)\, dx$ *exists and is finite. Suppose that* $0 \leqslant g_{n+1} \leqslant g_n$ *and that* $g_n(x) \to 0$ *for each* $x \in [0,1]$. *Then*
>
> $$\lim_{n \to \infty} \int_0^1 g_n(x)\, dx = 0 \ .$$

At this point the reader should return to the examples in Section 5.3 to see that they are in accord with this theorem. Certainly if the g_n's are not decreasing the result is not true, as examples in that section show.

---

* For a more complete discussion of convergence theorems in the Riemann theory, see W. A. J. Luxemburg, *Arzelá's dominated convergence theorem for the Riemann integral*, Am. Math. Monthly, **78** (1971) 970–979.

Corollary 3. *Let f_n, $f\colon [0,1] \to \mathbb{R}$ be non-negative functions and suppose $0 \leqslant f_n \leqslant f_{n+1} \leqslant f$ and suppose $\int_0^1 f(x)\,dx$ exists as an improper integral. Furthermore, suppose $f_n(x) \to f(x)$ for all $x \in [0,1]$. Then*

$$\lim_{n\to\infty} \int_0^1 f_n(x)\,dx = \int_0^1 f(x)\,dx \ .$$

This result follows by applying Theorem 9 to the functions $g_n(x) = f(x) - f_n(x)$ which decrease monotonically to zero. The details shall be left to the reader.

We used the interval $[0,1]$ for definiteness but any other interval could be used as well and the results could also generalize to $\mathbb{R}^n$.

The result we will need in Chapter 10 follows as a corollary.

Corollary 4. *If $f\colon [a,b] \to \mathbb{R}$, $f \geqslant 0$ and the improper integral $\int_a^b f^2 < \infty$ exists, then $\int_a^b (f - f_M)^2 \to 0$ as $M \to \infty$.*

Here we use Theorem 9 with $g_n = f - f_n$ (f_M is defined in the previous section).

EXAMPLE 1. Prove:

$$\lim_{n\to\infty} \int_0^1 e^{-n^2}x^p\,dx = 0 \qquad \text{if } p > -1 \ .$$

Solution: Theorem 9 applies. The functions $g_n(x) = e^{-nx^2}x^p \leqslant x^p$ so are integrable and moreover, the g_n decrease pointwise to zero.

EXAMPLE 2. Let g_n be non-negative and integrable on $[0,1]$. Let $g(x) = \sum_{n=1}^\infty g_n(x)$ and assume $g(x)$ is integrable. Prove

$$\int_0^1 g(x)\,dx = \sum_{n=1}^\infty \int_0^1 g_n(x)\,dx \ .$$

Solution: Let

$$f_n(x) = \sum_{k=1}^n g_k(x)\,dx, \qquad \text{so} \qquad \int_0^1 f_n(x)\,dx = \sum_{k=1}^n \int_0^1 g_k(x)\,dx \ .$$

Now the $f_n(x)$ increase to g so by Corollary 3 their integrals converge to the integral of g.

Exercises for Section 8.8

1. Show that Theorem 8 can be proved using the methods of Chapter 5 if the g_n are continuous.

2. Evaluate $\displaystyle\lim_{n\to\infty} \int_0^1 \frac{e^x \sin(nx)}{n}\, dx$.

3. Evaluate $\displaystyle\lim_{n\to\infty} \int_0^1 \frac{(1 - e^{-nx})}{\sqrt{x}}\, dx$.

8.9 The Dirac δ-Function; Introduction to Distributions

Around 1930, in his famous book *The Principles of Quantum Mechanics,* Dirac emphasized the usefulness of the δ-function, which he defined by the following properties:

$$\delta(x) = \begin{cases} 0, & \text{if } x \neq 0, \\ \infty, & \text{if } x = 0, \end{cases}$$

and

$$\int_{-\infty}^{\infty} \delta(x)\, dx = 1 .$$

One can imagine approximations to δ where $f_n \to \delta$ in some sense (Figure 8-15), but δ itself is hard to visualize directly. Physicists quickly realized (as engineers had done independently) the usefulness of such ideas and proceeded to use the δ-function in their computations. For example, to describe the charge density σ of a point charge with charge e it is convenient to write $\sigma = e\delta$. One imagines $e\delta$ as a limit of well defined charge densities σ_n smeared out over small areas which concentrate down to a single point as $n \to \infty$.

At the same time as the physicists and engineers were computing, mathematicians sat back in quiet amusement, occasionally pointing out that this

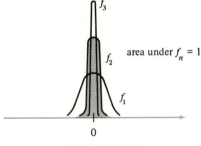

FIGURE 8-15

δ-function business was really all nonsense because no such function can exist. The definition does not really make sense, as anyone can plainly see. To add to the mathematician's enjoyment, Dirac proceeds to differentiate this function δ.

But the physicists turn out to have had a good idea after all. Today, distributions, of which δ is an example, are indispensible in the study of partial differential equations. But it took mathematicians almost 20 years to establish the theory of distributions satisfactorily. This was done by L. Schwartz and S. L. Sobolev around 1948, although hints of the theory had occurred in the works of earlier mathematicians as well. We will give only the briefest glimpse of the theory.

In actual computations, δ almost always appears under the integral sign in the form

$$\int \delta(x)f(x)\,dx = f(0)\,. \tag{1}$$

We can see the idea behind Eq. 1 from Dirac's definition because δ being zero away from $x = 0$ means that only $f(0)$ counts, so one ought to have

$$\int \delta(x)f(x)\,dx = f(0)\int \delta(x)\,dx = f(0)\,.$$

Eq. 1 is the clue as to how to proceed. Namely, consider the space $\mathscr{C}_b(\mathbb{R},\mathbb{R})$ of bounded continuous functions on $\mathbb{R}$ (Section 5.4). Then, regard

$$\delta\colon \mathscr{C}_b(\mathbb{R},\mathbb{R}) \to \mathbb{R}, f \mapsto f(0)\,.$$

Thus we do not regard δ as a function on $\mathbb{R}$ at all, but a function on $\mathscr{C}_b(\mathbb{R},\mathbb{R})$ which maps f to $f(0)$. This operation is well defined and δ is easily seen to be continuous (see Exercise 40, Chapter 5).

Thus it is possible to circumvent the difficulty with Dirac's definition by taking a whole new point of view; namely think of δ as being an assignment of a number to each function f. This assignment stands for the symbolic expression $\int \delta(x)f(x)\,dx$. Now any continuous function g also defines such an operation; it sends f to

$$\int_{-\infty}^{\infty} g(x)f(x)\,dx\,.$$

Thus distributions (of which linear maps from $\mathscr{C}_b$ to $\mathbb{R}$ are examples) include more than just ordinary functions.

How does one differentiate δ? For this, note that if g is differentiable, then

$$\int_{-\infty}^{\infty} \frac{dg}{dx}f(x)\,dx = -\int_{-\infty}^{\infty} g(x)\frac{df}{dx}(x)\,dx\,,$$

provided f is zero for large $|x|$, as can be seen by an integration by parts.

Thus dg/dx sends f to the same number that g sends $-df/dx$ to. Thus it is logical to *define* δ' by

$$\delta'(f) = \delta\left(-\frac{df}{dx}\right) = -\frac{df}{dx}(0).$$

This leads us to restrict $\mathscr{C}_b$ to the functions f which are C^1 and vanish for large $|x|$. Thus we might as well use C^∞ functions vanishing for large enough $|x|$. We are led to the next definition.

Definition 8. Let $\mathscr{D}$ denote the C^∞ functions which are identically zero outside some interval ($\mathscr{D}$ is called the *Schwartz space*). A *distribution* T is a linear map* $T: \mathscr{D} \to \mathbb{R}$. The derivative T' is defined by $T': \mathscr{D} \to \mathbb{R}, f \mapsto T(-df/dx)$.

If g is a continuous function it is customary to use the same symbol g for the distribution that maps $f \mapsto \int_{-\infty}^{\infty} g(x)f(x)\,dx$.

This is the elegant reformulation due to the founders of distribution theory. Much work remained to prove significant theorems about distributions, and this led to an important vitalization of the theory of partial differential equations. The physicists were pleased and everyone was happy.

Exercises for Section 8.9

1. Show $\delta''(f) = f''(0)$.

2. Let T_n and T be distributions. Say $T_n \to T$ if $T_n(f) \to T(f)$ for all $f \in \mathscr{D}$. Show that

$$\sqrt{\frac{n}{\pi}}\, e^{-nx^2} \to \delta.$$

3. If $T_n \to T$ (see Exercise 2), show that $T_n' \to T'$. Discuss and compare with Section 5.3.

4. Find a sequence of continuous functions g_n such that $g_n \to \delta'$.

Theorem Proofs for Chapter 8

Theorem 1. *(Darboux's Theorem.) Let $A \subset \mathbb{R}^n$ be bounded and lie in some rectangle S. Let $f: A \to \mathbb{R}$ be bounded and be extended to S by defining $f = 0$ outside A. Then f is integrable with integral I iff for any $\varepsilon > 0$ there is a $\delta > 0$ such that if P is any partition into rectangles $S_1, \ldots, S_N$ with sides $<\delta$ and if $x_1 \in S_1, \ldots, x_N \in S_N$, we have*

$$\left|\sum_{i=1}^{N} f(x_i)v(S_i) - I\right| < \varepsilon.$$

* Strictly speaking, one must require T to be continuous in the sense that if $f_n \to f$ uniformly on bounded sets and all derivatives of f_n converge uniformly to those of f on bounded sets, then $T(f_n) \to T(f)$. The actual topology on $\mathscr{D}$ is a bit complicated however.

278 INTEGRATION

Riemann's condition: f *is integrable iff for any* $\varepsilon > 0$ *there is a partition* P_ε *of* S *such that* $0 \leqslant U(f,P_\varepsilon) - L(f,P_\varepsilon) < \varepsilon$ *(and hence also for every partition* P *finer than* P_ε).

Proof: We shall show that the conditions "f integrable," "f satisfies Riemann's condition," and "f satisfies Darboux's condition" are equivalent. This will be done in four steps.

Step 1. *If* f *is integrable, then* f *satisfies Riemann's condition.*

Proof: Given $\varepsilon > 0$, there is a partition P'_ε such that

$$U(f,P'_\varepsilon) < I + \frac{\varepsilon}{2},$$

where $I = \int_A f$. We can do this since $I = \inf\{U(f,P) \,|\, P \text{ is a partition}\}$. If P is finer than P'_ε, then we know

$$U(f,P) \leqslant U(f,P'_\varepsilon) < I + \frac{\varepsilon}{2}.$$

Similarly, choose P''_ε such that for P finer than P''_ε we have

$$L(f,P) > I - \frac{\varepsilon}{2}.$$

Let $P_\varepsilon = P'_\varepsilon \cup P''_\varepsilon$. If P is finer than P_ε, we have

$$I - \frac{\varepsilon}{2} < L(f,P) \leqslant U(f,P) < I + \frac{\varepsilon}{2},$$

so

$$0 \leqslant U(f,P) - L(f,P) < \varepsilon,$$

which is Riemann's condition.

Step 2. *If* f *satisfies Riemann's condition, then* f *is integrable.*

Proof: For any $\varepsilon > 0$ there is a P_ε such that

$$0 \leqslant U(f,P_\varepsilon) - L(f,P_\varepsilon) < \varepsilon.$$

This implies that $S = s$. Indeed, for each P we have

$$L(f,P) \leqslant s \leqslant S \leqslant U(f,P)$$

so if $U(f,P_\varepsilon) - L(f,P_\varepsilon) < \varepsilon$, we also have $S - s < \varepsilon$ for every $\varepsilon > 0$ and hence $S = s$.

Step 3. *If* f *satisfies Darboux's condition, then* f *is integrable.*

Proof: We shall show that the I given in Darboux's condition will be the same as $S = \inf\{U(f,P) \,|\, P \text{ is a partition}\}$ and also the same as s. To accomplish this, given $\varepsilon > 0$, we shall produce a partition P such that

$$|U(f,P) - I| < \varepsilon,$$

which will show that $S \leqslant I$. Similarly, we will have $I \leqslant s$, and then $I \leqslant s \leqslant S \leqslant I$ will imply $s = S = I$.

For this, choose $\delta > 0$ such that if P is a partition with sides $<\delta$, then

$$|\sum f(x_i)v(S_i) - I| < \frac{\varepsilon}{2},$$

where $S_1, \ldots, S_N$ make up the partition P. Now we may choose x_i such that

$$|f(x_i) - \sup_{S_i}(f)| < \frac{\varepsilon}{(v(S_i)2N)}.$$

Then

$$|U(f,P) - I| \leqslant |U(f,P) - \sum f(x_i)v(S_i)| + |\sum f(x_i)v(S_i) - I|.$$

Now

$$|U(f,P) - \sum f(x_i)v(S_i)| < \sum \frac{\varepsilon v(S_i)}{(v(S_i)2N)} = \frac{\varepsilon}{2}$$

so that $|U(f,P) - I| < \varepsilon$ as required. The case for lower sums is similar.

Step 4. *If f is integrable then f satisfies Darboux's condition.*

Proof: Suppose that f is integrable with integral I. We will show, in two steps, that for any $\varepsilon > 0$, there is a $\delta > 0$ such that if P is any partition into rectangles $S_1, \ldots, S_N$, with sides $<\delta$ and if $x_1 \in S_1, \ldots, x_N \in S_N$, we have

$$\left| \sum_{i=1}^{N} f(x_i)v(S_i) - I \right| < \varepsilon.$$

Step 4.A. Let P be a partition of the rectangle $B \subset \mathbb{R}^n$. Given $\varepsilon > 0$, we shall show that there exists a $\delta > 0$ such that for any partition P' into subrectangles with sides less than δ, the sum of the volumes of the subrectangles of P' which are not entirely contained in some rectangle of P is less than ε.

In order to see this clearly, let us examine the cases $n = 1$ and $n > 1$ separately. First, suppose that we are working on the interval $[a,b]$; suppose that the partition P consists of N points. We assert that the δ that is needed is simply given by ε/N. Clearly then, the length of the intervals in P' which are not contained in an interval of P is $N \times \delta =$ (maximum number of intervals not contained entirely in an interval of P) $\times$ (maximum length of each such interval of P') $= \varepsilon$.

Next let us turn to the general case.

Let the partition P consist of rectangles $V_1, \ldots, V_M$. We denote the total "area" of the faces lying between any two rectangles by T. Let $\delta = \varepsilon/T$ and let P' be any partition of B into subrectangles of sides less than δ. Then for any rectangle $S \in P'$ such that S is not contained in one of the V_i, S intersects two adjacent rectangles. Now one can see that $v(S) \leqslant \delta A$, where A is the total area of faces between two subrectangles contained in S (see Figure 8-16). Thus $\sum_{S \in P'} v(S) < \delta T = \varepsilon$.

Step 4.B. Since f is bounded, there exists an $M > 0$ such that $|f(x)| < M$ for all $x \in S$. There exist partitions P_1 and P_2 of S such that $I - L(f,P_1) < \varepsilon/2$ and $U(f,P_2) - I < \varepsilon/2$. Choose a partition P which refines both P_1 and P_2. Then $U(f,P) - I < \varepsilon/2$ and $I - L(f,P) < \varepsilon/2$. By Step 1 there exists a $\delta > 0$ such that for any partition of P into rectangles of sides $<\delta$ the sum of the volumes of the subrectangles not contained in some subrectangle of P is less than $\varepsilon/2M$. Let $S_1, \ldots, S_N$ be a partition

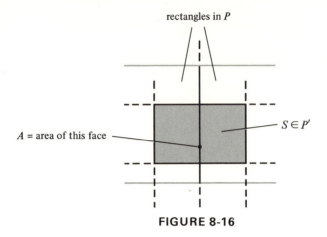

rectangles in P

A = area of this face

$S \in P'$

FIGURE 8-16

into subrectangles of side less than δ; let $S_1, \ldots, S_K$ be the subrectangles contained in some subrectangle of P, and let $S_{K+1}, \ldots, S_N$ be the remaining subrectangles.

Let $x_1 \in S_1, \ldots, x_N \in S_N$. Then

$$\sum_{i=1}^{N} f(x_i)v(S_i) = \sum_{i=1}^{K} f(x_i)v(S_i) + \sum_{i=K+1}^{N} f(x_i)v(S_i) \leqslant U(f,P) + M \cdot \frac{\varepsilon}{2M}$$

$$= U(f,P) + \frac{\varepsilon}{2} < I + \varepsilon .$$

Similarly,

$$\sum_{i=1}^{N} f(x_i)v(S_i) \geqslant L(f,P) - \frac{\varepsilon}{2} > I - \varepsilon .$$

Therefore

$$\left| \sum_{i=1}^{N} f(x_i)v(S_i) - I \right| < \varepsilon . \quad \blacksquare$$

In some later proofs it will be convenient to have the following technical fact at hand.

In the definition of measure zero, one can use either closed or open rectangles.

Proof: Let $A \subset \mathbb{R}^n$. First, suppose that given $\varepsilon > 0$ there are open rectangles $V_1, V_2, \ldots$ covering A of total volume $< \varepsilon$. Let $B_i = \mathrm{cl}(V_i)$. Then $B_1, B_2, \ldots$ are closed rectangles covering A with the same total volume $< \varepsilon$.

Conversely, given $\varepsilon > 0$, suppose we have a covering by closed rectangles $B_1, B_2, \ldots$ with total volume $< \varepsilon/2^n$. Then let V_i be the open rectangle containing B_i with twice the side. Then $v(V_i) = 2^n v(B_i)$, so

$$\sum_{i=1}^{\infty} v(V_i) = 2^n \sum_{i=1}^{\infty} v(B_i) < \varepsilon .$$

This same argument also works for content zero. See Exercise 11, p. 293. $\blacksquare$

Theorem 2.   *Suppose* $A_1, A_2, \ldots$ *have measure zero in* $\mathbb{R}^n$. *Then* $A_1 \cup A_2 \cup \cdots$ *has measure zero in* $\mathbb{R}^n$.

Proof: Since all of the A_i have measure zero, there is a covering of the A_i with rectangles $B_{i1}, B_{i2}, \ldots$ such that $\sum_{j=1}^{\infty} v(B_{ij}) < \varepsilon/2^i$. Since the collection $B_{i1}, B_{i2}, \ldots$ covers the A_i, the countable collection of all B_{ij} covers $A_1 \cup A_2 \cup \cdots$. But

$$\sum_{i,j=1}^{\infty} v(B_{ij}) = \sum_{i=1}^{\infty} \sum_{j=1}^{\infty} v(B_{ij}) < \sum_{i=1}^{\infty} \frac{\varepsilon}{2^i} = \varepsilon.$$

Since ε is arbitrary, $A_1 \cup A_2 \cup \cdots$ has measure zero. ∎

Note: It has not yet been justified that we can sum up the $v(B_{ij})$ by first j, then i. If this is true, then the terms can be rearranged in an absolutely convergent double series. That it is indeed true follows from Exercise 51, Chapter 5.

Theorem 3.   *Let* $A \subset \mathbb{R}^n$ *be bounded and let* $f: A \to \mathbb{R}$ *be a bounded function. Extend* f *to all of* $\mathbb{R}^n$ *by letting it be zero at points not contained in* A. *Then* f *is* (*Riemann*) *integrable iff the points at which the extended* f *is discontinuous form a set of measure zero.*

Proof: Consider some rectangle B which contains A. Then we must show that the function f is integrable on A iff the set of discontinuities of the function g, which equals f on A and zero elsewhere, has measure zero.

It is useful for the proof to have a measure of how "bad" a discontinuity is. In order to do this, we define the *oscillation of a function* h at x_0, written $O(h,x_0)$, to be $\inf \{\sup\{|h(x_1) - h(x_2)| \mid x_1, x_2 \in U\} \mid U \text{ is a neighborhood of } x_0\}$. Note that the inf is taken over all neighborhoods U of x_0. Thus $O(h,x_0) \geqslant 0$, and we claim that $O(h,x_0) = 0$ iff h is continuous at x_0. To see this, note that h is continuous at x_0 iff for any $\varepsilon > 0$, there is a neighborhood U of x_0 such that $\sup\{|h(x_0) - h(x_1)| \mid x_1 \in U\} < \varepsilon$, and this is equivalent to $O(h,x_0) = 0$. We are now ready to begin the proof—for convenience it is broken into two steps. Remember $g: B \to \mathbb{R}$, $g(x) = f(x)$ if $x \in A$ and $g(x) = 0$, $x \notin A$.

Step 1.   We assume the set of discontinuities of g has measure zero. Thus if we let $D_\varepsilon = \{x \mid O(g,x) \geqslant \varepsilon\}$ for $\varepsilon > 0$, and $D = \{\text{discontinuities of } g\}$, then $D_\varepsilon \subset D$. If y is an accumulation point of D_ε, every neighborhood of y contains a point of D_ε. Then every neighborhood U of y is a neighborhood of a point of D_ε, and by construction of D_ε, $\sup\{|f(x_1) - f(x_2)| \mid x_1, x_2 \in U\} \geqslant \varepsilon$. This implies $O(f,y) \geqslant \varepsilon$ and so $y \in D_\varepsilon$. This proves that D_ε is a closed set. Since $D_\varepsilon \subset B$, D_ε is bounded and therefore compact. Now D_ε has measure zero, since $D_\varepsilon \subset D$, so by definition there is a collection $B_1, B_2, \ldots$ of (open) rectangles which cover D_ε such that $\sum_{i=1}^{\infty} v(B_i) < \varepsilon$. We know that a finite number of the B_i cover D_ε, since D_ε is compact. Suppose $B_1, \ldots, B_N$ cover D_ε. Certainly, $\sum_{i=1}^{N} v(B_i) < \varepsilon$.

Now pick a partition of B. We can divide up the rectangles of the partition into two (not necessarily disjoint) collections, C1 and C2, defined as follows.

C1: Those rectangles which are contained in some B_i, $i = 1, \ldots, N$.

C2: Those rectangles which do not intersect D_ε.

For each rectangle S which does not intersect D_ε, the oscillation of g at each point of the rectangle is less than ε. Hence we can find a neighborhood U_x of each point x of the rectangle, such that $M_U(g) - m_U(g) < \varepsilon$, where $M_U(g) = \sup\{g(x) \mid x \in U\}$ and $m_U(g) = \inf\{g(x) \mid x \in U\}$. Now S is compact, so a finite collection of the open sets U_x covers S. Pick a refined partition such that each rectangle of the partition is contained in U_{x_i} for some U_{x_i} in the finite collection which covers S. If we do this for each S in C2, we get a partition P such that

$$U(g,P) - L(g,P) \leqslant \sum_{S \in C1} (M_S(g) - m_S(g))v(S) + \sum_{S \in C2} (M_S(g) - m_S(g))v(S)$$

$$\leqslant \varepsilon v(B) + \sum_{S \in C1} 2Mv(S), \quad \text{(where } |f(x)| < M \text{ on } A)$$

$$\leqslant \varepsilon v(B) + 2M\varepsilon, \quad \text{since } \sum_{S \in C1} v(S) < \sum_{i=1}^{N} v(B_i) < \varepsilon.$$

But ε is arbitrary, so by Riemann's condition, g and hence f is integrable.

Step 2. Suppose g is integrable. The set of discontinuities of g is the set of points of oscillation greater than zero. Hence $\{\text{discontinuities of } g\} = D_1 \cup D_{1/2} \cup D_{1/3} \cup \cdots$, where $D_{1/n} = \{x \in B \mid O(g,x) \geqslant 1/n\}$. Now by Theorem 1, we can find a partition P of B such that $U(g,P) - L(g,P) = \sum_{S \in P} (M_S(g) - m_S(g))v(S) < \varepsilon$. Now $D_{1/n} = \{x \in D_{1/n} \mid x$ lies on the boundary of some $S\} \cup \{x \in D_{1/n} \mid x \in \text{interior } (S) \text{ for some } S\} = S_1 \cup S_2$. The first of these sets, S_1, has measure zero, since we can cover the boundary of a rectangle with arbitrarily thin rectangles. Let C denote the collection of rectangles of the partition which have an element of $D_{1/n}$ in their interior. Then, if $S \in C$,

$$M_S(g) - m_S(g) \geqslant \frac{1}{n}$$

and

$$\frac{1}{n} \sum_{S \in C} v(S) \leqslant \sum_{S \in C} (M_S(g) - m_S(g))v(S) \leqslant \sum_{S \in P} (M_S(g) - m_S(g))v(S) < \varepsilon.$$

Hence C is a collection of rectangles which covers S_2 and $\sum_{S \in C} v(S) < n\varepsilon$. We can find a collection C' of rectangles which covers S_1 with $\sum_{S \in C'} v(S) < \varepsilon$. Then $C \cup C'$ covers $D_{1/n}$ and $\sum_{S \in C \cup C'} v(S) < (n + 1)\varepsilon$. But ε is arbitrary, so $D_{1/n}$ has measure zero. Finally, $\{\text{discontinuities of } g\} = D_1 \cup D_{1/2} \cup D_{1/3} \cup \cdots$ has measure zero by Theorem 2. ∎

Corollary 1. *A bounded set $A \subset \mathbb{R}^n$ has volume iff the boundary of A has measure zero.*

Proof: By Theorem 3 it suffices to show that the set of discontinuities of 1_A where

$$1_A(x) = \begin{cases} 0, & x \notin A, \\ 1, & x \in A, \end{cases}$$

is the boundary of A. But if $x \in \text{bd}(A)$, then any neighborhood of x intersects A and $\mathbb{R}^n \backslash A$. Hence there are points y in the neighborhood such that $1_A(x) - 1_A(y) = 1$. Thus 1_A is not continuous at x. If $x \notin \text{bd}(A)$, then there is a neighborhood of x which lies entirely in A or $\mathbb{R}^n \backslash A$. In either case 1_A is constant on this neighborhood so 1_A is continuous at x. ∎

Corollary 2.   *Let* $A \subset \mathbb{R}^n$ *be bounded and have volume. A bounded function* $f: A \to \mathbb{R}$ *with a finite or countable number of points of discontinuities is integrable.*

Proof: The discontinuities of the extended function g, which is equal to f on A and zero at points outside A are simply the discontinuities of f together with possibly some discontinuities of g on the boundary of A for the same reason as in the above proof. But bd(A) has measure zero by Corollary 1. Hence it is sufficient to show that a countable set has measure zero. But this follows at once from Theorem 2 and from the fact that a point has measure zero. ∎

Theorem 4.
(i) *Let* $A \subset \mathbb{R}^n$ *be bounded and have measure zero and let* $f: A \to \mathbb{R}$ *be any bounded integrable function. Then* $\int_A f(x)\, dx = 0$.
(ii) *If* $f: A \to \mathbb{R}$ *is integrable, and* $f(x) \geq 0$ *for all* x *and* $\int_A f(x)\, dx = 0$, *then the set* $\{x \in A \mid f(x) \neq 0\}$ *has measure zero.*

Proof: (i) We make the following observation about a set with measure zero, namely, that a set of measure zero cannot contain a non-trivial rectangle, that is, a rectangle $[a_1, b_1] \times \cdots \times [a_n, b_n]$ such that $a_i < b_i$ for each i. The reason is that a subset of a set of measure zero must be of measure zero and a non-trivial rectangle cannot have measure zero. This last assertion is intuitively clear; the details are given in Exercise 17. Let S be a rectangle enclosing A and extend f to S by setting it equal to 0 on $S \backslash A$; let P be any partition of S into subrectangles $S_1, \ldots, S_N$, and let M be such that $f(x) \leq M$ for all $x \in A$.
Then

$$L(f,P) = \sum_{i=1}^N m_{S_i}(f)v(S_i) \leq M \sum_{i=1}^N m_{S_i}(1_A)v(S_i) \ .$$

Suppose $m_{S_i}(1_A) \neq 0$ for some i, such that S_i is a (non-trivial) rectangle. This means that $S_i \subset A$, which contradicts the opening remarks of the proof. So for any non-trivial $S_i, m_{S_i}(1_A) = 0$. For any trivial S_i, $v(S_i) = 0$. Hence $\sum_{i=1}^N m_{S_i}(1_A)v(S_i) = 0$ or $L(f,P) \leq 0$. Now $\sup_{x \in S_i} f(x) = -\inf_{x \in S_i} - (f(x))$, so

$$U(f,P) = \sum_{S_i \in P} \sup_{x \in S_i} f(x)v(S_i) = -\sum_{S_i \in P} \inf_{x \in S_i} - f(x)v(S_i) = -L(-f,P),$$

and by the same arguments as above, $L(-f,P) \leq 0$. Hence $-L(-f,P) = U(f,P) \geq 0$. Since P was arbitrary, for any partition Q of S, $U(f,Q) \geq 0 \geq L(f,Q)$, hence

$$\overline{\int_A} f \geq 0 \geq \underline{\int_A} f \ ,$$

and so, since f is integrable,

$$\overline{\int_A} f = \underline{\int_A} f = \int_A f = 0 \ .$$

(ii) Consider the set $A_m = \{x \in A \mid f(x) > 1/m\}$; we shall first show that A_m has content zero.
Suppose we are given $\varepsilon > 0$. Let S be a rectangle enclosing A and extend f to S by

setting it equal to 0 on $S\backslash A$. Let P be a partition of the rectangle S such that $U(f,P) < \varepsilon/m$. Such a partition exists by the fact that $\int_A f = 0$. If $S_1, \ldots, S_K$ are the subrectangles of the partition P which have non-empty intersection with A_m, then if $M_{S_i}(f)$ is the sup of f on S_i,

$$\sum_{i=1}^{K} v(S_i) \leqslant \sum_{i=1}^{K} mM_{S_i}(f)v(S_i) < \varepsilon ,$$

since $mM_{S_i}(f) > 1$. Therefore $S_1, \ldots, S_K$ is a cover by closed rectangles of the set A_m such that $\sum_{i=1}^{K} v(S_i) < \varepsilon$. Hence A_m has content zero. Since A_m has content zero, it also has measure zero.

Finally, observe that

$$\{x \in A \mid f(x) \neq 0\} = \bigcup_{m=1}^{\infty} A_m .$$

Thus, by Theorem 2, this set has measure zero. ∎

Theorem 5.   *Let $A, B \subset \mathbb{R}^n$, $C \in \mathbb{R}$, $f, g: A \to \mathbb{R}$ be integrable. Then*
  (i) *$f + g$ is integrable and $\int_A f + g = \int_A f + \int_A g$.*
 (ii) *cf is integrable and $\int_A cf = c \int_A f$.*
(iii) *$|f|$ is integrable and $|\int_A f| \leqslant \int_A |f|$.*
 (iv) *If $f \leqslant g$, then $\int_A f \leqslant \int_A g$.*
  (v) *If A has volume, and $|f| \leqslant M$, then $|\int_A f| \leqslant Mv(A)$.*
 (vi) *(Mean-value Theorem for Integrals). If $f: A \to \mathbb{R}$ is continuous, A has volume and is compact and connected, then there is an $x_0 \in A$ such that $\int_A f(x)\,dx = f(x_0)v(A)$.*
(vii) *Let $f: A \cup B \to \mathbb{R}$. If A and B are such that $A \cap B$ has measure zero and $f \mid A \cap B$, $f \mid A$, and $f \mid B$ are integrable, then f is integrable and $\int_{A \cup B} f = \int_A f + \int_B f$.*

Proof: (i) Let S be a rectangle enclosing A and let f and g be extended to S by setting them equal to zero on $S\backslash A$. Suppose $\varepsilon > 0$ is given. By Theorem 1, there is a $\delta_1 > 0$ such that if P_1 is any partition of S into subrectangles $S_1, \ldots, S_N$ with sides less than δ_1 and if $x_1 \in S_1, \ldots, x_N \in S_N$, then

$$\left| \sum_{i=1}^{N} f(x_i)v(S_i) - \int_A f \right| < \frac{\varepsilon}{2} .$$

Similarly, there is a $\delta_2 > 0$ such that if P_2 is any partition of S into subrectangles $R_1, \ldots, R_M$ with sides less than δ_2 and if $x_1 \in R_1, \ldots, x_M \in R_M$, then

$$\left| \sum_{i=1}^{M} g(x_i)v(R_i) - \int_A g \right| < \frac{\varepsilon}{2} .$$

If we let $\delta = \min(\delta_1, \delta_2)$, then if P is any partition of S into subrectangles $T_1, \ldots, T_K$ with sides less than δ and if $x_1 \in T_1, \ldots, x_K \in T_K$, then

$$\left| \sum_{i=1}^{K} (g(x_i) + f(x_i))v(T_i) - \int_A f - \int_A g \right| < \varepsilon .$$

Hence by Theorem 1 we may conclude that $f + g$ is integrable and $\int_A (f + g) = \int_A f + \int_A g$.

(ii) Suppose $\varepsilon > 0$ is given. Let S be an enclosing rectangle for A with f extended to S as in (i). Let $\delta > 0$ be such that if P is any partition of S into subrectangles $S_1, \ldots, S_N$ with sides $< \delta$ and if $x_1 \in S_1, \ldots, x_N \in S_N$, then

$$\left| \sum_{i=1}^{N} f(x_i)v(S_i) - \int_A f \right| < \frac{\varepsilon}{|c|} .$$

This implies that

$$\left| \sum_{i=1}^{N} cf(x_i)v(S_i) - c\int_A f \right| < \varepsilon .$$

Hence cf is integrable and $\int_A cf = c \int_A f$.

(iii) This part is most easily proved as a corollary of (iv), which will therefore be proved first.

(iv) Let P be any partition of the enclosing rectangle S. As $f \leqslant g$, we see that

$$U(f,P) \leqslant U(g,P) .$$

So

$$\int_A f = \inf\{U(f,P) \mid P \text{ is any partition}\}$$

$$\leqslant \inf\{U(g,P) \mid P \text{ is any partition}\} = \int_A g .$$

Hence $\int_A f \leqslant \int_A g$.

(iii) We use the fact that if f is continuous at a point x in its domain, then $|f|$ is continuous at that point since $|f|$ is the composition of $y \mapsto |y|$ following f. Hence, if f is integrable over A, then by Theorem 3, $|f|$ is integrable over A. Now $-|f| \leqslant f \leqslant |f|$, so by (iv) $-\int_A |f| \leqslant \int_A f \leqslant \int_A |f|$, and therefore $|\int_A f| \leqslant \int_A |f|$.

(v) If P is any partition of the enclosing rectangle S into subrectangles $S_1, \ldots, S_N$, then

$$\int_A |f| \leqslant U(|f|,P) = \sum_{i=1}^{N} M_{S_i}(|f|)v(S_i)$$

$$\leqslant M \sum_{i=1}^{N} M_{S_i}(1_A)v(S_i) = MU(1_A,P) .$$

This implies that $\int_A |f| \leqslant Mv(A)$, and so

$$\left| \int_A f \right| \leqslant \int_A |f| \leqslant Mv(A) .$$

(vi) Let $m = \inf\{f(x) \mid x \in A\}$ and $M = \sup\{f(x) \mid x \in A\}$. By assumption, m and M are assumed values, as A is compact (Theorem 5, Chapter 4). Let $\lambda = (\int_A f)/v(A)$. (If $v(A) = 0$, the theorem follows from Theorem 4(i).) Then by (v), $m \leqslant \lambda \leqslant M$. Hence by the intermediate value theorem, there is an $x_0 \in A$ such that $f(x_0) = \lambda$, which proves the assertion.

Remark: More careful reasoning shows that compactness of A is not necessary; see Exercise 19.

(vii) Let $f_1 = f \cdot 1_A$, $f_2 = f \cdot 1_B$, and $f_3 = f \cdot 1_{A \cap B}$, so these represent the extensions of $f \mid A$, $f \mid B$, and $f \mid A \cap B$ required by the definition of integrability. By assumption, f_1, f_2, and f_3 are integrable and, for example, $\int_{A \cup B} f \cdot 1_A = \int_A f$ by definition. Now $f = f_1 + f_2 - f_3$ so by (i),

$$\int_{A \cup B} f = \int_{A \cup B} f_1 + \int_{A \cup B} f_2 - \int_{A \cup B} f_3$$

$$= \int_A f + \int_B f - \int_{A \cap B} f .$$

Since $A \cap B$ has measure zero, we have by Theorem 4(i) that $\int_{A \cap B} f = 0$. ∎

Theorem 6. (*Fundamental Theorem of Calculus*). *Let* $f : [a,b] \to \mathbb{R}$ *be continuous. Then* f *has an antiderivative* F *and*

$$\int_a^b f(x) \, dx = F(b) - F(a) .$$

This same formula holds for any antiderivative of f.

Proof: Define $F : [a,b] \to \mathbb{R}$ by $F(x) = \int_a^x f(y) \, dy$. We claim F is an antiderivative of f. Let $x \in \,]a,b[$ and let $h > 0$ be such that $]x, x + h[\,\subset\,]a,b[$. Then

$$\frac{F(x + h) - F(x)}{h} = \frac{\int_a^{x+h} f(y) \, dy - \int_a^x f(y) \, dy}{h} = \frac{\left(\int_x^{x+h} f(y) \, dy \right)}{h} .$$

Now given $\varepsilon > 0$, choose h such that $|f(y) - f(x)| < \varepsilon$ for all $y \in \,]x, x + h[$; such a choice is possible by the continuity of f. Therefore,

$$\left| \left(\int_x^{x+h} \frac{f(y)}{h} \, dy \right) - f(x) \right| = \left| \int_x^{x+h} \left(\frac{f(y) - f(x)}{h} \right) dy \right|$$

$$\leqslant \int_x^{x+h} \left| \frac{f(y) - f(x)}{h} \right| dy < \frac{\varepsilon h}{h} = \varepsilon .$$

Hence

$$\lim_{h \to 0+} \frac{F(x + h) - F(x)}{h} = f(x) \qquad \text{(the limit through } h > 0 \text{)} .$$

Similarly, we can show

$$\lim_{h \to 0-} \frac{F(x) - F(x - h)}{h} = f(x) \qquad \text{(the limit through } h < 0 \text{)} .$$

Hence F' exists and $F'(x) = f(x)$. From the definition of F, we get

$$\int_a^b f(y) \, dy = F(b) - F(a) .$$

The function F is easily seen to be continuous at a and b. Now let F_1 be any antiderivative of f. We shall show that $F_1 = F +$ constant. Since $F_1'(x) = F'(x) = f(x)$ for all $x \in \,]a,b[$, we have $(F_1 - F)'(x) = 0$ for all $x \in \,]a,b[$. By Example 1 Section 6.7, if a function has zero derivative on an interval, it must be constant on that interval. So $F_1 = F + \text{constant}$. Hence

$$\int_a^b f(y)\,dy = F_1(b) - F_1(a)$$

for any antiderivative F_1 of f. ∎

Theorem 7. *For $f \geq 0$ bounded and integrable on any cube $[-a,a]^n$, f is integrable iff the following condition holds: for any sequence B_k of bounded sets with volume such that (i) $B_k \subset B_{k+1}$ and (ii) for any cube, C, and we have $C \subset B_k$ for sufficiently large k, then* $\underset{k\to\infty}{\text{limit}} \int_{B_k} f$ *exists. In this case* $\underset{k\to\infty}{\text{limit}} \int_{B_k} f = \int_A f$.

Proof: Suppose f is integrable. Then if $[-a,a]^n \subset B_k \subset [-b,b]^n$, we have

$$\int_{[-a,a]^n} f \leq \int_{B_k} f \leq \int_{[-b,b]^n} f,$$

since $f \cdot 1_{[-a,a]^n} \leq f \cdot 1_{B_k} \leq f \cdot 1_{[-b,b]^n}$, where 1_A is the characteristic function of the set A. Hence as $\underset{a\to\infty}{\text{limit}} \int_{[a,a]^n} f$ exists, so does $\underset{k\to\infty}{\text{limit}} \int_{B_k} f$, and the limit is the same. We used hypothesis (ii) about B_k so that we could, by choosing k large, have $[-a,a]^n \subset B_k$ for any given a.

For the converse, $\int_{B_k} f$ is an increasing sequence in k, using (i), and has a limit. Say $\int_{B_k} f \to C$ as $k \to \infty$. Thus as $\int_{B_k} f \leq C$ for all k and for each a, $[-a,a]^n \subset B_k$ for some k, $\int_{[-a,a]^n} f \leq C$ for all a. Hence as $a \to \infty$ this is an increasing function of a bounded above so it converges (why?). ∎

Theorem 8.
(i) *Suppose $f: [a,\infty[\to \mathbb{R}$ is continuous and $f \geq 0$. Let F be an antiderivative of f. Then f is integrable iff* $\underset{x\to\infty}{\text{limit}}\, F(x)$ *exists. In this case*

$$\int_{[a,\infty]} f\,dx = \int_a^\infty f\,dx = \{\underset{x\to\infty}{\text{limit}}\, F(x)\} - F(a).$$

(ii) *Suppose $f: \,]a,b] \to \mathbb{R}$ is continuous and $f \geq 0$. Then f is integrable iff*

$$\underset{\varepsilon\to 0}{\text{limit}} \int_{a+\varepsilon}^b f(x)\,dx$$

exists. This limit equals $\int_a^b f(x)\,dx$.

Proof: (i) $\int_{[-b,b]} f \cdot 1_{[a,b]} = \int_a^b f$ for $b > a$, and $\int_a^b f = F(b) - F(a)$. Hence $\underset{b\to\infty}{\text{limit}} \int_{-b}^b f$ exists iff $\underset{b\to\infty}{\text{limit}}\, F(b)$ does, and $\int_a^\infty f = (\underset{b\to\infty}{\text{limit}}\, F(b)) - F(a)$.

(ii) The second part is a little trickier. For $\varepsilon > 0$ define f^ε to be 0 on $[a, a + \varepsilon]$ and f on $]a + \varepsilon, b]$. We proceed by first giving two preliminary steps.

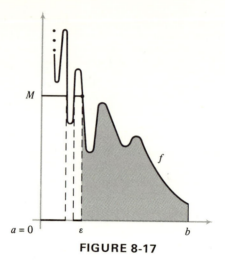

FIGURE 8-17

Step 1. If $M = \sup\{f(x) \mid a + \varepsilon \leqslant x \leqslant b\}$, then, recalling that $f_M(x) = f(x)$ if $f(x) \leqslant M$ and zero otherwise, we have

$$\int_a^b f^\varepsilon = \int_{a+\varepsilon}^b f \leqslant \int_a^b f_M$$

because $f \leqslant f_M$ on $[a,b]$. Note that f_M might not be zero on $[a,a + \varepsilon]$, as in Figure 8-17.

Step 2. For any ε and M,

$$\int_a^b f_M - \int_a^b f^\varepsilon \leqslant M\varepsilon .$$

This is because $f_M \leqslant M$ on $[a,a + \varepsilon]$, and for $x \in [a + \varepsilon,b]$, $f_M(x) = f(x)$, so

$$\int_a^b f_M - \int_a^b f^\varepsilon = \left(\int_a^{a+\varepsilon} f_M + \int_{a+\varepsilon}^b f \right) - \int_{a+\varepsilon}^b f = \int_a^{a+\varepsilon} f_M \leqslant \varepsilon M$$

since $f_M \leqslant M$.

Now, to demonstrate the theorem, first suppose $\int f_M \to I$ as $M \to \infty$. Notice that since $f \geqslant 0$, $\int f_M$ increases as M increases. We must show that $\int f^\varepsilon$ also converges to I as $\varepsilon \to 0$. It clearly increases to something $\leqslant I$ by Step 1. But, given $\delta > 0$, choose M such that $I - \int f_M < \delta/2$. Then if we let $\varepsilon = \delta/2M$, by Step 2, $\int f_M - \int f^\varepsilon < \delta/2$. Hence $I - \int f^\varepsilon < \delta$. Thus $\lim_{\varepsilon \to 0} \int f^\varepsilon = I$.

The converse follows in much the same way, again using Steps 1 and 2 to show that if $\int f^\varepsilon \to I$, then $\int f_M \to I$. ∎

In order to prepare for the proof of the monotone convergence theorem, we first prove the following lemma.*

---

* This proof of the monotone convergence theorem was pointed out by R. Gulliver.

Lemma.   *Suppose f is Riemann integrable, $f: [0,1] \to \mathbb{R}$, $|f| \leqslant M$, and $\int_0^1 f \geqslant \alpha > 0$. Then $E = \{x \in [0,1] \mid f(x) \geqslant \alpha/2\}$ contains a finite union of intervals of total length $l \geqslant \alpha/4M$.*

Proof: Let P be a partition of $[0,1]$ such that $0 \leqslant \int_0^1 f - L(f,P) \leqslant \alpha/4$. So $L(f,P) \geqslant 3\alpha/4$. We will now show that the intervals $R \in P$ with $R \subset E$ satisfy the conclusion; let l denote their total length. We have

$$\frac{3\alpha}{4} \leqslant L(f,P) = \sum_{\substack{R \in P \\ R \subset E}} \inf_{x \in R} f(x) v(R) + \sum_{\substack{R \in P \\ R \not\subset E}} \inf_{x \in R} f(x) v(R)$$

$$\leqslant Ml + \frac{\alpha}{2}(1 - l) \leqslant Ml + \frac{\alpha}{2}$$

So $l \geqslant \alpha/4M$, as claimed.   ∎

Theorem 9.   *(Lebesque's Monotone Convergence Theorem). Let $g_n: [0,1] \to \mathbb{R}$ be a sequence of non-negative functions such that each improper integral $\int_0^1 g_n < \infty$. Suppose that $0 \leqslant g_{n+1} \leqslant g_n$ and that $g_n(x) \to 0$ for each $x \in [0,1]$. Then $\lim\limits_{n \to \infty} \int_0^1 g_n = 0$.*

Proof: We have $0 \leqslant \int_0^1 g_{n+1} \leqslant \int_0^1 g_n$, that is, the integrals form a bounded decreasing sequence, so $\lambda = \lim\limits_{n \to \infty} \int_0^1 g_n$ exists, and $\lambda \geqslant 0$. We wish to show $\lambda = 0$, so assume $\lambda > 0$. Note that $\int_0^1 g_n \geqslant \lambda$ for all n. Consider $E_n = \{x \in [0,1] \mid g_n(x) \geqslant 2\lambda/5\}$. Observe that $E_{n+1} \subset E_n$. We want to apply the lemma, but g_n might not be bounded. However, $g_n \leqslant g_1$; since $\int_0^1 g_1$ exists as an improper integral, $\int_0^1 g_{1M} \to \int_0^1 g_1$ as $M \to \infty$. Here

$$g_{nM}(x) = \begin{cases} g_n(x), & g_n(x) < M, \\ M, & g_n(x) \geqslant M. \end{cases}$$

So we may choose $M > 2\lambda/5$ such that $0 \leqslant \int_0^1 (g_1 - g_{1M}) \leqslant \lambda/5$. Then for all n, $0 \leqslant \int_0^1 (g_n - g_{nM}) \leqslant \int_0^1 (g_1 - g_{1M}) \leqslant \lambda/5$, so that $\int_0^1 g_{nM} \geqslant 4\lambda/5 = \alpha$. Note that since $M > 2\lambda/5$, E_n may also be described as $\{x \in [0,1] \mid g_{nM}(x) \geqslant 2\lambda/5\}$. Therefore, by the lemma, E_n contains a finite union of intervals of total length $\geqslant \lambda/5M$. Now define

$$D = \bigcup_{n=1}^{\infty} \{x \in [0,1] \mid g_n \text{ is not continuous at } x\} ;$$

so D has measure 0. Thus D is contained in the union U of a countable number of disjoint open intervals with total length $< \lambda/5M$. It may be readily seen that E_n is not a subset of U. Observe that if x_0 is an accumulation point of E_n but is not in E_n, then g_n must be discontinuous at x_0, so $x_0 \in D \subset U$. That is, $\text{cl}(E_n) \subset E_n \cup U$. Define $F_n = \text{cl}(E_n) \backslash U$: by what we have just shown, $F_n \subset E_n$. But F_n is compact and $F_{n+1} \subset F_n$. Therefore, by the Cantor Intersection theorem, $\bigcap_{n=1}^{\infty} F_n \neq \varnothing$, and hence $\bigcap_{n=1}^{\infty} E_n \neq \varnothing$. But this means that for some $x \in [0,1]$, $g_n(x) \geqslant 2\lambda/5 > 0$ for all n, contradicting the hypothesis $g_n(x) \to 0$.   ∎

Corollary 4.   *If $f: [a,b] \to \mathbb{R}$, $f \geqslant 0$, and the improper integral $\int_a^b f^2 < \infty$, then $\int_a^b (f - f_M)^2 \to 0$ as $M \to \infty$.*

Proof: Apply the monotone convergence theorem with $g_n = (f - f_n)^2$. Since $(f - f_{n+1})^2 \leqslant (f - f_n)^2 \leqslant f^2$, the integrals $\int_a^b g_n$ exist, by the comparison theorem, and $g_{n+1} \leqslant g_n$. ∎

Worked Examples for Chapter 8

1. Show that a bounded set $A \subset \mathbb{R}^n$ has zero volume iff it can be covered by a finite number of rectangles of arbitrarily small total volume.

   Solution: Suppose A has zero volume and let $\varepsilon > 0$ be given. Let S be a closed rectangle containing A and let 1_A be the characteristic function of A (i.e. 1_A equals 1 on A and 0 on $S \backslash A$). Then by definition of zero volume, there is a partition P of S into subrectangles $S_1, \ldots, S_M$ such that $U(1_A, P) < \varepsilon$. Let P_0 be the collection of all those subrectangles S_i which intersect A. Then $U(1_A, P)$ is simply $\sum_{S \in P_0} v(S)$, and so P_0 is a collection of rectangles covering A with a total volume $< \varepsilon$.

   Suppose, on the other hand, that given $\varepsilon > 0$ A can be covered by a finite number of rectangles of total volume $< \varepsilon$. Let these rectangles be $V_1, \ldots, V_M$. Let S be a closed rectangle containing A and let P be a partition of S into subrectangles $S_1, \ldots, S_N$, such that each S_i is either contained in some V_j or has at most its boundary in common with some V_j's—the partition is defined by using all the edges of the V_j's. Then $U(1_A, P) = \sum_{i=1}^M v(V_i) < \varepsilon$. This implies that $\inf\{ U(1_A, P) \mid P$ is a partition$\} = 0$, and hence $L(1_A, P') = 0$ for any partition P', since $0 \leqslant L(1_A, P') \leqslant U(1_A, P)$ for all P. Therefore A has volume, and this volume is zero.

2. Let f_k be a sequence of bounded (Riemann) integrable functions defined on $[a,b]$. Suppose $f_k \to f$ uniformly. Then prove that f is integrable on $[a,b]$, and

$$\int_a^b f_k(x)\, dx \to \int_a^b f(x)\, dx .$$

   Solution: First, observe that f is bounded. Indeed, suppose we choose f_N such that $|f_N(x) - f(x)| < 1$ for all x. Then, using the triangle inequality,

$$|f(x)| \leqslant |f_N(x)| + 1 .$$

Therefore, since f_N is bounded, so is f.

   By Theorem 1, we must find a number I, such that for every $\varepsilon > 0$ there is a $\delta > 0$, such that

$$\left| \sum_{k=1}^n f(x'_k)(x_k - x_{k-1}) - I \right| < \varepsilon$$

for any subdivision $x_0, x_1, \ldots, x_n$ of $[a,b]$ with $|x_k - x_{k-1}| < \delta$ and $x_{k-1} \leqslant x'_k \leqslant x_k$. By Theorem 4, Chapter 5, we *expect* $I = \lim_{k \to \infty} \int_a^b f_k(x)\, dx$. Now $\int_a^b f_k(x)\, dx$ is a Cauchy sequence. To see this, note that

$$\left| \int_a^b f_k(x)\, dx - \int_a^b f_l(x)\, dx \right| < \varepsilon \qquad \text{if } |f_k(x) - f_l(x)| < \frac{\varepsilon}{(b-a)} .$$

Hence, the sequence converges to a value which we call I.

Given $\varepsilon > 0$, choose N such that $k \geqslant N$ implies $|\int_a^b f_k(x)\,dx - I| < \varepsilon/3$. Now choose N_1 such that $k \geqslant N_1$ implies $|f_k(x) - f(x)| < \varepsilon/(3(b - a))$ for all $x \in [a,b]$ and choose $N_2 = \max(N, N_1)$. Now since f_{N_2} is integrable, there is a $\delta > 0$ such that for $|x_k - x_{k-1}| < \delta$,

$$\left| \sum_{k=1}^{n} f_{N_2}(x_k')(x_k - x_{k-1}) - \int_a^b f_{N_2}(x)\,dx \right| < \frac{\varepsilon}{3}.$$

With this choice, we have by the triangle inequality,

$$\left| \sum_{k=1}^{n} f(x_k')(x_k - x_{k-1}) - I \right| \leqslant \left| \sum_{k=1}^{n} f(x_k')(x_k - x_{k-1}) - \sum_{k=1}^{n} f_{N_2}(x_k')(x_k - x_{k-1}) \right|$$

$$+ \left| \sum_{k=1}^{n} f_{N_2}(x_k')(x_k - x_{k-1}) - \int_a^b f_{N_2}(x)\,dx \right|$$

$$+ \left| \int_a^b f_{N_2}(x)\,dx - I \right| < \frac{\varepsilon}{3} + \frac{\varepsilon}{3} + \frac{\varepsilon}{3} = \varepsilon,$$

which proves the assertion.

Remark: In Theorem 4, Chapter 5, we established the more restrictive result that if f_k, f were continuous (and hence integrable), then

$$\int_a^b f(x)\,dx = \lim_{k \to \infty} \int_a^b f_k(x)\,dx.$$

This of course is also shown by the proof just completed. The above method also works on $A \subset \mathbb{R}^n$.

3. Show that $\int_0^a x^2\,dx = a^3/3$ by using the fundamental theorem. Verify this answer directly by showing that for any given $\varepsilon > 0$, there exists a partition P of $[0,a]$, such that $U(x^2,P) - L(x^2,P) < \varepsilon$ and that

$$\inf\{U(x^2,P)\,|\,P \text{ is any partition}\} = \sup\{L(x^2,P)\,|\,P \text{ is any partition}\} = \frac{a^3}{3}.$$

Solution: The function F defined by $F(x) = x^3/3$ is an antiderivative of $f(x) = x^2$, since $F'(x) = x^2$. Thus by the fundamental theorem

$$\int_0^a x^2\,dx = F(a) - F(0) = \frac{a^3}{3}.$$

In order to verify our answer using the upper and lower sums, we partition $[0,a]$ into the n subintervals $[0,a/n]$, $[a/n, 2a/n]$, $\ldots$, $[((n-1)/n, a]$. If we call this partition P, then

$$U(x^2,P) = \sum_{k=1}^{n} \left(\frac{ka}{n} \right)^2 \frac{a}{n} = \left(\frac{a^3}{n^3} \right)\left(\sum_{k=1}^{n} k^2 \right) = \left(\frac{a^3}{n^3} \right)\left(\frac{1}{6} \right) n(n + 1)(2n + 1);$$

(see Exercise 25) and

$$L(x^2,P) = \sum_{k=1}^{n} \left(\frac{(k-1)a}{n} \right)^2 \frac{a}{n} = \left(\frac{a^3}{n^3} \right)\left(\sum_{k=1}^{n-1} k^2 \right) = \left(\frac{a^3}{n^3} \right)\left(\frac{1}{6} \right)(n - 1)(n)(2n - 1).$$

Hence

$$U(x^2,P) = \left(\frac{a^3}{3}\right)\left(1 + \frac{1}{n}\right)\left(1 + \frac{1}{2n}\right)$$

and

$$L(x^2,P) = \left(\frac{a^3}{3}\right)\left(1 - \frac{1}{n}\right)\left(1 - \frac{1}{2n}\right).$$

From these expressions we see that by choosing n sufficiently large we get $U(x^2,P) - L(x^2,P) < \varepsilon$ and $\inf\{U(x^2,P) \mid P \text{ is any partition}\} = \sup\{L(x^2,P) \mid P \text{ is any partition}\} = a^3/3$.

4. Find $\displaystyle\int_{[0,\infty[} \frac{dx}{(1 + x)^2}$.

Solution: We are integrating a non-negative function defined on an unbounded set, so by definition

$$\int_{[0,\infty[} \frac{dx}{(x + 1)^2} = \lim_{a \to \infty} \int_0^a \frac{dx}{(x + 1)^2} .$$

Since

$$\frac{d(-1/(x + 1))}{dx} = \frac{1}{(x + 1)^2} ,$$

we can use the fundamental theorem to obtain

$$\int_0^a \frac{dx}{(x + 1)^2} = -\frac{1}{(a + 1)} + \frac{1}{(0 + 1)} = \frac{a}{(a + 1)} .$$

Hence

$$\lim_{a \to \infty} \int_0^a \frac{dx}{(x + 1)^2} = 1 = \int_{[0,\infty[} \frac{dx}{(x + 1)^2} .$$

Exercises for Chapter 8

1. (a) Let $f: A \subset \mathbb{R}^n \to \mathbb{R}$, where A is bounded and f is bounded and integrable over A. Consider another bounded integrable function $g: A \to \mathbb{R}$ such that $g(x) = f(x)$ except on a set $S \subset A$ of measure zero. Then assuming f and g are integrable on S and $A\backslash S$, prove $\int_A g = \int_A f$.
   (b) If $f: A \subset \mathbb{R}^n \to \mathbb{R}$ and $g: A \subset \mathbb{R}^n \to \mathbb{R}$ are bounded functions, integrable on the bounded set A, and $\int_A |f - g| = 0$, then prove $f(x) = g(x)$ for all $x \in A$, except possibly for a set of measure zero.

2. Give a proof of Theorem 5(iii) directly from Darboux's theorem and the triangle inequality for real numbers. Now go back and fill in the gap in Example 5, Chapter 6.

3. Prove that an increasing function $f: [a,b] \to \mathbb{R}$ is Riemann integrable. [Hint: at each discontinuity x_0 of f, $\displaystyle\lim_{x \to x_0-} f(x) < \lim_{x \to x_0+} f(x)$, and we can find a rational r_{x_0}

such that $(\underset{x \to x_0^-}{\text{limit}} f(x)) < r_{x_0} < (\underset{x \to x_0^+}{\text{limit}} f(x))$. Show that the discontinuities of f are countable.]

4. Show that $f(x) = x^{2n+1}$ for n an integer $\geqslant 0$, is not (absolutely) integrable over $\mathbb{R}$, even though $\underset{a \to \infty}{\text{limit}} \int_{-a}^{a} f$ exists.

5. In $\mathbb{R}^3$, prove that any subset of the xy-plane has measure zero.

6. If $f: A \subset \mathbb{R}^n \to \mathbb{R}$ and $g: A \subset \mathbb{R}^n \to \mathbb{R}$ are bounded integrable functions on the bounded set A such that $f(x) < g(x)$ for all $x \in A$ and $v(A) \neq 0$, then show $\int_A f < \int_A g$. [Hint: Employ Theorem 4(ii).]

7. Let $f: [a,b] \to \mathbb{R}$ be continuous and differentiable on $]a,b[$. Assume $f(a) = 0$, $f(b) = -1$ and $\int_a^b f(x)\,dx = 0$. Prove that there is a $c \in]a,b[$ such that $f'(c) = 0$.

8. Compute the following integrals.
    (a) $\int_0^{2\pi} \sin x\,dx$.
    (b) $\int_a^b x^2(\sin x)\,dx$.

9. Give a proof of Theorem 5(i), (ii), (iii), (iv), and (vii) for improper integrals. (The parts of Theorem 5 omitted here do not make sense if A has infinite volume.)

10. (a) If $A \subset A_1 \cup A_2 \cup \cdots \cup A_N$, all sets having volume, then show $v(A) \leqslant \sum_{i=1}^{N} v(A_i)$.
    (b) If A is compact, then show that A has measure zero iff A has content zero. Show that a bounded set B has volume iff $\text{bd}(B)$ has content zero.

11. If S is a closed or open rectangle show that the two definitions of volume coincide. That is, prove that $\int_S 1 = (b_1 - a_1)(b_2 - a_2)\cdots(b_n - a_n)$ where either $S = [a_1,b_1] \times \cdots \times [a_n,b_n]$ or $S =]a_1,b_1[\times \cdots \times]a_n,b_n[$. [Hint: For the closed rectangle use the partition consisting only of the rectangle S. This statement for open rectangles then follows from Exercise 1.]

12. Prove that A has measure zero iff for every $\varepsilon > 0$ there is a covering of A by sets $V_1, V_2, \ldots$ with volume such that $\sum_{i=1}^{\infty} v(V_i) < \varepsilon$.

13. Prove that a bounded function $f: S \to \mathbb{R}$ is integrable on the rectangle S iff there is a number I such that for any $\varepsilon > 0$ there is a partition P_ε such that for any refinement P of P_ε and any choice of $x_i \in S_i$ for $S_i \in P$, $|\sum_{S_i \in P} f(x_i)v(S_i) - I| < \varepsilon$.

14. Is $\int_0^\infty x^p\,dx$ convergent for any p? If so, which p?

15. Generalize Example 2 on page 290 to functions $f: A \subset \mathbb{R}^n \to \mathbb{R}$. .

16. (a) Suppose $f_k \to f$ uniformly on $A \subset \mathbb{R}^n$. Let A_k be the points of discontinuity of f_k (extended). Show that the discontinuities of f (extended) are contained in $A_1 \cup A_2 \cup \cdots$. [Hint: Study Theorem 1, Chapter 5.]
    (b) Use (a) to give another proof for Example 2.
    (c) Find functions $f_k: A \to \mathbb{R}$ which are integrable and such that $f_k \to f$ pointwise, but f is not integrable. [Hint: Consult Gelbaum and Olmsted, *Counterexamples in Analysis*, Example 5, Chapter 7.]

17. (a) Let P_n denote the division of $[a,b]$ into 2^n equal subintervals and form $L(f,P_n)$, $U(f,P_n)$ for $f: [a,b] \to \mathbb{R}$ bounded. Show that f is integrable iff $\lim_{n \to \infty} L(f,P_n) = \lim_{n \to \infty} U(f,P_n)$. Why do these limits always exist?

    (b) Generalize (a) for $A \subset \mathbb{R}^n$.

18. Let $f: B \to \mathbb{R}$ be integrable, $f \geq 0$. If $A \subset B$ and f is integrable on A, then $\int_A f \leq \int_B f$. Is this true if we do not assume $f \geq 0$?

19. (a) Generalize the mean-value theorem for integrals (Theorem 6(vi)) to the case where A is any bounded connected set. [Hint: If $M = \sup\{f(x) \mid x \in A\}$ and $m = \inf\{f(x) \mid x \in A\}$, then we do not necessarily have a and $b \in A$ such that $f(a) = m$ and $f(b) = M$. If $\lambda = \int_A f / v(A)$, then as usual $m \leq \lambda \leq M$. Consider separately the cases $\lambda = m$, $\lambda = M$, $m < \lambda < M$. For the first two cases use the fact that if $g \geq 0$ and $\int_B g = 0$, then $g(x) = 0$, except for a set of measure zero (Theorem 4(ii)). For the last case we can pick points x and $y \in A$, such that $f(x) < \lambda < f(y)$, and then we can apply the intermediate-value theorem as in Theorem 6(vi).]

    (b) If $\varphi(x) \geq 0$ for $x \in$ a connected bounded set $A \subset \mathbb{R}$, φ is continuous and increasing in x and f is integrable, then $f\varphi$ is integrable and $\int_A f\varphi = \varphi(c) \int_A f$ for some point $c \in A$. (This is the "second mean-value theorem.")

20. Suppose $f: \,]0,b] \to \mathbb{R}$ is continuous, positive, and integrable on $]0,b]$. Suppose further that as $x \to 0$ from the right, $f(x)$ increases monotonically to $+\infty$. Then prove that $\varepsilon f(\varepsilon) \to 0$ as $\varepsilon \to 0$.

21. Show that $\int_1^\infty x^{-p} \sin x \, dx$ converges if $p > 1$. Show that if $0 < p \leq 1$, then the convergence is conditional.

22. The *gamma function* is defined by the improper integral $\Gamma(p) = \int_0^\infty e^{-x} x^{p-1} \, dx$. Show that the integral is convergent for $p > 0$.

23. (a) If $\varphi: [a,b] \to \mathbb{R}^n$ is a continuous function, then show that the set $S = $ graph $\varphi = \{(x,\varphi(x)) \mid x \in [a,b]\} \subset \mathbb{R}^{n+1}$ has content and measure zero. [Hint: First, consider the case $n = 1$ and use the definition of continuity.]

    (b) For $\varphi: \mathbb{R} \to \mathbb{R}^n$ continuous, show that graph φ has measure zero. [Hint: Graph $\varphi = \bigcup_{n=1}^\infty$ graph $(\varphi \mid [-n,n])$.]

    (c) Let $f: [a,b] \to \mathbb{R}$ be integrable. Show the graph of f has volume zero by considering the difference of the upper and lower sums for f.

    (d) Show that the ellipse $x^2 + 3y^2 = 9$ in $\mathbb{R}^2$ has volume zero.

24. Give an example to show that the following is *not* equivalent to the integrability of f. For any $\varepsilon > 0$ there is a $\delta > 0$ such that if P is any partition into rectangles $S_1, \ldots, S_N$ with sides $< \delta$, there exist $x_1 \in S_1, \ldots, x_N \in S_N$ such that

$$\left| \sum_{i=1}^N f(x_i) v(S_i) - I \right| < \varepsilon.$$

25. Prove that

$$\sum_{k=1}^n k = \frac{n(n+1)}{2} \quad \text{and} \quad \sum_{k=1}^n k^2 = \frac{n(n+1)(2n+1)}{6}.$$

These formulas were used in Example 3. [Hint: Let $S = \sum_{k=1}^{n} k$. Write down S backwards and consider $S + S$. Consider the identity $(k + 1)^3 - k^3 = 3k^2 + 3k + 1$ and observe that

$$\sum_{k=1}^{n} \{(k + 1)^3 - k^3\} = (n + 1)^3 - 1$$

and

$$\sum_{k=1}^{n} \{(k + 1)^3 - k^3\} = 3\left(\sum_{k=1}^{n} k^2\right) + 3\left(\sum_{k=1}^{n} k\right) + n\ .$$

Alternatively, use induction.]

26. Consider a set $A \subset \mathbb{R}^n$, where A is bounded and has volume. Let $f: A \subset \mathbb{R}^n \to \mathbb{R}$, $f \geq 0$, but allow f to be unbounded. Suppose C_i is a sequence of compact sets with volume, $C_i \subset A$ with C_i increasing to A, and assume $v(C_i) \to v(A)$ (this is actually automatically true as shall be seen in Exercise 12, Chapter 9). Then f is integrable iff f is integrable on each C_i and $\lim\limits_{i \to \infty} \int_{C_i} f$ exists and in this case

$$\int_A f = \lim_{i \to \infty} \int_{C_i} f\ .$$

[Hint: Study Theorem 8(ii).]

27. Prove that if $f: A \subset \mathbb{R}^n \to \mathbb{R}$ is continuous, A is open with volume, and $\int_B f = 0$ for each $B \subset A$ with volume, then $f = 0$. [Hint: Use Theorem 5(vi).]

28. Let $f: [0,1] \to \mathbb{R}$ be integrable and be continuous at x_0. Show that the map $x \mapsto \int_0^x f(y)\, dy$ is differentiable with derivative $f(x_0)$. Give an example of a discontinuous integrable f for which this map is not differentiable. For bounded integrable f prove this map is always continuous, and in fact, Lipschitz.

29. Show that the Cantor set $C \subset [0,1]$ has measure zero (see Exercise 38 at the end of Chapter 3).

30. (a) Let $f: [a,b] \to \mathbb{R}$ be differentiable and assume that f' is integrable. Prove $\int_a^b f'(x)\, dx = f(b) - f(a)$.
    (b) Must f' always be integrable?

31. Prove the following analogues of the Weierstrass and Dirichlet tests for uniform convergence using the Cauchy criterion (Exercise 1, Section 8.7).
    (a) Let $f: [a,\infty[\times [c,d] \to \mathbb{R}$ and suppose there is a positive function $M(x)$, $x \in [a,\infty[$ such that $|f(x,s)| \leq M(x)$ for all $s \in [c,d]$, and $\int_a^\infty M(x)\, dx < \infty$. Then $F(x) = \int_a^\infty f(x,s)\, dx$ converges uniformly in s. If $f(x,s)$ is continuous in x, s, prove F is continuous.
    (b) Let $f: [a,\infty[\times [c,d] \to \mathbb{R}$ be continuous and suppose $|\int_a^r f(x,s)\, dx| \leq M$ for a constant M for all $r \geq a$, $s \in [c,d]$. Suppose $\varphi(x,s)$ is decreasing in x and $\varphi(x,s) \to 0$ as $x \to \infty$ uniformly in s. Prove $F(s) = \int_a^\infty \varphi(x,s) f(x,s)\, dx$ converges uniformly.

32. For $x > 0$ define $L(x) = \int_1^x 1/t \, dt$. Prove the following, using this definition.
    (a) L is increasing in x.
    (b) $L(xy) = L(x) + L(y)$.
    (c) $L'(x) = 1/x$.
    (d) $L(1) = 0$.
    (e) Properties (c) and (d) uniquely determine L. What is L?

33. Let $f: \mathbb{R} \to \mathbb{R}$ be continuous and set $F(x) = \int_0^{x^2} f(y) \, dy$. Prove $F'(x) = 2xf(x^2)$. Give a general theorem.

34. Let $f: [0,1] \to \mathbb{R}$ be Riemann integrable and suppose for every a, b with $0 \leqslant a < b \leqslant 1$ there is a $c, a < c < b$ with $f(c) = 0$. Prove $\int_0^1 f = 0$. Must f be zero? What if f is continuous?

35. Let $A_n = 1/n[(n + 1) + (n + 2) + \cdots + (n + n)]$. Prove $\lim_{n \to \infty}(1/n)A_n = 3/2$ using the Riemann integral.

36. Prove that $\lim_{n \to \infty} (n!)^{1/n}/n = e^{-1}$ by considering Riemann sums for $\int_0^1 \log x \, dx$ based on the partition $1/n < 2/n < \cdots < 1$.

37. (a) Under what conditions is $\int_a^b f(\varphi(t))\varphi'(t) \, dt = \int_{\varphi(a)}^{\varphi(b)} f(x) \, dx$?
    (b) Evaluate $\int dx/((1 - x)\sqrt{1 - x^2})$ using $x = \cos t$.

38. Let $f: [0,1] \to \mathbb{R}$,

$$f(x) = \begin{cases} 0, & \text{if } x \text{ is irrational}, \\ \dfrac{1}{q}, & \text{if } x = \dfrac{p}{q}, \end{cases}$$

where $p, q \geqslant 0$ with no common factor. Show f is integrable and compute $\int_0^1 f$.

39. Prove that

$$\log 2 = \lim_{n \to \infty} \left[\frac{1}{n + 1} + \frac{1}{n + 2} + \cdots + \frac{1}{2n} \right].$$

[Hint: Write the expression in brackets as

$$\frac{1}{n} \sum_{k=1}^{n} \left(\frac{1}{1 + \dfrac{k}{n}} \right)$$

and use Riemann sums.]

40. Let $R([a,b]) = \{f: [a,b] \to \mathbb{R} \mid f \text{ is Riemann integrable}\}$. Set

$$d(f,g) = \int_a^b |f(x) - g(x)| \, dx \,.$$

Is d a metric on the space $R([a,b])$?

41. Find an open subset of $\mathbb{R}$ contained in $]0,1[$ which does not have volume as follows.
    (a) Review the construction of the Cantor set (see Exercise 38, Chapter 3).

(b) Modify the Cantor set by letting C_k be obtained from C_{k-1} by removing the middle $1/2^k$th from each interval of C_{k-1} and letting $C_0 = [0,1]$. Set $C = \bigcap_{k=1}^{\infty} C_k$.

(c) Show that $v(C_k) = \prod_{i=1}^{k} (1 - 1/2^i) \geqslant 1/4$.

(d) Let U be the complement of C. Compute the boundary of U and using (c) show that it cannot have measure zero.

This exercise also produces an example of a compact set C with empty interior which does not have volume.

42. Find a subset A of $[0,1]$ such that $A = \text{cl}(\text{int } A)$ and yet $\text{bd}(A)$ does not have measure zero. (This exercise requires care and patience.)

43. It is a fact that*

$$\sin\left(\frac{\pi}{n}\right)\sin\left(\frac{2\pi}{n}\right)\cdots\sin\left(\frac{(n-1)\pi}{n}\right) = \frac{n}{2^{n-1}}.$$

Use this identity to evaluate $\int_0^\pi \log \sin x \, dx$.

44. Discuss generalizations of Theorem 9 to $\mathbb{R}^n$.

45. Discuss generalizations of Theorem 9 from $[0,1]$ to $[0,\infty[$.

46. (a) Suppose $U = \,]-1,1[\,\times\,]-1,1[\,\subset \mathbb{R}^2$, $f : U \to \mathbb{R}$. Assume that $\partial f/\partial x$ and $\partial f/\partial y$ exist at each point of U and are *bounded* on U, where (x,y) are the standard coordinates for $\mathbb{R}^2$. Show that f is *continuous* at $(0,0)$.

(b) Show by example that boundedness of the partial derivatives is necessary in part (a); mere existence is not enough.

47. For every $\alpha > 0$ compare $\int_0^N x^\alpha \, dx$ with $\sum_{n=1}^{N} n^\alpha$ and $\sum_{n=0}^{N-1} n^\alpha$, and hence determine

$$\underset{N \to \infty}{\text{limit}} \sum_{n=1}^{N} \frac{n^\alpha}{N^{1+\alpha}}.$$

48. For any function $f(x)$ continuous over the reals define the sequence $f_n(x) = n \int_x^{x+1/n} f(\xi) \, d\xi$ for $n = 1, 2, 3, \ldots$. Show that $df_n(x)/dx$ exists even if $df(x)/dx$ does not, and that $f(x) = \underset{n \to \infty}{\text{limit}} f_n(x)$, and that convergence to the limit is uniform when f is uniformly continuous.

49. Suppose $\{I_k\}$ is a collection of open intervals whose union covers a closed interval C on the real axis; show that some positive ε exists such that every subinterval of C no wider than ε lies entirely in at least one of the I_k's.

50. State whatever lemmas, theorems, and so forth are needed to justify each of the following assertions.

(a) $\underset{n \to \infty}{\text{limit}} \sum_{k=1}^{\infty} 2^{-k} \sin(k/n) = 0$.

(b) If $f(x)$ is given by a power series converging in $\,]-1,1[$, then the same is true for $f'(x)$.

(c) Let $f(x) = \tan(\pi x/2)$ and set $a_n = f^{(n)}(0)/n!$. Then $\sum_0^\infty a_n$ is *not* a convergent series. (Do not attempt to compute a_n.)

---

* See J. Marsden, *Basic Complex Analysis*, W. H. Freeman, San Francisco, p. 24.

(d) If $f_n(x)$ is differentiable on $[a,b]$ with $|f'_n(x)| < 10$ for all n, and $x \in [a,b]$ and $f_n(x) \to 0$ at each x, then $f_n(x) \to 0$ uniformly.

(e) $f(x) = \sum_1^\infty \dfrac{\cos(2^k x)}{3^k}$ has a continuous derivative.

(f) $\left| e^x - 1 - x - \dfrac{x^2}{2} - \cdots - \dfrac{x^{99}}{99!} \right| \leqslant e^x \dfrac{x^{100}}{100!}$ for $x > 0$.

(g) $f(x) = \sum_1^\infty \dfrac{x^n}{n^3}$ is continuous in the closed interval $[-1,1]$.

(h) $\displaystyle\lim_{x \to 0} \dfrac{e^{ax} - 1}{\sin bx} = \dfrac{a}{b}$

(i) If $\sum_1^\infty |a_n| < \infty$ then $\int_0^t \left(\sum_1^\infty a_n \cos nx \right) = \sum_1^\infty a_n \sin nt$.

(j) For some integer n, $n > 10^{100} (\log n)^{1000}$

51. Explain the following: A function defined on $[0,\infty[$ but not infinitely often differentiable cannot be expressed as the sum of a Dirichlet series

$$f(t) = \sum_{n=0}^\infty a_n e^{-nt} \, .$$

[Hint: let $t = -\log x$]

52. Let Ω be the set of points $(x,y) \in \mathbb{R}^2$ that can be expressed in the form $(\sin \theta + \sin \psi, \cos \theta - \cos \psi)$ for some (θ,ψ). Find an interior point of Ω (that is, a point which together with a disc around it belongs to Ω).

53. Show that the series

$$\sum_{k=1}^\infty \dfrac{1}{2^k - k \sin(kx)}$$

is uniformly convergent on $\mathbb{R}$.

54. Prove the "dominated convergence theorem for series": *If*
   (i) for each k, $a_k^n \to a_k$ as $n \to \infty$, a_k and $a_k^n \in \mathbb{R}^m$;
   (ii) for each n and k, $\|a_k^n\| \leqslant b_k$, $\|a_k\| \leqslant b_k$, some $b_k \in \mathbb{R}$; *and*
   (iii) $\sum_{k=1}^\infty b_k$ is convergent (so by the comparison test, $\sum_{k=1}^\infty a_k$ and $\sum_{k=1}^\infty a_k^n$ are convergent); *then*

$$\sum_{k=1}^\infty a_k^n \to \sum_{k=1}^\infty a_k \text{ as } n \to \infty \, .$$

Give an example to show that condition (iii) is necessary, even if it is assumed that $\sum_k a_k^n$ and $\sum_k a_k$ are convergent.

Chapter **9**

Fubini's Theorem and the Change of Variables Formula

9.1   Introduction

There are two fundamental integration theorems which help us to evaluate multiple integrals. The first result concerns the evaluation of multiple integrals by means of iterated integration. Using this method, we can calculate the value of a multiple integral by performing successive single integrations.

EXAMPLE 1.   If A is the square defined by $0 \leqslant x \leqslant 1$ and $0 \leqslant y \leqslant 1$, then

$$\int_A (x + y)x \, dx \, dy = \int_{x=0}^{1} \left(\int_{y=0}^{1} [x^2 + yx] \, dy \right) dx$$
$$= \int_0^1 \left(x^2 + \frac{1}{2} x \right) dx$$
$$= \frac{1}{3} + \frac{1}{4} = \frac{7}{12} \, .$$

If A is not a square but say a triangle then we extend the function to a square by letting it be zero outside A. Then, in the above process, the y integration becomes cut off at some point which depends on x, as indicated in Figure 9-1.

The intuition behind this method is as follows. For a given function $f(x,y)$, $0 \leqslant x \leqslant 1$, $0 \leqslant y \leqslant 1$, the number $\int_0^1 f(x,y) \, dy$ is the area under the graph of f on the line $x =$ constant. Integrating this area over x gives

299

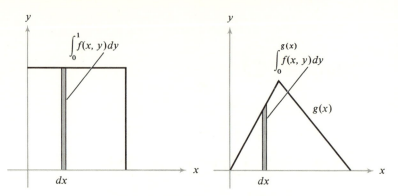

FIGURE 9-1   Iterated integrals.

the volume under the graph. This then suggests that $\int_A f(x,y)\,dx\,dy = \int_0^1 (\int_0^1 f(x,y)\,dy)\,dx$. A precise result will be given in Theorem 1, called Fubini's theorem.

The second basic theorem of this chapter is called the change of variables formula. This is used in conjunction with the above method to evaluate certain types of integrals. The following example is typical.

EXAMPLE 2.   Suppose we wish to evaluate $\int_A (x^2 + y^2)\,dx\,dy$ where A is the unit disc, $A = \{(x,y) \in \mathbb{R}^2 \mid x^2 + y^2 \leqslant 1\}$. This is easiest to evaluate if we change the variables to polar coordinates (r,θ). The change of coordinates is given by

$$x = r \cos \theta, \qquad y = r \sin \theta ,$$

for $r > 0$, $0 < \theta < 2\pi$. Then we have the "rule" $dx\,dy = r\,dr\,d\theta$. Since $x^2 + y^2 = r^2$, we have

$$\int_A (x^2 + y^2)\,dx\,dy = \int_A r^2 r\,dr\,d\theta$$

$$= \int_{r=0}^1 \int_{\theta=0}^{2\pi} r^3\,d\theta\,dr$$

$$= \int_{r=0}^1 2\pi r^3\,dr = \frac{\pi}{2}.$$

The justification of the "rule" $dx\,dy = r\,dr\,d\theta$ in Example 2 is given by the change of variables formula (Theorem 3 below). Notice that the extra factor r is just the Jacobian $\partial(x,y)/\partial(r,\theta) = r$. However it is easy to heuristically "justify" the rule by regarding dr and $d\theta$ as infinitesimals. Namely, dr represents a radial infinitesimal while $r\,d\theta$ represents an infinitesimal arc length. Thus $r\,dr\,d\theta$ is the area element in a sector bounded by r, $r + dr$ and θ, $\theta + d\theta$ (see Figure 9-2).

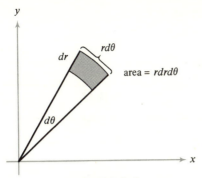

FIGURE 9-2

In the case of one dimension the change of variables formula is very easy. It states that if f is continuous on $[a,b]$ and we have a mapping $\varphi: [\alpha,\beta] \to [a,b]$ with $\varphi(\alpha) = a$, $\varphi(\beta) = b$, φ' exists and is continuous, then

$$\int_a^b f(x)\, dx = \int_\alpha^\beta f(\varphi(u))\varphi'(u)\, du \, .$$

To prove this, first find an F such that $F' = f$ which is possible by the fundamental theorem, and observe that

$$\int_a^b f(x)\, dx = F(b) - F(a) \, .$$

Now let $G(u) = F(\varphi(u))$. By the chain rule,

$$G'(u) = F'(\varphi(u))\varphi'(u) = f(\varphi(u))\varphi'(u) \, .$$

Hence, again by the fundamental theorem,

$$\int_\alpha^\beta f(\varphi(u))\varphi'(u)\, du = G(\beta) - G(\alpha) = F(b) - F(a)$$

as required. (This theorem is also true if f is not continuous, but is merely integrable as we shall see below.) This technique is often called "integration by substitution" and its power is well known to the student.

EXAMPLE 3.   To integrate $\int (1 + x^2)^{10}\, x\, dx$, let $y = 1 + x^2$ and note that $y' = 2x$, so $\int (1 + x^2)^{10} x\, dx = (1/2) \int y^{10} y'\, dx = (1/2) \int y^{10}\, dy = y^{11}/22 + C$ (C a constant).

The generalization of this result to higher dimensions is contained in the statement of the change of variables formula; while it is a good deal more subtle to prove, it is easy to understand and along with Fubini's theorem is the most powerful computational method we have.

Exercises for Section 9.1

1. Evaluate $\int_0^1 x^2 e^{x^3}\, dx$.

2. Evaluate $\displaystyle\int_0^1 \frac{\sin x}{\cos^2 x}\, dx$.

3. Evaluate $\int_A (x + y^2)\, dx\, dy$ where $A = [0,1] \times [0,1]$.

4. Evaluate $\int_A e^{-x^2 - y^2}\, dx\, dy$ where A is the unit disc in $\mathbb{R}^2$.

5. Evaluate $\int_A dx\, dy$ where A is the triangle in $\mathbb{R}^2$ bounded by the lines $x = 0$, $y = 0$, $x + y = 1$.

9.2 Fubini's Theorem

Let us now state the first of our two basic theorems. We start by giving Fubini's theorem for the case of the plane $\mathbb{R}^2$.

> **Theorem 1.** (i) *Let A be the rectangle described by $a \leqslant x \leqslant b$, $c \leqslant y \leqslant d$, and let $f\colon A \to \mathbb{R}$ be continuous. Then*
>
> $$\int_A f = \int_a^b \left(\int_c^d f(x,y)\, dy \right) dx$$
> $$= \int_c^d \left(\int_a^b f(x,y)\, dx \right) dy.$$
>
> *The expression*
>
> $$\int_a^b \left(\int_c^d f(x,y)\, dy \right) dx$$
>
> *means that the function*
>
> $$g(x) = \int_c^d f(x,y)\, dy.$$
>
> *is integrated from a to b.*
>
> (ii) *Suppose in (i) that f is integrable and the function $f_x\colon [c,d] \to \mathbb{R}$ defined by $f_x(y) = f(x,y)$ is integrable for each fixed $x \in [a,b]$. Then*
>
> $$\int_A f = \int_a^b \left(\int_c^d f(x,y)\, dy \right) dx \ .$$
>
> *One can similarly assume that*
>
> $$\int_a^b f(x,y)\, dx$$

exists for each y and obtain

$$\int_A f = \int_c^d \left(\int_a^b f(x,y)\, dx \right) dy.$$

As usual, we can apply this to a non-square region A by extending f to be zero outside A and applying the above to a containing rectangle. Examples are given below.

To be able to drop the assumptions of continuity or existence of the iterated integrals and replace them by just integrability of f is, unfortunately, not possible. To obtain such a general result the student will have to wait for more advanced courses in measure theory. However in actual practice, the above theorem is completely adequate. As mentioned in Section 9.1, the result is, intuitively, entirely reasonable.

The following corollary is a typical application of Theorem 1. This corollary can often be used effectively by breaking up a complicated region into smaller regions to each of which the corollary applies.

Corollary 1.   *Let* $\varphi, \psi \colon [a,b] \to \mathbb{R}$ *be continuous maps such that* $\varphi(x) \leqslant \psi(x)$ *for all* $x \in [a,b]$ *and let* $A = \{(x,y) \mid a \leqslant x \leqslant b,\ \varphi(x) \leqslant y \leqslant \psi(x)\}$. *Let* $f \colon A \to \mathbb{R}$ *be continuous. Then*

$$\int_A f = \int_a^b \left(\int_{\varphi(x)}^{\psi(x)} f(x,y)\, dy \right) dx$$

(see Figure 9-3).

There is an entirely analogous theorem with the roles of x and y interchanged. The corollary is an immediate consequence of the theorem if we remember that f is extended to be zero outside A. Theorem 1 and Corollary 1 are easily extended to multiple integrals, as shown in Theorem 2.

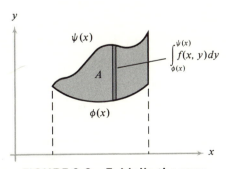

FIGURE 9-3   Fubini's theorem.

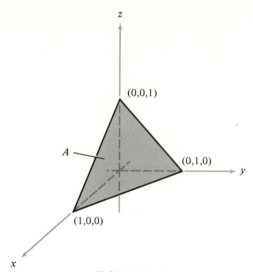

FIGURE 9-4

Theorem 2. (*i*) *Let* $A \subset \mathbb{R}^n$ *and* $B \subset \mathbb{R}^m$ *be rectangles and let* $f: A \times B \subset \mathbb{R}^n \times \mathbb{R}^m \to \mathbb{R}$ *be continuous. Define, for each* $x \in A$, $f_x: B \subset \mathbb{R}^m \to \mathbb{R}$ *by* $f_x(y) = f(x,y)$. *Then*

$$\int_{A \times B} f = \int_A \left(\int_B f(x,y) \, dy \right) dx = \int_A \left(\int_B f_x(y) \, dy \right) dx.$$

(*ii*) *If* f *is integrable and* f_x *is integrable for each fixed* x, *then again*

$$\int_{A \times B} f = \int_A \left(\int_B f(x,y) \, dy \right) dx \,.$$

Similarly, if $\int_A f(x,y) \, dx$ *exists for each* y, *then*

$$\int_{A \times B} f = \int_B \left(\int_A f(x,y) \, dx \right) dy \,.$$

In practice, this theorem may be used repeatedly to reduce a problem to iterated one-dimensional integrals.

EXAMPLE 1.   Evaluate

$$\int_A (x + y + z)^2 \, dx \, dy \, dz$$

where A is the three-dimensional volume sketched in Figure 9-4.

Solution: We proceed in the following manner. Here A is simply the set $\{(x,y,z) \in \mathbb{R}^3 \mid x \geq 0,\, y \geq 0,\, z \geq 0 \text{ and } x + y + z \leq 1\}$. Hence A consists of those points (x,y,z) for which $x \geqslant 0$, $y \geqslant 0$, $x + y \leqslant 1$, and $0 \leqslant z \leqslant 1 - (x + y)$. Let $B = \{(x,y) \mid x \geqslant 0,\ y \geqslant 0 \text{ and } x + y \leqslant 1\}$. Then by Theorem 2, and remembering that f is zero outside A,

$$\int_A (x + y + z)^2\, dx\, dy\, dz = \int_B \left(\int_0^{1-(x+y)} (x + y + z)^2\, dz \right) dx\, dy .$$

Similarly, B consists of those points (x,y) for which $x \in [0,1]$ and $y \in [0,1-x]$, so

$$\int_A (x + y + z)^2\, dx\, dy\, dz = \int_0^1 \int_0^{1-x} \int_0^{1-(x+y)} (x + y + z)^2\, dz\, dy\, dx .$$

Now use the fundamental theorem to evaluate these integrals. First note that

$$\frac{\partial(x + y + z)^3/3}{\partial z} = (x + y + z)^2 ,$$

so performing the z integration yields

$$\int_A (x + y + z)^2\, dx\, dy\, dz = \int_0^1 \int_0^{1-x} \left(\frac{(x + y + 1 - (x + y))^3}{3} \right.$$
$$\left. - \frac{(x + y + 0)^3}{3} \right) dy\, dx$$
$$= \int_0^1 \int_0^{1-x} \left(\frac{1}{3} - \frac{(x + y)^3}{3} \right) dy\, dx .$$

Again, using the fundamental theorem, the integration yields

$$= \int_0^1 \left(\frac{(1 - x)}{3} - \frac{(x + (1 - x))^4}{12} + \frac{(x)^4}{12} \right) dx$$
$$= \int_0^1 \left(\frac{1}{3} - \frac{x}{3} - \frac{1}{12} + \frac{x^4}{12} \right) dx$$
$$= \frac{1}{4} - \frac{1}{6} + \frac{1}{60} = \frac{15 - 10 + 1}{60} = \frac{1}{10} .$$

For improper integrals it is generally sufficient to apply the theorem first on a bounded region and then take limits. See Example 1, at the end of the chapter.

EXAMPLE 2. In the following integral change the order of integration and evaluate: $\int_0^1 \int_x^1 xy\, dy\, dx$.

Solution: The region in question is shown in Figure 9-5 (see Corollary 1). In the reverse order, we get

$$\int_0^1 \int_0^y xy\, dx\, dy = \int_0^1 \frac{y^3}{2}\, dy = \frac{1}{8} .$$

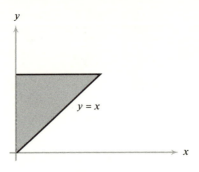

FIGURE 9-5

Exercises for Section 9.2

1. Show that the volume of region A in Example 1 is $1/6$.

2. Draw the region corresponding to the integral $\int_0^1 \int_1^{e^x} (x + y)\, dy\, dx$ and evaluate.

3. Interchange the order of integration in Exercise 2 and check that the answer is unaltered.

4. Let A be the region in $\mathbb{R}^3$ bounded by the planes $x = 0$, $y = 0$, $z = 2$ and the surface $z = x^2 + y^2$. Show that

$$\int_A x\, dx\, dy\, dz = 8\frac{\sqrt{2}}{15}.$$

9.3   Change of Variables Theorem

Next we turn to a rigorous statement of the change of variables formula for multiple integrals.

> **Theorem 3.**   *Let $A \subset \mathbb{R}^n$ be an open bounded set with volume and let $g: A \to \mathbb{R}^n$ be a C^1 mapping which is one-to-one, $Jg(x) \neq 0$ for all $x \in A$ and $|Jg(x)|$, $1/|Jg(x)|$ are bounded on A. Let $B = g(A)$ and assume B has volume. For $f: B \to \mathbb{R}$ bounded and integrable, $f \circ g\, |J(g)|$ is integrable on A and*
>
> $$\int_B f = \int_A (f \circ g)\, |Jg|,$$
>
> *that is,*
>
> $$\int_B f(y_1, \ldots, y_n)\, dy_1 \cdots dy_n = \int_A f(g(x_1, \ldots, x_n)) \frac{\partial(g_1, \ldots, g_n)}{\partial(x_1, \ldots, x_n)}\, dx_1 \cdots dx_n.$$

The proof of this theorem requires some subtle manipulations but we can give a fairly simple intuitive proof as follows.

To begin, let us suppose we isolate a small rectangle S in A. Then g is approximately affine near S, so $g(S)$ is approximately a parallelepiped. See Figure 9-6. If g were affine, the volume of $g(S)$ would be $|(\det \tilde{g})| \, v(S)$ where $\tilde{g}$ is the linear part of g. However, $g(x_0) + Dg(x_0)$ approximates g well near x_0 and is affine, so $v(g(S)) \approx |Jg| \, v(S)$. Thus

$$f(g(x)) \, |Jg(x)| \, dx \approx f(y) \, dy$$

where $y = g(x)$, and so "adding" these infinitesimal quantities gives the result.

EXAMPLE 1.   What is the volume of the parallelepiped spanned by the vectors $(1,1,1)$, $(2,3,1)$, $(0,1,1)$ in $\mathbb{R}^3$?

Solution: These vectors are the images of the standard basis under the linear mapping with matrix

$$\begin{pmatrix} 1 & 2 & 0 \\ 1 & 3 & 1 \\ 1 & 1 & 1 \end{pmatrix}.$$

By Theorem 3, the volume of the image set B is $|\det g|$ times one, the volume of A (the unit cube in $\mathbb{R}^3$). This determinant is easily seen to be 2, so the volume required is 2.

EXAMPLE 2.   Evaluate $\int_0^1 \int_0^1 (x^4 - y^4) \, dx \, dy$ using the change of variables $u = x^2 - y^2$, $v = 2xy$.

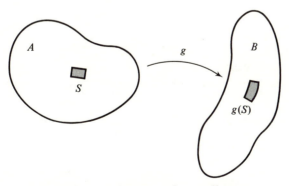

FIGURE 9-6   Change of coordinates.

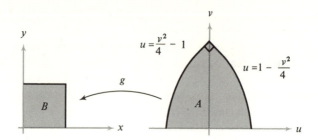

FIGURE 9-7

Solution: Take $B =]0,1[\times]0,1[$ and let g be the map taking (u,v) to the corresponding (x,y), so $g^{-1}(x,y) = (x^2 - y^2, 2xy)$. The region A corresponding to B is shown in Figure 9-7. We can check that $g: A \to B$ is one-to-one and

$$Jg(u,v) = \frac{1}{Jg^{-1}(x,y)} = \frac{1}{4(x^2 + y^2)}$$

since

$$\frac{\partial(u,v)}{\partial(x,y)} = \begin{vmatrix} 2x & -2y \\ 2y & 2x \end{vmatrix} = 4(x^2 + y^2),$$

which is non-zero. Thus by Theorem 3 our integral is*

$$\int_A u \cdot (x^2 + y^2) \frac{du\, dv}{4(x^2 + y^2)} = \frac{1}{4} \int_A u\, du\, dv.$$

Each integral can also be evaluated directly:

$$\int_0^1 \int_0^1 (x^4 - y^4)\, dx\, dy = \int_0^1 \left(\frac{1}{5} - y^4 \right) dy = 0.$$

Similarly

$$\int_A u\, du\, dv = \int_0^2 \int_{(v^2/4)-1}^{1-(v^2/4)} u\, du\, dv = 0.$$

Exercises for Section 9.3

1. Show that Theorem 3 contains the one variable theorem discussed in Section 9.1 as a special case.

2. What is the volume of the parallelepiped spanned by $(1,1,1,1)$, $(0,1,1,0)$, $(2,0,3,0)$, $(1,1,0,1)$ in $\mathbb{R}^4$?

---

* Strictly speaking, one should apply Theorem 3 only to the integral $\int_\varepsilon^1 \int_\varepsilon^1 (x^4 - y^4)\, dx\, dy$ and then let $\varepsilon \to 0$ in order to make $Jg(u,v)$ bounded (a refinement shows that this assumption is actually not necessary).

3. Transform $\int_0^1 \int_y^1 (x^2 + y^2)\, dx\, dy$ using the change of variables $x = u + v, y = u - v$.

4. Show that the volume of the parallelepiped spanned by vectors $v_1, \ldots, v_n$ in $\mathbb{R}^n$ is given by $|\det A_{ij}|^{1/2}$ where $A_{ij} = \langle v_i, v_j \rangle$.

5. Show that the area of the ellipse $x^2/a^2 + y^2/b^2 \leq 1$ is πab by making a change of variables and reducing the problem to one of finding the area of a circle.

9.4   Polar Coordinates

One standard application of the change of variables formula is to the evaluation of integrals using polar coordinates. The function which changes from polar coordinates to the standard rectangular coordinates is $g(r,\theta) = (r \cos \theta, r \sin \theta)$.

$$Jg(r,\theta) = \begin{vmatrix} \cos \theta & -r \sin \theta \\ \sin \theta & r \cos \theta \end{vmatrix} = r \cos^2 \theta + r \sin^2 \theta = r.$$

If we consider g to be defined on the set $\{(r,\theta) \mid r > 0, 0 < \theta < 2\pi\}$, then $Jg(r,\theta)$ is never zero and g is one-to-one on this set. We leave to the student the verification that g is one-to-one. Although the image of this function excludes the set of points on the x-axis with $x \geq 0$, this is a set of measure zero and therefore does not contribute to the value of an integral. (See Theorem 8.4, and Figure 9-8.)

EXAMPLE 1.   Consider a thin plate in the shape of an annulus with inner radius 1, outer radius 2, and mass density equal to $1/r^3$ at all points a distance r from the center. Compute the total mass. See Figure 9-9. If we let B denote the annulus, then B is the image (except for points on the positive x-axis) under the polar coordinate map g of $A = \{(r,\theta) \mid 0 < \theta < 2\pi, 1 < r < 2\}$.

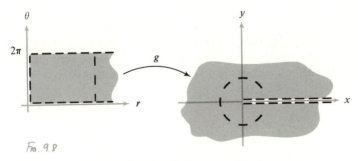

Fig. 9.8

FIGURE 9-8   Polar coordinates.

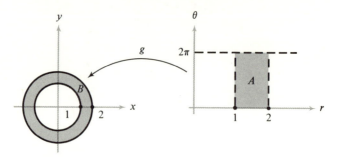

FIGURE 9-9

Hence the required mass is given by

$$\int_B (x^2 + y^2)^{-3/2}\, dx\, dy = \int_A (1/r^3)\, |Jg|\, dr\, d\theta = \int_A 1/r^2\, dr\, d\theta$$

$$= \int_0^{2\pi} \int_1^2 1/r^2\, dr\, d\theta = \int_0^{2\pi} (-1/2 + 1)\, d\theta = \pi.$$

Exercises for Section 9.4

Evaluate the integrals in Exercises 1 and 2 using polar coordinates.

1. $\int_D \exp(x^2 + y^2)\, dx\, dy$ where $D = \{(x,y) \mid x^2 + y^2 < 1\}$.

2. $\int_D \ln(x^2 + y^2)\, dx\, dy$ where $D = \{(x,y) \mid x \geqslant 0, y \geqslant 0, a^2 \leqslant x^2 + y^2 \leqslant b^2\}$.

3. Find the area of a circle of radius r using polar coordinates.

9.5   Spherical Coordinates

The same techniques which were applied to polar coordinates can also be applied to spherical coordinates. Here, let $g(r,\varphi,\theta) = (r \sin \varphi \cdot \cos \theta,\ r \sin \theta \cdot \sin \varphi, r \cos \varphi)$ and consider g to be defined on $\{(r,\varphi,\theta) \mid r > 0,\ 0 < \theta < 2\pi, 0 < \varphi < \pi\}$. See Figure 9-10. The image under g of this set is all of $\mathbb{R}^3$ except for the part of the xz-plane where $x \geqslant 0$. But we know (see Exercise 5, Chapter 8) that this is a set of measure zero and so can be safely neglected in integrals. The Jacobian determinant is given by

$$Jg(r,\varphi,\theta) = \begin{vmatrix} \sin \varphi \cos \theta & r \cos \varphi \cos \theta & -r \sin \varphi \sin \theta \\ \sin \varphi \sin \theta & r \cos \varphi \sin \theta & r \sin \varphi \cos \theta \\ \cos \varphi & -r \sin \varphi & 0 \end{vmatrix}$$

$$= r^2 \sin \varphi.$$

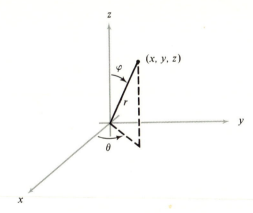

FIGURE 9-10 Spherical coordinates.

Hence $Jg(r,\varphi,\theta) \neq 0$ in the region specified above; g is also one-to-one on the region. Therefore the change of variables formula can be used to give

$$\int_{g(A)} f(x,y,z) \ dx \ dy \ dz$$

$$= \int_A f(r \cos \theta \cdot \sin \varphi, r \sin \theta \cdot \sin \varphi, r \cos \varphi) r^2 \sin \varphi \cdot dr \ d\varphi \ d\theta \ .$$

EXAMPLE 1.   Suppose we are given the function $f(x,y,z) = x^2 + y^2 + z^2$ and we want to integrate it over the set $B = \{(x,y,z) \mid x^2 + y^2 + z^2 < 1\}$. Then B is the image under g of $A = \{(r,\varphi,\theta) \mid 0 < r < 1, 0 < \theta < 2\pi, 0 < \varphi < \pi\}$ (except for the points of B on the xz-plane where $x \geqslant 0$). Hence

$$\int_B (x^2 + y^2 + z^2) \ dx \ dy \ dz = \int_A r^2 \ |Jg| \ dr \ d\varphi \ d\theta$$

$$= \int_A r^2 \cdot r^2 \sin \varphi \ dr \ d\varphi \ d\theta \qquad \text{(since } \sin \varphi > 0 \text{ in the relevant region)}$$

$$= \int_0^{2\pi} \int_0^{\pi} \int_0^1 r^4 \sin \varphi \ dr \ d\varphi \ d\theta$$

$$= \int_0^{2\pi} \int_0^{\pi} \frac{\sin \varphi}{5} \ d\varphi \ d\theta$$

$$= \frac{1}{5} \int_0^{2\pi} 2 \ d\theta = \left(\frac{4}{5}\right)\pi \ .$$

Exercises for Section 9.5

1. Show that $\int_D e^{[(x^2 + y^2 + z^2)^{3/2}]} \, dx \, dy \, dz = 4\pi(e - 1)/3$ where D is the unit ball in $\mathbb{R}^3$.

2. Let D be the unit ball in $\mathbb{R}^3$. Evaluate

$$\int_D \frac{dx \, dy \, dz}{\sqrt{2 + x^2 + y^2 + z^2}}$$

9.6 Cylindrical Coordinates

Cylindrical coordinates are treated much the same way as polar and spherical coordinates. The appropriate mapping is $g(r,\theta,z) = (r \cos \theta, r \sin \theta, z)$ on the set $\{(r,\theta,z) \mid r > 0, 0 < \theta < 2\pi\}$. See Figure 9-11. The Jacobian is $Jg(r,\theta,z) = r$, so the change of variables theorem becomes

$$\int_{g(A)} f(x,y,z) \, dx \, dy \, dz = \int_A f(r \cos \theta, r \sin \theta, z) r \, dr \, d\theta \, dz .$$

This is useful for triple integrals which have "cylindrical symmetry" as opposed to "spherical symmetry" problems for spherical coordinates.

EXAMPLE 1.   Evaluate $\int_R z e^{-x^2 - y^2} \, dx \, dy \, dz$ over the region

$$R = \{(x,y,z) \mid x^2 + y^2 \leqslant 1, 0 \leqslant z \leqslant 1\} .$$

Solution: Here we get

$$\int_{z=0}^{1} \int_{\theta=0}^{2\pi} \int_{r=0}^{1} z e^{-r^2} r \, dr \, d\theta \, dz = \frac{\pi}{2}(1 - e^{-1}) .$$

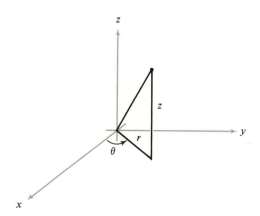

FIGURE 9-11

Exercises for Section 9.6

1. Evaluate $\int_D z \sqrt{x^2 + y^2}\, dx\, dy\, dz$ where $D = \{(x,y,z)\,|\,1 \leqslant x^2 + y^2 \leqslant 2, 1 \leqslant z \leqslant 2\}$.

2. Work out $\displaystyle\int_0^1 \int_{-1}^1 \int_{-\sqrt{1-y^2}}^{\sqrt{1-y^2}} z(x^2 + y^2)\, dx\, dy\, dz$ by using cylindrical coordinates.

3. Make a change of variables to evaluate $\int_D \exp\!\left(\dfrac{x + y}{x - y}\right) dx\, dy$ where $D = \{(x,y)\,|\,0 \leqslant y \leqslant x, 0 \leqslant x \leqslant 1\}$.

9.7 A Note on the Lebesgue Integral

Several times in this and the previous chapter, we have hinted of the existence of another theory of integration. There is such a theory called the Lebesgue theory. We shall now discuss the need for this theory and its basic underlying differences from the Riemann theory.

The need for the Lebesgue integral is largely a technical one. Namely, some functions might not be Riemann integrable and we might wish to integrate them anyway. For example, such a function can occur as a limit of Riemann integrable functions (a uniform limit of Riemann integrable functions is Riemann integrable, but a pointwise limit need not be). It is desireable, however, to work with "complete" spaces which contain limits of Cauchy sequences; for example $\mathbb{R}^n$ in Chapter 2 and $\mathscr{C}(A,\mathbb{R})$ in Chapter 5. In the next chapter on Fourier series we shall see a useful space of functions with the property that convergence in this space may be pointwise but not necessarily uniform. The Riemann theory is not sufficient to integrate such limit functions.

Henri Lebesgue's problem was to find a more general theory of integration than the Riemann theory and which, moreover was more useful technically. Actually, this is a simplification—the factual history is more complicated. For example, his work first received prominence when he used his ideas to give a characterization of Riemann integrable functions (see Theorem 3, Chapter 8).*

Here we can give only the briefest glimpse of the ideas. To develop them fully requires a course in itself (see for example H. L. Royden, *Real Analysis*).

The basis of Lebesgue's theory is as follows. For the Riemann integral, the area under the graph of a function was divided into vertical rectangles in order to define the integral. But why could it not be divided into horizontal rectangles just as well? See the illustration in Figure 9-12.

Intuitively this type of subdivision, where we take a partition on the y-axis rather than the x-axis, ought to yield the same area. But in fact, there is a

---

* For historical details see, for example, M. Kline, *Mathematical Thought From Ancient to Modern Times*, Oxford (1972).

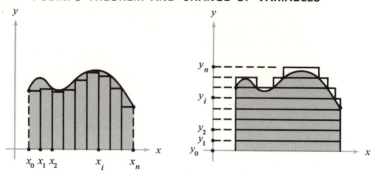

FIGURE 9-12 (a) Riemann approximations.
(b) Lebesgue approximations.

technical difference and Lebesgue's idea (horizontal rectangles) is what we want. This also leads to more complicated mathematics and let us now see why.

Let $f: [a,b] \to \mathbb{R}$ and suppose f is bounded and ≥ 0. Let $y_0 < y_1 < \cdots < y_n$ be a partition of the range of f. Our candidate for an approximating sum is (look at Figure 9-12 to see this):

$$\sum (y_{i+1} - y_i) \cdot (\text{length of the "interval"} \ \{x \mid f(x) \geq y_i\}) .$$

The technical point is that $\{x \mid f(x) \geq y_i\} = I_i$ might be a complicated set; after all the Riemann theory can already handle "decent" functions, so we have to be prepared to handle fairly complex ones. Thus our problem is that if I_i is a complicated set, how do we compute its length? This is where a main part of the Lebesgue theory comes in. One first has to develop the idea of the measure (or length) of a set. This turns out to be a bit complicated, but once this *measure theory* is developed, one can then proceed with the study of the Lebesgue approximating sums given above. Note that we already have seen one aspect of "measure" in Chapter 8 when we studied sets of measure zero. This concept is taken directly from the Lebesgue theory.

The conclusion is, therefore, that with considerably more effort, a more comprehensive theory of integration (which includes the Riemann theory as a special case) is possible. In more advanced areas of mathematics, the technical rewards are more than worth the extra effort involved.

Theorem Proofs for Chapter 9

Theorem 1. (i) *Let A be the rectangle described by $a \leq x \leq b$, $c \leq y \leq d$, and let $f: A \to \mathbb{R}$ be continuous. Then*

$$\int_A f = \int_a^b \left(\int_c^d f(x,y) \, dy \right) dx$$

$$= \int_c^d \left(\int_a^b f(x,y) \, dx \right) dy .$$

(*ii*) *Suppose in* (*i*) *that f is integrable and the function* $f_x : [c,d] \to \mathbb{R}$ *defined by* $f_x(y) = f(x,y)$ *is integrable for each fixed* $x \in [a,b]$. *Then*

$$\int_A f = \int_a^b \left(\int_c^d f(x,y)\, dy \right) dx .$$

One can similarly assume that

$$\int_a^b f(x,y)\, dx$$

exists for each y and obtain

$$\int_A f = \int_c^d \left(\int_a^b f(x,y)\, dx \right) dy .$$

Proof: As (i) is a special case of (ii), we only need to prove (ii). Let $g : [a,b] \subset \mathbb{R} \to \mathbb{R}$ be the function defined by

$$g(x) = \int_c^d f(x,y)\, dy .$$

We must show that g is integrable over $[a,b]$ and that

$$\int_A f = \int_a^b g(x)\, dx .$$

Suppose $a = x_0 < x_1 < \cdots < x_n = b$ and $c = y_0 < y_1 < \cdots < y_n = d$ are partitions of the intervals $[a,b]$ and $[c,d]$. Denote by $P_{[a,b]}$ the partition of $[a,b]$ given by the sets $V_i = [x_{i-1},x_i]$, denote by $P_{[c,d]}$ the partition of $[c,d]$ given by the sets $W_j = [y_{j-1},y_j]$, and denote by P_A the partition of A given by the sets

$$S_{ij} = [x_{i-1},x_i] \times [y_{j-1},y_j] .$$

Then

$$L(f,P_A) = \sum_{ij} m_{S_{ij}}(f)v(S_{ij})$$

$$= \sum_i \left(\sum_j m_{S_{ij}}(f)v(W_j) \right) v(V_i)$$

where $m_S(f)$ is the minimum (inf) of f on the set S. For $x \in V_i$ we have $m_{S_{ij}}(f) \leqslant m_{W_j}(f_x)$, where f_x is defined by $f_x(y) = f(x,y)$. Hence

$$\sum_j m_{S_{ij}}(f)v(W_j) \leqslant \sum_j m_{W_j}(f_x)v(W_j) \leqslant \int_c^d f_x(y)\, dy = g(x) .$$

As this inequality holds for any $x \in V_i$, we get

$$\sum_j m_{S_{ij}}(f)v(W_j) \leqslant m_{V_i}(g) .$$

Therefore

$$L(f,P_A) \leqslant \sum_i m_{V_i}(g)v(V_i) \leqslant L(g,P_{[a,b]}) .$$

From this and from a similar argument for upper bounds we obtain the inequalities

$$L(f,P_A) \leqslant L(g,P_{[a,b]}) \leqslant U(g,P_{[a,b]}) \leqslant U(f,P_A) .$$

Since f is integrable over A, the above inequalities show that g is integrable and

$$\int_a^b \left(\int_c^d f(x,y)\, dy \right) dx = \int_a^b g(x)\, dx = \int_A f \; .$$

Using the same argument as above, we can show that if we assume $\int_a^b f(x,y)\, dx$ exists for each y, we get

$$\int_A f = \int_c^d \left(\int_a^b f(x,y)\, dx \right) dy \; . \quad \blacksquare$$

Corollary 1. *Let* φ, $\psi\colon [a,b] \to \mathbb{R}$ *be continuous maps such that* $\varphi(x) \leqslant \psi(x)$ *for all* $x \in [a,b]$ *and let* $A = \{(x,y) \mid a \leqslant x \leqslant b, \varphi(x) \leqslant y \leqslant \psi(x)\}$. *Let* $f\colon A \to \mathbb{R}$ *be continuous or piecewise continuous. Then*

$$\int_A f = \int_a^b \left(\int_{\varphi(x)}^{\psi(x)} f(x,y)\, dy \right) dx \; .$$

Proof: Let $S = [a,b] \times [c,d]$ be a closed rectangle enclosing A and extend f to S by setting it equal to 0 on $S \backslash A$. The two sets graph$(\varphi) = \{(x,\varphi(x)) \mid x \in [a,b]\}$ and graph$(\psi) = \{(x,\psi(x)) \mid x \in [a,b]\}$ are '(by Exercise 23, Chapter 8) of measure zero. Thus the set of discontinuities of f defined on S is of measure zero, and thus f is integrable over S. Also for any x, f_x is continuous on $[c,d]$, except possibly at $\varphi(x)$ and $\psi(x)$, and so f_x is integrable for any $x \in [a,b]$. Thus we may apply Theorem 1 to get

$$\int_A f = \int_S f = \int_a^b \int_c^d f_x(y)\, dy\, dx = \int_a^b \left(\int_{\varphi(x)}^{\psi(x)} f_x(y)\, dy \right) dx \; . \quad \blacksquare$$

The proof of Theorem 2 is entirely analogous to the above proof, and so it will be left as an exercise.

We therefore turn to Theorem 3. The proof of this can be very laborious if not dealt with effectively. In particular the idea given in the text, p. 307 is hard to make precise if we interpret it too literally. The proof we give here is due to J. Schwartz in the "American Mathematical Monthly," *61* (1954) 81–85.

Theorem 3. *Let* $A \subset \mathbb{R}^n$ *be an open bounded set with volume and let* $g\colon A \to \mathbb{R}^n$ *be a* C^1 *mapping which is one-to-one,* $Jg(x) \neq 0$ *for all* $x \in A$ *and* $|Jg(x)|$, $1/|Jg(x)|$ *are bounded on* A. *Let* $B = g(A)$ *(an open set by the inverse function theorem) and assume* B *has volume. For* $f\colon B \to R$ *bounded and integrable,* $f \circ g\, |J(g)|$ *is integrable on* A *and*

$$\int_B f = \int_A (f \circ g)\, |Jg|$$

that is

$$\int_B f(y_1,\ldots,y_n)\, dy_1 \cdots dy_n = \int_A f(g(x_1,\ldots,x_n)) \frac{\partial(g_1,\ldots,g_n)}{\partial(x_1,\ldots,x_n)}\, dx_1,\ldots,dx_n \; .$$

Note: By a careful analysis of the proof one can show that f and $|Jg|$, $|Jg|^{-1}$ need not be assumed bounded (see Section 8.7 for improper integrals and see also the remark on p. 326).

The first stage of the proof consists of establishing our formula when $g = L$ is a linear map, in which case $JL = \det L$. This yields the geometric interpretation of $\det L$: *it is the factor by which volumes are changed under the transformation L.*

Since we do not want to assume this from linear algebra, we will go through the proof in some detail. We do need, however, to recall these two standard facts from linear algebra: (i) $\det TS = \det T \cdot \det S$, and, (ii) any matrix is a product of elementary matrices (see M. O'Nan, *Linear Algebra*, pages 91 and 241).

Lemma 1. *If $L: \mathbb{R}^n \to \mathbb{R}^n$ is a linear map and $A \subset \mathbb{R}^n$ is a set which has volume (that is, $\int_A 1_A$ exists), then the volume of $L(A)$ is $|\det L| \cdot v(A)$ i.e. $\int_{L(A)} 1_A = \int_A |\det L|$. (See Figure 9-13.)*

Proof: We will first consider the special case where A is a rectangle and L is a linear map whose matrix in terms of the standard basis is of one of the following two types:

$$
L_1 = \begin{pmatrix} 1 & 0 & & \cdots & & & 0 \\ 0 & 1 & & & & & \\ & & \ddots & & & & \\ & & & 1 & & & \\ \cdot & & & c & & & \\ & & & & \ddots & & \\ & & & & & 1 & 0 \\ 0 & & & \cdots & & 0 & 1 \end{pmatrix}
\quad \text{or} \quad
L_2 = \begin{pmatrix} 1 & 0 & & \cdots & & & 0 \\ 0 & 1 & & & & & 1 \\ & & \ddots & & & & \\ \cdot & & & 1 & & & \cdot \\ \cdot & & & & 1 & & \cdot \\ & & & & & \ddots & \\ & & & & & & 1 & 0 \\ 0 & & & \cdots & & & 0 & 1 \end{pmatrix}
$$

(The first matrix is obtained from the identity matrix by replacing a single diagonal 1 by a constant c. The second matrix is obtained from the identity matrix by writing a single one anywhere off the diagonal. These are called the elementary matrices.)

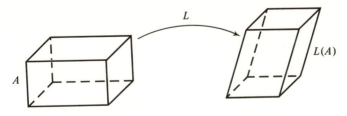

FIGURE 9-13   Image of a rectangle under a linear map.

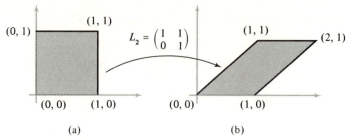

FIGURE 9-14 Image of a rectangle under an elementary matrix L_2.

If $A = [a_1,b_1] \times \cdots \times [a_n,b_n]$ and c is in the ith row, then

$$L_1(A) = [a_1,b_1] \times \cdots \times [ca_i,cb_i] \times \cdots \times [a_n,b_n],$$

so that obviously the volume of $L_1(A) = v(L_1(A)) = |c|\, v(A) = |\det L_1|\, v(A)$.

Now if the 1 off the diagonal is in the (i,j) position, then

$$L_2(A) = \{(x_1,\ldots,x_{i-1},x_i + x_j,x_{i+1},\ldots,x_n) \mid x_k \in [a_k,b_k]\}$$

for $1 \leqslant k \leqslant n$. (See Figure 9-14.) It is clear from the illustration that the volume in Figure 9-14b is the same as in Figure 9-14a, because they both have the same base and the same height. This fact can be verified analytically as follows. The set $L_2(A)$ can be broken up into three regions:

$$\{(x_1,\ldots,x_{i-1},x_i + x_j,x_{i+1},\ldots,x_n) \mid x_k \in [a_k,b_k]$$
$$\text{for } 1 \leqslant k \leqslant n \quad \text{and} \quad a_i + a_j \leqslant x_i + x_j \leqslant a_i + b_j\}$$

$$\{(x_1,\ldots,x_{i-1},x_i + x_j,x_{i+1},\ldots,x_n) \mid x_k \in [a_k,b_k]$$
$$\text{for } 1 \leqslant k \leqslant n \quad \text{and} \quad a_i + b_j \leqslant x_i + x_j \leqslant a_j + b_i\}$$

$$\{(x_i,\ldots,x_{i+1},x_i + x_j,x_{i+1},\ldots,x_n) \mid x_k \in [a_k,b_k]$$
$$\text{for } 1 \leqslant k \leqslant n \quad \text{and} \quad a_j + b_i \leqslant x_i + x_j \leqslant b_i + b_j\}.$$

By Fubini's theorem, the volume of the first set is

$$(b_1 - a_1) \cdots (b_{i-1} - a_{i-1})(b_{i+1} - a_{i+1}) \cdots (b_{j-1} - a_{j-1})(b_{j+1} - a_{j+1}) \cdots$$
$$(b_n - a_n) \int_{a_i+a_j}^{a_i+b_j} \left(\int_{a_j}^{a_j+a_i+b_j-x_i} 1\, dx_j \right) dx_i.$$

Now by the fundamental theorem

$$\int_{a_i+a_j}^{a_i+b_j} \left(\int_{a_j}^{a_j+a_i+b_j-x_i} 1\, dx_j \right) dx_i = \int_{a_i+a_j}^{a_i+b_j} (a_i + b_j - x_i)\, dx_i$$

$$= (a_i + b_j)(b_j - a_j) - \tfrac{1}{2}(a_i + b_j)^2 - \tfrac{1}{2}(a_i + a_j)^2$$

$$= \tfrac{1}{2}(b_j - a_j)^2.$$

Thus the volume of the first set is

$$(b_1 - a_1) \cdots (b_{i-1} - a_{i-1})(b_{i+1} - a_{i+1}) \cdots (b_{j-1} - a_{j-1})$$
$$(b_{j+1} - a_{j+1}) \cdots (b_n - a_n)\tfrac{1}{2}(b_j - a_j)^2 .$$

We can perform a similar integration over the third set to find that its volume is the same as that of the first. The second set is a rectangle with volume

$$(b_1 - a_1) \cdots (b_{i-1} - a_{i-1})(b_{i+1} - a_{i+1}) \cdots (b_{j-1} - a_{j-1})(b_{j+1} - a_{j+1}) \cdots$$
$$(b_n - a_n)(a_j + b_i - a_i - b_j)(b_j - a_j) .$$

The volume of $L_2(A)$ = sum of the above three volumes = $(b_1 - a_1) \cdots (b_n - a_n)$, that is, the same as that of the rectangle $[a_1, b_1] \times \cdots \times [a_n, b_n]$. Thus we have $v(L_2(A)) = |\det L_2|\, v(A)$, since $\det L_2 = 1$.

Now let A be an arbitrary set with volume and let L_i be one of the elementary matrices. Assume that $\det L_i \neq 0$ (that is, $c \neq 0$ if L_i is an elementary matrix of the first type). Let S be a rectangle enclosing A and let P be a partition of A into subrectangles $S_1, \ldots, S_N$, such that

$$U(1_A, P) - v(A) < \varepsilon(2\,|\det L_i|) \qquad \text{and} \qquad v(A) - L_i(1_A, P) < \varepsilon(2\,|\det L|) .$$

Then if we consider the sets $V_\varepsilon = \cup\, \{S_i \mid S_i \subset A\}$ and $W_\varepsilon = \cup\, \{S_i \mid S_i \cap A \neq \varnothing\}$, we have the result for rectangles that $v(L_i(V_\varepsilon)) = |\det L_i|\, L_i(1_A, P)$ and $v(L_i(W_\varepsilon)) = |\det L_i|\, U(1_A, P)$. So $v(L_i(W_\varepsilon)) - v(L_i(V_\varepsilon)) < \varepsilon$, and thus $L_i(A)$ has volume and $v(L_i(A)) = |\det L_i|\, v(A)$. If $\det L_i = 0$, that is L_i is a matrix of the first kind with $c = 0$, then $v(L_i(S)) = 0$ for any rectangles S, so $v(L_i(A)) = 0$ for any set A with volume.

Now let L be any linear map and let A be any set with volume. From fact (ii) mentioned earlier, we can write $L = L_1 L_2 \cdots L_k$, where the L_i are elementary matrices. By repeated application of what we have proved above, it follows that $L(A)$ has volume and

$$v(L(A)) = |\det L_1|\, |\det L_2| \cdots |\det L_n|\, v(A) = |\det L|\, v(A) . \quad \blacksquare$$

Lemma 2. *If the theorem is true for the function $f = 1$, then it is true for any integrable f.*

Proof: If the theorem is true for $f = 1$, it is true for any constant function (why?). Now let f be any integrable function on $g(A)$. Let S be a rectangle enclosing $g(A)$ and let P be a partition of $g(A)$ into rectangles $S_1, \ldots, S_N$. Recall that we define $m_{S_i}(f) = \inf\{f(x) \mid x \in S_i\}$. Denote by $m_{S_i}(f)$ the constant function on S_i with constant value $m_{S_i}(f)$. Then

$$L(f, P) = \sum_{i=1}^{N} m_{S_i}(f) v(S_i) = \sum_{i=1}^{N} \int_{S_i} m_{S_i}(f)$$

$$= \sum_{i=1}^{N} \int_{g^{-1}(S_i)} (m_{S_i}(f) \circ g)\, |Jg| \leqslant \sum_{i=1}^{N} \int_{g^{-1}(S_i)} (f \circ g)\, |Jg|$$

$$= \int_{g^{-1}(S)} (f \circ g)\, |Jg| = \int_A (f \circ g)\, |Jg| .$$

Hence

$$\int_{g(A)} f \leqslant \int_A (f \circ g)\, |Jg| .$$

A similar argument with $M_{S_i}(f) = \sup\{f(x) \mid x \in S_i\}$ shows that

$$\int_{g(A)} f \geq \int_A (f \circ g) \, |Jg|$$

so that

$$\int_{g(A)} f = \int_A (f \circ g) \, |Jg| \, . \quad \blacksquare$$

Lemma 3.   *The theorem is true if g is a linear transformation.*

Proof: By Lemma 1,

$$\int_{g(A)} 1 = \int_A |\det g| = \int_A |Jg| \, ,$$

since $g = Dg$. By Lemma 2,

$$\int_{g(A)} f = \int_A (f \circ g) \, |Jg| \, . \quad \blacksquare$$

One more key observation is given in Lemma 4.

Lemma 4.   *If the theorem is true for g: $A \to \mathbb{R}^n$ and for h: $B \to \mathbb{R}^n$, where $g(A) \subset B$, then the theorem holds for $h \circ g$: $A \to \mathbb{R}^n$.*

Proof: Make the following computation:

$$\int_{h \circ g(A)} f = \int_{h(g(A))} f = \int_{g(A)} (f \circ h) \, |Jh|$$

$$= \int_A (f \circ h \circ g)(|Jh| \circ g) \, |Jg|$$

$$= \int_A (f \circ (h \circ g)) \, |J(h \circ g)| \, . \quad \blacksquare$$

Some special notations will enable our proof to run more smoothly. If $x \in \mathbb{R}^n$, so that $x = (x_1, \ldots, x_n)$, we put $|x| = \max_{1 \leq i \leq n} |x_i|$. This "norm" has the convenient property that in terms of it, a cube with center p and side of length $2s$ can be characterized by the restriction $|x - p| \leq s$. If $A \colon \mathbb{R}^n \to \mathbb{R}^n$ is the linear transformation represented by the matrix a_{ij}, so that

$$A(x) = A(x_1, \ldots, x_n) = \left(\sum_{j=1}^n a_{1j} x_j, \ldots, \sum_{j=1}^n a_{nj} x_j \right),$$

we put

$$|A| = \max_{1 \leq i \leq n} \sum_{j=1}^n |a_{ij}| \, .$$

Thus, $|A(x)| \leq |A| \, |x|$. We also introduce the Jacobian matrix $j(x) = (j_{ik}(x))$ of the transformation $g(x) = (g_1(x) \cdots g_n(x))$ by putting

$$j_{ik}(x) = \frac{\partial g_i(x)}{\partial x_k}.$$

If C is a cube in the open set A, such that C is the set of all x characterized by a condition of the form $|x - p| \leqslant s$, then $v(C) = (2s)^n$. We have, by the mean-value theorem,

$$g_i(x) - g_i(p) = \sum_{k=1}^n j_{ik}[p + \theta_i(x)(x - p)](x_k - p_k),$$

where $0 \leqslant \theta_i(x) \leqslant 1$. It follows immediately that

$$|g(x) - g(p)| \leqslant s \max_{y \in C} |j(y)| ;$$

that is, $g(C)$ is entirely contained in the cube defined by

$$|z - g(p)| \leqslant s \max_{y \in C} |j(y)| ,$$

so we see that if $g(C)$ has volume, then

$$v(g(C)) \leqslant \left\{ \max_{y \in C} |j(y)|^n v(C) \right\}. \tag{1}$$

To ensure that $g(C)$ has volume, we prove another lemma.

Lemma 5. *If $h: U \subset \mathbb{R}^n \to \mathbb{R}^n$ is a C^1 map which is one-to-one, $Jh(x) \neq 0$, U is a bounded open set, and C is a set with volume such that $\mathrm{cl}(C) \subset U$, then $h(C)$ has volume.*

Proof: It is sufficient to show that the boundary $\mathrm{bd}(h(C))$ has content zero. First show that $\mathrm{bd}(h(C)) \subset h(\mathrm{bd}(C))$. Indeed, let $x \in \mathrm{bd}(h(C))$. Then to show $x \in h(\mathrm{bd}(C))$, let $y = h^{-1}(x)$. Then we must show $y \in \mathrm{bd}(C)$. Let V be a neighborhood of y, and suppose $V \subset U$. Then $h(V)$ is an open neighborhood of x, since h^{-1} is continuous. Thus, $h(V)$ contains points of $h(C)$ and $\mathbb{R}^n \backslash h(C)$, since $x \in \mathrm{bd}(h(C))$. Then as h is one-to-one, V contains points of A and $\mathbb{R}^n \backslash C$, so $y \in \mathrm{bd}(C)$. Applying this argument to h^{-1}, we see that in fact $\mathrm{bd}(h(C)) = h(\mathrm{bd}(C))$.

To show that $h(\mathrm{bd}(C))$ has volume zero, given $\varepsilon > 0$, cover $\mathrm{bd}(C)$ with rectangles $B_1, \ldots, B_N$ of total volume $\leqslant \varepsilon$. Equation 1 showed that $h(\mathrm{bd}(C))$ lies in a covering by rectangles with total volume $(\max |Jh(x)|)\varepsilon$. Here the maximum is over $B_1 \cup \cdots \cup B_N$. This shows that $h(\mathrm{bd}(C))$ has volume zero. ∎

By Lemma 3 we see that if A is a linear transformation and S has volume, then

$$v(A^{-1}(S)) = \det(A^{-1})v(S)$$

(take $f = 1$ on $A^{-1}(S)$, $f = 0$ on the complement, and apply Lemma 3). Now in Eq. 1 let $S = g(C)$ which we now know has volume; then since

$$|\det(A^{-1})| \, v(g(C)) \leqslant \left\{ \max_{y \in C} |A^{-1}j(y)| \right\}^n v(C) ,$$

we obtain

$$(g(C)) \leqslant |\det(A)| \left\{ \max_{y \in C} |A^{-1}j(y)| \right\}^n v(C) . \tag{2}$$

Now, let the cube C be subdivided into a finite set $C_1 \cdots C_M$ of non-overlapping cubes with centers $x_1 \cdots x_M$, and suppose that δ is greater than the length of a side of any of them. Apply Eq. 2 to each of $C_1 \cdots C_M$, taking, however, $A = j(x_i)$ in applying Eq. 2 to C_i; then add. This gives

$$v(g(C)) \leqslant \sum_{i=1}^{M} |\det(j(x_i))| \left\{ \max_{y \in C} |j^{-1}(x_i)j(y)| \right\}^n v(C_i) \,.$$

Now, since $j(x)$ is a continuous (matrix-valued) function, $j^{-1}(z)j(y)$ approaches the identity matrix δ_{ij} as z approaches y, and hence

$$\left\{ \max_{y \in C_i} |j^{-1}(x_i)j(y)| \right\}^n \leqslant 1 + \eta(\delta) \,,$$

where $\eta(\delta)$ approaches zero with δ. This gives

$$v(g(C)) \leqslant [1 + \eta(\delta)] \sum_{i=1}^{M} |\det(j(x_i))| \, v(C_i) \,;$$

as δ approaches zero, the sum on the right approaches $\int_C |Jg(x)| \, dx$, and the inequality becomes

$$v(g(C)) \leqslant \int_C |Jg(x)| \, dx \,. \tag{3}$$

If we examine the proof of Lemma 2 more closely and remember $f = 0$ outside B we get from Eq. 3 that

$$\int_{g(A)} f \leqslant \int_A (f \circ g) |Jg| \,. \tag{4}$$

Actually, Eq. 4 is enough for the theorem because Eq. 4 can be applied equally well to g^{-1} to get

$$\int_A (f \circ g) |Jg| \leqslant \int_{g(A)} (f \circ g \circ g^{-1}) |Jg \circ g^{-1}| \cdot |Jg^{-1}|$$

i.e.

$$\int_A f \circ g \, |Jg| \leqslant \int_{g(A)} f \,. \tag{5}$$

Combining Eqs. 4 and 5 gives the theorem. ∎

Worked Examples for Chapter 9

1. Use the change of variables formula and polar coordinates to show that

$$\int_{-\infty}^{\infty} e^{-x^2} \, dx = \sqrt{\pi}$$

(the function e^{-x^2} is called the Gaussian function; see Figure 9-15).

Solution: To use polar coordinates, we want to employ the expression $x^2 + y^2$.

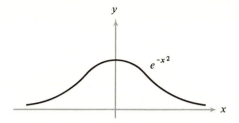

FIGURE 9-15   Gaussian function.

Therefore let us consider

$$I_b = \int_{A_b} e^{-x^2-y^2} \, dx \, dy$$

where A_b is the circle centered at the origin of radius b. Thus

$$I_b = \int_{r=0}^{b} \int_{\theta=0}^{2\pi} e^{-r^2} r \, d\theta \, dr$$

which we evaluate by iterated integrals to obtain

$$I_b = 2\pi \int_{r=0}^{b} r e^{-r^2} \, dr = \pi(1 - e^{-b^2})$$

since

$$\frac{d}{dr}(e^{-r^2}) = -2r e^{-r^2} \, .$$

On the other hand, we want to relate I_b to

$$\int_{-b}^{b} e^{-x^2} \, dx \, .$$

To see how this is accomplished, note that the improper integral

$$\int_{\mathbb{R}^2} e^{-x^2-y^2} \, dx \, dy = \operatorname*{limit}_{b \to \infty} \pi(1 - e^{-b^2}) = \pi \, .$$

Since the integral exists and $e^{-x^2-y^2} \geqslant 0$, we can evaluate the improper integral any way we please (Theorem 7, Chapter 8). Let us evaluate $\int_{\mathbb{R}^2} e^{-x^2-y^2} \, dx \, dy$ using the rectangle $[-b,b]^2 = [-b,b] \times [-b,b]$. Thus

$$\int_{\mathbb{R}^2} e^{-x^2-y^2} \, dx \, dy = \operatorname*{limit}_{b \to \infty} \int_{[-b,b]^2} e^{-x^2-y^2} \, dx \, dy$$

but

$$\int_{[-b,b]^2} e^{-x^2-y^2} \, dx \, dy = \left(\int_{-b}^{b} e^{-x^2} \, dx \right)\left(\int_{-b}^{b} e^{-y^2} \, dy \right) = \left(\int_{-b}^{b} e^{-x^2} \, dx \right)^2$$

by Fubini's theorem, and the fact that $e^{-x^2-y^2} = e^{-x^2}e^{-y^2}$. Hence

$$\operatorname*{limit}_{b \to \infty} \left(\int_{-b}^{b} e^{-x^2} \, dx \right)^2 = \pi \, .$$

Thus, as $e^{-x^2} \geqslant 0$, we have

$$\int_{-\infty}^{\infty} e^{-x^2}\, dx = \sqrt{\pi}$$

as required.

2. (Differentiation under the integral sign.) Suppose $f\colon [a,b] \times [c,d] \to \mathbb{R}$ is continuous and $\partial f/\partial y$ exists and is continuous on $[a,b] \times [c,d]$. Let

$$F(y) = \int_a^b f(x,y)\, dx \,.$$

Then prove F is differentiable and

$$F'(y) = \int_a^b \frac{\partial f}{\partial y}(x,y)\, dx \,.$$

Solution: Consider

$$\left| \frac{F(y+h) - F(y)}{h} - \int_a^b \frac{\partial f}{\partial y}(x,y)\, dx \right| = \left| \int_a^b \frac{f(x,y+h) - f(x,y)}{h} - \frac{\partial f}{\partial y}(x,y)\, dx \right|$$

$$= \left| \int_a^b \frac{\partial f}{\partial y}(x,c_{x,h}) - \frac{\partial f}{\partial y}(x,y)\, dx \right|$$

for some $c_{x,h}$ between y and $y + h$, by the mean-value theorem. However, $\partial f/\partial y$ is continuous and hence uniformly continuous on $[a,b] \times [c,d]$. Thus given $\varepsilon > 0$, choose $\delta > 0$ such that

$$\left| \frac{\partial f}{\partial y}(x_0,y_0) - \frac{\partial f}{\partial y}(x,y) \right| < \frac{\varepsilon}{(b-a)}$$

if $|x - x_0| < \delta$ and $|y - y_0| < \delta$. Therefore let $|h| < \delta$. Then

$$\left| \frac{F(y+h) - F(y)}{h} - \int_a^b \frac{\partial f}{\partial y}(x,y)\, dx \right|$$

$$\leqslant \int_a^b \left| \frac{\partial f}{\partial y}(x,c_{x,h}) - \frac{\partial f}{\partial y}(x,y) \right| dx < \frac{\varepsilon}{b-a}(b-a) = \varepsilon \,.$$

In part, this result justifies *differentiation under the integral sign*. There are analogous theorems for improper integrals.

3. Compute the volume of the ball of radius r in $\mathbb{R}^n$ and center the origin (that is, of the set $\{x \in \mathbb{R}^n \mid \|x\| < r\}$).

Solution: Use induction on the dimension n. Of course in $\mathbb{R}$ the ball is simply the open interval $]-r,r[$ and has volume $2r$. Suppose we have computed the volume of the $n - 1$ ball of radius r to be $a_{n-1}r^{n-1}$ (at first, one guesses the answer will be of this form; this is reasonable since the $n - 1$ ball is an $n - 1$ dimensional object). Then, since the boundary of the n-ball has measure zero, we may apply Fubini's theorem. For each fixed x_n, $0 \leqslant x_n < r$, the cross section of the n-ball of radius r, which is denoted as $B(n,r)$, is

$$\{(x_1,\ldots,x_{n-1},x_n) \mid x_1^2 + \cdots + x_{n-1}^2 < r^2 - x_n^2\} \,,$$

which is an $(n-1)$-ball of radius $(r^2 - x_n^2)^{1/2}$. Hence Fubini's theorem allows us to write

$$\int_{B(n,r)} 1 = \int_{-r}^{r} \left(\int_{B(n-1,(r^2-x_n^2)^{1/2})} dx_1 \cdots dx_{n-1} \right) dx_n$$

$$= \int_{-r}^{r} a_{n-1}((r^2 - x_n^2)^{1/2})^{n-1} \, dx_n \ .$$

Now let $x_n = r \sin \theta$ for $0 < \theta < \pi/2$, and hence $(r^2 - x_n^2)^{1/2} = r \cos \theta$ and $dx_n/d\theta = r \cos \theta > 0$ on $]0,\pi/2[$. Thus

$$\int_{B(n,r)} 1 = \int_{-r}^{r} a_{n-1}((r^2 - x_n^2)^{1/2})^{n-1} \, dx_n$$

$$= 2 \int_{0}^{r} a_{n-1}((r^2 - x_n^2)^{1/2})^{n-1} \, dx_n$$

$$= 2 \int_{0}^{\pi/2} a_{n-1}(r \cos \theta)^{n-1} r \cos \theta \, d\theta$$

$$= 2a_{n-1}r^n \int_{0}^{\pi/2} \cos^n \theta \, d\theta$$

$$= a_n r^n, \quad \text{where } a_n = 2a_{n-1} \int_{0}^{\pi/2} \cos^n \theta \, d\theta \ .$$

Using elementary calculus, one finds

$$\int_{0}^{\pi/2} \cos^n \theta \, d\theta = \begin{cases} \dfrac{(n-1)(n-3) \cdots}{n(n-2) \cdots} & n \text{ odd} \\[3mm] \dfrac{(n-1)(n-3) \cdots}{n(n-2) \cdots} \cdot \dfrac{\pi}{2} & n \text{ even} \ . \end{cases}$$

Hence we have

$$v(B(1,r)) = 2r$$

$$v(B(2,r)) = 2 \cdot a_1 \cdot r^2 \cdot \left(\frac{1}{2}\right) \cdot \left(\frac{\pi}{2}\right) = \pi r^2$$

$$v(B(3,r)) = 2 \cdot a_2 \cdot r^3 \cdot \left(\frac{2}{3}\right) = \left(\frac{4}{3}\right)\pi r^3$$

$$v(B(4,r)) = 2 \cdot a_3 \cdot r^4 \cdot \left(\frac{3}{8}\right) \cdot \left(\frac{\pi}{2}\right) = \frac{\pi^2}{2} r^4$$

$$v(B(5,r)) = 2 \cdot a_4 \cdot r^5 \cdot \left(\frac{8}{15}\right) = \left(\frac{8}{15}\right)\pi^2 r^5$$

$$v(B(6,r)) = 2 \cdot a_5 \cdot r^6 \cdot \left(\frac{15}{48}\right)\left(\frac{\pi}{2}\right) = \left(\frac{1}{6}\right)\pi^3 r^6 \ .$$

4. Improve the change of variables formula by replacing "A is open" with "A has volume."

Solution: There are two ways in which this can be done, and both shall be given as theorems.

Theorem.   *Let $g: D \subset \mathbb{R}^n \to \mathbb{R}^n$ be of class C^1, where D is open. Furthermore, let g be one-to-one and $Jg(x) \neq 0$ for all $x \in D$. Let $B = g(D)$. Suppose D and B have volume. Let $A \subset B$ have volume and $f: A \to R$ be integrable. Then*

$$\int_{g^{-1}(A)} (f \circ g)\, |Jg| = \int_A f.$$

Proof: Extend f to B by letting $f = 0$ outside A. Then by Theorem 3,

$$\int_D (f \circ g)\, |Jg| = \int_B f.$$

Now since $f = 0$ outside A, $f \circ g$ is zero outside $g^{-1}(A)$, and our conclusion follows. ∎

Theorem.   *Let B have volume and $f: B \to \mathbb{R}$ be integrable. Let A have volume and suppose $g: \text{int}(A) \subset \mathbb{R}^n \to \text{int}(B) \subset \mathbb{R}^n$ is C^1, one-to-one, onto, and $Jg(x) \neq 0$ for all $x \in A$. Then if $f: B \to \mathbb{R}$ is integrable,*

$$\int_B f = \int_A f \circ g\, |Jg|.$$

Proof: Since B has volume, and $\text{bd}(\text{int}(B)) \subset \text{bd}(B)$, $\text{int}(B)$ has volume (since $\text{bd}(B)$ has measure zero). Also, $\text{int}(B) \cup ((\text{bd}(B)) \cap B) = B$, and so by Theorem 8, Chapter 8,

$$\int_{\text{int}(B)} f = \int_B f.$$

Hence we get the result by Theorem 3. ∎

Notice that the conditions on g are equivalent to the existence of a C^1 inverse for g (by the inverse function theorem).

Remarks: In these two theorems one can show that $Jg(x) \neq 0$ can be dropped as a hypothesis (then g does not necessarily have a C^1 inverse). This is outlined in Exercise 5. Exercise 15 asks the reader to prove the change of variables formula for improper integrals. This becomes fairly easy using the usual change of variables formula and our discussion of improper integrals from Chapter 8.

Exercises for Chapter 9

1. Use cylindrical coordinates $g(r,\theta,z) = (r \cos \theta, r \sin \theta, z)$ on

$$\{(r,\theta,z) \mid r > 0, 0 < \theta < 2\pi\}$$

to calculate the integral over $A = \{(x,y,z) \mid x^2 + y^2 < 1, |z| < 1\}$ of $f(x,y,z) = (x^2 + y^2)z^2$.

2. Give a counter example to show that the change of variables formula does not hold if g is not one-to-one, even though $Jg(x) \neq 0$. [Hint: take $f = 1$ and $g(x,y) = (e^x \cos y, e^x \sin y)$.]

3. Evaluate the following integrals.
   (a) $\int_A x^2 y^2 \, dx \, dy$, where $A = \{(x,y) \mid 0 < x < y^2, 0 < y < 2 + x, x < 1\}$.
   (b) $\int_A \sin(x^2 + y^2) \, dx \, dy$, where A is the unit disc.
   (c) $\displaystyle\int_{\mathbb{R}^3} \frac{1}{x^2 + y^2 + z^2} \, dx \, dy \, dz$.
   (d) $\int_A y/\sqrt{x} \, dx \, dy$, where A is the unit square $= \{(x,y) \mid 0 < x < 1, 0 < y < 1\}$.
   (e) $\int_A x \, dx \, dy$, where $A = \{(x,y) \mid 0 < x < \sqrt{\pi}, 0 < y < \sin x^2\}$.
   (f) $\int_0^\pi \int_0^1 r^2 \, dr \, d\theta$.
   (g) $\int_{-1}^1 \int_0^{\sqrt{1-x^2}} (x^3 + 3y^2 x) \, dy \, dx$.

4. Compute the volume of the following sets:
   (a) A tetrahedron with the base area A and height h;
   (b) A cone with base radius r_0 and height h_0;
   (c) $\{(x,y) \mid x^2 < y < 1 - x^2\}$;
   (d) $\{(x,y,z) \mid x^2 + y^2 + z^2 < 1 \quad \text{and} \quad z < 1/2\}$.

5. *Sard's Theorem.* The purpose of this problem is to prove a simplified version of a fairly difficult theorem known as Sard's theorem.* In our case the statement of the theorem is as follows.

   Theorem. *Let* $g: A \subset \mathbb{R}^n \to \mathbb{R}^n$ *be of class* C^1, *where* A *is open. Let* $B = \{x \in A \mid Jg(x) = 0\}$. *Then* $g(B)$ *has measure zero.*

   Of course the set B in this theorem need not have measure zero (to see this take g to be a constant mapping). Before outlining the proof, we ask the reader to assume the result and show that in Theorem 3, the assumption that $Jg(x) \neq 0$ can be omitted (provided the (open) set of points where $Jg(x) \neq 0$ has volume).

   The theorem is proved as follows. First show that if U is a closed rectangle in A, it suffices to show that $g(U \cap B)$ has measure zero (show that $g(B)$ is the countable union of these intersections). In fact, we shall show $g(U \cap B)$ has content zero.

   Next, prove these two facts. For any $\varepsilon > 0$, there is a $\delta > 0$, such that for $x, y \in U$, $\|x - y\| < \delta$, we have

   $$\|g(x) - g(y) - Dg(x) \cdot (x - y)\| < \varepsilon \|x - y\|.$$

   Also there is an M such that

   $$\|g(x) - g(y)\| \leqslant M \|y - x\|.$$

   Let U have sides of length l. Choose N such that if U is divided into N^n rectangles (of sides l/N) and S is such a rectangle, then for $x, y \in S$ the above inequalities hold. (Choose $N \geqslant l/\delta$). Suppose $x \in S \cap B$. Then find a hyperplane H in $\mathbb{R}^n$ (thus H is

---

* A general treatment can be found in Milnor, *Topology from the Differentiable Viewpoint*, or in Sternberg, *Lectures on Differentiable Geometry*.

some $(n - 1)$-dimensional subspace) such that

$$\{Dg(x)(y - x) \mid y \in S\} \subset V.$$

Next show that $\{g(y) \mid y \in S\}$ lies in a cylinder of height $< 2\varepsilon\, n(l/N)$ and base $n - 1$ cube of side $< 2M\, n(l/N)$. Hence show that $g(U \cap B)$ lies in N^n rectangles of total volume $< \varepsilon\, K$, where $K = 2^n M^{n-1}\, (n)^n l^n$, a constant independent of N. This will prove the result.

6. Find $\int_A xy \sin(x^2 - y^2)\, dx\, dy$, where

$$A = \{(x,y) \mid 0 < y < 1, x > y \quad \text{and} \quad x^2 - y^2 < 1\}\,.$$

7. Prove:
   (a) If A has volume and λ is defined as

$$\lambda = \inf\left\{ \sum_{i=1}^{\infty} v(S_i) \mid S_1, S_2, \ldots \text{ is a countable cover of } A \text{ by open rectangles}\right\},$$

   then we have $v(A) = \lambda$.
   (b) Let A be a bounded set with volume and let A_i be a sequence of sets with volume such that the A_i are non-overlapping (that is, have non-intersecting interiors) and such that

$$\bigcup_{i=1}^{\infty} A_i = A\,.$$

   Then show that

$$v(A) = \sum_{i=1}^{\infty} v(A_i)\,.$$

8. (a) If $u: A \subset \mathbb{R}^2 \to\,]a,b[$ and $v: B \subset \mathbb{R}^2 \to\,]c,d[$ are two functions of class C^1 from the open sets A and B onto the intervals $]a,b[$ and $]c,d[$ such that $u(x,y) = u(x',y')$ and $v(x,y) = v(x',y')$ only when $(x,y) = (x',y')$, and

$$\frac{\partial u}{\partial x} \cdot \frac{\partial v}{\partial y} - \frac{\partial v}{\partial x}\frac{\partial u}{\partial y} \neq 0$$

   at any point $(x,y) \in A \cap B$, $W = \{(x,y) \mid a < u(x,y) < b, c < v(x,y) < d\}$, and if f is an integrable function on W, then show that

$$\int_W f = \int_c^d \int_a^b f(u,v) \cdot \left(\frac{\partial u}{\partial x} \cdot \frac{\partial v}{\partial y} - \frac{\partial v}{\partial x}\frac{\partial u}{\partial y}\right) du\, dv\,.$$

   (b) Use (a) to evaluate

$$\int_W (x^2 + y^2)\, dx\, dy\,,$$

   where $W = \{(x,y) \mid x > 0, y > 0, -1 < x^2 - y^2 < 1, xy < 1\}$.

9. Suppose $f:]a,b[\to \mathbb{R}$ and $g:]c,d[\to \mathbb{R}$ are two integrable functions and define $\tilde{f}(x,y) = f(x)$, $\tilde{g}(x,y) = g(y)$ (assume f and g are bounded). Then prove

$$\int_{[a,b] \times [c,d]} \tilde{f}(x,y)\tilde{g}(x,y)\, dx\, dy = \left(\int_a^b f(x)\, dx\right)\left(\int_c^d g(x)\, dx\right).$$

10. Use Fubini's theorem and the fundamental theorem of calculus to give an alternative proof of Theorem 9, Chapter 6. [Hint: If $\partial^2 f/\partial x_i\, \partial x_j > \partial^2 f/\partial x_j\, \partial x_i$, integrate the difference over a small rectangle.]

11. Let $S = \{(x,y) \in \mathbb{R}^2 \mid x \text{ rational}, 0 < x < 1, \text{ and write } x = p/m \text{ in lowest form}, y = k/m, k = 1,\ldots,m-1\}$. Then show that the interated integral

$$\int_0^1 \int_0^1 1_S \, dy\, dx = 0$$

but that

$$\int_{[0,1]\times[0,1]} 1_S$$

doesn't exist.

12. If A is a bounded set with volume and A_i is a sequence of sets with volume such that $A_{i+1} \supset A_i$ and $A_1 \cup A_2 \cup \cdots = A$, then $v(A_i) \to v(A)$ as $i \to \infty$. [Hint: Use Exercise 7 or the monotone convergence theorem].

13. Suppose $C \subset A \times B$, $v(C) = 0$, and

$$1_{C_x}(y) = \begin{cases} 1, & \text{if } (x,y) \in C, \\ 0, & \text{otherwise}, \end{cases}$$

is integrable over B for each $x \in A$. Let $C_x = \{y \in B \mid (x,y) \in C\}$. Then show C_x has volume zero in A for all x except possibly a set of measure zero. Give an example where $v(C_x) \neq 0$ for some x.

14. (a) Prove Theorem 2. [Hint: As in Theorem 1, it suffices to prove (ii). Do this in exactly the same way as for Theorem 1.]

(b) Prove the following generalization of Corollary 1. Let $A \subset \mathbb{R}^n$ be a closed rectangle and let $\varphi: A \subset \mathbb{R}^n \to \mathbb{R}^m$ and $\psi: A \subset \mathbb{R}^n \to \mathbb{R}^m$ be continuous functions, such that $\varphi_j(x) \leqslant \psi_j(x)$ for all $x \in A, 1 \leqslant j \leqslant m$. Let $D = \{(x,y) \in \mathbb{R}^n \times \mathbb{R}^m \mid x \in A, \varphi_j(x) \leqslant y_j \leqslant \psi_j(x), 1 \leqslant j \leqslant m\}$. See Figure 9-16. Let $f: A \to \mathbb{R}$

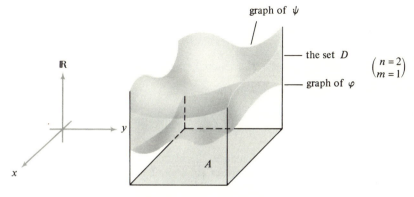

FIGURE 9-16   The set D.

be continuous, define $B_x \subset \mathbb{R}^m$ by

$$B_x = \{y \in \mathbb{R}^m \mid \varphi_j(x) \leqslant y_j \leqslant \psi_j(x), 1 \leqslant j \leqslant m\},$$

define $f_x: B_x \subset \mathbb{R}^m \to \mathbb{R}$ by $f_x(y)$, and define $g: A \subset \mathbb{R}^n \to \mathbb{R}$ by $g(x) = \int_B f_x$. Then g is integrable over A and $\int_A g = \int_D f$.

15. Investigate possible generalizations of Theorem 3 to unbounded regions.

16. For what values of p is r^p integrable in $\mathbb{R}^3$, where

$$r = \sqrt{x^2 + y^2 + z^2} \ ?$$

17. Let $f: [0,1] \to \mathbb{R}$,

$$f(x) = \begin{cases} n^2, & \text{if } x = 1/n, \\ 0, & \text{if } x \in [0,1], x \neq 1/n. \end{cases}$$

Prove f is integrable and $\int_0^1 f(x)\, dx = 0$. (*Warning: f is not bounded.*)

18. Let A be a closed rectangle in $\mathbb{R}^n$. Let $C \subset A$ have volume. Prove that for any $\varepsilon > 0$ there is a compact set $K \subset C$ such that $v(C\backslash K) < \varepsilon$, and a compact set $L \supset C$ such that $v(L\backslash C) < \varepsilon$.

19. Let $A \subset \mathbb{R}^n$, $B \subset \mathbb{R}^m$ have volume and $f: A \to \mathbb{R}$, $g: B \to \mathbb{R}$ integrable. Let $F(x,y) = f(x) + g(y)$. Show

$$\int_{A \times B} F(x,y)\, dx\, dy = \left(\int_A f \right) v(B) + \left(\int_B g \right) v(A).$$

20. Compute the area of the region $D = \{(x,y) \mid 1 \leqslant x < 3, x^2 \leqslant y \leqslant x^2 + 1\}$.

21. Suppose $g: \mathbb{R} \to \mathbb{R}$ is differentiable everywhere, and that $|g'(x)| \leqslant M$ for all $x \in \mathbb{R}$. Show that, if ε is small enough, the function f which is defined by $f(x) = x + \varepsilon g(x)$ is one-to-one.

22. Let f and g be two integrable real-valued functions on $[a,b]$. Let

$$h(x) = \max(f(x), g(x)) = \begin{cases} f(x), & \text{if } f(x) \geqslant g(x), \\ g(x), & \text{if } f(x) \leqslant g(x). \end{cases}$$

Show that $h(x)$ is integrable.

23. Let $f_n(x) = \sum_{m=1}^n (1/2^m)\sin mx$ be defined for all $x \in \mathbb{R}$.
   (a) Show $\lim_{n \to \infty} f_n(x)$ exists. [Hint: Show $f_n(x)$ is a Cauchy sequence.]

   (b) Show the sequence converges uniformly.

   (c) Show $\int_0^{2\pi} \left(\lim_{n \to \infty} f_n(t) \right) dt = 0$.

24. *True or False*
   (a) If f is a continuous function on $[0,1]$, then f is bounded on $[0,1]$.
   (b) If $f: S \to \mathbb{R}^n$ is a continuous function, where S is a compact subset of $\mathbb{R}$, then $f(S)$ is compact.
   (c) An integrable function on $[0,1]$ must be continuous on $[0,1]$.

(d) If U and V are open subsets of $\mathbb{R}$, then $U \times V = \{(x,y) \mid x \in U, y \in V\}$ is open in $\mathbb{R}^2$.

(e) If f and g are integrable functions on $[a,b]$, then $f - 2g$ is integrable on $[a,b]$.

(f) Any bounded sequence in $\mathbb{R}^n$ must have a convergent subsequence.

(g) If f is a continuous real-valued function on $[0,1]$ such that $f(x) \geqslant 0$ for all $x \in [0,1]$ and $\int_0^1 f(x)\,dx = 0$, then $f(x) = 0$ for all $x \in [0,1]$.

(h) If f is an infinitely differentiable real-valued function on $\mathbb{R}$, then f must have a power series expansion about each point of $\mathbb{R}$.

(i) If $f, f_1, f_2, f_3, \ldots$ are all continuous real-valued functions on $[0,1]$ such that $f(x) = \lim_{n \to \infty} f_n(x)$ for each $x \in [0,1]$, then $f_1, f_2, f_3, \ldots$ converges to f uniformly on $[0,1]$.

(j) If $a_0, a_1, a_2, \ldots$ is a sequence of real numbers and r is the radius of convergence of $\sum_{n=0}^{\infty} a_n x^n$, then $\sum_{n=0}^{\infty} a_n r^n$ converges.

(k) If $f_1, f_2, f_3, \ldots$ converges to f uniformly on $[a,b]$ and if f_n is integrable for each $n = 1, 2, 3, \ldots$, then f is integrable on $[a,b]$.

(l) If $f_1, f_2, f_3, \ldots$ converges to f uniformly on $[a,b]$ and if f_n is differentiable on $]a,b[$ for each $n = 1, 2, 3, \ldots$, then f is differentiable on $]a,b[$.

(m) An open connected subset of $\mathbb{R}^n$ is arcwise connected.

(n) If f is a differentiable real-valued function on $]0,1[$ and $f(1/2) \geqslant f(x)$ for all $x \in]0,1[$, then $f'(1/2) = 0$.

(o) If f is an integrable function on $[0,1]$ and $\varepsilon > 0$, then there exists a step function g on $[0,1]$ such that $\int_0^1 |f - g| < \varepsilon$.

(p) Every closed and bounded subset of $\mathbb{R}$ contains its least upper bound.

(q) If $f(x)$ has a power series expansion on $]-r,r[$, then f is differentiable on $]-r,r[$.

(r) If f is integrable on $[a,b]$ and on $[b,c]$, where $a < b < c$, then f is integrable on $[a,c]$.

(s) If $f \colon \mathbb{R}^n \to \mathbb{R}^m$ is continuous and S is closed in $\mathbb{R}^m$, then $f^{-1}(S)$ is closed in $\mathbb{R}^n$.

(t) If a series $\sum_{j=1}^{\infty} a_j$ converges conditionally, then the sequence of partial sums $\sum_{j=1}^{n} a_j$ is bounded.

(u) If f is a continuous real-valued function on $]a,b[$ then there exists a differentiable function F on $]a,b[$ such that $f = F'$ on $]a,b[$.

25. Suppose that $f, f_1, f_2, f_3, \ldots$ are continuous real-valued functions on $[0,1]$ and suppose $f_n \to f$ uniformly on $[0,1]$ as $n \to \infty$. Prove that each of $|f|, |f_1|, |f_2|, |f_3|, \ldots$ is integrable on $[0,1]$ and that $\int_0^1 |f_n| \to \int_0^1 |f|$ as $n \to \infty$.

26. Compute
(a) $\lim_{n \to \infty} (1 + 2^k + \cdots + n^k)/n^{k+1}$, where $k > 0$.

(b) $\lim_{n \to \infty} \left(\dfrac{1}{n + 1} + \dfrac{1}{n + 2} + \cdots + \dfrac{1}{2n} \right)$.

27. Show, if $f''(x) > 0$ for all x that f is *convex upward*, which means

$$f\left(\frac{x + y}{2}\right) \leqslant \frac{f(x) + f(y)}{2}.$$

[Hint: Consider the auxiliary function $G(x) = f\left(\dfrac{x+y}{2}\right) - \dfrac{f(x)+f(y)}{2}$ for fixed y.]

28. Suppose $f(x)$ is continuous on $]-1,1[$, $f(0) = 0$ and $f(x) \neq 0$ if $x \neq 0$. Prove: $\displaystyle\lim_{x \to 0} \dfrac{\log(1 + f(x))}{f(x)}$ exists. It equals what?

29. Let C be a cube in $\mathbb{R}^n$, f, $g \colon C \to \mathbb{R}$ bounded and integrable. Suppose $f(x) \leqslant g(x)$ for all x in a *dense* subset S of C (that is, cl$(S) = C$). Show that $\int_C f \leqslant \int_C g$.

30. Suppose $A \subset \mathbb{R}^n$ and A has zero volume. Suppose $f \colon A \to \mathbb{R}$ is a bounded function. Prove f is integrable and $\int_A f = 0$.

31. Consider the following theorem.

Theorem. *Let A be an open subset of $\mathbb{R}^n$, $\varphi \colon A \to \mathbb{R}^n$ a one-to-one continuously differentiable map whose jacobian $J\varphi$ is nowhere zero on A. Suppose that the function $f \colon \varphi(A) \to \mathbb{R}$ is continuous and is zero outside a compact subset of $\varphi(A)$ and that $\int_{\varphi(A)} f$ exists. Then $\int_{\varphi(A)} f = \int_A (f \circ \varphi) |J\varphi|$.*

Suppose A_i, $i = 1, 2, \ldots$ are open subsets of $\mathbb{R}^n$ such that the theorem is true for each A_i and the restriction of φ to A_i. Let $A = \bigcup A_i$. Show that the theorem is true for A and φ.

32. Suppose f is bounded, defined on $[0,b]$, $b > 0$ and suppose $\int_c^b f$ exists for all $0 < c < b$. Show $\int_0^b f$ exists.

33. Suppose $S \subset \mathbb{R}^n$ has volume and $t \in \mathbb{R}$. Define $tS = \{(tx_1,\ldots,tx_n) \mid (x_1,\ldots,x_n) \in S\}$. Show tS has volume and vol$(tS) = |t|^n$ vol(S). [Hint: define the map $\varphi \colon (y_1,\ldots,y_n) \to (y_1/t,\ldots,y_n/t)$.]

34. *True or False* If false give a counterexample; if true give a reason.
    (a) Suppose f is integrable on A and g is a function on A such that $g \leqslant f$. Then g is also integrable on A.
    (b) Suppose A has volume and f is continuous on A. Then $\int_A f$ exists.
    (c) Suppose A has volume and $\int_A f$ exists. Then f is continuous on A.
    (d) Any bounded subset of $\mathbb{R}^2$ has zero volume when considered as a subset of $\mathbb{R}^3$.
    (e) Suppose $I = [0,1]$, φ_1 and φ_2 are continuous functions: $I \to \mathbb{R}$ with $\varphi_1(x) < \varphi_2(x)$ for all $x \in I$. Then the set $S = \{(x,y) \mid x \in I, \varphi_1(x) < y < \varphi_2(x)\}$ has volume.

35. Compute $\displaystyle\lim_{n \to \infty} n \log\left(1 + \dfrac{1}{n}\right)$.

Chapter 10

Fourier Analysis

Fourier analysis arose historically in connection with problems of mathematical physics such as heat conduction and wave motion. This subject has now evolved into a vast theory with many applications, both mathematical and physical. This chapter is intended to give a brief but basic working knowledge of some Fourier methods, to introduce the student to the general theory,* and to delineate some fundamental applications.

Probably the best motivation for the study of Fourier analysis is obtained by examining a vibrating string. Although this topic is the subject of a detailed discussion later, here we are primarily interested in an heuristic approach. Further applications (for example, to quantum mechanics) are presented in later sections.

Consider then, a string of length l with clamped ends and which is free to vibrate when plucked. The position (vertical displacement) of the string is represented by a function $y(t,x)$, where t is the time and $x \in [0,l]$. See Figure 10-1. It is a fact from elementary physics that y obeys the *wave equation:*

$$\frac{\partial^2 y}{\partial t^2} = c^2 \frac{\partial^2 y}{\partial x^2}$$

where c is a constant determined by the nature of the string and the tension in it.† That the string has clamped ends entails that $y(t,0) = y(t,l) = 0$ for all t.

* A thorough treatment of the general theory requires concepts which are beyond the scope of this book. For this more advanced work, we refer the reader to books such as Zemanian, *Distribution Theory and Transform Analysis,* or to Rudin, *Real and Complex Analysis.*

† For a discussion of this point and other physical situations in which the wave equation arises (for example, sound waves, water waves) see R. P. Feynman, R. B. Leighton, and M. Sands, *The Feynman Lectures on Physics,* Addison-Wesley (1963), Ch. 47–51.

FIGURE 10-1

To simplify matters, let us first look at the case of special solutions, called *standing waves;* these are solutions of the form $y(t,x) = (\cos \omega t)u(x)$, where $y(t,x)$, as above, represents the vertical displacement at x at time t, and ω is the frequency. Thus $|u|$ represents the amplitude or wave shape. Physically, a standing wave is a synchronous up and down motion which repeats its shape periodically after time $t = 2\pi/\omega$, such as occurs when a string produces a pure note.

Certain solutions which correspond to *fundamental solutions* or *harmonics* are given by

$$y_n(t,x) = \sin\left(\frac{n\pi x}{l}\right)\cos(\omega_n t), \qquad n = 0, 1, 2, \ldots ,$$

where $\omega_n = n\pi c/l$ is the frequency. For $n = 2$ and $t = \pi/\omega_n$, we obtain the illustration in Figure 10-1.

It is both important and remarkable that any solution $y(x,t)$ describing the motion of the string can be decomposed into harmonics; that is, written as a series:

$$y(x,t) = \sum_{n=1}^{\infty} c_n y_n(x,t) = \sum_{n=1}^{\infty} c_n u_n(x)\cos(\omega_n t) ,$$

where y_n is as above, and $u_n(x) = \sin(n\pi x/l)$. We think of u_1 as the first harmonic component of y, u_2 the second, and so forth. Thus a complicated looking vibration (such as occurs on a violin string) is in reality an infinite combination of simple harmonics, where each harmonic component appears with weight c_n. In Figure 10-2, we illustrate how summing three sine curves of varying amplitudes can lead to a more complicated curve. For a general curve, one requires an infinite combination of sine curves.

The purpose of Fourier analysis is to carry out this procedure of decomposition using a general method. For finite regions (such as $[0,l]$ above) the appropriate method to use is *Fourier series*, while on an infinite region (the whole real line, for instance) *Fourier integrals* are required.

The series obtained from sin nx, cos nx, or e^{inx} are called the *classical Fourier series*. For other types of problems (the harmonic oscillator in quantum mechanics, for example), other types of basic solutions enter and arbitrary solutions need to be expanded in terms of these basic solutions

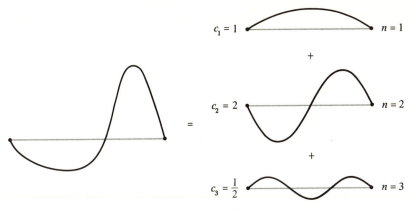

$c_1 = 1$ $n = 1$

+

$c_2 = 2$ $n = 2$

=

+

$c_3 = \frac{1}{2}$ $n = 3$

FIGURE 10-2

(for the quantum mechanical harmonic oscillator, for instance, Hermite functions are used). Therefore, it is useful to discuss the general theory of expansion, which will be done in Sections 10.1 and 10.2.

In Sections 10.3, 10.4, and 10.5 we will study the special case of trigonometric Fourier series, and show that the expansion procedure is justified. To justify this for other families of functions often requires an examination of special situations—for instance, the differential equation giving rise to the problem.

In this regard, there are two main theorems. The first deals with the important concept of mean convergence and states that any square integrable function has a Fourier series which converges in the mean. As we later explain, this must not be confused with convergence at each value x (called pointwise convergence). For the latter, one must use the basic theorem of Jordan (or Dirichlet-Jordan). The mean and pointwise convergence properties are proved in Theorems 8 and 9, respectively.

Some further theorems on convergence, such as justifying term-by-term differentiation, are given in Section 10.6.

A few simple but important applications of Fourier methods are presented in Section 10.7. There we study special cases of three problems—the wave equation, the Dirichlet problem (Laplace's equation) and the heat equation—all from the point of view of Fourier series. The discussion is quite rigorous and we are careful about differentiability properties, assumptions concerning initial or boundary values, and the manner in which these values are assumed.

In Section 10.8, an informal treatment of Fourier integrals is given, stating the basic properties and definitions without proofs (these are left for a more advanced course). Hopefully, however, the student will obtain some perspective of the role of Fourier integrals and their relationship to Fourier series.

Finally in Section 10.9 a brief glimpse is given of quantum mechanics and how the machinery of Sections 10.1–10.3 can be used to establish some of the basic results of the subject.

10.1   Inner Product Spaces

Before we begin our study of Fourier series themselves, we must learn certain concepts which will enable us to simplify the job. These ideas are really quite easy and reveal an important geometric aspect of Fourier series. In this chapter we also need some basic facts about complex numbers. The reader should therefore glance at the appendix to this section before proceeding if he is not already familiar with the elementary properties of complex numbers.

In Chapter 1 we studied the inner product $\langle , \rangle$ on $\mathbb{R}^n$. Now we want to extend these notions to an arbitrary vector space V. In the present chapter, V is no longer finite dimensional, but is a space of functions which is infinite dimensional—as was the space $\mathscr{C}(A,\mathbb{R}^m)$ studied in Chapter 5. For example, V might be a space of functions $f\colon [0,2\pi] \to \mathbb{R}$. This is a vector space if we take the usual definitions, $(f + g)(x) = f(x) + g(x)$ and $(\alpha f)(x) = \alpha f(x)$. The expression $\langle f,g \rangle = \int_0^{2\pi} f(x)g(x)\, dx$ is called the *inner product* of f and g.

It is important to allow complex values, for the simple reason that it is often more convenient to work with $e^{i\theta}$ than with $\sin\theta$ and $\cos\theta$. In the complex case, we let*

$$\langle f,g \rangle = \int_0^{2\pi} f(x)\overline{g(x)}\, dx\ ,$$

where $\overline{g(x)}$ is the complex conjugate of $g(x)$. The reason we use $\overline{g(x)}$ is so that we can (as in the real case) define the *length* or *norm* of f by

$$\|f\|^2 = \langle f,f \rangle = \int_0^{2\pi} |f(x)|^2\, dx$$

(for complex numbers z, we recall that $|z|^2 = z\bar{z}$ is a positive real number).

It is this sort of space V that the reader should keep in mind when studying the next two sections. Later on, we shall be explicitly dealing with this space, or spaces like it.

Our study begins with general spaces with inner products rather than the special one above (which is actually the one of most interest to us) because it is conceptually and notationally simpler to work with the notation $\langle , \rangle$ than with integrals. At this point only the following basic properties of $\langle , \rangle$ are significant.

* Physicists use the convention of putting the bar over the f.

Definition 1. Let V be a complex vector space (this is just a vector space where we allow complex numbers for scalar multiplication).

An *inner product* on V is a mapping $\langle \, , \, \rangle : V \times V \to \mathbb{C}$ (where $\mathbb{C}$ denotes the complex numbers) with the following properties.

(i) $\langle af + bg, h \rangle = a\langle f, h \rangle + b\langle g, h \rangle$, for all f, g, $h \in V$ and $a, b \in \mathbb{C}$.

(ii) $\langle f, h \rangle = \overline{\langle h, f \rangle}$.

(iii) $\langle f, f \rangle \geqslant 0$, and $\langle f, f \rangle = 0$ implies $f = 0$.

From (i) and (ii) we deduce that $\langle h, af + bg \rangle = \bar{a}\langle h, f \rangle + \bar{b}\langle h, g \rangle$. Notice that if all quantities were real, we would have the same properties as the usual inner product on $\mathbb{R}^n$ (Theorem 5, Chapter 1). As we have stressed, in general, V is not finite dimensional, so we must avoid using (finite) bases and matrices.

Theorem 1. *The space* V *of continuous functions* $f \colon [a,b] \to \mathbb{C}$ *forms an inner product space if we define*

$$\langle f, g \rangle = \int_a^b f(x)\overline{g(x)} \, dx \, .$$

The integral of a complex-valued function is defined as

$$\int_a^b f(x) \, dx = \int_a^b f_1(x) \, dx + i \int_a^b f_2(x) \, dx \, ,$$

where $f = f_1 + if_2$. The properties of complex integrals are similar to and may be deduced from real integrals. Some of these are listed in the appendix to this section.

The structure of an inner product space allows us to introduce many of the ideas considered in Chapter 1. The *norm* of f, denoted $\|f\|$, is defined by

$$\|f\|^2 = \langle f, f \rangle$$

and the *distance between f and g* by

$$d(f,g) = \|f - g\|$$

(see Theorem 5, Chapter 1).

We use the same language as in $\mathbb{R}^n$ by analogy. For example, we say f and g are *orthogonal* if $\langle f, g \rangle = 0$. Since V is a vector space, we can also talk about linear dependence and other related ideas we saw in $\mathbb{R}^n$. The following theorem develops the analogy with $\mathbb{R}^n$ further.

Theorem 2 (*The Cauchy Schwarz Inequality*). *Let f, g belong to the inner product space V; then*

$$|\langle f, g \rangle| \leqslant \|f\| \, \|g\| \, .$$

Furthermore, all the properties listed in Theorem 5 (II), (III), Chapter 1 hold (II (iii) also holds for α complex).

We can also introduce the notions of topology exactly as before (see Chapter 2). The main concept for us here is that of convergence of a sequence or series. A definition now follows.

> **Definition 2.** Let V be an inner product space and let f_n be a sequence in V. We say f_n *converges to* f and write $f_n \to f$ if $\|f_n - f\| \to 0$; that is, for every real number $\varepsilon > 0$ there is an N such that $n \geq N$ implies $\|f_n - f\| < \varepsilon$. Similarly, a series $\sum_{n=1}^{\infty} g_n$ *converges to* f if the sequence of partial sums $s_n = \sum_{k=1}^{n} g_k$ converges to f.

Suppose our space V consists of the functions $f: [a,b] \to \mathbb{C}$ (see Theorem 1). Then the Cauchy-Schwarz inequality reads

$$\left(\int_a^b f(x)\overline{g(x)} \, dx \right)^2 \leq \left(\int_a^b |f(x)|^2 \, dx \right)\left(\int_a^b |g(x)|^2 \, dx \right).$$

The triangle inequality ($\|f + g\| \leq \|f\| + \|g\|$) becomes what is called *Minkowski's inequality* and reads

$$\left\{ \int_a^b |f(x) + g(x)|^2 \, dx \right\}^{1/2} \leq \left\{ \int_a^b |f(x)|^2 \, dx \right\}^{1/2} + \left\{ \int_a^b |g(x)|^2 \, dx \right\}^{1/2}.$$

With this inner product on the space of functions V, convergence is called *convergence in the mean*. It is quite different from pointwise or uniform convergence and is generally much weaker. We write $f_n \to f$ (in mean), $f_n \to f$ (pointwise), and $f_n \to f$ (uniformly) to distinguish these types.

Thus $f_n \to f$ (in mean) is the same as

$$\left(\int_a^b |f_n(x) - f(x)|^2 \, dx \right) \to 0.$$

For example, consider the function f_n defined by $f_n(x) = 1 - nx$ for $0 \leq x \leq 1/n$ and $f_n(x) = 0$ for all other $x \in [0,1]$. Then $f_n \to 0$ (in mean), but f_n does not converge to $f = 0$ pointwise (at $x = 0$ specifically) and therefore not uniformly. See Figure 10-3. One could contemplate other types of convergence such as $\int |f_n - f| \to 0$, but the one above is the most appropriate for Fourier series because of that theory's close connection with inner product spaces.

Uniform convergence implies mean convergence (see Exercise 5, at the end of this chapter and Theorem 4, Chapter 5). However, pointwise convergence does not in general imply mean convergence (see Section 5.3).

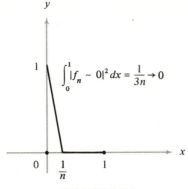

$$\int_0^1 |f_n - 0|^2\, dx = \frac{1}{3n} \to 0$$

FIGURE 10-3

The space V of Theorem 1 is easily extended to include other functions such as piecewise continuous functions (see Theorem 3). However, even if V is extended to all Riemann integrable functions it still suffers from a serious deficiency—it is not complete; that is, a Cauchy sequence may not converge. As in $\mathbb{R}^n$, a definition follows.

Definition 3. A sequence f_n in an inner product space V is a *Cauchy sequence* when for any $\varepsilon > 0$ there is an N such that $m, n \geqslant N$ implies $\|f_n - f_m\| < \varepsilon$. An inner product space is called *complete* if every Cauchy sequence converges. A complete inner product space is called a *Hilbert space*.

In order to make V in Theorem 1 complete, the concept of the Lebesgue integral must be used. Fortunately, our elementary discussion does not require this notion, but the student should be aware that a solution to this problem can be found. Of course, for the beginner in the subject, it is more important to get an intuitive grasp and a working knowledge, and this then is our goal.

Hence, the question is, can we work with more general functions and still have an inner product space? The answer is really quite simple. The only place in Theorem 1 where continuity was used was in the statement that

$$\int_a^b |f(x)|\, dx = 0 \qquad \text{implies } f = 0$$

(see the proof of Theorem 1). For a general f,

$$\int_a^b |f(x)|\, dx = 0 \qquad \text{implies } f(x) = 0$$

except possibly for those x in a set of measure zero (see Theorem 4, Chapter 8).

If we regard such an f as actually zero (modify f on this set of measure zero if necessary) then Theorem 1 carries over. We shall not try to make this any more precise because it is a technical point which tends to obscure what is going on. With that understanding, the next theorem follows.

Theorem 3. *Let* $V = \mathscr{L}^2$ *be the space of functions* $f\colon [a,b] \to \mathbb{C}$, *such that* $|f|^2$ *is integrable (that is,* $\int_a^b |f(x)|^2\, dx < \infty$*). Then the space* V *is an inner product space with inner product*

$$\langle f,g \rangle = \int_a^b f(x)\overline{g(x)}\, dx$$

and norm

$$\|f\| = \left(\int_a^b |f(x)|^2\, dx \right)^{1/2}.$$

Another convenient class of functions which forms an inner product space is the class of sectionally continuous (or piecewise continuous) ones. They are defined as follows.

Definition 4. A function $f\colon [a,b] \to \mathbb{C}$ is *sectionally continuous* if $[a,b]$ has a finite partition $a = x_0 < x_1 < \cdots < x_n = b$ such that f is continuous and bounded on each open section $]x_i, x_{i+1}[$, $i = 0, \ldots, n-1$.

APPENDIX TO SECTION 10.1:
Complex Numbers

The reader is possibly familiar with some aspects of complex numbers. We shall now quickly review the basic properties of these numbers.

We define the set $\mathbb{C}$ of complex numbers to be the set of ordered pairs of real numbers (a,b) (that is, elements of $\mathbb{R}^2$) which we shall write $a + bi$. For example $3 + 2i = (3,2)$, $2i = 0 + 2i = (0,2)$, $i = 0 + 1i = (0,1)$, and so forth. We define the operations $+$ and $\cdot$ as follows:

$$(a + bi) + (c + di) = (a + c) + (b + d)i \, ;$$

$$(a + bi) \cdot (c + di) = (ac - bd) + (ad + bc)i \, .$$

The reader may verify that the complex numbers with these operations do form a field (defined in Chapter 1); that is, all the usual algebraic rules for addition, subtraction, multiplication, and division hold.

Complex numbers are generally denoted z. Thus z stands for $z = a + bi = (a,b)$. Also, we usually just write $z_1 z_2$ for $z_1 \cdot z_2$.

Using the definition of $\cdot$, we see that the imaginary unit i satisfies

$$i \cdot i = (0 + 1i) \cdot (0 + 1i) = -1 \, .$$

Thus $i^2 = -1$, which is written as $i = \sqrt{-1}$.

Since we have defined the complex numbers to be ordered pairs of real numbers, they may also be naturally associated with points in $\mathbb{R}^2$. Thus, in particular, the metric on $\mathbb{R}^2$ induces a metric on the complex numbers. Also, from the norm on $\mathbb{R}^2$ we get a norm for $z = a + bi$ defined as $|a + bi| = \sqrt{a^2 + b^2} = \sqrt{(a + bi)(a - bi)}$. The relation between $a + bi$ and $a - bi$ is sufficiently important to be given a name. We call $a - bi$ the *complex conjugate*, or simply the conjugate of $a + bi$, and we write $\bar{z}$ for the conjugate of the complex number z. Thus using our definition of norm above, we have

$$|z|^2 = z \cdot \bar{z} \, .$$

Important properties of the complex conjugate are that $\overline{z_1 z_2} = \bar{z}_1 \bar{z}_2$, $\overline{z_1 + z_2} = \bar{z}_1 + \bar{z}_2$ and $\overline{z_1/z_2} = \bar{z}_1/\bar{z}_2$ (see Example 2).

We want to think of the complex numbers as an extension of the real numbers and therefore associate (or identify) the real number a with the complex number $a + 0i$. Then a complex number may be thought of as the sum of a real number a and a real multiple of $i = \sqrt{-1}$. In the number $a + bi$, a is called the *real part* and b is called the *imaginary part* to distinguish the two numbers in the ordered pair. Since we associated the real numbers with numbers of the form $a + 0i$, we see that a number is real iff its imaginary part is zero. That is, $a + bi$ is real iff $a + bi = a - bi$, which can be written as z is real iff $z = \bar{z}$.

It is useful to have a definition of e^z, where z may be a complex number. Since we want e^a to coincide with the usual definition for a real and we want $e^{a+bi} = e^a \cdot e^{bi}$, we only need to define $e^{i\theta}$ for θ real. We *define* $e^{i\theta} = \cos \theta + i \sin \theta$ and hence $e^{a+bi} = e^a(\cos b + i \sin b)$. Then since $\cos 0 + i \sin 0 = 1$, we have $e^{a+0i} = e^a$, and so our definition agrees with the usual definition in the case where a is real. The reader may also check (Exercise 1) that $e^{z_1 + z_2} = e^{z_1} \cdot e^{z_2}$.

Complex numbers are represented (as already stated) by points in $\mathbb{R}^2$. Using polar coordinates we can thus write

$$z = re^{i\theta} = r \cos \theta + ir \sin \theta \, ,$$

where $r = |z|$ and $\theta = \arg z$, the argument of z; see Figure 10-4.

Any function $f : [a,b] \to \mathbb{C}$ may be divided up into two real-valued functions, f_1 and f_2, such that $f(x) = f_1(x) + if_2(x)$ (that is, define $f_1(x)$ as the real part of $f(x)$ and $f_2(x)$ the imaginary part). Then we make the natural definition

$$\int_a^b f(x)\, dx = \int_a^b f_1(x)\, dx + i \int_a^b f_2(s)\, dx$$

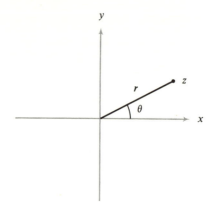

FIGURE 10-4

if both f_1 and f_2 are integrable over $[a,b]$. In this case f is called integrable over $[a,b]$. It is easily seen that

(i) $\int_a^b cf(x)\,dx = c\int_a^b f(x)\,dx$ for any complex number c and any integrable function f;

(ii) $\int_a^b [f(x) + g(x)]\,dx = \int_a^b f(x)\,dx + \int_a^b g(x)\,dx$ for integrable functions f and g; and

(iii) $\int_a^b \overline{f(x)}\,dx = \overline{\int_a^b f(x)\,dx}$.

One can also prove

(iv) $|\int_a^b f(x)\,dx| \leqslant \int_a^b |f(x)|\,dx$ with a little more effort.

Similarly, $f'(x) = f_1'(x) + if_2'(x)$ can be defined, and the usual rules for derivatives hold.

EXAMPLE 1. Show that it is *impossible* to define an order on the complex numbers satisfying all the order axioms (see Chapter 1).

Solution: We must have either $i \leqslant 0$ or $i \geqslant 0$. Suppose, first of all, that $i \leqslant 0$. Then $0 \leqslant -i$, and so $(-i)i \leqslant (-i)\cdot 0 = 0 \Rightarrow -(-1) \leqslant 0 \Rightarrow 1 \leqslant 0$. Then $-1 \geqslant 0$, so $(-1)(1) \leqslant 0 \Rightarrow -1 \leqslant 0$. But $-1 \geqslant 0$ and $-1 \leqslant 0 \Rightarrow -1 = 0$, which is not possible. On the other hand, suppose $i \geqslant 0$. Then $i(i) \geqslant 0 \Rightarrow -1 \geqslant 0$, which again leads to a contradiction. Hence such an ordering of the complex numbers is impossible.

EXAMPLE 2. For complex numbers z_1, z_2, prove that $\overline{z_1 z_2} = \bar{z}_1 \bar{z}_2$, $|z_1 z_2| = |z_1| \cdot |z_2|$, and $|z_1/z_2| = |z_1|/|z_2|$ if $z_2 \neq 0$. Also, show that $|z_1 + z_2| \leqslant |z_1| + |z_2|$.

Solution: First, we show that $\overline{z_1 z_2} = \bar{z}_1 \bar{z}_2$. Let $z_1 = x_1 + iy_1$ and $z_2 = x_2 + iy_2$. Then

$$z_1 z_2 = x_1 x_2 - y_1 y_2 + i(x_1 y_2 + x_2 y_1)$$

and so

$$\overline{z_1 z_2} = x_1 x_2 - y_1 y_2 - i(x_1 y_2 + x_2 y_1) \, .$$

Also,

$$\bar{z}_1 \bar{z}_2 = (x_1 - iy_1)(x_2 - iy_2) = x_1 x_2 - y_1 y_2 - i(x_1 y_2 + x_2 y_1)$$

and therefore $\overline{z_1 z_2} = \bar{z}_1 \bar{z}_2$.

To show that $|z_1 z_2| = |z_1| \, |z_2|$, note that $|z_1|^2 \, |z_2|^2 = z_1 \bar{z}_1 \cdot z_2 \bar{z}_2 = (z_1 z_2)(\overline{z_1 z_2}) = |z_1 z_2|^2$ by the above.

To prove that $|z_1/z_2| = |z_1|/|z_2|$, write $z_1 = z_2 \cdot z_1/z_2$. Then, by what has been shown, $|z_1| = |z_2| \, |z_1/z_2|$, which implies that $|z_1/z_2| = |z_1|/|z_2|$.

Finally, $|z_1 + z_2| \leqslant |z_1| + |z_2|$ is simply the triangle inequality for points in $\mathbb{R}^2$, which was proved in Chapter 1, Theorem 5.

EXAMPLE 3. If $f_1, \ldots, f_n$ are orthonormal vectors in the inner product space V, that is, $\langle f_i, f_j \rangle = 0$, if $i \neq j$, $\langle f_i, f_i \rangle = 1$. Prove that $f_1, \ldots, f_n$ are linearly independent.

Solution: Suppose $\sum_{i=1}^{n} c_i f_i = 0$. We must show that $c_i = 0$. Fix i, and let $g = \sum_{j=1}^{n} c_j f_j$; then form $\langle g, f_i \rangle$. We have

$$\langle g, f_i \rangle = \sum_{j=1}^{n} \langle c_j f_j, f_i \rangle$$

$$= \sum_{j=1}^{n} c_j \langle f_j, f_i \rangle$$

$$= \sum_{j=1}^{n} c_j \cdot \delta_{ji} = c_i$$

(where $\delta_{ji} = 1$ if $j = i$, and is zero if $j \neq i$). Since $g = 0$, we obtain $c_i = 0$, and so we have the desired result.

EXAMPLE 4. Let V be an inner product space and $f, g \in V$, $g \neq 0$. Define the *projection* of f on g as the vector $h = \langle f, g \rangle (g/\|g\|^2)$. Show that h and $f - h$ are orthogonal, and interpret this result geometrically.

Solution: We compute as follows.

$$\langle h, f - h \rangle = \langle h, f \rangle - \|h\|^2$$

$$= \frac{\langle g, f \rangle \langle f, g \rangle}{\|g\|^2} - \frac{\langle \langle f, g \rangle g, \langle f, g \rangle g \rangle}{\|g\|^4}$$

$$= \frac{\langle g, f \rangle \langle f, g \rangle}{\|g\|^2} - \frac{\langle f, g \rangle \overline{\langle f, g \rangle}}{\|g\|^2} = 0$$

since $\langle f, g \rangle = \overline{\langle g, f \rangle}$. Hence h and $f - h$ are orthogonal. The geometric significance of this is illustrated in Figure 10-5.

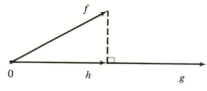

FIGURE 10-5

Exercises for Section 10.1

Exercises 1–7 deal with complex numbers; 8–12 deal with inner product spaces.

1. Prove that
   (a) $e^{z_1} \cdot e^{z_2} = e^{z_1 + z_2}$.
   (b) $e^z \neq 0$ for any complex number z.
   (c) $|e^{i\theta}| = 1$ for any real number θ.
   (d) $(\cos\theta + i\sin\theta)^n = \cos n\theta + i\sin n\theta$.

2. Show
   (a) $e^z = 1$ iff $z = k\, 2\pi i$ for some integer k.
   (b) $e^{z_1} = e^{z_2}$ iff $z_1 - z_2 = k\, 2\pi i$ for some integer k.

3. Use the power series for $\cos x$ and $\sin x$ to prove $e^z = \sum_{k=0}^{\infty} z^k/k!$ for $z = ix$. (As usual, $0! = 1$.)

4. Prove:
   (a) $\int_a^b cf(x)\, dx = c \int_a^b f(x)\, dx$
   for any complex number c and any integrable complex-valued function f.
   (b) $\int_a^b (f(x) + g(x))\, dx = \int_a^b f(x)\, dx + \int_a^b g(x)\, dx$
   for any integrable complex-valued functions f and g.

5. Compute $\int_0^1 e^{ix}\, dx$.

6. (a) For $z = re^{i\theta}$ and $w = \rho e^{i\varphi}$ show that $zw = r\rho e^{i(\theta + \varphi)}$. Interpret this result geometrically.
   (b) Interpret geometrically the process of multiplication by i.

7. For complex-valued functions (on an interval or an open subset of $\mathbb{R}^n$), discuss the sum, chain, product, and quotient rules for derivatives. Also, prove the fundamental theorem of calculus for complex functions on intervals.

8. Generalize Theorem 1 to functions defined on a set in $\mathbb{R}^n$.

9. Prove the following in an inner product space (compare Exercise 12, Chapter 1).
   (a) $\langle f,g \rangle = 0 \Rightarrow \|f + g\|^2 = \|f\|^2 + \|g\|^2$ (Pythagoras theorem).
   (b) $4\langle f,g \rangle = (\|f + g\|^2 - \|f - g\|^2) - i(\|f + ig\|^2 - \|f - ig\|^2)$.
   (c) $2\|f\|^2 + 2\|g\|^2 = \|f + g\|^2 + \|f - g\|^2$.
   (d) $\|f + g\| \cdot \|f - g\| \leq \|f\|^2 + \|g\|^2$.

10. Show that

$$\left| \int_a^b f(x)\,dx \right|^2 \leqslant (b-a) \int_a^b |f(x)|^2\,dx \,,$$

and deduce that a square integrable function on $[a,b]$ is also integrable. Is the converse true?

11. Let $\varphi_1, \ldots, \varphi_n$ be orthonormal vectors in V and $f \in V$. Define the *projection* of f on $\varphi_1, \ldots, \varphi_n$ by $g = \sum_{i=1}^{n} \langle f,\varphi_i \rangle \varphi_i$. Show that g and $f - g$ are orthogonal. Interpret geometrically.

12. (a) In an inner product space prove that $|\, \|f\| - \|g\| \,| \leqslant \|f - g\|$. In particular, $\|f\| \leqslant \|g\| + \|f - g\|$. [Hint: Write $f = (f - g) + g$ and apply the triangle inequality.]

   (b) If $f_n \to f$ (in mean), prove $\|f_n\|$ is a bounded sequence.

10.2 Orthogonal Families of Functions

In this section we study some general properties of orthogonal vectors in an inner product space. This is the basic general theory underlying Fourier analysis and is remarkably simple. The core of the problem, however, is treated in the next section. The main notions developed in this section are that of a general Fourier series, a complete orthonormal system, and the relations between these concepts.

Let V be a vector space with an inner product $\langle \,,\, \rangle$. A vector $\varphi \in V$ is called *normalized* if $\|\varphi\| = \langle \varphi, \varphi \rangle^{1/2} = 1$. For $g \in V$, if $g \neq 0$, then $g/\|g\|$ is normalized. Also, recall that f and g are called *orthogonal vectors* if $\langle f,g \rangle = 0$.

A sequence $\varphi_0, \varphi_1, \varphi_2, \ldots$ in V is called an *orthonormal family* if each φ_i is normalized and φ_i, φ_j are orthogonal if $i \neq j$. These conditions may be restated as

$$\langle \varphi_i, \varphi_j \rangle = \delta_{ij} \,,$$

where $\delta_{ij} = 1$ if $i = j$ and 0 if $i \neq j$.

Ultimately, we shall study the space $V = \mathscr{L}^2$ consisting of the square integrable functions $f: [a,b] \to \mathbb{C}$ with $\langle f,g \rangle = \int_a^b f(x)\overline{g(x)}\,dx$, as was stressed in Section 10.1, but for now, our discussion will be restrained to general inner product spaces.

The object of Fourier analysis is to write each $f \in V$ in the form

$$f = \sum_{k=0}^{\infty} c_k \varphi_k \,,$$

where $c_k \in \mathbb{C}$, and $\varphi_0, \varphi_1, \ldots$ is a given orthonormal family. In general, one cannot do this; however, if this can be done for each $f \in V$, the family

$\varphi_0, \varphi_1, \ldots$ is called *complete*. (This is not to be confused with the unrelated notion of complete in the sense that Cauchy sequences converge.)

The sum $\sum_{k=0}^{\infty} c_k\varphi_k$ is understood to be taken in the sense of "mean convergence," that is, if $s_n = \sum_{k=0}^{n} c_k\varphi_k$, then $\|\sum_{k=0}^{\infty} c_k\varphi_k - s_n\| \to 0$ as $n \to \infty$. The questions of pointwise or uniform convergence (in case V is a space of functions) are more subtle and will be dealt with in the ensuing sections.

Our first job is to determine the constants c_k in the expression for f. This is very easy and reasonable if we keep in mind the geometric intuition. We refer specifically to the fact that in $\mathbb{R}^n$, if $e_1, \ldots, e_n$ is an orthonormal basis, each $x \in \mathbb{R}^n$ is written

$$x = \sum_{i=1}^{n} x_i e_i ,$$

where $x_i = \langle x, e_i \rangle$. The latter is called the projection of x along e_i. The same is true in general.

Theorem 4. *Let V be an inner product space and suppose $f = \sum_{k=0}^{\infty} c_k\varphi_k$ for an orthonormal family, $\varphi_0, \varphi_1, \ldots$ in V (convergence in the mean) and $f \in V$. Then $c_k = \langle f, \varphi_k \rangle = \langle \varphi_k, f \rangle$.*

We gather some important terminology in the following definition.

Definition 5. An orthonormal family $\varphi_0, \varphi_1, \ldots$ in an inner product space V is called *complete* if every $f \in V$ can be written $f = \sum_{k=0}^{\infty} c_k\varphi_k$. We call $\sum_{k=0}^{\infty} \langle f, \varphi_k \rangle \varphi_k$ the *Fourier series* of f with respect to $\varphi_0, \varphi_1, \ldots$, and $\langle f, \varphi_k \rangle$ the *Fourier coefficients*.

Theorem 4 says that the only candidate for representing f in terms of the φ_k is the Fourier series with $c_k = \langle f, \varphi_k \rangle$. Also, note that saying $\varphi_0, \varphi_1, \ldots$ is complete can be stated as the condition that each f is "equal" to its Fourier series, that is, the Fourier series of f converges in the mean to f. Thus $\{\varphi_k\}$ is complete iff for every $f \in V$, $\|f - \sum_{k=0}^{n} \langle f, \varphi_k \rangle \varphi_k\| \to 0$ as $n \to \infty$.

Before proceeding with the theory, let us give some examples (which will be complete orthonormal families).

First, there are the *classical Fourier series* where φ_n are taken to be the functions

$$\varphi_n = \frac{e^{inx}}{\sqrt{2\pi}}, \qquad n = 0, \pm 1, \pm 2, \ldots, x \in [0, 2\pi] .$$

These will be studied in greater detail later and will be shown to be a complete orthonormal family in $V = \mathscr{L}^2$. For now, note that the Fourier series for

$f: [0,2\pi] \to \mathbb{C}$ for this family is given by

$$\sum_{k=-\infty}^{\infty} \frac{c_k e^{ikx}}{\sqrt{2\pi}}$$

where

$$c_k = \frac{1}{\sqrt{2\pi}} \int_0^{2\pi} f(x) e^{-ikx}\, dx = \langle f, \varphi_k \rangle$$

(the term is e^{-ikx} because we use the complex conjugate of g in $\langle f,g \rangle$). After we prove completeness (Section 10.3), we can assert that f equals its Fourier series in the sense of convergence in the mean.

Another family closely related to the above is

$$\frac{1}{\sqrt{2\pi}}, \frac{\cos mx}{\sqrt{\pi}}, \frac{\sin nx}{\sqrt{\pi}}, \qquad m, n = 1, 2, \ldots .$$

Accepting that this is orthonormal, the reader should write out the Fourier series of a function with respect to this family.

The above are really the only orthonormal families which are directly pertinent to our later discussions. However, for reference we give other classical examples which arise in practice. To describe these, the Gram-Schmidt process will be reviewed first.

Given an inner product space V and linearly independent vectors $g_0, g_1, g_2, \ldots$ in V, one can form a corresponding orthonormal system $\varphi_0, \varphi_1, \ldots$ by the Gram-Schmidt process. To do this, take

$$\varphi_0 = \frac{g_0}{\|g_0\|},$$

$$\varphi_1 = \frac{(g_1 - \langle g_1, \varphi_0 \rangle \varphi_0)}{\|g_1 - \langle g_1, \varphi_0 \rangle \varphi_0\|},$$

$$\varphi_2 = \frac{[g_2 - \langle g_2, \varphi_1 \rangle \varphi_1 - \langle g_2, \varphi_0 \rangle \varphi_0]}{\|g_2 - \langle g_2, \varphi_1 \rangle \varphi_1 - \langle g_2, \varphi_0 \rangle \varphi_0\|},$$

and so on. Geometrically this is the "obvious" thing to do. It is left to the reader to verify that the process leads to an orthonormal family; see Exercise 2.

The normalized *Legendre polynomials* are obtained by applying the Gram-Schmidt process to the polynomials, $1, x, x^2, \ldots, x^n, \ldots$ on $[-1,1]$. It can be shown by induction (a fairly tedious but straightforward proof) that the nth normalized Legendre polynomial is

$$P_n(x) = \frac{(2n+1)}{(\sqrt{2} \cdot 2^n n!)} \frac{d^n}{dx^n} (x^2 - 1)^n.$$

On $\mathbb{R} = \,]-\infty, \infty[$ the Gram-Schmidt process applied to the functions $x^n e^{-x^2/2}, n = 0, 1, 2, \ldots$ gives the normalized *Hermite functions*, and applied to the functions $x^n e^{-x}$, $n = 0, 1, 2, \ldots$ on $[0, \infty[$, it gives the normalized *Laguerre functions*.

These functions are more properly treated in the context of differential equations where they represent fundamental solutions of certain differential equations, just as $\sin nx$ was the fundamental solution for the vibrating string.

Let us continue with the general theory. The next result is called *Bessel's inequality*.

> **Theorem 5.** *Let* $\varphi_0, \varphi_1, \ldots$ *be an orthonormal system in an inner product space* V. *For each* $f \in V$, $\sum_{i=0}^{\infty} |\langle f, \varphi_i \rangle|^2$ *converges and we have the inequality*
>
> $$\sum_{i=0}^{\infty} |\langle f, \varphi_i \rangle|^2 \leqslant \|f\|^2 .$$

In particular, note that the Fourier coefficients $c_k = \langle f, \varphi_k \rangle$ converge to 0 as $k \to \infty$. Thus Bessel's inequality gives some control over the behavior of the Fourier coefficients. Some motivation for this result is given below.

Recall that $\varphi_0, \varphi_1, \ldots$ is a complete orthonormal system iff for every $f \in V$, we have

$$f = \sum_{k=0}^{\infty} \langle f, \varphi_k \rangle \varphi_k .$$

Parseval's theorem relates completeness of a system $\varphi_0, \varphi_1, \ldots$ to Bessel's inequality as follows.

> **Theorem 6.** *Let* V *be an inner product space and* $\varphi_0, \varphi_1, \ldots$ *an orthonormal system. Then* $\varphi_0, \varphi_1, \ldots$ *is complete iff for each* $f \in V$, *we have*
>
> $$\|f\|^2 = \sum_{n=0}^{\infty} |\langle f, \varphi_n \rangle|^2 .$$

Hence, we see under what conditions Bessel's inequality becomes an equality. This theorem gives many useful relations in Fourier series, but usually it is not very practical for telling when a given family $\varphi_0, \varphi_1, \ldots$ is complete (see, however, Exercises 7 and 75 at the end of the chapter). For this, one usually uses direct techniques, which are given in Section 10.3.

Geometrically, Parseval's relation may be regarded as a generalized Pythagoras' theorem. Recall that if g is perpendicular to h (that is, $\langle g, h \rangle = 0$), then $\|g + h\|^2 = \|g\|^2 + \|h\|^2$ (Exercise 9, Section 10.1). This is the ordinary Pythagoras theorem for right triangles. Now if $\sum_{n=0}^{\infty} \langle f, \varphi_n \rangle \varphi_n = f$, then f is a sum of orthogonal vectors $\langle f, \varphi_n \rangle \varphi_n$, so $\|f\|^2$ should equal the sum of the squares of the lengths of $\langle f, \varphi_n \rangle \varphi_n$. But, since φ_n is normalized, $\langle f, \varphi_n \rangle \varphi_n$

has square length $|\langle f,\varphi_n\rangle|^2$, so we should get $\|f\|^2 = \sum_{n=0}^{\infty}|\langle f,\varphi_n\rangle|^2$, which is Parseval's relation.

If we have an incomplete orthonormal system, then intuitively speaking, there are some terms missing on the right-hand side and so only an inequality prevails, namely, Bessel's inequality.

We have seen in Theorem 4 that it is natural and, indeed, obligatory to choose the Fourier coefficients $c_i = \langle f,\varphi_i\rangle$ when expanding $f = \sum_{i=0}^{\infty} c_i\varphi_i$. There is another reason for this choice which aids in the geometric understanding and is as follows.

The constants c_i, for which the length

$$\left\|f - \sum_{i=0}^{n} c_i\varphi_i\right\|$$

is smallest, are $c_i = \langle f,\varphi_i\rangle$, the Fourier coefficients (that is, the choice $c_i = \langle f,\varphi_i\rangle$ yields the *best mean approximation*). This is reasonable because $\sum_{i=0}^{n} \langle f,\varphi_i\rangle\varphi_i$ is just the projection of f on the space spanned by $\varphi_0, \ldots, \varphi_n$, and the shortest distance to a plane from a point is the perpendicular distance. See Figure 10-6. The precise statement is given in Theorem 7.

Theorem 7. *Let V be an inner product space and $\varphi_0, \varphi_1, \ldots, \varphi_n$ a set of orthonormal vectors in V. Then for each set of numbers $t_0, t_1, \ldots, t_n$,*

$$\left\|f - \sum_{k=0}^{n} t_k\varphi_k\right\| \geq \left\|f - \sum_{k=0}^{n} \langle f,\varphi_k\rangle\varphi_k\right\|$$

Equality holds iff $t_k = \langle f,\varphi_k\rangle$.

This concludes our brief treatment of the general theory. The remainder of the chapter is devoted to the study of the classical cases of the orthonormal

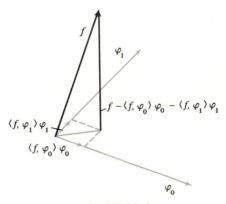

FIGURE 10-6

families

$$\left\{\frac{e^{inx}}{\sqrt{2\pi}}, n = 0, \pm 1, \ldots\right\} \quad \text{and} \quad \left\{\frac{1}{\sqrt{2\pi}}, \frac{\sin nx}{\sqrt{\pi}}, \frac{\cos mx}{\sqrt{\pi}}, n, m = 1, 2, \ldots\right\}$$

on $[0, 2\pi]$ or $[-\pi, \pi]$. The cases of corresponding orthonormal families on other intervals follow easily from this (see Exercise 3).

We cannot stress too strongly the fact that $f = \sum_{k=0}^{\infty} \langle f, \varphi_k \rangle \varphi_k$ means only that the sum converges in the mean to f and that this does not entail pointwise convergence without some additonal conditions. In the general situation (see Section 10.3) we usually do have mean convergence, but in order to obtain pointwise or uniform convergence, we require more careful hypotheses, such as continuity or differentiability assumptions on the function f.

EXAMPLE 1.   Let V be an inner product space and $\varphi_0, \varphi_1, \ldots$ a complete orthonormal system. Then show that $\varphi_1, \varphi_2, \ldots$ is not complete.

Solution: If $\varphi_1, \varphi_2, \ldots$ were complete, we could write $f = \sum_{i=1}^{\infty} \langle f, \varphi_i \rangle \varphi_i$ for each $f \in V$. Take $f = \varphi_0$; then we would have

$$\varphi_0 = \sum_{i=1}^{\infty} \langle \varphi_0, \varphi_i \rangle \varphi_i.$$

But $\langle \varphi_0, \varphi_i \rangle = 0$, $i = 1, 2, \ldots$, so $\varphi_0 = 0$, which is impossible since $\|\varphi_0\| = 1$. Hence $\varphi_1, \varphi_2, \ldots$ is not complete. As an alternative method of solving the problem observe that Parseval's relation

$$\|f\|^2 = \sum_{i=1}^{\infty} |\langle f, \varphi_i \rangle|^2$$

does not hold for $f = \varphi_0$, because the left side would be 1, while the right side would be 0. Similarly, $\varphi_N, \varphi_{N+1}, \ldots$ or any proper subcollection is not complete.

EXAMPLE 2.   If $\varphi_0, \varphi_1, \ldots$ is a complete orthonormal system in an inner product space V, and f is orthogonal to each φ_i, then $f = 0$.

Solution: Since the system is complete, we can write

$$f = \sum_{i=0}^{\infty} \langle f, \varphi_i \rangle \varphi_i.$$

By assumption, each $\langle f, \varphi_i \rangle = 0$, so that $f = 0$. If V were a Hilbert space, the converse to this would also be true. See Exercise 14, at the end of this chapter.

EXAMPLE 3. Show that the functions

$$\frac{1}{\sqrt{2\pi}}, \quad \frac{\cos mx}{\sqrt{\pi}}, \quad \frac{\sin nx}{\sqrt{\pi}}, \qquad m = 1, 2, \ldots; n = 1, 2, \ldots$$

are orthonormal on $[0, 2\pi]$.

Solution: In effect, this problem means that

$$\int_0^{2\pi} \left(\frac{1}{\sqrt{2\pi}}\right)^2 dx = 1, \qquad \int_0^{2\pi} \frac{\cos^2 mx}{\pi} dx = 1, \qquad \int_0^{2\pi} \frac{\sin^2 nx}{\pi} dx = 1$$

(normalization) and

$$\int_0^{2\pi} \frac{1}{\sqrt{2}\,\pi} \cos mx \, dx = 0, \qquad \int_0^{2\pi} \frac{1}{\sqrt{2}\,\pi} \sin nx \, dx = 0 \; ;$$

$$\int_0^{2\pi} \frac{1}{\pi} (\cos mx)(\sin nx) \, dx = 0, \qquad \text{all } m, n \; ;$$

$$\int_0^{2\pi} \frac{1}{\pi} (\cos mx)(\cos m'x) \, dx = 0, \qquad m \neq m' \; ;$$

$$\int_0^{2\pi} \frac{1}{\pi} (\sin nx)(\sin n'x) \, dx = 0, \qquad n \neq n' \; ;$$

(orthogonality). Each of these relations may be verified by elementary techniques. An easier way is to note that

$$\frac{1}{2\pi} \int_0^{2\pi} e^{inx} e^{-imx} \, dx = \delta_{nm} \, ,$$

because if $n \neq m$,

$$\int_0^{2\pi} e^{i(n-m)x} \, dx = \frac{1}{i(n - m)} e^{i(n-m)x} \bigg|_0^{2\pi} = 0 \; .$$

Taking the real and imaginary parts of this relation for all n, m gives the desired relations above.

This example also shows that

$$\left\{ \frac{e^{inx}}{\sqrt{2\pi}} \,\bigg|\, n = 0, \pm 1, \pm 2, \ldots \right\}$$

is an orthonormal system on $[0, 2\pi]$.

EXAMPLE 4. Let $f: [0, 2\pi] \to \mathbb{C}$ be such that

$$\int_0^{2\pi} |f(x)|^2 \, dx < \infty \; .$$

Then show that

$$\lim_{n \to \infty} \int_0^{2\pi} f(x)\sin nx \, dx = 0$$

and

$$\lim_{m \to \infty} \int_0^{2\pi} f(x)\cos mx \, dx = 0.$$

Solution: By Example 3, the sets

$$\left\{ \frac{\sin nx}{\sqrt{\pi}} \, \middle| \, n = 1,2,\ldots \right\} \quad \text{and} \quad \left\{ \frac{\cos mx}{\sqrt{\pi}} \, \middle| \, m = 1,2,\ldots \right\}$$

are orthonormal families. Hence, by Theorem 5, the Fourier coefficients of f with respect to these systems converge to zero, and the result follows immediately. As an exercise, a direct proof can be tried (see p. 416).

The reader may legitimately ask where the hypothesis

$$\|f\|^2 = \int_0^{2\pi} |f(x)|^2 \, dx < \infty$$

is used in this solution. This is required so that we can form the inner product space V of such functions, and obtain the upper bound $\|f\|^2 < \infty$ so the series of Theorem 5 will converge.

Exercises for Section 10.2

1. Take the case $V = \mathbb{R}^n$. Show that any n orthonormal vectors form a complete set.

2. Let $g_0, g_1, g_2, \ldots$ be linearly independent vectors in an inner product space. Inductively define

$$h_0 = g_0, \varphi_0 = \frac{h_0}{\|h_0\|}, \ldots, h_n = g_n - \sum_{k=0}^{n-1} \langle g_n, \varphi_k \rangle \varphi_k, \varphi_n = \frac{h_n}{\|h_n\|}.$$

Show that $\varphi_0, \varphi_1, \varphi_2, \ldots$ are orthonormal. Why must we assume that the g's are linearly independent?

3. (a) Suppose $\varphi_0(x), \varphi_1(x), \ldots$ are orthonormal functions on $[0,2\pi]$. Then show that the functions

$$\psi_n(x) = \sqrt{\frac{2\pi}{l}} \, \varphi_n\!\left(\frac{2\pi x}{l}\right)$$

are orthonormal on $[0,l]$.

(b) Write the family obtained by modifying $\dfrac{1}{\sqrt{2\pi}}, \dfrac{\sin nx}{\sqrt{\pi}}, \dfrac{\cos nx}{\sqrt{\pi}}$, or $\dfrac{e^{inx}}{\sqrt{2\pi}}$ to $[0,l]$ as in (a).

(c) Write the Fourier series of f for the families obtained in (b).

(d) Show that if the φ_n in (a) are complete, so are the ψ_n.

4. Assume for the moment that the functions $\dfrac{1}{\sqrt{2\pi}}, \dfrac{\sin nx}{\sqrt{\pi}}, \dfrac{\cos mx}{\sqrt{\pi}}$ are complete on the interval $[0, 2\pi]$ (this will be proved later).

(a) Apply this to the function x to show that

$$x = \pi - 2\sum_{n=1}^{\infty} \frac{\sin nx}{n}$$

(convergence in the mean).

(b) Using the Fourier coefficients found in (a), apply Parseval's relation to show that

$$\frac{\pi^2}{6} = \sum_{n=1}^{\infty} \frac{1}{n^2}.$$

(c) Use the same procedure on x^2 to get $\dfrac{\pi^4}{90} = \sum_{n=1}^{\infty} \dfrac{1}{n^4}.$

5. Prove that the Fourier series of a sum of two functions is the sum of the Fourier series.

6. Prove that

$$2\sum_{k=1}^{n} \cos k\theta = \left[\frac{\sin(n + 1/2)\theta}{\sin \theta/2}\right] - 1.$$

[Hint: First, note that

$$e^{i\theta} + e^{2i\theta} + \cdots + e^{ni\theta} = \frac{e^{i\theta}(1 - e^{in\theta})}{1 - e^{i\theta}}$$

and take the real and imaginary parts.] This result will be important to us later.

10.3 Completeness and Convergence Theorems

This section will investigate the problem of the convergence of the Fourier series of a function. We see that the Fourier series of a given function is completely determined by that function, but there is no prior guarantee that the series converges or, if it does converge, whether or not its sum is the given function. The type of convergence we obtain depends on the hypotheses we place on f. The important results are summarized in Table 10-1; further convergence theorems are given in Sections 10.4 and 10.6.

It is possible to weaken slightly the hypotheses of the pointwise convergence theorems presented, but this makes little difference in practice and requires lengthy expositions on topics such as functions of bounded variation. We shall discuss those slightly sharper results in the optional Section 10.4.*

* We should also like to mention that there is a deep result of L. Carleson which states that for $|f|^2$ integrable, the Fourier series of f converges pointwise to f, except possibly on a set of measure zero. However, this result is far beyond the scope of this book. See Acta. Math. *116* (1966) p. 135.

TABLE 10-1   Convergence Properties of Fourier Series

Hypotheses on the function f	Convergence of Fourier series
$\int_0^{2\pi} \lvert f(x)\rvert^2\, dx < \infty$	Converges in mean to f
f, f' both sectionally continuous	Converges pointwise (and in mean) to $$\frac{[f(x+) + f(x-)]}{2}$$
f continuous, f' sectionally continuous	Converges uniformly (Section 10.6), pointwise, and in mean to f

The practical aspects of Fourier series (that is, examples and computational methods) are given in Section 10.5.

From now on, we deal primarily with the following two orthonormal systems:

(a) Exponential system

$$\frac{e^{inx}}{\sqrt{2\pi}}, \qquad n = 0, \pm 1, \pm 2, \ldots$$

(b) Trigonometric system

$$\frac{1}{\sqrt{2\pi}}, \frac{\sin nx}{\sqrt{\pi}}, \frac{\cos mx}{\sqrt{\pi}}, \qquad n, m = 1, 2, \ldots$$

on the intervals $[0,2\pi]$ or $[-\pi,\pi]$.

These two systems are closely related. Indeed, the trigonometric system is obtained by taking the real and imaginary parts of the exponential system (see Exercise 1).

The Fourier series of a function $f: [0,2\pi] \to \mathbb{C}$ with respect to the exponential system is the series

$$\sum_{n=-\infty}^{\infty} c_n e^{inx} = \lim_{N \to \infty} \sum_{n=-N}^{N} c_n e^{inx},$$

where the Fourier coefficients are given by

$$c_n = \frac{1}{2\pi} \int_0^{2\pi} f(y)e^{-iny}\, dy$$

(we have gathered two $\sqrt{2\pi}$'s for convenience and for historical and conventional reasons).

The Fourier series of a function f with respect to the trigonometric system is

$$\frac{a_0}{2} + \sum_{n=1}^{\infty} a_n \cos nx + b_n \sin nx \,,$$

where the coefficients are given by

$$a_n = \frac{1}{\pi} \int_0^{2\pi} f(x) \cos nx \, dx, \qquad n = 0, 1, 2, \dots$$

and

$$b_n = \frac{1}{\pi} \int_0^{2\pi} f(x) \sin nx \, dx, \qquad n = 1, 2, \dots \, .$$

The reader should review Sections 10.1 and 10.2 if these statements are not clear.

The partial sums for the trigonometric series and the exponential series are the same (Exercise 1). Thus, if we can prove theorems for one system we will automatically obtain theorems for the other. The system used depends on the particular problem and, to some extent, personal taste. Examples of computational differences are given in the Section 10.5.

The primary goal here is to give theorems which enable us to say that a function "equals" its Fourier series. If we take the equality to be convergence in the mean, then this is a problem of completeness of the orthonormal system. Fortunately, and this is one of the main theorems of the subject, the above systems are complete. On the other hand, if we take the equality to mean pointwise or uniform convergence, then extra conditions must be put on f.

Let us first deal with completeness; the theorem is as follows.

Theorem 8.   *The exponential and trigonometric systems on* $[0, 2\pi]$ *(or* $[-\pi, \pi]$*) are complete in the space* $V = \mathscr{L}^2$ *of functions,* $f: [0, 2\pi] \to \mathbb{C}$ *with* $\int_0^{2\pi} |f(x)|^2 \, dx < \infty$ *(the integral may be improper).*

This means that for any function f with $|f|^2$ integrable (that is, f is square integrable), f equals the sum of its Fourier series in the sense of convergence in the mean.

The proof of this result is a little involved; completely different proofs of related results which will follow, especially Theorem 9, might be more easily understood. (See also Exercises 75 and 76 at the end of the chapter for alternative proofs of Theorem 8.)

It is false that we always get pointwise convergence. Indeed, one can

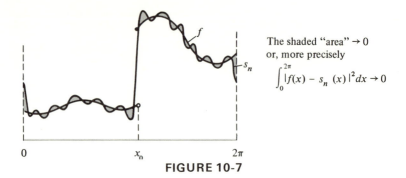

The shaded "area" $\to 0$
or, more precisely
$$\int_0^{2\pi} |f(x) - s_n(x)|^2 dx \to 0$$

FIGURE 10-7

even construct a continuous (periodic) function f whose Fourier series diverges at a given point.*

Let s_n be the nth partial sum of the trigonometric Fourier series for the real function f. Then the intuition behind Theorem 8 is illustrated in Figure 10-7. Notice that each s_n is a nice smooth function (a trigonometric polynomial) but as $n \to \infty$, s_n may still converge to something discontinuous as illustrated. If we demand uniform convergence, f must be continuous by Theorem 8, Chapter 5. Thus if f is discontinuous, we get mean convergence, but never uniform convergence.

The above theorem also follows, in special cases, from somewhat easier theorems given later in Section 10.6. The technique of the proof of Theorem 1 is important, for it shows that Fourier series in higher dimensions are also complete (see Exericse 18 at the end of the chapter). Theorem 8 has the advantage that it is valid for a wide class of functions f. However, it does not deal with the question of pointwise convergence. The next theorem does answer this question.

To state the theorem, we need some additional terminology. For this theorem we may use either real or complex functions, but it is often enough to consider real functions, for if $f = f_1 + if_2$, the Fourier series of f is that of $f_1 + i$ (that of f_2) (why?).

Suppose then, that $f\colon [0,2\pi] \to \mathbb{R}$ (or $[-\pi,\pi] \to \mathbb{R}$) has a possible discontinuity at $x_0 \in [0,2\pi]$. In case $x_0 = 0$ or 2π this shall mean that we are to take the function f *extended to be periodic;* that is, define $f(x + 2\pi) = f(x)$. This is reasonable because the Fourier series itself is periodic. This periodic extension is illustrated for two cases in Figure 10-8. Now recall that we define

$$f(x_0+) = \lim_{\substack{x \to x_0+}} f(x) = \lim_{\substack{x \to x_0 \\ x > x_0}} f(x)$$

if it exists (see p. 80). This means that for every $\varepsilon > 0$, there is a $\delta > 0$

* This uses more advanced methods; see, for example, Widom, Drasin, and Tromba, *Lectures on Measure and Integration Theory*, p. 153. Van Nostrand Mathematical Studies #20.

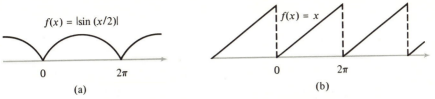

$f(x) = |\sin (x/2)|$

$f(x) = x$

0   2π

0   2π

(a)

(b)

FIGURE 10-8   (a) Continuous at 0. (b) Discontinuous at 0.

such that if $|x - x_0| < \delta$ and $x > x_0$, then $|f(x) - f(x_0+)| < \varepsilon$. Intuitively, $f(x_0+)$ means the value of f just to the right of x_0. See Figure 10-9. Of course, $f(x_0+)$ may not exist; look at Figure 10-9b. One defines $f(x_0-)$ in an entirely analogous way. A discontinuity, x_0, where both $f(x_0-)$ and $f(x_0+)$ exist, is called a *jump discontinuity* and $f(x_0+) - f(x_0-)$ is called the *jump* of f at x_0. The jump can, of course, be either positive or negative, and is zero iff f is continuous at x_0.

Suppose f is differentiable on some open interval $]x_0,x_0 + \varepsilon[$. Then we can form f' on this set, and hence can talk about $f'(x_0+)$ if it exists (by the above definition). Similarly, we can form $f'(x_0-)$. Intuitively, $f'(x_0+)$ is the slope of f just to the right of x_0. For instance, in Figure 10-8a, $f'(0+) = \lim_{x \to 0+}(d/dx)(\sin x/2) = 1/2$ and in Figure 10-8b, $f'(0+) = +1$.

There is a slightly weaker definition of $f'(x_0+)$ which is sometimes important. The above definition demanded that $f'(x)$ exists for $x > x_0$ and for $f'(x_0+) = \lim_{x \to x_0+} f'(x)$ to exist. It is easy to prove that if this is so, then

$$f'(x_0+) = \lim_{h \to 0+}\left\{\frac{f(x_0 + h) - f(x_0+)}{h}\right\}$$

(see Exercise 39 at the end of the chapter). For the following theorem the existence of this limit is sufficient, so we shall adopt it as our definition, with

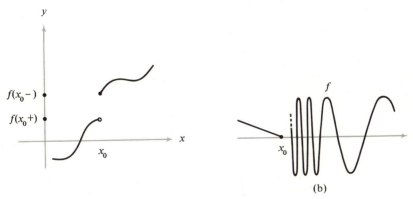

y

$f(x_0-)$

$f(x_0+)$

x_0

x

f

x_0

(b)

**FIGURE 10-9   (a) Jump discontinuity.
(b) $f(x_0 +)$ does not exist.**

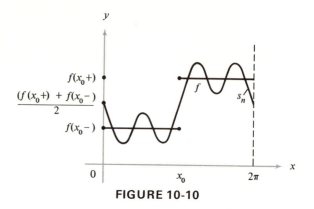

FIGURE 10-10

$f'(x_0-)$ similarly defined by

$$f'(x_0-) = \lim_{h \to 0+} \left\{ \frac{f(x_0-) - f(x_0 - h)}{h} \right\}.$$

The proof that this second method is actually a weaker assumption is left to the reader in the same exercise.

Observe that f is differentiable at x_0 iff $f(x_0+) = f(x_0-)$, $f'(x_0+)$, and $f'(x_0-)$ exist, and are equal.

The next theorem contains the principal result on pointwise convergence.

> **Theorem 9.** *Let $f: [0,2\pi] \to \mathbb{R}$ (or $f: [-\pi,\pi] \to \mathbb{R}$) be sectionally continuous, have a jump discontinuity at x_0, and assume that $f'(x_0+)$ and $f'(x_0-)$ both exist. Then the Fourier series of f (either in exponential or trigonometric form) evaluated at x_0 converges to $[f(x_0+) + f(x_0-)]/2$. In particular, if f is differentiable at x_0, the Fourier series of f converges at x_0 to $f(x_0)$.*

If x_0 is an endpoint of the interval, then as mentioned previously the numbers $f(x_0+)$ and $f(x_0-)$ are computed for the function after it is extended to be periodic (see Figure 10-8). In the section on theorem proofs, we give two proofs of this result. The first is quite short. The second, which is the classical proof, is longer but is also useful for other purposes required in Section 10.4, so it is included as well.

Notice that the Fourier series does not necessarily converge to $f(x_0)$ at a jump discontinuity but to the average of $f(x_0+)$ and $f(x_0-)$. A typical example is a step function (see Figure 10-10).

Theorem 9 is very nice because it gives us conditions which are easily verified in examples, and which do hold in most cases of interest. Furthermore, even in simple examples it is difficult to prove directly (without Theorem 9) that the Fourier series converges to the function.

In Theorem 9 we also have mean convergence of the Fourier series to f by Theorem 8. However, Theorem 9 tells us that in addition, the Fourier series converges at points where conditions of the theorem hold. Pointwise convergence is a more delicate and sometimes more useful condition.

EXAMPLE 1. Suppose $f: [0,2\pi] \to \mathbb{C}$ has $\int_0^{2\pi} |f(x)|^2\, dx < \infty$. Then show that

$$\int_0^{2\pi} |f(x)|^2\, dx = \sum_{n=-\infty}^{\infty} \frac{1}{2\pi} \left| \int_0^{2\pi} f(x) e^{-inx}\, dx \right|^2$$

$$= \frac{1}{2\pi} \left| \int_0^{2\pi} f(x)\, dx \right|^2$$

$$+ \sum_{n=1}^{\infty} \frac{1}{\pi} \left| \int_0^{2\pi} f(x)\cos nx\, dx \right|^2$$

$$+ \sum_{n=1}^{\infty} \frac{1}{\pi} \left| \int_0^{2\pi} f(x)\sin nx\, dx \right|^2 .$$

Solution: Let

$$V = \mathscr{L}^2 = \left\{ f: [0,2\pi] \to \mathbb{C} \ \middle| \ \|f\|^2 = \int_0^{2\pi} |f(x)|^2\, dx < \infty \right\}.$$

Then, by Theorem 8,

$$\left\{ \varphi_n(x) = \frac{e^{inx}}{\sqrt{2\pi}} \ \middle| \ n = 0, \pm 1, \pm 2, \ldots \right\}$$

is a complete orthonormal family in V. Thus, by Theorem 6, Parseval's relation holds, and so

$$\|f\|^2 = \sum_{n=-\infty}^{\infty} |\langle f, \varphi_n \rangle|^2 .$$

Here

$$\langle f, \varphi_n \rangle = \int_0^{2\pi} f(x)\overline{\varphi_n(x)}\, dx$$

$$= \int_0^{2\pi} \frac{f(x) e^{-inx}\, dx}{\sqrt{2\pi}}$$

so the first equality follows. Let us recall that

$$\sum_{n=-\infty}^{\infty} |\langle f, \varphi_n \rangle|^2 \qquad \text{means} \qquad \lim_{N \to \infty} \sum_{n=-N}^{N} |\langle f, \varphi_n \rangle|^2$$

for the exponential functions (that is, they are taken in the order $\varphi_0, \varphi_1, \varphi_{-1}, \varphi_2, \varphi_{-2}, \ldots$). (However, here the terms are positive, so the series can be rearranged arbitrarily, by Example 5 at the end of Chapter 5.)

The second equality follows by applying the same procedure to the complete orthonormal family,

$$\left\{\frac{1}{\sqrt{2\pi}}, \frac{\cos nx}{\sqrt{\pi}}, \frac{\sin mx}{\sqrt{\pi}} \,\middle|\, n,m = 1,2,\ldots\right\}.$$

One can also derive the second equality from the first by writing $e^{-inx} = \cos nx - i \sin nx$, squaring and gathering terms, and noting that the cross-terms from n and $-n$ cancel.

EXAMPLE 2. For the following functions on $[-\pi,\pi]$, state whether we have mean or pointwise convergence of the Fourier series and what the series converges to at $x_0 = 0$.

(a) $f(x) = \begin{cases} -2, & x \leqslant 0, \\ 2, & x > 0. \end{cases}$

(b) $f(x) = \begin{cases} x, & x \leqslant 1, \\ x + 1, & x > 1. \end{cases}$

(c) $f(x) = \sin x.$

(d) $f(x) = \begin{cases} 1 + x, & x \leqslant 0, \\ x \sin\left(\dfrac{1}{x}\right), & x > 0. \end{cases}$

Solution: The graphs of these functions are given in Figure 10-11. Each function is piecewise continuous, and the discontinuities are jump discontinuities. This is obvious except perhaps for (d). There, $f(x) = x \sin(1/x) \to 0$ as $x \to 0$, since $|x \sin(1/x)| \leqslant |x|$, so $f(0+) = 0$.

Also, at 0, $f'(0+)$ and $f'(0-)$ exist in cases (a), (b), and (c). All of these are fairly obvious. For instance, in (a), $f(x) = 2$ for $x > 0$, and so $\lim_{x\to 0+} f'(x) = 0$ exists. In case (d), this is not true. Here, for $h > 0$, we have

$$\frac{f(0 + h) - f(0+)}{h} = \sin\left(\frac{1}{h}\right),$$

which does not converge as $h \to 0$. Thus Theorem 9 does not apply in this case. However, in each case we do have mean convergence by Theorem 8. At $x = 0$, the Fourier series converges in (a) to $0 = [f(0+) + f(0-)]/2$, in (b) to 0, in (c) to 0, and in (d) our theorems fail. (One can show the convergence of the Fourier series in (d) to $1/2$ by a direct analysis.)

EXAMPLE 3. Find an example of a function f such that the Fourier series of f converges pointwise and in the mean to f, but does not converge uniformly.

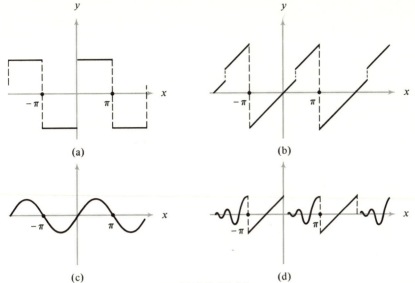

(a)　　　　　　　　　　　(b)

(c)　　　　　　　　　　　(d)

FIGURE 10-11

Solution: Let

$$f = \begin{cases} 0, & -\pi < x < 0, \\[2mm] \dfrac{1}{2}, & x = 0, \\[2mm] 1, & 0 < x < \pi, \\[2mm] \dfrac{1}{2}, & x = \pi. \end{cases}$$

The discontinuities of f are jump discontinuities (see Figure 10-12). From Theorem 8, the Fourier series of f converges to f in mean, and by Theorem 9, it converges pointwise, since $f(x_0) = [f(x_0+) + f(x_0-)]/2$ at each point.

However, the Fourier series cannot uniformly converge to f, because each $s_n(x)$ is continuous and if $s_n(x) \to f(x)$ uniformly, f would be continuous (Theorem 1, Chapter 5), which is not the case.

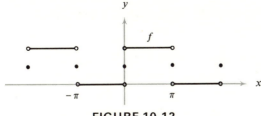

FIGURE 10-12

Exercises for Section 10.3

1. (a) Show that the nth partial sum of the trigonometric Fourier series of a (real or complex) function is equal to the nth partial sum of the exponential series. [Hint: prove this by writing $e^{inx} = \cos nx + i \sin nx$.]
   (b) Write the corresponding series on $[-\pi,\pi]$.
   (c) Show that if, on $[-\pi,\pi]$, f is even (that is, $f(x) = f(-x)$), then in the trigonometric Fourier series, all $b_n = 0$. The series is then called the *cosine series*.
   (d) Repeat question (c) for f odd; that is, if $f(-x) = -f(x)$ show all $a_n = 0$. The series is then called the *sine series*.

2. For $f: [0,2\pi] \to \mathbb{R}$, show that f is continuous at zero (in the sense of f being periodic) iff $f(0) = f(2\pi)$ and f is continuous in the usual sense at both points 0 and 2π in $[0,2\pi]$, that is, $\displaystyle\lim_{x \to 0+} f(x) = f(0)$, and $\displaystyle\lim_{x \to 2\pi-} f(x) = f(2\pi)$.

3. Suppose $f: [0,l] \to \mathbb{C}$ has $\int_0^l |f(x)|^2 \, dx < \infty$. Then show that

$$\int_0^l |f(x)|^2 \, dx = \frac{1}{l} \sum_{n=-\infty}^{\infty} \left| \int_0^l f(x) e^{in2\pi x/l} \, dx \right|^2$$

$$= \frac{1}{l} \left| \int_0^l f(x) \, dx \right|^2$$

$$+ 2 \sum_{n=1}^{\infty} \frac{1}{l} \left| \int_0^l f(x) \cos\left(\frac{2\pi nx}{l}\right) dx \right|^2$$

$$+ 2 \sum_{m=1}^{\infty} \frac{1}{l} \left| \int_0^l f(x) \sin\left(\frac{mx2\pi}{l}\right) dx \right|^2 .$$

4. For each of the following functions on $[-\pi,\pi]$ determine whether the Fourier series converges pointwise or in mean, and what the pointwise limit is if it exists.
   (a) $f(x) = x^n$ (consider all possible values of n; $n = \ldots, -3, -2, -1, 0, 1, 2, \ldots$).
   (b) $f(x) = \begin{cases} 0, & x < 0, \\ kx, & x \geqslant 0. \end{cases}$ for some $k \in \mathbb{R}$,
   (c) $f(x) = \tan x$.
   (d) $f(x) = e^{-x^2}$.
   (e) $f(x) = \begin{cases} e^{-1/x^2}, & x > 0, \\ 0, & x \leqslant 0. \end{cases}$

5. Use the theorems of this section to justify your manipulations in Exercise 4, Section 10.2.

10.4   Functions of Bounded Variation and Fejér Theory*

There is a theorem similar to the Jordan theorem (Theorem 9 above), but which holds under more general conditions and which also gives a criterion for uniform convergence. We shall just state this theorem without proof

---

* This section is optional and may be omitted without loss of continuity.

(although it is quite similar to the proof of Theorem 9, it is just a little more intricate). We shall be content to prove a weaker version in Section 10.6 and to prove a related theorem of Fejér.

To understand the theorem, the notion of a function of bounded variation is needed. Let $f: [a,b] \to \mathbb{R}$. Say that f is of *bounded variation* if there is a number M such that for all partitions $a = x_0 < x_1 < \cdots < x_n = b$ of $[a,b]$,

$$\sum_{k=1}^{n} |f(x_k) - f(x_{k-1})| \leq M .$$

Roughly, saying that f is of bounded variation means that the graph of f has finite arc length. The conditions of Theorem 9 imply that f is of bounded variation on some closed interval containing x_0, but being of bounded variation is generally a weaker condition. One can show that a function is of bounded variation iff it is the difference of two bounded monotone functions.* It follows (see Exercise 3, p. 292) that if f is of bounded variation, then its discontinuities are all jump discontinuities and are countable in number.

The Dirichlet-Jordan theorem is as follows (the proof is omitted).

Theorem 10.   *Let $f: [0,2\pi] \to \mathbb{R}$ be a bounded function.*

(i) *If f is of bounded variation on an interval $[x_0 - \varepsilon, x_0 + \varepsilon]$, (for some $\varepsilon > 0$) about x_0, then the Fourier series of f evaluated at x_0 converges to $[f(x_0+) + f(x_0-)]/2$.*

(ii) *If f is continuous and of bounded variation, then the Fourier series of f converges uniformly to f.*

Both Theorem 9 and the Dirichlet-Jordan theorem give sufficient conditions for the Fourier series to converge. Exercise 34 gives an example to show that the conditions are not necessary. Useful, necessary, and sufficient conditions are not known.

As we have remarked, the Fourier series of a continuous function need not converge pointwise. By the Dirichlet-Jordan theorem, such a function cannot be of bounded variation. Fejér's theory covers this case by weakening pointwise convergence of the series to Cesaro summability of the series. Let us recall from Section 5.9 that sequence $a_1, a_2, \ldots$ is said to converge *in the sense of Cesaro* or (C,1) if $\sigma_n = (a_1 + \cdots + a_n)/n$ converges. If $a_n \to x$, then $\sigma_n \to x$, but not necessarily conversely. For series, this criterion is applied to the partial sums.

In 1904 Fejér proved the remarkable fact that although the Fourier series of a continuous function need not converge pointwise, it is always (C,1) convergent.

---

* If f is of bounded variation, set $v(x) = \sup\{\sum_{k=1}^{n} |f(x_k) - f(x_{k-1})| \ \big| \ a = x_0 \leq x_1 \leq \cdots \leq x_n = x\}$, the variation of f. Write $f = p - q$, where $p = v + f/2$, $q = v - f/2$. One checks that p and q are increasing. The converse is easy to verify.

>   **Theorem 11** (*Fejér*).   *Let f be piecewise continuous on* $[0,2\pi]$ *and suppose* $f(x_0+)$ *and* $f(x_0-)$ *exist. Then the Fourier series of f converges* (C,1) *at* x_0 *to* $[f(x_0+) + f(x_0-)]/2$. *If f is continuous, the Fourier series converges* (C,1) *uniformly to f.*

Note that no assumption of bounded variation or differentiability is required. For practical applications, these refinements of Theorem 9 are not too important, but they are of considerable theoretical interest.

When one considers "distributions" or "generalized functions," such as the Dirac delta function (Section 8.9), Fourier series still make sense when suitably interpreted, and every distribution has a convergent Fourier series (convergence in an appropriate sense, see p. 277). These convergence facts are quite useful in practice, but space does not allow a treatment of them here.*

EXAMPLE 1.   Let us formally compute the Fourier series of the delta function, δ on $[-\pi,\pi]$. Recall that this function has the defining property:

$$\int_{-\pi}^{\pi} f(x)\,\delta(x - a)\,dx = f(a) \;.$$

Now

$$\int_{-\pi}^{\pi} \delta(x)e^{-inx}\,dx = 1 \;,$$

so the Fourier series of δ is

$$\sum_{n=-\infty}^{\infty} \frac{e^{inx}}{2\pi} \;.$$

Of course, this does not converge at $x = 0$, but we do not expect it to, since $\delta(0)$ is undefined.

What is true is that

$$\delta(x) = \sum_{n=-\infty}^{\infty} \frac{e^{inx}}{2\pi}$$

in the sense that it holds under the integral sign; that is, for any continuously differentiable function, f,

$$f(0) = \int_{-\pi}^{\pi} \delta(x)f(x)\,dx = \sum_{-\infty}^{\infty} \int_{-\pi}^{\pi} \frac{f(x)e^{inx}\,dx}{2\pi} \;.$$

The validity of this is quite obvious; in fact, from Theorem 9,

$$f(y) = \sum_{-\infty}^{\infty} \left(\int_{-\pi}^{\pi} f(x)e^{-inx}\,dx \right) \frac{e^{iny}}{2\pi}$$

for each y. (Since the sum is from $-\infty$ to $+\infty$, we can replace n by $-n$.)

---

* For a more complete discussion, see for example, Zemanian, *Distribution Theory and Transform Analysis*.

The situation for a general distribution and the proof of convergence of its Fourier series is analogous, that is, if T is a distribution, then as above,

$$T = \frac{1}{2\pi} \sum_{-\infty}^{\infty} a_n e^{inx},$$

where $a_n = T(e^{-inx})$.

Exercises for Section 10.4

1. Prove that the trigonometric series $\sum_{-\infty}^{\infty} e^{ikx}$ is (C,1) summable to 0 for x not a multiple of 2π.

2. Compute the Fourier series of δ', the derivative of the delta function.

10.5 Computation of Fourier Series

In this section we are mainly concerned with specific examples of Fourier series and methods that can be used to compute them. Included in our discussion is an interesting and important phenomenon which occurs in the behavior of a Fourier series at a jump discontinuity; this is known as the Gibbs' phenomenon.

The trigonometric and exponential forms of Fourier series are entirely equivalent as we have seen (Exercise 1, Section 10.3). For computations, the trigonometric form is often the most convenient. The various forms of Fourier series and their convergence properties are summarized in Tables 10-2 and 10-3. The functions can be real or complex, but we will work with real functions for simplicity.

There are several comments to be made on these formulas. The first two forms (Table 10-2) should be self-explanatory. The Fourier sine series arises when f is odd, because then we have $f(-x) = -f(x)$ and hence

$$a_n = \frac{1}{\pi} \int_{-\pi}^{\pi} f(x)\cos nx \, dx$$

$$= \frac{1}{\pi} \int_{\pi}^{-\pi} f(-x)\cos(-nx) \, d(-x)$$

$$= -\frac{1}{\pi} \int_{\pi}^{-\pi} f(x)\cos nx \, d(-x) = -a_n,$$

so $a_n = 0$.

Similarly, for f even, the Fourier series reduces to the cosine series. See Figure 10-13.

TABLE 10-2 Various Forms of Fourier Series

	The function f	Fourier series	Coefficients
Exponential Fourier series	f defined on $[0,2\pi]$ (or $[-\pi,\pi]$)	$\displaystyle\sum_{n=-\infty}^{\infty} c_n e^{inx}$	$\displaystyle c_n = \frac{1}{2\pi}\int_0^{2\pi} f(x)e^{-inx}\,dx \left(\text{or} \int_{-\pi}^{\pi}\right)$
Trigonometric Fourier series	f defined on $[-\pi,\pi]$ (or $[0,2\pi]$)	$\displaystyle\frac{a_0}{2} + \sum_{n=1}^{\infty}(a_n\cos nx + b_n\sin nx)$	$\displaystyle a_n = \frac{1}{\pi}\int_{-\pi}^{\pi} f(x)\cos nx\,dx, \quad n = 0,1,\ldots$ $\displaystyle b_n = \frac{1}{\pi}\int_{-\pi}^{\pi} f(x)\sin nx\,dx, \quad n = 1,2,\ldots$ $\displaystyle \left(\text{or} \int_0^{2\pi}\right)$
Fourier sine series	f defined on $[-\pi,\pi]$ and f is odd; $f(-x) = -f(x)$	$\displaystyle\sum_{n=1}^{\infty} b_n\sin nx$	$\displaystyle b_n = \frac{1}{\pi}\int_{-\pi}^{\pi} f(x)\sin nx\,dx, \quad n = 1,2,\ldots$ $\displaystyle = \frac{2}{\pi}\int_0^{\pi} f(x)\sin nx\,dx$

Fourier cosine series	f defined on $[-\pi,\pi]$ and f is even; $f(-x)=f(x)$	$\dfrac{a_0}{2}+\displaystyle\sum_{n=1}^{\infty}a_n\cos nx$	$a_n=\dfrac{1}{\pi}\displaystyle\int_{-\pi}^{\pi}f(x)\cos nx\,dx,\quad n=0,1,\ldots$ $=\dfrac{2}{\pi}\displaystyle\int_0^{\pi}f(x)\cos nx\,dx$
Exponential series on $[-l,l]$	f defined on $[-l,l]$	$\displaystyle\sum_{n=-\infty}^{\infty}c_n e^{inx\pi/l}$	$c_n=\dfrac{1}{2l}\displaystyle\int_{-l}^{l}f(x)e^{-inx\pi/l}\,dx$
Trigonometric series on $[-l,l]$	f defined on $[-l,l]$	$\dfrac{a_0}{2}+\displaystyle\sum_{n=1}^{\infty}\left(a_n\cos\left(\dfrac{n\pi x}{l}\right)+b_n\sin\left(\dfrac{n\pi x}{l}\right)\right)$	$a_n=\dfrac{1}{l}\displaystyle\int_{-l}^{l}f(x)\cos\left(\dfrac{n\pi x}{l}\right)dx,\quad n=0,1,\ldots$ $b_n=\dfrac{1}{l}\displaystyle\int_{-l}^{l}f(x)\sin\left(\dfrac{n\pi x}{l}\right)dx,\quad n=1,2,\ldots$
Half-interval cosine series	f defined on $[0,l]$	$\dfrac{a_0}{2}+\displaystyle\sum_{n=1}^{\infty}a_n\cos\left(\dfrac{n\pi x}{l}\right)$	$a_n=\dfrac{2}{l}\displaystyle\int_0^{l}f(x)\cos\left(\dfrac{n\pi x}{l}\right)dx,\quad n=0,1,\ldots$
Half-interval sine series	f defined on $[0,l]$	$\displaystyle\sum_{n=1}^{\infty}b_n\sin\left(\dfrac{n\pi x}{l}\right)$	$b_n=\dfrac{2}{l}\displaystyle\int_0^{l}f(x)\sin\left(\dfrac{n\pi x}{l}\right)dx,\quad n=1,2,\ldots$

TABLE 10-3 Convergence Properties

Properties of f	Convergence of Fourier series
$\displaystyle\int_0^{2\pi} \|f(x)\|^2\, dx < \infty$	Converges in mean to f
f has a jump discontinuity at x_0 and $f'(x_0 +)$, $f'(x_0 -)$ exist. If x_0 is an endpoint, we regard f as extended so it becomes periodic (see below for the half-interval forms)	Converges pointwise at x_0 to $$\frac{[f(x_0 +) + f(x_0 -)]}{2}$$
f continuous and f' sectionally continuous	Converges uniformly to f

For the interval $[-l,l]$, we replace orthonormal functions φ_k on $[-\pi,\pi]$ by $\psi_k(x) = \sqrt{\pi/l}\,\varphi_k(\pi x/l)$ which again are orthonormal on $[-l,l]$. This is just a change of scale and the same convergence properties also hold in this case. The reader should write down the sine and cosine series (for f odd or even) on a general interval $[-l,l]$.

The half-interval formulas are obtained as follows: for the cosine series extend f to $[-l,l]$ by defining

$$f(-x) = f(x) .$$

Then f becomes even and so has a cosine series. See Figure 10-14. For convergence at $x_0 = 0$ we must check this extended function, and not the original one. If $f(0+)$ exists, then the extended function evidently has no jump at 0. Similarly, we do not get a jump at l or $-l$. Thus the usual convergence criterion applies without modification for the cosine series.

The half-interval sine series is similar. On $[-l,0[$ define f by $f(-x) = -f(x)$, for $0 \leqslant x \leqslant l$ so that f is odd and so has a sine series for its Fourier

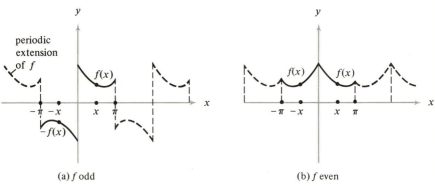

(a) f odd   (b) f even

FIGURE 10-13 (a) f is odd. (b) f is even.

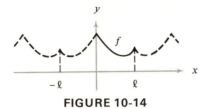

FIGURE 10-14

series. See Figure 10-15. In this case, at the point 0 the Fourier series is always zero (sin 0 = 0). Thus at 0, we do get an extra jump discontinuity introduced but the Fourier series is zero at these points. To ensure continuity of the extended f, we would have to impose the conditions that f be continuous and $f(0) = f(l)$. Thus the convergence criteria apply to the half-interval sine series without modification if we keep in mind that at 0, l the convergence is to zero.

From the general theory (Theorems 6 and 8) we know that Parseval's relation holds for each f with $\int_0^{2\pi} |f|^2 \, dx < \infty$; that is,

$$\int_0^{2\pi} |f|^2 \, dx = \sum_{n=-\infty}^{\infty} |c_n|^2 \, ,$$

where c_n are the Fourier coefficients in exponential form. Care must be taken in the above cases because c_n, a_n, b_n of the table are not the Fourier coefficients in the previous sense, as we have gathered factors of $\sqrt{2\pi}, \sqrt{\pi}$ for traditional reasons. But if we remember this, Parseval's relation is easy to find. The results are tabulated in Table 10-4.

If we know the Fourier series of $f(x)$, say, on $[-\pi, \pi]$, then we can get an expansion for the function $g(x) = \int_{-\pi}^{x} f(y) \, dy$ using the following theorem.

Theorem 12. *Suppose $\int_{-\pi}^{\pi} |f(x)|^2 \, dx < \infty$ and f has Fourier series*

$$\frac{a_0}{2} + \sum_{n=1}^{\infty} (a_n \cos nx + b_n \sin nx) \, .$$

Then letting $g(x) = \int_{-\pi}^{x} f(y) \, dy$, we have

$$g(x) = \frac{a_0(x + \pi)}{2} + \sum_{n=1}^{\infty} \left(a_n \int_{-\pi}^{x} \cos ny \, dy + b_n \int_{-\pi}^{x} \sin ny \, dy \right)$$

$$= \frac{a_0(x + \pi)}{2} + \sum_{n=1}^{\infty} \left\{ \left(\frac{a_n}{n} \right) \sin nx - \left(\frac{b_n}{n} \right) (\cos(nx) - (-1)^n) \right\}$$

and the convergence is uniform for $-\pi \leqslant x \leqslant \pi$.

Note that this expansion is not the Fourier series of g, but does give the Fourier expansion of $g(x) - a_0 x/2$. Also, the expression is obtained simply

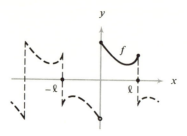

FIGURE 10-15

by integrating term by term the Fourier series of f. Similarly, any of the series in Table 10-2 may be integrated term by term to give a uniformly convergent series (the proof is the same in each case).

This is quite useful when the constants, a_n and b_n, have already been computed for f. Then to get the actual Fourier series for g, we can just

TABLE 10-4   Parseval's Relation

Type of series	Parseval's relation
Exponential series	$\dfrac{1}{2\pi}\displaystyle\int_0^{2\pi}|f(x)|^2\,dx = \sum_{n=-\infty}^{\infty}|c_n|^2\left(\text{or }\int_{-\pi}^{\pi}\right)$
Trigonometric series	$\dfrac{1}{\pi}\displaystyle\int_{-\pi}^{\pi}|f(x)|^2\,dx = \dfrac{a_0^2}{2}+\sum_{n=1}^{\infty}(a_n^2+b_n^2)\left(\text{or }\int_0^{2\pi}\right)$
Sine series	$\dfrac{1}{\pi}\displaystyle\int_{-\pi}^{\pi}|f(x)|^2\,dx = \sum_{n=1}^{\infty}b_n^2$
Cosine series	$\dfrac{1}{\pi}\displaystyle\int_{-\pi}^{\pi}|f(x)|^2\,dx = \dfrac{a_0^2}{2}+\sum_{n=1}^{\infty}a_n^2$
Exponential series on $[-l,l]$	$\dfrac{1}{2l}\displaystyle\int_{-l}^{l}|f(x)|^2\,dx = \sum_{n=-\infty}^{\infty}|c_n|^2$
Trigonometric series on $[-l,l]$	$\dfrac{1}{l}\displaystyle\int_{-l}^{l}|f(x)|^2\,dx = \dfrac{a_0^2}{2}+\sum_{n=1}^{\infty}(a_n^2+b_n^2)$
Half-interval cosine series	$\dfrac{2}{l}\displaystyle\int_0^{l}|f(x)|^2\,dx = \dfrac{a_0^2}{2}+\sum_{n=1}^{\infty}a_n^2$
Half-interval sine series	$\dfrac{2}{l}\displaystyle\int_0^{l}|f(x)|^2\,dx = \sum_{n=1}^{\infty}b_n^2$

substitute the series for x, and gather terms (the series for x is given below). See also Example 3 at the end of the chapter.

Recall that if we have any convergent expansion for a function in terms of $\cos nx$ and $\sin mx$, then it must be the Fourier series of the function (see Theorem 4). Differentiation of Fourier series requires more care and will be treated in Section 10.6; see for example Section 5.3.

In Table 10-5 are assembled some of the common Fourier expansions. In using this table, one should keep in mind that the Fourier series is linear. That is, the Fourier series of $af(x) + bg(x)$ is

$$a(\text{Fourier series of } f) + b(\text{Fourier series of } g).$$

TABLE 10-5   Some Fourier and Related Series

Function	Series	Valid pointwise on the interval
1. $f(x) = \begin{cases} 1, & 0 \leqslant x \leqslant \pi \\ 0, & -\pi \leqslant x < 0 \end{cases}$	$\dfrac{1}{2} + \left(\dfrac{2}{\pi}\right)\displaystyle\sum_{n=1}^{\infty} \dfrac{\sin(2n-1)x}{2n-1}$ $= \dfrac{1}{2} + \dfrac{1}{\pi}\displaystyle\sum_{n=1}^{\infty}(1-(-1)^n)\dfrac{\sin nx}{n}$	$]-\pi,\pi[,$ $\left[\tfrac{1}{2} \text{ at } x = 0, \pi, -\pi\right]$
1a. $f(x) = 1, \quad 0 \leqslant x \leqslant \pi$	$\dfrac{4}{\pi}\displaystyle\sum_{n=1}^{\infty}\dfrac{\sin(2n-1)x}{2n-1}$ (half-interval sine series)	$]0,\pi[$ $[0 \text{ at } x = 0, x = \pi]$
	1 (half-interval cosine series)	$[0,\pi]$
2. $f(x) = x$	$2\displaystyle\sum_{n=1}^{\infty}\dfrac{(-1)^{n+1}}{n}\sin nx$	$]-\pi,\pi[$ $[0 \text{ at } x = \pi, x = -\pi]$
	$\pi - 2\displaystyle\sum_{n=1}^{\infty}\dfrac{\sin nx}{n}$	$]0,2\pi[$ $[\pi \text{ at } x = 0, x = 2\pi]$
	$\dfrac{\pi}{2} - \dfrac{4}{\pi}\displaystyle\sum_{n=1}^{\infty}\dfrac{\cos(2n-1)x}{(2n-1)^2}$ (half-interval cosine series)	$[0,\pi]$
2a. $f(x) = \begin{cases} 0, & -\pi \leqslant x < 0 \\ x, & 0 \leqslant x \leqslant \pi \end{cases}$	$\dfrac{\pi}{4} - \dfrac{2}{\pi}\displaystyle\sum_{n=1}^{\infty}\dfrac{\cos(2n-1)x}{(2n-1)^2}$ $-\displaystyle\sum_{n=1}^{\infty}\dfrac{(-1)^n}{n}\sin(nx)$	 $]-\pi,\pi[$ $[\pi/2 \text{ at } x = \pi, x = -\pi]$

TABLE 10-5 (*continued*)

Function	Series	Valid pointwise on the interval
3. $f(x) = x^2$	$2\pi x - \dfrac{2\pi^2}{3} + 4\displaystyle\sum_{n=1}^{\infty} \dfrac{\cos nx}{n^2}$	$[0,2\pi]$
	$\dfrac{4}{3}\pi^2 + 4\displaystyle\sum_{1}^{\infty}\left(\dfrac{\cos nx}{n^2} - \pi\,\dfrac{\sin nx}{n}\right)$	$]0,2\pi[$ $[2\pi^2 \text{ at } 0,2\pi]$
	$\dfrac{\pi^2}{3} + 4\displaystyle\sum_{n=1}^{\infty}\dfrac{(-1)^n}{n}\cos nx$	$[-\pi,\pi]$
	$\pi x - \dfrac{8}{\pi}\displaystyle\sum_{n=1}^{\infty}\dfrac{\sin(2n-1)x}{(2n-1)^3}$	$[0,\pi]$
	$\displaystyle\sum_{n=1}^{\infty}\left(\dfrac{2\pi(-1)^{n+1}}{n}\right.$ $\left. - \dfrac{4}{\pi}\,\dfrac{(1-(-1)^n)}{n^3}\right)\sin nx$ (half-interval sine series)	$[0,\pi[$ (0 at $x = \pi$)
4. $f(x) = \sin x$	$\dfrac{2}{\pi} - \dfrac{4}{\pi}\displaystyle\sum_{n=1}^{\infty}\dfrac{\cos 2nx}{4n^2 - 1}$ (half-interval cosine series)	$[0,\pi[$
4a. $f(x) = \begin{cases} \sin x, & 0 \le x \le \pi \\ 0, & -\pi \le x < 0 \end{cases}$	$\dfrac{1}{\pi} + \dfrac{1}{2}\sin x - \dfrac{2}{\pi}\displaystyle\sum_{n=1}^{\infty}\dfrac{\cos 2nx}{4n^2 - 1}$	$[-\pi,\pi]$
4b. $f(x) = |\sin x|$	$\dfrac{2}{\pi} - \dfrac{4}{\pi}\displaystyle\sum_{n=1}^{\infty}\dfrac{\cos 2nx}{4n^2 - 1}$	all of $\mathbb{R}$
5. $f(x) = \cos x$	$\dfrac{8}{\pi}\displaystyle\sum_{n=1}^{\infty}\dfrac{n\sin 2nx}{4n^2 - 1}$ (half interval sine series)	$]0,\pi[$ [0 at $x = 0$, $x = \pi$]
6. $f(x) = e^x$	$\dfrac{\sinh \pi}{\pi}\displaystyle\sum_{-\infty}^{\infty}\dfrac{(-1)^n}{1 - in}e^{inx}$	$]-\pi,\pi[$ [$\cosh \pi$ at π, $-\pi$]

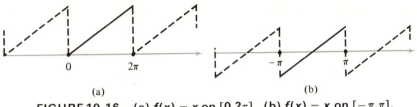

FIGURE 10-16   (a) $f(x) = x$ on $[0,2\pi]$. (b) $f(x) = x$ on $[-\pi,\pi]$.

Also, Theorem 12 can be used effectively to build further series successively by integrations, for example, $x, x^2, x^3, \ldots$. Also, note that if f is modified at a finite number (or even a countable number) of points, the Fourier series is unchanged (why?). Specific illustrations will be given shortly.

In these formulas, care should be taken with regard to the domain. For example, $f(x) = x$ on $[0,2\pi]$ is quite different from $f(x) = x$ on $[-\pi,\pi]$ as a periodic function (Figure 10-16).

Of course, on $]0,\pi[$ the functions and their series agree. A comparison of these series leads to many interesting identities. For the function x above, for example, we deduce that

$$x = \pi - 2\sum_{n=1}^{\infty} \frac{\sin nx}{n} = 2\sum_{n=1}^{\infty} \frac{(-1)^{n+1}}{n} \sin nx$$

for $0 < x < \pi$. However, off $]0,\pi[$ they will not agree. See Figure 10-17. We have sketched roughly what the above series look like up to the nth term.

In Table 10-5, $[-\pi,\pi]$, $[0,2\pi]$, and $[0,\pi]$ are presented for convenience. These can be changed to $[-l,l]$, $[0,l]$ by introducing constants and new variables as indicated in Table 10-2. Some further expansions are found in the exercises and examples.

We now turn to what is referred to as the *Gibbs' phenomenon*.* The Gibbs' phenomenon generally occurs when f has a jump discontinuity. The idea

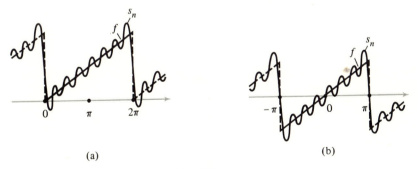

FIGURE 10-17

---

* Named after J. W. Gibbs, a mathematical physicist and physical chemist who discovered it. Gibbs is usually credited with inventing current vector notation around 1880.

is illustrated in Figure 10-18. This shows that if s_n is the nth partial sum of the Fourier series, then the maxima and minima of s_n near the jump always differ by more than the jump of f, and this excess remains as $n \to \infty$. Roughly, the Fourier series "overshoots" the jump and this overshoot persists in the limit. Another way of saying this is that as $n \to \infty$, $s_n(x)$ tends to approximate a vertical line longer than the jump.

The general case is a bit delicate so instead we consider just one special case of a jump discontinuity and determine the overshoot exactly.

Theorem 13. *Consider*

$$f(x) = \begin{cases} a, & -\pi \leqslant x < 0, \\ b, & 0 \leqslant x \leqslant \pi, \end{cases}$$

and suppose $a < b$. Let $s_n(x)$ be the nth partial sum of the trigonometric Fourier series. Then the maximum of s_n occurs at $\pi/2n$ and the minimum at $-(\pi/2n)$ and

$$\lim_{n \to \infty} s_n\left(\frac{\pi}{2n}\right) = \left(\frac{b-a}{2}\right)\left(\frac{2}{\pi}\int_0^\pi \frac{\sin t}{t}\,dt + 1\right) + b$$
$$\approx (b-a)(.089) + b\,.$$

Similarly,

$$\lim_{n \to \infty} s_n\left(-\frac{\pi}{2n}\right) = \left(\frac{b-a}{2}\right)\left(-\frac{2}{\pi}\int_0^\pi \frac{\sin t}{t}\,dt + 1\right) + a$$
$$\approx a - (b-a)(.089)$$

and the difference of these limits is

$$\left(\frac{b-a}{2}\right)\left(\frac{4}{\pi}\int_0^\pi \frac{\sin t}{t}\,dt\right) \approx (b-a)(1.179)\,.$$

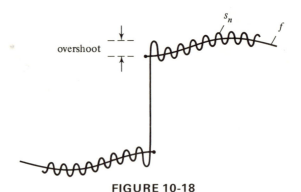

FIGURE 10-18

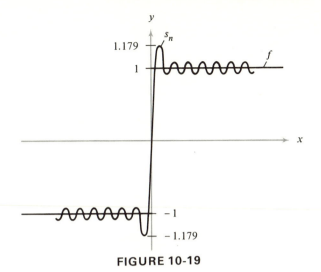

FIGURE 10-19

For $a = -1$ and $b = 1$, this is illustrated in Figure 10-19. Thus the overshoot of the maxima and minima is each about 9 percent of the jump in f.

EXAMPLE 1. Let us show that formula 1 of Table 10-5 is obtained by evaluating the Fourier coefficients by direct integration on $[-\pi,\pi]$ (see Table 10-2). We obtain

$$a_n = \frac{1}{\pi} \int_0^\pi \cos nx \, dx = \begin{cases} 1, & n = 0, \\ 0, & n = 1, 2, \dots . \end{cases}$$

$$b_n = \frac{1}{\pi} \int_0^\pi \sin nx \, dx = \frac{1}{\pi} \left(-\frac{\cos nx}{n} \right)\Big|_0^\pi$$

$$= \frac{1}{\pi} \left(\frac{1 - (-1)^n}{n} \right)$$

$$= \begin{cases} \dfrac{2}{\pi n}, & n \text{ odd}, \\ 0, & n \text{ even}. \end{cases}$$

This establishes formula 1.

EXAMPLE 2. Use the table to find the series for

$$g(x) = \begin{cases} -1, & -\pi \leqslant x < 0, \\ 1, & 0 \leqslant x \leqslant \pi. \end{cases}$$

Solution: Let f be defined as in formula 1. Then

$$g(x) = 2f(x) - 1,$$

so the expansion for g is

$$2\left(\frac{1}{2} + \frac{2}{\pi} \sum_{n=1}^{\infty} \frac{\sin[(2n-1)x]}{2n-1}\right) - 1 = \frac{4}{\pi} \sum_{n=1}^{\infty} \frac{\sin[(2n-1)x]}{2n-1}$$

since the Fourier expansion of 1 on $[-\pi,\pi]$ is 1. Of course, one could also obtain this directly.

Note that the half-interval sine expansion of 1 is not 1 itself but is the same as that of g above (explain).

EXAMPLE 3.   For each $0 < x < \pi$, prove that

$$\frac{\pi}{2} - \sum_{n=1}^{\infty} \frac{2}{\pi n^2}(1 - (-1)^n)\cos nx = \sum_{n=1}^{\infty} \frac{2(-1)^{n+1}}{n} \sin nx .$$

Solution: The left side is the cosine series for $f(x) = x$ on $[0,\pi]$, while the right side is the sine series. For each $0 < x < \pi$ we have convergence to the value x. At $\pm\pi$ the right side is zero.

EXAMPLE 4.   Establish the first formula in Table 10-5 for $f(x) = x$ and state how one obtains those for x^2.

Solution: Since $f(x) = x$ is odd, we use the sine series. Then

$$b_n = \frac{2}{\pi} \int_0^\pi x \sin nx \, dx .$$

Integrating by parts gives

$$b_n = \frac{2}{\pi}\left(-\frac{x \cos nx}{n}\right)\Big|_0^\pi + \frac{2}{\pi} \int_0^\pi \frac{\cos nx}{n} \, dx$$

$$= \frac{2}{\pi}\frac{\pi(-1)^{n+1}}{n} + \frac{2}{\pi}\cdot 0 .$$

Hence the series is

$$\sum_{n=1}^{\infty} b_n \sin nx = 2 \sum_{n=1}^{\infty} \frac{(-1)^{n+1}}{n} \sin nx .$$

The series for x^2 is obtained as follows. The first formula is the integral of the series for x on $[0,2\pi]$ (with a factor of 2, since $\int_0^x y \, dy = x^2/2$). The term $2\pi^2/3$ comes from the cosine term at 0 using $\sum_{n=1}^{\infty} 1/n^2 = \pi^2/6$ (Exercise 4, Section 10.2). Here Theorem 12 is used. The second formula uses the first and the expansion for x on $]0,2\pi[$ from formula 2 of Table 10.5. These can also be done directly. The third formula is the fourier series (= cosine series in this case) for x^2, and the remaining formulas are the integral of the cosine expansion of x on $[0,\pi]$, and this along with the sine expansion of x on $]0,\pi[$ are substituted, respectively.

EXAMPLE 5.   Find the Fourier series on $[-\pi,\pi]$ for

$$f(x) = \begin{cases} 0, & -\pi \leqslant x \leqslant 0, \\ x^2, & 0 \leqslant x \leqslant \pi. \end{cases}$$

Solution: If we integrate function 2a of Table 10-5, we get 1/2 of the f given here. Hence by Theorem 12,

$$\tfrac{1}{2}f(x) = \frac{\pi x}{4} + \frac{\pi^2}{4} - \frac{2}{\pi}\sum_1^\infty \int_{-\pi}^x \frac{\cos[(2n-1)x]}{(2n-1)^2}\,dx - \sum_1^\infty \int_{-\pi}^x \frac{(-1)^n}{n}\sin nx\,dx$$

$$= \frac{\pi x}{4} + \frac{\pi^2}{4} - \frac{2}{\pi}\sum_1^\infty \frac{\sin[(2n-1)x]}{(2n-1)^3}$$

$$+ \sum_1^\infty (-1)^n\left(\frac{\cos nx}{n^2}\right) - \frac{(-1)^n}{n^2}\cdot(-1)^n.$$

Inserting the first series for x from formula 2 gives

$$\frac{\pi}{4}\left(2\sum_1^\infty \frac{(-1)^{n+1}}{n}\sin nx\right)$$

$$+ \frac{\pi^2}{4} - \frac{2}{\pi}\sum_1^\infty \frac{\sin[(2n-1)x]}{(2n-1)^3} + \sum_1^\infty \frac{(-1)^n\cos nx}{n^2} - \sum_1^\infty \frac{1}{n^2}.$$

This is the desired series but in a slightly awkward form. Note that from Exercise 4, Section 10.2

$$\sum_1^\infty \frac{1}{n^2} = \frac{\pi^2}{6}.$$

The resulting series is thus

$$f(x) = 2\left(\frac{\pi^2}{12} + \sum_{n=1}^\infty \left\{\left[\frac{\pi(-1)^{n+1}}{2n} - \frac{1-(-1)^n}{\pi n^3}\right]\sin nx + \frac{(-1)^n}{n^2}\cos nx\right\}\right).$$

Exercises for Section 10.5

1. Establish the following in Table 10-5.
   (a) Formulas 2, 2a.
   (b) Formulas 4, 4a, 4b.

2. Find the half-interval sine series for x^2.

3. Establish the following:

$$x\cos x = -(\tfrac{1}{2})\sin x + 2\sum_{n=2}^\infty \frac{(-1)^n n\sin nx}{n^2-1}, \qquad -\pi < x < \pi.$$

4. Compute the Fourier series on $[-\pi,\pi]$ for each of the following functions.

(a) $f(x) = \begin{cases} 10, & x > 0, \\ -11, & x < 0. \end{cases}$

(b) $f(x) = x^2 + x + 3$.

(c) $f(x) = \begin{cases} x, & x > 0, \\ -x^2, & x < 0. \end{cases}$

5. Discuss the Gibbs' phenomenon for the function

$$f(x) = \begin{cases} 8, & x > 0, \\ -4, & x < 0, \end{cases}$$

on the interval $[-\pi,\pi]$.

6. By considering

$$f(x) = \begin{cases} \dfrac{\pi}{4}, & 0 \leqslant x \leqslant \pi, \\[3mm] -\dfrac{\pi}{4}, & -\pi \leqslant x < 0, \end{cases}$$

and the point $x = \pi/2$, prove *Leibnitz' formula:*

$$1 - \frac{1}{3} + \frac{1}{5} - \frac{1}{7} + \cdots = \frac{\pi}{4}.$$

10.6   Some Further Convergence Theorems

In this section, we give some additional convergence theorems concerned mainly with uniform convergence, differentiability, and integration of Fourier series.

We have already stated in Section 10.4 that if f is continuous and of bounded variation, then the Fourier series of f converges uniformly to f. Let us now give a slightly weaker version which is easier to prove and is almost as good in practice. The reader should be sure to fully understand the notion of uniform convergence (see Chapter 5). For example, in Figure 10-18, why is s_n not uniformly convergent to f?

> **Theorem 14.**   *Suppose f is continuous on $[-\pi,\pi]$, $f(-\pi) = f(\pi)$, and f' is sectionally continuous with jump discontinuities. Then the (trigonometric or exponential) Fourier series of f converges to f absolutely and uniformly. A similar statement holds for f on $[0,2\pi]$.*

In particular, this implies that the Fourier series converges in the mean and pointwise, which we knew already (see Theorems 8 and 9).

For example, consider $f(x) = |x|$ on $[-\pi,\pi]$. Here the conditions of Theorem 14 are satisfied (but they are not satisfied on $[0,2\pi]$ since this is not the same function), so the Fourier series of f, namely,

$$\frac{\pi}{2} - \frac{4}{\pi} \sum_1^{\infty} \frac{\cos[(2n-1)x]}{(2n-1)^2}$$

converges uniformly. See Figure 10-20.

Thus, by the definition of uniform convergence, there is, for every $\varepsilon > 0$, an N such that $n \geq N$ implies

$$\left| |x| - \left(\frac{\pi}{2} - \frac{4}{\pi} \sum_1^n \frac{\cos[(2n-1)x]}{(2n-1)^2} \right) \right| < \varepsilon$$

for all $x \in [-\pi,\pi]$.

One might think that if

$$f(x) = \frac{a_0}{2} + \sum_{n=1}^{\infty} (a_n \cos nx + b_n \sin nx)$$

then

$$f'(x) = \sum_{n=1}^{\infty} \{(-na_n)\sin nx + nb_n \cos nx\}$$

at each point where $f'(x)$ exists. Unfortunately, this is not true. For example, let

$$f(x) = \begin{cases} 1, & 0 < x \leq \pi, \\ -1, & -\pi \leq x \leq 0. \end{cases}$$

Then

$$f(x) = \frac{4}{\pi} \sum_{n=1}^{\infty} \frac{\sin(2n-1)x}{2n-1}.$$

So, for $x > 0$, we would expect

$$0 = \frac{4}{\pi} \sum_1^{\infty} \cos(2n-1)x.$$

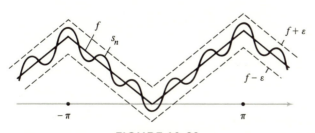

FIGURE 10-20

But this series does not converge since $\cos(2n - 1)x$ does not $\rightarrow 0$. (To make sense out of this, distribution theory must be used. Then it is possible to differentiate at will under all conditions when suitably interpreted.)

To get a differentiation theorem one naturally thinks of using Theorem 5, Chapter 5. However, we can get a better theorem in this case by arguing directly and the result is as follows.

> **Theorem 15.** *Let f be continuous on $[-\pi,\pi]$, $f(-\pi) = f(\pi)$, and let f' be sectionally continuous with jump discontinuities. Suppose f'' exists at $x \in [-\pi,\pi]$. Then the Fourier series for*
>
> $$f(x) = \frac{a_0}{2} + \sum_{n=1}^{\infty} (a_n \cos nx + b_n \sin nx)$$
>
> *may be differentiated term by term at x,*
>
> $$f'(x) = \sum_{n=1}^{\infty} (-na_n \sin nx + nb_n \cos nx) \ .$$
>
> *Furthermore, this is the Fourier series of f'.*

Thus, just as in Theorem 5, Chapter 5, one must be careful when differentiating series; certain conditions must hold to justify the operations. The result should be compared with Theorem 12 above.

EXAMPLE 1.   Consider $f(x) = |x| \ x \in [-\pi,\pi[$, $x \neq 0$, which satisfies the conditions of Theorem 15. It has the Fourier series

$$\frac{\pi}{2} - \frac{4}{\pi} \sum_{1}^{\infty} \frac{\cos(2n - 1)x}{(2n - 1)^2} \ .$$

Hence

$$f'(x) = \begin{cases} -1, & -\pi \leqslant x < 0 \ , \\ 1, & 0 < x \leqslant \pi \ , \end{cases}$$

has the Fourier series

$$\frac{4}{\pi} \sum_{1}^{\infty} \frac{\sin(2n - 1)x}{2n - 1} \ ,$$

which agrees with what we know.

EXAMPLE 2.   Give the version of Theorem 14 which is valid on $[-l,l]$.

Solution: We want to show that if f is continuous on $[-l,l]$, $f(-l) = f(l)$ and if $f'(x)$ is sectionally continuous with jump discontinuities, then the Fourier series

$$\frac{a_0}{2} + \sum_1^\infty \left\{ a_n \cos\left(\frac{n\pi x}{l}\right) + b_n \sin\left(\frac{n\pi x}{l}\right) \right\}$$

converges uniformly and absolutely to f, where

$$a_n = \frac{1}{l} \int_{-l}^{l} f(x)\cos\left(\frac{n\pi x}{l}\right) dx$$

$$b_n = \frac{1}{l} \int_{-l}^{l} f(x)\sin\left(\frac{n\pi x}{l}\right) dx$$

(see Table 10-2). The proof could be accomplished following the method of Theorem 14, but we can also deduce the result directly from this theorem as follows. Let $g(x): [-\pi,\pi] \to \mathbb{R}$ be defined by $g(x) = f(lx/\pi)$. Then

$$a_n = \frac{1}{l} \int_{-l}^{l} f(x)\cos\left(\frac{n\pi x}{l}\right) dx$$

$$= \frac{1}{\pi} \int_{-\pi}^{\pi} f\left(\frac{ly}{\pi}\right)\cos(ny) \, dy$$

(using $x = ly/\pi$). Thus a_n is also the Fourier coefficient of g and, similarly, for b_n.

Now g satisfies the conditions of Theorem 14, so

$$\frac{a_0}{2} + \sum_{n=1}^\infty (a_n \cos ny + b_n \sin ny)$$

converges uniformly and absolutely to g on $[-\pi,\pi]$. Replacing y by $\pi x/l$, we see that the same is true of f.

EXAMPLE 3. For each of the following functions, explain whether the Fourier series converges in the mean, pointwise, or uniformly. Determine if we can differentiate the Fourier series.
(a) $f: [0,2\pi] \to \mathbb{R}$,

$$f(x) = \begin{cases} \dfrac{1}{n}, & \dfrac{1}{(n+1)} \leqslant x < \dfrac{1}{n}, \\ & \qquad\qquad n = 1, 2, \ldots \\ 1, & \dfrac{1}{2} \leqslant x \leqslant 2, \end{cases}$$

(b) $f: [-\pi,\pi] \to \mathbb{R}$,
    $f(x) = \pi - |x|$.
(c) $f: [-\pi,\pi] \to \mathbb{R}$,

$$f(x) = \begin{cases} x^2 + 1, & -\pi \leqslant x < 0, \\ x + 1, & 0 < x \leqslant \pi. \end{cases}$$

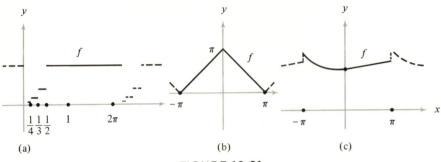

FIGURE 10-21

Solution: The three functions are sketched in Figure 10-21. In all three cases, f is bounded and hence square integrable (the function in (a) is integrable because its discontinuities form a countable set, see Theorem 3, Chapter 8. Hence the Fourier series converges in mean in all cases.

The Fourier series in (a) converges pointwise to the function, midway between the jumps at a discontinuity and to $1/2$ at the origin by Theorem 9.

In cases (a) and (c) the convergence is not uniform because f is not continuous (for continuity at the end points, one must look at the periodic extension; then (c) develops a discontinuity).

The function in (b) has a uniformly convergent Fourier series since it satisfies the conditions of Theorem 14.

The Fourier series of (c) converges to f at each x such that $-\pi < x < \pi$, and at $-\pi$ and π it converges to

$$\frac{1}{2}[f(-\pi) + f(\pi)] = \frac{1}{2}[(\pi^2 + 1) + (\pi + 1)] = \left(\frac{\pi^2 + \pi}{2}\right) + 1 \,.$$

Only the series of (b) may be differentiated to give the Fourier series of

$$f'(x) = \begin{cases} 1, & -\pi \leqslant x < 0 \,, \\ -1, & 0 < x \leqslant \pi \,, \end{cases}$$

which converges to f' for $x \neq 0$ and to 0 at $x = 0$.

Exercises for Section 10.6

For each of Exercises 1–3, determine what type of convergence the Fourier series will have and if we can differentiate the series.

1. $f(x) = x^2$ on $[-\pi,\pi]$.

2. $f(x) = \begin{cases} 3, & -\pi \leqslant x < -\frac{1}{2}, \\ 0, & -\frac{1}{2} \leqslant x < \frac{1}{2}, \\ 3, & \frac{1}{2} \leqslant x \leqslant \pi. \end{cases}$

3. $f(x) = \pi - x^2$ on $[-\pi,\pi]$.

4. Use Theorem 12 to find the Fourier series of x^3 on $[-\pi,\pi]$ using that of x^2 from Table 10-5.

5. (a) Suppose $f: [-\pi,\pi] \rightarrow \mathbb{R}$ is differentiable on $[-\pi,\pi]$, $f(-\pi) = f(\pi)$ and f', f'' are sectionally continuous, with jump discontinuities. Then show that

$$\frac{1}{\pi} \int_{-\pi}^{\pi} |f'(x)|^2 \, dx = \sum_{n=1}^{\infty} n^2(a_n^2 + b_n^2),$$

where a_n, b_n are the Fourier coefficients of f.

(b) Use (a) and Schwarz' inequality to deduce $\sum_{n=1}^{\infty} (a_n^2 + b_n^2)^{1/2} < \infty$.

6. Consider the half-interval cosine series for $\sin x$ on $[-\pi,\pi]$. Verify Theorem 15 directly in this case.

10.7   Applications

In this section, we briefly describe some applications of Fourier methods to simple boundary value problems which occur in mathematical physics. These examples are fairly easy, yet serve to illustrate some basic techniques. This material is intended solely for illustration and as a link with other courses in mathematics or physics which the student may be taking. It is by no means a complete course in boundary value problems. For example, we use only rectangular coordinates, when in fact polar and spherical coordinates are also very useful.

The problems we consider are standard ones—the vibrating string, heat conduction, and Laplace's equation. Some further applications to boundary value problems for ordinary differential equations are given in Exercises 19 and 71 at the end of the chapter. We begin then by considering the vibrating string.

From standard physical arguments we find that a good approximating mathematical model for a vibrating string with uniform density and (small) vertical displacement $y(x,t)$ at x at time t is that $y(x,t)$ should satisfy the *wave equation*,

$$\frac{\partial^2 y}{\partial t^2} = c^2 \frac{\partial^2 y}{\partial x^2}, \qquad 0 \leqslant x \leqslant l.$$

Here c is a constant determined by the physics of the string and represents the velocity of wave propagation along the string (as will be seen below). See Figure 10-22.

In order to completely specify the problem it is necessary to give the configuration of the string at $t = 0$; that is, how it is initially "plucked."

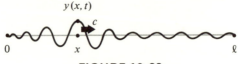

$y(x,t)$

FIGURE 10-22

This data will be given by giving the *initial condition* $y(x,0)$. Also note that it is physically reasonable to assume that dy/dt is zero at $t = 0$ (that is, the string is instantaneously motionless at the instant of plucking). It is also necessary to specify what happens at the ends of the string. Typically they are held fixed; for example, $y(0,t) = 0$, $y(l,t) = 0$ although other choices are possible. Such a specification is called the *boundary conditions*.

Once we have selected this model of the vibrating string, we have a purely mathematical problem, and the physics does not re-enter until one wishes to interpret the answer which the mathematics provides. There is a basic method used in these problems called *separation of variables* which yields special solutions, and from these one can build up general solutions.

Let us consider the case of a given initial displacement. To be more precise, let us call the *initial displacement problem* the problem of finding $y(x,t)$ for $0 < x < l$, such that

(1) (equation of motion) $\dfrac{\partial^2 y}{\partial t^2} = c^2 \dfrac{\partial^2 y}{\partial x^2}$.

(2) (initial conditions at $t = 0$) $\begin{cases} y(x,0) = f(x) & \text{(for given } f) \\ \dfrac{\partial y}{\partial t}(x,0) = 0 & \text{(no initial velocity)} \end{cases}$

(3) (boundary conditions) $y(0,t) = 0$, $y(l,t) = 0$ (for all t)

Thus we seek the motion of the string for future (or past) time when it is initially "plucked" in shape $f(x)$. For (2) and (3) to be consistent, we assume also that $f(0) = 0 = f(l)$. Other types of initial conditions are considered in Exercise 3.

Separation of variables means that we first seek solutions to the equation of motion of the form

$$y(x,t) = h(x)g(t) .$$

Thus substituting in the equation of motion, we obtain

$$h(x)g''(t) = c^2 h''(x)g(t) .$$

This will be satisfied if

$$h''(x) + \lambda h(x) = 0 \quad \text{and} \quad g''(t) + \lambda c^2 g(t) = 0$$

for a constant λ (why?). A solution of these equations with $h(0) = h(l) = 0$

and $g'(0) = 0$ is

$$h(x) = \sin\left(\frac{n\pi x}{l}\right) \quad \text{and} \quad g(t) = \cos\left(\frac{n\pi ct}{l}\right)$$

where

$$\lambda = \frac{n^2\pi^2}{l^2}, \quad n = 1, 2, 3, \dots .$$

Thus, for each n, a solution of the equations of motion satisfying conditions (1) and (3) above is given by

$$y_n(x,t) = \sin\left(\frac{n\pi x}{l}\right)\cos\left(\frac{n\pi ct}{l}\right), \quad n = 1, 2, \dots .$$

The initial conditions for this solution are

$$y(x,0) = \sin\left(\frac{n\pi x}{l}\right) \quad \text{and} \quad \frac{\partial y}{\partial t}(x,0) = 0 .$$

Thus we have the solution for a *particular* initial condition $\sin(n\pi x/l)$. However, we know that any f can be expanded in a half-interval sine series, and since all the conditions are linear, we should be able to add up the solutions corresponding to the terms in this expansion. This is done more precisely as follows.

Theorem 16. *In the initial displacement problem, suppose that f is twice differentiable. Then the solution to the initial displacement problem is*

$$y(x,t) = \frac{1}{2}\left[f(x - ct) + f(x + ct)\right]$$

$$= \sum_{n=1}^{\infty} b_n \sin\left(\frac{n\pi x}{l}\right)\cos\left(\frac{n\pi ct}{l}\right),$$

where the b_n are the half-interval sine coefficients,

$$b_n = \frac{2}{l}\int_0^l f(x)\sin\left(\frac{n\pi x}{l}\right) dx$$

and f is to be extended so that it is odd periodic. (Twice differentiable means we are assuming that the extended f is twice differentiable.) See Figure 10-23.

It happens in this case that the Fourier series solution could be simplified to a more easily handled and explicit form. Often, however, one must deal directly with the Fourier series itself.

Before generalizing, let us note the simple physical interpretation of the result. We see that the graph of $f(x - ct)$ is just that of f moved over to

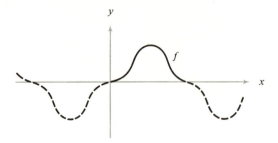

FIGURE 10-23

the right by ct, so we can interpret the function $g_t(x) = f(x - ct)$ as f moving to the right with velocity c after time t. Similarly, $h_t(x) = f(x + ct)$ is f moving to the left with velocity c. See Figure 10-24.

Thus, in Theorem 16, it can be seen that the initial shape of the string merely propagates away to the left and right with velocity c, each with 1/2 the initial amplitude, and reflections with sign-change at the endpoints.

To use the half-interval sine series, recall that we made f odd periodic. If we look only on the interval $[0,l]$, we see that when f moves to l it reflects from the wall; see Figure 10-25. Since the solution is the sum, there will be complicated cancelling (or "interference").

To keep track of this, it is useful to visualize a simpler situation first. Suppose f were concentrated near a point (possibly a δ-function) and we

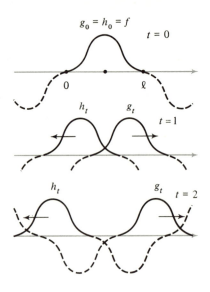

FIGURE 10-24

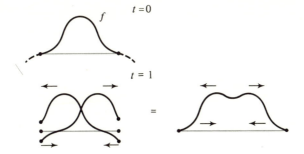

FIGURE 10-25

watch it move. These motions are called the *characteristics of the problem.* They should be visualized as if one were watching a movie. See Figure 10-26.

If we want to use genuine delta functions or functions f which are continuous but not twice differentiable, then we must generalize the scope of Theorem 1 and also generalize what we mean by a solution of $\partial^2 y/\partial t^2 = c^2(\partial^2 y/\partial x^2)$ for y which are not differentiable. This is done using the theory of distributions. Admitting them, Theorem 1 still holds for f a distribution (that is, the formal manipulations can be justified when properly interpreted). We shall then regard $(f(x - ct) + f(x + ct))/2$ as the solution for any f, differentiable or not.

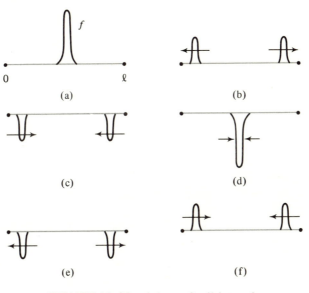

FIGURE 10-26 (a) $t = 0$. (b) $t = 1$.
(c) $t = 2$. (d) $t = 3$. (e) $t = 4$.
(f) $t = 5$. (g) return to (a).

In 2-dimensional problems (such as a vibrating drum) the wave equation reads

$$\frac{\partial^2 y}{\partial t^2} = c^2\left(\frac{\partial^2 y}{\partial x_1^2} + \frac{\partial^2 y}{\partial x_2^2}\right).$$

In this case the general solution can be written in Fourier series but does not have a simple explicit expression as in Theorem 14. The solution (for the similar initial displacement problem) on the rectangle $[0,l] \times [0,l']$, is given by

$$y(x_1,x_2,t) = \sum_{n,m=1}^{\infty} b_{nm} \sin\left(\frac{nx_1\pi}{l}\right)\sin\left(\frac{mx_2\pi}{l'}\right)\cos\left[\pi c t\sqrt{\left(\frac{n}{l}\right)^2 + \left(\frac{m}{l'}\right)^2}\right]$$

where

$$b_{nm} = \frac{4}{ll'}\int_0^l \int_0^{l'} f(x_1,x_2)\sin\left(\frac{nx_1\pi}{l}\right)\sin\left(\frac{mx_2\pi}{l'}\right) dx_1\, dx_2\,.$$

The reader is asked to go through the derivation of this in Exercise 68.

We now turn our attention to the problem of heat conduction. Consider a bar whose temperature is $T(x,t)$ at the point x at time t. Interpret $-(\partial T/\partial x)$ as the rate of heat flow. Thus the condition of "insulation" at $x = 0$ is $\partial T/\partial x = 0$ (evaluated at $x = 0$). The law of heat conduction asserts that

$$\frac{\partial T}{\partial t}(x,t) = k\frac{\partial^2 T}{\partial x^2}(x,t)\,,$$

where k is a constant determined by the conductivity of the material. This is called the *heat equation*.*

Notice that the above equation differs from the wave equation in that we have $\partial T/\partial t$ instead of $\partial^2 T/\partial t^2$. This difference is very important, for solutions to the heat conduction problem are very different in their behavior from those of the wave equation. For example, in the heat equation one obtains solutions only for $t \geqslant 0$. Intuitively, for the wave equation the graph of the solution "bounces around" like water waves. For the heat equation the solution diffuses out and becomes steady as $t \to \infty$ (as temperature tends to become evened out).

Thus, to study this simple situation, let us make the following model for heat conduction of a bar with insulated ends (for simplicity, take $k = 1$). Hence, we wish to solve for $T(x,t)$ satisfying

(1) (heat equation) $\dfrac{\partial T}{\partial t}(x,t) = \dfrac{\partial^2 T}{\partial x^2}(x,t),$ $0 < x < l, t \geqslant 0$

---

* For a derivation see Marsden-Tromba, *Vector Calculus*, Chapter 7, W. H. Freeman (1975).

(2) (initial conditions) $T(x,0) = f(x)$, $0 < x < l$

and

(3) (boundary conditions)
$$\begin{cases} \dfrac{\partial T}{\partial x}(0,t) = 0, \\[2mm] \dfrac{\partial T}{\partial x}(l,t) = 0, \end{cases} \qquad t \geqslant 0.$$

First, let us find some special solutions for special f by separation of variables. Let us try $T(x,t) = g(x)h(t)$; then we must have

$$g(x)h'(t) = g''(x)h(t) \, .$$

These equations are true if, for a constant λ,

$$g(x) + \lambda g''(x) = 0 \qquad \text{and} \qquad h(t) + \lambda h'(t) = 0 \, .$$

Solutions of these equations satisfying the boundary conditions are clearly given by

$$g(x) = \cos\!\left(\frac{n\pi x}{l}\right),$$

and

$$h(t) = e^{-n^2\pi^2 t/l^2}, \qquad n = 0, 1, 2, \ldots \, ,$$

where $\lambda = n^2\pi^2/l^2$. We use cosine and not sine so that the third boundary condition will hold. The reader can also see that these boundary conditions can't be met if we try to solve for g and h with $\lambda < 0$.

Thus a solution with $f(x) = \cos(n\pi x/l)$ is given by $e^{-n^2\pi^2 t/l^2}\cos(n\pi x/l)$, $n = 0, 1, 2, \ldots$. Since all expressions are linear and

$$f(x) = \sum_{n=1}^{\infty} a_n \cos\!\left(\frac{n\pi x}{l}\right) + \frac{a_0}{2}$$

(half-interval cosine series), we expect that the general solution with initial condition f is given by

$$\frac{a_0}{2} + \sum_{n=1}^{\infty} a_n e^{-n^2\pi^2 t/l^2} \cos\!\left(\frac{n\pi x}{l}\right) .$$

The relevant theorem is Theorem 17.

Theorem 17. *If f is square integrable, then for each $t > 0$*

$$T(x,t) = \frac{a_0}{2} + \sum_{n=1}^{\infty} a_n e^{-n^2\pi^2 t/l^2} \cos\!\left(\frac{n\pi x}{l}\right)$$

converges uniformly, is differentiable, and satisfies the heat equation and boundary conditions. At $t = 0$, *it equals* f *in the sense of convergence in the mean, and pointwise if* f *is of class* C^1. *As usual,*

$$a_n = \frac{2}{l} \int_0^l f(x)\cos\left(\frac{n\pi x}{l}\right) dx .$$

Thus this theorem gives the general solution to our problem. The exponential term makes the convergence rapid for $t > 0$. For $t < 0$, divergence usually prevails.

As $t \to \infty$ all the terms in the series $\to 0$, and also the sum $\to 0$, leaving

$$\lim_{t \to \infty} T(x,t) = \frac{1}{2}a_0 ,$$

so T becomes a uniform constant temperature in accordance with our intuition. The proof of this is simple; see Exercise 69 at the end of the chapter.

What happens as $t \to 0$ is answered by the following more delicate result.

Theorem 18.   *In Theorem 17,*

$$\lim_{\substack{t \to 0 \\ t > 0}} T(x,t) = f(x)$$

in the sense of convergence in mean, and converges uniformly (and pointwise) if f *is continuous, with* f' *sectionally continuous. More generally, for any* f, *if the Fourier series of* f *converges at* x *to* $f(x)$, *then* $T(x,t) \to f(x)$ *as* $t \to 0$.

This is an important result, for it tells us in what sense we recover the initial value $f(x)$ from those for $t > 0$. This is not derivable from differentiability of $T(x,t)$ for $t > 0$.

Our final application will be the consideration of Laplace's equation on a square. *Laplace's equation* in $\mathbb{R}^n$ is

$$\nabla^2 \varphi = 0$$

or, written out,

$$\sum_{k=1}^{n} \frac{\partial^2 \varphi}{\partial x_k^2} = 0 .$$

Such a function φ is called *harmonic*. This equation arises in many problems of electrostatics, fluid flow, and heat conduction.

The basic problem, called *Dirichlet's problem*, is the following. Given values of φ on some closed curve in the plane, find φ inside. This seemingly simple problem is at the core of the vast and deep subject of potential theory.*

---

* The terminology arises from electrostatics, in which φ represents the electric potential; again see Marsden-Tromba, *Vector Calculus*, Chapter 7 for details. This problem can also be attacked by methods of complex variables; see, for example, J. Marsden, *Basic Complex Analysis*, Chapter 5.

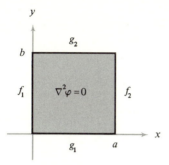

FIGURE 10-27

Let us use Fourier series to solve this problem for a square in $\mathbb{R}^2$. Cubes in $\mathbb{R}^3$ are similar. The problem is summarized as follows.

In $\mathbb{R}^2$, on $[0,a] \times [0,b]$ find a function φ such that

(1) (Laplace's equation) $\nabla^2\varphi = 0$.

(2) (boundary conditions) $\varphi(x,0) = g_1(x)$

$$\varphi(x,b) = g_2(x)$$
$$\varphi(0,y) = f_1(y)$$
$$\varphi(a,y) = f_2(y)$$

where f_i and g_j are given functions. See Figure 10-27.

First, let us get special solutions by separation of variables. Try

$$\varphi(x,y) = \varphi_1(x)\varphi_2(y) .$$

Then

$$\varphi_1''(x)\varphi_2(y) + \varphi_1(x)\varphi_2''(y) = 0 ,$$

if, for a constant λ, φ_1 and φ_2 satisfy the equations

$$\varphi_1''(x) + \lambda\varphi_1(x) = 0 \quad \text{and} \quad \varphi_2''(y) - \lambda\varphi_2(y) = 0 .$$

Solutions to this are

$$\varphi_1(x) = \sin\left(\frac{n\pi x}{a}\right), \qquad n = 1, 2, \ldots$$

$$\varphi_2(y) = \sinh\left[\frac{n\pi(b - y)}{a}\right]$$

where $\lambda = n^2\pi^2/a^2$. We choose $\sinh(z) = (e^z - e^{-z})/2$, $z = n\pi(b - y)/a$, rather than e^z or e^{-z}, because it will vanish when $y = b$. Similarly, we choose sine rather than cosine.

Thus

$$\varphi(x,y) = \sinh\left(\frac{n\pi(b - y)}{a}\right)\sin\left(\frac{n\pi x}{a}\right)$$

satisfies the boundary conditions

$$g_1 = \sinh\left(\frac{n\pi b}{a}\right)\sin\left(\frac{n\pi x}{a}\right) ,$$

$f_1 = f_2 = g_2 = 0$. Similarly, we obtain other basic solutions. It can be expected, therefore, that when $f_1 = f_2 = g_2 = 0$, the solution of the problem is

$$\varphi(x,y) = \sum_{n=1}^{\infty} b_n \sinh\left(\frac{n\pi(b-y)}{a}\right)\frac{\sin(n\pi x/a)}{\sinh(n\pi b/a)} \qquad (1)$$

where $g_1(x) = \sum_{n=1}^{\infty} b_n \sin(n\pi x/a)$ (half-interval sine series). Similar solutions hold for the other sides, and the sum is the solution for all sides.

Theorem 19 summarizes the conclusions.

Theorem 19.

(i) *Given g_1, let $\varphi(x,y)$ be defined as above. Suppose g_1 is of class C^2 and $g_1(0) = g_1(a) = 0$. Then φ converges uniformly, is the solution to the Dirichlet problem above with $f_1 = f_2 = g_2 = 0$, is continuous on the whole square, and $\nabla^2\varphi = 0$ on the interior.*

(ii) *If each of f_1, f_2, g_1, g_2 is of class C^2 and vanishes at the corners of the rectangle, then the solution $\varphi(x,y)$ is given as the sum of four series like Eq. (1) above, $\nabla^2\varphi = 0$ on the interior, ∇ is continuous on the whole rectangle and assumes the given boundary values. Furthermore, φ is C^{∞} on the interior.*

(iii) *If f_1, f_2, g_1, g_2 are only square integrable, the series for φ converges on the interior, $\nabla^2\varphi = 0$ and φ is C^{∞}. Also, φ takes on the boundary values in the sense of convergence in mean. This means, for example, $\lim_{y\to 0} \varphi(x,y) = \varphi(x,0) = g_1(x)$ with convergence in mean.*

The results (i) and (ii) are still true if we only assume that f_i and g_i are continuous, but they require a different method of proof. The present procedure is good, however, because it gives the solution explicitly in terms of Fourier series. The conditions (iii) are probably the most important in practice. See Example 3 below and Figure 10-28.*

EXAMPLE 1. In the initial displacement problem define the *total energy* of the string at time t to be

$$E(t) = \frac{1}{2}\int_0^l \left(\frac{\partial y}{\partial t}\right)^2 dx + \frac{c^2}{2}\int_0^l \left(\frac{\partial y}{\partial x}\right)^2 dx$$

(kinetic plus potential energy).

Show that $E(t)$ is constant in t.

---

* For further applications to problems in mathematical physics, we recommend Duff and Naylor, *Partial Differential Equations of Applied Mathematics,* Churchill, *Fourier Series and Boundary Value Problems,* and for a more exhaustive treatment, Courant and Hilbert, *Methods of Mathematical Physics.*

Solution: It suffices to show that $dE/dt = 0$. Now

$$\frac{dE}{dt} = \frac{1}{2} \int_0^l \frac{\partial}{\partial t} \left(\frac{\partial y}{\partial t}\right)^2 dx + \frac{c^2}{2} \int_0^l \frac{\partial}{\partial t} \left(\frac{\partial y}{\partial x}\right)^2 dx \ .$$

This is justified if y is twice continuously differentiable (see Example 2, Chapter 9). Then

$$\frac{c^2}{2} \int_0^l \frac{\partial}{\partial t} \left(\frac{\partial y}{\partial x}\right)^2 dx = \frac{c^2}{2} \int_0^l \frac{\partial y}{\partial x} \frac{\partial^2 y}{\partial t \, \partial x} dx \ .$$

Integrating the right-hand side by parts and using the fact that $\partial y/\partial t = 0$ at $x = 0, \ l$ gives

$$-\frac{c^2}{2} \int_0^l \frac{\partial y}{\partial t} \frac{\partial^2 y}{\partial x^2} dx$$

which equals, in view of the equation of motion,

$$-\frac{1}{2} \int_0^l \frac{\partial y}{\partial t} \frac{\partial^2 y}{\partial t^2} dx \ .$$

Thus $dE/dt = 0$, since the first term in dE/dt is

$$\frac{1}{2} \int_0^l \frac{\partial y}{\partial t} \frac{\partial^2 y}{\partial t^2} dx \ .$$

In case y is not twice differentiable more care is needed. (For example, if y is a δ-function, E is not even defined.)

For the heat equation, we do not have conservation of energy because, roughly speaking, the energy diffuses away. (See exercise 6).

EXAMPLE 2.   A bar with insulated ends has a temperature distribution given by $f(x) = x$, $0 < x < l$ at $t = 0$. Find the temperature distribution for $t > 0$.

Solution: According to Theorem 17, we simply take the half-interval cosine series for x and insert factors $e^{-(n^2 \pi^2 t)/l^2}$. Now the series for x is given by

$$x = \frac{l}{2} - \frac{4l}{\pi^2} \sum_{n=1}^{\infty} \frac{\cos[(2n-1)\pi x/l]}{(2n-1)^2} \ ,$$

so the required solution is

$$T(x,t) = \frac{l}{2} - \frac{4l}{\pi^2} \sum_{n=1}^{\infty} e^{-((2n-1)^2 \pi^2 t)/l^2} \frac{\cos[(2n-1)\pi x/l]}{(2n-1)^2} \ .$$

In general, one cannot reduce this expression to a compact form but must instead work with this series expansion.

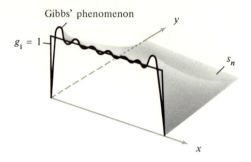

FIGURE 10-28

EXAMPLE 3.   Solve the Dirichlet problem on $[0,\pi] \times [0,\pi]$ with boundary values $g_1 = 1$, $f_1 = 0$, $f_2 = 0$, $g_2 = 0$. How are the boundary values assumed?

Solution: Here the sine series for 1 is

$$\frac{4}{\pi} \sum_{n=1}^{\infty} \frac{\sin(2n - 1)x}{2n - 1}.$$

According to Theorem 19, φ is obtained by inserting factors

$$\frac{\sinh[n(\pi - y)]}{\sinh(n\pi)}$$

into this. Thus we obtain

$$\varphi(x,y) = \frac{4}{\pi} \sum_{n=1}^{\infty} \frac{\sinh[(2n - 1)(\pi - y)]}{\sinh(2n - 1)\pi} \frac{\sin(2n - 1)x}{2n - 1}$$

as the solution. By Theorem 19, φ is C^{∞} and satisfies $\nabla^2 \varphi = 0$ on the interior of the square. Here $\varphi(x,0) = 1$ in the sense of convergence in mean, that is, $\varphi(x,y) \to 1$ as $y \to 0$ (in mean). The partial sum s_n of φ is roughly sketched in Figure 10-28.

Exercises for Section 10.7

1. For the initial displacement problem of a string, consider the "plucked string" with

$$f(x) = \begin{cases} hx\,, & 0 \leqslant x \leqslant \dfrac{l}{2}, \\[2ex] lh - hx\,, & \dfrac{l}{2} \leqslant x \leqslant l. \end{cases}$$

Show that f does not satisfy the hypotheses of Theorem 16. Find an expression for the solution and sketch it after time $t = l/c,\ 3l/2c$.

2. Suppose in the initial displacement problem, f has a maximum or a discontinuity at x_0. Show that, after time t, this feature is propagated like a characteristic.

3. (a) State clearly what the initial velocity problem for a string would be (no displacement). If the initial velocity is $g(x)$, then show that the solution is

$$y(x,t) = \frac{1}{2c} \left\{ \int_0^{x+ct} g(z)\, dz - \int_0^{x-ct} g(z)\, dz \right\}.$$

   Try to interpret physically.

   (b) Combine (a) with the initial displacement problem to get a solution for the problem with both initial displacement and velocity. (This is called d'Alembert's solution.)

4. In Theorem 17 prove that for any $t \geqslant 0$,

$$\frac{2}{l} \int_0^l T(x,t)\, dx = a_0.$$

   [Hint: What are the Fourier coefficients of $T(x,t)$ for t fixed?]

5. A bar with insulated ends has at $t = 0$ the temperature distribution $f(x,0) = x^2$. Find the temperature at $t > 0$ and the limit as $t \to \infty$.

6. Let $T(x,t)$ be a solution of the heat equation (Theorem 17) and set $L(t) = \int_0^l |T(x,t)|^2\, dx$. Show that $L(t)$ is non-increasing.

10.8 Fourier Integrals

This section is a short informal discussion of Fourier integrals. We shall just sketch the main results so that the reader can see a preview of some of this material and its place in Fourier analysis.

As we have seen in the previous sections, Fourier series are a very useful tool for analyzing functions on a finite interval. Since many functions are given on the whole real line $\mathbb{R}$, it would be nice to have an analogous theory on $\mathbb{R}$. Fourier integrals provide this theory.

Let us first argue heuristically. Consider $f: [-l,l] \to \mathbb{R}$. Then we can write f in terms of its Fourier series, using the exponential form,

$$f(x) = \sum_{-\infty}^{\infty} c_n e^{in\pi x/l},$$

where

$$c_n = \frac{1}{2l} \int_{-l}^{l} f(y) e^{-in\pi y/l}\, dy.$$

Let $\alpha = n\pi/l$ and introduce

$$c(\alpha) = \frac{1}{2\pi} \int_{-l}^{l} f(y)e^{-i\alpha y}\, dy \ .$$

Then

$$f(x) = \sum_{-\infty}^{\infty} c(\alpha)e^{i\alpha x}\, \frac{\pi}{l} \ .$$

For l large, α approximates a continuous variable, and this sum is roughly a Riemann sum with $\Delta\alpha = \pi/l$. This suggests that

$$f(x) = \int_{-\infty}^{\infty} c(\alpha)e^{i\alpha x}\, d\alpha \ ,$$

where

$$c(\alpha) = \frac{1}{2\pi} \int_{-\infty}^{\infty} f(y)e^{-i\alpha y}\, dy \ .$$

In short, when we extend our intervals to infinite ones, the Fourier series goes over into an integral.

Exactly the same steps as above can also be used in trigonometric form, except that integrals are taken from 0 to ∞ as are the corresponding sums.

The relevant theorem states that if f is sectionally continuous with jump discontinuities and $f'(x_0+)$ and $f'(x_0-)$ exist there, and $\int_{-\infty}^{\infty} |f(x)|\, dx < \infty$ (f is integrable), then

$$\frac{1}{2}\left[f(x+) + f(x-)\right] = \int_{-\infty}^{\infty} c(\alpha)e^{i\alpha x}\, d\alpha \ ,$$

where

$$c(\alpha) = \frac{1}{2\pi} \int_{-\infty}^{\infty} f(y)e^{-i\alpha y}\, dy \ .$$

One proves this in a way similar to the Jordan theorem (Theorem 9).

The above formula is called the *Fourier inversion formula*. In trigonometric form the formula is

$$\frac{1}{2}\left[f(x+) + f(x-)\right] = \int_{0}^{\infty} \left[A(\alpha)\cos \alpha x + B(\alpha)\sin \alpha x\right] d\alpha \ ,$$

where

$$A(\alpha) = \frac{1}{\pi} \int_{0}^{\infty} f(y)\cos(\alpha y)\, dy$$

and

$$B(\alpha) = \frac{1}{\pi} \int_{0}^{\infty} f(y)\sin(\alpha y)\, dy \ .$$

This form is especially convenient if f is even or odd.

In view of the inversion theorem, the *Fourier transform* of f is defined as

$$\hat{f}(\alpha) = \frac{1}{2\pi} \int_{-\infty}^{\infty} f(x)e^{-ix\alpha} \, dx .$$

Then if f is continuous, differentiable, and integrable on $\mathbb{R}$,

$$f(x) = \int_{-\infty}^{\infty} \hat{f}(\alpha)e^{i\alpha x} \, d\alpha .$$

There is a similar formula on $\mathbb{R}^n$; that is,

$$f(x) = \int_{\mathbb{R}^n} \hat{f}(\alpha)e^{i\langle \alpha, x \rangle} \, d\alpha ,$$

where

$$\hat{f}(\alpha) = \frac{1}{(2\pi)^n} \int_{\mathbb{R}^n} f(x)e^{-i\langle x, \alpha \rangle} \, dx ,$$

$x, \alpha \in \mathbb{R}^n$ and $\langle x, \alpha \rangle$ is the usual inner product in $\mathbb{R}^n$.

Suppose $f: [0,\infty[\to \mathbb{R}$. Then we can extend f to all of $\mathbb{R}$ by making it even or odd. Just as with the cosine and sine series, we can then introduce the *Fourier cosine transform* by extending f to be even, and setting

$$\hat{f}_c(\alpha) = \frac{2}{\pi} \int_0^{\infty} f(y)\cos(\alpha y) \, dy .$$

The inversion theorem becomes

$$f(x) = \int_0^{\infty} \hat{f}_c(\alpha)\cos(\alpha x) \, d\alpha .$$

Similarly, extending f to be odd leads to the *Fourier sine transform*,

$$\hat{f}_s(\alpha) = \frac{2}{\pi} \int_0^{\infty} f(y)\sin(\alpha y) \, dy ,$$

and the inversion formula becomes

$$f(x) = \int_0^{\infty} \hat{f}_s(\alpha)\sin(x\alpha) \, d\alpha .$$

A standard fact one should know (using Example 1, Chapter 9) is that the Fourier transform of $e^{-x^2/2}$ (Gaussian function) on $\mathbb{R}$ is $e^{-\alpha^2/2}/\sqrt{2\pi}$. Note that this is consistent with the inversion theorem.

In general, an *integral transform* is an association of the function

$$g(x) = \int_A k(x,y)f(y) \, dy$$

with the function f for some fixed function k called the *kernel*, and some fixed range A of integration. Such operations are common in mathematical physics.

Thus the Fourier transform is an integral transform with kernel

$$k(x,y) = \frac{1}{2\pi} e^{ix \cdot y} .$$

Here we come to an important, general problem. The transform maps f to g; that is, given f we get g. Can we invert this? In other words, given g can we invert the transformation to find f?

The Fourier inversion theorem solves this problem in the case of $k(x,y) = (1/2\pi)e^{ixy}$. That is, knowing the Fourier transform we can recover the function using the inversion formula.

Another common integral transform is the *Laplace transform* with kernel $k(x,y) = e^{-xy}$, and range $[0,\infty]$. Thus the Laplace transform of f is

$$L(f)(x) = \int_0^\infty e^{-xy} f(y)\, dy .$$

The inversion problem for Laplace transforms has a solution analogous to, but quite distinct from the Fourier transform.*

For $f: [-\pi,\pi] \to \mathbb{R}$ we have seen that (see Table 10-4)

$$\|f\|^2 = \int_{-\pi}^{\pi} |f(x)|^2\, dx = 2\pi \sum_{-\infty}^{\infty} |c_n|^2$$

for the Fourier coefficients c_n in exponential form. Since c_n is analogous to the Fourier transforms we might expect that something similar holds in terms of $\hat{f}$.

In fact this is true. If $|f|^2$ and $|f|$ are integrable, then letting

$$\|f\|^2 = \int_{-\infty}^{\infty} |f(x)|^2\, dx ,$$

we have

$$\|f\|^2 = 2\pi \|\hat{f}\|^2 .$$

More generally, $\langle f,g \rangle = 2\pi \langle \hat{f},\hat{g} \rangle$. Here the Fourier transform can be any of the types—exponential, trigonometric, sine, or cosine.

This result is variously known as *Parseval's relation* and *Plancherel's theorem*. To deal effectively with the technicalities involved requires the Lebesgue integral.

---

* See, for example, Marsden, *Basic Complex Analysis*, Chapter 7 for details.

For f and g integrable on $\mathbb{R}$ (or $\mathbb{R}^n$) we define the *convolution* of f and g by

$$(f * g)(x) = \int_{-\infty}^{\infty} f(x - y)g(y)\, dy \, .$$

This operation enters naturally in many problems. One of its main properties is that

$$\widehat{f * g} = 2\pi\hat{f} \cdot \hat{g}$$

(see Example 4 at the end of this chapter).

Fourier transforms have important applications in both pure and applied mathematics, but especially important in partial differential equations, such as the wave equation on all of $\mathbb{R}^n$. The reason is that in terms of Fourier transforms the equations become much simpler, often algebraic, and when these equations are solved the answer is obtained using the inversion formula. Convolutions are then encountered when we invert.

Using Fourier transforms many problems solved above on finite intervals can be translated easily to problems on the whole real line $\mathbb{R}$. The following exercises outline how to do this.

Exercises for Section 10.8

These problems can be done informally with little attention to rigor since this section has been so presented.

1. Show that if $\hat{f}(\alpha)$ is the Fourier transform of f, then $\hat{f'}(\alpha) = i\alpha\hat{f}(\alpha)$ and that $2\pi \int_{-\infty}^{\infty} \alpha^2 |\hat{f}|^2 \, d\alpha = \int_{-\infty}^{\infty} |f'|^2 \, dx$.

2. Let $f(x,y)$ satisfy $\partial^2 f/\partial x^2 + \partial^2 f/\partial y^2 = 0$. Suppose that $f(x,0) = g(x)$ and $\lim_{y \to \infty} f(x,y) = 0$ for all x.
   (a) Let $\hat{f}(\alpha,y)$ be the Fourier transform of $f(x,y)$, with y regarded as a constant. Show that $\hat{f}(\alpha,y) = \hat{g}(\alpha)e^{-|\alpha|y}$.
   (b) Show that $e^{-|\alpha|y}$ is the Fourier transform, with respect to x, of

   $$\frac{2y}{x^2 + y^2} \, .$$

   (c) Deduce that the solution of Laplace's equation is

   $$f(x,y) = \frac{1}{\pi} \int_{-\infty}^{\infty} \frac{y}{y^2 + (x - x')^2} g(x')\, dx' \, .$$

3. Suppose that $f(x,t)$ is a function for which

   $$\frac{\partial^2 f}{\partial t^2} = \frac{\partial^2 f}{\partial x^2} \, , \qquad f(x,0) = g(x) \, , \qquad \text{and} \qquad \frac{\partial f}{\partial t}(x,0) = h(x) \, .$$

Let $\hat{f}(\alpha,t)$ be the Fourier transform of $f(x,t)$, with t regarded as a constant.

(a) Show that $\hat{f}(\alpha,t) = \hat{g}(\alpha) \cos \alpha t + \hat{h}(\alpha)(\sin \alpha t/\alpha)$.

(b) Deduce that the solution of the wave equation is

$$f(x,t) = \frac{1}{2} [g(x - t) + g(x + t)] + \frac{1}{2} \int_0^t [h(x - \tau) + h(x + \tau)] \, d\tau \ .$$

4. Suppose that $f(x,t)$ is a function for which

$$\frac{\partial f}{\partial t} = k^2 \frac{\partial^2 f}{\partial x^2} \qquad -\infty < x < \infty , \qquad t \geqslant 0$$

and $f(x,0) = g(x)$. Let $\hat{f}(\alpha,t)$ be the Fourier transform of f.

(a) Show that $\hat{f}(\alpha,t) = \hat{g}(\alpha)e^{-k^2\alpha^2 t}$.

(b) Deduce that the solution of the heat equation is

$$f(x,t) = \frac{1}{2k\sqrt{\pi t}} \int_{-\infty}^{\infty} g(x')e^{-(x - x')^2/4k^2 t} \, dx' .$$

10.9 Quantum Mechanical Formalism

There is a close connection between the theory of Fourier series in an inner product space (developed in Sections 10.1 and 10.2) and some aspects of the formalism in quantum mechanics. Our purpose is to explain some of the aspects of this connection.

First, let us give a brief indication of the difference between classical and quantum mechanics. In classical mechanics, a particle's motion is described by a definite path with its associated definite velocity. In quantum mechanics, however, there is always some "uncertainty" about the position or velocity (or both). For atomic phenomenon this uncertainty is necessary and these effects are outside the domain of applicability of classical mechanics. For example, if an atomic particle with a definite initial velocity is prepared and projected at a screen, we do not know precisely what its future path will be. Instead, when we look for the particle we can only determine the probability of finding it in a given region.

In Figure 10-29, we consider* projecting particles through a screen with two slits onto a detection plate. Only the probability of location can be determined, not the exact location. For repeated trials, light and dark areas on the screen corresponding to high and low probability are obtained. This is represented by the curve in Figure 10-29. Other physical phenomena,

---

* This is an imaginary experiment which is used for illustrative purposes only. In real experiments, the "screen" might be, for instance, a crystal, and the slits might correspond to interatomic spacings.

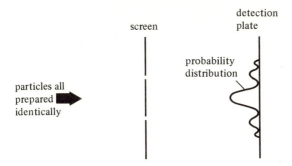

FIGURE 10-29   Double slit experiment.

like the discrete spectra for atoms, also require a quantum mechanical description.*

The above should not be taken to mean that classical mechanics is "false," while quantum mechanics is "true." Both are mathematical models which approximate nature well in their own limited circumstances. Quantum mechanics is, however, a more "refined theory" than classical mechanics.

The question then is how does one describe the behavior of a quantum mechanical particle? A single quantum mechanical particle is described by a complex-valued function, $\psi(x)$, where $x \in \mathbb{R}^3$. For more particles one must change $\mathbb{R}^3$ to another space (for N particles, $\mathbb{R}^{3N}$ is used). If the system depends on time, then we use $\psi(t,x)$. The probability density for locating the particle in space is given by $\psi(x)\overline{\psi(x)}$, so if the total probability is one, we should have

$$\|\psi\|^2 = \int_{\mathbb{R}^3} \psi(x)\overline{\psi(x)}\, dx = 1\,,$$

that is, ψ should be normalized. This last sentence provides one of the links between studying the mathematical model (that is, studying the inner product space of square integrable functions ψ and operations on it discussed below) and the physical interpretation of this model.

If one is measuring a definite quantity, such as the x coordinate, the momentum, or angular momentum, then, as above, these cannot be measured with certainty. The aspect of quantum mechanics we wish to explore is the mathematical structure of the wave functions ψ and the mathematical objects corresponding to physically measurable quantities. Of course, the subject goes much deeper than this and our discussion has only just begun to scratch the surface.

* Consult R. P. Feynman, R. B. Leighton, and M. Sands, *The Feynman Lectures on Physics*, Volume III for additional background.

Let us introduce some rigor into our discussion. First, suppose $V = \mathscr{L}^2$ is the space of functions, $\psi \colon \mathbb{R}^3 \to \mathbb{C}$, which are square integrable;

$$\|\psi\|^2 = \langle \psi, \psi \rangle = \int \psi(x) \overline{\psi(x)} \, dx < \infty \, .$$

As we have seen before, this is an inner product space with inner product

$$\langle f, g \rangle = \int_{\mathbb{R}^3} f(x) \overline{g(x)} \, dx \, .$$

Thus, all the discussion relating to an inner product space (see Sections 10.1 and 10.2) is relevant here.

A *quantum mechanical state* is (by definition) a function, $\psi \in V$ such that $\|\psi\| = 1$; that is, ψ is normalized. An *observable* is an operator A on V which is *symmetric* (or self-adjoint, or Hermitian); this means that $A \colon V \to V$ is a linear map satisfying

$$\langle Af, g \rangle = \langle f, Ag \rangle$$

for all $f, g \in V$. Actually, A may be defined only on some* elements of V. For example, if A is the Laplacian

$$Af = \frac{\partial^2 f}{\partial x^2} + \frac{\partial^2 f}{\partial y^2} + \frac{\partial^2 f}{\partial z^2} = \nabla^2 f,$$

then A is defined on those $f \in V$ whose second derivatives also lie in V. One can check formally that ∇^2 is symmetric using integration by parts twice. On the other hand, $\partial/\partial x$ is not symmetric, but $i(\partial/\partial x)$ is (the reader can prove this without difficulty).

An *eigenfunction* of an operator A is an $f \in V$, $f \neq 0$, such that $Af = \lambda f$ for some complex number λ called the *eigenvalue*. Observe that if f is an eigenfunction, then so is $f/\|f\|$, so we can assume our eigenfunctions are normalized.

There are two important remarks to be made concerning eigenfunctions of symmetric operators. First, if A is symmetric and f is an eigenfunction with eigenvalue λ, then λ is real. (*Proof:* $\langle Af, f \rangle = \langle \lambda f, f \rangle = \lambda \|f\|^2$. On the other hand, $\langle Af, f \rangle = \langle f, Af \rangle = \langle f, \lambda f \rangle = \bar{\lambda} \|f\|^2$, so $\lambda = \bar{\lambda}$, that is, λ is real.) Second, if f and g are eigenfunctions with eigenvalues λ and μ, and $\lambda \neq \mu$, then f and g are orthogonal. (*Proof:* Consider $\langle Af, g \rangle - \langle f, Ag \rangle = 0 = \langle \lambda f, g \rangle - \langle f, \mu g \rangle = (\lambda - \mu) \langle f, g \rangle$. Since $\mu \neq \lambda$, $\langle f, g \rangle = 0$.)

If f and g are independent eigenfunctions with the same eigenvalue λ, then we can obtain two new orthogonal eigenfunctions by the Gram-Schmidt

---

* Such operators are called *unbounded* and for them one has to distinguish between symmetric and self-adjoint. For further information see Reed-Simon, *Methods of Modern Mathematics Physics, Vol. I, Functional Analysis*, Academic Press (1972).

process. A similar operation can be performed for any number of eigenfunctions. Thus follows an important result. The eigenfunctions of a symmetric operator A form an orthonormal set. Hence, we have the connection with Fourier series; for, if φ_0, φ_1, φ_2, ... are the orthonormal eigenfunctions of A and if they are complete, then we can write $\psi = \sum_{n=0}^{\infty} \langle \psi, \varphi_n \rangle \varphi_n$ for any $\psi \in V$.

Unfortunately, the eigenfunctions are often not complete. For example, the Laplacian operator on $\mathbb{R}^n$ has no (square integrable) eigenfunctions. In many important problems ("bound-state problems"), however, the eigenfunctions are complete. This is proved in advanced courses on functional analysis by means of a theorem called the *spectral theorem*.

Let us return to the physical interpretation of observables. Again, let V be as above and let A be a symmetric operator. The main physical assumption is the following.

> **Physical Interpretation.** If A is "measured" in a state ψ, only the eigenvalues of A are observed. The value λ_n is observed with probability $|\langle \psi, \varphi_n \rangle|^2$, where $A\varphi_n = \lambda_n \varphi_n$.

This interpretation is consistent because (see Exercise 22 at the end of the chapter)

$$1 = \langle \psi, \psi \rangle = \sum \langle \psi, \varphi_n \rangle \langle \varphi_n, \psi \rangle = \sum |\langle \psi, \varphi_n \rangle|^2 ,$$

so the total probability is one. Furthermore, if the system is already in state φ_n (which it need not be generally), then we observe λ_n with probability one (that is, with certainty).

Thus the Fourier expansion of ψ exhibits ψ as a "mixture" of the eigenstates φ_n, and the squares of the absolute values of the Fourier coefficients are the probabilities of observing the particular eigenvalues.

As we have seen before, in a given state ψ and given an observable A, one cannot generally predict with certainty the observed value of A. But the average value observed, after many trials, is $\sum_{n=0}^{\infty} |\langle \psi, \varphi_n \rangle|^2 \lambda_n = \langle A\psi, \psi \rangle$ (see Exercise 3). This quantity $\langle A\psi, \psi \rangle$ is also called the *expectation value* of A in the state ψ. What is this for an eigenstate?

Let us now give some simple examples of observables. Probably the most important example is the *energy operator*, denoted H, also called the *Hamiltonian*. For a single particle in a potential U, it is given by

$$H\psi = -\frac{\hbar^2}{2m} \nabla^2 \psi + U\psi .$$

The justification of this choice depends on a more detailed analysis of the foundations of the subject; again, refer to Feynman's book for details. Here U is just a given real-valued function representing the potential, m is the particle's mass and $\hbar$ is a certain constant which depends on units of measurement ($\hbar = 1.05 \times 10^{-27}$ erg sec) and is called *Planck's constant.**

For example, in the hydrogen atom we can observe only discrete energy levels, which are eigenvalues of the operator

$$H\psi = -\frac{\hbar^2}{2m} \nabla^2 \psi - \frac{\psi}{r},$$

where $r(x,y,z) = (x^2 + y^2 + z^2)^{1/2}$, and m is the mass of the electron.

A word of caution—any ψ is an admissible state, not just the eigenstates. But the eigenstates are particularly important states because their eigenvalues give the values which are observable.

The reason the energy operator is so important is two-fold. First, its eigenvalues give the possible energies we can observe. Second, this operator governs the time dependence of ψ by means of the celebrated *Schrödinger equation* which reads as follows:

$$i\hbar \frac{\partial \psi}{\partial t} = H\psi$$

(the solution ψ of this equation using Fourier series is given in Exercise 23, at the end of the chapter).

Other operators are

(1) The *position operator* (in the x direction),

$$Q_x(\psi) = x\psi(x,y,z) .$$

(2) The *momentum operator* (in the x direction),

$$P_x(\psi) = \frac{\hbar}{i} \frac{\partial \psi}{\partial x} .$$

(3) The *angular momentum* operator (about the z axis),

$$J_z(\psi) = \frac{\hbar}{i}\left(y \frac{\partial \psi}{\partial x} - x \frac{\partial \psi}{\partial y}\right).$$

Similar definitions can be made for Q_y, Q_z, and so on.

The eigenfunctions of J_z are complete and the eigenfunctions and eigenvalues are computed in any quantum mechanics book. The operators Q_x, P_x do not have square integrable eigenfunctions.

Finally, before looking at a specific example, we examine the important notion of the commutator. The *commutator* of two operators, A, B, is the operator $[A,B]$ defined by

$$[A,B] = AB - BA ,$$

where AB means $A \circ B$; that is, $(AB)(f) = A(B(f))$.

---

* Actually $h = 2\pi\hbar$ is usually called Planck's constant.

Suppose f is an eigenfunction of both A and B. If $Af = \lambda f$, and $Bf = \mu f$, then

$$[A,B]f = (AB - BA)f = A(\mu f) - B(\lambda f)$$
$$= \mu \lambda f - \lambda \mu f = 0 \ .$$

Thus if A and B have the same eigenfunctions, then $[A,B]f = 0$ for such eigenfunctions. If the eigenfunctions are complete, we expect $[A,B] = 0$ for all f, but this requires more assumptions than we can go into here.

Conversely, if $[A,B] = 0$, we can select the eigenfunctions to be simultaneous eigenfunctions of A and B. This is easy to see if there are no repeated eigenvalues (the more general case requires a bit more argument). To illustrate, suppose $Af = \lambda f$. Then $A(Bf) - B(Af) = 0$, so $A(Bf) = \lambda(Bf)$. Thus Bf is an eigenfunction of A, so $Bf = \mu f$ for some μ (since by assumption, λ is a simple eigenvalue). Thus f is an eigenfunction of both A and B.

In summary, $[A,B] = 0$ iff A and B have simultaneous eigenfunctions. Physically this means that these eigenfunctions give exact observables for both A and B at once, or as we say, A and B can be *measured simultaneously.* Further justification of this statement is given by the famous *uncertainty principle* (Exercise 5), which states that the product of the "errors" in measuring A and B for the same ψ (that is, measuring A, B simultaneously) is at least $2 |\langle C\psi, \psi \rangle|$, where $C = [A,B]$. The definition of "error" is also given in Exercise 5.

Finally, let us look at a simple example. Other important examples such as the harmonic oscillator ($H = -(\hbar^2/2m)\nabla^2 + kr$) and the hydrogen atom ($H = -(\hbar^2/2m)\nabla^2 - 1/r$) are found in standard texts, and are a little more laborious to perform fully.

The example to be studied is that of a particle in an "infinitely deep well." We want to find the eigenfunctions and see if they are a complete orthonormal set. Here the problem is in one dimension for simplicity. We have

$$H = -\frac{\hbar^2}{2m}\frac{\partial^2 \psi}{\partial x^2} + V\psi \ ,$$

$V = 0$ on $[0,l]$, and $V = \infty$ outside. Since this is not workable within the space of square integrable functions, let us reformulate H by demanding

$$\begin{cases} H\psi = -\dfrac{\hbar^2}{2m}\dfrac{\partial^2 \psi}{\partial x^2}, & \text{on } [0,l] \ , \\[2mm] \psi \text{ and } H\psi = 0, & \text{outside } [0,l] \ . \end{cases}$$

Thus H is really an operator on the functions $\psi(x)$, $0 \leqslant x \leqslant l$ with $\psi(0) = \psi(l) = 0$.

It then follows that ψ is an eigenfunction iff there is a constant E, such that

$$-\frac{\hbar^2}{2m}\frac{\partial^2 \psi}{\partial x^2} = E\psi \ .$$

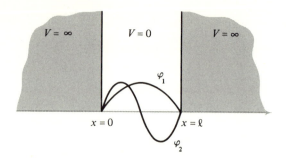

FIGURE 10-30

The general solution to this equation is easily seen to be

$$\psi(x) = A \sin \lambda x + B \cos \lambda x ,$$

where $\lambda^2 = 2mE/\hbar^2$. If $\psi = 0$ at $0, l$, we must have $B = 0$ and also, $\lambda = n\pi/l$, $n = 1, 2, 3, \ldots$.

Thus, for the problem, the eigenfunctions are (normalized),

$$\varphi_n(x) = \sqrt{\frac{2}{l}} \sin \lambda_n x$$

(see Figure 10-30) where $\lambda_n = n\pi/l$, $n = 1, 2, \ldots$, and the eigenvalues are

$$E_n = \frac{\hbar^2 n^2 \pi^2}{2ml^2} .$$

Thus these E_n are the only possible energy values we can observe.

Here these functions are complete, as has been proved in Section 10.3. Thus if a particle is in a state ψ, the probability we will observe energy E_n is given by

$$|\langle \psi, \varphi_n \rangle|^2 = \frac{2}{l} \left| \int_0^l \psi(x) \sin\left(\frac{n\pi x}{l}\right) dx \right|^2 .$$

Exercises for Section 10.9

1. Let A be an operator on V. Define its *adjoint* A^* by $\langle A^*x, y \rangle = \langle x, Ay \rangle$ (assume such an A^* exists). Prove that $(AB)^* = B^*A^*$.

2. Let V be $\mathbb{R}^n$ and $A: V \to V$ linear. Prove that A is symmetric in our sense if its matrix with respect to any orthonormal basis is symmetric in the usual sense of matrices, that is, $a_{ij} = a_{ji}$.

3. Let A be a symmetric operator with a complete set of eigenfunctions $\varphi_0, \varphi_1, \varphi_2, \ldots$

and eigenvalues $\lambda_0, \lambda_1, \ldots$. If $\psi \in V$, argue that the expectation of A is given by

$$\langle A\psi,\psi\rangle = \sum_{n=0}^{\infty} |\langle\psi,\varphi_n\rangle|^2 \, \lambda_n \, .$$

Interpret this quantum mechanically (that is, probabilistically).

4. Suppose $\langle Af,g\rangle = -\langle f,Ag\rangle$ for all $f, g \in V$. Show that A^2 is symmetric and that it has only negative eigenvalues.

5. (Uncertainty principle.) Let A be a symmetric operator. The *uncertainty* (or variance) in observing A in a state ψ is given by

$$\Delta^2(A,\psi) = \langle (A - \langle A\psi,\psi\rangle)^2\psi,\psi\rangle \, .$$

(a) Show that this equals $\langle A^2\psi,\psi\rangle - \langle A\psi,\psi\rangle^2$.
(b) Let A, B be two symmetric operators and let $C = [A,B]$. Show that

$$\Delta^2(A,\psi)\,\Delta^2(B,\psi) \geqslant 4\,|\langle C\psi,\psi\rangle|^2 \, .$$

Note the special case $[A,B] = 0$ and interpret it. [Hint: Show that for any two symmetric operators A and B, 2 (imaginary part of $\langle B\psi,A\psi\rangle$) $= \langle C\psi,\psi\rangle$. Apply the Schwarz inequality and replace A by $A - \langle A\psi,\psi\rangle$ and replace B similarly.]

(c) For the case $A\psi = x\psi$ and $B\psi = (\hbar/i)\,\partial\psi/\partial x$ (position and momentum), show that

$$\Delta^2(A,\psi)\,\Delta^2(B,\psi) \geqslant 4\hbar^2$$

(for $\|\psi\| = 1$). This is called the *Heisenberg uncertainty principle*.

Theorem Proofs for Chapter 10

Theorem 1. *The space V of continuous functions $f: [a,b] \to \mathbb{C}$ forms an inner product space if we define*

$$\langle f,g\rangle = \int_a^b f(x)\overline{g(x)}\, dx \, .$$

Proof: The properties of the inner product follow from these computations:

(i) $\langle af + bg,h\rangle = \displaystyle\int_a^b [af(x) + bg(x)]\overline{h(x)}\, dx$

$$= a\int_a^b f(x)\overline{h(x)}\, dx + b\int_a^b g(x)\overline{h(x)}\, dx$$

$$= a\langle f,h\rangle + b\langle g,h\rangle \, .$$

(ii) $\overline{\langle f,g\rangle} = \overline{\displaystyle\int_a^b f(x)\overline{g(x)}\, dx} = \int_a^b \overline{f(x)g(x)}\, dx$

$$= \int_a^b \overline{f(x)}g(x)\, dx$$

$$= \langle g,f\rangle \, .$$

(iii) Note from (ii) that $\langle f,f \rangle = \overline{\langle f,f \rangle}$, so $\langle f,f \rangle$ is real; thus $\langle f,f \rangle \geqslant 0$ makes sense. Here

$$\langle f,f \rangle = \int_a^b f(x)\overline{f(x)}\, dx = \int_a^b |f(x)|^2\, dx \geqslant 0\, ,$$

since $|f(x)|^2 \geqslant 0$.

Finally, suppose $\langle f,f \rangle = 0$. Use the fact that if h is continuous and $h \geqslant 0$, then

$$\int_a^b h(x)\, dx = 0 \qquad \text{implies } h = 0$$

(see Section 8.4), so $\langle f,f \rangle = 0$ implies $\int_a^b |f|^2\, dx = 0$ and hence $f = 0$. ∎

Theorem 2. (*The Cauchy Schwarz Inequality.*) *Let f, g belong to the inner product space V; then*

$$|\langle f,g \rangle| \leqslant \|f\|\, \|g\|\, .$$

Furthermore, all the properties listed in Theorem 5 (II), (III), Chapter 1 hold (II (iii) also holds for α complex).

Proof: We shall only prove that $|\langle f,g \rangle| \leqslant \|f\|\, \|g\|$, the rest being routine as in Theorem 5, Chapter 1.

First, let us prove the inequality when $\|g\| = 1$. Now

$$
\begin{aligned}
0 \leqslant \|f - \langle f,g \rangle g\|^2 &= \langle f - \langle f,g \rangle g, f - \langle f,g \rangle g \rangle \\
&= \langle f,f \rangle - \langle f,g \rangle \overline{\langle f,g \rangle} - \overline{\langle f,g \rangle}\langle f,g \rangle + \langle f,g \rangle \overline{\langle f,g \rangle}\langle g,g \rangle \\
&= \langle f,f \rangle - \langle f,g \rangle \overline{\langle f,g \rangle} \\
&= \|f\|^2 - |\langle f,g \rangle|^2\, .
\end{aligned}
$$

Thus $|\langle f,g \rangle|^2 \leqslant \|f\|^2$.

For the general case $|\langle f,g \rangle| \leqslant \|f\|\, \|g\|$, we can suppose $g \neq 0$, so $\|g\| \neq 0$. Let $h = g/\|g\|$ so $\|h\| = 1$. Then

$$|\langle f,h \rangle| \leqslant \|f\|\, .$$

But

$$|\langle f,h \rangle| = \frac{|\langle f,g \rangle|}{\|g\|} \leqslant \|f\|\, ,$$

so we obtain the result. ∎

This method is similar to that used to prove Theorem 1, Chapter 5, except now a bit more care was needed to keep track of complex conjugates. The reader should derive the other properties, taking special care with the triangle inequality $\|f + g\| \leqslant \|f\| + \|g\|$.

Theorem 3. *Let V be the space of functions $f: [a,b] \to \mathbb{C}$, such that $|f|^2$ is integrable (that is, $\int_a^b |f(x)|^2\, dx < \infty$). Then the space V is an inner product space with*

$$\langle f,g \rangle = \int_a^b f(x)\overline{g(x)}\, dx$$

and norm

$$\|f\|^2 = \left(\int_a^b |f(x)|^2 \, dx \right)^{1/2} .$$

The space of sectionally continuous functions is also an inner product space.

Proof: First, if $\|f\| = 0$, we have $\int_a^b |f(x)|^2 \, dx = 0$, so by Theorem 4(ii), Chapter 8, f is zero except possibly on a set of measure zero. Since we are identifying functions which agree except on a set of measure zero, $f = 0$. Finally, $\langle f,g \rangle$ satisfies all the other rules of an inner product space as in Theorem 1. We need only show that $\langle f,g \rangle$ is finite (that is, fg is integrable).

If we work only with bounded functions, it is clear that $f\bar{g}$ is integrable and bounded, as are both f and g (see Chapter 8). However, we also wish to allow improper integrals, so f and g need not be bounded. If we split f and g into real and imaginary parts, and into positive and negative parts, this easily reduces to the case of f and g real and positive (the reader is asked to carry out the details as an exercise). Define, for each $M \geqslant 0$, $(fg)_M$ as in Chapter 8. We want to show

$$\underset{M \to \infty}{\text{limit}} \int_a^b (fg)_M < \infty .$$

However, for $M \geqslant 1$, one easily sees that

$$0 \leqslant (fg)_M \leqslant f_M g_M$$

so

$$\int_a^b (fg)_M \leqslant \langle f_M, g_M \rangle \leqslant \|f_M\| \, \|g_M\|$$

by the Schwarz inequality. But $\|f_M\| \leqslant \|f\|$ and $\|g_M\| \leqslant \|g\|$, so

$$\int_a^b (fg)_M \leqslant \|f\| \, \|g\| < \infty .$$

Hence we obtain the result (the limit exists as the integral increases with M; we only needed to show it was bounded above).

Finally, for sectionally continuous functions, observe that they form a vector space (Exercise 9 at the end of the chapter) and are bounded (Exercise 11). Hence both functions, f and $|f|^2$, are integrable, since the set of discontinuities is finite (Theorem 3, Chapter 8). ∎

Theorem 4. *Let V be an inner product space and suppose $f = \sum_{k=0}^{\infty} c_k \varphi_k$ for an orthonormal family, $\varphi_0, \varphi_1, \ldots$ in V (convergence in the mean) and $f \in V$. Then $c_k = \langle f, \varphi_k \rangle = \overline{\langle \varphi_k, f \rangle}$.*

Proof: Let $s_n = \sum_{k=0}^{n} c_k \varphi_k$, so that $\|f - s_n\| \to 0$. Fix i and choose $n \geqslant i$. Form

$$\langle f - s_n, \varphi_i \rangle = \langle f, \varphi_i \rangle - \langle s_n, \varphi_i \rangle .$$

This expression approaches zero as $n \to \infty$, since $|\langle f - s_n, \varphi_i \rangle| \leqslant \|f - s_n\|$. But for $n \geqslant i$, we have

$$\langle s_n, \varphi_i \rangle = \sum_{k=0}^{n} \langle c_k \varphi_k, \varphi_i \rangle = \sum_{k=0}^{n} c_k \langle \varphi_k, \varphi_i \rangle = \sum_{k=0}^{n} c_k \delta_{ki} = c_i .$$

Thus

$$\langle f, \varphi_i \rangle - c_i \to 0$$

as $n \to \infty$. Since this expression is independent of n, we have $\langle f, \varphi_i \rangle = c_i$. ∎

Theorem 5. *Let* $\varphi_0, \varphi_1, \dots$ *be an orthonormal system in an inner product space V. For each* $f \in V$, $\sum_{i=0}^{\infty} |\langle f, \varphi_i \rangle|^2$ *converges and we have the inequality*

$$\sum_{i=0}^{\infty} |\langle f, \varphi_i \rangle|^2 \leq \|f\|^2 .$$

Proof: Let $s_n = \sum_{i=0}^{n} \langle f, \varphi_i \rangle \varphi_i$. We first show that $f - s_n$ and s_n are orthogonal. To see this, it is enough to show that $f - s_n$ and φ_i, $1 \leq i \leq n$ are orthogonal (why?). Indeed,

$$\langle f - s_n, \varphi_i \rangle = \langle f, \varphi_i \rangle - \langle s_n, \varphi_i \rangle$$

and

$$\langle s_n, \varphi_i \rangle = \langle f, \varphi_i \rangle ,$$

since

$$\langle s_n, \varphi_i \rangle = \sum_{j=0}^{n} \langle \langle f, \varphi_j \rangle \varphi_j, \varphi_i \rangle = \sum_{j=0}^{n} \langle f, \varphi_j \rangle \, \delta_{ij} = \langle f, \varphi_i \rangle$$

(this is the same computation as in Theorem 4). Now if g and h are orthogonal, $\|g + h\|^2 = \|g\|^2 + \|h\|^2$ (Pythagoras' relation, Exercise 9, Section 10.1), so, by the above,

$$\|f\|^2 = \|f - s_n + s_n\|^2 = \|s_n\|^2 + \|f - s_n\|^2 ;$$

hence

$$\|s_n\|^2 \leq \|f\|^2 .$$

Now

$$\|s_n\|^2 = \left\| \sum_{i=0}^{n} \langle f, \varphi_i \rangle \varphi_i \right\|^2 = \sum_{i=0}^{n} |\langle f, \varphi_i \rangle|^2 \|\varphi_i\|^2 ,$$

since the φ_i are orthogonal and therefore

$$\|s_n\|^2 = \sum_{i=0}^{n} |\langle f, \varphi_i \rangle|^2 \leq \|f\|^2 .$$

Thus the series $\sum_{i=0}^{\infty} |\langle f, \varphi_i \rangle|^2$ has partial sum $\|s_n\|^2$, which is an increasing sequence, since the terms of the series are ≥ 0 and the series is bounded above by $\|f\|^2$; hence the series converges with sum $\leq \|f\|^2$. ∎

Theorem 6. *Let V be an inner product space and* $\varphi_0, \varphi_1, \dots$ *an orthonormal system. Then* $\varphi_0, \varphi_1, \dots$ *is complete iff for each* $f \in V$, *we have*

$$\|f\|^2 = \sum_{n=0}^{\infty} |\langle f, \varphi_n \rangle|^2 .$$

Proof: Let

$$s_n = \sum_{i=0}^{n} \langle f, \varphi_i \rangle \varphi_i .$$

In the proof of Theorem 5 it was shown that

$$\|f\|^2 = \|f - s_n\|^2 + \|s_n\|^2 .$$

Now suppose $\varphi_0, \varphi_1, \ldots$ is complete. Then $s_n \to f$, and so $\|f - s_n\|^2 \to 0$. Therefore, since

$$\|s_n\|^2 = \sum_{i=0}^{n} |\langle f,\varphi_i \rangle|^2 ,$$

we have, letting $n \to \infty$,

$$\|f\|^2 = \sum_{i=0}^{\infty} |\langle f,\varphi_i \rangle|^2 .$$

Conversely, if this relation holds, then $\|f\|^2 - \|s_n\|^2 \to 0$ as $n \to \infty$. Hence $\|f - s_n\|^2 \to 0$, that is, $s_n \to f$, which means that

$$f = \sum_{i=0}^{\infty} \langle f,\varphi_i \rangle \varphi_i . \quad \blacksquare$$

Theorem 7. *Let V be an inner product space and $\varphi_0, \varphi_1, \ldots, \varphi_n$ a set of orthonormal vectors in V. Then for each set of numbers $t_0, t_1, \ldots, t_n$*

$$\left\| f - \sum_{k=0}^{n} t_k \varphi_k \right\| \geq \left\| f - \sum_{k=0}^{n} \langle f,\varphi_k \rangle \varphi_k \right\|$$

with equality iff $t_k = \langle f,\varphi_k \rangle$.

Proof: Let $c_k = \langle f,\varphi_k \rangle$, $s_n = \sum_{i=0}^{n} c_i \varphi_i$, and $h_n = \sum_{i=0}^{n} t_i \varphi_i$. Then it is required to show that

$$\|f - s_n\|^2 \leq \|f - h_n\|^2$$

with equality iff $t_k = c_k$. For this, it shall be shown that

$$\|f - h_n\|^2 = \|f\|^2 - \sum_{k=0}^{n} |c_k|^2 + \sum_{k=0}^{n} |c_k - t_k|^2 ,$$

which evidently suffices to prove the theorem. Now to prove this equality, note that

$$\|f - h_n\|^2 = \langle f - h_n, f - h_n \rangle$$

$$= \langle f,f \rangle - \langle f,h_n \rangle - \langle h_n,f \rangle + \langle h_n,h_n \rangle .$$

First,

$$\langle h_n,h_n \rangle = \sum_{i,j} \langle t_i \varphi_i, t_j \varphi_j \rangle$$

$$= \sum_{i,j} t_i \bar{t}_j \delta_{ij} = \sum_{i=0}^{n} |t_i|^2 .$$

Second,

$$\langle f,h_n \rangle = \left\langle f, \sum_{k=0}^{n} t_k \varphi_k \right\rangle = \sum_{k=0}^{n} c_k \bar{t}_k .$$

Thus

$$\| f - h_n \|^2 = \| f^2 \| - \sum_{k=0}^{n} c_k \bar{t}_k - \sum_{k=0}^{n} t_k \bar{c}_k + \sum_{k=0}^{n} |t_k|^2$$

$$= \| f \|^2 - \sum_{k=0}^{n} |c_k|^2 + \sum_{k=0}^{n} |c_k - t_k|^2$$

as required. ∎

Theorem 8.   *The exponential and trigonometric systems on* $[0,2\pi]$ *(or* $[-\pi,\pi]$ *are complete in the space of functions* $f: [0,2\pi] \to \mathbb{C}$ *with* $\int_0^{2\pi} |f(x)|^2 \, dx < \infty$ *(the integral may be improper).*

Proof*:   By our remarks in the text and Exercise 1, Section 10.3, it suffices to consider the exponential case. Two necessary facts are contained in the following lemmas.

Lemma 1.   *(Stone-Weierstrass theorem in a special case.) Let* $f: [0,2\pi] \to \mathbb{C}$ *be continuous and let* $f(0) = f(2\pi)$ *(periodicity). Then for any* $\varepsilon > 0$ *there is an n and constants* $c_i, i = -n, \ldots, -1, 0, 1, \ldots, n$, *such that if we form the function*

$$p_n(x) = c_0 + c_1 e^{ix} + c_2 e^{2ix} + \cdots + c_n e^{nix}$$

$$+ c_{-1} e^{-ix} + c_{-2} e^{-2ix} + \cdots + c_{-n} e^{-nix}$$

then

$$|f(x) - p_n(x)| < \varepsilon$$

for all $x \in [0,2\pi]$.

The Stone-Weierstrass theorem was proved in Chapter 5. See also, Exercise 44(b), Chapter 5. The proof of the next lemma is technical; it may be omitted on a first reading of the proof.

Lemma 2.   *Let* $f: [0,2\pi] \to \mathbb{C}$ *be square integrable, and* $\varepsilon > 0$. *Then there is a continuous function* $g: [0,2\pi] \to \mathbb{C}$ *with* $g(0) = g(2\pi)$ *such that*

$$\| f - g \|^2 = \int_0^{2\pi} |f(x) - g(x)|^2 \, dx < \varepsilon .$$

Proof: First suppose that f is ≥ 0 and bounded by M. Given $\varepsilon > 0$ choose a partition P of $[0,2\pi]$ such that, setting $h = f^2$,

$$\left| \int h - \sum_{i=1}^{n} h(c_i)(x_{i+1} - x_i) \right| < \frac{\varepsilon}{2} ,$$

and a similar estimate for f. We can, by drawing straight lines, construct a continuous g such that g is constant $= f(c_i)$ on $[y_i, z_i]$, where $[y_i, z_i] \subset [x_i, x_{i+1}]$ and $|y_i - x_i| <$

---

* A proof due to Luxemburg and not relying on the Stone-Weierstrass theorem is outlined in Exercise 75. Another proof due to Lebesgue is given in Exercise 76. Both proofs, however, rely on the converse of Example 2, Section 10.2 (see Exercise 14), which uses completeness of $\mathscr{L}^2$, that is, the Lebesgue integral.

$\varepsilon/8M^2n$, $|x_{i+1} - z_i| < \varepsilon/8M^2n$, and g is bounded by M. It is then easy to see that

$$\int |f - g|^2 = \int (f^2 + g^2 - 2fg) < \frac{\varepsilon}{2} + 4M^2 \times \frac{\varepsilon}{8M^2n} \times n = \varepsilon$$

by adding and subtracting the approximations for $\int f^2 = \int h$ and $\int f$ and using the definition of g. The details are left to the reader.

The general case may be dealt with by writing f as $f = f^+ - f^-$ (see Chapter 8), so we can assume $f \geqslant 0$. Then we can form f_M as in Chapter 8 and choose M large such that $\int |f - f_M|^2 < \varepsilon/4$ which is possible by Corollary 4 of the monotone convergence theorem (Chapter 8). By the above we can choose g such that $\int |g - f_M|^2 < \varepsilon/4$, and thus $\int |f - g|^2 < \varepsilon$, since

$$\|f - g\| \leqslant \|f - f_M\| + \|g - f_M\| \leqslant \frac{\sqrt{\varepsilon}}{2} + \frac{\sqrt{\varepsilon}}{2} = \sqrt{\varepsilon} . \quad \blacksquare$$

To prove the theorem from these lemmas requires two steps.

Step 1:   *Proof of the theorem for f continuous and periodic.*
Let

$$s_n = \sum_{-n}^{n} \langle f, \varphi_k \rangle \varphi_k, \qquad \text{where } \varphi_k(x) = \frac{e^{ixk}}{\sqrt{2\pi}} .$$

Then for $\varepsilon > 0$ we must show there is an N such that $n \geqslant N$ implies $\|f - s_n\| < \varepsilon$. It suffices to produce a single n, because by Theorem 7, $\|f - s_{n+k}\| \leqslant \|f - s_n\|$ (we get a better approximation by taking more terms—see also, Exercise 21, p. 436). Now choose p_n as in Lemma 1, so $|f(x) - p_n(x)| < \varepsilon/\sqrt{2\pi}$ and form the corresponding s_n.
Now

$$\|f - p_n\|^2 = \int_0^{2\pi} |f(x) - p_n(x)|^2 \, dx$$

$$\leqslant \int_0^{2\pi} \left(\frac{\varepsilon}{\sqrt{2\pi}} \right)^2 dx = \varepsilon^2 .$$

Thus $\|f - p_n\| < \varepsilon$. However, by Theorem 7,

$$\|f - s_n\| \leqslant \|f - p_n\| < \varepsilon ,$$

since the Fourier series gives the best mean approximation to f. This proves Step 1.

Step 2:   *General case.*
In view of Lemma 2 and Step 1, it suffices to prove the following fact. Here V is the space of square integrable functions, but the lemma is stated in general terms.

Lemma 3.   *Let V be an inner product space and let $\varphi_0, \varphi_1, \ldots$ be an orthonormal family. Suppose $f \in V$ and $f_n \to f$. If we have*

$$f_n = \sum_{k=0}^{\infty} \langle f_n, \varphi_k \rangle \varphi_k$$

for each n, then

$$f = \sum_{k=0}^{\infty} \langle f, \varphi_k \rangle \varphi_k .$$

Proof: Given $\varepsilon > 0$, choose N such that $k \geqslant N$ implies $\| f_k - f \| = \varepsilon/3$. Choose M such that $n \geqslant M$ implies

$$\left\| \sum_{j=0}^{n} \langle f_N, \varphi_j \rangle \varphi_j - f_N \right\| < \frac{\varepsilon}{3}.$$

Then using the triangle inequality,

$$\left\| \sum_{j=0}^{n} \langle f, \varphi_j \rangle \varphi_j - f \right\|$$

$$\leqslant \left\| \sum_{j=0}^{n} \langle f, \varphi_j \rangle \varphi_j - \sum_{j=0}^{n} \langle f_N, \varphi_j \rangle \varphi_j \right\| + \left\| \sum_{j=0}^{n} \langle f_N, \varphi_j \rangle \varphi_j - f_N \right\| + \| f_N - f \| .$$

By Bessel's inequality, the first term is $\leqslant \| f - f_N \|$ (see the proof of Theorem 5). Thus $n \geqslant M$ implies

$$\left\| \sum_{j=0}^{n} \langle f, \varphi_j \rangle \varphi_j - f \right\| < \frac{\varepsilon}{3} + \frac{\varepsilon}{3} + \frac{\varepsilon}{3} = \varepsilon ,$$

which proves our assertion.   ∎

Theorem 9.   Let $f: [0, 2\pi] \rightarrow \mathbb{R}$ (or $f: [-\pi, \pi] \rightarrow \mathbb{R}$) be sectionally continuous, have a jump discontinuity at x_0, and assume that $f'(x_0+)$ and $f'(x_0-)$ both exist. Then the Fourier series of f (either in exponential or trigonometric form) evaluated at x_0 converges to $[f(x_0+) + f(x_0-)]/2$. In particular, if f is differentiable at x_0, then the Fourier series of f converges at x_0 to $f(x_0)$.

It is convenient to first prove the following special case:

Lemma 4.   Let $f: [-\pi, \pi] \rightarrow \mathbb{C}$ be square integrable and differentiable at x_0 (as usual, extend f so it is periodic). Then the Fourier series of f at x_0 converges to $f(x_0)$.

Proof: (The proof of Lemma 4 was pointed out by P. Chernoff.) By translating and adding a constant we can assume $x_0 = 0$ and $f(x_0) = 0$ (why?). Define a new function $g(x)$ by setting

$$g(x) = \begin{cases} \dfrac{f(x)}{e^{ix} - 1} , & x \neq 0 , \\[2mm] \dfrac{f'(0)}{i} , & x = 0 . \end{cases}$$

By the quotient rule of calculus it follows that g is continuous at 0. Since $1/(e^{ix} - 1)$ is bounded in absolute value outside a neighborhood of 0, it follows that g is square integrable (why?).

Now $f(x) = (e^{ix} - 1)g(x)$. Let $c_n(f)$ be the nth Fourier coefficient of f and $c_n(g)$ that for g. Then from the definition

$$c_n(f) = c_{n-1}(g) - c_n(g_1) .$$

So

$$\sum_{n=-N}^{N} c_n(f) = c_{-N-1}(g) - c_N(g) ,$$

since we have a telescoping sum. Since $x_0 = 0$, $\sum_{-N}^{N} c_n(f)$ is the Nth partial sum at $x = 0$ of the Fourier series of f. But $c_N(g) \to 0$ by Bessel's inequality. Hence $\sum_{-N}^{N} c_n(f) \to 0 = f(x_0)$. ∎

Actually, we do not need the fact that f is differentiable at x_0. If f is Lipschitz at x_0 (that is, there is a constant M such that $|(f(x) - f(x_0))/(x - x_0)| \leqslant M$ for $|x - x_0| < \delta$, $x \neq x_0$) we could obtain the same result by a similar proof (we only need g in the proof to be square integrable—or even just integrable). For example, if f is continuous and $f'(x_0+)$ and $f'(x_0-)$ exist, then this condition is satisfied (why?).

It is now quite easy to prove Theorem 9 from Lemma 4 and the above remarks. Now, consider

$$h(x) = \begin{cases} f(x_0-), & x < x_0, \\ f(x_0), & x = x_0, \\ f(x_0+), & x > x_0. \end{cases}$$

Then h is a step function and we can easily compute its Fourier series directly (see Section 10.5). We know this series converges to $[f(x_0-) + f(x_0+)]/2$ at x_0. Now consider

$$k(x) = f(x) - h(x).$$

Then $k(x_0) = 0 = k(x_0+) = k(x_0-)$ and $k'(x_0+)$, $k'(x_0-)$ exist. Hence, by Lemma 4, the Fourier series of k converges to 0 at x_0. Therefore, the Fourier series for f converges to $[f(x_0+) + f(x_0-)]/2$ at x_0. This proves the assertion. ∎

Now we turn to the longer classical proof of Theorem 9. Later it will be convenient to have this longer proof at hand, despite the fact that it is more complex than the one just given. First, let us explain the basic idea behind this proof. Let $s_n(x)$ be the nth partial sum of the trigonometric Fourier series. We shall write

$$s_n(x) = \int_0^{2\pi} f(\xi) D_n(x - \xi)\, d\xi$$

for some function D_n specified later (Lemma 9); we say s_n is the *convolution* of f and D_n. Then we show that D_n has unit area and "concentrates" around 0; that is, behaves like a Dirac delta function. As $n \to \infty$, the convolution will then pick off the value of f at x. See Figure 10-31. For this reason, D_n is also called an *approximate identity*.

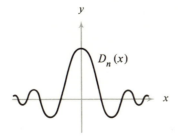

FIGURE 10-31

Before we can formalize these ideas, we need some preliminary results. The first lemma is a generalization of Example 4, Section 10.2, and is called the *Riemann-Lebesgue lemma*.

Lemma 5. *Suppose f is bounded and (Riemann) integrable on* $[a,b]$. *Then*

$$\lim_{\alpha \to \infty} \int_a^b f(x) \sin(\alpha x)\, dx = 0$$

(where the limit is taken through all real $\alpha > 0$*).*

Proof: First, suppose f is a constant M. Then

$$\left| \int_a^b f(x) \sin \alpha x\, dx \right| = |M| \left| \int_a^b \sin \alpha x\, dx \right|$$

$$= |M| \frac{|\cos(\alpha a) - \cos(\alpha b)|}{\alpha}$$

$$\leq \left| \frac{2M}{\alpha} \right| \to 0 \qquad \text{as } \alpha \to \infty .$$

Thus the result is true for constant f.

Now, for the general case, given $\varepsilon > 0$ choose a partition $P = \{x_0, x_1, \ldots, x_n\}$ of $[a,b]$ such that $U(f,P) - L(f,P) < \varepsilon/2$. Then

$$U(f,P) = \sum_{i=1}^n M_i(x_i - x_{i-1})$$

and

$$L(f,P) = \sum_{i=1}^n m_i(x_i - x_{i-1}) ,$$

where M_i is the maximum of f on $[x_{i-1}, x_i]$ and m_i is the minimum. Let m be the step function equal to m_i on $]x_{i-1}, x_i]$. Then choose N so that

$$\int_a^b m(x) \sin(\alpha x)\, dx = \sum_{i=1}^n \int_{x_{i-1}}^{x_i} m_i \sin(\alpha x)\, dx < \frac{\varepsilon}{2}$$

if $\alpha \geq N$, which is possible because m_i is constant and n is fixed and finite. Then, by the triangle inequality, for $\alpha \geq N$

$$\left| \int_a^b f(x) \sin(\alpha x)\, dx \right| \leq \left| \int_a^b m(x) \sin \alpha x\, dx \right| + \left| \int_a^b [f(x) - m(x)] \sin \alpha x\, dx \right|$$

$$< \frac{\varepsilon}{2} + \int_a^b |M(x) - m(x)|\, dx ,$$

where M equals M_i on $]x_{i-1}, x_i]$. (Here we have used the fact that $|\sin \alpha x| \leq 1$.) But $M(x) - m(x) \geq 0$ and

$$\int_a^b M(x) - m(x)\, dx = U(f,P) - L(f,P) < \frac{\varepsilon}{2} ,$$

so, for $\alpha > N$,

$$\left| \int_a^b f(x)\sin(\alpha x)\, dx \right| < \varepsilon$$

by the above. ∎

Lemma 6. *Suppose* $g: [0,a] \rightarrow \mathbb{R}$ *is sectionally continuous, and* $g'(0+)$ *exists. Then*

$$\lim_{k \rightarrow \infty} \int_0^a g(x)\, \frac{\sin kx\, dx}{x} = \frac{\pi}{2}\, g(0+)\, .$$

Proof: Since

$$\int_0^a g(x)\, \frac{\sin kx\, dx}{x} = g(0+)\int_0^a \frac{\sin kx\, dx}{x} + \int_0^a \left[\frac{g(x) - g(0+)}{x} \right]\sin kx\, dx\, ,$$

it suffices to show that

$$\int_0^a \frac{\sin kx}{x}\, dx \rightarrow \frac{\pi}{2} \qquad \text{as } k \rightarrow \infty, \qquad \text{and} \tag{1}$$

$$\int_0^a \left[\frac{g(x) - g(0+)}{x} \right]\sin kx\, dx \rightarrow 0 \qquad \text{as } k \rightarrow \infty\, . \tag{2}$$

To clarify Eq. 1,

$$\int_0^a \frac{\sin kx}{x}\, dx = \int_0^{ka} \frac{\sin t}{t}\, dt\, ,$$

which converges to $\pi/2$, as $k \rightarrow \infty$, since $\int_0^\infty (\sin t)/t\, dt = \pi/2$; see Example 1, p. 271 and Exercise 29, p. 437.

To prove Eq. 2, observe that $[g(x) - g(0+)]/x$ is bounded and integrable (since, as $x \rightarrow 0$, this approaches a limit $g'(0+)$). Therefore

$$\int_0^a \left[\frac{g(x) - g(0)}{x} \right]\sin kx\, dx \rightarrow 0$$

as $k \rightarrow \infty$ by Lemma 5. ∎

Note that Lemma 5 is needed for α real and an arbitrary interval $[a,b]$. This case does not follow at once from Example 4, Section 10.2, but requires the direct argument we gave.

Lemma 7. *Let* g *be sectionally continuous on* $]a,b[$ *and have a jump discontinuity at* x_0. *Suppose* $g'(x_0+)$ *and* $g'(x_0-)$ *exist. Then*

$$\lim_{k \rightarrow \infty} \int_a^b g(x)\left(\frac{\sin k(x - x_0)}{(x - x_0)} \right) dx = \frac{\pi[g(x_0+) + g(x_0-)]}{2}\, .$$

Proof: Write the above integral as a sum,

$$\int_a^b = \int_a^{x_0} + \int_{x_0}^b$$

and note that

$$\int_a^{x_0} g(x)\frac{\sin k(x - x_0)}{(x - x_0)}\,dx = \int_0^{x_0-a} g(x_0 - t)\frac{\sin kt\,dt}{t}$$

and

$$\int_{x_0}^b g(x)\frac{\sin k(x - x_0)}{(x - x_0)}\,dx = \int_0^{b-x_0} g(x_0 + t)\frac{\sin kt}{t}\,dt\,.$$

Now let $k \to \infty$ and employ Lemma 6. We then get $(\pi g(x_0-))/2$ and $\pi g(x_0+)/2$ for the limit of these two integrals, respectively, as $k \to \infty$. (See also, Exercise 40, p. 440). ∎

Lemma 8. *Let* $f\colon [0,2\pi] \to \mathbb{R}$. *Then the nth partial sum of the Fourier series of f may be written as*

$$s_n(x) = \frac{1}{2\pi}\int_0^{2\pi} f(t)\,dt + \frac{1}{\pi}\sum_{k=1}^n \int_0^{2\pi} f(t)\cos[k(t - x)]\,dt\,.$$

Proof: This is clear if we remember that

$$\cos[k(t - x)] = \cos(kt)\cos(kx) + \sin(kt)\sin(kx)\,. ∎$$

Lemma 9. *Let* $s_n(x)$ *be the nth partial sum of the Fourier series of f. Then*

$$s_n(x) = \frac{1}{2\pi}\int_0^{2\pi} f(t)D_n(t - x)\,dt\,,$$

where

$$D_n(u) = \frac{\{\sin[(n + 1/2)u]\}}{[\sin(u/2)]}\,.$$

Proof: This follows from Lemma 8 and the identity

$$\sum_{k=-n}^n e^{iku} = \frac{\sin[(n + 1/2)u]}{\sin(u/2)}$$

(Exercise 6, Section 10.2). ∎

We are now ready to prove Theorem 9. We must show that

$$s_n(x_0) \to \frac{[f(x_0+) + f(x_0-)]}{2}$$

as $n \to \infty$. We shall assume $0 < x_0 < 2\pi$. The reader is asked to consider the cases $x_0 = 0$, $x_0 = 2\pi$ separately. By Lemma 9,

$$s_n(x_0) = \frac{1}{\pi}\int_0^{2\pi} g(t)\left\{\frac{\sin(n + 1/2)(t - x_0)}{t - x_0}\right\}dt\,,$$

where

$$g(t) = f(t)\left\{\frac{(t - x_0)/2}{\sin[(t - x_0)/2]}\right\}\,,\qquad x_0 \neq t\,.$$

By Lemma 7 (which is applicable by Exercise 41 at the end of the chapter), we have

$$s_n(x_0) \rightarrow \frac{[g(x_0+) + g(x_0-)]}{2} .$$

Now it is a simple matter to see that

$$g(x_0+) = f(x_0+) \quad \text{and} \quad g(x_0-) = f(x_0-),$$

and so the theorem is obtained. ∎

Theorem 11 (*Fejér.*) *Let f be piecewise continuous on* $[0,2\pi]$ *and suppose* $f(x_0+)$ *and* $f(x_0-)$ *exist. Then the Fourier series of f converges* (C,1) *at* x_0 *to* $[f(x_0+) + f(x_0-)]/2$. *If f is continuous, the Fourier series converges* (C,1) *uniformly to f.*

Proof: For notational reasons it is slightly more convenient to use $[-\pi,\pi]$ rather than $[0,2\pi]$ for this proof. Of course, this does not effect the conclusions.

With our usual notation,

$$s_n(x) = \sum_{k=-n}^{n} c_k e^{ikx} ,$$

the *n*th partial sum of the Fourier series of f. To discuss (C,1) summability, we must consider

$$\sigma_n(x) = \frac{1}{n+1} \sum_{k=0}^{n} s_k(x) .$$

Using Lemma 9 we obtain

$$\sigma_n(x) = \frac{1}{n+1} \sum_{k=0}^{n} \frac{1}{2\pi} \int_{-\pi}^{\pi} f(x - t) D_k(t) \, dt$$

that is,

$$\sigma_n(x) = \frac{1}{2\pi} \int_{-\pi}^{\pi} f(x - t) F_n(t) \, dt ,$$

where by definition the *Fejér kernel* is

$$F_n(t) = \frac{1}{n+1} \sum_{k=0}^{n} D_k(t) .$$

We shall need the following lemmas.

Lemma 10. $F_n(t) = \dfrac{1}{n+1} \dfrac{\sin^2[(n+1)t/2]}{\sin^2[t/2]} .$

Proof: By the formula for D_n (Lemma 9), we have

$$(n+1)F_n(t) = \sum_{k=0}^{n} \frac{\sin(k+1/2)t}{\sin 1/2t}$$

$$= \frac{1}{\sin 1/2t} \operatorname{Im} \left\{ \sum_{k=0}^{n} e^{i(k+1/2)t} \right\} \qquad (\text{Im} = \text{imaginary part})$$

$$= \frac{1}{\sin 1/2t} \, \mathrm{Im}\left\{ e^{it/2} \cdot \frac{e^{i(n+1)t} - 1}{e^{it} - 1} \right\}$$

$$= \frac{1}{\sin 1/2t} \, \mathrm{Im}\left\{ \frac{e^{i(n+1)t} - 1}{e^{it/2} - e^{-it/2}} \right\}$$

$$= \frac{1 - \cos(n+1)t}{2 \sin^2 1/2t} = \frac{\sin^2 1/2(n+1)t}{\sin^2 1/2t} \cdot \quad \blacksquare$$

Lemma 11.   *The Fejér kernel has the following properties:*
  (i) $F_n(t)$ *is* 2π-*periodic*

  (ii) $\dfrac{1}{2\pi} \int_{-\pi}^{\pi} F_n(t)\, dt = 1$

  (iii) $F_n(t) \geqslant 0$
  (iv) *For each fixed* $\delta > 0$, $\displaystyle\lim_{n \to \infty} \int_{\delta \leqslant |t| \leqslant \pi} F_n(t)\, dt = 0$.

  Proof: (i), (ii) follow from the definition of F_n; (iii) follows from Lemma 10. (iv) for $\delta \leqslant |t| \leqslant \pi$ we have $1/(\sin^2 t/2) \leqslant 1/(\sin^2 \delta/2)$. Hence

$$0 \leqslant F_n(t) \leqslant \frac{1}{n+1} \frac{1}{\sin^2 \delta/2}, \qquad \delta \leqslant |t| \leqslant \pi \; .$$

Since this $\to 0$ uniformly as $n \to \infty$, the integral $\int_{\delta \leqslant |t| \leqslant \pi} F_n(t)\, dt \to 0$. $\blacksquare$

Let us now prove Theorem 11. By using the same technique that was used in the proof of Theorem 9 (see the arguments following Lemma 4) it suffices to prove the last part of the theorem. Thus assume f is continuous. We have

$$\sigma_n(x) = \frac{1}{2\pi} \int_{-\pi}^{\pi} f(x - t) F_n(t)\, dt \; .$$

Hence by (ii) of Lemma 11,

$$f(x) - \sigma_n(x) = \frac{1}{2\pi} \int_{-\pi}^{\pi} (f(x) - f(x - t)) F_n(t)\, dt \; .$$

Accordingly, by (iii) of Lemma 11 (positivity of F_n)

$$|f(x) - \sigma_n(x)| \leqslant \frac{1}{2\pi} \int_{-\pi}^{\pi} |f(x) - f(x - t)| \, F_n(t)\, dt \; .$$

Given $\varepsilon > 0$ we can, by uniform continuity of f find $\delta > 0$ so that $|f(x) - f(y)| \leqslant \varepsilon$ if $|x - y| \leqslant \delta$. Then

$$|f(x) - \sigma_n(x)| \leqslant \frac{1}{2\pi} \int_{|t| \leqslant \delta} |f(x) - f(x - t)| \, F_n(t)\, dt$$

$$+ \frac{1}{2\pi} \int_{\delta \leqslant |t| < \pi} |f(x) - f(x - t)| \, F_n(t)\, dt \; .$$

The first integral is

$$\leqslant \frac{1}{2\pi} \int_{|t| \leqslant \delta} \varepsilon F_n(t) \, dt$$

$$\leqslant \frac{1}{2\pi} \int_{-\pi}^{\pi} \varepsilon F_n(t) \, dt = \varepsilon \ .$$

The second integral is

$$\leqslant \frac{1}{2\pi} \int_{\delta \leqslant |t| \leqslant \pi} 2M \, F_n(t) \, dt$$

$$= \frac{M}{\pi} \int_{\delta \leqslant |t| \leqslant \pi} F_n(t) \, dt$$

where $M = \sup_t |f(t)|$. Now, by property (iv) of Lemma 11, we may choose N so that if $n \geqslant N$ this last integral is $\leqslant \varepsilon$. Accordingly, if $n \geqslant N$, $|f(x) - \sigma_n(x)| \leqslant \varepsilon + \varepsilon = 2\varepsilon$. ∎

For an integrable function one can prove that the Cesaro sums converge to the function except possibly on a set of measure zero (see Hewitt and Stromberg, *Real and Abstract Analysis*, p. 294). This result is not as deep as that of L. Carleson mentioned on p. 353.

Theorem 12. *Suppose $\int_{-\pi}^{\pi} |f(x)|^2 \, dx < \infty$, and f has Fourier series*

$$\frac{a_0}{2} + \sum_{n=1}^{\infty} (a_n \cos nx + b_n \sin nx) \ .$$

Then, letting $g(x) = \int_{-\pi}^{x} f(y) \, dy$, we have

$$g(x) = \frac{a_0}{2}(x + \pi) + \sum_{n=1}^{\infty} \left(a_n \int_{-\pi}^{x} \cos ny \, dy + b_n \int_{-\pi}^{x} \sin ny \, dy \right)$$

$$= \frac{a_0}{2}(x + \pi) + \sum_{n=1}^{\infty} \left\{ \frac{a_n}{n} \sin nx - \frac{b_n}{n} (\cos(nx) - (-1)^n) \right\}$$

and the convergence is uniform.

Proof: It is enough to prove the following lemma, as we shall see below.

Lemma 12. *Suppose $f_n \colon [a,b] \to \mathbb{R}$ is such that $\int_a^b |f_n(x)|^2 \, dx < \infty$ and $f_n \to f$ in mean. Let*

$$g_n(x) = \int_a^x f_n(y) \, dy \qquad and \qquad g(x) = \int_a^x f(y) \, dy \ .$$

Then $g_n \to g$ uniformly on $[a,b]$.

Proof: We have

$$|g_n(x) - g(x)|^2 \leqslant \left(\int_a^x |f_n(x) - f(x)| \, dx \right)^2$$

$$\leqslant \left(\int_a^x |f_n(x) - f(x)|^2 \, dx \right) (x - a)$$

by the Schwarz inequality. This is bounded by

$$\| f_n - f \|^2 (b - a)$$

from which the result is obvious. ∎

For the theorem, let $s_n(x)$ be the nth partial sum of the Fourier series and take $f_n = s_n$ in the lemma. We know $f_n \to f$ in mean (Theorem 8), so $g_n \to g$ uniformly. Here g_n is the partial sum of the integrated Fourier series, so we have the result. ∎

Theorem 13.   *Consider*

$$f(x) = \begin{cases} a, & -\pi \leqslant x < 0, \\ b, & 0 \leqslant x \leqslant \pi, \end{cases}$$

and suppose $a < b$. Let $s_n(x)$ be the nth partial sum of the trigonometric Fourier series. Then the maximum of s_n occurs at $\pi/2n$ and the minimum at $-\pi/2n$ and

$$\lim_{n \to \infty} s_n\left(\frac{\pi}{2n}\right) = \left(\frac{b-a}{2}\right)\left(\frac{2}{\pi}\int_0^\pi \frac{\sin t}{t} dt + 1\right) + b$$

$$\approx (b - a)(.089) + b.$$

Similarly,

$$\lim_{n \to \infty} s_n\left(-\frac{\pi}{2n}\right) = \left(\frac{b-a}{2}\right)\left(-\frac{2}{\pi}\int_0^\pi \frac{\sin t}{t} dt + 1\right) + a$$

$$\approx a - (b - a)(.089)$$

and the difference of these limits is

$$\left(\frac{b-a}{2}\right)\left(\frac{4}{\pi}\int_0^\pi \frac{\sin t}{t} dt\right) \approx (b - a)(1.179).$$

Proof: Let us first prove this for the special case $a = -1$, $b = 1$. We have seen that if

$$g(x) = \begin{cases} -1, & -\pi \leqslant x < 0, \\ 1, & 0 \leqslant x \leqslant \pi, \end{cases}$$

the Fourier series of g is

$$\frac{4}{\pi}\sum_{n=1}^\infty \frac{\sin[(2n-1)x]}{2n-1}.$$

Let

$$s_n(x) = \frac{4}{\pi}\sum_{k=1}^n \frac{\sin[(2k-1)x]}{2k-1}.$$

By differentiating, we see that s_n has its maximum at $x_n = \pi/2n$ (some details here are left to the reader). The value here is

$$s_n\left(\frac{\pi}{2n}\right) = \frac{4}{\pi}\sum_{k=1}^n \frac{\sin[(2k-1)\pi/2n]}{2k-1}$$

$$= \frac{2}{\pi}\sum_{k=1}^n \frac{\sin[(2k-1)\pi/2n]}{((2k-1)\pi/2n)}\left(\frac{\pi}{n}\right).$$

This sum is a Riemann sum for the function $\sin y/y$ on $[0,\pi]$ with partition $\{0,\pi/n,2\pi/n,\ldots,\pi\}$. Hence if we choose n even and let $n \to \infty$, this converges (by Theorems 1 and 3, Chapter 8) to

$$\frac{2}{\pi} \int_0^\pi \frac{\sin y}{y}\, dy \;.$$

The case of the minimum of f for $x < 0$ holds as f and s_n are both odd.

The numerical value of the integral is approximately 1.179 and is computed by numerical methods such as the trapezoid rule (we omit the details).

The general case for f follows by observing that its Fourier series has nth partial sum, $\frac{1}{2}(b - a)(s_n + 1) + a$ (why?). ∎

Theorem 14. *Suppose f is continuous on $[-\pi,\pi]$, $f(-\pi) = f(\pi)$, and f' is sectionally continuous with jump discontinuities. Then the (trigonometric or exponential) Fourier series of f converges to f absolutely and uniformly.*

Proof: We can write

$$f(x) = \frac{a_0}{2} + \sum_{n=1}^{\infty}(a_n \cos nx + b_n \sin nx)$$

by Theorem 9. Also, the Fourier coefficients of f' are

$$\alpha_n = \frac{1}{\pi}\int_{-\pi}^{\pi} f'(x) \cos nx\, dx\;, \qquad \beta_n = \frac{1}{\pi}\int_{-\pi}^{\pi} f'(x)\sin nx\, dx\;,$$

and we have

$$\alpha_n = \frac{1}{\pi} f(x)\cos nx \bigg|_{-\pi}^{\pi} + \frac{n}{\pi}\int_{-\pi}^{\pi} f(x)\sin nx\, dx$$

$$= nb_n\;,$$

since we can integrate by parts, and $f(\pi) = f(-\pi)$. Similarly, $\beta_n = -na_n$.

Some care is needed above in justifying integration by parts, since f' exists only in sections. But if it is applied on each section using the fact that it has jump discontinuities only, and noting the continuity of f, then we get the above results. The reader should write out the details if they are not clear.

Now, a lemma is stated.

Lemma 13. *Under the conditions of Theorem 14, we have*

$$\sum_{n=1}^{\infty} |a_n| < \infty\;, \qquad \sum_{n=1}^{\infty} |b_n| < \infty\;,$$

and $na_n \to 0$, $nb_n \to 0$.

Proof: We know that $\sum_{n=1}^{\infty} \beta_n^2$ converges (by Bessel's inequality for f'). Now let $s_n = \sum_1^n |a_k|$. Then

$$s_n = \sum_1^n \frac{|\beta_k|}{k} = \sum_1^n \sqrt{\frac{\beta_k^2}{k^2}}$$

$$\leqslant \left\{ \left(\sum_1^n \beta_k^2\right)\left(\sum_1^n \frac{1}{k^2}\right) \right\}^{1/2}$$

by the Schwarz inequality. Since this is bounded, so is s_n, and therefore it converges (an increasing sequence converges iff it is bounded). Thus $\sum_{n=0}^{\infty} |a_n|$ converges. Since $\beta_n \to 0$, we also have $na_n \to 0$. The case of the b_n's is similar.

To prove the theorem, we need only show that $a_0/2 + \sum_{n=1}^{\infty} (a_n \cos nx + b_n \sin nx)$ converges uniformly, since we know the limit must be $f(x)$. Thus it suffices to show that $\sum_{n=1}^{\infty} (a_n \cos nx + b_n \sin nx)$ is uniformly convergent.

This is simple using the lemma. Note that

$$|a_n \cos nx + b_n \sin nx| \leqslant |a_n| + |b_n| = M_n$$

and by the lemma, $\sum M_n$ converges. Hence by the Weierstrass M-test (Theorem 3, Chapter 5), the series converges uniformly and absolutely. ∎

Theorem 15. *Let f be continuous on $[-\pi,\pi]$, $f(-\pi) = f(\pi)$, and let f' be sectionally continuous with jump discontinuities. Suppose f'' exists at $x \in [-\pi,\pi]$. Then the Fourier series for f*

$$f(x) = \frac{a_0}{2} + \sum_{n=1}^{\infty} (a_n \cos nx + b_n \sin nx)$$

may be differentiated term by term at x:

$$f'(x) = \sum_{n=1}^{\infty} (-na_n \sin nx + nb_n \cos nx) .$$

Furthermore, this is the Fourier series of f'.

Proof: The proof of Theorem 14 showed that the Fourier coefficients of f' are given by

$$\alpha_n = nb_n , \qquad \beta_n = -na_n .$$

This remark suffices to prove the theorem, since if f'' exists, $f'(x)$ will be the sum of Fourier series (Theorem 9). ∎

Theorem 16. *In the initial displacement problem, suppose that f is twice differentiable. Then the solution to the initial displacement problem is*

$$y(x,t) = \frac{1}{2} [f(x - ct) + f(x + ct)]$$

$$= \sum_{n=1}^{\infty} b_n \sin\left(\frac{n\pi x}{l}\right) \cos\left(\frac{n\pi ct}{l}\right)$$

where b_n are the half-interval sine coefficients of f.

Proof: First, note that the series for $y(x,t)$ converges because $\sum_{1}^{\infty} b_n \sin(n\pi x/l)$ converges uniformly and absolutely to f (Theorem 14).

Let us now show

$$\sum_{n=1}^{\infty} b_n \sin\left(\frac{n\pi x}{l}\right) \cos\left(\frac{n\pi ct}{l}\right) = \frac{1}{2} [f(x - ct) + f(x + ct)].$$

For this, note

$$2 \sin\left(\frac{n\pi x}{l}\right) \cos\left(\frac{n\pi ct}{l}\right) = \sin\left[\frac{n\pi(x - ct)}{l}\right] + \sin\left[\frac{n\pi(x + ct)}{l}\right]$$

so that

$$\sum_1^\infty b_n \sin\left(\frac{n\pi x}{l}\right)\cos\left(\frac{n\pi ct}{l}\right) = 1/2 \sum_1^\infty b_n \sin\left[\frac{n\pi(x-ct)}{l}\right] + \sum_1^\infty b_n \sin\left[\frac{n\pi(x+ct)}{l}\right]$$

$$= 1/2\,[f(x-ct) + f(x+ct)].$$

Now we verify that

$$y(x,t) = 1/2\,[f(x-ct) + f(x+ct)]$$

satisfies all the conditions. First,

$$\frac{\partial^2 y}{\partial t^2} = 1/2c^2[f''(x-ct) + f''(x+ct)] = c^2\frac{\partial^2 y}{\partial x^2}.$$

Second, at $t = 0$, $y(x,0) = f(x)$, and

$$\frac{\partial y}{\partial t}(x,0) = 1/2c[-f'(x) + f'(x)] = 0.$$

Third, $y(0,t) = 1/2[f(-ct) + f(ct)] = 0$, because f is odd (when extended) and

$$y(l,t) = 1/2[f(l-ct) + f(l+ct)] = 0$$

because $f(l-ct) = -f(ct-l) = -f(ct+l)$, since $f(x) = f(x+2l)$ by periodicity. ∎

Theorem 17. *If f is square integrable, then for each $t > 0$*

$$T(x,t) = \frac{a_0}{2} + \sum_{n=1}^\infty a_n e^{-n^2\pi^2 t/l^2}\cos\left(\frac{n\pi x}{l}\right)$$

converges uniformly, is differentiable, and satisfies the heat equation and boundary conditions. At $t = 0$ it equals f, in the sense of convergence in the mean, or pointwise if f is of class C^1. As usual,

$$a_n = \frac{2}{l}\int_0^l f(x)\cos\left(\frac{n\pi x}{l}\right)dx.$$

Proof: To show that $T(x,t)$ satisfies the heat equation, what we must do is justify term-by-term differentiation in both x and t. For this we use Theorem 5, Chapter 5. What we must show is that the series of derivatives

$$-\sum_{n=1}^\infty \frac{a_n\pi^2 n^2}{l^2}e^{-n^2\pi^2 t/l^2}\cos\left(\frac{n\pi x}{l}\right)$$

(which represents both $\partial T/\partial t$ and $\partial^2 T/\partial x^2$) converges uniformly in t and in x. For this we use the Weierstrass M-test in each case. Since $|a_n|$ is bounded ($a_n \to 0$, in fact), we can omit the terms $a_n\pi^2/l^2$. Now in x, let $M_n = n^2 e^{-n^2\pi^2 t/l^2}$. By using the ratio test, we see that $\sum M_n < \infty$, so the series will converge uniformly in x.

Uniformly in t means uniformly for all $t \geqslant \varepsilon$, where $\varepsilon > 0$ is arbitrary but fixed. In this case we let $M_n = n^2 e^{-n^2\pi^2\varepsilon/l^2}$ and note that $\sum M_n$ converges. (We cannot allow $t = 0$.) The rest of the theorem is obvious. ∎

Theorem 18. *In Theorem 17*

$$\lim_{\substack{t\to 0 \\ t>0}} T(x,t) = f(x)$$

in the sense of convergence in mean, and converges uniformly (and pointwise) if f is continuous, with f' sectionally continuous. (We do not require $f(0) = f(l)$ here.) More generally, for any f, if the Fourier series of f converges at x to $f(x)$, then $T(x,t) \to f(x)$ as $t \to 0$.

Proof: For the first part, it will suffice to show the following.

Lemma 14.   *For each $t > 0$ suppose $f_t \in V$, an inner product space and $\varphi_0, \varphi_1, \ldots$ is a complete orthonormal basis. Let*

$$f_t = \sum_{n=0}^{\infty} c_n(t)\varphi_n, \qquad f = \sum_{n=0}^{\infty} c_n\varphi_n.$$

If

$$\lim_{t \to 0} \sum_{n=0}^{\infty} |c_n(t) - c_n|^2 = 0,$$

then $f_t \to f$ (in mean).

Proof: The result is clear, since by Parseval's relation

$$\| f_t - f \|^2 = \sum_{n=0}^{\infty} |c_n(t) - c_n|^2. \quad \blacksquare$$

In the case of Theorem 18, we must show that

$$\lim_{t \to 0} \sum_{n=1}^{\infty} |a_n|^2 (1 - e^{-2n\pi^2 t/l^2})^2 = 0.$$

To do this, it is enough to show that the function

$$g(t) = \sum_{n=1}^{\infty} |a_n|^2 (1 - e^{-n^2\pi^2 t/l^2})^2$$

is continuous in t, for $g(0) = 0$, and hence we would have $\lim\limits_{t \to 0} g(t) = 0$. To show that $g(t)$ is continuous, we shall show that the series converges uniformly in t. To do this, Abel's test will be used. The form we needed is the following.

Lemma 15.   *Let $\sum_{n=1}^{\infty} c_n$ be a convergent series and $\varphi_n(t)$ a uniformly bounded, decreasing (respectively, increasing) sequence; $t \geq 0$. Then $g(t) = \sum_{n=1}^{\infty} c_n\varphi_n(t)$ converges uniformly in t. In particular, g is continuous and $g(0) = \lim\limits_{t \to 0} g(t)$.*

See Theorem 13, Chapter 5 for the proof. One deduces the increasing case from the decreasing case by considering $-g(t)$, instead of $g(t)$. In our case $c_n = |a_n|^2$ and $\varphi_n(t) = (1 - e^{-n^2\pi^2 t/l^2})^2$. Now $\varphi_n \leq \varphi_m$ if $n \leq m$, and $|\varphi_n(t)| \leq 1$. Thus from the lemma and the fact that $\sum c_n$ converges, we have our result.

Now suppose f' is sectionally continuous. Then, from the proof of Theorem 12, $\sum_{n=1}^{\infty} |a_n| < \infty$. Thus for a given x,

$$|f(x) - T(x,t)| \leq \sum_{n=1}^{\infty} |a_n| (1 - e^{-n^2\pi^2 t/l^2}).$$

By an argument like the above, the series on the right converges uniformly, so we can let $t \to 0$ in each term to conclude that

$$T(x,t) \to f(x) \text{ as } t \to 0 .$$

Indeed, note that the convergence is uniform in x because we have the bound

$$\sum_{n=1}^{\infty} |a_n| \, (1 - e^{-n^2 \pi^2 t / l^2}) ,$$

which $\to 0$ as $t \to 0$, is independent of x.

Finally, suppose

$$\sum_{n=1}^{\infty} a_n \cos\left(\frac{n\pi x}{l}\right)$$

converges for some fixed x. Then we wish to show that (for this x fixed)

$$\lim_{t \to 0} g(t) = \lim_{t \to 0} \sum_{n=1}^{\infty} a_n e^{-n^2 \pi^2 t / l} \cos\left(\frac{n\pi x}{l}\right) = 0 .$$

Here we cannot make the same estimate as above because the factor $\cos(n\pi x/l)$ is essential for $\sum a_n \cos(n\pi x/l)$ to converge. However, Lemma 15 can again be applied with $c_n = a_n \cos(n\pi x/l)$ and $\varphi_n(t) = e^{-n^2 \pi^2 t / l^2}$ to yield the desired conclusion, since the φ_n are decreasing and are bounded by 1. ∎

Notice that from this we also conclude that

$$\lim_{t \to t_0} T(x,t) = T(x,t_0)$$

that is, T is continuous in t, in each of the three cases of Theorem 18. Indeed, we already know that for $t > 0$, $T(x,t)$ is differentiable and hence continuous. However, $T(x,t)$ may not be differentiable at $t = 0$, but the above theorem does show that we have continuity at $t = 0$.

These same methods using Abel's and Dirchlet's test are important for establishing convergence in other problems (such as Laplace's equation) as we shall see below.

Theorem 19

(i) *Given g_1, let $\varphi(x,y)$ be defined as on p. 392. Suppose g_1 is of class C^2 and $g_1(0) = g_1(a) = 0$. Then φ converges uniformly, is the solution to the Dirichlet problem with $f_1 = f_2 = g_2 = 0$, is continuous on the whole square, and $\nabla^2 \varphi = 0$ on the interior.*

(ii) *If each of f_1, f_2, g_1, g_2 is of class C^2 and vanishes at the corners of the rectangle, then the solution $\varphi(x,y)$ is given as the sum of four series like those in (i), $\nabla^2 \varphi = 0$ on the interior, φ is continuous on the whole rectangle and assumes the given boundary values. Furthermore, φ is C^{∞} on the interior.*

(iii) *If f_1, f_2, g_1, g_2 are only square integrable, the series for φ converges on the interior, $\nabla^2 \varphi = 0$ and φ is C^{∞}. Also, φ takes on the boundary values in the sense of convergence in mean. This means, for example, $\lim_{y \to 0} \varphi(x,y) = \varphi(x,0) = g_1(x)$ with convergence in mean.*

Proof: For simplicity, let us take the case $a = b = \pi$, the general case being just a change of coordinates. To prove parts (i) and (ii) of the theorem, we show that $\varphi(x,y)$

converges uniformly in x and y and that we can differentiate twice, term by term, on the interior. In view of preceding remarks, this suffices to prove the theorem. Part (ii) is an immediate consequence of (i) and linearity; the boundary values are assumed simply because g_1 is represented by its Fourier series.

Now by Theorems 14 and 15,

$$g_1(x) = \sum_1^\infty b_n \sin nx \,, \qquad g_1'(x) = \sum_1^\infty nb_n \cos nx$$

and these series converge uniformly and absolutely. Here we use the fact that $g_1(0) = g_1(\pi) = 0$.

To show that φ converges uniformly, as in Theorem 18, we use Abel's test (Theorem 13, Chapter 5) on the square $[0,\pi] \times [0,\pi]$. Thus we must show, since the series for g_1 converges uniformly, that $\varphi_n = (\sin n(\pi - y))/\sin n\pi$ is decreasing with n and these functions are uniformly bounded. If we can show that they are decreasing, then uniform boundedness follows easily because $0 \leqslant \varphi_n(y) \leqslant \varphi_1(y)$ and φ_1 is bounded, since it is continuous. In fact, $\varphi_1 \leqslant 1$, in this case.

To show that $\varphi_{n+1} \leqslant \varphi_n$, let us fix y and consider $\psi(t) = (\sinh t(\pi - y))/\sinh t\pi$, $t > 0$. It suffices to show that $\psi'(t) \leqslant 0$, for then ψ decreases as t increases, and, in particular, $\psi(n + 1) \leqslant \psi(n)$. This is a special case of the following lemma.

Lemma 16. *For constants* α, β *if* $\beta > 0$, $\beta \geqslant \alpha$, *and* $\psi(t) = \sinh(\alpha t)/\sinh(\beta t)$, *then* $\psi'(t) \leqslant 0$, *for* $t \geqslant 0$.

Proof:

$$\sinh^2(\beta t)\psi'(t) = \alpha \sinh(\beta t)\cosh(\alpha t) - \beta \sinh(\alpha t)\cosh(\beta t)$$

$$= -\frac{\beta^2 - \alpha^2}{2}\left[\frac{\sinh(\alpha + \beta)t}{\alpha + \beta} - \frac{\sinh(\beta - \alpha)t}{\beta - \alpha}\right]$$

using the identity $\sinh(u + v) = \sinh u \cosh v + \sinh v \cosh u$. If the term in brackets is $\geqslant 0$ we are finished, since $\beta^2 - \alpha^2 > 0$. This is in fact true. To see it, let

$$\rho(t) = \frac{\sinh(\alpha + \beta)t}{\alpha + \beta} - \frac{\sinh(\beta - \alpha)t}{\beta - \alpha}.$$

Now $\rho(0) = 0$ and $\rho'(t) = \sinh((\alpha + \beta)t) - \sinh((\beta - \alpha)t) \geqslant 0$, since $\sinh$ is increasing. Hence $\rho(t) \geqslant 0$ for all $t \geqslant 0$.

This establishes the first part of the proof, which says that the series for $\varphi(x,y)$ converges uniformly. ∎

For the differentiability part, Theorem 5, Chapter 5 is employed. Thus we must show that

$$\lambda(x,y) = \sum_1^\infty n^2 b_n \sinh n(\pi - y)\frac{\sin nx}{\sinh(n\pi)}$$

converges uniformly (this is the second formal y derivative; the second x derivative is its negative).

Here it is important to realize that we can get uniform convergence only if we stay away from the boundary; in fact, for any $\varepsilon > 0$ we shall establish uniform convergence on $0 < \varepsilon \leqslant y \leqslant \pi$ and x arbitrary. With this extra restriction the delicacy of Abel's test is no longer needed; the Weierstrass M-test will do. We have $|b_n| \leqslant M$. Let

$$M_n = n^2 M \frac{\sinh[n(\pi - \varepsilon)]}{\sinh(\pi n)}.$$

Then M_n bounds the terms in λ. But $2 \sinh[n(\pi - \varepsilon)] < e^{n(\pi - \varepsilon)}$ and $2 \sinh(n\pi) \geqslant e^{n\pi}(1 - e^{n\pi})$ from the definition of sinh. Thus

$$M_n \leqslant Mn^2 \frac{e^{-n\varepsilon}}{[1 - e^{-2\pi}]}.$$

Since $\varepsilon > 0$, $\sum M_n$ converges, so we have uniform convergence.

Note that we could use n^k instead of n^2 here and still have convergence; in fact, we can differentiate any number of times, that is, φ is C^∞ (a little thought shows that φ is analytic—see Example 2, p. 181).

The proof of part (iii) is now routine. To show $\nabla^2 \varphi = 0$ and φ is a C^∞ function on the interior, the proof is the same as that above (all that was used was that the b_n are bounded). For convergence in mean we proceed exactly as with the proof of Theorem 18, using Lemma 15. ∎

Worked Examples for Chapter 10

1. Let $f: [0,\pi] \to \mathbb{C}$ be a continuous function. Prove that the following inequality holds.

$$\left| \int_0^\pi f(x)\sin x \, dx \right|^2 + \cdots + \left| \int_0^\pi f(x)\sin nx \, dx \right|^2 \leqslant \frac{\pi}{2} \int_0^\pi |f(x)|^2 \, dx .$$

Solution: This follows immediately from Bessel's inequality applied to the following (incomplete) orthonormal system on $[0,\pi]$:

$$\sqrt{\frac{2}{\pi}} \sin x, \ldots, \sqrt{\frac{2}{\pi}} \sin nx .$$

Notice that if we had used an infinite sum, we would have equality by Parseval's theorem (see Table 10-4).

2. Let V be an inner product space. Show that if $f_n \to f$ (in mean) and $g_n \to g$ (in mean), then

$$\langle f_n, g_n \rangle \to \langle f, g \rangle .$$

Solution: First, make an estimate using the Schwarz inequality and the triangle inequality:

$$|\langle f_n, g_n \rangle - \langle f, g \rangle| \leqslant |\langle f_n, g_n \rangle - \langle f_n, g \rangle| + |\langle f_n, g \rangle - \langle f, g \rangle|$$
$$= |\langle f_n, g_n - g \rangle| + |\langle f_n - f, g \rangle|$$
$$\leqslant \|f_n\| \|g_n - g\| + \|f_n - f\| \|g\| .$$

The result follows from this. Given $\varepsilon > 0$ choose N so that $n \geqslant N$ implies each of the following estimates.

(1) $\|f_n - f\| < \dfrac{\varepsilon}{2}\|g\|$;

(2) $\|f_n - f\| \leqslant 1$; and

(3) $\|g_n - g\| < \dfrac{\varepsilon}{2(\|f\| + 1)}$.

(Why is this possible?)

Then $\|f_n - f\| \leqslant 1$ implies $\|f_n\| \leqslant \|f\| + 1$ (why?), and for $n \geqslant N$

$$|\langle f_n, g_n \rangle - \langle f, g \rangle| \leqslant (\|f\| + 1)\|g_n - g\| + \|f_n - f\|\|g\|$$

$$\leqslant (\|f\| + 1)\frac{\varepsilon}{2}(\|f\| + 1)^{-1} + \frac{\varepsilon\,\|g\|}{2\,\|g\|}$$

$$= \varepsilon .$$

This proves that $\langle f_n, g_n \rangle \to \langle f, g \rangle$.

3. Let $f: [0, 2\pi] \to \mathbb{R}$ be square integrable and define $g(x) = \int_0^x f(y)\,dy$. Find the Fourier expansion of g and state where it is valid.

Solution: From Theorem 12,

$$g(x) = \frac{a_0 x}{2} + \sum_1^\infty \frac{1}{n}[a_n \sin nx - b_n(\cos nx - (-1)^n)],$$

which converges uniformly (and hence, pointwise) and in mean. Now

$$x = \pi - 2\sum_1^\infty \frac{\sin nx}{n}$$

(Table 10-5), which converges in mean and pointwise for $x \neq 0, 2\pi$. Thus

$$g(x) = \frac{a_0 \pi}{2} + \left(\sum_1^\infty \frac{b_n(-1)^n}{n} \right) + \sum_1^\infty \frac{1}{n}[(a_0 + a_n)\sin nx - b_n \cos nx],$$

which converges in mean and pointwise if $x \neq 0, 2\pi$. Since the Fourier coefficients are unique (Theorem 4) this is the Fourier expansion of $g(x)$.

Since g is bounded by $\int_0^{2\pi} |f(x)|\,dx$, which is finite, g is square integrable, so it certainly has a Fourier series. Note also that g is continuous but need not be differentiable (see Exercise 61).

4. Let $f: [0, 2\pi] \to \mathbb{R}$, $g: [0, 2\pi] \to \mathbb{R}$ and extend by periodicity. Define the *convolution* of f and g by

$$(f * g)(x) = \int_0^{2\pi} f(y)g(x - y)\,dy .$$

Compute the Fourier series of $f * g$ in terms of that of f and g, using the exponential form of Fourier series.

Solution: The Fourier coefficients of $f * g$ are given by

$$c_n = \frac{1}{2\pi} \int_0^{2\pi} (f * g)(x)e^{-inx}\, dx$$

$$= \frac{1}{2\pi} \int_0^{2\pi} \int_0^{2\pi} f(y)g(x - y)e^{-inx}\, dy\, dx$$

$$= \frac{1}{2\pi} \int_0^{2\pi} \int_0^{2\pi} f(y)e^{-iny}g(x - y)e^{-in(x-y)}\, dy\, dx$$

since $e^{-inx} = e^{-iny} \cdot e^{-inx+iny}$. Changing variables ($y \mapsto y$ and $x - y \mapsto t$) and using periodicity (we may interchange the order of integration by Fubini's theorem, Section 9.2) leads to

$$\frac{1}{2\pi} \int_0^{2\pi} f(y)e^{-iny}\, dy \int_0^{2\pi} g(t)e^{-int}\, dt \ .$$

Let $f(x) = \sum_{-\infty}^{\infty} a_n e^{inx}$ and $g(x) = \sum_{-\infty}^{\infty} b_n e^{inx}$ be the Fourier series of f and g, so

$$a_n = \frac{1}{2\pi} \int_0^{2\pi} f(x)e^{-inx}\, dx$$

and

$$b_n = \frac{1}{2\pi} \int_0^{2\pi} g(x)e^{-inx}\, dx \ .$$

Then the above computation shows that we have

$$c_n = 2\pi a_n b_n \ ,$$

and thus

$$(f * g)(x) = 2\pi \sum_{-\infty}^{\infty} a_n b_n e^{inx} \ .$$

A similar operation could be done with the trigonometric series, although the computations and the results are much more awkward. Sufficient conditions on f and g for the above to be valid are that f and g be sectionally continuous, or more generally, square integrable. If we want the series to converge pointwise we must add the hypotheses of Theorem 9.

5. Let us consider $f(x) = \cos \lambda x$, $-\pi \leqslant x \leqslant \pi$, where λ is real and non-integral. Compute

$$c_n = \frac{1}{2\pi} \int_{-\pi}^{\pi} \cos \lambda x\, e^{-inx}\, dx$$

$$= \frac{1}{4\pi} \int_{-\pi}^{\pi} (e^{i\lambda x} + e^{-i\lambda x})e^{-inx}\, dx$$

$$= \frac{1}{4\pi}\left[\frac{e^{i(\lambda - n)x}}{i(\lambda - n)} + \frac{e^{-i(\lambda + n)x}}{-i(\lambda + n)} \right]_{x=-\pi}^{\pi}$$

$$= \frac{(-1)^n}{4\pi i} \cdot 2i \sin \lambda \pi \left(\frac{1}{\lambda - n} + \frac{1}{\lambda + n} \right)$$

$$= \frac{(-1)^n}{2\pi} \sin \lambda \pi \cdot \frac{2\lambda}{\lambda^2 - n^2} \ .$$

By Theorem 9 the fourier series converges to $f(x)$ at all points. Hence, for $|x| \leqslant \pi$,

$$\cos \lambda x = \frac{\sin \pi\lambda}{2} \sum_{k=-\infty}^{\infty} (-1)^k \frac{2\lambda}{\lambda^2 - k^2} e^{ikx} \ .$$

In particular, if we set $x = \pi$ then

$$\cos \pi\lambda = \frac{\sin \pi\lambda}{2\pi} \sum_{k=-\infty}^{\infty} (-1)^k \frac{2\lambda}{\lambda^2 - k^2} (-1)^k$$

$$= \frac{\sin \pi\lambda}{\pi} \left\{ \frac{1}{\lambda} + \sum_{k=1}^{\infty} \frac{2\lambda}{\lambda^2 - k^2} \right\} \ .$$

Hence

$$\pi \frac{\cos \pi\lambda}{\sin \pi\lambda} - \frac{1}{\lambda} = \sum_{k=1}^{\infty} \frac{2\lambda}{\lambda^2 - k^2} \ , \qquad \lambda \neq \text{integer} \ .$$

Note that the series on the right converges uniformly for $0 \leqslant \lambda \leqslant \lambda_0 < 1$. Note also that $\pi(\cos \pi\lambda/\sin \pi\lambda) - (1/\lambda) \to 0$ as $\lambda \to 0$, and so is Riemann integrable. By integrating,

$$\log\left(\frac{\sin \pi\lambda}{\pi\lambda} \right) = \sum_{k=1}^{\infty} \log\left(1 - \frac{\lambda^2}{k^2} \right), \qquad |\lambda| < 1 \ .$$

By exponentiating,

$$\frac{\sin \pi\lambda}{\pi\lambda} = \prod_{k=1}^{\infty} \left(1 - \frac{\lambda^2}{k^2} \right)$$

or

$$\frac{\sin \pi\lambda}{\pi\lambda} = \lambda \prod_{k=1}^{\infty} \left(1 - \frac{\lambda^2}{k^2} \right), \qquad |\lambda| < 1 \ .$$

Actually the product on the right defines a function of λ of period 2 (see Exercise 75) as does the left side, so the above formula holds for *all* real values of λ. This product formula for the sine was discovered (though not rigorously proved) by Euler.*

If we take $\lambda = 1/2$ then

$$1 = \frac{\pi}{2} \prod_{k=1}^{\infty} \left(1 - \frac{1}{4k^2} \right) = \frac{\pi}{2} \prod_{k=1}^{\infty} \frac{(2k-1)(2k+1)}{2k\,2k}$$

or

$$\frac{\pi}{2} = \prod_{k=1}^{\infty} \frac{2k\,2k}{(2k-1)(2k+1)}$$

$$\frac{\pi}{2} = \lim_{n\to\infty} \frac{(2\cdot2)(4\cdot4)(6\cdot6)\cdots(2n\cdot2n)}{(1\cdot3)(3\cdot5)(5\cdot7)\cdots(2n-1\cdot2n+1)}$$

which is called *Wallis' product formula* for $\pi/2$.

6. An interesting application of the Parseval relation to the isoperimetric problem is as follows: show that among all plane curves of a given perimeter, the largest area is enclosed by the circle.

---

* For another method of proof, see J. Marsden, *Basic Complex Analysis*, Chapter 7.

Solution: Let $(x(t), y(t))$, $0 \leqslant t \leqslant 2\pi$, be a parametric representation of a simple closed curve. We assume that $x(t)$, $y(t)$ are C^1 functions of t, and that the parameter t is arc length. Thus the total length is 2π, and $\dot{x}(t)^2 + \dot{y}(t)^2 = 1$, ($\cdot = d/dt$) so $\int_0^{2\pi} (\dot{x}(t)^2 + \dot{y}(t)^2)\, dt = 2\pi$.

The enclosed area is given by*

$$A = \int_0^{2\pi} x(t)\dot{y}(t)\, dt \ .$$

We claim that $A \leqslant \pi$, and $A = \pi$ only if the curve is a circle. To prove this we will express A in terms of the Fourier coefficients of x and y. Write

$$x(t) = \frac{a_0}{2} + \sum_{k=1}^{\infty} (a_k \cos kt + b_k \sin kt)$$

$$y(t) = \frac{\alpha_0}{2} + \sum_{k=1}^{\infty} (\alpha_k \cos kt + \beta_k \sin kt) \ .$$

All coefficients are real. By a change of origin in the plane, we may assume $a_0 = \alpha_0 = 0$.

The Fourier series of the derivatives $\dot{x}(t)$, $\dot{y}(t)$ are then

$$\dot{x}(t) = \sum_{k=1}^{\infty} (kb_k \cos kt - ba_k \sin kt)$$

and

$$\dot{y}(t) = \sum_{k=1}^{\infty} (k\beta_k \cos kt - b\alpha_k \sin kt) \ .$$

Accordingly, by Parseval's relation,

$$2\pi = \int_0^{2\pi} (\dot{x}^2 + \dot{y}^2)\, dt = \pi \sum_{k=1}^{\infty} k^2(a_k^2 + b_k^2 + \alpha_k^2 + \beta_k^2)$$

and the area is

$$A = \int_0^{2\pi} x\dot{y}\, dt = \pi \sum_{k=1}^{\infty} k(a_k\beta_k - b_k\alpha_k) \ .$$

Hence

$$2\pi - 2A = \pi \sum_{1}^{\infty} \{k^2(a_k^2 + b_k^2 + \alpha_k^2 + \beta_k^2) - 2k(a_k\beta_k - b_k\alpha_k)\}$$

$$= \pi \sum_{1}^{\infty} (k^2 - k)(a_k^2 + b_k^2 + \alpha_k^2 + \beta_k^2) + \pi \sum_{1}^{\infty} k\{(a_k - \beta_k)^2 + (\alpha_k + b_k)^2\} \ .$$

Thus $\pi - A \geqslant 0$, and $\pi - A = 0 \Leftrightarrow$
(i) $a_k = b_k = \alpha_k = \beta_k = 0$ for $k \geqslant 2$
(ii) $a_1 = \beta_1$, $\alpha_1 = -b_1$.
In this case,

$$x(t) = a_1 \cos t + b_1 \sin t$$

$$y(t) = -b_1 \cos t + a_1 \sin t = -x(t + \pi/2) \ .$$

---

* This is a standard calculus formula; see, for example, Marsden–Tromba, *Vector Calculus*, Chapter 17.

Equivalently, for some R and δ

$$x(t) = R\cos(t + \delta)$$
$$y(t) = R\sin(t + \delta).$$

The condition $\dot{x}^2 + \dot{y}^2 = 1$ implies $R = 1$ and therefore we have a circle of radius 1.

Exercises for Chapter 10

1. Let V be an inner product space and $M \subset V$ a vector subspace. Define the *orthogonal complement* of M by

$$M^\perp = \{f \in V \mid \langle f,g \rangle = 0 \text{ for all } g \in M\}.$$

Show that M is a vector subspace of V and is closed (that is, if $f_n \in M$, and $f_n \to f$ (in mean), then $f \in M$). [Hint: Make use of Example 2 above.]

2. Prove that the Legendre polynomials (see Section 10.2) are complete in $\mathscr{L}^2$ of $[-1,1]$. [Hint: First show that any polynomial can be expanded in Legendre polynomials, and then employ the method of proof of Theorem 8.]

3. (a) Use the Fourier series for e^x on $[-\pi,\pi]$ in order to prove the following identity:

$$(\pi \coth \pi - 1)/2 = \sum_1^\infty \frac{1}{n^2 + 1}.$$

(b) Use the half interval cosine series for $\cos ax$ where a is not an integer in order to prove the following identity:

$$\pi \cot \pi a = \frac{1}{a} + \sum_1^\infty \frac{2a}{a^2 - n^2}.$$

4. Prove that if $f_n \to f$ (in mean), then $\|f_n\| \to \|f\|$. Is the converse true?

5. Prove that uniform convergence implies mean convergence (on finite intervals).

6. Consider the space l_2 of all sequences $x = (x_1, x_2, \ldots)$ of real numbers with $\sum_{i=1}^\infty x_i^2 < \infty$. Show that l_2 is an inner product space with $\langle x,y \rangle = \sum_{i=1}^\infty x_i y_i$. In addition, show that this space is complete (Cauchy sequences converge).

7. Let

$$l_2 = \left\{(x_1, x_2, \ldots) \;\middle|\; \sum_{i=1}^\infty x_i^2 < \infty\right\}$$

which, by Exercise 6, is a Hilbert space. Let $\varphi_n \in l_2$, $n = 1, 2, \ldots$ be defined as $\varphi_n = (0,0,\ldots,1,0,\ldots)$ with the 1 in the nth spot. Show in two ways that $\varphi_1, \varphi_2, \ldots$ is a complete orthonormal set: (a) directly, (b) using Theorem 6 (see also Exercise 14(c)).

8. Find functions f_n and f on $[-1,1]$ such that $f_n \to f$ (pointwise) but not in mean.

9. Verify that the sectionally continuous functions $f: [a,b] \to \mathbb{C}$ form a vector space.

10. Find a sequence f_n of sectionally continuous functions with $f_n \to f$ pointwise (respectively, in mean) such that f is not sectionally continuous.

11. Prove that if $f: [a,b] \to \mathbb{C}$ is sectionally continuous, then so is $|f|$. Show also that $|f|$ is bounded.

12. Prove that if f and g are square integrable on $[a,b]$, $\|f - g\| = 0$ iff $f = g$ except on a set of measure zero.

13. In the proof of Theorem 7 we showed that

$$\left\| f - \sum_{k=0}^{n} t_k \varphi_k \right\|^2 = \|f\|^2 - \sum_{k=0}^{n} |\langle f, \varphi_k \rangle|^2 + \sum_{k=0}^{n} |\langle f, \varphi_k \rangle - t_k|^2 .$$

Use this equality with $t_k = \langle f, \varphi_k \rangle$ to prove Bessel's inequality.

14. Suppose V is a Hilbert space (that is, is a complete inner product space). Let φ_0, $\varphi_1, \ldots$ be an orthonormal set in V.
    (a) For each $f \in V$, show that

$$s_n = \sum_{k=0}^{n} \langle f, \varphi_k \rangle \varphi_k$$

converges to some element of V. [Hint: Show that

$$\|s_n - s_m\|^2 = \sum_{i=n+1}^{m} |\langle f, \varphi_i \rangle|^2$$

and use Bessel's inequality to show s_n is a Cauchy sequence.]
    (b) For each $f \in V$ show that, if $s = \sum_{i=0}^{\infty} \langle f, \varphi_i \rangle \varphi_i$, $f - s$ is orthogonal to each φ_i and $f - s$ is orthogonal to s.
    (c) Show that if whenever f is orthogonal to each φ_i we have $f = 0$, then φ_0, $\varphi_1, \ldots$ is complete. [Hint: By (b), $f - s$ is orthogonal to each φ_i, so $f = s$.]

15. Show that in V of Theorem 1, we do *not* have the Bolzano-Weierstrass theorem; that is, in a closed bounded set, a sequence need not have any convergent subsequences. [Hint: Consider the elements $\varphi_n(x) = (\sin nx)/\sqrt{\pi}$ on $[0, 2\pi]$ and show that $d(\varphi_n, \varphi_m) = \sqrt{2}$, $n \neq m$.]

16. Compute the Fourier series of $f(x) = \dfrac{x + x^2}{2}$, $0 \leqslant x \leqslant 2\pi$.

17. (a) Let $\varphi_0, \varphi_1, \ldots$ be orthonormal vectors in an inner product space, V. If

$$\sum_{k=0}^{\infty} c_k \varphi_k = 0$$

then show $c_k = 0$, $k = 0, 1, 2, \ldots$.
    (b) If $\varphi_0, \varphi_1, \ldots$ is a complete set, and $\langle f, \varphi_i \rangle = \langle g, \varphi_i \rangle$ for all i, then prove $f = g$.
    (c) If V is a space of integrable functions, prove that (b) implies $f(x) = g(x)$ except possibly on a set of measure zero.

18. This is an exercise on Fourier series in several variables.

(a) We have seen that $e^{inx}/\sqrt{2\pi}$ $n = 0, \pm 1, \ldots$ are orthonormal functions on $[0,2\pi]$. Consider now the functions

$$\varphi_{n,m}(x,y) = \frac{e^{inx+imy}}{2\pi} = \frac{e^{inx}}{\sqrt{2\pi}} \frac{e^{imy}}{\sqrt{2\pi}} .$$

Show that $\varphi_{n,m}$ are orthonormal on $[0,2\pi] \times [0,2\pi]$.

(b) Generalize (a) to construct $\varphi_{n,m}$, given general φ_n which are orthonormal on $[a,b]$, rather than just the case $e^{inx}/\sqrt{2\pi}$.

Remark: The $e^{inx}/\sqrt{2\pi}$ are complete and so are the $\varphi_{n,m}$. This is proved in Theorem 8 for $e^{inx}/\sqrt{2\pi}$ and the proof for $\varphi_{n,m}$ is similar.

(c) For $f: [0,2\pi] \times [0,2\pi] \to \mathbb{C}$, write the Fourier series for f (with respect to $\varphi_{n,m}$ above).

19. (Sturm-Liouville problems.) Consider the differential equation

$$\frac{d^2f}{dx^2} + [q(x) + \lambda p(x)]f(x) = 0$$

to be solved for $f(x)$, with boundary conditions $f(a) = f(b) = 0$ and $a \leqslant x \leqslant b$. The functions q, p are fixed and assume $p(x) > 0$. The λ for which solutions f exist are called *eigenvalues*.

(a) Show that if f and g are solutions with eigenvalues λ and μ and $\lambda \neq \mu$, then

$$\int_a^b p(x)f(x)g(x) \, dx = 0 .$$

[Hint: Use the differential equations to show that

$$(\lambda - \mu)p(x)f(x)g(x) = \frac{d}{dx}[g'(x)f(x) - f'(x)g(x)]$$

and then integrate.]

(b) Interpret (a) as orthogonality of f and g with

$$\langle f,g \rangle = \int_a^b p(x)f(x)g(x) \, dx .$$

Show that this is an inner product.*

20. Show that $\{\sqrt{2/\pi}\, \sin nx \mid n = 1,2,\ldots\}$ is an orthonormal family on $[0,\pi]$. What is a Fourier series for this family? Is it complete?

21. Let $\varphi_0, \varphi_1, \ldots$ be an orthonormal system in an inner product space V and $f \in V$. Let $s_n = \sum_{k=0}^{n} \langle f,\varphi_k \rangle \varphi_k$, the nth partial sum of the Fourier series. Show that for any integer, $p \geqslant 0$,

$$\| f - s_{n+p} \| \leqslant \| f - s_n \| .$$

[Hint: Use Theorem 7, or a direct argument. Deduce that $\lim_{n \to \infty} \| f - s_n \|$ always exists.]

---

* Many orthonormal systems arise this way. The trigonometric system arises with $p = 1, q = 0$. There is an advanced theorem which asserts that such systems are complete. See, for example, Coddington and Levinson, *Theory of Ordinary Differential Equations*.

22. Let V be an inner product space and $\varphi_0, \varphi_1, \ldots$ a complete orthonormal system. For $f, g \in V$, show that

$$\langle f, g \rangle = \sum_{k=0}^{\infty} \langle f, \varphi_k \rangle \langle \varphi_k, g \rangle .$$

Of course, part of this problem is to show that the sum converges. [Hint: Write f and g as limits of Fourier series and apply Example 2 above.]

Exercises 23–28 refer to quantum mechanical systems (Section 10.9).

23. Show that the solution of $i\hbar(\partial\psi/\partial t) = H\psi$, if $\psi = \psi_0$ at $t = 0$, is given by

$$\psi = \sum_{n=0}^{\infty} \langle \psi_0, \varphi_n \rangle e^{-iE_nt/\hbar} \varphi_n ,$$

where $\varphi_n \in V$ are the eigenfunctions of H with eigenvalues E_n (you may assume the series can be differentiated term by term and that the eigenfunctions are complete). What happens if ψ is already an eigenfunction?

24. Suppose $i\hbar(\partial\psi/\partial t) = H\psi$. Prove that $\langle \psi, H\psi \rangle$ is constant in time. This result is called *conservation of energy*.

25. Compute the commutators of the operators, Q_x, Q_y, Q_z, P_x, P_y, P_z, and J_x, J_y, J_z given in the text. What is the uncertainty principle for these operators (Exercise 5, Section 10.9)?

26. If A and B are symmetric, is $[A,B]$ symmetric? What about $i[A,B]$?

27. Solve for the eigenfunctions in a deep box if we replace $[0,l]$ by $[-l,l]$.

28. If A is symmetric, then show $\langle A\psi, \psi \rangle$ is real for any $\psi \in V$. Interpret this result using Exercise 3, Section 10.9.

29. In the second proof of Theorem 9 we used the fact that

$$\int_0^{\infty} \frac{\sin x}{x} dx = \frac{\pi}{2}$$

(recall that this integral is conditionally convergent; see p. 271). Prove this fact as follows. Let

$$F(t) = \int_0^{\infty} e^{-tx} \frac{\sin x}{x} dx .$$

Then show that

$$F'(t) = -\int_0^{\infty} e^{-tx} \sin x \, dx = -(t^2 + 1)^{-1}$$

(see Example 2 at the end of Chapter 9). Hence $F(t) = -\tan^{-1} t + C$. Show that $F(t) \to 0$ as $t \to \infty$, so $C = \pi/2$. Then look at $F(0)$ for the result. (The main difficulty here is the justification of these steps.)*

---

* This integral can also be evaluated using complex variables methods; see J. Marsden, *Basic Complex Analysis*, Chapter 4.

30. Convolutions were defined in Example 4 on p. 430. Show that $f * g = g * f$. Use Example 4 to write out Parseval's relation for $f * g$.

31. The derivative of the delta function is defined by $\int_{-\infty}^{\infty} \delta'(x)f(x)\,dx = -f'(0)$ (see Section 8.9). Compute the Fourier transform of δ'; how is it related to that of δ?

32. (The Riesz-Fischer theorem.) The Riesz-Fischer theorem represents one of the early and most important successes of the Lebesgue integral. For this problem we do not assume a knowledge of the Legesgue integral, but take for granted that the set of all square integrable functions forms a Hilbert space. Assuming this, the Riesz-Fischer theorem is quite easy. Sometimes, depending on how you read the history, the fact we just took for granted is called the Riesz-Fischer theorem!
    (a) Prove the following theorem.

    Riesz–Fischer theorem. *Let V be a Hilbert space and $\varphi_0, \varphi_1, \ldots$ a complete orthonormal set. Let $c_0, c_1, \ldots$ be complex numbers and suppose $\sum_{n=0}^{\infty} |c_n|^2 < \infty$. Then there exists an $f \in V$ with*

    $$\langle f, \varphi_i \rangle = c_i .$$

    Thus every series

    $$\sum_{n=0}^{\infty} c_n \varphi_n$$

    with

    $$\sum_{n=0}^{\infty} |c_n|^2 < \infty$$

    is the Fourier series of some f.

    [Hint: Let $f_n = \sum_{i=0}^{n} c_i \varphi_i$ and show that f_n is a Cauchy sequence, by showing that $\|f_m - f_n\|^2 = \sum_{i=n+1}^{m} |c_i|^2$.]
    (b) Use (a) to prove that for every sequence, $c_k, k = 0, \pm 1, \pm 2, \ldots$ with

    $$\sum_{-\infty}^{\infty} |c_k|^2 = \lim_{N \to \infty} \sum_{-N}^{N} |c_k|^2 < \infty ,$$

    there is a square integrable f on $[0,2\pi]$ (or $[-\pi,\pi]$) such that

    $$f = \sum_{-\infty}^{\infty} c_k \varphi_k ,$$

    where $\varphi_k = e^{ikx}$ and the convergence is convergence in the mean.
    (c) Is $\sum_{\substack{n=-\infty \\ n\neq 0}}^{\infty} (1/n)e^{inx}$ the Fourier series of some function? Is $\sum_{\substack{n=-\infty \\ n\neq 0}}^{\infty} (1/\sqrt{n})e^{inx}$?

33. If $f: [a,b] \to \mathbb{R}$ has a discontinuity only at $x_0 \in \,]a,b[$ and f' is bounded on $]x_0,b[$ and on $]a,x_0[$, then prove f is a function of bounded variation. [Hint: Use the mean-value theorem and arrange the partition P so $x_0 \in P$.] Show that one can apply the Jordan-Dirichlet theorem.

34. Find a function which is continuous and periodic on $[0,2\pi]$, and whose Fourier series converges at each point, but for which the hypotheses of both the Jordan and the Jordan-Dirichlet theorems fail. [Hint: Consider the function $x \sin(1/x)$ for

$x > 0$ and extend this function to be odd periodic. Make use of Exercise 1(d), Section 10.4.]

35. Investigate the nature of convergence of the Fourier series of each of the following functions on $]-\pi,\pi]$.

    (a) $f(x) = x^3$.

    (b) $f(x) = (\sin x)^2$.

    (c) $f(x) = \begin{cases} x^3, & x > 0, \\ -x^2, & x \leqslant 0. \end{cases}$

    (d) $f(x) = \begin{cases} x^2 + 1, & x \geqslant 0, \\ 0, & x < 0. \end{cases}$

    (e) $f(x) = \begin{cases} 1, & x \leqslant 0, \\ x^2 \sin \dfrac{1}{x}, & x > 0. \end{cases}$

    [Hint: Use Exercise 33.]

36. Suppose f is real, square integrable on $[-l,l]$, and

$$a_n = \frac{1}{l} \int_{-l}^{l} f(x)\cos\left(\frac{n\pi x}{l}\right) dx , \qquad n = 0, 1, 2, \ldots$$

$$b_n = \frac{1}{l} \int_{-l}^{l} f(x)\sin\left(\frac{n\pi x}{l}\right) dx , \qquad n = 1, 2, \ldots .$$

Show that

$$\frac{a_0^2}{2} + \sum_{n=1}^{\infty} a_n^2 + \sum_{n=1}^{\infty} b_n^2 = \frac{1}{l} \int_{-l}^{l} f(x)^2 \, dx .$$

37. Form the function

$$\varphi(x) = a \sin x + b \sin 2x + c \sin 3x$$

on $[0,\pi]$. For what values of a, b, c is φ closest in mean to the constant function 1? What about on the interval $[-\pi,\pi]$?

38. Let $g: \,]a,b[\, \to \mathbb{R}$ be continuous and suppose $g(a+)$ and $g(b-)$ exist. Then prove g is bounded. [Hint: Define $h: [a,b] \to \mathbb{R}$ by $h(a) = g(a+)$, $h(b) = g(b-)$ and $h = g$ on $]a,b[$. Show that h is continuous.] What does this say about the definition of a sectionally continuous function?

39. (a) Suppose f is differentiable for $x > x_0$ and $\underset{x \to x_0 +}{\text{limit}}\ f'(x)$ exists. Then show that $f(x_0 +)$ exists as well.

    (b) If $\underset{x \to x_0 +}{\text{limit}}\ f'(x)$ exists, show that it equals

$$\underset{h \to 0+}{\text{limit}}\left\{ \frac{f(x_0 + h) - f(x_0 +)}{h} \right\} .$$

    [Hint: Extend f so it is continuous on an interval $[x_0,x]$. Then apply the mean-value theorem to the above difference quotient.]

    (c) Consider the functions $f_1(x) = x \sin(1/x)$, $f_2(x) = x^2 \sin(1/x)$, $f_3(x) = x^3 \sin(1/x)$, for $x > 0$. Which of $\underset{x \to 0+}{\text{limit}}\ f_i'(x)$ and $f_i'(0+)$ exist?

40. Let x_0 be a jump discontinuity of f. Define $h(x) = f(x_0 - x)$ and $g(x) = f(x_0 + x)$. Show that $h(0+) = f(x_0-)$ and $g(0+) = f(x_0+)$.

41. Let $f: [a,b] \to \mathbb{R}$ have a jump discontinuity at $x_0 \in \,]a,b[$ with $f'(x_0+), f'(x_0-)$ existing. Let $\varphi: \,]a,b[\to \mathbb{R}$ be differentiable and suppose

$$g(x) = f(x)\varphi(x) .$$

Then prove g has a jump discontinuity at x_0, $g'(x_0+)$ and $g'(x_0-)$ exist, and we have

$$g(x_0+) = f(x_0+)\varphi(x_0)$$
$$g(x_0-) = f(x_0-)\varphi(x_0) .$$

Apply this when

$$\varphi(x) = \begin{cases} \dfrac{(x - x_0)/2}{\sin[(x - x_0)/2]} , & x \neq x_0 , \\ 1 , & x = x_0 , \end{cases}$$

to complete the proof of Theorem 9.

42. If $f_n \to f$ in mean on $[a,b]$, then prove $f_n \to f$ in mean on any subinterval.

43. Suppose $f_n \to f$ and $g_n \to g$ in mean on $[a,b]$. Let $h(x) = \int_a^x f(y)g(y)\,dy$ and let h_n be defined similarly by $h_n(x) = \int_a^x f_n(y)g_n(y)\,dy$. Then prove that $h_n \to h$ uniformly. [Hint: Modify Lemma 12.]

44. Establish the following formulas.

(a) $x \sin x = 1 - \dfrac{\cos x}{2} - 2 \sum_1^\infty \dfrac{(-1)^n \cos nx}{n^2 - 1} , \qquad -\pi \leqslant x \leqslant \pi .$

(b) $\log[\sin \tfrac{1}{2}x] = -\log 2 - \sum_1^\infty \dfrac{\cos nx}{n} , \qquad \text{on }]0,2\pi[.$

45. Establish formulas 5 and 6 in Table 10-5.

46. Apply Parseval's relation to formulas 4a and 6 in Table 10-5 to obtain some arithmetical identities. What justifies reading off a_n and b_n as the coefficients of $\cos nx$ and $\sin nx$?

47. Discuss the Gibbs' phenomenon for $f(x) = 2, x \geqslant 0$; $f(x) = 0, x < 0$.

48. Let f have a jump discontinuity at x_0 and let $f'(x_0+), f'(x_0-)$ exist and $f'(x)$ exist and be continuous for $x \in \,]x_0 - \varepsilon, x_0[$ and $x \in \,]x_0, x_0 + \varepsilon]$. Show that f "exhibits a Gibbs' phenomenon at x_0" and the "overshoot" is $\approx (f(x_0+) - f(x_0-)) \cdot (1.179)$. [Hint: Let

$$h(x) = \begin{cases} f(x_0-) , & x \leqslant x_0 , \\ f(x_0+) , & x > x_0 , \end{cases}$$

and consider $k(x) = f(x) - h(x)$. Use Theorem 9 and the fact that near x_0, the Fourier series of k is uniformly small.]

49. Use the Fourier series of $|\sin x|$ in Table 10-5 to show that

$$\sum_1^\infty \frac{1}{4n^2 - 1} = \frac{1}{2} \qquad \text{and} \qquad \sum_1^\infty \frac{(-1)^n}{4n^2 - 1} = \frac{1}{2} - \frac{\pi}{4} .$$

50. Use the Fourier sine series on $[0,\pi]$ to show that

$$\cos \pi x = \frac{8}{\pi} \sum_{n=1}^{\infty} \frac{n}{4n^2 - 1} \sin(2\pi n x) , \qquad 0 < x < 1 .$$

51. In Table 10-5, state the discontinuities of all functions (including endpoints of intervals) and the values of the Fourier series at those points.

52. Derive formula 2a of Table 10-5 by noting that $f(x) = (x + |x|)/2$ and observing that the series for $|x|$ is just the half-interval cosine series for x.

53. Use the Jordan theorem at $x = 0$ in Example 3, Section 10.5 to prove that $\pi^2/8 = \sum_{n=1}^{\infty} 1/(2n - 1)^2$.

54. (a) Let f be smooth (C^∞) on $[-\pi,\pi]$, and suppose $f(-\pi) = f(\pi)$, $f^{(k)}(-\pi) = f^{(k)}(\pi)$, $k = 1, 2, \ldots$. Then prove that the Fourier series of f may be differentiated any number of times and will still converge uniformly.
    (b) Show that for any integer p, $n^p a_n \to 0$, $n^p b_n \to 0$, where a_n, b_n are the Fourier coefficients of f.

55. (a) Let $f: [0,\pi] \to \mathbb{R}$ be continuous and let f' be sectionally continuous with jump discontinuities. Then show that the half-interval cosine series of f converges uniformly and absolutely to f.
    (b) Justify the same conclusion for the half-interval sine series if we assume also that $f(0) = f(\pi) = 0$. Explain.
    (c) Show that without the condition $f(-\pi) = f(\pi)$ in Theorem 14, the conclusion is false.

56. Let

$$f(x) = \begin{cases} 0 , & -\pi \leqslant x < 0 , \\ 1 , & 0 \leqslant x \leqslant \pi . \end{cases}$$

Then show that if we differentiate the Fourier series of f, we get the Fourier series of $\delta(x) - \delta(x + \pi)$ (δ is the Dirac δ function). Can you explain in what sense $f' = \delta$?

57. Give the theorem that is obtained by combining Theorem 14 and Corollary 3 (p. 109) and show why Theorem 15 is better.

58. Use Theorem 15 to derive a differentiation theorem for half-interval cosine series.

59. If f is square integrable on $[-\pi,\pi]$ with Fourier coefficients, a_n, b_n, then prove that $\sum_{n=1}^{\infty} a_n/n$ and $\sum_{n=1}^{\infty} b_n/n$ converge absolutely. [Hint: Use the method of Lemma 13.]

60. (a) Let $f: [-\pi,\pi] \to \mathbb{R}$ be square integrable, and $g(x) = \int_{-\pi}^{x} f(y)\, dy$. Find the Fourier expansion of g and state where it is valid.
    (b) Repeat part (a) for the half-interval cosine series of $f: [0,\pi] \to \mathbb{R}$.

61. In Example 3, p. 430 show that g is of bounded variation (hence the Jordan-Dirichlet theorem applies).

62. Use Example 3 to find the Fourier series of x^3 on $[0,2\pi]$ using that of x^2 from Table 10-5.

63. For each of the following, determine what type of convergence the Fourier series has and whether or not the series can be differentiated termwise.

   (a) $f(x) = x^3$ on $[-\pi,\pi]$.

   (b) $f(x) = \begin{cases} 3, & -\pi \leqslant x < -\frac{1}{2}, \\ x, & -\frac{1}{2} \leqslant x < 1, \\ 1, & 1 \leqslant x \leqslant \pi. \end{cases}$

   (c) $f(x) = \pi^3 - |x|^3$ on $[-\pi,\pi]$.

   (d) $f(x) = \begin{cases} x^3 + 8, & -\pi \leqslant x \leqslant 0, \\ x^2 + 8, & 0 \leqslant x \leqslant \pi. \end{cases}$

   (e) $f(x) = \begin{cases} 0, & -\pi \leqslant x \leqslant 0, \\ x^3 \sin\left(\dfrac{1}{x}\right), & 0 < x \leqslant \pi. \end{cases}$

   (f) $f(x) = \begin{cases} -2n(x+1)x + 2n + 1, & \dfrac{1}{n+1} \leqslant x \leqslant \dfrac{1}{n}, \\ 0, & -\pi \leqslant x \leqslant 0, \\ 1, & \text{otherwise on } [-\pi,\pi]. \end{cases}$

   (g) $\log\left|\sin\left(\dfrac{x}{2}\right)\right|$ on $]-\pi,\pi[$.

64. Suppose f is square integrable on $[0,2\pi]$ with Fourier coefficients a_n, b_n. If $f(0+)$, $f(0-)$, $f'(0+)$, and $f'(0-)$ all exist, then prove $\sum_{n=0}^{\infty} a_n$ converges. What assumptions guarantee that $\sum_{n=1}^{\infty} b_n$ converges?

Exercises 65–71 are based on Section 10.7.

65. If the temperature at the ends of a bar is kept constant at zero, show that the temperature T after time t, if T is equal to f at $t = 0$, is given by

$$T(x,t) = \sum_{n=1}^{\infty} b_n e^{-(n^2\pi^2 t/l)} \sin\left(\frac{n\pi x}{l}\right),$$

where the b_n are the half-interval sine series coefficients of f.

66. Show that the solution to the heat equation is always C^∞ for $t > 0$ (see Theorem 19).

67. (a) On $[0,\pi] \times [0,\pi]$ find a function φ such that $\nabla^2 \varphi = 0$ and

$$\varphi(x,0) = \left(\frac{x-\pi}{2}\right)^2 - \frac{\pi^2}{4}, \quad \varphi(0,y) = 0, \quad \varphi(\pi,y) = 0, \quad \varphi(x,\pi) = 0 .$$

   Explain in what way the boundary values are assumed.

   (b) Repeat the problem with $\varphi(x,0) = x$.

68. Show that the solution in the text is correct for the 2-dimensional wave equation. Derive the fundamental solutions by separation of variables. You may wish to use Exercise 18.

69. In Theorem 17, prove that

$$\lim_{t \to \infty} \sum_{n=1}^{\infty} a_n e^{-(n^2\pi^2 t/l^2)}\left(\cos \frac{n\pi x}{l}\right) = 0$$

(uniformly in x). [Hint: Let $|a_n| \leqslant M$. Then the sum is, in absolute value,

$$\leqslant M \sum_{n=1}^{\infty} e^{-(n^2\pi^2 t/l^2)} \leqslant M \cdot \frac{e^{-\pi^2 t/l^2}}{[1 - e^{-\pi^2 t/l^2}]} \cdot]$$

70. In Theorem 19(iii), show that φ converges uniformly on any compact set in the interior of the square. [Hint: The distance from a side is >0.]

71. (Boundary-value problems for ordinary differential equations.) Suppose f on $[-\pi,\pi]$ has $f(-\pi) = f(\pi)$ and Fourier coefficients a_n, b_n. The fact that f' has coefficients $\alpha_n = nb_n$ and $\beta_n = -na_n$ (see Theorem 14) is useful in solving certain boundary-value problems.
   (a) Solve the equation $f''(x) + kf(x) = g(x)$ for given g on $[-\pi,\pi]$ if we require $f'(-\pi) = f'(\pi)$, and $f(-\pi) = f(\pi)$ by noting that $-n^2 a_n + ka_n = \tilde{a}_n$ and $-n^2 b_n + kb_n = \tilde{b}_n$, where $\tilde{a}_n$ and $\tilde{b}_n$ are the coefficients for g. Hence show that

$$f(x) = \frac{\tilde{a}_0}{2k} + \sum_{n=1}^{\infty} \frac{\tilde{a}_n}{k - n^2} \cos nx + \frac{\tilde{b}_n}{k - n^2} \sin nx.$$

   (b) Solve the equation corresponding to (a) for $[-l,l]$.

72. Let δ be a given real number, $0 < \delta < \pi$. Define

$$f(x) = \begin{cases} 1, & |x| \leqslant \delta, \\ 0, & \delta < |x| \leqslant \pi. \end{cases}$$

   (a) Calculate the Fourier series of f.
   (b) By evaluating at $x = \pi$, show that

$$\frac{\delta}{2} = \sum_{k=1}^{\infty} (-1)^{k+1} \frac{\sin k\delta}{k}.$$

   (c) What does the Parseval relation say in the case of f?

73. Evaluate

$$\lim_{k \to \infty} \int_0^{\pi} \sqrt{x} \sin^2 kx \, dx.$$

74. Verify that the infinite product in Example 5

$$f(\lambda) = \lambda \prod_{n=1}^{\infty}\left(1 - \frac{\lambda^2}{n^2}\right)$$

is periodic with period 2, $f(\lambda + 2) = f(\lambda)$. First show $f(\lambda + 1) = -f(\lambda)$.

75. (From notes of W. A. J. Luxemburg.) Let φ_n be a set of orthonormal functions in $\mathscr{L}^2$ of the interval $[a,b]$. Show that (a) φ_n is complete iff $x - a = \sum_{n=1}^{\infty} |\int_a^x \varphi_n(t) \, dt|^2$

for all $x \in [a,b]$ and (b) φ_n is complete iff

$$(b-a)^2/2 = \sum_{n=1}^{\infty} \int_a^b \left| \int_a^x \varphi_n(t) \, dt \right|^2 dx \, .$$

[Hint: (a) For $\Rightarrow$, apply Parsevals relation to the characteristic function of $[g,x]$. For $\Leftarrow$, assume $\langle g,\varphi_n \rangle = 0$ for all n, $\|g\| = 1$. Apply Bessel's inequality to the characteristic function of $[a,x]$ using the orthonormal system $\{\varphi_n\} \cup \{g\}$. Conclude that $\int_a^x g(t) \, dt = 0$ for all $x \in [a,b]$ and hence that g is zero (except on a set of measure zero). Now use Exercise 14(c) to conclude that $\{\varphi_n\}$ are complete. (You may assume that $\mathscr{L}^2$ is complete, that is, is a Hilbert space, for this problem). (b) For $\Rightarrow$, integrate the result in (a) term by term (using the monotone convergence theorem). For $\Leftarrow$, show that $\int_a^b \{(x-a) - \sum_{n=1}^{\infty} |\int_a^x \varphi_n(t) \, dt|^2\} \, dx = 0$ and use Bessel's inequality to show that the integrand is $\geqslant 0$. Hence apply (a).

Now verify that (b) holds for the exponential system on $[0,2\pi]$ and deduce its completeness.]

76. (Lebesgue's proof of completeness of the trigonometric system from A. Zygmund, *Trigonometric Series*.) Give another proof of Theorem 8 as follows. Let $f: [-\pi,\pi] \rightarrow \mathbb{R}$ be continuous and be orthogonal to $\cos nx$, $\sin mx$. Prove $f = 0$ as follows: assume $f(x) > \varepsilon$ for $x \in I =]x_0 - \delta, x_0 + \delta[$. Let $T_n(x) = [t(x)]^n$, $t(x) = 1 + \cos(x - x_0) - \cos \delta$ and show $T_n(x) \geqslant 0$, on I, $T_n(x) \rightarrow \infty$ uniformly on every closed subinterval of I, and T_n are uniformly bounded outside I. Use this to show $\langle f,T_n \rangle = 0$ is impossible for n large. For general f, attempt to apply the results just obtained to $F(x) = \int_{-\pi}^x f(t) \, dt$.

Appendices

Appendix A

Notes on the Axioms of Set Theory

By István Fáry

A.1 Introduction

There is no rigorous mathematics today which does not use concepts of set theory. For this reason we started with set theory in this text. The purpose of this appendix is to help bridge the gap between the approach in this text and that in more formal set theory courses using a book like Halmos [18].* Any introduction to set theory has to take into account the following facts.

(a) The concept of set is so basic that it is impossible to define it in terms of more basic notions.

(b) Because of (a), we specify the concept of set with axioms, but the axiomatic method may not be familiar to the student.

(c) Axiomatic set theory involves logic, but some concepts of logic may not be familiar either.

In view of these circumstances, the most effective approach and the one used in this text, is to start working with the intuitive concept of set (Introductory chapter) and come back to foundations later on. When this method is used, the question arises whether to take up logic first, or else to treat the axiomatic set theory without formal logic, like any other chapter of axiomatized mathematics. We chose the second approach.

This plan corresponds to the historical development: set theory, based on intuitive concepts came first, then criticism of this inspired the axiomatic foundations, and finally an intensive discussion of this method heralded new developments in logic. It may be useful, therefore, to say something about the history of our subject.

A.2 On the History of Set Theory

Set theory is possibly the most important chapter of mathematics. It includes facts about finite sets, but the importance of the theory comes from the fact that it can deal

* See the references listed on p. 473.

with infinite sets. The theory dates from the moment when characteristic properties of infinite sets were recognized and the mathematical consequences pursued. In this sense the founder of the theory was Georg Cantor (1845–1918). He published his important papers just before the turn of the century. There was a heated debate of his work, and famous mathematicians disagreed about fundamental questions. In the recent history of mathematics this is rather unusual.

Cantor was led to discover facts about infinite sets in connection with his work on so-called trigonometric series. Let us mention that a trigonometric series is of the form (see Chapter 10)

$$\sum_{k=0}^{\infty} (a_k \cos kx + b_k \sin kx) .$$

Convergence properties of these series are delicate questions, and distinguishing points according to the behavior of the series leads to very general types of sets of numbers. For this reason Cantor dealt with sets of real numbers first, but discovered soon that he had to deal with infinite sets in general.

In one of his papers, he gave the following "definition" or "description" of the concept of set:

> We understand by "set" any gathering M of well-defined, distinguishable objects m (which will be called "elements" of M) of our intuition or our ideas (1) into a whole.*

It is customary today to be "ashamed" of the original definition of Cantor, and to say that it is not a definition. As a point of fact, there are many so-called "definitions" in other fields which do not come close to the clarity and precision of (1). Nevertheless, the concept of set being so important, we will not accept ultimately Cantor's definition. However, for the moment we will use (1) to clarify our ideas about sets.

The first point is that we "gather together" objects, and we do not care in which *order* they are taken. For example, if we talk about "the set of natural numbers" we do not imply that the elements of this set are given in some "order," even though there is a "natural order" for integers. For practical purposes we may give the elements in some order, but this has nothing to do with the set itself. Better yet, we define "order" in terms of sets.

The words "well-defined, distinguishable objects" in (1) point out another aspect of the concept of set. That is, the *elements* of the set "do not appear twice," thus, for example a set consisting of 2, 2, 2, 3 contains 2 and 3 and nothing else. Hence a "set" "contains" some objects which "belong" to the set; some other objects may not belong to the set. For example, 1003 belongs to the set of natural numbers (positive integers), 3.14159 does not belong to it.

Finally the "whole" at the end of (1) refers to the fact that sets themselves are treated as objects, in the sense that they may be elements of other sets. Thus we may consider sets whose elements are sets. As a point of fact, these are the most important sets in set theory.

---

* The original German text is (Collected Papers, p. 282): "Unter einer "Menge" verstehen wir jede Zusammenfassung *M* von bestimmten wohlunterschiedenen Objekten *m* unserer Anschauung oder unseres Denkens (welche die "Elemente" von *M* genannt werden) zu einem Ganzen."

Let us now criticize Cantor's definition—take the following definition. An integer p is a prime number if $p \neq 1$, and ± 1, $\pm p$ are the only divisors of p. In this definition the concept of "prime number" is defined in terms of other concepts (integers, divisor, $+1$, -1, $-p$), and we suppose that the latter concepts are known or were defined without the use of the concept of prime number. The definition thus reduces the concept of prime number to these other concepts. This definition also tells us what to do in order to test whether or not 1003 is a prime number (it is not; it is divisible by 17). Let us see whether (1) can stand such criteria. We have in this sentence a number of other concepts: "gathering," "well-defined," "distinguishable," "whole" (not to mention our "intuition," our "ideas"). It is only fair to ask which concept is simpler: "set" or "gathering." (As a point of fact the German word "Zusammenfassung" sounds better, but does not escape the criticism.) Similarly, we can question every one of the other concepts, and wonder if it is simpler than the concept of set, and could be conceived prior to it.

In a later paper Cantor came back to the question and discovered a germ of the axiomatic description. Let us add that Cantor's definition was also criticized on the grounds that it does not exclude contradictory sets, as we will see below, and his second approach was motivated by this criticism.

A.3 Remarks on Logic

We want to handle logic in an uncritical and unsophisticated way; nevertheless, we want to say a few words about conventions of mathematical language. It is probably fair to say that the basis of our rational thinking is the following belief: if we start with true premises, and make correct deductions from them, then we reach a true conclusion. We could refuse to accept this but would not get far in mathematics. If we take this belief seriously (as we do in mathematics), rather sophisticated results can be reached. For example, suppose that $2, 3, 5, \ldots, 17$ were the only prime numbers ≥ 2. Then form the number $n = 2 \cdot 3 \cdot 5 \cdots 17 + 1$ (where the points indicate that we have to write all the seven primes from 2 to 17). Then n is not divisible by a prime ≤ 17, hence it is a prime or it has a divisor which is a prime ≥ 18. As the conclusion plainly contradicts the premise, both cannot be true, and *as our reasoning was correct*, the premise must be false—there is a prime number ≥ 18. This is not surprising as 19 happens to be a prime number, but we reached the conclusion by reasoning and not by experience. This reasoning, sometimes called reductio ad absurdum, is used frequently.

There are English sentences in which we can erase a word, write x in its place, and still get a meaningful sentence. For example, in the sentence "two is smaller than five" erasing two and writing x gives the sentence "x is smaller than five." Such a combination of words is called a *propositional function* or *condition* and could be denoted $S(x)$. Now writing "seven" in place of x we get a false sentence and writing "three" in place of x we get a true sentence. Then $S(x)$ is meaningful if x is an integer, and is true for some integers and false for other integers. Given now an arbitrary condition $S(x)$, we may

$$\text{take all objects whose name, substituted in} \atop \text{the place of } x \text{ in } S(x), \text{ gives a true sentence.} \qquad (2)$$

It is understood that x may occur several times, and substitution must be done consistently (thus x is just a sort of "place holder" in this case). On the basis of definition (1) we thus obtain a set. There will be a standard notation for this set:

$$\{x \mid S(x)\}. \tag{3}$$

In spite of the fact that (2) is consistent with (1) and with the usual concept of set, we run into contradictions if we use (2) indiscriminately. Take the following example.

> *The set that does not contain itself as an element.* (4)

This sentence seems to be all right; after all, who ever saw a set which contained itself as an element. Erase "The set," and write x:

$$S(x) = x \text{ does not contain itself as an element.} \tag{5}$$

Then take the corresponding set (2), and call it M as Cantor does (M for "Menge"). Let us ask the question: does M contain M? If it does not, then it should, by the sentence which defines it. If it does, then it should not, by virtue of the same sentence.

This property of construction (2), first noticed by Bertrand Russell, is shocking, and discouraging. When we were inspecting Cantor's definition, we suggested that it was not really bad and actually helped clarify ideas. Now we find that, at the same time, it allows forming the impossible set M.

The example of the set M may suggest that there is something inherently wrong with the concept of set, or at least with the concept of "big" sets. In fact M is as big as they come—it contains *every* single "decent" set. However, the kind of contradiction we have in connection with (5) is well known in classical logic. Let us mention first an example, which can be formulated in terms of "small" sets. Let N be the set of men living in a small village. Suppose that the barber of the village declares: I will shave $x \in N$ if x does not shave himself. It seems then that this sentence defines a subset $P \subset N$. However, the question whether the barber belongs to P leads to the following dilemma: "I will shave myself, if I do not shave myself."

The dilemma above was extensively discussed by Greek logicians who did not use the concept of set. Hence, the contradiction may be independent of this concept. This seems to be confirmed by the following paradox.

Suppose that during one of my lectures a student in the class says,

> *The last sentence on the blackboard is false.* (6)

This can happen, unfortunately. If it does, I normally do the following: I again read the sentence. If I find that the student is right, I apologize, erase the sentence, and write down the sentence corrected. If I find that the student was mistaken, I say so aloud, and leave the sentence on the blackboard. To make this concrete, suppose now that I lecture on set theory, and reach the point up to and including sentence (6); sentences (1) through (6) are on the blackboard (in order), and nothing else. If a student says now "The last sentence on the blackboard is false", I am at a loss what to do. If he is right, then (6) is false, which means that it is true, hence the student was wrong, but in this case the sentence is right, which means that it is false.

It would be interesting to pursue further these questions of logic, but our aim was simply to indicate why it is advisable to restrict the form of sentences when defining subsets of a set in our axiomatic set theory.

A.4   Language of our Axioms

In a complete, advanced presentation of the axioms of set theory, formalized logic must be used. Thus at least part of the language of the theory is formalized. We turn now to describe this part of the language, without effectively carrying out a formalization. In this description, we follow [18].

There will be two basic types of sentences, namely assertions of belonging

$$x \in A \tag{7}$$

and assertions of equality

$$A = B \; ; \tag{7'}$$

all other sentences are to be obtained from such *atomic* sentences by repeated applications of the usual logical operators, subjected to the rules of grammar and unambiguity.

To make the definition explicit, it is necessary to append to it a list of the "usual logical operators", and the rules of syntax. Our list of "logical operators" will be

$$
\left\{
\begin{array}{l}
\left\{
\begin{array}{l}
\textit{not} \\
\textit{and} \\
\textit{or (in the non-exclusive sense)} \\
\textit{if—then—(meaning implies)} \\
\textit{if and only if (abbreviated iff)}
\end{array}
\right. \\
\left\{
\begin{array}{l}
\textit{for some (there exists)} \\
\textit{for all}
\end{array}
\right.
\end{array}
\right. \tag{8}
$$

Notice that "not" operates on a single sentence, the next four operators act on two sentences (S and T, . . . , S iff T) and the last two act on conditions (for some x, $S(x)$ holds, and so forth.)

This list is redundant: it is proved in logic that the first five can be replaced by a single operator, hence everything really comes to the concept of implication, or some very closely connected concept. [*Example:* instead of the sentence "S and T," where S and T are sentences, we can say "not (not S or not T)." This is clumsy in colloquial English but very simple with appropriate logical symbolism. As we do not want to use formalized logic, we use the longer list (8).] In our list (8) the first five operators are called *logical connectives*, and the last two are called *quantifiers*. In the usual formalism "for some x" is sometimes written $\exists x$ and "for all x" is denoted $\forall x$. The connection between these two quantifiers is as follows: the negation of "for some x, $S(x)$ holds" is "for all x, not $S(x)$ holds." The negation of "for all x, $S(x)$ holds" is "there is an x, such that not $S(x)$ holds." This is very important; in fact, possibly the *main* idea to be learned here. Very often the connection between the two quantifiers appears in the following form. We want to prove a statement:

$$\textit{for every } \varepsilon > 0 \cdots \textit{holds true} \; . \tag{9}$$

The negation of this is:

$$\textit{there exists } \varepsilon_0 > 0, \textit{ such that } \cdots \textit{ does not hold true .} \tag{10}$$

If we can now deduce a contradiction from (10), we have a proof of (9).

As for the rules of sentence construction, we make the following agreements:

(i) Put "not" before a sentence and enclose the result between parentheses. (The reason for parentheses, here and below, is to guarantee unambiguity. Note, incidentally, that they make all other punctuation marks unnecessary. The complete parenthetical equipment that the definition of sentences calls for is rarely needed. We shall always omit as many parentheses as it seems safe to omit without leading to confusion. In normal mathematical practice, to be followed here, several different sizes and shapes of parentheses are used, but that is for visual convenience only.)

(ii) Put "and" or "or" or "if and only if" between two sentences and enclose the result between parentheses.

(iii) Replace the dashes in "if—then—" by sentences and enclose the result in parentheses.

(iv) Replace the dash in "for some—" or in "for all—" by a letter, follow the result by a sentence, and enclose the whole in parentheses. (If the letter used does not occur in the sentence, no harm is done. According to the usual and natural convention "for some $y(x \in A)$" just means "$x \in A$." It is equally harmless if the letter used has already been used with "for some—" or "for all—". Recall that "for some $x(x \in A)$" means the same as "for some $y(y \in A)$"; it follows that a judicious change of notation will always avert alphabetic collisions.)

This is about all we need to know on logic. The axiomatics of set theory depend heavily on the logical apparatus used, but we believe that the axioms and their immediate corollaries can be understood on this modest basis, and the rest of the text is but an exercise on the use of quantifiers, mostly in the form of (9) and (10) above.

A.5   The Axioms

Instead of giving a definition of the concept of "set A" and that of "belonging to a set," denoted $a \in A$, we will give properties of these concepts. Enumerating properties is the main feature of the axiomatic method.

We will state now the axioms in the wording of [18], accompanying them with a few remarks.

1. Axiom of Extension.   *Two sets are equal if and only if they have the same elements.* ([18], *p.* 2.)

This axiom means, in particular, if we want to prove $A = B$ we have to prove that $x \in A$ implies $x \in B$ and that $x \in B$ implies $x \in A$. This fact is so important, that it is worthwhile to have a notation for the case when only half of it, say the first half is satisfied. We then write $A \subset B$. This will be a relation between sets; it is not an undefined concept but it was *defined* in terms of "set" and "belonging." See Example 1, p. 6 for a concrete application of the axiom of extension.

2. Axiom of Specification.   To every set A and to every condition S(x) there corresponds a set B whose elements are exactly those elements x of A for which S(x) holds. ([18], *p.* 6.)

We introduce here the important notation

$$\{x \in A \mid S(x)\} \tag{11}$$

to denote the set B. Notice that (11) is the same as our set (3) except that we do not form now the set of all objects satisfying a certain condition, but only those which are already elements of some set (set A in (11)). This allows us, for example, to form sets of real numbers quite arbitrarily, like

$$I = \{x \in \mathbb{R} \mid a \leqslant x \leqslant b\} \tag{12}$$

where our sentence $S(x)$ is $a \leqslant x \leqslant b$, provided we know that $\mathbb{R}$ is a set. (This is not yet implied by Axioms 1 and 2.) The simplest set (11) can be formed with the *atomic* sentence (7) then we get $A = \{x \in A \mid x \in A\}$, hence A is a subset of A. If our sentence $S(x)$ is not satisfied by *any* element of A, (11) describes the *empty* set $\varnothing$. We can always write an impossible condition, for example $x \notin A$. Then $\varnothing = \{x \in A \mid x \notin A\}$. Conclusion: if there is any set, there is an empty set containing no elements (our axioms do not say yet that there are sets at all; we have to postulate this later).

On the basis of Axiom 2, we introduce the important set theoretical operation of intersection. Given sets A and B, we write $\{x \in A \mid x \in B\}$; this set is denoted $A \cap B$ as you know. $B \cap A$ would be $\{x \in B \mid x \in A\}$; this is clearly the same set. The most general operation is the intersection of a collection of sets C (instead of a set of sets we sometimes say collection of sets, but, for us, "collection" shall be synonymous with "set"): suppose C is a set, and, if $A \in C$, then A is also a set. We define:

$$\bigcap \{A \mid A \in C\} = \{x \in A_0 \mid A_0 \in C \text{ and } x \in A \text{ for all } A \in C\}. \tag{13}$$

Hence x is an element of the intersection if it belongs to all sets that belong to C. $A \cap B$ corresponds to the case when C contains two elements, one being A the other being B. If all elements of C are indexed with integers so that $C = \{A_n\}$, we write

$$\bigcap_{n=1}^{\infty} A_n = \{x \in A_1 \mid x \in A_n \text{ for all } n\}. \tag{14}$$

Clearly (14) is the set (13) in this special case (in some cases the elements of C cannot be indexed this way).

3. Axiom of Pairing.   For any two sets there exists a set that they both belong to. ([18], *p.* 9.)

4. Axiom of Unions.   For every collection of sets there exists a set that contains all the elements that belong to at least one of the sets of the given collection. ([18], *p.* 12.)

5. Axiom of Powers.   For each set there exists a collection of sets that contains among its elements all the subsets of the given set. ([18], *p.* 19.)

If C is as in (13), we write

$$\bigcup \{A \mid A \in C\} \tag{15}$$

to denote the set postulated in Axiom 4. The notations $A \cup B$ and $\bigcup A_n$ are used in special cases similar to (14). If we want to prove that $x \in \bigcup A_n$ we must prove $x \in A_n$ for at *least one n;* if we want to prove $x \in \bigcap A_n$, we must prove $x \in A_n$ *for all n.* The logical quantifiers $\exists x$, $\forall x$ are thus closely connected to the set theoretical operations $\cup, \cap$.

We must carefully distinguish the pairing and the union: the set $\{A,B\}$, postulated in Axiom 3, has two elements A, B if $A \neq B$ and a single element A if $B = A$ (this is not excluded). For example, given the set $\varnothing$, we can form the set $\{\varnothing,\varnothing\} = \{\varnothing\}$ which is a non-empty set; it has one element. Axiom 4 postulates the existence of $A \cup B$. This set does not contain, in general, A or B as elements; its elements are either elements of A or elements of B. For example, $\varnothing \cup \varnothing = \varnothing$ has no element, hence it is different from $\{\varnothing\}$.

Axioms 3 and 4 also imply the existence of the set $\{A,B,C\}$ with three elements. *Proof:* Form $\{A,B\}$ and $\{C,C\} = \{C\}$. Then form the pair $\{\{A,B\},\{C\}\} = D$. Take the union of the elements of D. Similarly, given n sets $A_1, \ldots, A_n$, we can form the set $\{A_1,\ldots,A_n\}$ containing these elements.

The Axiom of Powers is a very important tool of set theory. We know already what countable sets are. We have proved in Example 4 of the introductory chapter that if A is countable, the power set $\mathscr{P}(A)$ is not countable, and more generally, we have shown that there is no bijection from A to $\mathscr{P}(A)$. This was discovered by Cantor; set theory, as we understand today, was launched by this discovery. If A is countable, then there is a bijection from $\mathscr{P}(A)$ to $\mathbb{R}$, that is, the set of real numbers. Hence, if we accepted the existence of the integers as a set, Axioms 1–5 would imply the existence of the set $\mathbb{R}$ (or something akin to it, which can be used in place of $\mathbb{R}$). But these axioms do not postulate the existence of any set, yet alone the existence of infinite sets.

Before formulating the last group of axioms, we want to examine the question of existence of sets more closely. If we understand sets in the sense of Cantor's definition (1), all our axioms are clearly satisfied. From the axioms we can deduce, however, that some sets which can be formed in virtue of Cantor's definition are not sets in the sense of the axioms. Specifically, given a set A we can form $B = \{x \in A \mid x \notin x\}$. Suppose now that $B \in B$. Then $B \notin B$, hence this is not possible. In conclusion, $B \notin B$, and in particular $B \notin A$. Summing up, to any set A, a set B can be constructed which is not an element of A. Hence the axioms *exclude* the existence of a set which would contain *all* sets. On the other hand Cantor's definition would admit such a set. Similarly, the contradiction concerning the set M of (5) shows presently that M is *not* a set. The axiomatic system thus accomplished our purpose: on the basis of the axioms we can introduce a part of Cantor's set theory, which is indispensable in mathematics, and at the same time we exclude the known contradictions of Cantor's theory.

If we replace the word "set" in Axioms 1 through 5 by the words "finite set," we have consistent statements. As we want to introduce the concept of "set" with these axioms, we must accept any interpretation consistent with them. Hence, there is a need for an axiom of infinity.

Definition.   If x is a set, we define $x^+ = x \cup \{x\}$, and call it the *successor* of x.

6. Axiom of Infinity.   *There exists a set containing $\varnothing$ and containing the successor of each of its elements.* ([18], p. 44.)

This is the sort of axiom needed to introduce the integers. The next axiom, which has been a point of controversy in the history of set theory, asserts that from any collection of sets, we can "pick" out one representative from each set in the collection. This is stated more precisely in the next axiom.

7. Axiom of Choice. *If A is a collection of non-empty sets, then there exists a choice set C, such that $x \cap C$ contains a single element for any x in A.* ([18], p. 59.)

We will use the concept of "ordered pair (a,b) of elements a, b"; an ordered pair contains a first element (coordinate) a and a second element (coordinate) b; in case $a = b$ these coordinates are equal. The concept of ordered pair could be reduced to the concept of set by defining $(a,b) = \{\{a\},\{a,b\}\}$ (see [18], pp. 22–25); we will not give details of this here.

If A and B are given sets we can form the set of *all* ordered pairs (a,b); this set is denoted $A \times B$. By definition a map $f : A \rightarrow B$ is a subset of $A \times B$ such that: (1) given $a \in A$ there is a $b \in B$ such that $(a,b) \in f$; (2) if $(a,b_1) \in f$ and $(a,b_2) \in f$ then $b_2 = b_1$. *Note:* Instead of $(a,b) \in f$ we write $b = f(a)$. You may then proceed to define the following terms, notations, and concepts in connection with functions: injection, surjection, bijection, restriction, extension, $f(X)$ if $X \subset A$, $f^{-1}(Y)$ if $Y \subset B$, composition of maps. If $B \subset \mathbb{R}$, f is usually called a *real-valued* function.

8. Axiom of Substitution. *If $S(a,b)$ is a sentence such that for each a in a set A the set $\{b \mid S(a,b)\}$ can be formed, then there exists a function F with domain A such that $F(a) = \{b \mid S(a,b)\}$ for each a in A.* ([18], p. 75.)

Remark. By definition, a function F has a range, hence the axiom requires the existence of a set B such that $F \subset A \times B$.

We can easily remember these axioms, if we summarize them in suggestive form as follows. The axiom of extension gives a criterion for the equality of two sets. The axioms of specification, pairing, unions, and powers allow us to specify subsets, form pairs, and finite sets in general, intersections and unions, and the collection of all subsets of a given set (called the power set of the given set). We postulate the existence of infinite sets. The axiom of choice insures that we can choose a single element from each (non-empty) set of a collection of sets and form a set with the chosen elements. The axiom of substitution shows that we can substitute for each element of a given set some set depending on this element.

If we give completely detailed proofs in mathematics, we have to go back to these axioms, and first principles of logic. In actual practice we mainly use the set theoretical operations of union, intersection, complement, difference, power set, and choice set (the latter usually implicitly).

Appendix B

Miscellaneous Problems

1. Find a set $A \subset \mathbb{R}^2$, such that
$$\text{int}(\text{cl}(A)) \neq \text{int}(A) .$$

2. For two sets $A, B \subset \mathbb{R}^n$, define the distance between them by
$$d(A,B) = \inf\{d(x,y) \mid x \in A, y \in B\} .$$

   (a) Find two closed sets such that $A \cap B = \varnothing$ and yet $d(A,B) = 0$.
   (b) For A compact and $x \notin A$, show that $d(A,x) = d(y,x) > 0$ for some $y \in A$. [Hint: Use Theorem 5, Chapter 5.]
   (c) For any set $A \subset \mathbb{R}^n$ show that $\text{cl}(A) = \{x \in \mathbb{R}^n \mid d(x,A) = 0\}$.

3. Let $A \subset \mathbb{R}^n$. Show that A is compact iff every continuous map $f \colon A \to \mathbb{R}$ is bounded above and assumes its maximum at some point of A.

4. Let x_n be a sequence in $\mathbb{R}^n$, such that there is a constant M with $\|x_n\| \leqslant M$ for all n. Then prove x_n has a convergent subsequence.

5. Show that $f \colon \mathbb{R} \to \mathbb{R}$ has a continuous derivative iff the double limit
$$\underset{(x,y)\to(x_0,x_0)}{\text{limit}} \frac{f(x) - f(y)}{x - y}$$
exists for every $x_0 \in \mathbb{R}$.

6. For continuous functions $f, g \colon [a,b] \to \mathbb{R}$, define
$$\langle f,g \rangle = \int_a^b f(x)g(x)\, dx$$

   Show that $\langle\, ,\, \rangle$ has all the properties of an inner product (Theorem 5(I), Chapter 1). Hence deduce the inequality
$$\left(\int_a^b f(x)g(x)\, dx \right)^2 \leqslant \left(\int_a^b f(x)^2\, dx \right)\left(\int_a^b g(x)^2\, dx \right) .$$

7. Show that $\sum_{n=1}^{\infty} (x + 1)^2/n^2$ converges uniformly on $[0,1]$. Also show that $\sum_{n=1}^{\infty} (x + 1)^n/n!$ converges uniformly on the same interval.

8. Given that $\sum_{n=0}^{\infty} x^n = 1/(1 - x)$ for $|x| < 1$, differentiate both sides to obtain

$$\sum_{n=1}^{\infty} nx^{n-1} = \left(\frac{1}{1 - x}\right)^2, \qquad \text{for } |x| < 1 .$$

Can you justify this?

9. Let $f(x) = 1/(1 + x)^2$, so $f > 0$. Now $d(-1/1 + x)/dx = f(x)$, so

$$\int_{-\infty}^{\infty} f(x)\, dx = \lim_{a \to \infty} \int_{-a}^{a} f(x)\, dx = \lim_{a \to \infty}\left[\frac{-1}{1 + a} + \frac{-1}{1 - a}\right] = 0 .$$

What is wrong with this argument?

10. Let $f: \mathbb{R}^2 \to \mathbb{R}$ and $c \in \mathbb{R}$. What conditions on f will guarantee that $f(x,y) = c$ defines a smooth curve in the plane? (Say $y = g(x)$ or $x = b(y)$). Interpret geometrically.

11. Answer *true* or *false*.
    (a) The rationals are an ordered field.
    (b) A continuous function $f: \mathbb{R} \to \mathbb{R}$ is uniformly continuous.
    (c) A closed bounded subset of a metric space is compact.
    (d) The real numbers are connected.
    (e) An open set is bounded.
    (f) A compact set is closed.
    (g) A differentiable function is continuous.
    (h) $[0,\infty[$ is closed.
    (i) $]0,1[$ is compact.

    (j) $\left\{\left.\frac{1}{n}\right| n \in N\right\}$ is bounded.

12. Give an example of a continuous function $f: \mathbb{R} \to \mathbb{R}$ such that $f(\mathbb{R})$ is not closed. If $A \subset \mathbb{R}$ is a closed bounded interval, must $f(A)$ be closed?

13. At first, one thinks the intervals $]0,1[$ and $[0,1]$ are very similar. State at least five significant differences between them in terms of topology and continuous functions.

14. (a) Find an example of a closed set $A \subset \mathbb{R}^n$ such that $A = \text{bd}(A)$.
    (b) If $A = \text{bd}(A)$, then show that A is closed.
    (c) Prove: If $\text{int}(A) = \varnothing$, then $A = \text{bd}(A)$ iff A is closed.
    (d) Prove: If A is closed and $A \subset \text{bd}(A)$, then $A = \text{bd}(A)$ and $\text{int}(A) = \varnothing$.
    (e) Find a set A such that $A \subset \text{bd}(A)$, but $A \neq \text{bd}(A)$.

15. (a) Let f be integrable and suppose that for every partition P, $b \leqslant U(f,P)$. Then show $b \leqslant \int_A f$.
    (b) Suppose $U(f,P) \leqslant U(g,P)$ for every P. Then prove that $\int_A f \leqslant \int_A g$.
    (c) Is it true that $\int_A f \leqslant \int_A f^2$? Distinguish the cases $|f| \geqslant 1$ and $|f| \leqslant 1$.

16. Are the following statements true or false? (All sets have volume and all functions are bounded and integrable.)

(a) If $A \supset B$, and $A \backslash B$ has measure zero, then $\int_A f = \int_B f$.

(b) If $\{x \mid f(x) \neq g(x)\}$ has measure zero, then $\int_A f = \int_A g$.

(c) If $f \geqslant 0$, $g \geqslant 0$, and $\int_A f = \int_A g$, then $f = g$ on A except possibly on a set of measure zero.

(d) The same question as (c) except $f \geqslant g$.

17. Let $f: [a,b] \to \mathbb{R}$ be integrable. (a) Prove $F(x) = \int_a^x f(t)\, dt$ is uniformly continuous. (b) Show that F has a derivative at x_0 if f is continuous at x_0. (c) Show F is differentiable except possibly on a set of measure zero.

18. (a) Let $T: \mathbb{R}^n \to \mathbb{R}^n$ be a linear mapping. Prove that T is norm preserving (that is $\|Tx\| = \|x\|$) iff T preserves the inner product $\langle Tx, Ty \rangle = \langle x, y \rangle$. [Hint: See Exercise 12, Chapter 1.]

(b) If T preserves the norm (or inner product), then T is an isomorphism.

19. Prove Cavalieri's Principle: If $A, B \subset \mathbb{R}^3$ have volume, and every plane parallel to the xy-plane intersects A and B in equal area, then A and B have the same volume. [Hint: Make use of Fubini's theorem.]

   Remark: In connection with this problem, see Gelbaum and Olmsted, *Counterexamples in Analysis*, Example 6, Chapter 11. For applications see McAloon-Tromba *Calculus*, Chapter 6.

20. Let $f: A \subset \mathbb{R}^n \to \mathbb{R}$. If f is continuous at x, show that $|f|$ is as well.

21. Show that a set $A \subset \mathbb{R}^n$ has volume iff for any $\varepsilon > 0$ there exists a set $V_\varepsilon \subset A$ and a set $W_\varepsilon \supset A$ such that V_ε and W_ε have volume and $v(W_\varepsilon \backslash V_\varepsilon) = v(W_\varepsilon) - v(V_\varepsilon) < \varepsilon$. Show that if the latter condition holds, the volume of A is

$$\inf\{v(W_\varepsilon) \mid \varepsilon > 0\} = \sup\{v(V_\varepsilon) \mid \varepsilon > 0\} .$$

22. Let $f: A \subset \mathbb{R}^n \to \mathbb{R}$ be integrable and $f \geqslant 0$. Let $S = \{(x,y) \in \mathbb{R}^n \times \mathbb{R} \mid x \in A$ and $0 \leqslant y \leqslant f(x)\}$. Show that $S \subset \mathbb{R}^{n+1}$ has volume $\int_A f$. Interpret geometrically.

23. Show that the volume of the figure obtained by rotating the area under the graph of a non-negative function $f: [a,b] \to \mathbb{R}$ is given by $\int_a^b \pi f(x)^2\, dx$. See Figure A-1.

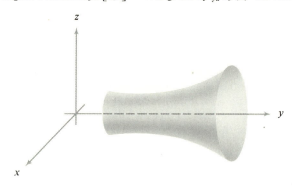

FIGURE A-1   Volume of rotation.

Use this formula to compute the volume of

$$\{(x,y,z) \in \mathbb{R}^3 \mid 1 < x < 2, y^2 + z^2 < x^4\}.$$

24. Evaluate the following integrals.
    (a) $\int_A (1 - x^2 - y^2) \, dx \, dy$; A is the unit disc.
    (b) $\int_A y \, dx \, dy$; $A = \{(x,y) \mid 0 < x < y, \, y < \pi/2 \sin x\}$.
    (c) $\int_A \sqrt{x^2 + y^2} \, dx \, dy$; $A = \{(x,y) \mid (x - 1)^2 + y^2 < 1\}$.
    (d) $\int_A |z| \, dx \, dy \, dz$; $A = \{(x,y,z) \mid x^2 + y^2 + z^2 < 4x^2 + y^2 > 1\}$.
    (e) $\int_A 1 \, dx \, dy \, dz$; $A = \{(x,y,z) \mid x^2 + y^2 + z^2 > 1z < x^2 + y^2 < 1\}$.
    (f) $\int_A (x^2 + y^2) \, dx \, dy$; $A = \{(x,y) \mid x^2 + y^2 < 1 - x\}$.
    (g) $\int_A xyz \, dx \, dy \, dz$; $A = [a,b] \times [c,d] \times [e,f]$.
    (h) $\int_A (x^2 - y^2)\sin^2(x + y) \, dx \, dy$; $A = \{(x,y) \mid \pi < x + y < 2\pi \text{ and } -\pi < x - y < \pi\}$. [Hint: Use the substitution $u = x - y, v = x + y$.]

25. Let $f: [a,b] \to \mathbb{R}$ be continuous and suppose $f(a)f(b) < 0$. Then show there is an $x \in \,]a,b[$ such that $f(x) = 0$.

26. Let $f: A \subset \mathbb{R}^n \to \mathbb{R}$ be continuous. Suppose $B \subset A$ is such that B is bounded and $\mathrm{cl}(B) \subset A$. Then show there are points $x_0, y_0 \in \mathrm{cl}(B)$, such that

$$f(x_0) = \inf\{f(x) \mid x \in B\} \quad \text{and} \quad f(y_0) = \sup\{f(x) \mid x \in B\}.$$

27. Let $f: \mathbb{R}^n \to \mathbb{R}^n$ be of class C^1 and suppose $Jf(x) \neq 0$ for all x. Let $x_0 \in \mathbb{R}^n$ and $B = \{x \in \mathbb{R}^n \mid f(x) = x_0\}$. Show that B has no accumulation points.

28. If $f: A \subset \mathbb{R}^n \to B \subset \mathbb{R}^n$ (where A and B are open sets) is a one-to-one function of class C^1 with $Jf(x) \neq 0$ for each $x \in A$, then prove $f^{-1}: B \to A$ is also of class C^1.

29. Give an example of a function f of class C^1 which has derivative equal to zero at a point x but is one-to-one in a neighborhood of x. Show that f^{-1} cannot be differentiable at $f(x)$.

30. Let the sets A and B have volume. Then show $A \cup B$ has volume. If $A \cap B$ and $A \backslash B$ have volume as well, then prove
    (a) $v(A \cup B) = v(A) + v(B) - v(A \cap B)$;
    (b) $v(A \backslash B) = v(A) - v(B)$ if $A \supset B$.

31. Suppose $f: A \subset \mathbb{R}^n \to \mathbb{R}$ is integrable and $f = g$ except on a set of content zero. Then show that g is integrable. Show that this is false if we replace "content zero" by "measure zero."

32. Show that the bounded integrable functions $f: A \to \mathbb{R}$ on a bounded set A form a vector space. Also show that if f and g are bounded and integrable, so is fg. If f and g are integrable but unbounded, need fg be integrable?

33. Let $f(x,y) = x - y^2$. Is there a real valued function $g(x)$ defined near $x = 0$ such that $f(x,g(x)) = 0$? Show that g is not unique. How does this tie up with the implicit function theorem?

34. Let $f: A \subset \mathbb{R}^n \to \mathbb{R}^m$ be a function such that for any open set $V \subset A, f(V)$ is open. For a set $B \subset \mathbb{R}^n$ such that $\mathrm{cl}(B) \subset f(A)$, show that

$$f(\mathrm{int}(f^{-1}(B))) \subset \mathrm{int}(B) \quad \text{and} \quad \mathrm{bd}(B) \subset f(\mathrm{bd}(f^{-1}(B))).$$

35. Let $f: A \subset \mathbb{R}^n \to \mathbb{R}^m$ be a map. Show that f is continuous iff for every compact set $K \subset A$, the restriction $f \mid K: K \to \mathbb{R}^m$ is continuous.

36. (a) Show that if U_i is a family of disjoint open sets in $\mathbb{R}^n$, then the family is countable. [Hint: Pick a point with rational coordinates in each set and use the fact that the set of such points is countable.]
    (b) If $A \subset \mathbb{R}^n$ is open, show that the components of A are open and are countable.
    (c) Prove that any open set in $\mathbb{R}$ is the countable union of intervals.

37. Let $A \subset \mathbb{R}^n$ be closed. Show that A is compact iff for every $\varepsilon > 0$ there is a finite covering of A by sets with diameter $< \varepsilon$.

38. Let $f: [a,b] \to [\alpha,\beta]$ be strictly increasing and onto. Then show that f and f^{-1} are continuous.

39. Prove the Lebesque Covering lemma: Let $A \subset \mathbb{R}^n$ be a compact subset of $\mathbb{R}^n$, and let $\{V_\alpha\}$ be an open cover of A. Then there exists an $\varepsilon > 0$, such that if S is any rectangle contained in A having sides less than ε, then S is contained in some open set of the cover.

40. Let $f: A \subset \mathbb{R}^n \to \mathbb{R}$ be a mapping. Let $M \subset A$ be the set of (strict) local maxima of f. Then show $f(M)$ is finite or countable. Give examples.

41. Find and classify the critical points of the following functions.
    (a) $f(x,y) = y^4 + 2x^2y$,
    (b) $f(x,y,z) = xy + xz + zy$,
    (c) $f(x,y) = (\sin x)(\sin y)$.

42. Let S be an open connected set in $\mathbb{R}^n$. Let A be a component of $\mathbb{R}^n \backslash S$. Then show $\mathbb{R}^n \backslash A$ is connected.

43. Find a non-constant continuous function $f: [0,1] \to \mathbb{R}$ which has its maximum at $x_0 \in \,]0,1[$ but $f'(x_0)$ does not exist.

44. A set $B \subset A$ is said to be *dense* in A if $\text{cl}(B) \supset A$. Show that this is equivalent to the condition that for every open set U with $A \cap U \neq \varnothing$, we have $B \cap U \neq \varnothing$. Is A dense in $\text{cl}(A)$? Show that $\mathbb{R}^n$ has a countable dense subset.

45. (a) Let $A \subset \mathbb{R}^n$ have volume. Then show $\text{int}(A)$ and $\text{cl}(A)$ have volume, and $v(A) = v(\text{int}(A)) = v(\text{cl}(A))$.
    (b) Prove that if A is a set and $\text{cl}(A)$ has volume, then we cannot conclude that A has volume.
    (c) Prove that if $\text{int } A = \varnothing$, we cannot conclude that A has content or is of measure zero.

46. Let $g: A \subset \mathbb{R}^n \to B \subset \mathbb{R}^n$ be C^1 on the open set A and $B = g(A)$. We say g is *volume preserving* if for every set $D \subset A$ with $g(D)$ and D having volume, $v(g(D)) = v(D)$. Suppose g is one-to-one and $Jg(x) \neq 0$ at each $x \in A$. Then prove that g is volume preserving iff $|Jg(x)| = 1$ for all $x \in A$.

47. Show that if A has content zero, then $\text{cl}(A)$ has content zero. Is this true for measure zero?

48. Let $f: A \subset \mathbb{R}^n \to \mathbb{R}^m$ be of class C^1 with A an open set. Let $A_0 \subset A$ and $\text{cl}(A_0) \subset A$ and suppose A_0 is compact. If A_0 has content or measure zero, then prove that $f(A_0)$ does as well. [Hint: Consider the case where $Jf(x) = 0$ separately and use Sard's theorem (Exercise 5, Chapter 9).]

49. Let $A \subset \mathbb{R}^n$ and let B denote the set of accumulation points of A. Show that B is a closed set. Find an example where B consists of a single point.

50. A set of $A \subset \mathbb{R}^n$ is called *homeomorphic* to $B \subset \mathbb{R}^m$ if there is a continuous map $\varphi: A \to B$ with a continuous inverse φ^{-1}. We call φ a *homeomorphism*.
    (a) Find an example of a bijection $\varphi: A \to B$ which is continuous but is not a homeomorphism.
    (b) Let $f: A \subset \mathbb{R}^n \to \mathbb{R}^m$ be continuous, with Γ the graph of f ($\Gamma = \{(x, f(x)) \in \mathbb{R}^n \times \mathbb{R}^m \mid x \in A\}$). Show that A and Γ are homeomorphic.

51. Let $f, g: \mathbb{R}^n \to \mathbb{R}^m$ be continuous and $B = \{x \in A \mid f(x) = g(x)\}$. Show that B is a closed set.

52. Let $f: A \subset \mathbb{R}^n \to \mathbb{R}$ be bounded and integrable and A have volume. Let $B \subset A$ have volume. Then show that the restriction of f to B is integrable.

53. Find a function $f: \mathbb{R}^2 \to \mathbb{R}^2$ which has a Jacobian equal to 1 everywhere, but is not onto.

54. Let f be a monotone function; $f: [a,b] \to \mathbb{R}$, say f is non-decreasing: $f(x) \leqslant f(y)$ if $x \leqslant y$.
    (a) For any $x \in [a,b]$, show that the left and right limits

    $$f(x+) = \lim_{h \to 0+} f(x + h)$$

    $$f(x-) = \lim_{h \to 0+} f(x - h)$$

    exist.
    (b) Show that f has at most a countable set of discontinuities. [Hint: Let P_n be the set of points where the jump of f exceeds $1/n$. Show P_n is finite and consider the union of all the P_n, $n = 1, 2, 3, \ldots$.]
    (c) It is a famous theorem of Lebesgue that for such f, the derivative of f exists except possibly for points in a set of measure zero. Consider some examples to verify the validity of the results. Look up a proof in, for example, Hewitt and Stromberg, *Real and Abstract Analysis*, and write a brief essay on the essential features of the proof.

55. Prove that the transformation

$$\begin{cases} x_1 = u_1 \\ x_2 = u_1 + u_2 \\ x_3 = u_1 + u_2 + u_3 \\ \quad \cdot \\ \quad \cdot \\ \quad \cdot \\ x_n = u_1 + \cdots + u_n \end{cases}$$

leaves volumes unchanged.

56. Let $g: [0,1] \to \mathbb{R}$ be integrable. Prove that

$$\int_0^1 \left[\int_x^1 g(t)\, dt \right] dx = \int_0^1 t g(t)\, dt \ .$$

57. Reverse the order of integration in

$$\int_0^1 \int_0^{(8-8x^2)^{1/3}} \int_0^{(16-16x^2-2y^3)^{1/4}} f(x,y,z)\, dz\, dy\, dx .$$

58. Let K be a compact set. If $\{f_n\}$ is a uniformly convergent sequence of continuous real valued functions on K, prove $\{f_n\}$ is equicontinuous. The converse is not true. Give a counter example.

59. Let B be the open region bounded by the curves $x = -y^2$, $x = 2y - y^2$ and $x = 2 - y^2 - 2y$. Introducing the change of variables $x = u - (u+v)^2/4$, $y = (u+v)/2$ evaluate $\iint\limits_B x\, dx\, dy$.

60. Let $S \subset \mathbb{R}^n$ have volume and $t > 0$. Let R be the set of points

$$\{(tx_1, \ldots, tx_n) \mid (x_1, \ldots, x_n) \in S\} \ .$$

Show $v(R) = t^n v(S)$. What if $t < 0$?

61. Explain how the Gibbs' phenomenon is possible and yet the Fourier series still converges in the mean and pointwise.

62. (a) Let $f(x)$ on $[-\pi,\pi]$ have Fourier series

$$\frac{a_0}{2} + \sum_{n=1}^{\infty} [a_n \cos nx + b_n \sin nx] \ .$$

Define the *reflection* of f by $g(x) = f(-x)$. Show that the Fourier series of g is

$$\frac{a_0}{2} + \sum_{n=1}^{\infty} [a_n \cos nx - b_n \sin nx] \ .$$

(b) Recall that the Fourier series of

$$f(x) = \begin{cases} 0, & -\pi \leqslant x \leqslant 0, \\ x, & 0 \leqslant x \leqslant \pi, \end{cases}$$

is

$$\frac{\pi}{4} + \sum_{n=1}^{\infty} \left(\frac{(-1)^n - 1}{\pi n^2} \cos nx - \frac{(-1)^n}{n} \sin nx \right).$$

Use (a) to show that the Fourier series of $|x|$ on $[-\pi,\pi]$ is

$$\frac{\pi}{2} + 2 \sum_{n=1}^{\infty} \frac{(-1)^n - 1}{\pi n^2} \cos nx = \frac{\pi}{2} - 4 \sum_{n=1}^{\infty} \frac{\cos[(2n-1)x]}{\pi(2n-1)^2} \ .$$

(c) Use (b) to show that

$$\frac{\pi^2}{8} = 1 + \frac{1}{3^2} + \frac{1}{5^2} + \frac{1}{7^2} + \cdots .$$

(d) Use (b) to obtain the Fourier cosine series of x on $[0,\pi]$ and conversely.

63. Using Table 10-5, find the Fourier series of each of the following functions:
(a) $f(x) = a + bx$, on $[-\pi,\pi]$,
(b) $f(x) = a - bx$, sine series on $[0,\pi]$,
(c) $f(x) = x^2 + \sin x$, on $[0,2\pi]$.
To what values do the series converge at each point?

64. (a) Let V be an inner product space and $\varphi_0, \varphi_1, \ldots$ a complete orthonormal basis. Suppose W is a subspace of V and $f \in W$ if and only if $\langle f, \varphi_0 \rangle = 0$. Then prove $\varphi_1, \varphi_2, \ldots$ is a complete orthonormal system for W. Generalize.
(b) Apply (a) to the trigonometric system and
   (i) $W = \{f : [-\pi,\pi] \to \mathbb{R} \mid \int_{-\pi}^{\pi} f(x)\, dx = 0\}$,
   (ii) $W = \{f : [-\pi,\pi] \to \mathbb{R}, \text{ which are even}\}$,
   (iii) $W = \{f : [-\pi,\pi] \to \mathbb{R}, \text{ which are odd}\}$.

65. Let
$$f(x) = \begin{cases} -1, & -l < x < 0, \\ 1, & 0 < x < l, \\ 0, & x = 0, x = l, \end{cases}$$

and extend so f is periodic. Then for all x, show that

$$f(x) = \frac{4}{\pi} \sum_{n=1}^{\infty} \frac{1}{2n-1} \sin\left(\frac{(2n-1)\pi x}{l}\right).$$

66. If $f : [a,b] \to \mathbb{R}$ is square integrable, then prove that f is integrable, that is, $\int_a^b |f|^2\, dx < \infty$ implies $\int_a^b |f|\, dx < \infty$. [Hint: Use the Schwarz inequality.]

67. Let $f : [-\pi,\pi] \to \mathbb{R}$ be
$$f(x) = \begin{cases} 0, & -\pi \leqslant x < 0, \\ 1, & 0 \leqslant x \leqslant \pi. \end{cases}$$

The Fourier series of f in exponential form is

$$\sum_{-\infty}^{\infty} \frac{e^{inx}}{2\pi i n}.$$

What kind of convergence do we have?

68. (a) Suppose that $f : [-\pi,\pi] \to \mathbb{R}$ is sectionally continuous with jump discontinuities. Then show that the sum of the Fourier series of f at x depends only on the values of f in any neighborhood of x. This property is called *Riemann's localization property*. [Hint: Apply Theorem 9, Chapter 10.]
(b) The Fourier coefficients of f depend on f throughout $[-\pi,\pi]$. How do you reconcile this with (a)? [Hint: Study the proof of Theorem 9, Chapter 10.]

69. Suppose we have $f : [-\pi,\pi] \times [-\pi,\pi] \to \mathbb{R}$; consider its Fourier series

$$\sum_{n,m=-\infty}^{\infty} c_{n,m} e^{inx} e^{imy}$$

(see Exercise 18, Chapter 10).

(a) Write out the Fourier series of f in trigonometric form.

(b) For y fixed, let $g(x) = f(x,y)$. Show that the exponential Fourier coefficients of g are

$$c_n = \sum_{m=-\infty}^{\infty} c_{nm} e^{imy} .$$

(c) If f is square integrable, we know that its Fourier series converges to f in mean (see Theorem 8, Chapter 10). The purpose here is to give a pointwise convergence theorem. Hence, show that if f is of class C^1 and $f(x,\pi) = f(x,-\pi)$, $f(\pi,y) = f(-\pi,y)$, then prove the Fourier series of f converges to f pointwise. [Hint: Use (b) and Theorem 9, Chapter 10.]

70. What types of convergence hold for the Fourier series of the following functions?

(a) $f(x) = \begin{cases} -3, & 0 \leqslant x \leqslant \pi , \\ 2, & -\pi \leqslant x < 0 , \end{cases}$

(b) $f(x) = \begin{cases} x^2 + 1, & -\pi \leqslant x \leqslant 0 , \\ -\pi x + 1, & 0 \leqslant x \leqslant \pi , \end{cases}$

(c) $f(x) = x^2 + 3, \qquad -\pi \leqslant x \leqslant \pi,$

(d) $f(x) = \sin x \qquad$ on $[0,l]$,

(e) $f(x) = 1 \qquad$ on $[0,\pi]$ (both sine and cosine series).

71. Discuss the Gibbs' phenomenon for the function

$$f(x) = \begin{cases} -3, & 0 \leqslant x \leqslant \pi , \\ 2, & -\pi \leqslant x < 0 . \end{cases}$$

72. For what values of p is $\sum_{n=1}^{\infty} (\sin nx)/n^p$ the Fourier series of a square integrable function (see Exercise 32, Chapter 10).

73. (a) Show that

$$\left| \sum_{k=0}^{n} \cos kx \right| \leqslant \left| \frac{1}{\sin(x/2)} \right| \qquad x \neq 0 .$$

[Hint: See Exercise 6, Section 10.2.]

(b) Consider the Fourier series for the step function

$$f(x) = \sum_{n=1}^{\infty} \frac{\cos nx}{n} .$$

Show that for any $\delta > 0$, this converges uniformly on $[\delta,\pi]$. [Hint: Use (a) and the Dirichlet test.]

(c) Generalize (b) to any Fourier series

$$f(x) = \sum_{n=1}^{\infty} b_n \cos nx$$

with b_n decreasing. Conclude that f must be continuous on $]0,\pi]$.

(d) Deduce from (c) that if f has a discontinuity at x_0 and $0 < x_0 < \pi$, then the Fourier coefficients of f are not decreasing.

74. A string on $[0,l]$ is initially displaced at $t = 0$ by $f(x) = (x - l/2)^2 - l^2/4$. Find a formula for the displacement after time t.

75. (a) If a bar with insulated ends has temperature $T = $ constant at $t = 0$, then show that $T = $ constant for all $t > 0$.
    (b) If a bar on $[0,\pi]$ has temperature at $t = 0$ given by $\sin x$, find the temperature for $t > 0$.
    (c) Same as (b) except $T = \cos x$ at $t = 0$.

76. (a) Find a function φ on $[0,\pi] \times [0,\pi]$ such that $\nabla^2\varphi = 0$ and $\varphi(x,0) = \cos x$, $\varphi(x,\pi) = 0 = \varphi(0,y) = \varphi(\pi,y)$.
    (b) In (a) replace $\varphi(0,y) = 0$ by $\varphi(0,y) = 1$ and find the function.
    (c) In what sense are the boundary values in (a) and (b) assumed?

77. Let V be an inner product space. Usually, $\|f_n\| \to \|f\|$ does not imply $f_n \to f$ (Exercise 16, Chapter 3). However, show that $\|f_n\| \to \|f\|$ does imply $f_n \to f$ in mean if f_n is the nth partial sum of the Fourier series with respect to an orthonormal family.

78. Let $f: \mathbb{R} \to \mathbb{R}$ be twice differentiable, then show that
    (a) if $F(x,y) = f(xy)$, then $x\,\partial F/\partial x = y\,\partial F/\partial y$,
    (b) if $F(x,y) = f(ax + by)$, then $b\,\partial F/\partial x = a\,\partial F/\partial y$,
    (c) if $F(x,y) = f(x^2 + y^2)$, then $y\,\partial F/\partial x = x\,\partial F/\partial y$, and
    (d) if $F(x,y) = f(x + cy) + f(x - cy)$, then $c^2\,\partial^2 F/\partial x^2 = \partial^2 F/\partial y^2$.

79. Let $f:]0,1[\to \mathbb{R}$ be continuous and bounded. Prove that
$$\Gamma = \{(x,f(x)) \in \mathbb{R}^2 \mid x \in \,]0,1[\,\}$$
    is not closed.

80. Prove Kronecker's lemma: if $\sum_{n=1}^{\infty} x_n/n$ converges, then $(x_1 + \cdots + x_n)/n \to 0$ as $n \to \infty$ (that is, $x_n \to 0$ in the Cesaro sense).

81. Let
$$f(x) = \begin{cases} \sin\left(\dfrac{1}{x}\right), & x \neq 0, \\ 0, & x = 0. \end{cases}$$
    Prove f has an antiderivative $F: \mathbb{R} \to \mathbb{R}$.

82. (a) Let $I \subset \mathbb{R}$ be an open interval and let $f: I \to \mathbb{R}^n$ be continuous. Assume there are two maps $g_1, g_2: I \subset \mathbb{R} \to \mathbb{R}^n$ such that
$$\frac{1}{n^2}\left(f(x + h) - f(x) - hg_1(x) - \frac{h^2}{2}g_2(x)\right) \to 0$$
    uniformly on every compact $K \subset I$ as $h \to 0$. Set $\Delta_h f(x) = f(x + h) - f(x)$ and $\Delta_h \Delta_h f(x) = \Delta_h f(x + h) + \Delta_h f(x)$. Then prove that
$$\frac{\Delta_h \Delta_h f}{h^2} \to g_2(x) \qquad \text{as } h \to 0$$

uniformly on every compact $K \subset I$. Deduce g_2 is continuous and

$$\frac{\Delta_h f(x)}{h} \to g_1(x) \qquad \text{as } h \to 0 .$$

Is f of class C^2?

(b) Examine (a) for

$$f(x) = \begin{cases} x^2 \sin \dfrac{1}{x}, & x \neq 0 , \\ 0, & x = 0 . \end{cases}$$

83. Let M be a compact metric space and let $\varphi : M \to M$ satisfy $d(\varphi(x), \varphi(y)) < d(x,y)$ for $x, y \in M$, $x \neq y$. Prove φ has a unique fixed point. Give a counterexample if $M = \mathbb{R}$.

84. Let $f : [a,b] \to \mathbb{R}$ be a bounded integrable function, $f(x) \geqslant m > 0$ for all $x \in [a,b]$. Show that

$$\left(\int_a^b \frac{1}{f} \right) \left(\int_a^b f \right) \geqslant (b - a)^2 .$$

85. Suppose $f : [0,1] \to \mathbb{R}$ is integrable, $\int_0^1 f(x)\, dx \geqslant 7$, $0 \leqslant f(x) \leqslant 10$ for all $x \in [0,1]$. Define the set $E = \{x \in [0,1] \mid f(x) \geqslant 1\}$, and assume E has volume. Show that $v(E) \geqslant 1/2$.

86. Suppose $f : [0,2\pi] \to \mathbb{R}$ is continuous and $f(0) = f(2\pi)$. Let $s_N = \sum_{k=-N}^{N} \langle f, \varphi_k \rangle \varphi_k$ be the Nth partial sum of the Fourier series for f, and define

$$\Phi(y) = \int_0^y f(x)\, dx \qquad \Sigma_N(y) = \int_0^y s_N(x)\, dx .$$

State whether each of the following "Must Be True" (MBT) or "Could Be False" (CBF).

(a) $\int_0^{2\pi} s_N(x) e^{ix}\, dx \to \int_0^{2\pi} f(x) e^{ix}\, dx$ as $N \to \infty$.
(b) $\int_0^{2\pi} x^2 s_N(x)\, dx \to \int_0^{2\pi} x^2 f(x)\, dx$.
(c) $\| s_N - f \| \to 0$.
(d) $s_N(2) \to f(2)$.
(e) Σ_N is the Nth partial sum of the Fourier series for Φ.
(f) $\| \Sigma_N - \Phi \| \to 0$.
(g) $\Sigma_N(2) \to \Phi(2)$.
(h) $\Sigma_N \to \Phi$ uniformly on $[0,2\pi]$.

87. *The Poisson kernel and harmonic functions.* Let $f(\theta)$ be continuous and periodic, $-\pi \leqslant \theta \leqslant \pi$. (We can think of $f(\theta)$ as a function defined on the circumference of the unit circle in the plane.) By Fejér's theorem, we know that the Fourier series of f converges to f in the (C,1) sense. Deduce that

$$\lim_{r \to 1^-} \sum_{k=-\infty}^{\infty} c_k r^{|k|} e^{ik\theta} = f(\theta) .$$

(Note the exponent $|k|$ for the negative indices). Define

$$u(r,\theta) = \sum_{k=-\infty}^{\infty} c_k r^{|k|} e^{ik\theta}$$

for $0 \leqslant r < 1$. We regard u as a function in the interior of the unit disk in the plane. In *rectangular* coordinates x, y we have

$$u(x,y) = c_0 + \sum_{k=1}^{\infty} (c_k(x + iy)^k + c_{-k}(x - iy)^k).$$

Prove that this series converges *uniformly* in any disk of radius < 1.

Show that (by the general theory of power series) we can differentiate u term by term any number of times. In this way prove that

$$\frac{\partial^2 u}{\partial x^2} + \frac{\partial^2 u}{\partial y^2} = 0 \, ;$$

that is, u is a solution of Laplace's equation—a so-called *harmonic function*. We have already seen that $u(r,\theta) \to f(\theta)$ as $r \to 1-$, so we have solved the "Dirichlet problem": to find a harmonic function in the unit disk which has a given function for its boundary values.

For $0 \leqslant r < 1$ prove that

$$u(r,\theta) = \frac{1}{2\pi} \int_{-\pi}^{\pi} f(t)P_r(\theta - t)\, dt$$

where

$$P_r(\theta - t) = \sum_{-\infty}^{\infty} r^{|k|} e^{ik(\theta - t)}.$$

The function $P_r(y) = \sum_{-\infty}^{\infty} r^{|k|} e^{iky}$ is called the *Poisson kernel*. Sum this series explicitly to prove that

$$P_r(y) = \frac{1 - r^2}{1 + r^2 - 2r\cos y}.$$

Show that this kernel has the same crucial properties that the Féjér kernel has (see p. 420), namely
(a) 2π − periodicity,
(b) $\dfrac{1}{2\pi}\displaystyle\int_{-\pi}^{\pi} P_r(t)\, dt = 1$,
(c) $P_r(t) \geqslant 0$,
(d) For each fixed $\delta > 0$, $\displaystyle\lim_{r \to 1-} \int_{\delta \leqslant |t| \leqslant \pi} P_r(t)\, dt = 0$.
Deduce that $u(r,\theta)$ discussed above converges to $f(\theta)$ *uniformly* as $r \to 1-$.

88. Let $f : \mathbb{R}^n \to \mathbb{R}^n$ be a diffeomorphism of $\mathbb{R}^n$ with positive Jacobian and with $f(0) = 0$. Prove that there is a curve f_t, $0 \leqslant t \leqslant 1$ joining f continuously to the identity where each f_t is a diffeomorphism. (One says that f is *isotopic* to the identity.) [Hint: Consider the map $g_t(x) = f(xt)/t$. Show that this joins f to $Df(0)$. Now show that a non-singular matrix with positive determinant can be joined to the identity through matrices of this class. You may consult outside texts for this last part.]

Exercises 89–96 are "examination style" based on Chapters 1–6, and 8.

89. (a) Define the least upper bound of a set S.
    (b) Find $\sup\{x \in \mathbb{R} \mid x^2 + x < 3\}$.
    (c) What is meant by saying that $\mathbb{R}$ is complete?
    (d) Let x_n be a convergent sequence in $\mathbb{R}$. Prove that x_n is a Cauchy sequence.
    (e) Define $x_0 = \alpha$, and inductively $x_n = (x_{n-1} + 1)/2$ where $0 < \alpha < 1$. Prove x_n converges to 1 as $n \to \infty$.

90. (a) Define the phrase "$A \subset \mathbb{R}^n$ is open."
    (b) Define the phrase "$A \subset \mathbb{R}^n$ is compact."
    (c) State the Heine-Borel theorem.
    (d) Find the closure of $\{(x,y) \in \mathbb{R}^2 \mid x^2 < y\}$. Prove your assertion.
    (e) Give an example of a set $B \subset \mathbb{R}^2$ such that (i) int $B = \varnothing$ but (ii) int(cl(B)) $\neq \varnothing$.
    (f) Let $B \subset \mathbb{R}^n$ be a set satisfying (i), (ii) of part (e). Prove that bd(B) = cl(B).

91. (a) Define the term "connected set."
    (b) Define the term "path-connected set."
    (c) State and prove a general version of the intermediate value theorem.
    (d) Prove that $\{(x,y) \in \mathbb{R}^2 \mid x = 0,\ y \geqslant 0\} \cup \{(x,y) \in \mathbb{R}^2 \mid x = y,\ x > 0\}$ is connected.
    (e) If A and B are connected sets in $\mathbb{R}^n$ and $A \cap B \neq \varnothing$, prove that $A \cup B$ is connected.

92. (a) Define what is meant by "$F: A \subset \mathbb{R}^n \to \mathbb{R}$ is continuous on A."
    (b) Give an equivalent reformulation of your definition in (a).
    (c) Explain the difference between continuity and uniform continuity; give illustrative examples.
    (d) Prove that the continuous image of a compact set is compact.
    (e) Let A be compact, $A \subset \mathbb{R}^n$ and $f: A \to \mathbb{R}$ continuous. Prove f achieves its maximum at some point of A.

93. (a) Define what it means for a sequence of functions $f_n: A \subset \mathbb{R}^n \to \mathbb{R}$ to converge uniformly.
    (b) Prove that $\displaystyle\sum_{k=1}^{\infty} \frac{(\sin kx)^2}{k^{3/2}}$ converges uniformly for $x \in \mathbb{R}$.
    (c) Is $f(x) = \displaystyle\sum_{k=1}^{\infty} \frac{(\sin kx)^2}{k^{3/2}}$ a continuous function of x? Justify your answer.
    (d) Let $f_k(x) = \dfrac{1}{kx} + 1$ for $k = 1, 2, 3, \ldots, x \in\ ]0,1[$. Prove $f_k \to 0$ pointwise.
    (e) Does f_k in part (d) converge uniformly?

94. (a) Let $f_n: [0,b] \to \mathbb{R}$ be continuous functions, differentiable on $]a,b[$, with $f_n'(x)$ continuous. Suppose f_n converges uniformly to f, f_n' converges uniformly to g. State a theorem concerning differentiability of f.
    (b) Prove your theorem in (a); clearly state any results used.
    (c) Let $f_k(x) = \sin kx/k^2$. Does your theorem work?
    (d) State a result which would guarantee that the following operation would be valid:
$$\sum_{k=1}^{\infty} \int_a^b g_k(x)\, dx = \int_a^b \sum_{k=1}^{\infty} g_k(x)\, dx .$$

(e) Define $e^x = \sum_{n=0}^{\infty} \dfrac{x^n}{n!}$. Use (d) to prove $\int_0^x e^y \, dy = e^x - 1$.

95. (a) Let $f: A \subset \mathbb{R}^n \to \mathbb{R}^n$ where A is open. Give a definition of the derivative of f.
    (b) For $f: A \subset \mathbb{R}^n \to \mathbb{R}$, define the gradient of f and discuss the geometrical meaning of $\langle \operatorname{grad} f(x), e \rangle$.
    (c) If S is a surface $f = $ constant, argue that $\operatorname{grad} f(x)$ is perpendicular to S if $x \in S$.
    (d) Find the equation of the plane tangent to the surface $x^2 + y^3 + z^4 = 3$ at $(1,1,1)$.
    (e) Argue that the two surfaces $x^2 + y^2 + z^2 = 3$ and $x^3 + y^3 + z^3 = 3$ are tangent at the point $(1,1,1)$.

96. Define the phrase "$f: [a,b] \to \mathbb{R}$ is Riemann integrable" by
    (a) defining upper and lower sums, and defining upper and lower integrals.
    (b) Is $f(x) = \sin x / (x^2 + 3x + 1)$ Riemann integrable on $[0,3]$?
    (c) State the fundamental theorem of calculus.
    (d) Let $f: [a,b] \to \mathbb{R}$ be Riemann integrable. Define $F(x) = \int_a^x f(t)\,dt$. If f is continuous at x_0, prove $F'(x_0) = f(x_0)$. Does F' exist if f is not continuous at x_0?

Exercises 97–101 are "examination style" based on Chapters 6–10.

97. (a) Let $f: \mathbb{R}^n \to \mathbb{R}^m$. Define what it means for f to be differentiable at $x \in \mathbb{R}^n$.
    (b) Is it true that existence of the partial derivatives implies that f is differentiable? Discuss.
    (c) Let $f: \mathbb{R}^2 \to \mathbb{R}^3$, $f(x,y) = (xy, e^y, \cos x)$. Compute $Df(1,0)$.
    (d) Write down a formula for $\partial h / \partial x$ if $h(x,y) = f(g(x,y), k(y), p(x))$. Justify this in terms of the chain rule.
    (e) Let $f: \mathbb{R} \to \mathbb{R}$ be differentiable. Assume f and f' have no common zeros. Prove that f has only finitely many zeros in $[0,1]$.

98. (a) What does the inverse function theorem state for functions $f: \mathbb{R} \to \mathbb{R}$?
    (b) Consider the equations
$$\begin{cases} x^3 + y^{40} \quad\quad = 2, \\ xz + y^2 + y = 3. \end{cases}$$

    Show that they are solvable for $y(x)$, $z(x)$ near $x = 1$, $y = 1$, $z = 1$. Compute dy/dx at $x = 1$.
    (c) Let $\varphi: [0,1] \to [0,1]$ be continuous. Prove that φ has a fixed point.
    (d) Let $F: \mathbb{R}^n \to \mathbb{R}^n$ be C^1 and have non-zero Jacobian at every point. Prove $F(\mathbb{R}^n)$ is open.
    (e) Let $f: \mathbb{R}^2 \to \mathbb{R}$ be continuous. Show f is not one-to-one. [Hint: If f was one-to-one then the images of the x and y axes would both be intervals in $\mathbb{R}$.]

99. (a) Define the term "$A \subset \mathbb{R}$ has measure zero."
    (b) Give an example of a set in $\mathbb{R}$ which has measure zero but does not have volume.

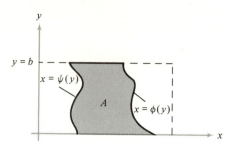

FIGURE A-2

(c) State general conditions under which a function $f: [0,1] \to \mathbb{R}$ which is bounded, is Riemann integrable.

(d) State one criterion for Riemann integrability and use it to prove that a continuous function $f: [0,1] \to \mathbb{R}$ is Riemann integrable. (You may use any relevant theorems about continuous functions, if they are clearly stated.)

(e) Briefly outline the steps that are used to prove that a continuous function $f: D \to \mathbb{R}$ is Riemann integrable, when $D = \{(x,y) \in \mathbb{R}^2 \mid x^2 + y^2 \leqslant 1\}$. (No proofs are required here, just a brief essay describing the relevant facts.)

100. (a) Evaluate $\int_A e^{-x^2-y^2} \, dx \, dy$ where $A = \{(x,y) \in \mathbb{R}^2 \mid x^2 + y^2 \leqslant 1\}$.

(b) Evaluate $\int_B x \, dx \, dy$ where B is the region in the plane bounded by $x = 0$, $y = 0$ and $x + y = 1$.

(c) State one version of Fubini's theorem.

(d) Use (c) to write a formula for $\int_A f(x,y) \, dx \, dy$ where A is as shown in Figure A-2.

(e) Let $\varphi: \mathbb{R}^2 \to \mathbb{R}^2$ be C^1 and bijective with $J\varphi \neq 0$. Assume

$$\int_A dx \, dy = \int_{\varphi(A)} dx \, dy$$

for all open discs A. Prove $J\varphi = 1$.

101. (a) Let V be an inner product space and $\varphi_0, \varphi_1, \varphi_2, \dots$ on orthonormal set in V. Write the Fourier series of $f \in V$ relative to φ_i. What if the φ_i are complete?

(b) Explain how (a) is related to the formula

$$f = \frac{a_0}{2} + \sum_{n=1}^{\infty} a_n \cos nx + \sum_{n=1}^{\infty} b_n \sin nx$$

where

$$a_k = \frac{1}{\pi} \int_{-\pi}^{\pi} f(x)\cos kx \, dx \qquad k = 0, 1, 2, \dots$$

$$b_k = \frac{1}{\pi} \int_{-\pi}^{\pi} f(x)\sin kx \, dx \qquad k = 1, 2, \dots$$

(c) Compute the Fourier series of $f(x) = x$, $-\pi < x < \pi$.

(d) What is the pointwise limit of the series in (c)? Does the series converge in the mean? Discuss.

(e) Assuming the completeness theorems in the text, prove that $\{\sin nx \mid n = 1,2,\ldots\}$ is complete in $\mathscr{L}^2([0,\pi],\mathbb{C})$ (square integrable functions on $[0,\pi]$).

Exercises 102–109 are "examination style" based on Chapters 1–7.

102. Let $B_0(1)$ be the open unit ball in $\mathbb{R}^n$ centered around 0, and let $f: B_0(1) \to \mathbb{R}$ be continuous. Assume there exists a continuous map $g: B_0(1) \to L(\mathbb{R}^n,\mathbb{R})$ (linear maps from $\mathbb{R}^n$ to $\mathbb{R}$), such that for each pair of points $x, y \in B_0(1)$,

$$f(y) - f(x) = \int_0^1 \{(g(ty + (1-t)x))(y-x)\} \, dt \ .$$

Show that f is C^1, and that $Df = g$.

103. Let $D^2 = \{x \in \mathbb{R}^2 \mid \|x\| \leqslant 1\}$ (that is, D^2 is the closed unit ball, centered at 0, in $\mathbb{R}^2$). For each integer $n \in \mathbb{N}$, let $f_n: D^2 \to \mathbb{R}$ be continuous, and assume that $f: D^2 \to \mathbb{R}$ is a continuous function such that the sequence $\{f_n\}_{n\in\mathbb{N}}$ converges uniformly to f. Is the set of functions $\{f_n\}_{n\in\mathbb{N}}$ equicontinuous and/or bounded? Justify your answer.

104. Let $f, g: \mathbb{R} \to \mathbb{R}$ be C^2 functions, and let $h: \mathbb{R}^2 \to \mathbb{R}$ be a C^2 function. Define $\alpha: \mathbb{R}^2 \to \mathbb{R}$ by $\alpha(x,y) = h(x,f(x) - g(y))$. Compute the following partial derivatives of α (in terms of the partial derivatives of h, f, and g):

(a) $\dfrac{\partial \alpha}{\partial x}$ and $\dfrac{\partial \alpha}{\partial y}$,

(b) $\dfrac{\partial^2 \alpha}{\partial x^2}$,

(c) $\dfrac{\partial^2 \alpha}{\partial x \, \partial y}$,

(d) $\dfrac{\partial^2 \alpha}{\partial y \, \partial x}$,

(e) $\dfrac{\partial^2 \alpha}{\partial y^2}$.

105. Let $M(n,\mathbb{R})$ be the vector space of $n \times n$ matrices with real-valued entries. Define a map $\varphi: M(n,\mathbb{R}) \to M(n,\mathbb{R})$ by $\varphi(A) = A^2$ for each $A \in M(n,\mathbb{R})$. Show that φ is a C^∞ map. For each $A, B \in M(n,\mathbb{R})$, calculate $(D\varphi(A))(B)$ (that is, calculate the derivative of φ at the "point" A and in the direction determined by B).

106. For each pair of functions $f, g: I \to \mathbb{R}$, define $f \vee g: I \to \mathbb{R}$ by $(f \vee g)(x) = \max\{f(x),g(x)\}$.
(a) If f and g are continuous, show that $f \vee g$ is continuous.
(b) Define a map $\psi: \mathscr{C}(I) \times \mathscr{C}(I) \to \mathscr{C}(I)$ by $\psi(f,g) = f \vee g$ for each pair $f, g \in \mathscr{C}(I)$. Show that ψ is continuous. ($\mathscr{C}(I)$ denotes the space of all continuous real valued functions on $I = [0,1]$).

107. Define $f: \mathbb{R}^2 \to \mathbb{R}$ by $f(x,y) = e^{(x^2)} \cos(xy) - 1$. Does there exist a sufficiently small positive number ε such that, for $|x|, |y| < \varepsilon$, the equation $f(x,y) = 0$ can be solved for y uniquely and differentiably in terms of x? Or, rephrased: does there exist $\varepsilon > 0$ and $g: \,]-\varepsilon,\varepsilon[\, \to \mathbb{R}$ such that
(a) g is C^1,
(b) $g(0) = 0$,
(c) $f(x,g(x)) = 0$,
(d) for each $x \in \,]-\varepsilon,\varepsilon[$, the point $(x,g(x))$ is the only point in $\mathbb{R}^2$ whose first

coordinate is x, such that the absolute value of both coordinates is less than ε, which solves the equation $f(x,y) = 0$. Justify your answer.

108. Define a map $\alpha: \mathscr{C}(I) \to \mathscr{C}(I)$ as follows: for each $f \in \mathscr{C}(I)$, define $\alpha(f) \in \mathscr{C}(I)$ by $(\alpha(f))(x) = \int_0^x f(t)\, dt$.
    (a) Show that $\alpha: \mathscr{C}(I) \to \mathscr{C}(I)$ is a continuous linear map.
    (b) Is α a compact linear map? Justify your answer. (A linear map is called *compact* if the closure of the image of the unit ball is compact).

109. Define $f: \mathbb{R}^2 \to \mathbb{R}^2$ by $f(x,y) = (e^x \sin(y), e^y \cos(x))$. Does there exist an open neighborhood U of $0 \in \mathbb{R}^2$ such that $f(U)$ is open in $\mathbb{R}^2$, $f \mid U$ (f restricted to U) is injective, and such that $(f \mid U)^{-1}$ is C^∞? Justify your answer.

Appendix C

Suggestions for Further Study

The number of books on advanced calculus and introductory analysis is overwhelming. Despite the large number of recent texts, some of the older books remain the best. Some favorites are:

[1] Carslaw, H. S., 1930. *Theory of Fourier's Series and Integrals*. 3rd. ed. New York: Dover.

[2] Hardy, G. H., 1947. *Pure Mathematics*. 9th ed. New York: Cambridge Univ. Press.

[3] Hobson, E. W., 1921. *The Theory of Functions of a Real Variable and the Theory of Fourier's Series*, Cambridge, Eng: Cambridge Univ. Press.

[4] Titchmarsh, E. C., 1937. *Theory of Fourier Integrals*. London: Oxford Univ. Press.

[5] Whittaker, E. T. and Watson, G. N., 1926. *A Course of Modern Analysis*. Cambridge, Eng: Cambridge Univ. Press.

Of the more recent texts on roughly the same level as this one, the following have been popular. Of these, [6, 7, 8, 9, 12, 13, 15, 16] are fairly classical, while [10, 11, 14] tend to be a bit more abstract.

[6] Apostol, T. M., 1957. *Mathematical Analysis*. Reading, Mass: Addison-Wesley.

[7] Bartle, R. G., 1964. *The Elements of Real Analysis*. New York: Wiley.

[8] Buck, R. C., 1965. *Advanced Calculus*. 2nd ed. New York: McGraw-Hill.

[9] Graves, L. M., 1956. *Theory of Functions of Real Variables*. 2nd ed. New York: McGraw-Hill.

[10] Lang, S., 1968. *Analysis I*. Reading, Mass: Addison-Wesley.

[11] Loomis, L. H. and Sternberg, S., 1968. *Advanced Calculus*. Reading, Mass: Addison-Wesley.

473

[12] Olmsted, J. M. A., 1961. *Advanced Calculus.* New York: Appleton-Century-Crofts.

[13] ——, 1956. *Real Variables.* New York: Appleton-Century-Crofts.

[14] Rosenlicht, M., 1968. *Introduction to Analysis.* Glenview, Ill: Scott, Foresman and Co.

[15] Rudin, W., 1964. *Principles of Mathematical Analysis.* 2nd. ed. New York: McGraw-Hill.

[16] Widder, D. V., 1965. *Advanced Calculus.* 2nd. ed. Englewood, Cliffs, New Jersey: Prentice-Hall.

For more information on the foundations of set theory, consult [17].

[17] Dieudonné, Jean, 1966. *Foundations of Modern Analysis.* New Jersey: Prentice-Hall.

In [17] there is not much material on logic and the axioms of set theory, but you will find concisely all the facts on set theory which are of practical importance in the course. In addition, [17] develops thoroughly the abstract differential calculus (see our Chapters 6, 7) in the context of Banach spaces.

The axioms in Appendix A of this text are taken verbatim from [18].

[18] Halmos, Paul R, 1960. *Naive Set Theory.* New York: D. Van Nostrand Co.

The following are some general references for more advanced work in real analysis including Lebesgue integration and abstract analysis in general Banach and Hilbert spaces.

[19] Burkhill, J. C., 1951. *The Lebesgue Integral.* Cambridge, Eng: Cambridge Univ. Press.

[20] Halmos, P. R., 1950. *Measure Theory.* New York: D. Van Nostrand.

[21] Hewitt, E. and Stromberg, K., 1969. *Real and Abstract Analysis.* New York: Springer Verlag.

[22] Gleason, A. M., 1966. *Fundamentals of Abstract Analysis.* Reading, Mass: Addison-Wesley.

[23] Lang, S., 1969. *Analysis II.* Reading, Mass: Addison-Wesley.

[24] Royden, H. L., 1963. *Real Analysis.* New York: Macmillan.

[25] Rudin, W., 1966. *Real and Complex Analysis.* New York: McGraw-Hill.

[26] ——, 1973. *Functional Analysis.* New York: McGraw-Hill.

[27] Simmons, G., 1963. *Introduction to Topology and Modern Analysis.* New York: McGraw-Hill.

A handy book to use for finding counterexamples to theorems with missing hypotheses is [28].

[28] Gelbaum, B. R. and Olmsted, J. M. H., 1964. *Counterexamples in Analysis.* San Francisco: Holden Day.

Our text studied quite a bit about series. The classical references are [29, 30].

[29] Hardy, G. H., 1949. *Divergent Series*. London: Oxford Univ. Press.

[30] Knopp, K., 1951. *Theory and Application of Infinite Series*. New York: Hafner.

Those wishing to pursue distribution theory can consult the following, in addition to [26].

[31] Gelfand, I. M. and Shilov, G. E., 1964. *Generalized Functions*. New York: Academic Press.

[32] Schwartz, L., 1966. *Théorie des distributions*. Paris: Hermann.

[33] Zemanian, A., 1965. *Distribution Theory and Transform Analysis*. New York: McGraw-Hill.

The following texts develop the theory of ordinary differential equations and integral equations. Of these [34] and [35] are comprehensive treatises.

[34] Coddington, E. A. and Levinson, N., 1955. *Theory of Ordinary Differential Equations*. New York: McGraw-Hill.

[35] Hartman, P., 1964. *Ordinary Differential Equations*. New York: Wiley.

[36] Hurewicz, W., 1958. *Lectures on Ordinary Differential Equations*. Cambridge, Mass: M.I.T. Press.

[37] Roxin, E. O., 1972. *Ordinary Differential Equations*. Belmont, Cal: Wadsworth.

[38] Widom, H., 1969. *Lectures on Integral Equations*. New York: Van Nostrand Mathematical Studies #17.

Advanced calculus can be elegantly applied to study problems in geometry and vector analysis. Besides [10, 11, 23], consult

[39] Flemming, W., 1965. *Functions of Several Variables*. Reading, Mass: Addison-Wesley.

[40] Spivak, M., 1965. *Calculus on Manifolds*. New York: Benjamin.

We have already cited several texts which deal with Fourier series [1, 3, 4, 21, 23, 25, 32, 33]. Others, somewhat more advanced, are:

[41] Stein, M. and Weiss, G., 1971. *Introduction to Fourier Analysis on Euclidean Spaces*. Princeton, New Jersey: Princeton Univ. Press.

[42] Widom, H., 1969. *Lectures on Measures and Integration*. New York: Van Nostrand Mathematical Studies #20.

[43] Zygmund, Z., 1959. *Trigonometric Series*. 2nd. ed. Cambridge, Eng: Cambridge Univ. Press.

Our chapter on Fourier series gave an introduction to partial differential equations. Further information can be found in the following texts. The last two texts use distribution theory with [47] being advanced.

[44] Churchill, R. V., 1963. *Fourier Series and Boundary Value Problems*. 2nd ed. New York: McGraw-Hill.

[45] Courant, R. and Hilbert, D., 1962. *Methods of Mathematical Physics*. (2 volumes), New York: Wiley-Interscience.

[46] Duff, G. F. D. and Naylor, D., 1966. *Differential Equations of Applied Mathematics*. New York: Wiley.

[47] Sobolev, S. L., 1963. *Applications of Functional Analysis in Mathematical Physics*. Providence, Rhode Island: American Mathematical Society Translations, Vol. 7.

A few references on quantum mechanics follow. [50] is a standard elementary text while [48, 49] are more advanced and more mathematically oriented.

[48] Jauch, J. M., 1968. *Foundations of Quantum Mechanics*. Reading, Mass: Addison-Wesley.

[49] Mackey, G. W., 1963. *The Mathematical Foundations of Quantum Mechanics*. New York: Benjamin.

[50] Merzbacher, E., 1970. *Quantum Mechanics*. 2nd. ed. New York: Wiley.

There are a number of important topics in classical analysis which we did not cover. For example, we could have studied the gamma function following [16] or [51].

[51] Artin, E., 1964. *The Gamma Function*. New York: Holt, Rinehart and Winston.

(This topic is often covered in courses in complex variables as well).

There are a large number of excellent texts which are not in English. For example:

[52] Bourbaki, N., 1961. *Élements de Mathématique; Fonctions d'une variable réelle*. Paris: Hermann.

[53] Dieudonné, J., 1971. *Calcul Infinitésimal*. Paris: Hermann.

A rigorous treatment of elementary analysis did not evolve rapidly or smoothly. The creators of this area of mathematics traveled over cobblestones and encountered numerous blind alleys before experiencing their brilliant insights. An appreciation of this history is important to the student's education in mathematics. A recommended text is

[54] Kline, M., 1972. *Mathematical Thought from Ancient to Modern Times*. New York: Oxford Univ. Press.

Appendix D

Answers to
Selected Exercises

Introduction
Prerequisites: Sets and Functions

1. (a) $f(A_0) = \{1\}, f^{-1}(B_0) = A$.
   (b) $f(A_0) = A_0, f^{-1}(B_0) = B_0$.
   (c) $f(A_0) = \{1, 0, -1\}, f^{-1}(B_0) = \{x \mid x \leqslant 0\}$.
2. (a) and (c) are neither one-to-one nor onto, (b) one-to-one and onto.
3. (a) $x \in f^{-1}(C_1 \cup C_2) \Leftrightarrow f(x) \in C_1 \cup C_2$,
   $\Leftrightarrow f(x) \in C_1 \quad$ or $f(x) \in C_2$,
   $\Leftrightarrow x \in f^{-1}(C_1) \quad$ or $x \in f^{-1}(C_2)$,
   $\Leftrightarrow x \in f^{-1}(C_1) \cup f^{-1}(C_2)$,
   hence $f^{-1}(C_1 \cup C_2) = f^{-1}(C_1) \cup f^{-1}(C_2)$.
   (d) $y \in f(D_1 \cap D_2)$ implies that there exists $x \in D_1 \cap D_2$ such that $y = f(x)$. Since $x \in D_1$ and $x \in D_2$, then $y \in f(D_1)$ and $y \in f(D_2)$, hence $y \in f(D_1) \cap f(D_2)$.
4. (a) To verify Exercises 3(a) and (d) for the function in Exercise 1(c),

$$f^{-1}(C_1 \cup C_2) = \{1, 0, -1\} = \{1\} \cup \{0, -1\} = f^{-1}(C_1) \cup f^{-1}(C_2)$$

verifying 3(a), and

$$f(D_1 \cap D_2) = f(\{1\}) = \{1\} = \{1, -1\} \cap \{1\} = f(D_1) \cap f(D_2)$$

verifying

$$f(D_1 \cap D_2) \subset f(D_1) \cap f(D_2).$$

6. Define $f:]0,1[\to \mathbb{R}$ by

$$f(x) = \begin{cases} (x - \tfrac{1}{2})/x, & \text{if } 0 < x \leqslant \tfrac{1}{2}, \\ (x - \tfrac{1}{2})/(1 - x), & \text{if } \tfrac{1}{2} < x < 1. \end{cases}$$

Verify that this is a bijection.

8. Define φ: $\{\ldots, -2, -1, 0, 1, 2, 3, \ldots\} \to \{1, 2, 3, \ldots\}$ by

$$\varphi(n) = \begin{cases} 2n, & \text{if } n > 0, \\ 1, & \text{if } n = 0, \\ -2n + 1, & \text{if } n < 0. \end{cases}$$

Verify that this is a bijection.

9. Let $A_i = \{a_{i1}, a_{i2}, \ldots\}$, and define $f \colon \bigcup A_i \to \mathbb{N}$, $a_{ij} \mapsto i + (k - 1)(k - 2)/2$ where $k = i + j$. Then f maps $\bigcup A_i$ one-to-one and onto $\mathbb{N}$.

10. To show $\bigcup \mathscr{A} \subset \bigcup \mathscr{B}$, note that $x \in \bigcup \mathscr{A}$ implies there exists $A \in \mathscr{A} \subset \mathscr{B}$ with $x \in A$, hence $x \in \bigcup \mathscr{B}$.

11. $f \circ (g \circ h)(x) = f(g \circ h(x)) = f(g(h(x))) = (f \circ g)(h(x)) = (f \circ g) \circ h(x)$.

12. (i) Assume $f \colon A \to B$ is a bijection. Define $g \colon B \to A$ as follows: for $y \in B$, let $g(y) = x$ where $f(x) = y$ (x exists by onto-ness and x is unique by one-to-oneness).
(ii) Assume there exists $g \colon B \to A$ such that $f \circ g = $ identity and $g \circ f = $ identity. To show f is onto, let $y \in B$ and let $x = g(y)$. Then $f(x) = y$. To show f is one-to-one, if $f(x_1) = f(x_2)$ then $x_1 = g(f(x_1)) = g(f(x_2)) = x_2$. (Verify that $g = f^{-1}$ and is unique.)

13. $(f^{-1} \circ g^{-1}) \circ (g \circ f) = f^{-1} \circ (g^{-1} \circ g) \circ f = f^{-1} \circ f = $ identity, and similarly $(g \circ f) \circ (f^{-1} \circ g^{-1}) = $ identity. Thus by Exercise 12, $f^{-1} \circ g^{-1} = (g \circ f)^{-1}$ and $g \circ f$ is a bijection.

Chapter 1
The Real Line and Euclidean n-Space

1.1   The Real Line $\mathbb{R}^n$

1. $\text{Sup}(S) = 1$; S is not bounded below.

3. $\dfrac{3^n}{n!} = \dfrac{3 \cdot 3 \cdots 3}{1 \cdot 2 \cdots n} \leqslant \dfrac{9}{2} \cdot \dfrac{3}{n} = \dfrac{27}{2n}$ so pick $N > \dfrac{27}{2\varepsilon}$.

5. $x_n = (\sqrt{n^2 + 1} - n)\left(\dfrac{\sqrt{n^2 + 1} + n}{\sqrt{n^2 + 1} + n}\right) = \dfrac{1}{\sqrt{n^2 + 1} + n} \to 0$ as $n \to \infty$.

6. No; let $x_n = 1 + \frac{1}{2} + \frac{1}{3} + \cdots + \dfrac{1}{n - 1}$.

7. $\text{Sup}(Q)$ is an upper bound of P, hence $\text{sup}(Q) \geqslant \text{sup}(P)$.

1.2   Euclidean n-Space $\mathbb{R}^n$

2. $\cos^{-1}\left(\dfrac{2}{\sqrt{17}}\right)$.

3. We have $\left.\begin{array}{r} 3x + 2y + 2z = 0 \\ y = 0 \end{array}\right\}$ and so $\{(-2, 0, 3)\}$ spans the solution space of this system of equations.

Exercises for Chapter 1 (at end of chapter)

1. (a) $\sup(S) = \sqrt{5}$; $\inf(S) = -\sqrt{5}$.
   (b) Neither $\sup(S)$ nor $\inf(S)$ exist.
   (c) $\sup(S) = 1$; $\inf(S) = 0$.
   (d) $\sup(S) = 0$; $\inf(S) = -1$.
   (e) $\sup(S) = 1/3$; $\inf(S) = .3$.
   (f) $\sup(S) = b$; $\inf(S) = a$ for each case.

2. Let k be square free, that is, such that for no prime p does $p^2 \mid k$, (p^2 divides k); suppose $\sqrt{k} = a/b$ for some integers a, b, and that a and b have no common factors. Then $k = a^2/b^2$ implies $b^2 k = a^2$ implying $k \mid a^2$ (k divides a^2). But k square free implies $k^2 \mid a^2$. (This is a consequence of the fact that any integer has a unique prime factorization.) Then $b^2 k = a'^2 k^2$ implies $b^2 = a'^2 k$ implying $k \mid b$, contradicting the assumption that a and b have no common factors.

3. (a) Suppose $x > 0$. Let $\varepsilon = x/2$. Then $x < x/2$ implies $0 < x/2 < 0$, a contradiction. Hence $x = 0$.
   (b) Let $x = \min\{\varepsilon/2, 1/2\}$.

5. By the completeness axiom, $\sup(S) \in \mathbb{R}$ exists. By Theorem 2, there exists a point $x_{n_0} \in S$ such that $\sup(S) - x_{n_0} < \varepsilon$. x_n increasing implies $\sup(S) \geqslant x_n \geqslant x_{n_0}$ for all $n > n_0$, hence for all $n > n_0$, $0 < \sup(S) - x_n < \varepsilon$. Thus $\lim_{n \to \infty} x_n = \sup(S)$.

7. Let $a = \sup(A)$, $b = \sup(B)$, and $z = x + y \in A + B$. Then $z = x + y \leqslant a + y \leqslant a + b$, hence $a + b$ is an upper bound for $A + B$. If $\varepsilon > 0$, there exists a $x \in A$, $y \in B$ such that $a < x + \varepsilon/2$, and $b < y + \varepsilon/2$ implying $(a + b) < (x + y) + \varepsilon/2 + \varepsilon/2 = (x + y) + \varepsilon$. Thus by Theorem 2, $a + b = \sup(A + B)$.

12. (a) $\|x + y\|^2 = \langle x + y, x + y \rangle = \langle x,x \rangle + \langle x,y \rangle + \langle y,x \rangle + \langle y,y \rangle$
    $$= \|x\|^2 + 2\langle x,y \rangle + \|y\|^2,$$

    and similarly

    $$\|x - y\|^2 = \|x\|^2 - 2\langle x,y \rangle + \|y\|^2.$$

    Adding gives the result.

    This proves that the sum of the squares of the diagonals of a parallelogram is twice the sum of the squares of the sides.

    (b) $\|x + y\|^2 \|x - y\|^2 = [\|x\|^2 + 2\langle x,y \rangle + \|y\|^2] \cdot [\|x\|^2$
    $$- 2\langle x,y \rangle + \|y\|^2]$$
    $$= (\|x\|^2 + \|y\|^2)^2 - 4\langle x,y \rangle^2 \leqslant (\|x\|^2 + \|y\|^2)^2.$$

    (c) Similar to (a).

14. (a) Use induction on n. The Schwarz inequality follows because

    $$\sum_{1 \leqslant i < j \leqslant n} (x_i y_j - x_j y_i)^2 \geqslant 0$$

    and thus

    $$\left(\sum_{i=1}^{n} x_i y_i \right)^2 = \left(\sum_{i=1}^{n} x_i^2 \right)\left(\sum_{i=1}^{n} y_i^2 \right) - \sum_{1 \leqslant i < j \leqslant n} (x_i y_j - x_j y_i)^2 \leqslant \left(\sum_{i=1}^{n} x_i^2 \right)\left(\sum_{i=1}^{n} y_i^2 \right).$$

(b) $(x + y)^2 = x(x + y) + y(x + y)$ and by (a), $\sum x_j(x_j + y_j) + \sum y_j(x_j + y_j) \leqslant$ $(\sum x_j^2)^{1/2}(\sum (x_j + y_j)^2)^{1/2} + (\sum y_j^2)^{1/2}(\sum (x_j + y_j)^2)^{1/2}$. Combining terms and dividing by $(\sum (x_j + y_j)^2)^{1/2}$ gives the result.

15. Let $d(x_1,x_2) = r$. Then by induction $d(x_n,x_{n+1}) \leqslant r/2^{n-1}$, and so by the triangle inequality

$$d(x_n,x_{n+k}) \leqslant d(x_n,x_{n+1}) + d(x_{n+1},x_{n+2}) + \cdots + d(x_{n+k-1},x_{n+k})$$

$$\leqslant \frac{r}{2^{n-1}} + \frac{r}{2^n} + \cdots + \frac{r}{2^{n+k-2}}$$

$$= r\sum_{i=0}^{k-1} \frac{1}{2^{n+i-1}} = \frac{r}{2^{n-1}} \sum_{i=0}^{k-1} \frac{1}{2^i}$$

$$\leqslant \frac{r}{2^{n-1}} \sum_{i=0}^{\infty} \frac{1}{2^i} = \frac{r}{2^{n-1}} \cdot 2 = \frac{r}{2^{n-2}} \, .$$

Thus, if we pick N large enough so that $r/2^{N-2} < \varepsilon$, then $n > N$ implies $d(x_n,x_{n+k}) < \varepsilon$. Therefore, x_n is Cauchy.

17. Let $L = \{x \in \mathbb{R} \mid x \text{ is a lower bound for } S\}$. Then $\inf(S) \geqslant y$ for all $y \in L$, hence $\inf(S) \geqslant \sup(L)$. Also $\inf(S) \in L$ implies $\sup(L) \geqslant \inf(S)$, giving the equality.

18. (a) $|x_n - x| = |x - x_n|$ hence $x_n \to x$ iff for all $\varepsilon > 0$ there exists a N such that $n \geqslant N$ implies $|x_n - x| = |x - x_n| < \varepsilon$ iff $-x_n \to -x$.

(b) Assume every increasing sequence which is bounded above converges (that is, assume the completeness axiom). Let x_n be a decreasing sequence which is bounded below; we must show that x_n converges. $\{-x_n\}$ is an increasing sequence which is bounded above, so $-x_n$ converges, say to a. Thus by (a), x_n converges to $-a$. The other direction proceeds the same way.

(c) Use Exercise 5(a) and the fact that $\sup\{-x_1,-x_2,\ldots\} = -\inf\{x_1,x_2,\ldots\}$.

19. $y_1 = y_2 = y_3 = \frac{1}{2}$.

22. (a) Given $\varepsilon > 0$, let N be such that $n \geqslant N$ implies $|x_n - x| < \varepsilon/|a|$. Then $n \geqslant N$ implies $|ax_n - ax| = |a| \cdot |x_n - x| < |a| \cdot \varepsilon/|a| = \varepsilon$, so $ax_n \to ax$.

23. $x \geqslant 0$ for all $x \in P$ so 0 is a lower bound for P; also, given $\varepsilon > 0$ there is $x \in P$ with $x < 0 + \varepsilon$, namely an $x_k \in P$ such that $kx_k \leqslant 1$ where $k > 1/\varepsilon$. Thus by Exercise 4, $0 = \inf(P)$.

24. No; let $P =]0,1[$ and $Q = [0,1]$, then $\sup(P) = \sup(Q) = 1$ and $\inf(P) = \inf(Q) = 0$ but $P \neq Q$.

27. Pick each b_n such that $b_n = |b_n| < \varepsilon/2^n$ (this is possible because $a_n \to 0$). Then

$$\sum_{n=1}^{\infty} b_n < \sum_{n=1}^{\infty} \frac{\varepsilon}{2^n} = \varepsilon \cdot \sum_{n=1}^{\infty} \frac{1}{2^n} = \varepsilon \cdot 1 = \varepsilon \, .$$

32. $x_{n+1} = \dfrac{x_1 + \cdots + x_{n-1}}{2} + \dfrac{x_n}{2} = x_n + \dfrac{x_n}{2} = \dfrac{3}{2}x_n$, so for any n, $x_n = (3/2)^{n-1}$ (prove

this by induction). Let $M > 0$. Now $(3/2)^n = (1 + 1/2)^n \geqslant 1 + n/2$, and by the Archimedean principle there exists a N such that $N > 2M - 1$, so $x_N = (3/2)^{N-1} \geqslant 1 + (N - 1)/2 > M$ proving that $x_n \to \infty$.

33. (a) By l'Hopital's rule, $\lim\limits_{x\to\infty} \dfrac{\log x}{x} = \lim\limits_{x\to\infty} \dfrac{1/x}{1} = \lim\limits_{x\to\infty} \dfrac{1}{x} = 0.$

(b) Use (a) and continuity of e^x to show that $x^{1/x}\,(= e^{(1/x)\log x}) \to 1$ for all real x.

Chapter 2
Topology of $\mathbb{R}^n$

2.1 Open Sets

1. Let $x \in \mathbb{R}^2\backslash\{(0,0)\}$. Since $x \neq (0,0)$, $d(x,(0,0)) = r > 0$; then $D(x,r) \subset \mathbb{R}^2\backslash\{(0,0)\}$, for $(0,0) \in D(x,r)$ implies $d(x,(0,0)) < r = d(x,(0,0))$, which is impossible. Hence $\mathbb{R}^2\backslash\{(0,0)\}$ is open.
3. Let $(x_0,y_0) \in B$. Then $x_0 \in A$. Hence there exists a $\delta > 0$ such that $]x_0 - \delta, x_0 + \delta[\subset A$. Claim, $D((x_0,y_0),\delta) \subset B$. For $(x,y) \in D((x_0,y_0),\delta)$ implies $d(x,x_0) \leqslant d((x,y), (x_0,y_0)) < \delta$, hence $x \in A$.
4. Let $A = \bigcup\limits_{y\in B} D(y,1)$. Then $x \in A \Leftrightarrow$ there exists a $y \in B$ such that $x \in D(y,1)$ (that is, $d(x,y) < 1$) for some $y \in B \Leftrightarrow y \in C$. C is open, being the union of open sets.
5. No; let A be any open subset of $\mathbb{R}$ and $B = \{0\}$. Then $A \cdot B = \{0\}$ which is not open. *Note:* If B is also open then $A \cdot B$ is open.

2.2 Interior of a Set

1. $\text{int}(S) = \{(x,y) \in \mathbb{R}^2 \mid xy > 1\}$.
3. Yes, $x \in \text{int}(A)$ implies there exists an open set U with $a \in U \subset A \subset B$, hence $x \in \text{int}(B)$.
4. Yes. If $x \in \text{int}(A) \cap \text{int}(B)$, then there exist open sets U, V with $x \in U \subset A$ and $x \in V \subset B$. Now $x \in U \cap V \subset A \cap B$ and $U \cap V$ is open, so $x \in \text{int}(A \cap B)$. If $x \in \text{int}(A \cap B)$, then there exists an open set U with $x \in U \subset A \cap B \subset A$ and B; so $x \in \text{int}(A) \cap \text{int}(B)$.

2.3 Closed Sets

1. Yes.
2. No; $(0,1) \in \mathbb{R}^2\backslash S$ and any neighborhood about $(0,1)$ will contain points of S.
5. No. If $x \in \mathbb{R}\backslash S = \{x \in \mathbb{R} \mid x \text{ is rational}\}$ there is no neighborhood of x not containing irrational points, hence $\mathbb{R}\backslash S$ is not open, and S is not closed.

2.4 Accumulation Points

1. $\{(x,y) \in \mathbb{R}^2 \mid y = 0 \text{ and } 0 \leqslant x \leqslant 1\}$.
2. Yes; since any open set N containing x contains points of A other than x, which are also points of B.
3. (a) No accumulation points (a ball of radius $1/2$ around any (m,n) contains only (m,n)).

(b) All of $\mathbb{R}^2$ (for any point in $\mathbb{R}^2$ there is a point arbitrarily close with rational coordinates).

(c) $\{(x,0) \in \mathbb{R}^2 \mid x \in \mathbb{R}\}$ = the x-axis.

(d) $\{(1/n,0) \mid n$ an integer, $n \neq 0\}$ (see (c) and (d) by graphing the sets.)

4. No (but yes if $x \notin A$ by Theorem 2, Chapter 1); for instance if $A = \{1\}$ then $\sup(A) = 1$ but 1 is not an accumulation point of A (A has no accumulation points).

2.5   Closure of a Set

1. $\mathrm{cl}(S) = \{(x,y) \in \mathbb{R}^2 \mid x \geqslant y^2\}$.

2. $\{0\} \cup \{1/n \mid n = 1,2,3,\ldots\}$.

3. $\mathbb{R}^2$.

4. (a) $\mathrm{cl}(A)\backslash A = (A \cup \{\text{accumulation points of } A\})\backslash A = (A\backslash A) \cup \{\text{accumulation points of } A\}\backslash A = \{\text{accumulation points of } A\}\backslash A \subset \{\text{accumulation points of } A\}$.

   (b) Not necessarily, let $A =]0,1[$. Then every point of A is an accumulation point of A so $\mathrm{cl}(A)\backslash A = \{0,1\}$ misses all the accumulation points which are points of A.

5. If $x \in A$ then $x \in \mathrm{cl}(A)$, If $x \notin A$ use Theorem 2, Chapter 1, to show x is an accumulation point of A.

2.6   Boundary of a Set

1. $\mathrm{bd}(A) = \{0\} \cup A$.

2. (a) Suppose $\mathrm{cl}(A)\backslash A \neq \varnothing$, otherwise the statement is vacuously true. Let $x \in \mathrm{cl}(A)\backslash A$, and N be a neighborhood of x. $x \in \mathbb{R}\backslash A$ implies $N \cap \mathbb{R}\backslash A \neq \varnothing$, and x an accumulation point of A implies that there exists a $y \in A$ such that $y \in N$. Hence $N \cap A \neq \varnothing$, and by Theorem 6, $x \in \mathrm{bd}(A)$.

   (b) The converse is not true; let $A =$ the rationals in $[0,1]$. Then $\mathrm{bd}(A) = [0,1]$ so $1/2 \in \mathrm{bd}(A)$, but $1/2 \in A$ so $1/2 \notin \mathrm{cl}(A)\backslash A$.

3. $\mathrm{bd}(A) = \{(x,y) \in \mathbb{R}^2 \mid x = y\}$.

4. No, for if $A = \{x \mid x \in [0,1]$ and x is rational$\}$ then $\mathrm{int}\,A = \varnothing$, $\mathrm{bd}(\mathrm{int}\,A) = \varnothing$ but, $\mathrm{bd}(A) = [0,1]$.

5. Yes.

2.7   Sequences

1. $(0,0)$.

2. It contains limits of all its sequences (since a subsequence of a convergent sequence converges to the same limit as the whole sequence) so use Theorem 9(i).

3. Use Theorem 9(ii).

5. $\mathrm{cl}(S) = \{x \in \mathbb{R} \mid x^2 \leqslant 2\} = [-\sqrt{2},\sqrt{2}]$.

2.8   Series in $\mathbb{R}$ and $\mathbb{R}^n$

1. For all k,

$$\sum_{n=1}^{k} x_n = \left(\sum_{n=1}^{k} \frac{(\sin n)^n}{n^2}, \sum_{n=1}^{k} \frac{1}{n^2} \right).$$

Hence $\sum\limits_{n=1}^{\infty} x_n$ converges iff $\sum\limits_{n=1}^{\infty} \dfrac{(\sin n)^n}{n^2}$ and $\sum\limits_{n=1}^{\infty} \dfrac{1}{n^2}$ converge. By Theorem 13(iii),

$\sum\limits_{n=1}^{\infty} \dfrac{1}{n^2}$ converges, and since $\left| \dfrac{(\sin n)^n}{n^2} \right| = \dfrac{|\sin n|^n}{n^2} \leqslant \dfrac{1}{n^2}$, by the comparison test

$\sum\limits_{n=1}^{\infty} (\sin n)^n / n^2$ converges. Thus $\sum\limits_{n=1}^{\infty} x_n$ converges.

4. If $n \geqslant 4$, then $\dfrac{2^n + n}{3^n - n} \leqslant \left(\dfrac{3}{4}\right)^n$, so by Theorems 13(i) and 13(ii) $\sum\limits_{n=0}^{\infty} (2^n + n)/(3^n - n)$
converges. Alternatively the ratio test may be used. ($\lim\limits_{n\to\infty} |a_{n+1}/a_n| = 2/3$.)

5. $\lim\limits_{n\to\infty} |a_{n+1}/a_n| = \lim\limits_{n\to\infty}(n + 1)/3 \to \infty$, hence $\sum\limits_{n=0}^{\infty} a_n$ does not converge.

Exercises for Chapter 2 (at end of chapter)

1. (a) Let $x \in \,]1,2[$ and $\delta = \min\{2 - x, x - 1\}$, then $]x - \delta, x + \delta[= D(x,\delta) \subset \,]1,2[$, so $]1,2[$ is open.
   (b) Show $\mathbb{R}\backslash[2,3]$ is open.
   (c) $\bigcap\limits_{n=1}^{\infty} [-1,1/n[= [-1,0]$ is closed.
   (d) $\mathbb{R}^n$ is open in $\mathbb{R}^n$.
   (e) Closed.
   (f) Neither open nor closed. See Exercise 5 of Section 2.3.
   (g) Neither open nor closed.
   (h) Let $\{x_n\}$ be a convergent sequence in $S = \{x \in \mathbb{R}^n \mid \|x\| = 1\}$, say $x_n \to x$. Now for any $x, y \in \mathbb{R}^n$, $\left| \|x\| - \|y\| \right| \leqslant \|x - y\|$, hence $x_n \to x$ implies $\|x_n\| \to \|x\|$. But for all $n, \|x_n\| = 1$, hence $\|x\| = 1$, so $x \in S$, proving by Theorem 9(i) that S is closed.

2. (a) $\mathrm{int}(A) = A$, $\mathrm{cl}(A) = [1,2]$, $\mathrm{bd}(A) = \{1,2\}$.
   (b) $\mathrm{int}(A) = \,]2,3[$, $\mathrm{cl}(A) = A$, $\mathrm{bd}(A) = \{2,3\}$.
   (c) $\mathrm{int}(A) = \,]-1,0[$, $\mathrm{cl}(A) = A$, $\mathrm{bd}(A) = \{-1,0\}$.
   (d) $\mathrm{int}(A) = A$, $\mathrm{cl}(A) = A$, $\mathrm{bd}(A) = \varnothing$.
   (e) $\mathrm{int}(A) = \varnothing$, $\mathrm{cl}(A) = A$, $\mathrm{bd}(A) = A$.
   (f) $\mathrm{int}(A) = \varnothing$, $\mathrm{cl}(A) = [0,1]$, $\mathrm{bd}(A) = [0,1]$.
   (g) $\mathrm{int}(A) = \{(x,y) \in \mathbb{R}^2 \mid 0 < x < 1\}$, $\mathrm{cl}(A) = \{(x,y) \in \mathbb{R}^2 \mid 0 \leqslant x \leqslant 1\}$, $\mathrm{bd}(A) = \{(x,y) \in \mathbb{R}^2 \mid x = 0 \text{ or } x = 1\}$.
   (h) $\mathrm{int}(A) = \varnothing$, $\mathrm{cl}(A) = A$, $\mathrm{bd}(A) = A$.

5. Let $x \in \mathrm{int}(A)$; then there exists an open set U with $x \in U \subset A$, and U open implies there exists an $\varepsilon > 0$ such that $D(x,\varepsilon) \subset U \subset A$. Conversely, if there exists an $\varepsilon > 0$ with $D(x,\varepsilon) \subset A$, then since $D(x,\varepsilon)$ is open, there exists an open set $U = D(x,\varepsilon)$ such that $x \in U \subset A$, hence $x \in \mathrm{int}(A)$.

6. (a) $x_n = (-1)^n$ has no limit.
   (b) $(1,0)$.
   (c) $(0,0)$.
   (d) $(0,0)$ ($1/n^n \leqslant 1/n \to 0$).

7. It is required to show that $U = \text{cl}(U)\backslash(\text{cl}(U) \cap \text{cl}(\mathbb{R}^n\backslash U))$. Since U is open, $\mathbb{R}^n\backslash U$ is closed, so $\text{cl}(\mathbb{R}^n\backslash U) = \mathbb{R}^n\backslash U$. Thus

$$\text{cl}(U)\backslash(\text{cl}(U) \cap \text{cl}(\mathbb{R}^n\backslash U)) = \text{cl}(U) \cap [\mathbb{R}^n\backslash\text{cl}(U) \cap \text{cl}(\mathbb{R}^n\backslash U)]$$
$$= \text{cl}(U) \cap [\mathbb{R}^n\backslash\text{cl}(U)] \cup \{\text{cl}(U) \cap [\mathbb{R}^n\backslash(\mathbb{R}^n\backslash U)]\}$$
$$= \varnothing \cup \{\text{cl}(U) \cap U\}$$
$$= U$$

This is not true for every set in $\mathbb{R}^n$; for example let $U = [0,1]$. Then $\text{cl}(U)\backslash\text{bd}(U) = [0,1]\backslash\{0,1\} = \,]0,1[\, \neq U$.

8. $S \subset \mathbb{R}$ bounded above implies S has a supremum in $\mathbb{R}$. Then $\sup(S) \in \text{cl}(S)$ by Exercise 5, Section 2.5 and S closed implies $S = \text{cl}(S)$.

10. (a) False (let $A = $ rationals; then $\text{int}(A) = \varnothing$, $\text{cl}(A) = \mathbb{R}$, $\text{int}(\text{cl}(A)) = \mathbb{R}$).
    (b) True (since $A \subset \text{cl}(A)$).
    (c) False (let $A = \,]0,1[\,$; then $\text{cl}(\text{int } A) = \text{cl}(A) = [0,1] \neq A$).
    (d) False (let $A = $ rationals in $[0,1]$; then $\text{bd}(A) = [0,1]$, $\text{cl}(A) = [0,1]$, $\text{bd}(\text{cl}(A)) = \{0,1\}$).
    (e) True (A open implies $\text{bd}(A) = \text{cl}(A) \cap (\mathbb{R}^n\backslash A) \subset \mathbb{R}^n\backslash A$).

12. (a) Clearly $\text{int}(\text{int } A) \subset \text{int}(A)$. Conversely, let $x \in \text{int}(A)$, then there is an open set U with $x \in U \subset A$. Let $V = U \cap \text{int}(A) \neq \varnothing$, then V is an open set such that $x \in V \subset \text{int}(A)$, so $x \in \text{int}(\text{int } A)$.
    (b) Let $x \in \text{int}(A) \cup \text{int}(B)$ so either $x \in \text{int}(A)$ or $x \in \text{int}(B)$. If $x \in \text{int}(A)$ then there exists an open set U with $x \in U \subset A \subset A \cup B$ so $x \in \text{int}(A \cup B)$. If $x \in \text{int}(B)$, by the same argument, $x \in \text{int}(A \cup B)$.
    (c) See solution to Exercise 4, Section 2.2.

16. $\{a_n\}$ is an increasing sequence ($x < (\sqrt{2})^x \Leftrightarrow (\ln x)/x < \ln \sqrt{2}$, which is true for all $x > 0$) and is bounded above by 2, for if $a_n \leqslant 2$, then $a_{n+1} = (\sqrt{2})^{a_n} \leqslant (\sqrt{2})^2 = 2$. limit $a_n = 2$, computed as in Exercise 43.
    $\underset{n \to \infty}{}$

17. For all m, $|x_m \sin m| \leqslant |x_m|$ so since $\sum |x_m|$ converges, $\sum |x_m \sin m|$ converges by the comparison test. Therefore $\sum x_m \sin m$ converges absolutely and thus converges by Theorem 12.

18. Let $\varepsilon = d(x,y)$, $U = D(x,\varepsilon/2)$, $V = D(y,\varepsilon/2)$.

21. If x_k is Cauchy and U is a neighborhood of 0, find $\varepsilon > 0$ such that $D(0,\varepsilon) \subset U$. Then find N such that $k, l \geqslant N$ implies $\|x_k - x_l\| < \varepsilon$. Then $k, l \geqslant N$ implies $x_k - x_l \in U$. For the converse, given $\varepsilon > 0$ choose $U = D(0,\varepsilon)$.

24. Let $A \subset \mathbb{R}^n \times \mathbb{R}^m$ be open and let $(x,y) \in A$. Pick $\varepsilon > 0$ such that $D((x,y),\varepsilon) \subset A$. Let $\varepsilon' = \varepsilon/\sqrt{2}$; then $D(x,\varepsilon') \times D(y,\varepsilon') \subset A$. For the converse, let $(x,y) \in A$ and let $U \subset \mathbb{R}^n$ and $V \subset \mathbb{R}^m$ be open sets with $(x,y) \in U \times V \subset A$. Pick $\varepsilon > 0$ such that $D(x,\varepsilon) \subset U$ and $D(y,\varepsilon) \subset V$; then $D((x,y),\varepsilon) \subset U \times V \subset A$, so A is open.

26.
$$a_n^2 - 2 = \left(\frac{2 + a_{n-1}}{1 + a_{n-1}}\right)^2 - 2 = \frac{2 - a_{n-1}^2}{1 + 2a_{n-1} + a_{n-1}^2} = k_n(2 - a_{n-1}^2)$$

where $k_n = 1/(1 + 2a_{n-1} + a_{n-1}^2)$ is a positive number less than 1. It follows that $a_n^2 - 2$ is alternately positive and negative, and hence that a_n is alternately above and below $\sqrt{2}$. Further, since $k_n < 1$, the even terms a_{2n} are increasing and the odd terms a_{2n+1} are decreasing. The sequence is bounded above and below by 2 and 1 respectively so the odd sequence and the even sequence (being decreasing and increasing respectively) have limits. By writing

$$a_n = 1 + \frac{1}{1 + a_{n-1}}, a_{n+1} = 1 + \frac{1}{2 + \dfrac{1}{1 + a_{n-1}}}, \text{ and setting } \alpha = 1 + \frac{1}{2 + \dfrac{1}{1 + \alpha}},$$

we find $\alpha = \sqrt{2}$. Thus the limit of both of the "every other" sequences is $\sqrt{2}$ and an easy argument shows therefore that $\lim_{n \to \infty} a_n = \sqrt{2}$.

27. $\inf(B) = \sqrt{2}$.

28. (a) The integers.

(b) Any open interval.

(c) $\left\{ \dfrac{1}{2}, \dfrac{1}{3}, \dfrac{1}{4}, \ldots, \dfrac{1}{n}, \ldots, 1\dfrac{1}{2}, 1\dfrac{1}{3}, 1\dfrac{1}{4}, \ldots, 1\dfrac{1}{n}, \ldots, 2\dfrac{1}{2}, 2\dfrac{1}{3}, \ldots, 2\dfrac{1}{n}, \ldots, m + \dfrac{1}{2}, \ldots, m + \dfrac{1}{n}, \ldots \right\}$

(d) A point in $\mathbb{R}$, the unit circle in $\mathbb{R}^2$, a line segment (including endpoints) in $\mathbb{R}^2$.

29. Yes.

30. Let $U \subset \mathbb{R}$ be open and bounded. If $U = \varnothing$, then $U = \,]1,1[$. Now suppose $U \neq \varnothing$, and $x \in U$. U open implies there exists a $y, z \in \mathbb{R}$ such that $[x,y[,]z,x] \subset U$, hence $H = \{y \mid [x,y[\subset U\}$ and $L = \{z \mid]z,x] \subset U\} \neq \varnothing$ and are both bounded so $\sup(H) = h$, $\inf(L) = l \in \mathbb{R}$. Let $I_x = \,]l,h[$, and $I = \{I_x \mid x \in U\}$. Then $U = \bigcup I$ and $I_x \cap I_y = \varnothing$ if $I_x \neq I_y$. Since $x \in U$ implies $x \in I_x \subset \bigcup I$, $U \subset \bigcup I$. Now let $y \in I_x = \,]a,b[$, so if $x < y < b$, there exists a z such that $y \in [x,z] \subset U$, hence $y \in U$, $I_x \subset U$ and $\bigcup I \subset U$. Now let $I_x = \,]a,b[$, $I_y = \,]c,d[\subset U$ such that $I_x \cap I_y \neq \varnothing$. $c \notin U$ otherwise there exists an $\varepsilon > 0$ such that $]c - \varepsilon, y] \subset U$ contradicting the definition of c. Thus $c \notin \,]a,b[$, hence $c \leqslant a$. Similarly $a \leqslant c$ implies $a = c$, and $b = d$, hence $I_x = I_y$. This is not true in $\mathbb{R}^n$; for example consider the set $\{(x,y) \mid x^2 + y^2 < 1\}$.

32. Immediate from Theorems 9 and 10.

33. Subtract $(s_n + s_{n-1})$ from both sides of $s_{n+1} + s_{n-1} \geqslant 2s_n$ to get $s_{n+1} - s_n \geqslant s_n - s_{n-1}$; let $\alpha_n = s_{n+1} - s_n$, so α_n is increasing. Furthermore α_n is bounded, since $|\alpha_n| = |s_{n+1} - s_n| \leqslant |s_{n+1}| + |-s_n| = |s_{n+1}| + |s_n| \leqslant 2M$ where M is a bound for s_n. Thus α_n converges, $\alpha_n \to \alpha$. Suppose $\alpha \neq 0$, say $\alpha > 0$. Since the α_i's are increasing to α, there exists a N such that $n > N$ implies $\alpha_n > \alpha/2$. We thus have

$$s_n = s_0 + (s_1 - s_0) + \cdots + (s_n - s_{n-1}) = s_0 + \sum_{i=1}^{n} \alpha_i$$

$$= s_0 + \sum_{i=1}^{N} \alpha_i + \sum_{i=N+1}^{n} \alpha_i \leqslant s_0 + \sum_{i=1}^{N} \alpha_i + (n - N)\frac{\alpha}{2} \to \infty$$

as $n \to \infty$, a contradiction since s_n is bounded. We get a similar contradiction assuming $\alpha < 0$. Thus $\alpha = 0$.

34. $d(x_{n+p}, x_n) \leqslant d(x_{n+p}, x_{n+p-1}) + \cdots + d(x_{n+1}, x_n)$

$\leqslant (r^{n+p-1} + \cdots + r^n)\, d(x_0, x_1)$.

Now $r < 1$ implies $\sum r^n$ converges, hence for any $\varepsilon > 0$ there exists a M such that

$n > M$ implies $|r^{n+p-1} + \cdots + r^n| < \varepsilon/d(x_0,x_1)$ implying $d(x_{n+p},x_n) < \varepsilon$, and so $\{x_n\}$ is a Cauchy sequence.

38. Given $\varepsilon > 0$ choose n large enough so that $k > n$ implies $1/k < \varepsilon/2$. Then $k, l > n$ implies $\|x_k - x_l\| \leqslant (1/k) + (1/l) < (\varepsilon/2) + (\varepsilon/2) = \varepsilon$, so x_k is Cauchy and thus converges.

39. For all x, $y \in S$, $\sup(S) \geqslant x$, $-\inf(S) \geqslant -y$ implies $x - y \leqslant \sup(S) - \inf(S)$, hence $\sup(S) - \inf(S)$ is an upper bound for the set. If $\varepsilon > 0$, there exists a $v, w \in S$ such that $v + \varepsilon/2 > \sup(A)$, and $\varepsilon/2 - w > -\inf(A)$ which implies $(v - w) + \varepsilon > \sup(A) - \inf(A)$, and hence $\sup(A) - \inf(A)$ is the sup of the set.

41. Let U be a neighborhood of x; we must show U contains some point of A_1, other than x. Since $\bigcap_{n=1}^{\infty} A_n = \varnothing$, there exists a n such that $x \notin A_n$. Then by Theorem 5, since $x \in \mathrm{cl}(A_n)$, x is an accumulation point of A_n, so U contains a point y of A_n, $y \neq x$. But $A_n \subset A_{n-1} \subset \cdots \subset A_1$ so $y \in A_1$.

42. No; let $A = \,]0,1]$ and $x = 0$. Then $d(x,A) = 0$ but there is no point $z \in A$ with $d(z,0) = 0$, for $0 \notin A$. As another example let A be the open unit disc in $\mathbb{R}^2$ and $x = (1,0)$. Then $d(x,A) = 0$ but there is no $z \in A$ with $d(x,z) = 0$. If A is closed, however, the assertion is always true (see Exercise 17 at the end of Chapter 3).

43. x_n is clearly increasing and we prove by induction that x_n is bounded above by 3: $x_1 = \sqrt{3} < 3$. Now assume $x_{n-1} < 3$. Then $x_n = \sqrt{3 + x_{n-1}} < \sqrt{3 + 3} = \sqrt{6} < 3$. Thus x_n has a limit; call it x. x satisfies $x = \sqrt{3 + x}$ (by taking limits on both sides of $x_n = \sqrt{3 + x_{n-1}}$) and so $x = (1 \pm \sqrt{13})/2$. Since all the x_n's are positive the limit must be $\geqslant 0$, so $x = (1 + \sqrt{13})/2$.

Chapter 3
Compact and Connected Sets

3.1 Compact Sets: the Heine-Borel and Bolzano-Weierstrass Theorems

1. (a) Not compact because it is not closed.
   (b) Not compact because it is not bounded.
   (c) Not compact because it is not closed.

2. $[0,1]$ is compact so any sequence in it has a convergent subsequence by Theorem 1.

4. If A is bounded, then $\mathrm{cl}(A)$ is bounded. Suppose there exists a M such that for all $x \in A$, $\|x\| < M$. Then $A \subset \mathrm{cl}(D(0,M))$ implies $\mathrm{cl}(A) \subset \mathrm{cl}(D(0,M))$. Since $\mathrm{cl}(A)$ is also closed, $\mathrm{cl}(A)$ is compact.

5. No; let $A = \{0, 1/2,2/3,3/4,4/5,\ldots,1,2,3,4,5,6,\ldots\}$. Then A has the single accumulation point 1 and A is infinite, but A is not compact since it is not bounded.

3.2 Nested Set Property

2. No; let $F_k = \,]0,1/k[$.

3. If $F_k = \{x_l \mid l \geqslant k\}$, then $\bigcap F_k = \varnothing$. None of the sets F_k are compact.

3.3   Path-Connected Sets

1. (a) Not path-connected, since any path between two rationals must contain an irrational.
   (b) Path-connected.
   (c) Path-connected.
   (d) Not path-connected. If the point $(1,0)$ were added, it would be.
3. No. For instance, let $\varphi: [0,4] \to \mathbb{R}^3$ be the curve which wraps around the unit circle in the x-y plane twice in such a way that $[0,1]$ gets sent to the first half of the circle, $[1,2]$ to the second half, $[2,3]$ to the first half again, and $[3,4]$ to the second half again. Let $c = \varphi([2,3])$. Then $\varphi^{-1}(c) = [0,1] \cup [2,3]$ is not connected. (If φ is one-to-one, then $\varphi^{-1}(c) = [c,d]$ is connected.)

3.4   Connected Sets

1. No. $]-1/2, 1\,1/2[$ and $]2, 3\,1/2[$ are two open sets which are disjoint and whose union contains A.
2. Yes, it is path-connected.
4. (a) The components are $[0,1]$ and $[2,3]$.
   (b) The components are $\ldots \{-2\}, \{-1\}, \{0\}, \{1\}, \{2\}, \ldots$.
   (c) Each rational is a component.

Exercises for Chapter 3 (at end of chapter)

1. (a) Connected, not compact.
   (b) Connected and compact.
   (c) Connected and compact.
   (d) Neither connected nor compact.
   (e) Compact, not connected if it contains more than 1 point.
   (f) $n = 1$, compact and not connected; $n \geqslant 2$, compact and connected.
   (g) Connected and compact.
   (h) Compact, not necessarily connected.
   (i) Neither compact nor connected.
   (j) Compact, not necessarily connected.
3. (a) If a set has an accumulation point x, then we can find a sequence of points in the set which converges to x. Hence if every infinite subset has an accumulation point in A, one sees that A satisfies the Bolzano-Weierstrass property (Theorem 1(iii)) and is thus compact (distinguish the cases of a repeating sequence and a sequence with infinitely many distinct points). For the converse, suppose A is compact. Given an infinite subset of A we may pick a sequence of distinct points of A. Since A is compact this sequence has a subsequence converging to a point in A, which must be an accumulation point of the subset.
   (b) Let B be the bounded infinite set. Then $B \subset D(0,M)$ for some M and hence $B \subset \mathrm{cl}(D(0,M))$. Since $\mathrm{cl}(D(0,M))$ is compact, every infinite subset of it has an accumulation point by (a). Hence B has an accumulation point.
5. (a) $F_k = \{x \in \mathbb{R}^2 \mid \|x\| < k/(k+1)\}, \qquad k = 1, 2, \ldots$.
   (b) $F_k =]k - 1/3, k + 1/3[, \qquad k = \cdots -3, -2, -1, 0, 1, 2, \ldots$.

6. By Theorem 2 there exists a $x \in \bigcap F_k$. Now suppose there exists a $y \in \bigcap F_k$, $y \neq x$. Then $d(x,y) \neq 0$. By hypothesis there exists a N such that $n \geqslant N$ implies $\text{diam}(F_n) < d(x,y)$. Then, since $x, y \in F_N$, $d(x,y) \leqslant \text{diam}(F_n) < d(x,y)$, a contradiction.

7. For all k, $\text{cl}(A_k) = \{x_k, x_{k+1}, \ldots\} \cup \{x\}$, hence $x \in \text{cl}(A_k)$ for all k implies $x \in \bigcap \text{cl}(A_k)$. Now suppose $y \in \bigcap \text{cl}(A_k)$, $y \neq x$. There exists a N such that $n \geqslant N$ implies $\|x_n - x\| < d(x,y)$. But $y \in A_n$, so $y = x_j$, $j \geqslant N$ and thus $\|y - x\| < d(x,y)$, a contradiction.

9. (a) False; $[0,1]$ is compact but $\mathbb{R}\backslash[0,1]$ is not connected. For $\mathbb{R}^n$, $A = \{x \in \mathbb{R}^n \mid 1 \leqslant \|x\| \leqslant 2\}$ is compact but $\mathbb{R}^n\backslash A$ is not connected.
   (b) False; same examples as in (a).
   (c) False; $]a,b]$ is connected but neither open nor closed.
   (d) False for $n = 1$, true for $n \geqslant 2$. ($\mathbb{R}^n\backslash A$ is path connected for $n \geqslant 2$.)

11. (a) Suppose $B \subset U \cup V$ where $B \cap U \neq \varnothing$, $B \cap V \neq \varnothing$, $B \cap U \cap V = \varnothing$, and U, V open. Then $A \subset U \cup V$ and $A \cap U \cap V = \varnothing$, and it remains to show $U \cap A \neq \varnothing$ and $V \cap A \neq \varnothing$ (for then we will have shown A is not connected, a contradiction). $B \cap U \neq \varnothing$, so let $x \in B \cap U$. If $x \in A$ the exercise is complete; if $x \notin A$, then since $x \in B \subset \text{cl}(A)$ we have x is an accumulation point of A. Thus every neighborhood of x contains points of A, so in particular U contains points of A, so $U \cap A \neq \varnothing$. Similarly $V \cap A \neq \varnothing$.

13. Let x_n be a sequence of points with $x_n \in F_n$. Then x_n is clearly a Cauchy sequence (since $\text{diam}(F_n) \to 0$ and $F_{n+1} \subset F_n$) and thus converges, say to x, since M is complete. For all n, x is a limit of a sequence of elements of F_n, so since all F_n's are closed, $x \in F_n$ for all n, that is, $x \in \bigcap F_n$. To see that x is the only element in $\bigcap F_n$ use an argument similar to that of Exercise 6.

16. $|\|x_k\| - \|x\|| \leqslant \|x_k - x\|$. Hence given $\varepsilon > 0$ there is an N such that $k \geqslant N$ implies $\|x_k - x\| < \varepsilon$ implies $|\|x_k\| - \|x\|| < \varepsilon$, so $\|x_k\| \to \|x\|$. The converse is false. Let $x_k = (-1)^k$. Let $\{x_k\}$ be a sequence in $D = \{x \in \mathbb{R}^n \mid \|x\| \leqslant 1\}$ with $x_k \to x$. We must show that $\|x\| \leqslant 1$, that is, that any convergent sequence in D converges to a point in D. By the above, $\|x_k\| \to \|x\|$. Now $\{\|x_k\|\}$ is a sequence in $[0,1]$, which is closed, hence $\|x\| \in [0,1]$, and $\|x\| \leqslant 1$.

17. There is a sequence $z_n \in A$ such that $d(x,z_n) \to d(x,A)$ (for the proof, imitate Example 2 at the end of Chapter 1). There exists a N such that $n \geqslant N$ implies $d(x,z_n) - d(x,A) < 1$, that is, $d(x,z_n) < 1 + d(x,A)$. Thus the sequence z_n with the first N terms chopped off lies in the closed ball of radius $1 + d(x,A)$ about x; this ball is compact so it follows that z_n has a convergent subsequence, say z_{n_i}, $z_{n_i} \to z$. Since $d(x,z_{n_i})$ is a subsequence of $d(x,z_n)$ and $d(x,z_n) \to d(x,A)$ it follows that $d(x,z_{n_i}) \to d(x,A)$. We will prove $d(x,z_{n_i}) \to d(x,z)$ and thus by uniqueness of limits $d(x,A) = d(x,z)$. By the triangle inequality we have $|d(z_{n_i},x) - d(z,x)| \leqslant d(z,z_{n_i}) \to 0$ as $n_i \to \infty$. It remains to show that $z \in A$; this is true because $z \in \text{cl}(A)$ and A is closed.

18. The sets F_n satisfy the hypotheses of Theorem 2, hence $\bigcap F_n \neq \varnothing$. Furthermore, $\text{diam}(F_n) \to 0$ so by Exercise 6, $\bigcap F_n$ has exactly one point x. $x^2 \leqslant 2$ and $2 \leqslant x^2$, hence $x^2 = 2$.

21. (a) First note that $\varnothing$ and A are open and closed relative to A, since $\varnothing = \varnothing \cap A$ and $\varnothing$ is open and closed in $\mathbb{R}^n$, and $A = \mathbb{R}^n \cap A$ and $\mathbb{R}^n$ is open and closed in

$\mathbb{R}^n$. Now assume A is not connected, that is $A \subset U \cup V$ where U and V are open, $A \cap U \cap V = \emptyset$, $A \cap U \neq \emptyset$, and $A \cap V \neq \emptyset$. Then $U \cap A$ is open and closed relative to A, since $U \cap A = U \cap A$ where U is open in $\mathbb{R}^n$, and $U \cap A = (\mathbb{R}^n \setminus V) \cap A$ where $\mathbb{R}^n \setminus V$ is closed in $\mathbb{R}^n$ ($U \cap A = (\mathbb{R}^n \setminus V) \cap A$ because $U \cap V \cap A = \emptyset$). For the other direction, assume there is subset W of A such that $W \neq \emptyset$, $W \neq A$, and $W = V \cap A = U \cap A$ with V open, U closed in $\mathbb{R}^n$. Let $\mathbb{R} = V$ and $S = \mathbb{R}^n \setminus V$; then $A \subset \mathbb{R} \cup S$, $\mathbb{R}$ and S are open, $A \cap \mathbb{R} \cap S = \emptyset$, $A \cap \mathbb{R} \neq \emptyset$, and $A \cap S \neq \emptyset$; thus A is not connected.

(b) $\mathbb{R}^n$ is path-connected and therefore connected so the result follows by (a).

23. $\mathbb{Q} \subset]-\infty,\sqrt{2}[\cup]\sqrt{2},\infty[$; both intervals are open, disjoint, and so forth. $\mathbb{R} \setminus \mathbb{Q} \subset]-\infty,1/2[\cup]1/2,\infty[$ (where, for example, $]-\infty,\sqrt{2}[$ is defined to be $\{x \in \mathbb{R} \mid x < \sqrt{2}\}$).

25. The sequence $\sin(n)$, $n = 1, 2, \ldots$, is contained in the compact set $[-1,1]$ and hence has a convergent subsequence $\sin(n_k)$.

26. Assume the nested set property. Let x_n be a Cauchy sequence. To show it converges, let $A_k = \{x_k, x_{k+1}, \ldots\}$ and take $F_k = \text{cl}(A_k)$ in the nested set property. (For the special definition of completeness of $\mathbb{R}$, that is, that every increasing sequence which is bounded above converges, do the same thing.)

28. Let $x \in A$ and assume x is not an accumulation point of A; let U be a neighborhood of x such that $U \cap A = x$. Let ε be such that $D(x,2\varepsilon) \subset U$. Let $W = D(x,\varepsilon)$ and $V = \mathbb{R}^n \setminus \text{cl}(W)$. Then V and W are open, $A \subset V \cup W$, $A \cap V \cap W = \emptyset$, $A \cap V \neq \emptyset$ (since A contains points other than x), and $A \cap W \neq \emptyset$ (it contains x). Thus A is not connected, a contradiction.

29. A is both compact and connected.

30. (a) True; use Theorem 2, Chapter 2.

(b) False; let $U_k =]-1/k,1/k[$ in $\mathbb{R}$. Then $\bigcap_{k=1}^{\infty} U_k = \{0\}$.

33. $\|x_{n+p} - x_n\| \leqslant \|x_{n+p} - x_{n+p-1}\| + \cdots + \|x_{n+1} - x_n\|$

$$\leqslant \frac{1}{(n+p-1)^2 + (n+p-1)} + \cdots + \frac{1}{n^2+n}$$

$$\leqslant \frac{1}{(n+p-1)^2} + \cdots + \frac{1}{n^2} \leqslant \sum_{j=n}^{\infty} \frac{1}{j^2} \to 0 \text{ as } n \to \infty$$

because $\sum_{j=1}^{\infty} (1/j^2)$ converges; thus x_n is Cauchy so it converges. *Note:* the problem also works if we are given just $\|x_{n+1} - x_n\| \leqslant a_n$ where $\sum_{n=1}^{\infty} a_n$ is any convergent series.

35. (a) If $a = 0, 1$ then $a_n = 1$ for all n so $\{a_n\}$ is constant. Now suppose $a \neq 0, 1$. Then $a_n - a_{n-1} = 1 - a_{n-1} + a_{n-1}^2 - a_{n-1} = (1 - a_{n-1})^2 > 0$, hence a_n is monotone increasing.

(b) Let $0 < a < 1$, then for all n, $a_n < 1$. Suppose $a_{n-1} < 1$, then $0 < a_{n-1} - a_{n-1}^2 < 1$ and $1 > 1 - (a_{n-1} - a_{n-1}^2) = a_n > 0$. Hence if $0 < a < 1$, then $\{a_n\}$ is bounded. Now suppose $a = 1 + h$, $h > 0$. It can be shown then that $a_n \geqslant 1 + (n-1)h^2 \to \infty$ as $n \to \infty$, hence a_n is unbounded. Finally, if $a < 0$, then $|a_n| > 1 + (n-1)a^2 \to \infty$ as $n \to \infty$. Therefore $\{a_n\}$ is bounded only when $0 \leqslant a \leqslant 1$.

(c) From (a) and (b), if $0 \leqslant a \leqslant 1$, then $\{a_n\}$ is bounded and non-decreasing, hence converges, and if $a = 0, 1$, then $a_n \to \infty$.

36. Divide $\mathbb{R}^n$ into n-dimensional cubes of side 1; thus we get a countable number of cubes. There must be some cubes with an uncountable, hence infinite number of points of A in it, otherwise A would have only a countable number of points. So take an infinite sequence of distinct points a_n in $A \cap S$. Since cl(S) is compact, a_n has a convergent subsequence, $a_{n_i} \to a$, and a is an accumulation point of A.

38. (a) $C \subset [0,1]$ and so is bounded. Also, each F_n is closed, being the union of a finite number of closed sets; hence the Cantor set is closed, being the intersection of the collection of closed sets $\{F_n\}$. Thus C is compact.
    (b) The endpoints of each interval of F_n are elements of *every* F_n and hence elements of $\bigcap F_n$. There are 2^n intervals in F_n, and there are an infinite number of F_n's.
    (c) Suppose $]a,b[\subset Ca, \neq b$, then C contains an interval of length $(b - a)$. But the intervals in F_n have length $1/3^n$ and there exists a N such that $1/3^N < b - a$, so $]a,b[\not\subset F_N$; hence $]a,b[\not\subset \bigcap F_n$. (Provided by Nancy Hildreth.)

40. Suppose $\bigcap F_k$ is not connected, then by Exercise 39 there exist open sets U, V such that $\bigcap F_k \subset U \cup V$, $U \cap V = \varnothing$, $\bigcap F_k \cap U \neq \varnothing$, $\bigcap F_k \cap V \neq \varnothing$. We claim $U \cup V$ contains some F_k, which will be a contradiction since all the F_k's are connected. Suppose $U \cup V$ contains no F_k; then for all k there exists a $x_k \in F_k$ such that $x_k \notin U \cup V$. Since $x_k \in F_1$ for all k and F_1 is compact there exists a convergent subsequence $x_{k_i} \to x$, and we have $x \notin U \cup V$ since $U \cup V$ is open and $x_{k_i} \notin U \cup V$ for all i. But since x is the limit of a sequence in each closed set F_k, $x \in F_k$ for all k implies $x \in \bigcap F_k$ a contradiction since $\bigcap F_k \subset U \cup V$. Thus $U \cup V$ must contain some F_k, the desired contradiction. An example showing compactness is necessary; let $F_n = \{(x,y) \in \mathbb{R}^2 \mid |y| \geqslant 1\} \cup \{(x,y) \in \mathbb{R}^2 \mid |x| \geqslant n\}$. Then $\{F_n\}_{n=1}^\infty$ is a nest of closed connected sets but $\bigcap_{n=1}^\infty F_n = \{(x,y) \in \mathbb{R}^2 \mid |y| \geqslant 1\}$ is not connected.

Chapter 4
Continuous Mappings

4.1   Continuity

1. (a) Let $\delta = \min\left\{1, \dfrac{\varepsilon}{1 + 2\,|x_0|}\right\}$, then $|x - x_0| < \delta$ implies $|x^2 - x_0^2| = |x - x_0|\,|x + x_0| \leqslant \delta(|x| + |x_0|) \leqslant \delta(\delta + 2\,|x_0|)$ since $|x| - |x_0| < |x - x_0| < \delta$, so $|x| < \delta + |x_0|$. Finally $|x^2 - x_0^2| \leqslant \delta(1 + 2\,|x_0|) < \varepsilon$.
   (b) Let $(x_n, y_n) \to (x_0, y_0)$; then (as proved in Chapter 1) $x_n \to x_0$, so by Theorem 1(ii) f is continuous.

2. Let $f \colon \mathbb{R}^2 \to \mathbb{R}, (x,y) \mapsto x$. Then $A = f^{-1}(U)$, and since f is continuous by Exercise 1(b), and U is open, A is open.

3. $A = f^{-1}([0,1])$, and f is continuous so $[0,1]$ closed implies A is closed.

4. (a) $f(x) = 1$, $U = $ any open set;

   (b) $f(x) = \begin{cases} 0 & \text{if } x \leqslant 0 \\ x & \text{if } x > 0, x < 1, U = \,]-1,2[\\ 1 & \text{if } x \geqslant 1 \,. \end{cases}$

   $f(U) = [0,1]$, closed.

4.2   Images of Compact and Connected Sets

1. (a) Closed, not necessarily compact or connected.
   (b) Open, not necessarily compact or connected.
   (c) Connected, not necessarily compact, open or closed.
   (d) Compact and connected; not necessarily open or closed.

3. If $f(x) = x/(1 + x)$ if $x > 0$, $x/(1 - x)$ if $x < 0$, $B = \mathbb{R}$ then $f(B) = \,]-1,1[$. If B is also bounded then B is compact, so $f(B)$ is compact and $f(B)$ is closed.

4. $A = f(A \times B)$ where $f \colon \mathbb{R}^2 \to \mathbb{R}$, $f(x,y) = x$. Thus A is connected if $A \times B$ is, since f is continuous (see Exercise 1(b) of Section 4.1).

5. Yes. Let $x \in A$, and $y \in B$. Then there exists a $\delta > 0$ such that $D((x,y),\delta) \subset A \times B$. Then $\{(z,y) \mid z \in \,]x - \delta, x + \delta[\,\} \subset D((x,y),\delta)$ implies $]x - \delta, x + \delta[\subset A$, hence A is open.

4.3   Operations on Continuous Functions

1. (a) Everywhere.
   (b) f is continuous on $\mathbb{R}\backslash\{1,-1\}$.
   (c) Everywhere.

2. Let $p_1 \colon \mathbb{R}^2 \to \mathbb{R}$, $(x,y) \mapsto x$, $p_2 \colon \mathbb{R}^2 \to \mathbb{R}$, $(x,y) \mapsto y$, $f \colon \mathbb{R} \to \mathbb{R}$, $x \mapsto x$. By earlier exercises, p_1 and p_2 are continuous, so by Theorem 3, $f \circ p_1$, $f \circ p_2$ are continuous. Let $h \colon \mathbb{R}^2 \to \mathbb{R}, (x,y) \mapsto (f \circ p)(x,y)\,(f \circ p_2)(x,y)$, then by Theorem 4, h is continuous. Then if $a_k \to a$, $b_k \to b$, $(a_k, b_k) \to (a,b)$ and $h(a_k, b_k) = (f \circ p_1)(a_k, b_k) \cdot (f \circ p_2)(a_k, b_k) = a_k \cdot b_k \to h(a,b) = a \cdot b$.

3. Use the fact that $\{.56\}$ is closed, and $\sin x$ is continuous. A is not compact.

4. It is sufficient to show $g(x) = |x|$, and $h(x) = \sqrt{x}$ are continuous, for $f = g \circ h$.

5. $f = g \circ h$, where $g(x) = \sqrt{x}$, $h(x) = x^2 + 1$, and g, h are continuous.

4.4   The Boundedness of Continuous Functions on Compact Sets

1. Let $f(x) = x/(1 + |x|)$, then f is bounded, $\sup f(\mathbb{R}) = 1$, $\inf f(\mathbb{R}) = -1$, but f does not attain either value on $\mathbb{R}$.

3. M is bounded since $M \subset K$ and K is bounded. M is closed since $M = f^{-1}\{\sup f(K)\}$, f is continuous and $\{\sup f(K)\}$ is closed. Hence M is compact.

4. $f \circ c$ is continuous and $[0,1]$ is compact. Less briefly, c is continuous and $[0,1]$ is compact, so $c([0,1])$ is compact. Since f is continuous, f attains its maximum and minimum on $c([0,1])$.

5. Let $A = \,]0,\infty[$, then $\sup f(A) = 1$, which f does not attain on $]0,\infty[$. (For all $x \in \,]0,\infty[, x > |\sin x|$, and $\lim\limits_{x \to 0} \sin x/x = 1$).

4.5   The Intermediate Value Theorem

1. Quadratic polynomials need not be negative anywhere so the method fails; the method works for quintic polynomials, and in general, for all odd degree polynomials.

2. Let $\{x_n, f(x_n)\}$ be any convergent sequence in Γ, $(x_n, f(x_n)) \to (x,y)$. If $y = f(x)$, then $(x,y) \in \Gamma$, and we have shown Γ closed. f is continuous, hence $x_n \to x$ implies $f(x_n) \to f(x)$, so $(x_n, f(x_n)) \to (x, f(x))$. Thus $y = f(x)$.

5. $f([0,1])$ would have to be closed (since $[0,1]$ is compact), and $]0,1[$ is not closed.

4.6 Uniform Continuity

1. $\left|\dfrac{1}{x} - \dfrac{1}{y}\right| = \left|\dfrac{x-y}{xy}\right| \leqslant \left|\dfrac{x-y}{a^2}\right|$. Let $\delta = a^2\varepsilon$, then $|x-y| < \delta$ implies $\left|\dfrac{1}{x} - \dfrac{1}{y}\right| < \dfrac{\delta}{a^2} = \varepsilon$.

2. See solution to Exercise 1 or use the fact that $f'(x)$ is bounded.

4. No; let $f(x) = g(x) = x$. If f and g are bounded, yes; let M be such that $|f(x)| < M$ and $|g(x)| < M$ for all x, and let $\varepsilon > 0$ be given. Pick δ such that $|x - y| < \delta$ implies $|f(x) - f(y)| < \varepsilon/2M$ and $|g(x) - g(y)| < \varepsilon/2M$. Then $|x - y| < \delta$ implies $|f(x)g(x) - f(y)g(y)| \leqslant |f(x)|\,|g(x) - g(y)| + |g(y)|\,|f(x) - f(y)| < M(\varepsilon/2M) + M(\varepsilon/2M) = \varepsilon$.

Exercises for Chapter 4 (at end of chapter)

1. (a) It is sufficient to show that f is continuous on $]a,\infty[$, for every $a > 0$. Let $x_0 \in \,]a,\infty[$, suppose $x_0 = a + \eta$. Let $\delta = \inf\{1, \eta, a^4\varepsilon/(1 + 2x_0)\}$. Then

$$|x - x_0| < \delta \text{ implies } \left|\frac{1}{x^2} - \frac{1}{x_0^2}\right| = \left|\frac{x^2 - x_0^2}{x^2 x_0^2}\right| \leqslant \left|\frac{x^2 - x_0^2}{a^4}\right| \text{ since } x_0, x > a,$$

and $\left|\dfrac{1}{x^2} - \dfrac{1}{x_0^2}\right| \leqslant \dfrac{|x + x_0|\cdot|x - x_0|}{a^4} \leqslant \dfrac{|x - x_0|\,(|x| + |x_0|)}{a^4} < \dfrac{\delta(\delta + 2\,|x_0|)}{a^4} \leqslant$

$\dfrac{\delta(1 + 2\,|x_0|)}{a^4} = \varepsilon.$

(b) Given $\varepsilon > 0$, let $\delta = $ anything > 0.

(c) Yes; it is a composition of continuous functions.

2. (a) f continuous at every point of A implies f continuous at every point of B.

3. (a) No, let $f(x) = \sin x$, $k = \{1\}$.

(b) f is continuous on all of $\mathbb{R}^n$, so f is continuous on $\mathrm{cl}(B)$ which is compact. $f(\mathrm{cl}(B))$ is compact and thus bounded; so since $f(B) \subset f(\mathrm{cl}(B))$, $f(B)$ is also bounded.

6. (a) If c_k converges then every subsequence converges to the same limit, so one direction is clear. For the other direction, suppose $x_n \nrightarrow c$; we will find a subsequence of x_n which has no subsequence converging to c. Since $x_n \nrightarrow c$, there exists a $\varepsilon > 0$ such that for all N there exists a $n > N$ with $|x_n - c| > \varepsilon$. So let n_i be such that $n_i > i$ and $|x_{n_i} - c| > \varepsilon$. Then $\{x_{n_i}\}$ is a subsequence which has no subsequence converging to c.

(b) If f is continuous, then the graph of f is closed (see solution to Exercise 2, Section 4.5). For the other direction, suppose the graph of f is closed and f is bounded. Let $x_n \to x$; we want to show $f(x_n) \to f(x)$. By (a) it suffices to show that every subsequence of $f(x_n)$ has a further subsequence which converges to $f(x)$. Let $f(x_{n_i})$ be a subsequence of $f(x_n)$; since the set of values of f is bounded, $f(x_{n_i})$ has a convergent subsequence $f(x_{n_{i_j}}) \to y$. Thus $(x_{n_{i_j}}, f(x_{n_{i_j}})) \to (x,y)$; but then since the graph of f is closed, (x,y) must be in the graph, that is $y = f(x)$. Thus every subsequence of $f(x_n)$ has a further subsequence which converges to $f(x)$, so $f(x_n) \to f(x)$, and therefore f is continuous. If f is unbounded the theorem fails; for example let $f(x) = 1/x$ if $x \neq 0$, 0 if $x = 0$. Then the graph of f is closed but f is not continuous.

7. We will show that $(f^{-1})^{-1}(C)$ is closed for every closed subset C of B. Let C be a closed subset of B. Then C is bounded so C is compact. Hence $f(C)$ is compact so $f(C)$ is closed. An example where the conclusion fails with B not compact: let $B =]0,2\pi]$, $f: B \to \mathbb{R}^2$, $f(\theta) = (\cos\theta, \sin\theta)$. Then f^{-1} is not continuous since when δ is small $(\cos\delta,\sin\delta)$ is close to $(\cos 2\pi,\sin 2\pi)$ but δ is not close to 2π (this needs to be made precise).

9. Let $A = [a,b]$, $B = [b,c]$. Let V be closed in $\mathbb{R}^m$; we show $h^{-1}(V)$ is closed. $h^{-1}(V) = h^{-1}(V) \cap (A \cup B) = (h^{-1}(V) \cap A) \cup (h^{-1}(V) \cap B) = f^{-1}(V) \cup g^{-1}(V)$, a union of two closed sets and therefore closed. A generalization to $A, B \subset \mathbb{R}^n$: Let $f: A \to \mathbb{R}^m$ and $g: B \to \mathbb{R}^m$ be continuous, and suppose $f = g$ on $A \cap B$. Let $h: A \cup B \to \mathbb{R}^m$ be defined by $h(x) = \begin{cases} f(x) \text{ if } x \in A \\ g(x) \text{ if } x \in B \end{cases}$; then h is continuous. (The proof is exactly the same.)

12. (a) Given $\varepsilon > 0$, let $\delta < \varepsilon/L$. Then $\|x - y\| < \delta$ implies $\|f(x) - f(y)\| \leqslant L\|x - y\| < L\delta < L\varepsilon/L = \varepsilon$.

(b) Let $f(x) = \sin x^2$.

(c) The sum of two Lipschitz, functions f, g is Lipschitz, for if L_1, L_2 are their Lipschitz constants respectively, then $\|f(x) + g(x) - f(y) - g(y)\| \leqslant \|f(x) - f(y)\| + \|g(x) - g(y)\| \leqslant L_1\|x - y\| + L_2\|x - y\| = (L_1 + L_2) \cdot \|x - y\|$. The product of two Lipschitz functions is not necessarily Lipschitz, for example, if $f(x) = x$, then $f(x) \cdot f(x) = x^2$ is not even uniformly continuous.

(d) The sum of two uniformly continuous functions is uniformly continuous, but the product is not necessarily uniformly continuous.

14. (a) Let $f(x,y) = \begin{cases} 0 & \text{if } y > x \\ 1 & \text{if } x \geqslant y \end{cases}$.

Then $\lim\limits_{x \to 0} \lim\limits_{y \to 0} f(x,y) = 1$ and $\lim\limits_{y \to 0} \lim\limits_{x \to 0} f(x,y) = 0$.

15. We must show

$$\sup\{f_1(x) + \cdots + f_N(x) \mid x \in A\} \leqslant \sup\{f_1(x) \mid x \in A\} + \cdots + \sup\{f_N(x) \mid x \in A\}.$$

First note that the right side equals $\sup\{f_1(x_1) + \cdots + f_N(x_N) \mid x_1,\ldots,x_N \in A\}$ (see Exercise 7, Chapter 1). Then since $\{f_1(x) + \cdots + f_N(x) \mid x \in A\} \subset \{f_1(x_1) + \cdots + f_N(x_N) \mid x_1,\ldots,x_N \in A\}$, the result follows. And as an example where equality fails, let $A = [0,1]$, $f_1: [0,1] \to \mathbb{R}$ be defined by

$$f_1(x) = \begin{cases} 0 & \text{if } x \leqslant 1/2 \\ 1 & \text{if } x > 1/2 \end{cases},$$

and $f_2: [0,1] \to \mathbb{R}$ be defined by

$$f_2(x) = \begin{cases} 1 & \text{if } x \leqslant 1/2 \\ 0 & \text{if } x > 1/2 \end{cases}.$$

Then $m = 1$, $m_1 + m_2 = 1 + 1 = 2$.

16. Use the estimate $\|f(x,y) - f(x_0,y_0)\| \leqslant \|f(x,y) - f(x_0,y)\| + \|f(x_0,y) - f(x_0,y_0)\|$.

18. Use the intermediate value theorem (Theorem 6).

19. Let $A \subset \mathbb{R}^2$ be the graph of $\tan x$, $-\pi/2 < x < \pi/2$. Then A is closed since $\tan x$

is continuous (see Exercise 6 or Exercise 2, Section 4.5). If $f(x,y) = x$, then $f(A) =]-\pi/2,\pi/2[$, which is not closed.

21. (a) Yes, $f'(x)$ is bounded.
    (b) Yes, $f'(x)$ is bounded.
    (c) Yes, $f'(x)$ is bounded.
    (d) No; we must find $\varepsilon > 0$ such that for all $\delta > 0$ there exists a x, y with $|x - y| < \delta$ and $|x \sin x - y \sin y| > \varepsilon$. Let $\varepsilon = 1$ and take any $\delta > 0$. Pick $n > 1/\pi \sin(\delta/2)$. Let $x = n\pi + \delta/2, y = n\pi$. Then $|x - y| = \delta/2 < \delta$, but

$$|(n\pi + \delta/2)\sin(n\pi + \delta/2) - n\pi \sin(n\pi)| = |(n\pi + \delta/2)\sin(n\pi + \delta/2)|$$
$$= |(n\pi + \delta/2)\sin(\delta/2)|$$
$$> |n\pi \sin(\delta/2)| \ .$$

25. (a) Directly: We show $\lim\limits_{x \to 0+} f(x)$ exists. We have $|f'(x)| \leqslant M$ for all $x \in]0,1[$. Hence by the mean value theorem $|f(x) - f(y)|/|x - y| \leqslant M$ for all $x, y \in]0,1[$, so $|f(x) - f(y)| \leqslant M |x - y|$ for all $x, y \in]0,1[$. Suppose $x_n \to 0+$, that is, $x_n \to 0, x_n \in]0,1[$. Then since $|f(x_n) - f(x_m)| \leqslant M |x_n - x_m|, f(x_n)$ is Cauchy as given ε, pick N such that $n, m \geqslant N$ implies $|x_n - x_m| < \varepsilon/M$; then $n, m \geqslant N$ implies $|f(x_n) - f(x_m)| < \varepsilon$. Thus $f(x_n)$ converges, say to a. It remains to show that for any other sequence $y_n \to 0+$ we also have $f(y_n) \to a$. We know $f(y_n)$ converges (as $f(x_n)$ did), say to b. Let $\varepsilon > 0$ be given. Pick N_1 such that $n \geqslant N_1$ implies $|x_n| < \varepsilon/6M, N_2$ such that $n \geqslant N_2$ implies $|y_n| < \varepsilon/6M, N_3$ such that $n \geqslant N_3$ implies $|b - f(y_n)| < \varepsilon/3$, and N_4 such that $n \geqslant N_4$ implies $|f(x_n) - a| < \varepsilon/3$. Let $N = \max\{N_1, N_2, N_3, N_4\}$. Then $n \geqslant N$ implies

$$|b - a| \leqslant |b - f(y_n)| + |f(y_n) - f(x_n)| + |f(x_n) - a|$$
$$< \varepsilon/3 + M |x_n - y_n| + \varepsilon/3$$
$$\leqslant \varepsilon/3 + M(|x_n| + |y_n|) + \varepsilon/3$$
$$< \varepsilon/3 + M(\varepsilon/6M + \varepsilon/6M) + \varepsilon/3 = \varepsilon \ .$$

Thus since ε was arbitrary, we have $b = a$.

(b) Indirectly: We have $|f'(x)| \leqslant M$ for all x, so by Example 2, Section 4.6, f is uniformly continuous. Thus by Exercise 24(c), f has a unique continuous extension f^* to $[0,1]$, so by definition of $\lim\limits_{x \to 0+} f(x)$ and definition of continuity of f^* at 0, $\lim\limits_{x \to 0+} f(x)$ exists and is equal to $f^*(0)$.

26. If f' is continuous, then $f'([a,b])$ is compact, since $[a,b]$ is compact. Thus f' is bounded on $[a,b]$, so f is uniformly continuous on $[a,b]$.

27. $\dfrac{81}{64}$.

28. Yes.

29. We have $|f(x) - f(y)|/|x - y| \leqslant |x - y|$ for all $x, y \in \mathbb{R}$. We will show for all $x_0 \in \mathbb{R}, f'(x_0) = \lim\limits_{x \to x_0}(f(x) - f(x_0))/(x - x_0)$ exists and is equal to 0. Let $\varepsilon > 0$ be given and $\delta = \varepsilon$. Then $|x - x_0| < \delta$ implies $|(f(x) - f(x_0))/(x - x_0) - 0| \leqslant |x - x_0| < \varepsilon$; thus $f'(x_0)$ exists and equals 0, so by elementary calculus f is constant.

30. (a) Let $\varepsilon > 0$ be given and $\delta = \varepsilon^2$. To show that $|x - y| < \delta$ implies $(\sqrt{x} - \sqrt{y}) < \varepsilon$ for all $x, y \geq 0$, or in other words $|x^2 - y^2| < \delta$ implies $|x - y| < \varepsilon$ for all $x, y \geq 0$. Now, $|x^2 - y^2| < \varepsilon^2$ implies $|x - y| \, |x + y| < \varepsilon^2$ implies $|x - y| \, |x - y| < \varepsilon^2$, (since for $x, y \geq 0$ we have $|x - y| \leq |x + y|$) implies $|x - y| < \varepsilon$. Thus $\sqrt{x}$ is uniformly continuous on $[0, \infty[$.

   (b) We know $\dfrac{x - x^k}{\log x}$ is continuous on $]0,1[$; it remains to show continuity of f at 0 and 1; that is, that $\displaystyle\lim_{x \to 0} \frac{x - x^k}{\log x} = 0$ and $\displaystyle\lim_{x \to 1} \frac{x - x^k}{\log x} = 1$. This is easily accomplished by use of l'Hopital's rule. f is uniformly continuous, being continuous on a compact set.

33. First assume that A is relatively compact, that is, $\mathrm{cl}(A)$ is compact. By the Bolzano-Weierstrass theorem every sequence in $A \subset \mathrm{cl}(A)$ has a subsequence which converges to a point in $\mathrm{cl}(A) \subset \mathbb{R}^n$. For the converse, assume every sequence in A has a subsequence which converges to a point in $\mathbb{R}^n$. To show $\mathrm{cl}(A)$ is compact, we take a sequence y_n in $\mathrm{cl}(A)$ and show it has a convergent subsequence. Let $x_n \in A$ be such that $d(x_n, y_n) < 1/n$. x_n has a convergent subsequence, $x_{n_i} \to x \in \mathbb{R}^n$. Claim $y_{n_i} \to x$. For the proof, given $\varepsilon > 0$ pick N_1 such that $n_i \geq N_1$ implies $d(x_{n_i}, x) < \varepsilon/2$, and pick $N_2 > 2/\varepsilon$. Let $N = \max\{N_1, N_2\}$. Then $n_i \geq N$ implies $d(y_{n_i}, x) \leq d(y_{n_i}, x_{n_i}) + d(x_{n_i}, x) \leq 1/n_i + \varepsilon/2 < \varepsilon/2 + \varepsilon/2 = \varepsilon$. Since $\{y_{n_i}\}$ is a sequence in the closed set $\mathrm{cl}(B)$, $x \in \mathrm{cl}(B)$. Thus $\{y_n\}$ has a convergent subsequence, and $\mathrm{cl}(B)$ is compact.

Chapter 5
Uniform Convergence

5.1   Pointwise and Uniform Convergence

1. Yes, for if $\varepsilon > 0$ and $N > 3/\varepsilon$, then $n > N$ implies that for all $x \in [0,1]$,

$$|f_n(x) - f(x)| = \left| x^2 - \frac{2x}{n} + \frac{1}{n^2} - x^2 \right|$$

$$= \left| \frac{1}{n^2} - \frac{2x}{n} \right| \leq \frac{1}{n^2} + \left| \frac{2}{n} x \right| \leq \frac{1}{n^2} + \frac{2}{n} \leq \frac{3}{n} < \varepsilon,$$

since $|x| \leq 1$, independently of x.

2. No, since the limit function $f(x) = \begin{cases} x & \text{if } x \in [0,1[\\ 0 & \text{if } x = 1 \end{cases}$ is not continuous but each f_n is.

4. Yes, $\{f_n\}$ converges uniformly to $f = 0$ on $[0, .999]$ for $|f_n(x) - f(x)| = |x^n| \leq |.999^n| \to 0$ as $n \to \infty$, independently of x.

5. $f_k(x) = \displaystyle\sum_0^k \frac{x^{n/2}}{n(n!)^2}$ converges uniformly to $f(x)$, since $|f_k(x) - f(x)| = \displaystyle\sum_{n=k+1}^{\infty} \frac{x^{n/2}}{n(n!)^2} \leq$

$$\sum_{n=k+1}^{\infty} \frac{1}{n(n!)^2} < \sum_{n=k+1}^{\infty} \frac{1}{n^2} \to 0 \text{ independently of } x \text{ since } \sum_{n=0}^{\infty} \frac{1}{n^2} \text{ is a convergent series.}$$

Thus since the f_k's are continuous, f is.

5.2   The Weierstrass M-Test

1. (a) Converges pointwise, not uniformly.
   (b) $f_n(x) = e^{-x^2/n}/n \to f(x) = 0$ uniformly. To show uniformity, $|f_n(x) - f(x)| = |f_n(x)| = 1/ne^{x^2/n} < 1/n \to 0$, independently of x.

2. $|x^n/n^2| \leq 1/n^2 = M_n$, and since $\sum_{n=1}^{\infty} M_n$ converges, $f_k(x) = \sum_{n=1}^{k} x^n/n^2$ converges uniformly, by the M-test.

4. The series converges uniformly everywhere on $\mathbb{R}$ by the Weierstrass M-test, since

$$\frac{1}{x^2 + n^2} \leq \frac{1}{n^2} = M_n.$$

5. Use the Weierstrass M-test with $M_k = |a_k|$.

5.3   Integration and Differentiation of Series

1. The limit function is $f(x) = \begin{cases} 1/x & \text{if } x > 0 \\ 0 & \text{if } x = 0 \end{cases}$ which is not continuous, hence the convergence is not uniform and Theorem 4 cannot be applied.

2. For $x = 0, 1$, $f_n(x) = 0 \to 0$. For $x < 1$, $\sum_0^{\infty} n^3 x^n$ converges by the ratio test, so $n^3 x^n \to 0$ and $f_n(x) = n^3 x^n(1 - x) \to 0$. Thus $f_n \to f = 0$ pointwise on $[0,1]$. However, the convergence is not uniform, since

$$\int_0^1 f_n(x)\, dx = n^3 \left(\frac{1}{n+1} - \frac{1}{n+2} \right) = \frac{n^3}{(n+1)(n+2)} \to \infty$$

but

$$\int_0^1 f(x)\, dx = \int_0^1 0\, dx = 0.$$

3. $f_n \to 0$ uniformly since by locating the maximum of f_n at $\frac{n}{(n+1)}$, $|f_n(x)| \leq \sqrt{n}$.

$$\left(\frac{n}{n+1} \right)^n \cdot \left(\frac{1}{n+1} \right) \leq \frac{\sqrt{n}}{n+1} \to 0.$$ Thus Theorem 4 is valid. The derivatives converge to zero pointwise but not uniformly, so the hypotheses of Theorem 5 fail. Is the conclusion valid?

5.4   The Space of Continuous Functions

1. No. Let $f(x) = \begin{cases} 1, & x \leq 1 \\ 1/x, & x > 1 \end{cases}$.

Then $f \in B$. Let $\varepsilon > 0$, and $g(x) = \begin{cases} 1, & x \leq 1 \\ 1/x - \varepsilon/2, & x > 1 \end{cases}$.

Then $g \notin B$, since if $x > 2/\varepsilon$, $g(x) = 1/x - \varepsilon/2 < \varepsilon/2 - \varepsilon/2 = 0$. But $\|f - g\| < \varepsilon$, hence $D(f,\varepsilon) \not\subset B$, so B is not open. Also, $\text{int}(B) = \{f \in \mathscr{C}_b(\mathbb{R},\mathbb{R}) \mid \text{there exists a } \delta > 0 \text{ such that } f > \delta\}$.

2. $\text{cl}(B) = \{f \in \mathscr{C}_b(\mathbb{R},\mathbb{R}) \mid f(x) \geqslant 0 \text{ for all } x \in \mathbb{R}\}$.

4. $f_n(x) = \dfrac{1}{n} \cdot \dfrac{nx}{1 + nx} \leqslant \dfrac{1}{n} \cdot 1 \to 0$ as $n \to \infty$ independently of x. Hence $f_n \to 0$ uniformly, that is, $f_n \to 0$ in $\mathscr{C}([0,1],\mathbb{R})$.

5. Pick N such that $n \geqslant N$ implies $\|f_n - f\| < 1$. Then

$$M = \max\{\|f_1\|,\ldots,\|f_N\|,1 + \|f\|\}$$

is a bound for $\{\|f_n\|\}$. It is not closed unless f is an element of it, that is, unless $f_n = f$ for some n.

5.5   The Arzela-Ascoli Theorem

1. $f_n(0) = 0$ implies f_n bounded, for let M be such that $|f_n'(x)| \leqslant M$ for all n and for all $x \in \,]0,1[$, then by the mean value theorem $|f_n(x) - f_n(0)| = |f_n(x)| \leqslant M\,|x - 0| = M\,|x| \leqslant M$.

2. No, let $f_n(x) = 1$ if n is even, 2 if n is odd.

4. B is compact by the remark after Theorem 9, and I is continuous (see Example 3). Hence I is a continuous function on a compact set, so it assumes its maximum at a point $f_0 \in B$.

5.6   Fixed Points and Integral Equations

1. $|\alpha| < 1$.

2. $f(x) = \displaystyle\sum_{k=0}^{\infty} x^{2k}$.

3. $r < \frac{1}{2}$.

5. $f(x) = 1 + \int_0^x 3xf(y)\,dy$. Let $T(f)(x) = 1 + \int_0^x 3xf(y)\,dy$ and calculate $T(0)(x)$, $T^2(0)(x)$.

5.7   The Stone-Weierstrass Theorem

1. By Example 2, the polynomials on $[0,2\pi]$ are dense, so since $\sin x$ is continuous on $[0,2\pi]$, there is a polynomial p with $|p(x) - \sin x| < \varepsilon$ for all $x \in [0,2\pi]$. Let $\varepsilon = 1/100$.

3. The answer to the second part is yes.

4. Use Theorem 12.

5. Yes, by Theorem 12.

5.8   The Dirichlet and Abel Tests

1. $\displaystyle\sum_{n=1}^{\infty} \frac{x^n}{n!} e^{-nx}$ converges uniformly by the M-test with $M_n = \dfrac{1}{n!}$ where $\displaystyle\sum_{n=0}^{\infty} \frac{1}{n!} = e < \infty$.

2. $\displaystyle\sum_{n=1}^{\infty} \frac{(-1)^n x^n}{n}$ converges uniformly, by Dirichlet's test. The partial sums of $\sum f_n(x) = \sum (-1)^n$ are bounded by 1, and $g_n(x) = \dfrac{x^n}{n}$ are non-negative, decreasing with n

and $\to 0$ uniformly since $|g_n(x)| = \left|\dfrac{x^n}{n}\right| \leqslant \dfrac{1}{n} \to 0$ independently of x.

4. $\displaystyle\sum_{n=1}^{\infty} \frac{\sin nx}{n} e^{-nx}$ converges uniformly by Abel's test, since $\displaystyle\sum_{n=1}^{\infty} f_n(x) = \sum_{n=1}^{\infty} \frac{\sin nx}{n}$ converges uniformly by Example 1, and $\varphi_n(x) = e^{-nx}$ are decreasing with n and bounded by 1.

5.9 Power Series and Cesaro and Abel Summability

1. $R = 1, R = 0$.

2. Differentiate $\sum x^k = \dfrac{1}{(1-x)}$ using Corollary 4.

3. $S_n = 1, 1, 0, 1, 1, 0, \ldots$ so $\sigma_n \to 2/3$.

4. Use $1 - x^2 + x^3 - x^5 + \cdots = \dfrac{1-x^2}{1-x^3}$ or else use Theorem 17.

Exercises for Chapter 5 (at end of chapter)

1. (a) Let $\varepsilon > 0$ be given. Pick K such that $k \geqslant K$ implies $m_k < \varepsilon$. Then $k \geqslant K$ implies $\|f_k(x) - f(x)\| < \varepsilon$ for all $x \in A$, that is, $f_k \to f$ uniformly on A.

   (b) Let $\varepsilon > 0$ be given. Since $m_k \to m$, $\{m_k\}$ is Cauchy. So pick K such that $k, l \geqslant K$ implies $|m_k - m_l| < \varepsilon$. Then $k, l \geqslant K$ implies $\|f_k(x) - f_l(x)\| < \varepsilon$ for all $x \in A$, so by Theorem 2 (the Cauchy Criterion) f_k converges uniformly on A.

2. (a) $\left|\dfrac{\sin x}{k}\right| \leqslant \dfrac{1}{k} \to 0$ independently of x, thus $\dfrac{\sin x}{k} \to f = 0$ uniformly. Clearly the limit function $f = 0$ is continuous.

   (b) $\dfrac{1}{kx + 1} \to 0$ which is continuous. The convergence is not uniform since $f_k(1/k) = 1/2$ for all k.

   (c) $\dfrac{x}{kx + 1} \to 0$ which is continuous. The convergence is uniform since $\dfrac{x}{kx + 1} = \dfrac{1}{k + 1/x} \leqslant \dfrac{1}{k} \to 0$ independently of x.

   (d) $f_k'(x) = \dfrac{1 - kx^2}{(1 + kx^2)^2}$ so the maximum of f_k occurs at $x = 1/\sqrt{k}$ where its value is $1/2\sqrt{k}$; thus given $\varepsilon > 0$ pick $K > 1/4\varepsilon^2$. Then $k > K$ implies $|f_k(x)| < \varepsilon$ for all x so $f_k \to 0$ uniformly.

   (e) $1 \to 1$ uniformly and $\dfrac{\cos x}{k^2} \to 0$ uniformly since $\left|\dfrac{\cos x}{k^2}\right| \leqslant \dfrac{1}{k^2} \to 0$ independently of x, so $\left(1, \dfrac{\cos x}{k^2}\right) \to (1,0)$ uniformly. It remains to be verified that the component functions converging uniformly implies that the function converges uniformly. This can be done in a way similar to the proof for plain convergence of components.

3. (a) Does not converge anywhere, since $\sum_{k=1}^{\infty} g_k(x) = \sum_{k=K}^{\infty} (-1)^k$ where K is the smallest integer bigger than x, which does not converge.

(b) Converges uniformly on $\mathbb{R}$ by the M-test with $M_k = 1/k^2$. Thus the function
$$g(x) = \sum_{k=1}^{\infty} g_k(x) \text{ is continuous.}$$

(c) Converges uniformly by Dirichlet's test with $f_n(x) = (-1)^n$, and $g_n(x) = \dfrac{\cos(nx)}{\sqrt{n}}$, where $g_n \to 0$ uniformly since $\left| \dfrac{\cos(nx)}{\sqrt{n}} \right| \leqslant \dfrac{1}{\sqrt{n}} \to 0$ independently of x. Thus the limit function $g(x) = \sum_{k=1}^{\infty} g_k(x)$ is continuous.

(d) Converges to the continuous function $g(x) = x/(1 - x)$ (see the geometric series test, Chapter 2). However, the convergence is not uniform, since if it were, we would have $\sum_{1}^{n} g_k(x) = \dfrac{x - x^{n+1}}{1 - x} \to \dfrac{x}{1 - x}$ uniformly, that is, $\dfrac{x^{n+1}}{1 - x} \to 0$ uniformly. But that would imply $\dfrac{x^{n+1}}{1 - x}$ is uniformly bounded, a contradiction since near $x = 1$ the denominator goes to 0 and the numerator is bounded below by $1/2$, hence the quantity increases without bound.

7. Let $S_1 = \{|f(x)| \cdot |g(x)| \mid x \in A\}$ and $S_2 = \{|f(x)| \cdot |g(y)| \mid x,y \in A\}$. Then $S_1 \subset S_2$ so $\sup(S_1) \leqslant \sup(S_2)$. Clearly $\|fg\| = \sup(S_1)$ and $\|f\| \cdot \|g\| = \sup(S_2)$. An example where equality holds is $A = [0,1]$, $f(x) = g(x) = x$ and an example where strict inequality holds is $A = [0,1]$, $f(x) = x + 1$, $g(x) = 1/(x + 1)$. Then $\|f\| = 2$, $\|g\| = 1$, $\|f\| \cdot \|g\| = 2$, but $\|fg\| = 1$ since $f \cdot g = 1$.

8. No.

11. (a) No, completeness is necessary. For example let $f(x) = x^2$ on the non-complete metric space $]0,1/3]$ (not complete since the Cauchy sequence $\{1/3,1/4,\ldots,1/n,\ldots\}$ does not converge). f is a contraction since $|x^2 - y^2| = |x - y| |x + y| \leqslant 2/3 |x - y|$. Yet there is no fixed point, since $f(x) = x$ implies $x^2 = x$ implies $x = 0$ or 1, and 0, 1 are not in the metric space.

(b) No. Let $X = [2,\infty[$, $f(x) = x + 1/x$. If X is compact this cannot happen. Consider $g: X \to \mathbb{R}$, $g(x) = d(f(x),x)$. g is continuous as f is continuous and the distance function is continuous. So since X is compact, g assumes its minimum on X, say at $x_0 \in X$. We claim x_0 is a fixed point of f. Assume x_0 is not a fixed point; then $d(f(x_0),x_0) > 0$ so $d(f(x_0),x_0) > d(f(f(x_0)),f(x_0))$, a contradiction since g assumes its minimum at x_0.

13. We know $f_k \to f$ pointwise. Pick $x_0 \in]a,b[$ and pick N_1 such that $k \geqslant N_1$ implies $|f_k(x_0) - f(x_0)| < \varepsilon/2$. $f_k' \to f'$ uniformly, so there exists a N_2 such that $k \geqslant N_2$ implies $|f_k'(x) - f'(x)| < \varepsilon/2(b - a)$ for all $x \in]a,b[$. Applying the mean value theorem to the function $(f_k - f)$, $|(f_k(x) - f(x)) - (f_k(x_0) - f(x_0))| \leqslant M |x - x_0|$. Thus $|f_k(x) - f(x)| < (\varepsilon/2(b - a)) |x - x_0| + |f_k(x_0) - f(x_0)| < \varepsilon/2 + \varepsilon/2 = \varepsilon$.

14. Let $S = \bigcap_{n=1}^{\infty} f^n(X)$. $S \neq \varnothing$ because $x_0 \in S$ where x_0 is the fixed point of f. To show x_0 is the only point in S, suppose $x \in S$. Then for all n there exists a x_n such that $x = f^n(x_n)$. Since X is compact, x_n has a convergent subsequence $x_{n_k} \to y$. We have $d(f^{n_k}(y),x) = d(f^{n_k}(y),f^{n_k}(x_{n_k})) \leqslant \lambda^{n_k} d(y,x_{n_k}) \to 0 \ (\lambda < 1)$. Thus $f^{n_k}(y) \to x$. But $f^{n_k}(y) \to x_0$ (see the proof of the contraction mapping theorem). Thus $x = x_0$ since limits are unique.

15. Use Theorem 11, Chapter 2. For a counterexample if $\sum g_k$ is just convergent, let $g_n = (-1)^n/n$. $\sum g_k$ converges by Dirichlet's test with $f_k(x) = (-1)^k$, $g_k(x) = 1/k$. But the subseries of even terms $\sum 1/2n = (1/2) \sum 1/n$ doesn't converge.

18. Let $f_k \colon [0,1] \to [0,1]$ be $f_k(x) = \begin{cases} 1/k, & \text{if } x < 1/k, \\ 0, & \text{if } x \geqslant 1/k. \end{cases}$

Then $f_k \to 0$ uniformly, since for $\varepsilon > 0$ and $K > 1/\varepsilon$, $k \geqslant K$ implies

$$|f_k(x) - 0| = |f_k(x)| = \begin{cases} 0, & x \geqslant 1/k \\ 1/k, & 0 < x < 1/k \end{cases} < \varepsilon .$$

22. Let $\varepsilon > 0$ be given. For $x \in A$ let δ_x be as in the problem. Consider the open cover $\{D(x,\delta_x/2) \mid x \in A\}$ and let $\{D(x_n,\delta_n/2) \mid n = 1,\ldots,N\}$ be a finite subcover. Let $x \in A$; then there exists a n such that $d(x,x_n) < \delta_n/2$. Let $\delta = \min\{\delta_1/2,\ldots,\delta_N/2\}$. Then $d(x,y) < \delta$ implies $d(y,x_n) \leqslant d(y,x) + d(x,x_n) < \delta_n/2 + \delta_n/2 = \delta_n$ implies $d(f(x),f(y)) < \varepsilon$ for all $f \in B$.

23. No. Let $f(x) = \begin{cases} 0, & \text{if } x < -1, \\ 1, & \text{if } x \geqslant -1. \end{cases}$ Then $f \circ f \equiv 1$.

25. Use the intermediate value theorem. (If $f(0) < f(1)$ show f is increasing; if $f(0) > f(1)$ show f is decreasing.) Use the intermediate value theorem to show that if $x < y < z$ and $f(x) < f(z) < f(y)$, then f is not one-to-one.

26. Let $T \colon \mathscr{C}[0,1] \to \mathscr{C}[0,1]$ be the function $T(f)(x) = A(x) + \int_0^1 k(x,y)f(y)\,dy$. We will show T is a contraction, and thus that T has a fixed point since $\mathscr{C}[0,1]$ is a complete metric space. Let $M = \max_{(x,y)\in[0,1]\times[0,1]} |k(x,y)|$; we have $M < 1$. Then

$$\|T(f) - T(g)\| = \sup_{x\in[0,1]} \left| A(x) + \int_0^1 k(x,y)f(y)\,dy - A(x) - \int_0^1 k(x,y)g(y)\,dy \right|$$

$$= \sup_{x\in[0,1]} \left| \int_0^1 k(x,y)[f(y) - g(y)]\,dy \right|$$

$$\leqslant \sup_{x\in[0,1]} \left| \int_0^1 M(f(y) - g(y))\,dy \right| \leqslant M \sup_{x\in[0,1]} \int_0^1 |f(y)g(y)|\,dy$$

$$= M \|f - g\|. \text{ Thus } T \text{ is a contraction with } \lambda = M .$$

27. Use the method of Exercise 25, Chapter 4 (or use the exercise itself, parts (b) or (c)).

28. Yes on $[0,396]$, since $f_n(x) = x/n \leqslant 396/n \to 0$ independently of x. But f_n is not uniformly convergent on $\mathbb{R}$. Let $\varepsilon = 1$, then for all n there exists a x such that $f_n(x) > 1$, namely any $x > n$.

29. (a) f is uniformly continuous on $[-1,1]$, so if $\varepsilon > 0$, there exists a δ such that for

all $x, y \in]-1,1[\subset [-1,1], |f(x) - f(y)| < \varepsilon$, hence f is uniformly continuous on $]-1,1[$.

(b) Yes; it is uniformly continuous on the compact set $[0,1]$, and its derivative is bounded on $[1,\infty[$ so it is uniformly continuous there. Thus it is uniformly continuous on $[0,\infty[$.

(c) Yes, for the derivative of f is bounded.

(d) Yes. f is continuous on $[0,1]$ so it is uniformly continuous on $[0,1]$, and so it is uniformly continuous on $]0,1]$.

(e) No. As $x \to -1$, $\ln(1 + x^3)$ decreases to $-\infty$. Thus $\sin(\ln(1 + x^3))$ oscillates between $+1$ and -1 infinitely many times in any neighborhood of $x = -1$, so $\lim_{x \to -1} f(x)$ does not exist, and f is not continuous on $[-1,1]$.

30. See proof of Theorem 7, Chapter 4. The proof given applies to any $f: K \to B$ where K is a compact metric space, and B is a metric space.

31. Let $\varepsilon > 0$ be given. Pick N such that $n \geq N$ implies $|a_n - a| < \varepsilon/2$. Pick M such that $n \geq M$ implies $\left| \dfrac{a_1 + a_2 + \cdots + a_N - Na}{n} \right| < \dfrac{\varepsilon}{2}$. Then $n \geq M$ implies

$$
\begin{aligned}
|b_n - a| &= \left| \frac{a_1 + \cdots + a_n - na}{n} \right| = \left| \frac{(a_1 - a) + \cdots + (a_n - a)}{n} \right| \\
&= \left| \frac{(a_1 - a)}{n} + \cdots + \frac{(a_N - a)}{n} + \frac{(a_{N+1} - a)}{n} + \cdots + \frac{(a_n - a)}{n} \right| \\
&\leq \left| \frac{(a_1 - a)}{n} + \cdots + \frac{(a_N - a)}{n} \right| + \left| \frac{(a_{N+1} - a)}{n} + \cdots + \frac{(a_n - a)}{n} \right| \\
&= \left| \frac{a_1 + \cdots + a_N - Na}{n} \right| + \left| \frac{(a_{N+1} - a)}{n} + \cdots + \frac{(a_n - a)}{n} \right| \\
&< \varepsilon/2 + (n\varepsilon/2)/n = \varepsilon/2 = \varepsilon/2 + \varepsilon.
\end{aligned}
$$

33. (a) Yes. Given $\varepsilon > 0$, choose N such that $n \geq N$ implies $|f_n(0)| < \varepsilon$ and $|f_n(1)| < \varepsilon$. Then $n \geq N$ implies for all $x \in [0,1]$, $-\varepsilon < f_n(0) \leq f_n(x) \leq f_n(1) < \varepsilon$, since each f_n is increasing implies $|f_n(x)| < \varepsilon$ for all x, so $f_n \to 0$ uniformly.

(b) No. Let $f_n(x) = x^n$. The limit $f(x) = \begin{cases} 0, & \text{if } x < 1, \\ 1, & \text{if } x = 1, \end{cases}$ is not continuous so the convergence is not uniform, but all the f_n's are increasing.

36. (a) $\text{limit}_{x \to 0}\left(\lim_{y \to 0} \dfrac{x^2 y}{x^4 + y^2} \right) = \text{limit}_{x \to 0}(0) = 0.$

(b) $\text{limit}_{y \to 0}\left(\lim_{x \to 0} \dfrac{x^2 y}{x^4 + y^2} \right) = \text{limit}_{y \to 0}(0) = 0.$

(c) $\text{limit}_{(x,y) \to (0,0)}\left(\dfrac{x^2 y}{x^4 + y^2} \right)$ does not exist.

Let $(x,y) \to (0,0)$ along the path (x,cx^2) for some constant c. Then $f(x,cx^2) = \dfrac{cx^4}{x^4 + cx^4} = \dfrac{c}{1 + c^2}$, and the limit as $(x,y) \to (0,0)$ along this path is $\dfrac{c}{1 + c^2}$, which is a different value for each c. Thus the limit does not exist.

38. $1 + \dfrac{1}{2} + \dfrac{1}{4} + \dfrac{1}{8} + \cdots = \sum\limits_{n=1}^{\infty} \dfrac{1}{2^n}$ converges by the geometric series test, $1 - \dfrac{1}{2} +$

$\dfrac{1}{3} - \dfrac{1}{4} + \cdots = \sum\limits_{n=1}^{\infty} \dfrac{(-1)^{n=1}}{n}$ converges by Dirichlet's test with $f_n(x) = (-1)^n, g_n(x) =$

$1/n$, and $1 + \dfrac{1}{2} + \dfrac{1}{3} + \dfrac{1}{4} + \cdots = \sum\limits_{n=0}^{\infty} \dfrac{1}{n}$ does not converge, by the p-series test. (See Chapter 2.)

39. Let $\varepsilon > 0$; then since f is continuous on a compact set there exists a $\delta > 0$ such that for all $x, y \in [0,1] \mid x - y \mid < \delta$ implies $|f(x) - f(y)| < \varepsilon$. Let N be such that $1/N < \delta$, and divide $[0,1]$ into intervals $[j/n,(j+1)/n]$, $j = 0, \ldots, n$. Define $g(x) = f(j/n)$, if $x \in [j/n,(j+1)/n[$, and $g(1) = f(1)$. Then for any $x \in [0,1[$, there exists a j such that $x \in [j/n,(j+1)/n[$ implies $x - j/n < \delta$ implies $|f(x) - f(j/n)| = |f(x) - g(x)| < \varepsilon$, and if $x = 1$, $f(x) - g(x) = 0$. Thus $\|f - g\| < \varepsilon$, and g is simple.

40. Let $\varepsilon > 0$ and $f_0 \in \mathscr{C}([0,1],\mathbb{R})$; since f_0 is continuous $f_0([0,1])$ must be a closed interval, say $[a,b]$. Since g is continuous on $\mathbb{R}$, g is uniformly continuous on $[a',b'] = [a-1,b+1]$, so there exists a $\delta > 0$ such that $\delta < 1$ and for all $x, y \in [a,b]$, $|x - y| < \delta$ implies $|g(x) - g(y)| < \varepsilon/2$. Let $f \in \mathscr{C}([0,1],\mathbb{R})$ be such that $\|f - f_0\| < \delta$. Then for all $x \in [0,1]$, $|g(f(x)) - g(f_0(x))| < \varepsilon/2$ since $|f(x) - f_0(x)| < \delta$. Hence $\|g \circ f - g \circ f_0\| < \varepsilon$ and F is continuous.

   Now suppose g is uniformly continuous, and $\varepsilon > 0$. Then there exists a δ such that $|x - y| < \delta$ implies $|g(x) - g(y)| < \varepsilon/2$, for all $x, y \in \mathbb{R}$. Let $f_1, f_2 \in \mathscr{C}([0,1],\mathbb{R})$, $\|f_1 - f_2\| < \delta$, then for all $x, y \in [0,1]$, $|f_1(x) - f_2(x)| < \delta$ implies $|g \circ f_1(x) - g \circ f_2(x)| < \varepsilon/2$ implies $\|g \circ f_1 - g \circ f_2\| < \varepsilon$, hence F is uniformly continuous.

41. By Example 2, Section 5.7, the polynomials are dense in $\mathscr{C}([-1000,1000],\mathbb{R})$. Since $f(x) = |x|^3 \in \mathscr{C}([-1000,1000],\mathbb{R})$ then there exists a polynomial p such that $|p(x) - |x|^3| < 1/10$ for all $x \in [-1000,1000]$.

46. (a) We first show that the limit function f is uniformly continuous. Let $\varepsilon > 0$ be given. Pick $\delta > 0$ such that $\|x - y\| < \delta$ implies $\|f_n(x) - f_n(y)\| < \varepsilon/3$ for all n. Let $\|x - y\| < \delta$; pick N such that $\|f(x) - f_N(x)\|$ and $\|f_N(y) - f(y)\| < \varepsilon/3$. We have $\|f(x) - f(y)\| \leq \|f(x) - f_N(x)\| + \|f_N(x) - f_N(y)\| + \|f_N(y) - f(y)\| < \varepsilon/3 + \varepsilon/3 + \varepsilon/3 = \varepsilon$. Thus $\|x - y\| < \delta$ implies $\|f(x) - f(y)\| < \varepsilon$ so f is uniformly continuous. We now show the convergence is uniform. Let $\varepsilon > 0$ be given. Pick $\delta_1 > 0$ such that $\|x - y\| < \delta_1$ implies $\|f_n(x) - f_n(y)\| < \varepsilon/3$ for all n; and $\delta_2 > 0$ such that $\|x - y\| < \delta_2$ implies $\|f(x) - f(y)\| < \varepsilon/3$. Let $\delta = \min\{\delta_1,\delta_2\}$. For $x \in A$ pick N_x such that $n \geq N_x$ implies $\|f_n(x) - f(x)\| < \varepsilon/3$. Consider the open cover $\{D(x,\delta) \mid x \in A\}$ of A and let $\{D(x_n,\delta) \mid n = 1,2,\ldots,M\}$ be a finite subcover. Let $N = \max\{N_{x_1},\ldots,N_{x_M}\}$. Now let $x \in A$; let $\|x - x_i\| < \delta$. Then $n \geq N$ implies $\|f_n(x) - f(x)\| \leq \|f_n(x) - f_n(x_i)\| + \|f_n(x_i) - f(x_i)\| + \|f(x_i) - f(x)\|$. The first term is $< \varepsilon/3$ because $\|x - x_i\| < \delta \leq \delta_1$; the second term is $< \varepsilon/3$ because $n \geq N \geq N_{x_i}$; and the third term is $< \varepsilon/3$ because $\|x - x_i\| < \delta \leq \delta_2$. Thus we have $n \geq N$ implies $\|f_n(x) - f(x)\| < \varepsilon$ for all $x \in A$, so the convergence is uniform.

   (b) $f_n \to 0$ pointwise (this is clear). But $f_n \not\to 0$ uniformly, since $f_n(1/n) =$

$\dfrac{1/n^2}{(1/n^2) + 0} = 1 \not\to 0$ (that is, for $\varepsilon > 1$, no matter how big n is there is always an x, namely $x = 1/n$, such that $|f_n(x)| \geqslant \varepsilon$). We conclude from (a) that the f_n's are not equicontinuous.

48. Let $f(x) = \begin{cases} 1/2, & \text{if } x < 1/2, \\ x, & \text{if } x \geqslant 1/2, \end{cases}$ and $g(x) = \begin{cases} x, & \text{if } x < 1/2, \\ 1/2, & \text{if } x \geqslant 1/2. \end{cases}$

Then $(f + g)(x) = x + 1/2$ and $(f - g)(x) = |x - 1/2|$. We get $2\|f\|^2 + 2\|g\|^2 = 2 \cdot 1^2 + 2 \cdot 1^2 = 4$ but $\|f + g\|^2 + \|f - g\|^2 = (3/2)^2 + (1/2)^2 = 10/4 \neq 4$.

Chapter 6
Differentiable Mappings

6.1   Definition of the Derivative

1. $Df(x) = \sin x + x \cos x$

2. $\displaystyle \lim_{x \to x_0} \frac{\|[f(x) + g(x)] - [f(x_0) + g(x_0)] - [Df(x_0) + Dg(x_0)](x - x_0)\|}{\|x - x_0\|}$

$\displaystyle = \lim_{x \to x_0} \frac{\|f(x) - f(x_0) - Df(x_0)(x - x_0)\|}{\|x - x_0\|}$

$\displaystyle + \lim_{x \to x_0} \frac{\|g(x) - g(x_0) + Dg(x_0)(x - x_0)\|}{\|x - x_0\|} = 0.$

Thus by the definition of the derivative $D(f + g) = Df + Dg$.

4. First $f(0) = 0$, since $\|f(0)\| \leqslant M \cdot 0^2 = 0$. Now let $\varepsilon > 0$ be given and $\delta = \varepsilon/M$. Then $\|x\| < \delta$ implies $\dfrac{\|f(x) - f(0) - 0\|}{\|x - 0\|} = \dfrac{\|f(x)\|}{\|x\|} \leqslant M\|x\| < M \cdot \varepsilon/M = \varepsilon,$

hence $\displaystyle \lim_{x \to 0} \frac{\|f(x) - f(0) - 0\|}{\|x - 0\|} = 0.$

5. No. Let $f(x) = x$, then $Df(x) = 1$ for all x.

6. For $f(x) = \sqrt{x}$ on $[0,1]$, yes but for $g(x) = \sqrt{|x|}$ on $[-1,1]$; no, since g is not differentiable at 0.

6.2   Matrix Representation

1. $Df(x,y,z) = \begin{pmatrix} 4yx^3 & x^4 & 0 \\ e^z & 0 & xe^z \end{pmatrix}.$

2. $Df(x,y,z) = \operatorname{grad} f(x,y,z) = (2xe^{x^2+y^2+z^2}, 2ye^{x^2+y^2+z^2}, 2ze^{x^2+y^2+z^2}).$

3. By Exercise 2, Section 6.1, $D(L + g)(0) = DL(0) + Dg(0)$, and by Exercise 4, Section 6.1 $Dg(0) = 0$, hence $Df(0) = DL(0) + 0 = L$, by Example 2.

6.3 Continuity of Differentiable Mappings; Differentiable Paths

1. We show $f'(0) = 0$. Let $\varepsilon > 0$ be given, and $|x - 0| = |x| < \varepsilon$, then $\left| \dfrac{f(x) - f(0)}{x - 0} \right| =$ $\left| \dfrac{f(x)}{x} \right| \leqslant \left| \dfrac{x^2}{x} \right| = |x| < \varepsilon$. f is continuous at 0 since it is differentiable there.

2. No, for $f(x) = |x|$, where $M = 1$ for all x, is not differentiable at $x = 0$.

3. No; let $f(x) = -|x|$. Maximum of f occurs at $x = 0$ but f is not differentiable there.

4. f is continuous but not differentiable at $x = 0$.

5. $c'(1) = (6,e,3)$.

6.4 Conditions for Differentiability

1. Show that $\partial f/\partial x$ and $\partial f/\partial y$ are continuous at $(0,0)$.

2. By computing limits of difference quotients we find $\partial f(0,0)/\partial x = \partial f(0,0)/\partial y = 0$. Thus if f were differentiable, $Df(0,0)$ would have to be the constant function 0 (by Theorem 2). But

$$\lim_{(x,y) \to (0,0)} \frac{|f(x,y) - f(0,0) - 0|}{\|(x,y) - (0,0)\|} = \lim_{(x,y) \to (0,0)} \frac{|f(x,y)|}{\|(x,y)\|} = \lim_{(x,y) \to (0,0)} \frac{|xy|}{x^2 + y^2}$$

does not exist, since if we go along the path $y = Mx$ we get

$$\lim_{(x,y) \to 0} \frac{|xy|}{x^2 + y^2} = \lim_{x \to 0} \frac{Mx^2}{x^2 + M^2 x^2} = \lim_{x \to 0} \frac{M}{1 + M^2} = \frac{M}{1 + M^2},$$

which is different for every M. This gives an example of a function all of whose directional derivatives exist at every point, but which itself is not differentiable.

3. $z = 0$.

4. $f(x,y) = x^3 + y^4$ and $Df(x,y) = (3x^2, 4y^3)$, so $Df(1,3) = (3,108)$. Thus the tangent plane is $z = 82 + (3,108) \begin{pmatrix} x - 1 \\ y - 3 \end{pmatrix} = -245 + 3x + 108y$.

6.5 The Chain Rule or Composite Mapping Theorem

1.
$$\frac{\partial h}{\partial x} = \frac{\partial f}{\partial u} \frac{\partial u}{\partial x} + \frac{\partial f}{\partial v} \frac{\partial v}{\partial x}$$

$$\frac{\partial h}{\partial y} = \frac{\partial f}{\partial u} \frac{\partial u}{\partial y} + \frac{\partial f}{\partial v} \frac{\partial v}{\partial y} + \frac{\partial f}{\partial w} \frac{\partial w}{\partial y}$$

$$\frac{\partial h}{\partial z} = \frac{\partial f}{\partial u} \frac{\partial u}{\partial z} + \frac{\partial f}{\partial w} \frac{\partial w}{\partial z}$$

where $\dfrac{\partial f}{\partial u}, \dfrac{\partial f}{\partial v}$, and $\dfrac{\partial f}{\partial w}$ are evaluated at $g(x,y,z)$ and denote the partials of f with respect to the 1st, 2nd, and 3rd variables of f respectively, and $\dfrac{\partial u}{\partial x}, \dfrac{\partial u}{\partial y}$, and so forth are evaluated at (x,y,z).

3. $\dfrac{\partial F}{\partial y} = 2yf'(x^2 + y^2)$ so $x\dfrac{\partial F}{\partial y} = 2xyf'(x^2 + y^2)$ and

$\dfrac{\partial F}{\partial x} = 2xf'(x^2 + y^2)$ so $y\dfrac{\partial F}{\partial x} = 2xyf'(x^2 + y^2)$.

4. If $h(r,\theta,\varphi) = f(r\cos\theta\sin\varphi, r\sin\theta\sin\varphi, r\cos\varphi)$ where $f: \mathbb{R}^3 \to \mathbb{R}$ then

$$\frac{\partial h}{\partial r} = \frac{\partial f}{\partial x}\cos\theta\sin\varphi + \frac{\partial f}{\partial y}\sin\theta\sin\varphi + \frac{\partial f}{\partial z}\cos\varphi$$

$$\frac{\partial h}{\partial \theta} = -\frac{\partial f}{\partial x}r\sin\theta\sin\varphi + \frac{\partial f}{\partial y}r\cos\theta\sin\varphi$$

$$\frac{\partial h}{\partial \varphi} = \frac{\partial f}{\partial x}r\cos\theta\cos\varphi + \frac{\partial f}{\partial y}r\sin\theta\cos\varphi - \frac{\partial f}{\partial z}r\sin\varphi$$

where $\dfrac{\partial f}{\partial x}, \dfrac{\partial f}{\partial y},$ and $\dfrac{\partial f}{\partial z}$ are evaluated at $(r\cos\theta\sin\varphi, r\sin\theta\sin\varphi, r\cos\varphi)$.

5. Since $F(x,f(x)) = 0 = $ constant, we have $\dfrac{\partial F(x,f(x))}{\partial x} = 0$. Thus

$$0 = \frac{\partial F(x,f(x))}{\partial x} = \left(\frac{\partial F}{\partial x}\bigg|_{(x,f(x))} \cdot \frac{\partial x}{\partial x}\right) + \left(\frac{\partial F}{\partial y}\bigg|_{(x,f(x))} \cdot f'(x)\right)$$

$$= \frac{\partial F}{\partial x}\bigg|_{(x,f(x))} + \left(\frac{\partial F}{\partial y}\bigg|_{(x,f(x))} \cdot f'(x)\right)$$

and the result follows.

6.6   Product Rule and Gradients

1. Let $g(t) = x_0 + th$. Then $Dg(t) = h$ and $\dfrac{d}{dt}f(x_0 + th)\bigg|_{t=0} = \dfrac{d}{dt}f\circ g(t)\bigg|_{t=0} = $
$Df(g(0)) \cdot Dg(0) = Df(x_0) \cdot h$.

2. Let $F(x,y,z) = x^2 - y^2 + xyz - 1$, then grad $F(x,y,z) = (2x + yz, -2y + xz, xy)$ and grad $F(1,0,1)/\|$grad $F(1,0,1)\| = (2,1,0)/\sqrt{5}$.

3. The equation is

$$\langle \text{grad } F(1,0,1),(x,y,z)\rangle = \langle (2,1,0),(x - 1,y,z - 1)\rangle = 2x + y - 2 = 0.$$

4. In the direction of grad $f(x,y) = (2xye^{x^2}, e^{x^2})$.

6. The surface $z = f(x_1,\ldots,x_n)$ in $\mathbb{R}^{n+1}$ may be written as the set of those points $(x_1,\ldots,x_n,z)$ satisfying $F(x_1,\ldots,x_n,z) = 0$ where $F(x_1,\ldots,x_n,z) = f(x_1,\ldots,x_n) - z$. The tangent plane at (x_0,z_0) is $\langle (x - x_0, z - z_0),\text{grad } F(x_0,z_0)\rangle = 0$ which becomes $z = z_0 + Df(x_0) \cdot (x - x_0)$. The unit sphere $x^2 + y^2 + z^2 = 1$ in $\mathbb{R}^3$ is a surface of the form $F(x,y,z) = c$ which is not the graph of a function so the analysis of p. 165 does not apply.

6.7   Mean-Value Theorem

1. Let $x, y \in \mathbb{R}, x < y$. Then there exists a $c \in]x,y[$ such that $f(y) - f(x) = f'(c)(y - x)$, and since $f'(c) > 0, f(y) - f(x) = f'(c)(y - x) > 0$; hence $f(y) > f(x)$.

2.

$$\frac{f'(x_0)}{g'(x_0)} = \frac{\underset{x \to x_0}{\text{limit}} \dfrac{f(x) - f(x_0)}{x - x_0}}{\underset{x \to x_0}{\text{limit}} \dfrac{g(x) - g(x_0)}{x - x_0}} = \underset{x \to x_0}{\text{limit}} \left(\frac{f(x) - 0}{g(x) - 0} \right) = \underset{x \to x_0}{\text{limit}} \frac{f(x)}{g(x)}.$$

3. (a) $\underset{x \to 0}{\text{limit}} \dfrac{\sin x}{x} = \underset{x \to 0}{\text{limit}} \dfrac{\cos x}{1} = \cos 0 = 1.$

   (b) $\underset{x \to 0}{\text{limit}} \dfrac{e^x - 1}{x} = \underset{x \to 0}{\text{limit}} \dfrac{e^x}{1} = e^0 = 1.$

5. This is an immediate consequence of Theorem 7(i). If A is not convex this is not necessarily true. Let $A = \{x \in \mathbb{R} \mid x < 0 \text{ or } x > 1\}$ and define $f: A \to \mathbb{R}$ by
$f(x) = \begin{cases} 1, & x > 1 \\ 0, & x < 0 \end{cases}$. Then f is differentiable on A with $f'(x) = 0$ for all $x \in A$, so
for all $x \in A, |f'(x)| < 1/10$, but if $x = -y = 2, |f(x) - f(y)| = 1 > (1/10)|x - y| = 4/10$.

6.8   Taylor's Theorem and Higher Derivatives

2. Verify the conditions of Example 2.

3. f is not C^1 but is only differentiable. However Taylor's theorem for $r = 1$ in the form $f(0 + h) = f(0) + f'(0) \cdot h + R_1(0,h)$ where $R_1(0,h)/h \to 0$ as $h \to 0$ is valid.

4. The Taylor series representation is $-x - (1/2)x^2 - (1/3)x^3 - \cdots = \sum_{k=1}^{\infty} (1/k)x^k$. Now for $k > 1, |f^{(k)}(0)| = |(-1)(k - 1)| = k - 1 < 2^k$, so by Example 2, $\log(1 - x) = \sum_{k=1}^{\infty} -(1/k)x^k$ for $x \in]-1,1[$. Finally, let δ be such that $0 < \delta < 1$. Then for any $x \in [-\delta,\delta], |a_n| = |(-1)x^n/n| < \delta^n$, and since $\sum_{k=1}^{\infty} \delta^k$ converges, by the Weierstrass M-test, $\sum_{k=1}^{\infty} (-1)x^k/k$ converges uniformly on $[-\delta,\delta]$.

6. $f(h,k) = 1 + h + h^2/2 - k^2/2 + R_2((h,k),0)$,   where $R_2((h,k),0)/|(h,k)|^2 \to 0$ as $(h,k) \to (0,0)$.

6.9   Maxima and Minima

2. $Df(x,y) = (2x + 2y, 2x + 2y) = 0$ iff $x = -y$. Now $-D^2 f(x,y) = \begin{pmatrix} -2 & -2 \\ -2 & -2 \end{pmatrix}$
and so $\Delta_1 = -2$ and $\Delta_2 = 0$. Thus the test fails. However, $f(x,y) = (x + y)^2 + 6$, so $(0,0)$ is a minimum.

3. Local minimum.

4. Assume A is positive definite, and suppose $Ax = \lambda x$. Then $\langle x,Ax \rangle = \langle x,\lambda x \rangle = \lambda \langle x,x \rangle$ is positive and since $\langle x,x \rangle$ is positive, λ is positive. *Note:* The converse, that is, eigenvalues of A positive implies A positive definite, is also true and is not hard to prove using the fact that a symmetric matrix can be diagonalized by an orthogonal matrix.

Exercises for Chapter 6 (at end of chapter)

2. f_i differentiable implies there exists a $\delta_i > 0$ such that if $|x - x_0|$ then
$$\left| f_i(x) - f_i(x_0) - \frac{df_1}{dx}(x_0)(x - x_0) \right| < \frac{\varepsilon}{m} |x - x_0|.$$

Let $\delta = \min\{\delta_i \mid i = 1, \ldots, m\}$, then $|x - x_0| < \delta$ implies

$$\left\| f(x) - f(x_0) - \left(\frac{df_1}{dx}(x_0), \ldots, \frac{df_m}{dx}(x_0) \right)(x - x_0) \right\|$$

$$= \left\| f_1(x) - f_1(x_0) - \frac{df_1}{dx}(x_0)(x - x_0), \ldots, f_m(x) - f_m(x_0) - \frac{df_m}{dx}(x_0)(x - x_0) \right\|$$

$$\leqslant \left| f_1(x) - f_1(x_0) - \frac{df_1}{dx}(x_0)(x - x_0) \right| + \cdots + \left| f_m(x) - f_m(x_0) - \frac{df_m}{dx}(x_0)(x - x_0) \right|$$

$$\leqslant \left(\frac{\varepsilon}{m} + \cdots + \frac{\varepsilon}{m} \right) |x - x_0| = \varepsilon |x - x_0| .$$

Hence f is differentiable at x_0.

3. If $f = 0$ the exercise is complete, so suppose there exists a $x_0 \in [0, \infty[$ such that $f(x_0) \neq 0$ say $f(x_0) > 0$. (The argument if $f(x_0) < 0$ is similar.) By the intermediate value theorem there exists a $x_1 \in]0, x_0[$ such that $f(x_1) = f(x_0)/2$. Since $f(x) \to 0$, there exists a $y > x_0$ such that $f(y) < f(x_0)/2$, so again by the intermediate value theorem there exists a $x_2 \in]x_0, y[$ such that $f(x_2) = f(x_0)/2$. Then if $g(x) = f(x) - f(x_0)/2$, $g(x_1) = g(x_2) = 0$, and therefore by Rolle's theorem there exists a $x_3 \in]x_1, x_2[$ such that $g'(x_3) = f'(x_3) = 0$.

5. (a) $(2x \cos(x^2 + y^3) \quad 3y^2 \cos(x^2 + y^3))$.

   (b) $\begin{pmatrix} z\cos x & 0 & \sin x \\ 0 & z\cos y & \sin y \end{pmatrix}$.

   (c) $(y \quad x)$.

   (d) $(2x \quad 2y)$.

   (e) $\begin{pmatrix} y\cos(xy) & x\cos(xy) \\ -y\sin(xy) & -x\sin(xy) \\ 2y^2x & 2x^2y \end{pmatrix}$.

   (f) $((y + z)x^{y+z-1} \quad (\ln x)x^{y+z} \quad (\ln x)x^{y+z})$.

   (g) $(yz \quad xz \quad xy)$.

   (h) $\begin{pmatrix} (y\ln z)z^{xy} & (x\ln z)z^{xy} & (xy)z^{xy-1} \\ 2x & 0 & 0 \\ yz/\cos^2(xyz) & xz/\cos^2(xyz) & xy/\cos^2(xyz) \end{pmatrix}$.

7. (a) $(3,6)$ is a local minimum and $(1,2)$ is a saddle.

   (b) $(\pm n\pi + \pi/2, 1)$ for n even are saddle points; $(\pm n\pi + \pi/2, 1)$ for n odd are local minima.

   (d) The critical points are the plane $z = -x, -y$. They are all local minima since $f(x,y,z) = 0$ there and by inspection $f(x,y,z)$ is always $\geqslant 0$. (The theorems on the Hessian fail since the Hessian has $\Delta_3 = 0$.)

8. (a), (b), and (c) are immediate consequences of Theorem 12, the definition of $H_{x_0}(f)$, and the conditions for positive and negative definiteness of a matrix given on page 185.

12. Let $h_x: \mathbb{R} \to \mathbb{R}^n$, $t \mapsto tx$, for $x \in \mathbb{R}$, and $g = f \circ h_x: \mathbb{R} \to \mathbb{R}$, so $g(t) = f(tx) = t^m f(x)$. Then differentiating,

$$Dg(t) = Df(h_x(t)) \circ Dh_x(t) = Df(tx)(x) = \frac{d}{dt}(tf(x)) = mt^{m-1}f(x),$$

so, setting $t = 1$, $Df(x)(x) = mf(x)$. Now let $L: \mathbb{R}^{n_1} + \cdots + \mathbb{R}^{n_k} \to \mathbb{R}^m$ be multi-linear. Then $L(tx) = L(tx_1,\ldots,tx_k) = tL(x_1,tx_2,\ldots,tx_k) = \cdots = t^k L(x_1,\ldots,x_k)$, and therefore L is homogeneous of degree t.

13. (a) Let $T: \mathbb{R}^3 \to \mathbb{R}^3$, $(x,y,z) \mapsto (h(x),g(x,y),z)$, then $F = f \circ T$ and $DF(x,y,z) = Df(T(x,y,z)) \circ DT(x,y,z)$

$$= \left(\frac{\partial f}{\partial h} \quad \frac{\partial f}{\partial g} \quad \frac{\partial f}{\partial z}\right)_{(T(x,y,z))} \begin{pmatrix} \dfrac{\partial h}{\partial x} & 0 & 0 \\[2mm] \dfrac{\partial g}{\partial x} & \dfrac{\partial g}{\partial y} & 0 \\[2mm] 0 & 0 & 1 \end{pmatrix}_{(x,y,z)}$$

$$= \left(\frac{\partial f}{\partial h} \cdot \frac{\partial h}{\partial x} + \frac{\partial f}{\partial g} \cdot \frac{\partial g}{\partial x}, \frac{\partial f}{\partial g} \cdot \frac{\partial g}{\partial y}, \frac{\partial f}{\partial z}\right)_{(x,y,z)}$$

is the general formula for $DF(x,y,z)$.

(b)
$$\frac{\partial G(x,y,z)}{\partial x} = \frac{dh}{dw}(f(x,y,z) \cdot g(x,y)) \cdot \left[f(x,y,z)\frac{\partial g(x,y)}{\partial x} + g(x,y)\frac{\partial f(x,y,z)}{\partial x}\right],$$

$$\frac{\partial G(x,y,z)}{\partial y} = \frac{dh}{dw}(f(x,y,z) \cdot g(x,y)) \cdot \left[f(x,y,z)\frac{\partial g(x,y)}{\partial y} + g(x,y)\frac{\partial f(x,y,z)}{\partial y}\right],$$

$$\frac{\partial G(x,y,z)}{\partial z} = \frac{dh}{dw}(f(x,y,z) \cdot g(x,y)) \cdot g(x,y) \cdot \frac{\partial f(x,y,z)}{\partial z},$$

are the general formulas. For the specific f, g, and h in the problem we have

$$\frac{\partial G}{\partial x} = \cos((x^2 + yz) \cdot (y^3 + xy)) \cdot (2xy^3 + 3x^2y + y^2z),$$

$$\frac{\partial G}{\partial y} = \cos((x^2 + yz)(y^3 + xy)) \cdot (3x^2y^2 + x^3 + 4y^3z + 2xyz),$$

$$\frac{\partial G}{\partial z} = \cos((x^2 + yz)(y^3 + xy)) \cdot (y^4 + y^2x).$$

15. $S = f^{-1}(\{0\})$ is closed since f is continuous and $S \subset [0,1]$ so it is bounded, hence S is compact. If S is infinite, by the Bolzano-Weierstrass theorem, S has an accumulation point $x_0 \in S$, so $f(x_0) = 0$. Choose $\{x_n\} \subset B$ such that $x_n \to x_0$ and for all n, $x_n \neq x_0$. Then $f'(x_0) = \lim_{n \to \infty} \frac{f(x_n) - f(x_0)}{x_n - x_0} = \lim_{n \to \infty} \frac{0 - 0}{x_n - x_0} = 0$, contradicting the hypothesis that there is no $x \in \mathbb{R}$ such that $f(x) = 0 = f'(x)$. Thus S is finite.

16. Let $g(x) = f(x) - Df(0)(x)$. Then since $Df(0) = Df(x_0)$,

$$\|g(x_0 + h) - g(x_0)\| = \|f(x_0 + h) - Df(0)(x_0 + h) - f(x_0) + Df(0)(x_0)\|$$

$$= \|f(x_0 + h) - f(x_0) - Df(0)(h)\| < \varepsilon \|h\|$$

for $\|h\| < \delta(\varepsilon)$, hence $Dg(x_0) = 0$, and since x_0 was arbitrary, for all $x \in \mathbb{R}^n$, $Dg(x) = 0$ implies g is a constant function and thus $f = Df(0) + c, c \in \mathbb{R}^m$.

17. Imitate the proof of Theorem 12.

18. By the intermediate value theorem $f(x) = x^3 + bx + c$ has at least one root since $f \to \infty$ as $x \to \infty$ and $f \to -\infty$ as $x \to -\infty$. Now suppose $x_1 < x_2$ and $f(x_1) = f(x_2) = 0$, then there exists a $x_3 \in]x_1, x_2[$ such that $f'(x_3) = 3x_3^2 + b = 0$, that is, $3x_3^2 = -b$; but $b > 0$, a contradiction.

19. (a) $f(x,y) = x^2 + 2xy + y^2 + 0$.
    (b) $f(x,y) = 1 + x + y + \frac{1}{2}(x^2 + 2xy + y^2) + R_2(x,y)$.

20. (a) Clearly $\|0\| = 0$. Conversely, assume $\|L\| = 0$. Then for all $\varepsilon > 0$ there exists a $M < \varepsilon$ with $\|Lx\| \leqslant M \|x\|$ for all $x \in \mathbb{R}^n$. Let $x \in \mathbb{R}^n$ and $\varepsilon > 0$, so there exists a $M < \varepsilon/\|x\|$ such that $\|Lx\| \leqslant M \|x\| < \varepsilon$. Since x and ε were arbitrary, $L(x) = 0$, and $L = 0$.
    (b) Let $a \in \mathbb{R}$. $\|aL\| = \inf\{M \mid \|a(Lx)\| \leqslant M \|x\|$ for all $x\} = \inf\{M \mid |a| \|Lx\| \leqslant M \|x\|$ for all $x\} = |a| \inf\{M \mid \|Lx\| \leqslant M \|x\|\} = |a| \|L\|$.
    (c) Clearly $\|L\| \geqslant 0$ for all L.
    (d) $\|L_1 + L_2\| = \inf\{M \mid \|(L_1 + L_2)x\| \leqslant M \|x\|$ for all $x\}$. We have $A = \{M \mid \|(L_1 + L_2)x\| \leqslant M \|x\|$ for all $x\} \supset \{M \mid \|L_1(x)\| + \|L_2(x)\| \leqslant M \|x\|$ for all $x\} \supset \{M \mid \|L_1\| + \|L_2\| \leqslant M\} = B$.

21. This is a direct consequence of Theorem 12 and the discussion on page 185.

22. $f:]0,1[\to \mathbb{R}$, $f(x) = x$. For $f: \mathbb{R} \to \mathbb{R}$, no (see Exercises 2 of Section 4.5 and Exercise 6 at the end of Chapter 4). In fact, any *bounded* continuous function $f:]0,1[\to \mathbb{R}$ will have a graph which is not closed. If A is closed then the graph of f must be closed. If $\{(x_k, f(x_k))\}$ is a convergent sequence in the graph G of f, then $\lim_{k \to \infty} x_k = x \in A$, since A is closed. By the continuity of f, $f(x_k) \to f(x)$, hence $(x_n, f(x_n)) \to (x, f(x)) \in G$. (Provided by Dave Nishball.)

23. $0 + 0 - 1/2x^2 + 0 - 2/41x^4$.

25. Work through the proof of Theorem 4 and notice that continuity of $\partial f/\partial x^n$ is not necessary.

26. (a) $f'(a) = \lim_{h \to 0} \dfrac{f(a + h) - f(a)}{h} = \lim_{h \to 0^+} \dfrac{f(a + h) - f(a)}{h} = \lim_{h \to 0^+} f'(x_h)$, for some $x_h \in]a, h[$, by the mean value theorem, so $f'(a) = \lim_{h \to 0^+} f'(x_h) = l$.
    (b) No, since $\lim_{x \to 0^+} f(x) = 1 \neq f(0)$.

28. **Lemma.** Let $f: [a,b] \to \mathbb{R}$ be differentiable on $[a,b]$ and suppose $f'(b) > 0$ and $f'(a) < 0$. Then there exists a $x_0 \in]a,b[$ such that $f'(x_0) = 0$.
    Proof: Since $f'(a) < 0$, f has a local maximum at a. Similarly $f'(b) > 0$ implies f has a local maximum at b. By the compactness of $[a,b]$, $\inf f([a,b]) = f(x_0)$ for some $x_0 \in [a,b]$. Now by the above, $x \neq a$ and $x_0 \neq b$, hence $x_0 \in]a,b[$, so $f(x_0)$ is a local minimum of f on some open interval and hence $f'(x_0) = 0$. Now suppose

$f'(b) > c > f'(a)$. To show there exists a $x_0 \in \,]a,b[$ with $f'(x_0) = c$. Let $g(x) = f(x) - cx$. Then $g'(x) = f'(x) - c$ and $f'(b) > c > f'(a)$ implies $f'(b) - c = g'(b) > 0 > f'(a) - c = g'(a)$. Hence by the lemma there exists a $x_0 \in \,]a,b[$ such that $g'(x_0) = 0 = f'(x_0) - c$, that is, $f'(x_0) = c$. (Provided by Cindy Fleming.)

29. (a) Geometric series test. $f(x) = \begin{cases} xe^x & \text{if } x > 0 \\ e^x - 1 & \text{if } x = 0 \,. \\ 0 & \end{cases}$

(b) No; use l'Hopital's rule.

(c) No on $[0,\infty[$, yes on $[\delta,\infty[$ for all $\delta > 0$.

(d) No on $[0,\infty[$, yes on $]0,\infty[$ (apply the M-test on each interval $[\delta,\infty[$, $\delta > 0$).

30. f differentiable does imply f continuous. f may not assume its maximum, hence T may be empty. $f'(x) = 0$ does not imply f has a maximum or minimum. $f(x)$ a maximum does not imply $f(x) \geqslant 0$ there (for example, $f(x) = -3$). $T \neq S \cap \{x \mid f(x) \geqslant 0\}$. ($\{x \mid f(x) \geqslant 0\}$ is closed and S is closed since f and f' are continuous.) T really is closed, since $T = f^{-1}(a)$ ($a = \sup(f)$), $\{a\}$ is closed and f is continuous.

32. $\dfrac{\partial f}{\partial x}(0,0) = \underset{h \to 0}{\text{limit}} \dfrac{f(h,0) - f(0,0)}{h} = \underset{h \to 0}{\text{limit}} \dfrac{0 - 0}{h} = 0$ and if $(x,y) \neq 0$,

$$\frac{\partial f}{\partial x}(x,y) = \frac{x^4 y + 4x^2 y^3 - y^5}{x^4 + 2x^2 y^2 + y^4},$$

so

$$\frac{\partial^2 f}{\partial y \partial x}(0,0) = \underset{k \to 0}{\text{limit}} \frac{\dfrac{\partial f}{\partial x}(0,k) - \dfrac{\partial f}{\partial x}(0,0)}{k} = \underset{k \to 0}{\text{limit}} \frac{-k^5}{k \cdot k^4} = -1 \,.$$

Similarly,

$$\frac{\partial f}{\partial y}(0,0) = 0 \,, \qquad \frac{\partial f}{\partial y}(x,y) = \frac{x^5 - 4x^3 y^2 - xy^4}{x^4 + 2x^2 y^2 + y^4},$$

and

$$\frac{\partial^2 f}{\partial x \, \partial y}(0,0) = \underset{h \to 0}{\text{limit}} \frac{\dfrac{\partial f}{\partial y}(h,0) - \dfrac{\partial f}{\partial y}(0,0)}{h} = \underset{h \to 0}{\text{limit}} \frac{h^5}{h^4 h} = +1 \neq \frac{\partial^2 f}{\partial y \, \partial x} \,.$$

34. $x_n = 1/2 + x_{n-1}$; $\underset{n \to \infty}{\text{limit}}\, x_n = \sqrt{2} - 1$. (See solution to Exercise 26, Chapter 2 for similar methods.)

35. Suppose x_1, x_2, $x_3 \in \,]a,b[$ are such that $x_1 < x_2 < x_3$ and $f(x_1) = f(x_2) = f(x_3) = 0$. Then there exists a $x_4 \in \,]x_1,x_2[$ and $x_5 \in \,]x_2,x_3[$ with $f'(x_4) = f'(x_5) = 0$, by Rolle's Theorem. Now apply Rolle's Theorem to f' so that there exists a $c \in \,]x_4,x_5[$ with $f''(c) = 0$.

Chapter 7
The Inverse and Implicit Function Theorems and Related Topics

7.1   Inverse Function Theorem

1. $\begin{vmatrix} \dfrac{\partial u}{\partial x} & \dfrac{\partial u}{\partial y} \\[2mm] \dfrac{\partial v}{\partial x} & \dfrac{\partial v}{\partial y} \end{vmatrix} = \begin{vmatrix} 2x & -2y \\ 2y & 2x \end{vmatrix} = 4x^2 + 4y^2 = 0$ iff $(x,y) = (0,0)$.

3. This does not contradict Theorem 1 since f is not C^1 at $x = 0$.

5. $\begin{vmatrix} \dfrac{\partial u}{\partial x} & \dfrac{\partial u}{\partial y} & \dfrac{\partial u}{\partial z} \\[2mm] \dfrac{\partial v}{\partial x} & \dfrac{\partial v}{\partial y} & \dfrac{\partial v}{\partial z} \\[2mm] \dfrac{\partial w}{\partial x} & \dfrac{\partial w}{\partial y} & \dfrac{\partial w}{\partial z} \end{vmatrix}_{(0,0)} = \begin{vmatrix} 1 + yz & xz & xy \\ y & 1 + x & 0 \\ 2 & 0 & 1 + 6z \end{vmatrix}_{(0,0)} = \begin{vmatrix} 1 & 0 & 0 \\ 0 & 1 & 0 \\ 2 & 0 & 1 \end{vmatrix} = 1 ,$

so the system is invertible in a neighborhood of $(0,0,0)$.

7.2   Implicit Function Theorem

2. $\dfrac{\partial F}{\partial y} = 2y + 1 = 0$ iff $y = -1/2$.

4. $\begin{vmatrix} \dfrac{\partial F_1}{\partial u} & \dfrac{\partial F_1}{\partial v} & \dfrac{\partial F_1}{\partial w} \\[2mm] \dfrac{\partial F_2}{\partial u} & \dfrac{\partial F_2}{\partial v} & \dfrac{\partial F_2}{\partial w} \\[2mm] \dfrac{\partial F_3}{\partial u} & \dfrac{\partial F_3}{\partial v} & \dfrac{\partial F_3}{\partial w} \end{vmatrix}_{(0,0,0,0,0,-2)} = \begin{vmatrix} 1 & 0 & 0 \\ 0 & 1 & 1 \\ 0 & 0 & 1 \end{vmatrix} = 1 \neq 0$

so u, v, w can be expressed in terms of x, y, z, for (x,y,z) in some neighborhood of $(0,0,0)$.

7.3   Straightening-Out Theorem

1. $x \neq 0$ and $y \neq 0$.

3. The theorem does not apply near $(0,0)$ since $Df(0,0) = (3x^2, 2y)_{(0,0)} = (0,0)$, but f can be straightened out near $(0,1)$ as $Df(0,1) = (0,2) \neq 0$.

7.4 Further Consequences of the Implicit Function Theorem

1. Yes, near $(0,1)$.

7.5 An Existence Theorem for Ordinary Differential Equations

3. Clearly $x = 0$ is a solution; and since

$$\frac{d}{dt}\left(\frac{t^2}{4}\right) = t/2 = \sqrt{t^2/4} \text{ for } t \geq 0, \ x(t) = \begin{cases} 0, & t \leq 0 \\ t^2/4, & t > 0 \end{cases}$$

is a solution. By Theorem 6, then, $f(x,t) = \sqrt{x}$ cannot be Lipschitz on a neighborhood of $x = 0$.

5. (a) The series $e^{tA} = \sum_{n=0}^{\infty} (t^n/n!)A^n$ is absolutely convergent; it can be differentiated term by term; and $(d/dt)(e^{tA}x(0)) = A \sum_{n=0}^{\infty} (t^n/n!)A^n x(0) = Ae^{tA}x(0)$.

   (b) Yes, it can be extended to ∞ by shifting the origin of time, $e^{tA} = e^{(t-b)A}e^{bA}$. The various times are $b, 2b, \ldots, nb, \ldots$.

7.6 The Morse Lemma

1. Index $= 1$

3. $x^2 - 2xy + y^2 = (x - y)^2$, $(0,0)$ is a degenerate critical point.

5. (a) Use Theorem 7 and the fact that critical points are "preserved" by a change of coordinates. You could also use Taylor's theorem to prove this.

7.7 Constrained Extrema and Lagrange Multipliers

1. $(\sqrt{2/3}, -\sqrt{2/3}, \sqrt{2/3})$ is a maximum and $(-\sqrt{2/3}, \sqrt{2/3}, \sqrt{-2/3})$ is a minimum.

2. No extrema.

3. $(\pm\sqrt{3}, 0)$.

4. $(9/\sqrt{70}, 4/\sqrt{70})$ (max) and $(-9\sqrt{70}, -4\sqrt{70})$ (min).

Exercises for Chapter 7 (at end of chapter)

1. $\dfrac{\partial f}{\partial x} = \dfrac{\partial g}{\partial u} + \dfrac{\partial g}{\partial v}\dfrac{\partial h}{\partial x}$.

4. Let $L: \mathbb{R}^n \to \mathbb{R}^n$, $x \mapsto (g_1(x_1), \ldots, g_n(x_n))$, then $h = f \circ L$ and in matrices

$$Dh(x) = Df(L(x)) \circ DL(x) = Df(L(x)) \begin{pmatrix} \dfrac{\partial L_1}{\partial x_1} & \cdots & \dfrac{\partial L_1}{\partial x_n} \\ & & \\ \cdot & & \cdot \\ \cdot & & \cdot \\ \cdot & & \cdot \\ \dfrac{\partial L_n}{\partial x_1} & \cdots & \dfrac{\partial L_n}{\partial x_n} \end{pmatrix} = Df(L(x)) \begin{pmatrix} g_1'(x_1) & & 0 \\ & \cdot & \\ & & \cdot \\ & & \cdot \\ 0 & & g_n'(x_n) \end{pmatrix}$$

6. From linear algebra since $Jf(x) \neq 0$ we have

$$Dg(y_0) = (Df(x_0))^{-1} = \frac{1}{Jf(x_0)} \operatorname{adj}(Df(x_0))$$

and

$$\operatorname{adj}(Df(x_0)) = \begin{vmatrix} \dfrac{\partial(f_2,f_3)}{\partial(x_2,x_3)} & -\dfrac{\partial(f_1,f_3)}{\partial(x_2,x_3)} & \dfrac{\partial(f_1,f_2)}{\partial(x_2,x_3)} \\[2mm] \dfrac{\partial(f_2,f_3)}{\partial(x_1,x_3)} & \dfrac{\partial(f_1,f_3)}{\partial(x_1,x_3)} & -\dfrac{\partial(f_1,f_2)}{\partial(x_1,x_3)} \\[2mm] \dfrac{\partial(f_2,f_3)}{\partial(x_1,x_2)} & -\dfrac{\partial(f_1,f_3)}{\partial(x_1,x_2)} & \dfrac{\partial(f_1,f_2)}{\partial(x_1,x_2)} \end{vmatrix}_{(x_0)} .$$

Hence

$$D_1 g_1(y_0) = \frac{1}{Jf(x_0)} \cdot \frac{\partial(f_2,f_3)}{\partial(x_2,x_3)} ,$$

and so on, and since

$$\frac{\partial(f_2,f_3)}{\partial(x_2,x_3)} = \begin{vmatrix} D_2 f_2(x_0) & D_2 f_3(x_0) \\ D_3 f_2(x_0) & D_3 f_3(x_0) \end{vmatrix} ,$$

and so on, combining we get

$$Jf(x_0)D_1 g_i(y_0) = \begin{vmatrix} \delta_{i,1} & D_1 f_2(x_0) & D_1 f_3(x_0) \\ \delta_{i,2} & D_2 f_2(x_0) & D_2 f_3(x_0) \\ \delta_{i,3} & D_3 f_2(x_0) & D_3 f_3(x_0) \end{vmatrix} .$$

10. (a) $Jf(x,y) = \dfrac{\partial f_1}{\partial x}\dfrac{\partial f_2}{\partial y} - \dfrac{\partial f_1}{\partial y}\dfrac{\partial f_2}{\partial x} = \dfrac{\partial f_1}{\partial x}\dfrac{\partial f_1}{\partial x} - \dfrac{\partial f_1}{\partial y}\left(-\dfrac{\partial f_1}{\partial y}\right)$

$= \left(\dfrac{\partial f_1}{\partial x}\right)^2 + \left(\dfrac{\partial f_1}{\partial y}\right)^2 = 0$ iff $\dfrac{\partial f_1}{\partial x} = \dfrac{\partial f_1}{\partial y} = 0$ iff $\dfrac{\partial f_2}{\partial y} = \dfrac{\partial f_2}{\partial x} = 0$.

So the implicit function theorem says that if f satisfies the Cauchy Riemann equations and $Df(x,y) \neq 0$ then f is locally invertible.

11. (a) Use Exercise 3.
    (b) Use Theorem 4.

16. No.

18. $\begin{vmatrix} \dfrac{\partial F_1}{\partial u} & \dfrac{\partial F_1}{\partial v} \\[2mm] \dfrac{\partial F_2}{\partial u} & \dfrac{\partial F_2}{\partial v} \end{vmatrix} = \begin{vmatrix} 3u^2 & x \\ y & 3v^2 \end{vmatrix} = 9u^2 v^2 - xy$, so if $9u_0^2 v_0^2 \neq x_0 y_0$;

u and v can be expressed as functions of x and y near (x_0, y_0). And

$$
\begin{pmatrix} \dfrac{\partial u}{\partial x} & \dfrac{\partial u}{\partial y} \\[2ex] \dfrac{\partial v}{\partial x} & \dfrac{\partial v}{\partial y} \end{pmatrix} = - \begin{pmatrix} \dfrac{\partial F_1}{\partial u} & \dfrac{\partial F_1}{\partial v} \\[2ex] \dfrac{\partial F_2}{\partial u} & \dfrac{\partial F_2}{\partial v} \end{pmatrix}^{-1} \begin{pmatrix} \dfrac{\partial F_1}{\partial x} & \dfrac{\partial F_1}{\partial y} \\[2ex] \dfrac{\partial F_2}{\partial x} & \dfrac{\partial F_2}{\partial y} \end{pmatrix}
$$

so in particular $\dfrac{\partial u}{\partial x} = \left(\dfrac{\partial F_2}{\partial v} \dfrac{\partial F_1}{\partial x} - \dfrac{\partial F_1}{\partial v} \dfrac{\partial F_2}{\partial x} \right) \Big/ \left(\dfrac{\partial F_1}{\partial u} \dfrac{\partial F_2}{\partial v} - \dfrac{\partial F_1}{\partial v} \dfrac{\partial F_2}{\partial u} \right)$.

23. (a) In $\mathbb{R}^2$ an example of such a C is when C consists of rays from the origin, and areas between them.
    (b) Let $I = \{x \mid x \in C \text{ and } \|x\| = 1\}$. C is closed, hence I is compact and since f is continuous on I, there exists a $x_0 \in I$ such that $f(x_0) = \sup\{\|f(x)\| \mid x \in I\}$. Then for any $x \in C$, $x \neq 0$, $x/\|x\| \in I$ and $\|f(x)\| = \|x\| \|f(x/\|x\|)\| \leqslant \|x\| \|f(x_0)\| \|(x/\|x\|)\| = \|f(x_0)\| \|x\|$, so let $M = \|f(x_0)\|$. If $x = 0$, $\|f(x)\| = \|f(0 \cdot y)\| = 0 \|f(y)\| = 0 = M = 0$, for any $y \in \mathbb{R}^n$.

25. Show that f maps $\text{cl}(D(0,r))$ (a compact set) into itself, and satisfies the hypotheses of the contraction mapping principle.

29. Consult Section 5.8 and 5.9.

31.
$$ x(0) = 0, \qquad x_1(t) = 0, \qquad x_2(t) = 0 + \int_0^t (1 + 0)\, ds = t\,, $$

$$ x_3(t) = \int_0^t (1 + s^2)\, ds = t - \frac{t^3}{3}, \dots\,, $$

$$ x_n(t) = t + \frac{1}{3} t^3 + \frac{1}{3 \cdot 5} t^5 + \cdots + \frac{1}{3 \cdot 5 \cdots (2k - 3)} t^{2k-3}, \dots\,, $$

so

$$ x(t) = \sum_{k=2}^{\infty} \frac{1}{3 \cdot 5 \cdots (2k - 3)} t^{2k-3}\,. $$

The radius of convergence is given by

$$ \frac{1}{R} = \lim_{n \to \infty} \frac{|a_n + 1|}{|a_n|} = \lim_{k \to \infty} \frac{1}{(2k - 1)(2k - 2)} = 0 $$

implies $R = \infty$.

32. The index is 0.

Chapter 8
Integration

8.1 Review of Integration in $\mathbb{R}$ and $\mathbb{R}^2$

1. For $f = 1$, and $P = \{a = x_0, x_1, \dots, b = x_N\}$ any partition of $[a,b]$,

$$ U(f,P) = \sum_{n=0}^{N-1} 1(x_{n+1} - x_n) $$

$$= (b - x_{N-1}) + (x_{N-1} - x_{N-2}) + \cdots + (x_2 - x_1) + (x_1 - a)$$
$$= b - a = L(f,P) \,,$$

so $\sup\{U(f,P)\} = \inf\{L(f,P)\} = b - a$.

2. Let
$$P = \{x_0 = a, x_1, \ldots, x_N = b\}$$

be a partition of $[a,b]$. For any

$$n = 1, \ldots, N, \quad \sup_{x \in [x_{n-1}, x_n]} f(x) = S$$

is an upper bound for $\{g(x) \mid x \in [a,b]\}$, for $S \geqslant f(x) \geqslant g(x)$ for all $x \in [a,b]$. Thus for all n, $S_n \geqslant \sup_{x \in [x_{n-1}, x_n]} g(x) = t_n$, so

$$U(f,P) = \sum_{n=0}^{N-1} S_n(x_n - x_{n-1}) \geqslant U(g,P) = \sum_{n=0}^{N-1} t_n(x_n - x_{n-1}) \,.$$

Thus $\int_a^b g = \inf\{U(g,P)\}$ is a lower bound for $\{U(f,P)\}$, so $\int_a^b f = \inf\{U(f,P)\} \geqslant \int_a^b g$.

8.2   Integrable Functions

1. Let $m_i = \inf\{f(x) \mid x \in [x_i, x_{i+1}]\}$ and $M_i = \sup\{f(x) \mid x \in [x_i, x_{i+1}]\}$. We have $R = \sum f(c_i)(x_{i+1} - x_i)$ where $c_i \in [x_i, x_{i+1}]$, and since $m_i \leqslant f(c_i) \leqslant M_i$ we have $\sum m_i(x_{i+1} - x_{i-1}) \leqslant \sum f(c_i)(x_{i+1} - x_i) \leqslant \sum M_i(x_{i+1} - x_i)$, the desired result.

2. We use Riemann's condition. Let $\varepsilon > 0$ be given and let P_ε be the partition of $[0,1]$ $P_\varepsilon = \{0, 1/2 - \varepsilon/4, 1/2 + \varepsilon/4, 1\}$. Then clearly $L(f,P_\varepsilon) = 0$ and $U(f,P_\varepsilon) = \varepsilon/2$, so $U(f,P_\varepsilon) - L(f,P_\varepsilon) = \varepsilon/2 < \varepsilon$. Thus f is integrable. We know $\int_0^1 f(x)\,dx = \sup_P L(f,P)$ and $L(f,P) = 0$ for all partitions P, so $\int_0^1 f(x)\,dx = 0$.

8.3   Volume and Sets of Measure Zero

1. We show that given $\varepsilon > 0$ the upper half of the unit circle can be covered with rectangles whose total volume is $< \varepsilon/2$. We use Riemann's condition on the function $y = \sqrt{1 - x^2}$. This function f is integrable on $[0,1]$ (since it is continuous) so there is a partition P such that $U(f,P) - L(f,P) < \varepsilon/2$. However, $U(f,P) - L(f,P)$ is just the sum $\sum (M_i - m_i)[x_{i+1} - x_i]$ where $M_i = \sup\{f(x) \mid x \in [x_{i+1}, x_i]\}$ and $m_i = \inf\{f(x) \mid x \in [x_{i+1}, x_i]\}$. So let v_i be the rectangle $[x_i, x_{i+1}] \times [m_i, M_i]$. Then the v_i's cover the upper half of the unit circle and their total volume is $< \varepsilon/2$. Similarly we can cover the lower half with a finite number of rectangles whose total volume is $< \varepsilon/2$. Thus the whole unit circle can be covered with a finite number of rectangles whose total volume is $< \varepsilon$.

2. The answer to both parts is no. It does have measure zero.

5. No. The boundary of the rationals in $[0,1]$ is the whole interval $[0,1]$ which does not have measure zero.

8.4   Lebesgue's Theorem

1. f is bounded by 1 on $A = [-1,1]$, A is bounded and has volume by Corollary 1, and f has no discontinuities, so by Corollary 2 f is integrable on A.

3. $\int_A f = 1$.

4. Since f is continuous, A is open, and $f(x_0) > 0$, there is a neighborhood $D(x_0, \varepsilon)$ of x_0 on which f is >0. By Theorem 4(ii), if $\int_A f = 0$ then $\{x \in A \mid f(x) \neq 0\}$ has measure zero. But $D(x_0, \varepsilon) \subset \{x \in A \mid f(x) \neq 0\}$, so if $\{x \in A \mid f(x) \neq 0\}$ had measure zero then so would $D(x_0, \varepsilon)$ (see remarks after Example 1). But $D(x_0, \varepsilon)$ clearly does not have measure zero, so $\{x \in A \mid f(x) \neq 0\}$ does not have measure zero, and therefore $\int_A f \neq 0$. But $\int_A f \geq 0$ since $f \geq 0$, so $\int_A f > 0$.

8.5 Properties of the Integral

1. Not necessarily. Let $r_1, r_2, \ldots, r_n, \ldots$ be an enumeration of the rationals in $[0,1]$ and let $A_1 = \{r_1\}$, $A_2 = \{r_2\}, \ldots$. Each A_i has volume but $A = \bigcup_{i=1}^{\infty} A_i =$ rationals in $[0,1]$ does not have volume.

4. $A \cap B$ has zero volume implies $A \cap B$ has measure zero implies $\int_{A \cup B} 1 = \int_A 1 + \int_B 1$ (by Theorem 5(vii)) implies $v(A \cup B) = v(A) + v(B)$.

8.6 Fundamental Theorem of Calculus

3. $\frac{1}{6}(e^{75} - 1)$.

5. (a) $[0,1]$ has volume and f is bounded and the set of discontinuities is countable and so has measure zero.

   (b) $L(f,P) = 0$ for all P since in every open set there is an irrational so $\int_0^1 f(x)\,dx = \sup\{L(f,P)\} = 0$.

8.7 Improper Integrals

2. Let $n \in N$ be such that $n > p + 2$, then since $e^x = 1 + x + x^2/2! + x^3/3! + \cdots \geq x^{n+1}/(n+1)!$, if $x \geq 1$ then $(n+1)!/x \geq e^{-x}x^n > e^{-x}x^{p+2}$, and $(n+1)!/x \to 0$ implies $e^{-x}x^p/x^2 \to 0$. In particular there exists a N such that $x > N$ implies $e^{-x}x^{p+2} < 1$, that is, $e^{-x}x^p < 1/x^2$. Then by the comparison test, $\int_1^\infty e^{-x}x^p < \infty$.

3. $e^{-x}x^p \leq x^r$ so use the comparison test and Example 2(b).

5. $\displaystyle\int_1^\infty \frac{x^\alpha}{1 + x^\alpha}$ diverges for $\alpha \geq 0$ by the comparison test since $\dfrac{x^\alpha}{1 + x^\alpha} \geq 1/2$ if $x \geq 1$.

   It diverges also for $-1 \leq \alpha < 0$ because $\dfrac{x^\alpha}{1 + x^\alpha} = \dfrac{1}{1 + x^{-\alpha}} \geq \dfrac{1}{2x^{-\alpha}}$ if $x \geq 1$, and

   $\displaystyle\int_1^\infty \frac{1}{2} x^\alpha\,dx$ diverges by Example 2(a). Finally $\displaystyle\int_1^\infty \frac{x^\alpha}{1 + x^\alpha}\,dx$ converges for $\alpha < -1$

   because $\dfrac{1}{1 + x^{-\alpha}} \leq \dfrac{1}{x^{-\alpha}} = x^\alpha$ and $\displaystyle\int_1^\infty x^\alpha\,dx$ converges by Example 2(a).

8.8 Some Convergence Theorems

1. Use Dini's theorem.

3. 1.

Exercises for Chapter 8 (at end of chapter)

1. (a) Let $B = \{x \in A \mid g(x) \neq f(x)\}$, then B has measure zero, so, assuming that the function $f - g$ is integrable on B and $A \backslash B$, $\int_A (f - g) = \int_{A \backslash B} (f - g) +$

$\int_B (f - g) = 0 + 0 = 0$, since $f - g = 0$ on $A \backslash B$, and by Theorem 4(i), since B has measure zero, $\int_B (f - g) = 0$. Thus $\int_A (f - g) = 0$, that is, $\int_A f = \int_A g$.

(b) $|f - g| \geqslant 0$ so apply Theorem 4(ii) to obtain that $\{x \mid f(x) - g(x) \neq 0\}$ has measure zero.

5. We show that the xy plane has measure zero in $\mathbb{R}^3$ and hence any subset of the xy plane has measure zero. Let $\varepsilon > 0$ and

$$S_n = [-n,n] \times [-n,n] \times \left[-\frac{\varepsilon}{n^2 2^{n+4}}, \frac{\varepsilon}{n^2 2^{n+4}} \right].$$

Then xy plane $\subset \bigcup_{i=1}^{\infty} S_i$ and $\sum_{i=1}^{\infty} v(S_i) = \sum_{i=1}^{\infty} \frac{2n \cdot 2n \cdot 2\varepsilon}{n^2 2^{n+4}} = \sum_{i=1}^{\infty} \frac{\varepsilon}{2^{n+1}} = \frac{\varepsilon}{2} < \varepsilon.$

6. $\int_A g - \int_A f = \int_A (g - f) \geqslant 0$ since $g - f \geqslant 0$ on A. But $\int_A (g - f) \neq 0$, since otherwise $v(A) = 0$ by Theorem 4(ii). Thus $\int_A g - \int_A f = \int_A (g - f) > 0$.

7. Since f is continuous and $f(b) = -1$, there exists an $\varepsilon > 0$ such that f is negative on $[b - \varepsilon, b]$. Now $\int_a^b f(x)\,dx = \int_a^{b-\varepsilon} f(x)\,dx + \int_{b-\varepsilon}^b f(x)\,dx = \int_a^{b-\varepsilon} f(x)\,dx + l$ where $l < 0$. Thus $\int_a^{b-\varepsilon} f(x)\,dx > 0$, so there exists a $x_1 \in \,]a,b - \varepsilon[$ such that $f(x_1) > 0$ (otherwise $\int_a^{b-\varepsilon} f(x)\,dx \leqslant 0$). So by the intermediate value theorem there exists a $x_2 \in \,]x_1, b[$ such that $f(x_2) = 0$, and hence by Rolle's theorem there exists a $c \in \,]a, x_2[$ such that $f'(c) = 0$.

10. (b) If A has zero content, then A has measure zero. Conversely suppose A is compact and has measure zero. A compact implies A closed implies $\text{bd}(A) \subset A$, hence $\text{bd}(A)$ has measure zero. By the lemma on page 280, there exist open rectangles $P_1, P_2, \ldots, Q_1, Q_2, \ldots$ such that $A \subset \bigcup_{i=1}^{\infty} P_i$, $\text{bd}(A) \subset \bigcup_{i=1}^{\infty} Q_i$, $\sum_{i=1}^{\infty} v(P_i) < \varepsilon/2$ and $\sum_{i=1}^{\infty} v(Q_i) < \varepsilon/2$. Then $A \cup \text{bd}(A) = \text{cl}(A) \subset \left(\bigcup_{i=1}^{\infty} P_i \right) \cup \left(\bigcup_{i=1}^{\infty} Q_i \right)$ and $\sum_{i=1}^{\infty} v(Q_i) + \sum_{i=1}^{\infty} v(P_i) < \varepsilon$. A compact implies there exists a finite set $\{P_{1'}, \ldots, P_{n'}, Q_{1'}, \ldots, Q_{m'}\} \subset \{P_1, P_2, \ldots, Q_1, Q_2, \ldots\}$ which covers A. Now

$$\sum_{i=1}^{n'} v(\text{cl}(P_i)) + \sum_{i=1'}^{m'} v(\text{cl}(Q_i)) = \sum_{i=1'}^{n'} v(P_i) + \sum_{i=1'}^{m'} v(Q_i)$$

$$\leqslant \sum_{i=1}^{\infty} v(P_i) + \sum_{i=1}^{\infty} v(Q_i) < \varepsilon$$

(see Exercise 11). Let B be a closed rectangle containing A with a partition T containing the rectangles $\text{cl}(P_i)$ and $\text{cl}(Q_j)$. By Theorem 3, since $\text{bd}(A)$ has measure zero, A has volume, that is $\int_A 1_A$ exists. Then

$$0 \leqslant \int_A 1_A \leqslant \sum_{S \in T} \sup_{x \in S} 1_A(x) v(s) \leqslant \sum_{i=1'}^{n'} v(\text{cl}(P_i)) + \sum_{j=1'}^{m'} v(\text{cl}(Q_j)) < \varepsilon .$$

Since ε was arbitrary, $\int_A 1_A = 0$. For the second part, $\text{bd}(B)$ is compact (being closed and bounded) so B has volume iff $\text{bd}(B)$ has measure zero iff $\text{bd}(B)$ has content zero.

14. Case (a), $p < -1$. $\lim_{a \to 0} \int_a^1 x^p\,dx = \frac{1}{p+1} \lim_{a \to 0} (1 - a^{p+1}) = -\infty$, so $\int_0^{\infty} x^p\,dx$

does not exist. Case (b), $p = -1$. $\displaystyle\lim_{a \to \infty} \int_1^a x^{-1}\,dx = \lim_{a \to \infty} \log a = +\infty$ hence

$\displaystyle\int_0^\infty x^p\,dx$ does not exist. Case (c), $p > -1$. $\displaystyle\lim_{a \to \infty} \int_1^a x^p\,dx = \frac{1}{p+1}\lim_{a \to \infty}(a^{p+1} - 1) =$

$+\infty$, and again $\displaystyle\int_0^\infty x^p\,dx$ does not exist. Thus for no $p \in \mathbb{R}$ does $\displaystyle\int_0^\infty x^p\,dx$ exist.

16. (a) Let g denote f extended, then {discontinuities of g_k} $\subset$ bd(A) $\cup$ {discontinuities of f_k}, and {discontinuities of g} $\subset$ bd(A) $\cup$ {discontinuities of f}. It is sufficient to show that {discontinuities of f} $\subset \bigcup_{k=1}^\infty$ {discontinuities of f_k}, to imply that {discontinuities of g} $\subset \bigcup_{i=1}^\infty$ {discontinuities of g_k}. If $x_0 \in$ {discontinuities of f}, then since $f_k \to f$ uniformly, for all $N > 0$, there exists a $M > N$ such that f_M is discontinuous at x_0, hence $x_0 \in \bigcup_{k=1}^\infty$ {discontinuities of f_k}.

(b) Let $\{f_k\}$ be a sequence of bounded integrable functions on A, such that $f_k \to f$ uniformly on A. Let N be such that $n > N$ implies for all $x \in A$, $|f_n(x) - f(x)| < 1$, then for all $x \in A$, $n > N$ implies $|f(x)| < 1 + |f_n(x)| < 1 + M_n$, where M_n bounds f_n, thus f is bounded. To show f Riemann integrable it is sufficient to show that {discontinuities of f} $= D$ has measure zero. Now for all k, f_k Riemann integrable implies {discontinuities of f_k} $= D_k$ has measure zero, and $\bigcup_{k=1}^\infty D_k$, by Theorem 2 has measure zero. Since by (a) $D \subset \bigcup_{k=1}^\infty D_k$, D has measure zero, and hence f is Riemann integrable.

18. For all $x \in B$, $1_A(x) \leqslant 1_B(x)$ and $f(x) \geqslant 0$ implies $f(x)1_A(x) \leqslant f(x) \cdot 1_B(x)$ which implies $\int_B f(x)1_A(x) = \int_A f(x)\,dx \leqslant \int_B f(x)1_B(x)\,dx = \int_B f(x)\,dx$. In general if there exists a x such that $f(x) < 0$ this is not true, for example, let $f: [0,1] \to \mathbb{R}$, $x \mapsto -1$ and $A = [0,1/2]$. Then $\int_0^{1/2} f(x)\,dx = -1/2 \nleqslant \int_0^1 f(x)\,dx = -1$.

20. $f \geqslant 0$, continuous, and increasing monotonically as $x \to 0$, and $\int_a^b f(x)\,dx$ convergent implies that for any $x \in \,]a,b]$, $\int_a^x f(x)\,dx$ exists for $\displaystyle\lim_{\varepsilon \to 0} \int_{a+\varepsilon}^b f(x)\,dx =$
$\displaystyle\lim_{\varepsilon \to 0} \int_{a+\varepsilon}^x f(x)\,dx + \int_x^b f(x)\,dx$ and $\displaystyle\lim_{\varepsilon \to 0} \int_{a+\varepsilon}^b f(x)\,dx$ convergent implies $\displaystyle\lim_{\varepsilon \to 0} \int_{a+\varepsilon}^x f(x)\,dx$ convergent. Now for all $y \in \,]a,x]$, $f(y) \geqslant f(x)$, since f is monotonically increasing; hence $\int_a^{a+x} f(t)\,dt \geqslant \int_a^{a+x} f(x)\,dt = f(x)x$. Then for all $\varepsilon > 0$ there exists a δ such that $0 < x < \delta$ implies $0 < f(x)x \leqslant \int_a^{a+x} f(t)\,dt = \int_a^b f(t)\,dt - \int_{a+x}^b f(t)\,dt < \varepsilon$; hence $x \to 0$ implies $xf(x) \to 0$.

22. If $0 < p$, then $-1 < p - 1 < \alpha$ and on $[0,1]$, $x^{p-1} > e^{-x}x^{p-1}$, hence since $\int_0^1 x^{p-1}\,dx$ converges, $\int_0^1 e^{-x}x^{p-1}\,dx$ converges. Claim that for all α there exists a M such that $x \geqslant 0$ implies $e^{-x}x^\alpha < x^{-2}$, that is, $e^{-x}x^{\alpha+2} < 1$. It suffices to show that $e^{-x}x^n \to 0$ as $x \to \infty$ for any $n \in \mathbb{N}$. Now $e^x = 1 + x + x^2/2! + \cdots > x^{n+1}/(n+1)!$ implies $(n+1)/x > e^{-x}x^n \to 0$. Then since $\int_1^\infty x^{-2}\,dx$ converges, so does $\int_1^\infty e^{-x}x^{p-1}\,dx$ and so the Γ function $\int_0^\infty e^{-x}x^{p-1}\,dx$ converges for $p > 0$.

24. Let $A = [0,1]$, $f(x) = \begin{cases} 0, & x \in [0,1]\backslash\mathbb{Q}, \\ 1, & x \in [0,1] \cap \mathbb{Q}, \end{cases}$ then if $\varepsilon > 0$, taking any δ, and

if $|P| < \delta$, then there exists a $x_i \in S_i \cap \mathbb{Q}$ for all i, so $|\sum_{i=1}^N f(x_i)v(S_i) - 1| = 0 < \varepsilon$, but f is not integrable.

27. Suppose there exists a $x \in A$ such that $f(x) \neq 0$. Then there exists a $\delta > 0$ such

that $f > 0$ on $D(x,\delta)$, since A is open and f is continuous. Now by Theorem 5(vi), there exists a $y \in D(x,\delta)$ such that $f(y)v(D) = \int_D f$. By $f(y) > 0$ implies $\int_D f \neq 0$, a contradiction. Hence $f = 0$ on A.

29. For each k, the total length of the intervals in $F_k = \bigcap_{n=1}^{k} F_n$ is $(2/3)^k$. So for $\varepsilon > 0$, let k be such that $(2/3)^k < \varepsilon$, then the union of the 2^k intervals of F_k cover the Cantor set C and have volume $(2/3)^k < \varepsilon$. Thus C has zero volume and hence measure zero.

33. Let $g(x) = x^2$, $h(x) = \int_0^x f(y)\,dy$, then $F = h \circ g$ and so $F'(x) = h'(g(x)) \circ g'(x) = 2xf(x^2)$.

35. $\dfrac{1}{n} \cdot A_n$ is just an upper sum for $f(x) = x$ on $[1,2]$, as

$$\frac{1}{n}A_n = \frac{1}{n}\left[\left(1 + \frac{1}{n}\right) + \left(1 + \frac{2}{n}\right) + \cdots + \left(1 + \frac{n-1}{n}\right) + 2\right] = \sum_{k=1}^{n}\left(1 + \frac{k}{n}\right)\left(\frac{1}{n}\right),$$

where the partition is

$$P = \left\{1,1 + \frac{1}{n},\ldots,1 + \frac{n-1}{n},2\right\}.$$

Since

$$\int_1^2 x\,dx = 2 - \frac{1}{2} = \frac{3}{2}, \qquad \lim_{n\to\infty}\frac{1}{n}A_n = \frac{3}{2}.$$

36.
$$\int_0^1 \log x\,dx = -1 = \lim_{n\to\infty}\sum_{k=1}^{n}\log\frac{\left(\frac{k}{n}\right)}{n}$$

$$= \lim_{n\to\infty}\frac{1}{n}\left(\log\frac{1}{n} + \log\frac{2}{n} + \cdots + \log 1\right)$$

$$= \lim_{n\to\infty}\frac{1}{n}\log\frac{n!}{n^n} \qquad \text{implies}$$

$$e^{-1} = e^{\lim_{n\to\infty}(1/n)\log(n!/n^n)} = \lim_{n\to\infty}\frac{1}{n}(n!)^{1/n}.$$

39. By definition, $\log 2 = \int_1^2 \dfrac{dt}{t}$. Let $P_k = \left\{1,1 + \dfrac{1}{k},\ldots,1 + \dfrac{j}{k},\ldots,2\right\}$ be a partition

of $[1,2]$. Since $f(t) = \dfrac{1}{t}$ is decreasing,

$$\inf_{t\in[1+j-1/k,\,1+j/k]} f(t) = f\left(1 + \frac{j}{k}\right) = \frac{k}{(k+j)}.$$

Then

$$L\left(\frac{1}{t},P_k\right) = \sum_{j=1}^{k}\frac{k}{k+j}\cdot\frac{1}{k} = \frac{1}{k+1} + \cdots + \frac{1}{2k},$$

and

$$\lim_{k\to\infty}L\left(\frac{1}{t},P_k\right) = \lim_{k\to\infty}\left(\frac{1}{k+1} + \frac{1}{k+2} + \cdots + \frac{1}{2k}\right) = \int_1^2\frac{dt}{t} = \log 2.$$

40. (a) $d(f,g) = \int_a^b |f(x) - g(x)| \, dx \geqslant 0,$ by Theorem 5(iv).

    (b) $d(f,g) = \int_a^b |f(x) - g(x)| \, dx = \int_a^b |g(x) - f(x)| \, dx = d(g,f).$

    (c) $d(f,g) = \int_a^b |f(x) - g(x)| \, dx = 0$ implies by Theorem 4
       and since $|f(x) - g(x)| \geqslant 0$ that $|f(x) - g(x)| = 0$, that is, $f(x) = g(x)$, except
       possibly on a set of measure zero.

    (d) $d(f,g) = \int_a^b |f(x) - g(x)| \, dx \leqslant \int_a^b |f(x) - h(x)| \, dx + \int_a^b |h(x) - g(x)| \, dx = d(f,h) +$
       $d(h,g)$. Thus d does not satisfy the criterion $d(f,g) = 0$ iff $f = g$. But as all the
       other properties are satisfied d is called a semi-distance.

Chapter 9
Fubini's Theorem and the Change of Variables Formula

9.1   Introduction

3. $5/6$.

4. $-\pi(e^{-1} - 1)$.

5. $1/2$.

9.2   Fubini's Theorem

2. $e^2/4 - 1/4$.

9.3   Change of Variables Theorem

2. 2.

3. $\displaystyle\int_0^{1/2} \int_v^{1-v} (2u^2 + 2v^2) \cdot 2 \, du \, dv = \frac{1}{3}$.

9.4   Polar Coordinates

1. $\pi(e - 1)$.

2. $\dfrac{\pi}{2} [a^2 \log a - b^2 \log b - 1/2(a^2 - b^2)]$.

9.5   Spherical Coordinates

1. $\displaystyle\int_D e^{(x^2 + y^2 + z^2)^{3/2}} \, dx \, dy \, dz = \int_0^{2\pi} \int_0^{\pi} \int_0^1 e^{r^3} r^2 \sin \varphi \, dr \, d\varphi \, d\theta = \frac{4\pi}{3}(e - 1)$.

2. $\dfrac{2}{9} \pi((1 - a^3)^{1/2} + (1 + a^3)^{3/2})$.

9.6   Cylindrical Coordinates

1. $(2\sqrt{2} - 1)\pi$

2. $\pi/4$.

Exercises for Chapter 9 (at end of chapter)

1. $\pi/3$.
3. (b) $\pi(1 - \cos 1)$.
   (c) ∞.
   (d) 2.
   (e) 1.

4. (b) $v(A) = \displaystyle\int_A 1_A = \int_0^{2\pi} \int_0^{h_0} \int_0^{\left(\frac{h_0-z}{h_0}\right)r_0} r\, dr\, dz\, d\theta$

$$= \int_0^{2\pi} \int_0^{h_0} \frac{r_0^2}{2h_0^2} (h_0^2 - 2h_0 z + z^2)\, dz\, d\theta$$

$$= \int_0^{2\pi} \frac{r_0^2}{2h_0^2}\left(h_0^3 - h_0^3 + \frac{h_0^3}{3}\right) d\theta = \int_0^{2\pi} \frac{r_0^2 h_0}{6}\, d\theta$$

$$= \frac{\pi r_0^2 h_0}{3}.$$

   (c) $v(A) = \displaystyle\int_0^{1/\sqrt{2}} \int_{x^2}^{1-x^2} dy\, dx = \int_0^{1/\sqrt{2}} (1 - 2x^2)\, dx$

$$= \frac{2}{3\sqrt{2}}.$$

   (d) $v(A) = \displaystyle\int_0^{2\pi} \int_{-1}^{1/2} \int_0^{\sqrt{1-z^2}} r\, dr\, dz\, d\theta$

$$= \frac{1}{2}\int_0^{2\pi} \int_{-1}^{1/2} (1 - z^2)\, dz\, d\theta$$

$$= \frac{1}{2}\int_0^{2\pi} \frac{9}{8}\, d\theta = \frac{9\pi}{8}.$$

6. $\displaystyle\int_0^1 \int_y^{\sqrt{1+y^2}} xy \sin(x^2 - y^2)\, dx\, dy$

$$= -\frac{1}{2}\int_0^1 y(\cos 1 - 1)\, dy$$

$$= \frac{1}{4}(1 - \cos 1)$$

9. $\displaystyle\int_{[a,b]\times[c,d]} \tilde{f}(x,y)\tilde{g}(x,y)\, dx\, dy = \int_a^b \left(\int_c^d f(x)g(y)\, dy\right) dx$

$$= \int_a^b f(x)\left(\int_c^d g(y)\, dy\right) dx$$

$$= \left(\int_a^b f(x)\, dx\right)\left(\int_c^d g(y)\, dy\right).$$

(You must show $\tilde{f}\tilde{g}$ is integrable.)

11. For any $x \in [0,1]$, $\int_0^1 1_s(x,y) \, dy = 0$ for if x is irrational, then $1_s(x,y) = 0$, and if x is rational, $1_s(x,y) \neq 0$ only for a finite number of $y \in [0,1]$. But $\int_{[0,1] \times [0,1]} 1_s$ does not exist. For, if P is any partition of $[0,1] \times [0,1]$, then $L(1_s, P) = 0$ since there is an irrational in every open subset of $\mathbb{R}^2$.

12. $A = A_1 \cup (A_2 \backslash A_1) \cup (A_3 \backslash A_2) \cup \ldots$, and the $(A_{i+1} \backslash A_i)$'s are pairwise disjoint, hence by Exercise 7b, $v(A) = v(A_1) + \sum_{i=2}^{\infty} v(A_i \backslash A_{i-1})$. Now for all j, $v(A_j) = v(A_1) + \sum_{i=2}^{j} v(A_i \backslash A_{i-1})$, which is a consequence of Theorem 5, Chapter 8, hence $v(A) = \lim_{j \to \infty} v(A_j)$.

13. $v(C_x) = \int_B 1_{C_x}(y) \, dy$, and $v(C) = \int_{A \times B} 1_C = \int_A (\int_B 1_{C_x}(y) \, dy) \, dx = 0$ implies $\int_B 1_{C_x}(y) \, dy = 0$ except possibly on a set of measure zero, since $\int_B 1_{C_x}(y) \, dy \geq 0$. If $C = \{(1/2, y) \mid y \in [0,1]\}$, then $v(C) = 0$, but $v(C_{1/2}) = 1 \neq 0$.

16. $\int_{\mathbb{R}^3} (x^2 + y^2 + z^2)^{p/2} \, dx \, dy \, dz = \int_{\mathbb{R}^3} r^p \, dx \, dy \, dz = \int_0^\infty \int_0^{2\pi} \int_0^\pi r^p \cdot r^2 \sin \varphi \, d\varphi \, d\theta \, dr = 4\pi \int_0^\infty r^{p+2} \, dr$. Now by Exercise 14, Chapter 8, $\int_0^\infty r^{p+2} \, dr$ does not exist for any p, hence $\int_{\mathbb{R}^3} r^p \, dx \, dy \, dz$ does not exist for any p.

17. Let $f_M(x) = \begin{cases} f(x), & f(x) \leq M, \ x \in [0,1] \\ 0, & f(x) > M \end{cases}$. Then f is integrable iff $\lim_{M \to \infty} \int f_M$ exists. But for all M, $\int_0^1 f_M = 0$, since $f_M \neq 0$ for only a finite number of points, that is, $1, 1/2, \ldots, 1/n$, where n is the greatest natural number such that $n^2 \leq M$. Thus $\lim_{M \to \infty} \int_0^1 f_M = 0$, and $\int_0^1 f(x) \, dx = \lim_{M \to \infty} \int_0^1 f_M(x) \, dx = 0$.

19. $F(x,y) = f(x) + g(y)$ implies $\int_{A \times B} F(x,y) \, dx \, dy = \int_{A \times B} f(x) \, dx \, dy + \int_{A \times B} g(y) \, dx \, dy$. Using Fubini's theorem: $\int_{A \times B} f(x) \, dx \, dy = \int_A (\int_B f(x) \, dy) \, dx$ and $f(x)$ being constant with respect to y gives $\int_B f(x) \, dy = v(B) \cdot f(x)$ implies $\int_{A \times B} f(x) \, dx \, dy = \int_A v(B) f(x) \, dx = v(B) \int_A f(x) \, dx$. Similarly $\int_{A \times B} g(y) \, dx \, dy = v(A) \int_B g(y) \, dy$. Hence summing we get $\int_{A \times B} F(x,y) \, dx \, dy = v(B) \int_A f + v(A) \int_B g$, which shows that F is integrable.

20. $\int_D 1_D = \int_1^3 \int_{x^2}^{1+x^2} dy \, dx = \int_1^3 dx = 2$.

35. Consider $\dfrac{d}{dx} \log x$ at $x = 1$.

Chapter 10
Fourier Analysis

10.1 Inner Product Spaces

1. (a) Let $z_1 = x_1 + iy_1$, $\qquad z_2 = x_2 + iy_2$.
   Then

$$e^{z_1} \cdot e^{z_2} = [e^{x_1}(\cos(y_1) + i \sin(y_1))][e^{x_2}(\cos(y_2) + i \sin(y_2))]$$
$$= e^{x_1 + x_2}[\cos(y_1)\cos(y_2) - \sin(y_1)\sin(y_2)]$$
$$+ i[\sin(y_1)\cos(y_2) + \cos(y_1)\sin(y_2)]$$
$$\times e^{x_1 + x_2}(\cos(y_1 + y_2) + i \sin(y_1 + y_2)) = e^{z_1 + z_2}.$$

(b) $e^z \cdot e^{-z} = e^0 = 1$, so $e^z = 0$ is impossible.

(c) $|e^{i\theta}|^2 = \cos^2\theta + \sin^2\theta = 1$.

5. $-i(e^i - 1)$.

9. (a) $\langle f,g \rangle = 0$ implies $\langle g,f \rangle = \overline{\langle f,g \rangle} = 0$ and so $\|f + g\|^2 = \langle f + g, f + g \rangle = \langle f,f \rangle + \langle f,g \rangle + \langle g,f \rangle + \langle g,g \rangle = \|f\|^2 + \|g\|^2$.

(b)
$$\|f + g\|^2 - \|f - g\|^2 = \|f\|^2 + \|g\|^2 + \langle f,g \rangle + \langle g,f \rangle$$
$$- (\|f\|^2 + \|g\|^2 - \langle f,-g \rangle - \langle -g,f \rangle)$$
$$= \langle f,g \rangle + \langle g,f \rangle - \langle f,-g \rangle - \langle -g,f \rangle$$
$$= 2\langle f,g \rangle + 2\langle g,f \rangle$$
$$i(\|f + ig\|^2 - \|f - ig\|^2) = i[2\langle f,ig \rangle + 2\langle ig,f \rangle]$$
$$= 2\langle f,g \rangle - 2\langle g,f \rangle$$

and adding, we get (b).

(c)
$$\|f + g\|^2 + \|f - g\|^2 = 2\|f\|^2 + 2\|g\|^2 + \langle f,g \rangle + \langle g,f \rangle$$
$$+ \langle f,-g \rangle + \langle -g,f \rangle$$
$$= 2\|f\|^2 + 2\|g\|^2.$$

(d)
$$\|f + g\|^2 \cdot \|f - g\|^2 = [(\|f\|^2 + \|g\|^2)$$
$$+ (\langle f,g \rangle + \langle g,f \rangle)][(\|f\|^2 + \|g\|^2)$$
$$-(\langle f,g \rangle + \langle g,f \rangle)]$$
$$= (\|f\|^2 + \|g\|^2)^2 - (\langle f,g \rangle + \langle g,f \rangle)^2$$
$$\leqslant (\|f\|^2 + \|g\|^2)^2.$$

10. By Schwartz's inequality

$$\left| \int_a^b f(x)\,dx \right|^2 = |\langle 1,f \rangle|^2 \leqslant \|1\|^2 \|f\|^2 = (b - a) \int_a^b |f|^2\,dx\ .$$

The converse is not true. Let $f(x) = \begin{cases} x^{-1/2}, & x \neq 0 \\ 0, & x = 0 \end{cases}$ then f is integrable on

$[0,1]$ but f^2 is not.

11. $\langle f - g,g \rangle = \langle f,g \rangle - \langle g,g \rangle$

$$= \left\langle f, \sum_{i=1}^n \langle f,\varphi_i \rangle \varphi_i \right\rangle - \left\langle \sum_{i=1}^n \langle f,\varphi_i \rangle \varphi_i, \sum_{j=1}^n \langle f,\varphi_j \rangle \varphi_j \right\rangle$$

$$= \sum_{i=1}^n \overline{\langle f,\varphi_i \rangle} \langle f,\varphi_i \rangle - \sum_{i=1}^n \sum_{j=1}^n \langle f,\varphi_i \rangle \overline{\langle f,\varphi_j \rangle} \langle \varphi_i,\varphi_j \rangle.$$

Now since $\langle \varphi_i,\varphi_j \rangle = \delta_{ij}, \displaystyle\sum_{i=1}^n \sum_{j=1}^n \langle f,\varphi_i \rangle \overline{\langle f,\varphi_j \rangle} \langle \varphi_i,\varphi_j \rangle$

$$= \sum_{i=1}^n \langle f,\varphi_i \rangle \overline{\langle f,\varphi_i \rangle}, \text{ and hence } \langle f - g,g \rangle = 0\ .$$

Geometrically, we have resolved $f = g + (f - g)$ into two components: g and $f - g$: g lies in the plane P generated by $\varphi_1, \ldots, \varphi_n$ and $w = f - g$ is orthogonal to P; g is the projection of f onto P along w. The term $\langle f,\varphi_i\rangle$ is the component of f in the direction of φ_i.

10.2 Orthogonal Families of Functions

1. Any n-orthonormal vectors are linearly independent for if $c_1\varphi_1 + \cdots + c_n\varphi_n = 0$ then $\langle c_1\varphi_1 + \cdots + c_n\varphi_n, \varphi_i\rangle = c_i = 0$ for all $i = 1, \ldots, n$, and hence $\varphi_1, \ldots, \varphi_n$ form a basis for $\mathbb{R}^n$.

2. Since $g_0, g_1, \ldots$ are linearly independent $h_0 = g_0 \neq 0$. Suppose we have constructed $\varphi_0, \ldots, \varphi_{n-1}$ from $g_0, \ldots, g_{n-1}$ and that the $\{\varphi_k\}_{k=0}^{n-1}$ are orthonormal. Then $h_n = g_n - \sum_{k=0}^{n-1} \langle g_n,\varphi_k\rangle\varphi_k$ is non-zero for $0 = g_n - \sum_{k=0}^{n-1} \langle g_n,\varphi_k\rangle\varphi_k$ implies $g_n = \sum_{k=0}^{n-1} \langle g_n,\varphi_k\rangle\varphi_k$ implies g_n is a linear combination of $g_0, \ldots, g_{n-1}$, a contradiction. For $j < n$ we have

$$\langle h_n,\varphi_j\rangle = \left\langle g_n - \sum_{k=0}^{n-1} \langle g_n,\varphi_k\rangle\varphi_k,\varphi_j \right\rangle$$

$$= \langle g_n,\varphi_j\rangle - \sum_{k=0}^{n-1} \langle g_n,\varphi_k\rangle\langle\varphi_k,\varphi_j\rangle$$

$$= \langle g_n,\varphi_j\rangle - \langle g_n,\varphi_j\rangle = 0 .$$

Thus $\varphi_n = h_n/\|h_n\|$ is orthogonal to each φ_i, $i = 1, \ldots, n-1$, and since $\|\varphi_n\| = 1$, we have by induction that $\{\varphi_i \mid i = 0,1,\ldots\}$ is an orthonormal family.

3. (a) $\langle\psi_n,\psi_n\rangle = \displaystyle\int_0^l \left[\sqrt{\frac{2\pi}{l}}\,\varphi_n\!\left(\frac{2\pi x}{l}\right)\right]\left[\sqrt{\frac{2\pi}{l}}\,\varphi_m\!\left(\frac{2\pi x}{l}\right)\right] dx$

$\qquad = \dfrac{2\pi}{l}\displaystyle\int_0^l \varphi_n\!\left(\frac{2\pi x}{l}\right)\varphi_m\!\left(\frac{2\pi x}{l}\right) dx.$

Let $u = \dfrac{2\pi x}{l}$ then $\langle\psi_n,\psi_m\rangle = \dfrac{2\pi}{l}\displaystyle\int_0^{2\pi} \varphi_n(u)\varphi_m(u)\,\frac{l}{2\pi}\,du = \delta_{nm}.$

(b) $\dfrac{1}{\sqrt{l}}, \sqrt{\dfrac{2}{l}}\sin\!\left(\dfrac{2\pi n x}{l}\right), \sqrt{\dfrac{2}{l}}\cos\!\left(\dfrac{2\pi n x}{l}\right); \dfrac{1}{\sqrt{l}}e^{\left(\frac{2\pi i n x}{l}\right)}$

(c) $f(x) = \left\langle f(x),\dfrac{1}{\sqrt{l}}\right\rangle\dfrac{1}{\sqrt{l}}$

$\qquad + \dfrac{2}{l}\displaystyle\sum_{n=1}^{\infty}\left[\left\langle f(x),\sin\!\left(\dfrac{2\pi n x}{l}\right)\right\rangle\sin\!\left(\dfrac{2\pi n x}{l}\right) + \left\langle f(x),\cos\!\left(\dfrac{2\pi n x}{l}\right)\right\rangle\cos\!\left(\dfrac{2\pi n x}{l}\right)\right]$

$\quad f(x) = \left\langle f(x),\dfrac{1}{\sqrt{l}}\right\rangle\dfrac{1}{\sqrt{l}} + \dfrac{1}{l}\displaystyle\sum_{n=1}^{\infty}\left\langle f(x),e^{\left(\frac{2\pi i n x}{l}\right)}\right\rangle.$

6. $\displaystyle\sum_{k=1}^{n} x^k = x(1 + x + \cdots + x^{n-1}) = \dfrac{x(1-x)(1 + x + \cdots + x^{n-1})}{1-x} = \dfrac{x(1-x^n)}{1-x}$

implies

$$\sum_{k=1}^{n} e^{ik\theta} = \frac{e^{i\theta}(1 - e^{in\theta})}{1 - e^{i\theta}} = \frac{e^{i\theta}(1 - e^{in\theta})}{(1 - e^{i\theta})} \frac{(1 - e^{-i\theta})}{(1 - e^{-i\theta})}$$

$$= \frac{(e^{i(\theta/2)} - e^{i(n+1/2)\theta})(e^{i(\theta/2)} - e^{-i(\theta/2)})}{2 - (e^{i\theta} + e^{-i\theta})}$$

$$= \frac{2i(e^{i(\theta/2)} - e^{i(n+1/2)\theta})\sin(\theta/2)}{2 - 2\cos\theta}$$

$$= \frac{i(e^{i(\theta/2)} - e^{i(n+1/2)\theta})\sin(\theta/2)}{2\sin^2(\theta/2)} = \frac{i(e^{i(\theta/2)} - e^{i(n+1/2)\theta})}{2\sin(\theta/2)}$$

So $\displaystyle\sum_{k=1}^{n}\cos(k\theta) = \text{real}\left(\sum_{k=1}^{n} e^{ik\theta}\right) = \frac{\sin(n + 1/2)\theta}{2\sin(\theta/2)} - 1/2$.

10.3 Completeness and Convergence Theorems

1. (a) $\displaystyle s_N = \sum_{n=-N}^{N}\left\langle f(x), \frac{e^{inx}}{\sqrt{2\pi}}\right\rangle\frac{e^{inx}}{\sqrt{2\pi}}$

$$= \langle f(x),1\rangle\frac{1}{2\pi} + \frac{1}{2\pi}\sum_{n=1}^{N}\langle f(x),e^{inx}\rangle e^{inx} + \langle f(x),e^{-inx}\rangle e^{-inx}.$$

But

$$\langle f(x),e^{inx}\rangle e^{inx} = \langle f(x),\cos(nx)\rangle\cos(nx) + \langle f(x),\sin(nx)\rangle\sin(nx)$$
$$+ i(\langle f(x),\cos(nx)\rangle\sin(nx) - \langle f(x),\sin(nx)\rangle\cos(nx))$$

and

$$\langle f(x),e^{-inx}\rangle e^{-inx} = \langle f(x),\cos(nx)\rangle\cos(nx) + \langle f(x),\sin(nx)\rangle\sin(nx)$$
$$+ i(\langle f(x),\sin(nx)\rangle\cos(nx) - \langle f(x),\cos(nx)\rangle\sin(nx))$$

implies

$$s_N = \langle f(x),1\rangle\frac{1}{2\pi} + \frac{1}{\pi}\sum_{n=1}^{N}\{\langle f(x),\cos(nx)\rangle\cos(nx) + \langle f(x),\sin(nx)\rangle\sin(nx)\}.$$

(b) $\displaystyle f(x) = \sum_{-\infty}^{\infty}\left\langle f(x),\frac{e^{inx}}{\sqrt{2\pi}}\right\rangle\frac{e^{inx}}{\sqrt{2\pi}}$

$$f(x) = \left\langle f(x),\frac{1}{\sqrt{2\pi}}\right\rangle\frac{1}{\sqrt{2\pi}} + \sum_{n=1}^{\infty}\left\{\left\langle f(x),\frac{\cos(nx)}{\sqrt{\pi}}\right\rangle\frac{\cos(nx)}{\sqrt{\pi}} + \left\langle f(x),\frac{\sin(nx)}{\sqrt{\pi}}\right\rangle\frac{\sin(nx)}{\sqrt{\pi}}\right\}.$$

(c) $\displaystyle \langle f(x),\sin(nx)\rangle = \int_{-\pi}^{\pi} f(x)\sin(nx)\,dx$

$$= \int_{-\pi}^{0} f(x)\sin(nx)\,dx + \int_{0}^{\pi} f(x)\sin(nx)\,dx$$

$$= -\int_\pi^0 f(-x)\sin(-nx)\,dx + \int_0^\pi f(x)\sin(nx)\,dx$$

$$= \int_\pi^0 f(x)\sin(nx)\,dx + \int_0^\pi f(x)\sin(nx)\,dx = 0.$$

(d) $\langle f(x),\cos(nx)\rangle = \int_{-\pi}^0 f(x)\cos(nx) + \int_0^\pi f(x)\cos(nx)$

$$= \int_\pi^0 f(x)\cos(nx) + \int_0^\pi f(x)\cos(nx) = 0.$$

4. (a) *Case 1: $n \geqslant 0$ even.* Then the Fourier series converges uniformly to $f(x)$, since f is continuous and f' is sectionally continuous.

Case 2: $n > 0$ odd. Then the Fourier series converges uniformly to $f(x)$ for all $x \in (-\pi,\pi)$ and to zero for $x = -\pi$ or π.

Case 3: $n \leqslant -1$ odd. Then $f(x) = x^n$ is not square integrable and a Fourier series does not exist.

(b) Fourier series converges pointwise to $f(x)$ except at $x = \pi$.

(c) $f(x)$ is not square integrable so the Fourier series is not defined.

10.4   Functions of Bounded Variation and Fejér Theory

2. $\delta' = \displaystyle\sum_{-\infty}^{\infty} \frac{in}{2\pi} e^{inx}$.

10.5   Computation of Fourier Series

1. (a) $f_1(x) = x$, $-\pi < x < \pi$ is an odd function, so $a_n = 0$.

$$b_n = \frac{1}{\pi}\int_{-\pi}^\pi x\,\sin(nx) = \frac{1}{\pi}\left[-x\,\frac{\cos nx}{n} + \frac{\sin nx}{n^2}\right]_{-\pi}^\pi$$

$$= -\frac{2\pi}{n\pi}\cos \pi n = -\frac{2}{n}(-1)^n.$$

$$f_1(x) = \sum_{n=1}^\infty \frac{2(-1)^{n+1}}{n}\sin nx.$$

(b) $f_2(x) = \begin{cases} x & 0 < x < 2\pi \\ f_2(x) = f_2(2\pi + x) \end{cases}$

$f_2(x) = f_1(x - \pi) + \pi$ implies

$$f_2(x) = \pi + 2\sum_{n=1}^\infty \frac{(-1)^{n+1}}{n}\sin n(x - \pi)$$

$$= \pi + 2\sum_{n=1}^\infty \frac{(-1)^{n+1}}{n} \cdot (-1)^n \sin nx = \pi - 2\sum_{n=1}^\infty \frac{\sin nx}{n}.$$

(c) $f_3(x) = \begin{cases} |x| & -\pi < x < \pi \\ f_3(x) = f_3(x + 2\pi) \end{cases}$

$$\frac{a_0}{2} = \frac{1}{\pi}\int_0^\pi x\,dx = \frac{1}{\pi}\frac{x^2}{2}\bigg|_0^\pi = \frac{\pi}{2}$$

$$a_n = \frac{2}{\pi}\int_0^\pi x\cos nx\,dx = \frac{2}{\pi}\left[x\,\frac{\sin nx}{n} + \frac{\cos nx}{n^2}\right]_0^\pi$$

$$= \frac{2}{\pi}\left[\frac{\cos \pi n}{n^2} - 1\right] = \begin{cases} 0 & n \text{ even} \\ \dfrac{4}{\pi n^2} & n \text{ odd} \end{cases}$$

$$\text{implies } f_3(x) = \frac{\pi}{2} - \frac{4}{\pi}\sum_{n=1}^\infty \frac{\cos(2n-1)}{(2n-1)^2}.$$

2. $\displaystyle\sum_{n=1}^\infty \left\{\frac{2\pi(-1)^{n+1}}{n} - \frac{4}{\pi}\frac{(1-(-1)^n)}{n^3}\right\}\sin(nx).$

4. (a) $\displaystyle -\frac{1}{2} + \frac{42}{\pi}\sum_{n=1}^\infty \frac{\sin(2n-1)x}{2n-1}.$

(b) $\displaystyle 3 + \left(\frac{\pi^2}{3} + 4\sum_{n=1}^\infty \frac{(-1)^n}{n^2}\cos(nx)\right) + 2\sum_{n=1}^\infty \frac{(-1)^{n+1}}{n}\sin(nx).$

10.6   Some Further Convergence Theorems

1. The Fourier series for f converges absolutely and uniformly and may be differentiated term by term to get the absolutely and uniformly convergent Fourier series for f'.

2. The Fourier series converges in mean to f, and by Theorem 9 for $x \neq \pm\tfrac{1}{2}$, the Fourier series converges to $f(x)$. The series may not be differentiated term by term.

4. $\displaystyle\sum_{n=1}^\infty \left(12\frac{(-1)^n}{n^3} - \frac{2\pi^2(-1)^n}{n}\right)\sin(nx).$

5. (b) $\displaystyle\sum_{n=1}^\infty (a_n^2 + b_n^2)^{1/2} = \sum_{n=1}^\infty n(a_n^2 + b_n^2)^{1/2} \cdot \frac{1}{n}$

$$\leqslant \left(\sum_{n=1}^\infty [n(a_n^2 + b_n^2)^{1/2}]^2\right)^{1/2}\left(\sum_{n=1}^\infty \left(\frac{1}{n}\right)^2\right)^{1/2}$$

$$\leqslant \left[\sum_{n=1}^\infty n^2(a_n^2 + b_n^2)\right]^{1/2}\left[\sum_{n=1}^\infty \frac{1}{n^2}\right]^{1/2}.$$

Both of these series converge.

10.7   Applications

1. f is continuous but not differentiable at $x = 1/2$. Let $g(x)$ be the half-interval cosine

series expansion for x. Then $f(x) = h\left[g\left(\dfrac{\pi x}{l} + \dfrac{\pi}{2}\right) - \dfrac{\pi}{2}\right]$. So from Table 10-5 we have

$$f(x) = h\left[-\frac{4}{\pi}\sum_{n=1}^{\infty}\frac{\cos[(2n-1)(\pi x/l + \pi/2)]}{(2n-1)^2}\right]$$

$$= \sum_{n=1}^{\infty}\frac{4h\sin[(2n-1)(\pi x/l)]}{\pi(2n-1)^2}.$$

So $y(x,t) = \displaystyle\sum_{n=1}^{\infty}\frac{4h\sin[(2n-1)(\pi x/l)]}{\pi(2n-1)^2}\cos\frac{(n\pi ct)}{l}$.

At time $t = \dfrac{l}{c}$, $y\left(x,\dfrac{l}{c}\right) = \displaystyle\sum_{n=1}^{\infty}\frac{4h\sin[(2n-1)(\pi x/l)]}{\pi(2n-1)^2}\cos n\pi$

$$= \sum_{n=1}^{\infty}\frac{(-1)^n}{\pi(2n-1)^2}\,4h\sin\left[(2n-1)\left(\frac{\pi x}{l}\right)\right], \text{ and}$$

$$y\left(x,\frac{3l}{2c}\right) = \frac{1}{2}\left[f\left(x + \frac{3l}{2c}\right) + f\left(x - \frac{3l}{2c}\right)\right]$$

$$= -f(x).$$

2. At time t, $y(x,t)$ is the sum of two functions $1/2f(x + ct), 1/2f(x - ct)$ which have max, min or discontinuities at $x = x_0 - ct + lm$ and $x = x_0 + ct + ln$ respectively where m, n are integers chosen such that $0 < x_0 - ct + lm \leqslant l$ and $0 < x_0 + ct + ln \leqslant l$.

4. For fixed τ, let $T_\tau(x) = T(x,\tau)$. Then

$$T_\tau(x) = \frac{a_0}{2} + \sum_{n=1}^{\infty}[a_n e^{-n^2\pi^2\tau/l^2}]\cos\left(\frac{n\pi x}{l}\right).$$

So $a_0 = \dfrac{2}{l}\displaystyle\int_0^l T_\tau(x)\,dx = \dfrac{2}{l}\int_0^l T(x,\tau)\,dx$.

5. $f(x,\tau) = \dfrac{l^2}{3} + \dfrac{4l^2}{\pi^2}\displaystyle\sum_{n=1}^{\infty}\frac{(-1)^n}{n^2}e^{-n^2\pi^2\tau/l^2}\cos\left(\frac{n\pi x}{l}\right)$ and $\displaystyle\lim_{\tau\to\infty} f(x,\tau) = \dfrac{l^2}{3}$.

10.8   Fourier Integrals

1. Differentiate under the integral sign.

4. (a) From Exercise 1, $\partial\hat{f}/\partial t = -k^2\alpha^2\hat{f}(\alpha,t)$; so integrating, using $\hat{f}(\alpha,0) = \hat{g}(\alpha)$ we get $\hat{f}(\alpha,t) = \hat{g}(\alpha)e^{-k^2\alpha^2 t}$.

  (b) Use the theorem stated in the text on convolutions to find the inverse Fourier transform of $\hat{f}(\alpha,t)$, together with the fact about the Fourier transform of the Gaussian stated on p. 397.

10.9   Quantum Mechanical Formalism

1. $\langle(AB)^*x,y\rangle = \langle x,ABy\rangle = \langle A^*x,By\rangle = \langle B^*A^*x,y\rangle$ implies $\langle((AB)^* - B^*A^*)x,y\rangle = 0$ for all x, y implies $(AB)^* = B^*A^*$.

2. It is enough to show $\langle Av_i, v_j \rangle = \langle v_i, Av_j \rangle$ where $\{v_i, \ldots, v_n\}$ is an orthonormal basis. But $\langle Av_i, v_j \rangle = \langle \sum_{k=1}^n a_{ki}v_k, v_j \rangle = \sum_{k=1}^n a_{ki}\langle v_k, v_j \rangle = a_{ji} = a_{ij} = \sum_{k=1}^n \langle v_i, v_k \rangle a_{kj} = \langle v_i, \sum_{k=1}^n a_{kj}v_k \rangle = \langle v_i, Av_j \rangle.$

3. $\langle A\psi, \psi \rangle = \left\langle A\left(\sum_{n=1}^\infty \langle \psi, \varphi_n \rangle \varphi_n \right), \sum_{k=1}^\infty \langle \psi, \varphi_k \rangle \varphi_k \right\rangle$

$$= \sum_{n,k} \langle \psi, \varphi_n \rangle \overline{\langle \psi, \varphi_k \rangle} \langle A(\varphi_n), \varphi_k \rangle = \sum_{n,k} \langle \psi, \varphi_n \rangle \overline{\langle \psi, \varphi_k \rangle} \langle \lambda_n \varphi_n, \varphi_k \rangle$$

$$= \sum_{n,k} \langle \psi, \varphi_n \rangle \overline{\langle \psi, \varphi_k \rangle} \lambda_n \langle \varphi_n, \varphi_k \rangle = \sum_{n=1}^\infty |\langle \psi, \varphi_n \rangle|^2 \lambda_n .$$

The above would certainly be true in a finite dimensional space. In fact this is true in general. The expectation of A is just the sum of the observables $\{\lambda_n\}$ of A weighted by the probability with which they can be observed when A operates on the state ψ.

Exercises for Chapter 10 (at end of chapter)

1. If $f_1, f_2 \in M^\perp$ then for all $g \in M$ $\langle af_1 + bf_2, g \rangle = a\langle f_1, g \rangle + b\langle f_2, g \rangle = 0$ implies $af_1 + bf_2 \in M^\perp$. So $M^\perp$ is a subspace. Suppose $f_n \to f$ in mean, $\{f_n\} \in M^\perp$. Let $g \in M$, then $|\langle f, g \rangle| = |\langle f - f_n, g \rangle + \langle f_n, g \rangle| = |\langle f - f_n, g \rangle| \le \|f - f_n\| \|g\| \to 0$ as $n \to \infty$ so $\langle f, g \rangle = 0$ implies $f \in M^\perp$.

3. (a) $\cosh x = \dfrac{e^x + e^{-x}}{2} = \dfrac{\sinh(\pi)}{\pi} \sum_{-\infty}^\infty \dfrac{(-1)^n}{1 - in}\left(\dfrac{e^{inx} + e^{-inx}}{2} \right)$

$$= \dfrac{\sinh(\pi)}{\pi} \sum_{-\infty}^\infty \dfrac{(-1)^n}{1 - in} \cos(nx) .$$

Thus $\pi \coth(\pi) = \pi \dfrac{\cosh(\pi)}{\sinh(\pi)} = \sum_{-\infty}^\infty \dfrac{(-1)^n}{1 - in} \cos(n\pi)$

$$= 1 + \sum_{n=1}^\infty \left[\dfrac{(-1)^n}{1 - in} + \dfrac{(-1)^n}{1 + in} \right](-1)^n$$

and so $\pi \coth(\pi) - 1 = \sum_{n=1}^\infty \dfrac{2}{1 + n^2} .$

(b) $a_n = \dfrac{2}{\pi} \int_0^\pi \cos(ax)\cos(nx) \, dx$

$a_0 = \dfrac{2}{\pi} \int_0^\pi \cos(ax) \, dx = \dfrac{2}{\pi} \dfrac{\sin(ax)}{a} \Big|_0^\pi = \dfrac{2}{\pi} \dfrac{\sin(a\pi)}{a}$

$a_n = \dfrac{1}{\pi} \int_0^\pi \cos(a + n)x + \cos(a - n)x \, dx$

$$= \dfrac{1}{\pi} \left[\dfrac{\sin(a + n)\pi}{a + n} + \dfrac{\sin(a - n)\pi}{a - n} \right]$$

$$= \dfrac{1}{\pi} \left[\dfrac{2a(-1)^n \sin(a\pi)}{a^2 - n^2} \right] = \dfrac{2\sin(a\pi)}{\pi} \dfrac{(-1)^n}{a^2 - n^2}$$

implies $\cos ax = \dfrac{\sin(a\pi)}{\pi}\left[\dfrac{1}{a} + 2\sum_{n=1}^{\infty}\dfrac{(-1)^n}{a^2 - n^2}\cos(nx)\right]$

implies $\pi\cot(a\pi) = \dfrac{1}{a} + \sum_{n=1}^{\infty}\dfrac{2}{a^2 - n^2}$.

4. $(\|f_n\| - \|f\|)^2 = \|f_n\|^2 - 2\|f_n\|\,\|f\| + \|f\|^2$

$\qquad\qquad \leqslant \|f_n\|^2 - \langle f,f_n\rangle - \langle f_n,f\rangle + \|f\|^2$ (Schwarz inequality)

$\qquad\qquad = \langle f_n - f, f_n - f\rangle = \|f_n - f\|^2$

implies $\displaystyle\lim_{n\to\infty}|\,\|f_n\| - \|f\|\,| \leqslant \lim_{n\to\infty}\|f_n - f\| = 0$. The converse is false.

8. Let $f_n(x) = \begin{cases} 0, & -1 \leqslant x < -1 + 1/2^{n-2}, \\ 2^{n-1}, & -1 + 1/2^{n-2} \leqslant x \leqslant -1 + 1/2^{n-1}, \\ 0, & -1 + 1/2^{n-1} < x \leqslant 1. \end{cases}$

13. $\left\|f - \sum_{k=0}^{n}\langle f,\varphi_k\rangle\varphi_k\right\|^2 = \|f\|^2 - \sum_{k=0}^{n}|\langle f,\varphi_k\rangle|^2 + 0$ implies

$\|f\|^2 = \sum_{k=0}^{n}|\langle f,\varphi_k\rangle|^2 + \left\|f - \sum_{k=0}^{n}\langle f,\varphi_k\rangle\varphi_k\right\|^2 \geqslant \sum_{k=0}^{n}|\langle f,\varphi_k\rangle|^2$ implies

$\|f\|^2 \geqslant \displaystyle\lim_{n\to\infty}\sum_{k=0}^{n}|\langle f,\varphi_k\rangle|^2 = \sum_{k=0}^{\infty}|\langle f,\varphi_k\rangle|^2$.

15.
$$d^2(\varphi_n,\varphi_m) = \left\|\dfrac{\sin(nx)}{\sqrt{\pi}} - \dfrac{\sin(mx)}{\sqrt{\pi}}\right\|^2$$

$$= \left\|\dfrac{\sin(nx)}{\sqrt{\pi}}\right\|^2 - \left\langle\dfrac{\sin(nx)}{\sqrt{\pi}},\dfrac{\sin(mx)}{\sqrt{\pi}}\right\rangle$$

$$- \left\langle\dfrac{\sin(mx)}{\sqrt{\pi}},\dfrac{\sin(nx)}{\sqrt{\pi}}\right\rangle + \left\|\dfrac{\sin(mx)}{\sqrt{\pi}}\right\|^2$$

$$= 1 - 0 - 0 + 1 \qquad n \neq m$$

implies $d(\varphi_n,\varphi_m) = \sqrt{2}$, $n \neq m$ and $d(\varphi_n,\varphi_n) = 0$. So if $S = \{\varphi_n \mid n = 0,1,\ldots\}$ then S is bounded by $\sqrt{2}$, and if $\{\varphi_{n_k}\}_{k=0}^{\infty} \subset S$ and $\displaystyle\lim_{k,j\to\infty} d(\varphi_{n_k},\varphi_{n_j}) = 0$. Then there exists a K, J such that for all $k \geqslant K$, for all $j \geqslant J$, $d(\varphi_{n_k},\varphi_{n_j}) < 1$ implies $k = j = K$, so for all $k \geqslant K$, $\varphi_{n_k} = \varphi_{n_K}$. Thus $\displaystyle\lim_{k\to\infty}\varphi_{n_k} = \varphi_{n_K}$ implies S is closed. Let $\{\varphi_{n_k}\} \subset \{\varphi_n\}_{n=0}^{\infty}$ subsequence. Then if $i < j$, $d(\varphi_{n_i},\varphi_{n_j}) = \sqrt{2}$ implies $\displaystyle\lim_{i,k\to\infty} d(\varphi_{n_i},\varphi_{n_k}) \neq 0$. So no subsequence can converge.

16. $\dfrac{\pi}{2} + \dfrac{2\pi^2}{3} + 2\sum_{1}^{\infty}\left(\dfrac{\cos(nx)}{n^2}\right) - \left(\dfrac{2\pi + 1}{2^n}\right)\sin(nx)$.

20. $\dfrac{2}{\pi}\displaystyle\int_{0}^{\pi}\sin(nx)\sin(mx) =$

Case 1: $n = m$; $\dfrac{2}{\pi} \displaystyle\int_0^\pi \sin^2(nx) = \dfrac{2}{\pi}\left[\dfrac{x}{2} + \dfrac{\cos(2nx)}{4n}\right]_0^\pi = 1$

Case 2: $n \neq m$; $\dfrac{2}{\pi} \displaystyle\int_0^\pi \cos(n - m)x - \cos(n + m)x \, dx$

$$= \dfrac{2}{\pi}\,\dfrac{\sin(n - m)x}{n - m} - \dfrac{\sin(n + m)x}{n + m}\Bigg|_0^\pi = 0 \;.$$

$$f(x) = \sum_{n=1}^\infty b_n \sin(nx) \quad \text{where} \quad b_n = \dfrac{2}{\pi}\int_0^\pi f(x)\sin(nx) \;.$$

$S = \{\sqrt{2/\pi}\, \sin(nx) \mid n = 1,2,\ldots\}$ is complete for if $f:[0,\pi] \to \mathbb{R}$ is square integrable then we can extend f to a square integrable odd function on $[-\pi,\pi]$, say $\hat{f}$. Then $\hat{f}$ has a half-interval sine-series expansion namely $\hat{f}(x) = \sum_{n=1}^\infty b_n \sin(nx)$ where

$$b_n = \dfrac{2}{n}\int_0^\pi \hat{f}(x)\sin(nx) \, dx = \dfrac{2}{\pi}\int_0^\pi \hat{f}(x)\sin(nx) \, dx$$

and

$$\dfrac{1}{\pi}\int_{-\pi}^\pi |\hat{f}(x)|^2 \, dx = \sum_{i=1}^\infty b_n^2$$

implies

$$\dfrac{2}{\pi}\int_0^\pi |f(x)|^2 \, dx = \sum_{i=1}^\infty b_n^2$$

implies

$$\int_0^\pi |f(x)|^2 \, dx = \sum_{n=1}^\infty \left(\sqrt{\dfrac{\pi}{2}}\, b_n\right)^2 = \sum_{n=1}^\infty \left|\left\langle f, \sqrt{\dfrac{2}{\pi}}\,\sin(nx)\right\rangle\right|^2 \;.$$

Thus S is complete.

21. Let $A = f - s_n$, $B = s_{n+p} - s_n$. Then $\langle s_n, B\rangle = 0$ implies

$$\langle A,B\rangle = \sum_{i=n+1}^p \langle \overline{f,\varphi_i}\rangle\langle f,\varphi_i\rangle = \|B\|^2$$

$$\langle B,A\rangle = \sum_{i=n+1}^p \langle f,\varphi_i\rangle\langle \varphi_i,f\rangle = \sum_{i=n+1}^p |\langle f,\varphi_i\rangle|^2 = \langle B,B\rangle = \|B\|^2$$

So $\langle f - s_{n+p}, f - s_{n+p}\rangle = \langle A - B, A - B\rangle$
$$= \langle A,A\rangle - \langle A,B\rangle - \langle B,A\rangle + \|B\|^2$$
$$= \|A\|^2 - \|B\|^2 \leq \|A\|^2$$

Thus $\|f - s_{n+p}\| \leq \|f - s_n\|$. We therefore have $\{\|f - s_n\|\}_{n=0}^\infty$ a monotonically decreasing sequence bounded from below by zero and from above by $\|f\| < +\infty$. So $\displaystyle\lim_{n\to\infty} \|f - s_n\| = \text{glb}\{\|f - s_n\| \mid n = 0,1,2,\ldots\} < +\infty$.

23. $i\hbar\,\dfrac{\partial\psi}{\partial t} = i\hbar\left[\displaystyle\sum_{n=0}^\infty \langle \psi_0,\varphi_n\rangle e^{-iE_n t/\hbar} - \left(\dfrac{iE_n}{\hbar}\right)\varphi_n\right]$

$$= \sum_{n=0}^{\infty} \langle \psi_0, \varphi_n \rangle e^{-iE_n t/\hbar} E_n \varphi_n$$

$$= \sum_{n=0}^{\infty} \langle \psi_0, \varphi_n \rangle e^{-iE_n t/\hbar} H(\varphi_n)$$

$$= H\left(\sum_{n=0}^{\infty} \langle \psi_0, \varphi_n \rangle e^{-iE_n t/\hbar} \varphi_n \right)$$

$$= H(\psi) .$$

Also $\psi(0,x,y,z) = \sum_{n=0}^{\infty} \langle \psi_0, \varphi_n \rangle \varphi_n(x,y,z) = \psi_0$. If $H(\psi(t,x,y,z)) = E_n \psi(t,x,y,z) = i\hbar \dfrac{\partial \psi}{\partial t}$, then $\displaystyle\int \dfrac{1}{\psi(t,x,y,z)} \dfrac{\partial \psi}{\partial t} dt = \int \dfrac{E_n}{i\hbar} dt$ implies $\ln \psi(t,x,y,z) = \dfrac{E_n}{i\hbar} t + c(x,y,z)$, that is, $\psi(t,x,y,z) = e^{-iE_n t/\hbar} \cdot e^{c(x,y,z)}$ which implies $e^{-iE_n t/\hbar} H(e^{c(x,y,z)}) = H(\psi(t,x,y,z)) = E_n e^{-iE_n t/\hbar} e^{c(x,y,z)}$ and so $H(e^{c(x,y,z)}) = E_n e^{c(x,y,z)}$, that is, $e^{c(x,y,z)}$, is an eigenfunction of H implies $e^{c(x,y,z)} = k\varphi_n(x,y,z)$ where k is a constant. So

$$\psi(t,x,y,z) = ke^{-iE_n t/\hbar} \varphi_n(x,y,z) \qquad \text{and} \qquad \psi_0 = \psi(0,x,y,z) = k\varphi_n(x,y,z) .$$

24. Since H is symmetric,

$$\frac{d}{dt} \langle \psi, H(\psi) \rangle$$

$$= \left\langle \frac{\partial \psi}{\partial t}, H(\psi) \right\rangle + \left\langle H(\psi), \frac{\partial \psi}{\partial t} \right\rangle$$

$$= \frac{-i}{\hbar} \langle H(\psi), H(\psi) \rangle + \frac{i}{\hbar} \langle H(\psi), H(\psi) \rangle = 0 .$$

26. (a) No; $\langle [A,B](\psi), \sigma \rangle = \langle AB(\psi), \sigma \rangle - \langle BA(\psi), \sigma \rangle$

$$= \langle \psi, BA(\sigma) \rangle - \langle \psi, AB(\sigma) \rangle$$

$$= \langle \psi, [B,A](\sigma) \rangle .$$

So if $[A,B] = [A,B]^{\tau}$ then $[A,B] = [B,A] = -[A,B]$ implies $[A,B] = 0$ implies $AB = BA$. As an example let $A = J_x$, $B = P_y$. Then $J_x P_y \neq P_y J_x$ so $[J_x, P_y] = \hbar Pz/i$ is not symmetric.

(b) Yes; $\langle i[A,B](\psi), \sigma \rangle = i\langle [A,B](\psi), \sigma \rangle$

$$= i\langle \psi, [B,A](\sigma) \rangle$$

$$= \langle \psi, i[A,B](\sigma) \rangle .$$

30. $f * g(x) = \displaystyle\int_{0}^{2\pi} f(y)g(x-y) \, dy = \int_{x}^{x-2\pi} f(x-w)g(w)(-dw)$

$$= \int_{x-2\pi}^{x} f(x-w)g(w)\,dw = \int_{x}^{x+2\pi} g(w)f(x-w)\,dw$$

$$= \int_{x}^{0} g(w)f(x-w)\,dw + \int_{0}^{2\pi} g(w)f(x-w)\,dw + \int_{2\pi}^{x+2\pi} g(w)f(x-w)\,dw$$

$$= \int_{x}^{0} g(w)f(x-w)\,dw + \int_{0}^{2\pi} g(w)f(x-w)\,dw + \int_{0}^{x} g(w)f(x-w)\,dw$$

$$= \int_{0}^{2\pi} g(w)f(x-w)\,dw ,$$

and $f * g(x) = \sum_{-\infty}^{\infty} (2\pi a_n b_n \cdot \sqrt{2\pi}) \dfrac{e^{inx}}{\sqrt{2\pi}}$ implies $\|f * g\|^2 = 8\pi^3 \sum_{-\infty}^{\infty} a_n^2 b_n^2$ is Parseval's relation.

33. Choose M so that $|f'(x)| < M$ on $]x_0,b[$ and on $]a,x_0[$. Then

$$\sup_{P} \sum_{P} |f(\tau_{i+1}) - f(\tau_i)| = \sup_{x_0 \in P} \sum_{P} |f(\tau_{i+1}) - f(\tau_i)|$$

$$= \sup_{x_0 \in P} \left(\sum_{P \cap [a,x_0]} |f(\tau_{i+1}) - f(\tau_i)| + \sum_{P \cap [x_0,b]} |f(\tau_{i+1}) - f(\tau_i)| \right)$$

$$= \sup_{x_0 \in P} \left(\sum_{P \cap [a,x_0]} |f'(\xi_i)| (\tau_{i+1} - \tau_i) + \sum_{P \cap [x_0,b]} |f'(\eta_i)| (\tau_{i+1} - \tau_i) \right)$$

where $\tau_{i+1} < \xi_i < \tau_i$, $\tau_{i+1} < \eta_i < \tau_i$

$$\leqslant \sup_{x_0 \in P} \left(\sum_{P \cap [a,x_0]} M(\tau_{i+1} - \tau_i) + \sum_{P \cap [x_0,b]} M(\tau_{i+1} - \tau_i) \right) = M(b-a).$$

So f is of bounded variation on $[a,b]$. By the Dirichlet-Jordan theorem the Fourier series converges to f on $[a,b] \backslash \{x_0\}$ and to $\dfrac{f(x_0+) + f(x_0-)}{2}$ at x_0.

35. (a) Pointwise to $f(x)$ for $x \neq \pi, -\pi$ and to zero for $x = \pi, -\pi$, and in mean.
   (b) By Theorem 12, uniformly, pointwise, and in mean.
   (c) Pointwise to $\dfrac{f(x+) + f(x-)}{2}$ and in mean.
   (d) Pointwise to $\dfrac{f(x+) + f(x-)}{2}$ and in mean.
   (e) $f'(x) = \begin{cases} 0, & x \leqslant 0 \\ 2x \sin(1/x) - \cos(1/x), & x > 0 \end{cases}$ implies $|f'(x)| \leqslant 2\pi + 1$ and f is sectionally continuous. By Exercise 33 the Jordan-Dirichlet theorem applies. Thus convergence is pointwise to $\dfrac{f(x+) + f(x-)}{2}$.

36. Use Parseval's relation.

37. By Theorem 7, $a = 1/\pi \int_0^\pi \sin x \, dx = 4/\pi$, $b = 2/\pi \int_0^\pi \sin(2x) \, dx = 0$ and $c = 2/\pi \int_0^\pi \sin(3x) \, dx = +4/3\pi$. On $[-\pi,\pi]$, $f(x) = 1$ an even function implies $a = b = c = 0$.

43. $|h(x) - h_n(x)| = \left| \int_a^x [f(y)g(y) - f_n(y)g_n(y)] \, dy \right|$

$\leq \int_a^b |f(y)g(y) - f_n(y)g_n(y)| \, dy$

$\leq \int_a^b \{|f(y)g(y) - f(y)g_n(y)| + |f(y)g_n(y) - f_n(y)g_n(y)|\} \, dy$

$\leq \int_a^b |f(y)| \, |g(y) - g_n(y)| \, dy + \int_a^b |f(y) - f_n(y)| \, |g_n(y)| \, dy$

$= \langle |f|, |g - g_n| \rangle + \langle |f - f_n|, |g_n| \rangle$

$\leq \|f\| \, \|g - g_n\| + \|f - f_n\| \, \|g_n\|$

$\leq \|f\| \, \|g - g_n\| + \|f - f_n\|(\|g_n - g\| + \|g\|) \to 0 \text{ as } n \to \infty.$

44. (a) $x \sin x$ is an even function. So $b_n = 0$ and

$a_n = \dfrac{2}{\pi} \int_0^\pi x \sin(x) \cos(nx) \, dx = \dfrac{1}{\pi} \int_0^\pi x[\sin(n+1)x - \sin(n-1)x] \, dx$

$= \begin{cases} \dfrac{1}{\pi} \left[(x)\left(\dfrac{\cos(n-1)x}{n-1} - \dfrac{\cos(n+1)x}{n+1} \right) \right]_0^\pi = \dfrac{\cos(n-1)\pi}{n-1} - \dfrac{\cos(n+1)\pi}{n+1} & \\ & n \neq 1 \\ \dfrac{1}{\pi}(x)\left(-\dfrac{\cos(2x)}{2} \right) \Big|_0^\pi = -\dfrac{1}{2} & n = 1 \end{cases}$

$= \begin{cases} \dfrac{(-1)^{n+1}2}{n^2 - 1} & n \neq 1 \\ -\dfrac{1}{2} & n = 1 \end{cases}$

implies $a_0 = 2$, implies $x \sin x = \dfrac{2}{2} - \dfrac{\cos(x)}{2} - 2 \sum_1^\infty \dfrac{(-1)^n \cos(nx)}{n^2 - 1}$.

(b) $a_n = \dfrac{2}{\pi} \int_0^\pi \log(\sin 1/2 \, x) \cos(nx) \, dx$

$= \dfrac{4}{\pi} \int_0^{\pi/2} \log(\sin x) \cos(2nx) \, dx$.

If $n = 0$, $a_0 = \dfrac{4}{\pi} \int_0^{\pi/2} \log(\sin x) \, dx$.

Now $\int_0^\pi \log(\sin x) \, dx = 2 \int_0^{\pi/2} \log(\sin 2x) \, dx$

$= 2 \int_0^{\pi/2} \{\log(2) + \log(\sin x) + \log(\cos x)\} \, dx$

$= \pi \log(2) + 2 \int_0^{\pi/2} \log(\sin x) + 2 \int_0^{\pi/2} \log(\cos x) \, dx$.

But $\displaystyle\int_0^{\pi/2} \log(\cos x)\,dx = \int_{\pi/2}^{\pi} \log\!\left(\cos\!\left(x - \frac{\pi}{2}\right)\right) = \int_{\pi/2}^{\pi} \log(\sin x)\,dx$

$$= \int_{\pi/2}^0 \log(\sin(\pi - x))(-dx) = -\int_{\pi/2}^0 \log(\sin x)\,dx$$

$$= \int_0^{\pi/2} \log(\sin x)\,dx .$$

Thus

$$2\int_0^{\pi/2} \log(\sin x)\,dx = \pi \log(2) + 4\int_0^{\pi/2} \log(\sin x)\,dx$$

implies

$$\int_0^{\pi/2} \log(\sin x)\,dx = -\frac{\pi}{2}\log(2)$$

implies $a_0 = -2\log(2)$.

If $n \neq 0$,

$$a_n = \frac{4}{\pi}\left\{\frac{\log(\sin x)\sin(2nx)}{2n}\Bigg|_0^{\pi/2} - \frac{1}{2n}\int_0^{\pi/2} \frac{\cos(x)\sin(2nx)}{\sin(x)}\,dx\right\} .$$

The first term is zero because

$$\lim_{x\to 0} \frac{\log(\sin x)\sin(2nx)}{2n} = \lim_{x\to 0} \frac{\log(\sin x)}{2n/\sin(2nx)} = \lim_{x\to 0} \frac{\cos x/\sin x}{[2n/\sin^2(2nx)]\cdot 2n\cos(2nx)}$$

$$= \lim_{x\to 0} \frac{\sin^2(2nx)}{4n^2\sin x} = \lim_{x\to 0} \frac{2n\sin(2nx)\cos(2nx)}{4n^2\cos x} = 0 .$$

So

$$a_n = -\frac{1}{n\pi}\int_0^{\pi/2} \frac{\sin(2n+1)x + \sin(2n-1)x}{\sin x}\,dx .$$

However we can use Exercise 6, p. 353 to deduce $a_n = -1/\pi$, so

$$\log(\sin x/2) = -\log z - \sum_{n=1}^{\infty} \frac{1}{n}\cos nx .$$

49. $\displaystyle |\sin x| = \frac{2}{\pi} - \frac{4}{\pi}\sum_1^{\infty} \frac{\cos(2nx)}{4n^2 - 1} .$

$$0 = |\sin(0)| = \frac{2}{\pi} - \frac{4}{\pi}\sum_1^{\infty} \frac{\cos(0)}{4n^2 - 1} = \frac{2}{\pi} - \frac{4}{\pi}\sum_1^{\infty} \frac{1}{4n^2 - 1} .$$

So

$$\frac{1}{2} = \sum_1^{\infty} \frac{1}{4n^2 - 1} .$$

$$1 = \left|\sin\!\left(\frac{\pi}{2}\right)\right| = \frac{2}{\pi} - \frac{4}{\pi}\sum_1^{\infty} \frac{(-1)^n}{4n^2 - 1} .$$

52. $\displaystyle f(x) = \frac{x}{2} + \frac{|x|}{2} = \left(\sum_{n=1}^{\infty} \frac{(-1)^{n+1}}{n}\sin(nx)\right) + \left(\frac{\pi}{4} - \frac{2}{\pi}\sum_{n=1}^{\infty} \frac{\cos(2n-1)x}{(2n-1)^2}\right) .$

57. We obtain the following theorem.

    Theorem. *Let f be continuous on $[-\pi,\pi]$, $f(-\pi) = f(\pi)$ and let f' be continuous, f'' sectionally continuous. Then the Fourier series for f*

    $$f(x) = \frac{a_0}{2} + \sum_{n=1}^{\infty} (a_n \cos(nx) + b_n \sin(nx))$$

    may be differentiated term by term and we obtain

    $$f'(x) = \sum_{n=1}^{\infty} (-na_n \sin(nx) + nb_n \cos(nx)) \, .$$

    Furthermore this is the Fourier series of f'.

    Proof: Theorem 14 shows that the Fourier series for both f and f' converge absolutely and uniformly. Hence by Corollary 3, we may differentiate the series for f to get the series for f'. The advantage of Theorem 13 is that we need only to know that f'' exists at a particular point $x \in [-\pi,\pi]$. f'' need not be continuous.

59. Let

    $$S_k = \sum_{n=1}^{k} \frac{|a_n|}{n} \, .$$

    Then by Schwarz's inequality

    $$S_k = \sum_{n=1}^{k} |a_n| \cdot \frac{1}{n} \leqslant \left(\sum_{n=0}^{k} |a_n|^2 \right)^{1/2} \left(\sum_{n=1}^{k} \frac{1}{n^2} \right)^{1/2} < +\infty \, ,$$

    because

    $$\sum_{n=0}^{k} |a_n|^2 < \infty$$

    by Bessel's inequality.

62. $2\pi^3 + \sum_{n=1}^{\infty} \left\{ \frac{12\pi}{n^2} \cos(nx) + \left(\frac{12}{n^3} - \frac{8\pi^2}{n} \right) \sin(nx) \right\} .$

65. We want a function $T(x,t)$ such that

    (a) $\dfrac{\partial T}{\partial t}(x,t) = \dfrac{\partial^2 T}{\partial x^2}(x,t) \qquad 0 < x < l \qquad t \geqslant 0 \qquad$ (heat equation)

    (b) $T(x,0) = f(x) \qquad\qquad\quad 0 < x < l \qquad\qquad\qquad$ (initial condition)

    (c) $T(0,t) = T(l,t) = 0 \qquad\quad t \geqslant 0 \qquad\qquad\qquad$ (boundary condition)

    As usual we try $T(x,t) = g(x)h(t)$. Then we must have $g(x)h'(t) = g''(x)h(t)$. These equations are true if, for a constant λ, $g(x) + \lambda g''(x) = 0$, and $h(t) + \lambda h'(t) = 0$. Solutions of these equations satisfying the boundary conditions are $g_n(x) = \sin(n\pi x/l)$ and $h_n(t) = e^{-n^2\pi^2 t/l^2}$, $n = 0, 1, 2, \ldots$ and where $\lambda_n = n^2\pi^2/l^2$. We use sine and not cosine in order that $T_n(0,t) = T_n(l,t) = g_n(0)h_n(t) = 0$. Thus a solution with $f(x) = \sin(n\pi x/l)$ is given by $T_n(x,t) = \sin(n\pi x/l)e^{-n^2\pi^2 t/l^2}$. Since the equations are linear and $f(x) = \sum_{n=1}^{\infty} b_n \sin(n\pi x/l)$ (half-interval sine series) we expect that the general solution with initial condition f is given by

    $$T(x,t) = \sum_{n=1}^{\infty} b_n \sin\left(\frac{n\pi x}{l} \right) e^{-n^2\pi^2 t^2/l^2} \, .$$

69. From Theorem 17, there exists a M such that $|a_n| < M$. Using the identity $\sum_{n=1}^{\infty} x^n = x/(1 - x)$ for $|x| < 1$ we see for large t,

$$\left| \sum_{n=1}^{\infty} a_n e^{-n^2\pi^2 t/l^2} \cos\left(\frac{n\pi x}{l}\right) \right| \leqslant M \sum_{n=1}^{\infty} \{e^{-\pi^2 t/l^2}\}^n$$

$$= M \frac{e^{-\pi^2 t/l^2}}{1 - e^{-\pi^2 t/l^2}} \to 0 \text{ as } t \to \infty .$$

Index

The author gratefully thanks Jody Anderson and Barbara Komatsu for this index.